INTRODUCTION TO CLASSICAL REAL ANALYSIS

KARL R. STROMBERG
KANSAS STATE UNIVERSITY

WADSWORTH INTERNATIONAL GROUP
a division of WADSWORTH, INC.
BELMONT, CALIFORNIA

> **With utmost respect, I dedicate this work to my teacher, my collaborator, and my friend**
> **Edwin Hewitt**

ISBN 0-534-98012-0

© Copyright 1981 by Wadsworth, Inc., 10 Davis Drive, Belmont, California 94002

This book is a co-publishing project of Wadsworth International Group and Prindle, Weber & Schmidt.

Library of Congress Cataloging in Publication Data

Stromberg, Karl Robert
  An introduction to classical real analysis.

  Bibliography: p.
  Includes index.
  1. Mathematical analysis. I. Title.
QA300.S89   515   80-26939
ISBN 0-534-98012-0

Printed in the United States of America.

10 9 8 7 6 5 4 3

Jacket image "Manifestation" © Copyright 1981 (pending) by Michael Lasuchin. All rights reserved. Used by permission of the artist.

Jacket, cover and text design by David Foss, in collaboration with the Prindle, Weber & Schmidt production staff. Text composition in Times Roman and Baskerville by Computype, Inc. Jacket printed by New England Book Components, Inc. Text printed and bound by the Alpine Press, Inc.

PREFACE

This volume has evolved from lectures that I have given at the University of Oregon and at Kansas State University during the past twenty years. The subject is classical analysis. It is "real analysis" in the sense that none of the Cauchy theory of analytic functions is discussed. Complex numbers, however, do appear throughout. Infinite series and products are discussed in the setting of complex numbers. The elementary functions are defined as functions of a complex variable. I do depart from the classical theme in Chapter 3, where limits and continuity are presented in the contexts of abstract topological and metric spaces.

   The approach here is to begin with the axioms for a complete ordered field as the definition of the real number system. Based only upon that, an uncompromisingly rigorous Definition-Theorem-Proof style is followed to completely justify all else that is said. For better or for worse, I have scrupulously avoided any presumption at all that the reader has any knowledge of mathematical concepts until they are formally presented here. Thus, for example, the number π is not mentioned until it has been precisely defined in Chapter 5.

   I hope that this book will be found useful as a text for the sort of courses in analysis that are normally given nowadays in most American universities to advanced undergraduate and beginning graduate students. I have included every topic that I deem necessary as a preparation for learning complex and abstract analysis. I have also included a selection of optional topics. The table of contents is a brief guide to the topics included and to which ones may be safely omitted without disturbing the logical continuity of the presentation. I also hope that this book will be found useful as a reference tool for mature mathematicians and other scientific workers.

   One significant way in which this book differs from other texts at this level is that the integral which we first mention is the Lebesgue integral on the real line. There are at least three good reasons for doing this. First, the F. Riesz approach (after which mine is modelled) is no more difficult to understand than is the traditional theory of the Riemann integral as it currently appears in nearly every calculus book. Second, I feel that students profit from acquiring a thorough understanding of Lebesgue integration on Euclidean spaces before they enter into a study of abstract measure theory. Third, this is the integral that is most useful to current applied mathematicians and theoretical scientists whether or not they ever study abstract mathematics. Of course, it is clearly shown in Chapter 6 how the Riemann integral is a special case of the Lebesgue integral. Stieltjes integration is presented in a graded sequence of exercises. The proofs of these exercises are easy, but any instructor who wishes to include them in his lectures is obviously free to do so.

I sincerely hope that the exercise sets will prove to be a particularly attractive feature of this book. I spent at least three times as much effort in preparing them as I did on the main text itself. Most of the exercises take the form of simple assertions. The exercise is to prove the assertion. A great many of the exercises are projects of many parts which, when completed in the order given, lead the student by easy stages to important and interesting results. Many of the exercises are supplied with copious hints. I feel that the only way to truly learn mathematics is by just plain hard work. It does not suffice to simply read through a book and agree with the author. I do encourage all serious students to work diligently through the exercises provided here. Thomas Edison's dictum that genius is ten percent inspiration and ninety percent perspiration has never been truer than it is here.

I have found that for a two semester (or three quarter) course, it is easy to cover all the sections in Chapters 1 through 7 that are not marked with asterisks in the Table of Contents. I also find time to include some of the optional sections or part of Chapter 8. In doing this, I make it a practice of assigning a lot of the easier textual material as reading for the students, while I work through many of the harder exercises in class. I see no point in copying the text onto the blackboard.

If it is only possible to spend one semester (about fifteen weeks) on classical real analysis, then one can proceed as follows. Assign all of Chapter 0 and much of Chapter 1 as reading. Omit all sections which bear asterisks in the Table of Contents. Spend only one week on each of Chapters 1, 5, and 7 and only three weeks on each of Chapters 2, 3, 4, and 6 by making the following additional omissions. In Chapter 3, proceed only through "Uniform Convergence," omitting "Baire Category." In Chapter 4 omit "Differentiability Almost Everywhere." In Chapter 6 stop with "The Riemann Integral," but be sure to work through many of the exercises at the end of that section. In Chapter 7, stop with "Some Theorems of Abel."

I take great pleasure in offering my warmest thanks to my good friends Bob Burckel and Louis Pigno who gave me such valuable assistance in preparing this book through their constant encouragement, their proofreading and their many stimulating conversations. I also thank the four women who valiantly typed the technically complicated manuscript. They are Twila Peck, Judy Bernhart, Marie Davis, and Marlyn Logan. Finally, it is a pleasure to thank my publishers and editors John Martindale, Arthur Weber, Paul Prindle, John Kimmel, and David Foss for their excellent help and for their patience and understanding through this seemingly interminable project.

Karl Stromberg
Manhattan, Kansas
July, 1980

CONTENTS

---

    *Sections marked with a single asterisk are not actually needed in conjunction with any of the unmarked sections. These sections may be safely omitted if time permits only a short course.

    **Sections marked with a double asterisk are not actually needed in conjunction with any other section. They are included as interesting and useful applications of the theory.

0

PRELIMINARIES

The purpose of this brief chapter is to record some of the reasonably standard set theoretic notations and definitions that will be used throughout this book.

Sets and Subsets

A set is a collection of objects called the *members* [or *elements* or *points*] of the set. If A is a set and p is an object, we write $p \in A$ to mean that p is a member of A. If p is not a member of A, we write $p \notin A$.

Let A and B be any two sets. If each member of A is also a member of B, then we say that A is a *subset* of B or that B is a *superset* of A and we write $A \subset B$ or $B \supset A$. In case we have both $A \subset B$ and $B \subset A$, we say that A is *equal* to B and we write $A = B$. Thus, $A = B$ means that A and B have precisely the same members. To say that A is *not* a subset of B means that A has some member p which is not a member of B, and if this is the case, we write $A \not\subset B$ or $B \not\supset A$. The set that has no members is called the *void set* [or *empty set*] and is denoted by the symbol $\emptyset$. It is clear that $\emptyset \subset B$ (for any set B) since $\emptyset$ has no member p which fails to be a member of B. We write $A \neq B$ to mean that A and B are not equal. If $A \subset B$ and $A \neq B$, we write $A \subsetneq B$ and say that A is a *proper subset* of B.

Sometimes it is possible to specify a particular set by listing its members. Thus, for example, the set whose only members are a and b and c is denoted by $\{a,b,c\}$. Notice that $\{a,b,c\} = \{c,a,b\}$. The symbol {purple, white} denotes the set whose only members are "purple" and "white." A set that has exactly one member is called a *singleton*. The symbol $\{x\}$ denotes the singleton whose only member is x. Thus $A = \{x\}$ means that $x \in A$ *and* $y = x$ for each $y \in A$.

More frequently, however, we shall specify a particular set by stating some proposition $P(x)$ about objects x whose truth is both necessary and sufficient to qualify an object x for membership in the set being specified. When this is done, we

1

write $\{x : P(x)\}$ to denote the set of all objects x such that $P(x)$ is true. We also use $\{x \in S : P(x)\}$ in place of $\{x : x \in S$ and $P(x)\}$ to denote the subset of a given set S consisting of just those members x of S for which $P(x)$ is true. For example, anticipating Chapter 1, the symbol $\{x \in \mathbb{R} : 0 < x < 1\}$ denotes the set of all real numbers x such that $0 < x$ and $x < 1$.

Operations on Sets

Let A and B be given sets. We define three other sets by

$$A \cup B = \{x : x \in A \text{ or } x \in B\},$$

$$A \cap B = \{x : x \in A \text{ and } x \in B\},$$

$$A \backslash B = \{x : x \in A \text{ and } x \notin B\}.$$

These are called the *union* of A and B, the *intersection* of A and B, and the *difference* of A and B, respectively. If X is a set and $A \subset X$, then the set $X \backslash A$ is called the *complement* of A relative to X. We say that A and B are *disjoint* if $A \cap B = \varnothing$.

If $\mathcal{F}$ is a set and each member of $\mathcal{F}$ is a set, then we call $\mathcal{F}$ a *family of sets*. We define the *union* of $\mathcal{F}$ to be the set

$$\cup \mathcal{F} = \{x : x \in A \text{ for some } A \in \mathcal{F}\}$$

and we define the *intersection* of $\mathcal{F}$ to be the set

$$\cap \mathcal{F} = \{x : x \in A \text{ for every } A \in \mathcal{F}\}.$$

We say that $\mathcal{F}$ is *pairwise disjoint* [or simply *disjoint*] if $A \cap B = \varnothing$ whenever $A \in \mathcal{F}$, $B \in \mathcal{F}$, and $A \neq B$.

As examples of the above, let $A = \{1,2,3\}$, $B = \{2,3,4,5\}$, $C = \{3,5,4\}$, $D = \{6\}$, $\mathcal{E} = \{A,D,C\}$, and $\mathcal{F} = \{B,C,A\}$. Then $C \subset B$, $C \neq B$, $C \subsetneqq B$, $B \cup C = B$, $B \cap C = C$, $B \backslash C = \{2\}$, $C \backslash B = \varnothing$, $A \cup B = A \cup C = \cup \mathcal{F} = \{3,2,4,1,5\}$, $A \cap B = \{2,3\}$, $A \backslash B = \{1\}$, $B \backslash A = \{4,5\}$, $\cap \mathcal{F} = \{3\} = A \cap C$, $\cup \mathcal{E} = \{1,2,3,6,4,5\}$, $\cap \mathcal{E} = \varnothing$, $\mathcal{E}$ is not pairwise disjoint, $\mathcal{F} \cup \mathcal{E} = \{A,B,C,D\} \neq \mathcal{F} \cup D = \{A,B,C,6\}$, $\mathcal{F} \cap \mathcal{E} = \{A,C\} \neq A \cup C$, and $\{A \backslash B, B \backslash C, B \backslash A, D\}$ is pairwise disjoint.

If I is a set and if to each $i \in I$ there is assigned a set A_i, then the family $\{x : x = A_i$ for some $i \in I\}$ is more commonly denoted by $\{A_i : i \in I\}$ or $\{A_i\}_{i \in I}$ and this family is called an *indexed family of sets*. In this case, I is called the *index set* for the family and the members of I are called *indices*. We write $\bigcup_{i \in I} A_i$ and $\bigcap_{i \in I} A_i$ for the union and the intersection, respectively, of this indexed family. Notice that any family $\mathcal{F}$ can be made an indexed family by taking $\mathcal{F}$ for the index set and letting $A_F = F$ for each $F \in \mathcal{F}$ to obtain $\mathcal{F} = \{A_F : F \in \mathcal{F}\}$. The family $\mathcal{F}$ of the preceding paragraph may be indexed by taking $I = C = \{3,4,5\}$, $A_3 = B$, $A_4 = A$, and $A_5 = C$.

Ordered Pairs and Relations

An *ordered pair* is an object of the form (x, y) where x and y are objects called the *first coordinate* and the *second coordinate* of (x, y), respectively. The important feature of ordered pairs is the order in which the coordinates appear. Thus, we write $(x, y) = (u, v)$ if and only if $x = u$ and $y = v$. Notice that $\{x, y\} = \{y, x\}$ but that $(x, y) \neq (y, x)$ unless $x = y$. In axiomatic set theory, the ordered pair (x, y) is defined to be the set $\{\{x\}, \{x, y\}\}$.

A *relation* is any set R each of whose members is an ordered pair. We define the *domain* and the *range* of a relation R to be the sets

$$\operatorname{dom} R = \{ x : (x, y) \in R \text{ for some } y \}$$

and

$$\operatorname{rng} R = \{ y : (x, y) \in R \text{ for some } x \},$$

respectively. We define the *inverse* of R to be the relation

$$R^{-1} = \{ (y, x) : (x, y) \in R \}.$$

If R and S are relations, we define the *composite relation* R followed by S to be the relation

$$S \circ R = \{ (x, z) : (x, y) \in R \text{ and } (y, z) \in S \text{ for some } y \}.$$

Notice the inverted order of the notation. It is clear that $\operatorname{dom}(S \circ R) \subset \operatorname{dom} R$, $\operatorname{rng}(S \circ R) \subset \operatorname{rng} S$, and $S \circ R = \varnothing$ if and only if $(\operatorname{rng} R) \cap (\operatorname{dom} S) = \varnothing$. One checks that $(S \circ R)^{-1} = R^{-1} \circ S^{-1}$.

As examples of the above, let $R = \{(1, 2), (2, 2), (2, 3), (3, 4)\}$ and $S = \{(2, 4), (3, 4), (2, 1)\}$. Then $S \circ R = \{(1, 4), (1, 1), (2, 4), (2, 1)\}$, $\operatorname{dom} R = \{1, 2, 3\}$, $\operatorname{rng} S = \{1, 4\} = \operatorname{rng}(S \circ R)$, $\operatorname{dom}(S \circ R) = \{1, 2\}$, $R \circ S = \{(2, 2)\}$, $\operatorname{dom}(R \circ S) = \{2\} = \operatorname{rng}(R \circ S)$, $\operatorname{dom} S = \{2, 3\}$, $\operatorname{rng} R \{2, 3, 4\}$, and $R^{-1} \circ S^{-1} = \{(1, 1), (4, 1), (1, 2), (4, 2)\} = (S \circ R)^{-1}$.

For any relation R, we may write xRy to mean that $(x, y) \in R$.

Equivalence Relations

If X is a set and R is a relation, we call R an *equivalence relation on X* if $\operatorname{dom} R = X$ and, for all $x, y, z \in X$, we have

    (i) $(x, x) \in R$ (reflexive),

    (ii) $(x, y) \in R$ implies $(y, x) \in R$ (symmetric),

    (iii) $(x, y) \in R$ and $(y, z) \in R$ imply $(x, z) \in R$ (transitive).

Condition (ii) says $R^{-1} \subset R$ and (iii) says $R \circ R \subset R$. One checks that any such R satisfies $R^{-1} = R = R \circ R$ and that, conversely, any relation R satisfying these two equalities is an equivalence relation on its domain.

If R is an equivalence relation on X and $x \in X$, then the *equivalence class* containing x is the set $E_x = \{ y \in X : (x, y) \in R \}$. It is easy to check that the family $X/R = \{ E_x : x \in X \}$ of all such equivalence classes is a pairwise disjoint family of nonvoid sets and its union is X. [It may happen that $E_x \cap E_y \neq \emptyset$ even though $x \neq y$. However, this happens only if $E_x = E_y$.] Conversely, if X is any set and $\mathcal{F}$ is a pairwise disjoint family of nonvoid sets with $\cup \mathcal{F} = X$, then the relation R defined by

$$R = \{ (x, y) : \{x, y\} \subset E \text{ for some } E \in \mathcal{F} \}$$

is an equivalence relation on X and $X/R = \mathcal{F}$.

An example of an equivalence relation on the set $\mathbb{Z}$ of all integers is the relation $\sim$ defined by

$$\sim = \{ (x, y) : x \in \mathbb{Z}, y \in \mathbb{Z}, x - y \text{ is divisible by } 3 \}.$$

Thus, $x \sim y$ means that x and y are integers and that $x - y = 3n$ for some integer n. In this case, we have $\mathbb{Z}/\sim = \{ E_0, E_1, E_2 \}$ where $E_0 = \{3n : n \in \mathbb{Z}\}$, $E_1 = \{3n + 1 : n \in \mathbb{Z}\}$, and $E_2 = \{3n + 2 : n \in \mathbb{Z}\} = \{3m - 1 : m \in \mathbb{Z}\} = E_{-1} = E_5$. This relation is usually called congruence modulo 3.

Functions

A *function* is a relation f having the property that, for each $x \in \operatorname{dom} f$, the set $\{ y : (x, y) \in f \}$ is a singleton. That is, we call a relation f a function if and only if there do *not* exist $(x, y) \in f$ and $(x, z) \in f$ with $y \neq z$. For example, the relation $f = \{(1, 1), (1, 2)\}$ is not a function, while its inverse $f^{-1} = \{(1, 1), (2, 1)\}$ is a function. Notice that $\emptyset$ is a function by default.

If f is a given function and $x \in \operatorname{dom} f$ is given, then the set $\{ y : (x, y) \in f \}$ has exactly one member: we call this member the *value of f at x* and we denote it by the symbol $f(x)$ [read: f of x]. It is clear that, in order to specify a particular function f, it is sufficient to specify the domain X of f and to specify the value $f(x)$ of f at each point x of X, for then

$$f = \{ (x, y) : x \in X, y = f(x) \}.$$

We shall usually follow this procedure in defining particular functions. We shall often use an arrow $x \rightarrow f(x)$ to indicate which values are to be assigned to the points x in the domain of f. For instance, "the function f defined on $\mathbb{R}$ by $x \rightarrow x^2$" completely defines the function $f = \{ x, x^2 \} : x \in \mathbb{R} \}$ [see Chapter 1 where $\mathbb{R}$ is discussed].

Notations such as $f : X \rightarrow Y$ mean that f is some function with $\operatorname{dom} f = X$ and $\operatorname{rng} f \subset Y$. In such a case, we say that f *maps X into Y*. If $\operatorname{rng} f = Y$, we say that f maps X onto Y. Our terminology and notation seem to suggest that f, which is actually an immobile set of ordered pairs, is performing an act of transforming points of X into points of Y according to some rule. It is often useful or helpful to adopt this harmless way of thinking of functions. Of course, the specific function under consideration is itself the "rule of transformation." There are many words which have the same meaning as the word "function." Among these are "mapping," "map," "transformation," "operator," and "representation." One learns by experience the contexts in which one or another of these alternative terms is preferred.

Let f be a function and let A be a set. We define the *image of A under f* to be the set

$$f(A) = \{ y : y = f(x) \text{ for some } x \in A \cap (\text{dom } f) \}$$

and the *inverse image of A under f* to be the set

$$f^{-1}(A) = \{ x : x \in \text{dom } f, f(x) \in A \}.$$

Notice that $f(A) \neq \emptyset$ if and only if $A \cap (\text{dom } f) \neq \emptyset$ and that $f^{-1}(A) \neq \emptyset$ if and only if $A \cap (\text{rng} f) \neq \emptyset$.

If $f : X \to Y$ is a function and $E \subset X$, we define the *restriction of f to E* to be the function

$$f|_E = \{ (x, y) : x \in E, y = f(x) \}.$$

Thus $\text{dom}(f|_E) = E$ and $(f|_E)(x) = f(x)$ for every $x \in E$.

A function f is said to be *one to one* [or *biunique*] if the relation f^{-1} is also a function. That is, for each $y \in \text{rng} f$, the set $f^{-1}(\{ y \}) = \{ x : (x, y) \in f \}$ is a singleton. Equivalently said, a function f is one to one if and only if $f(x) \neq f(x')$ whenever $x \neq x'$ in dom f. To say that f is a *one to one correspondence from X onto Y* simply means that f is a one to one function with dom $f = X$ and rng $f = Y$. Functions that are *not* one to one are sometimes called *many to one correspondences*, while their inverses (which are *not* functions) are called *one to many correspondences* [or, somewhat archaically, *multiple valued functions*].

Let $f : X \to Y$ and $g : Y \to Z$ be functions. Then it is clear that the composite relation $g \circ f$ is a function, that $\text{dom}(g \circ f) = X$, that $(g \circ f)(x) = g(f(x))$ for all $x \in X$, and that $\text{rng}(g \circ f) = g(\text{rng } f)$. With the dynamic view of functions mentioned above, $g \circ f : X \to Z$ is the result of first applying f to get from X to Y and then applying g to move on to Z. If one is trying to build a function that has certain prescribed properties it should not be forgotten that a very useful method of construction is the formation of composite functions. Another method, to be studied extensively in this book, is to obtain the desired function as some kind of limit of functions that come "nearer and nearer" to doing the required job.

Products of Sets

If X and Y are sets, we define the *Cartesian product of X and Y* [named in honor of René Descartes (1596–1650), who introduced rectangular coordinates into the Euclidean plane and thereby founded analytic geometry as we know it] to be the set

$$X \times Y = \{ (x, y) : x \in X, y \in Y \}.$$

Observe that any function $f: X \to Y$ is a subset of $X \times Y$. Indeed, the function f itself is precisely what is ordinarily called *the graph of f*.

Let $\{X_i : i \in I\}$ be an indexed family of sets. We define the *Cartesian product* of this family to be the set

$$\underset{i \in I}{\times} X_i = \{x : x \text{ is a function, dom } x = I, x(i) \in X_i \text{ for each } i \in I\}.$$

For a point $x \in \times_{i \in I} X_i$, we usually write x_i instead of $x(i)$ and we call the point $x_i \in X_i$ the *i*th *coordinate of* x. An important special case of this is that in which $X_i = X$ for all $i \in I$, in which case we write X^I in place of $\times_{i \in I} X_i$. Thus X^I denotes the set of all functions having domain I and having all of their values in X.

In case n is a natural number (positive integer) [see Chapter 1 for our discussion of natural numbers], and I is the set of all natural numbers not exceeding n, the Cartesian product of a family $\{X_i : i \in I\}$ indexed by I is usually denoted by

$$\overset{n}{\underset{i=1}{\times}} X_i \quad \text{or} \quad X_1 \times X_2 \times \ldots \times X_n$$

and a point x in this product set is usually written out as an *ordered n tuple*

$$x = (x_1, x_2 \ldots x_n)$$

by listing the coordinates of x [values of the function x] in the order that they occur according to the "size" of the indices. For such I and the case that $X_i = X$ for all $i \in I$, we usually denote this product set by X^n instead of X^I.

Readers who are interested in a more thorough or more rigorous discussion of set theory than we have presented here should consult books on that subject. Some good ones are listed in our bibliography. We believe that what we have said will be adequate for understanding the present book.

1

NUMBERS

The foundation upon which mathematical analysis rests is the real number system $\mathbb{R}$. The fundamental property of $\mathbb{R}$ that makes analysis possible is completeness. Thus before we take up the main topics of this book [convergence, continuity, differentiation, integration, infinite series, etc.] it is necessary for us to discuss $\mathbb{R}$, as such, with particular attention to the notion of completeness.

The ancient Greek mathematicians, who were chiefly interested in geometry, already realized that the rational numbers [ratios of whole numbers] were inadequate in order to assign a numerical length to each line segment. Indeed, a square having sides of length 1 must, by the Pythagorean theorem, have diagonals of length d where $d^2 = 2$. But it is a familiar fact that there is no rational number d having this property.* This state of affairs made it necessary to invent some more numbers.

It is possible to lay the rational numbers off on a line, as it is done in analytic geometry courses, and then to *define* the real numbers to be the points of that line. The procedure is to first label two distinct points on the line as 0 and 1 and then, by compass and straight edge construction, one can construct any (directed) rational distance from 0. Next, one can define addition and multiplication of real numbers by such constructions and thence prove the usual rules of arithmetic. However, the proofs of such formulas as $x(y + z) = xy + xz$ are quite complicated and tend to offer the danger of error through too much dependence on geometric intuition rather than the axioms of geometry. We shall, and we advise the student to, use such geometrical pictures merely as guides to discovery and proof, not as proofs.

What then are the real numbers? The answer to this question is that the real

---

*If d is rational, then $d = a/b$ where a and b are whole numbers, not both even. Assuming $d^2 = 2$, we have $2b^2 = a^2$, from which it follows that a is even, say $a = 2c$. Then $b^2 = 2c^2$ so b is even—a contradiction.

number system is the unique complete* ordered field. The nature of a proof that this field $\mathbb{R}$ exists and is unique depends upon the starting assumptions. Presupposing a knowledge of the rational number system, a construction of $\mathbb{R}$, and a proof of its uniqueness, is given in §5 of [12] (see Bibliography). A different construction of $\mathbb{R}$, based only upon the five Peano postulates for the natural numbers, is given in [15]. It is, however, possible to deduce these postulates as theorems of axiomatic set theory (see [7] or [20]). For our part, we will simply set down as axioms the defining properties of a complete ordered field and assume the existence of such an object $\mathbb{R}$. All theorems that we prove about $\mathbb{R}$ will be based on this assumption alone. Our axioms all appear as theorems about real numbers in [7], [12], and [20].

Axioms for $\mathbb{R}$

The *real number system* consists of a set $\mathbb{R}$, a subset P of $\mathbb{R}$, and two binary operations $(x, y) \to x + y$ and $(x, y) \to xy = x \cdot y$ from $\mathbb{R} \times \mathbb{R}$ into $\mathbb{R}$ such that the following axioms are satisfied.

Axiom I [commutative laws]

$$x + y = y + x \quad \text{and} \quad xy = yx$$

for all $x, y \in \mathbb{R}$.

Axiom II [associative laws]

$$x + (y + z) = (x + y) + z \quad \text{and} \quad x(yz) = (xy)z$$

for all $x, y, z \in \mathbb{R}$.

Axiom III [distributive law]

$$x(y + z) = xy + xz$$

for all $x, y, z \in \mathbb{R}$.

Axiom IV [identity elements]   There exist two distinct elements 0 and 1 in $\mathbb{R}$ such that

$$0 + x = x \quad \text{and} \quad 1x = x$$

for all $x \in \mathbb{R}$.[†]

---

*The term *complete* as used here refers to order completeness as in Axiom VIII below. It is not equivalent to (but does imply) metric completeness: every Cauchy sequence converges (2.29).

[†]It is easy to see that the elements 0 and 1 that obey this axiom are unique since, for example, if 0 and $0'$ both obey it, then $0 = 0' + 0 = 0 + 0' = 0'$.

Axiom V [inverse elements] If $x \in \mathbb{R}$, then there exists a unique $-x \in \mathbb{R}$ such that

$$x + (-x) = 0.$$

If $x \in \mathbb{R}$ and $x \neq 0$, then there exists a unique $x^{-1} \in \mathbb{R}$ such that

$$xx^{-1} = 1.$$

(1.1) **Theorem** **[cancellation laws]** *If* $x, y, z, w \in \mathbb{R}$ *and* $w \neq 0$, *then*
(i) $x + z = y + z$ *implies* $x = y$ *and*
(ii) $xw = yw$ *implies* $x = y$.

Proof The proofs are similar so we prove only (ii). We have $x = 1x = x1 = x(ww^{-1}) = (xw)w^{-1} = (yw)w^{-1} = y(ww^{-1}) = y1 = 1y = y$. The reader should justify each of these equalities. $\square$

(1.2) **Theorem** *If* $x, y, z, w \in \mathbb{R}$ *with* $z \neq 0 \neq w$, *then*
(i) $x0 = 0$,
(ii) $-(-x) = x$,
(iii) $(w^{-1})^{-1} = w$,
(iv) $(-1)x = -x$,
(v) $x(-y) = -(xy) = (-x)y$,
(vi) $(-x) + (-y) = -(x + y)$,
(vii) $(-x)(-y) = xy$,
(viii) $(x/z)(y/w) = (xy)/(zw)$, *
(ix) $(x/z) + (y/w) = (wx + zy)/(zw)$.

Proof (i) Since $x0 + x0 = x(0 + 0) = x0 = 0 + x0$, just apply (1.1.i). (ii) We have $x + (-x) = 0 = (-x) + [-(-x)] = -(-x) + (-x)$ so again use (1.1.i). (iii) is similar to (ii). (iv) Observe that $(-1)x + x = (-1)x + 1x = x(-1 + 1) = x0 = 0 = -x + x$. For (v) use (iv), associativity, and commutativity. For (vi) use (iv) and the distributive law. To prove (vii) use (v) and (ii) to write $(-x)(-y) = -[x(-y)] = -[-(xy)] = xy$. We leave (viii) and (ix) as exercises for the reader. $\square$

Axioms I–V constitute the definition of a *field*; i.e., a field is any set F together with two binary operations $(x, y) \rightarrow x + y$ and $(x, y) \rightarrow xy$ that map $F \times F$ into F such that Axioms I–V [with $\mathbb{R}$ replaced by F] obtain. Theorems (1.1) and (1.2) hold in any field F; the proofs are the same as those given above. There are many examples of fields. The simplest is the two-element field: let $F = \{0, 1\}$ and define $0 \cdot 0 = 1 \cdot 0 = 0 \cdot 1 = 0 + 0 = 1 + 1 = 0$ and $0 + 1 = 1 + 0 = 1 \cdot 1 = 1$.

The remaining three axioms concern the given subset P of $\mathbb{R}$ which induces an ordering on $\mathbb{R}$; see (1.3) and (1.4) below.

---

*We often write a/b or $\frac{a}{b}$ to denote ab^{-1}.

Axiom VI   The three sets $\{0\}$, P and $-P = \{x \in \mathbb{R} : -x \in P\}$ are pairwise disjoint and their union is $\mathbb{R}$.

Axiom VII   If $x, y \in P$, then $x + y \in P$ and $xy \in P$.

(1.3)   **Definition**   The elements of P are called *positive numbers* and the elements of $-P$ are called *negative numbers*. For x, y in $\mathbb{R}$ we write $x < y$ and $y > x$ to mean that $y - x \in P$.* We also write $x \leq y$ and $y \geq x$ to mean that either $y - x \in P$ or $y = x$.

(1.4)   **Theorem**   *For all x, $y, z \in \mathbb{R}$ we have*
(i) [transitivity] $x < y$ and $y < z$ imply $x < z$,
(ii) [trichotomy] exactly one of the relations $x < y$, $x = y$, $x > y$ holds,
(iii) $x < y$ implies $x + z < y + z$,
(iv) $x < y$, $z > 0$ imply $xz < yz$,
(v) $x < y$, $z < 0$ imply $xz > yz$,
(vi) $1 > 0$ and $-1 < 0$,
(vii) $z > 0$ implies $1/z > 0$,
(viii) $0 < x < y$ implies $0 < 1/y < 1/x$.

Proof   (i) Since $y - x \in P$ and $z - y \in P$, their sum $z - x$ is also in P by Axiom VII; i.e., $x < z$. According to Axiom VI the number $y - x$ is in exactly one of the sets P, $\{0\}$, $-P$, and so (ii) follows. We leave the easy proofs of (iii), (iv), and (v) to the reader. To prove (vi) we need only show that $1 \in P$. Assume, to the contrary, that $1 \notin P$. Since $1 \notin \{0\}$, it follows that $1 \in -P$ and $-1 \in P$. Thus

$$1 = (-1)(-1) \in P$$

and the proof is complete. To prove (vii) assume that $z > 0$ and $1/z < 0$. Then (iv) gives $1 = (1/z)z < 0z = 0$, which contradicts (vi). Since $(1/x) - (1/y) = (y - x)[1/(xy)] \in P$, which is the product of two positive numbers, (viii) holds. $\square$

The reader can easily formulate and prove theorems similar to (1.4) in which some occurrences of $<$ are replaced by $\leq$. In particular, we have

(1.5)   **Theorem**   *If $a, b, c, d \in \mathbb{R}$ and $a \leq b$ and $c \leq d$, then $a + c \leq b + d$.*

Proof   Since $b - a$ and $d - c$ are both in $P \cup \{0\}$, their sum, $(b + d) - (a + c)$, is also in $P \cup \{0\}$. $\square$

Next we define the absolute value function on $\mathbb{R}$.

---

*We write $y - x$ to denote $y + (-x)$.

(1.6)   **Definition**   For each $x \in \mathbb{R}$ we define the *absolute value* of x to be the number $|x|$ given by

$$|x| = x \qquad \text{if } x \geqq 0,$$
$$|x| = -x \quad \text{if } x \leqq 0.$$

(1.7)   **Lemma**   *If $a, b \in \mathbb{R}$ and $b \geqq 0$, then $|a| \leqq b$ if and only if $-b \leqq a \leqq b$.*

Proof   This is obvious upon consideration of the two cases $a \geqq 0$ and $a < 0$. $\square$

(1.8)   **Theorem**   *For all real numbers x and y we have*
(i) $|xy| = |x| \, |y|$,
(ii) $|x + y| \leqq |x| + |y|$,
(iii) $\big||x| - |y|\big| \leqq |x - y|$.

Proof   We omit the easy proof of (i). To prove (ii), use (1.7) to write

$$-|x| \leqq x \leqq |x| \tag{1}$$

and

$$-|y| \leqq y \leqq |y|. \tag{2}$$

Using (1.5) to add (1) and (2), we obtain

$$-(|x| + |y|) \leqq x + y \leqq |x| + |y|$$

and then another application of (1.7) yields (ii).
        To get (iii) we use (ii) to write

$$|x| = |(x - y) + y| \leqq |x - y| + |y|,$$

from which it follows that

$$|x| - |y| \leqq |x - y|. \tag{3}$$

Interchanging the roles of x and y in (3) and using (i) gives

$$|y| - |x| \leqq |y - x| = |x - y|,$$

and so, multiplying by -1, we get

$$-|x - y| \leqq |x| - |y|. \tag{4}$$

Now (3), (4), and (1.7) imply (iii). $\square$

        Our first seven axioms define the notion of an *ordered field*. Thus (1.1)–(1.8) hold good in any ordered field. There are many examples of ordered fields that can be constructed, e.g., the rational numbers, the real algebraic numbers, and the field of rational functions over an ordered field. The thing that distinguishes the real field $\mathbb{R}$ from all other ordered fields is the following [completeness] axiom.

Axiom VIII [Dedekind completeness]   Let A and B be subsets of $\mathbb{R}$ such that
   (i) $A \neq \varnothing$ and $B \neq \varnothing$,
   (ii) $A \cup B = \mathbb{R}$,
   (iii) $a \in A$, $b \in B$ imply $a < b$.
Then there exists exactly one $x \in \mathbb{R}$ such that
   (iv) $u \in \mathbb{R}$, $u < x$ imply $u \in A$ and
   (v) $v \in \mathbb{R}$, $x < v$ imply $v \in B$.
[Plainly the number x must be in either A or B but not both. Thus either
$A = \{u \in \mathbb{R} : u \leqq x\}$, $B = \mathbb{R} \backslash A$ or $A = \{u \in \mathbb{R} : u < x\}$, $B = \mathbb{R} \backslash A$.]

Assumption   We now assume the existence of an object $\mathbb{R}$ that satisfies the above eight axioms. The reader who feels any qualms about this assumption may prove the existence [and uniqueness] of $\mathbb{R}$ from more primitive axioms, as mentioned in the introduction to this chapter.

The Supremum Principle

We next deduce a fundamental principle which will be used frequently in the sequel. First we need some definitions.

(1.9)   Definitions   Let E be a nonvoid set of real numbers. A number $b \in \mathbb{R}$ is called an *upper* [resp. *lower*] *bound* for E if $x \leqq b$ [resp. $b \leqq x$] for all $x \in E$. In case such a b exists, we say that E is *bounded above* [resp. *below*]. If E has both an upper bound and a lower bound, then we say that E is *bounded*. By a *supremum* [or *least upper bound*] of E we mean a number $s \in \mathbb{R}$ such that
   (i) s is an upper bound for E and
   (ii) if b is an upper bound for E, then $s \leqq b$.
[It is clear that E can have at most one supremum.] If E has a supremum s, we write

$$s = \sup E.$$

The notation $s = \text{lub}\, E$ is used in some books.
      The *infimum* [or *greatest lower bound*] of E is defined similarly. It is a lower bound for E which is larger than every other lower bound for E.

(1.10)   Examples   (a) Let $E = \{x \in \mathbb{R} : x \geqq 1\}$. Then any number $\leqq 1$ is a lower bound for E and $1 = \inf E$, but E has no upper bound.
      (b) Let $E = \{x \in \mathbb{R} : 0 < x \leqq 1\}$. Then $\sup E = 1 \in E$ while $\inf E = 0 \notin E$.

(1.11)   Supremum Principle   *Every nonvoid set of real numbers that is bounded above has a supremum in* $\mathbb{R}$.

      Proof   Let $\varnothing \neq E \subset \mathbb{R}$ be such that E has an upper bound. Let B be the set of all upper bounds for E and let $A = \mathbb{R} \backslash B$. Then $B \neq \varnothing$ and if $x \in E$, then $x - 1 \in A$ so $A \neq \emptyset$. Moreoever, if $a \in A$ and $b \in B$, then a is not an upper bound for E so there exists $x \in E$ such that $a < x \leq b$ and so $a < b$. It follows from Axiom VIII that there exists exactly one real number s such that

$$a \leq s \quad \text{for all } a \in A \tag{1}$$

and

$$s \leq b \quad \text{for all } b \in B. \tag{2}$$

It is impossible that $s \in A$, since if it were there would exist $x \in E$ with $s < x$. But then, writing $a = (s + x)/2$, we would have $a \in A$ [because $a < x \in E$] and $s < a$, contrary to (1). Therefore, s is an upper bound for E. Combining this fact with (2) shows that $s = \sup E$. □

(1.12)   Corollary   *Every nonvoid set of real numbers that is bounded below has an infimum in* $\mathbb{R}$.

Proof   Let F be such a set and let E be the set of all lower bounds for F. Then $E \neq \emptyset$ and each element of F is an upper bound for E. By (1.11), $s = \sup E$ exists in $\mathbb{R}$. We claim that $s = \inf F$. It is clear that $s \geq$ each lower bound for F; i.e., $s \geq x$ for all $x \in E$. Thus, we need only show that s is a lower bound for F. Assume that this is false. Then there exists $y \in F$ such that $y < s$. Then y is not an upper bound for E, so there exists $x \in E$ such that $y < x$. This last inequality contradicts the definition of E, and so the proof is complete. □

The following theorem gives a slightly different characterization of the supremum. We leave its proof to the reader.

(1.13)   Theorem   *Let* $\emptyset \neq E \subset \mathbb{R}$ *and let* $s \in \mathbb{R}$. *Then* $s = \sup E$ *if and only if*
(i) $x \leq s$ *for all* $x \in E$, *and*
(ii) *for each* $\epsilon > 0$, $\epsilon \in \mathbb{R}$, *there exists* $x \in E$ *such that* $x > s - \epsilon$.

The Natural Numbers

The presentation of the natural numbers in this section may seem a little strange to some readers since, in fact, the natural numbers are the numbers that are most familiar to everyone. However, having described the real numbers axiomatically, it is necessary for us to define and to study properties of the natural numbers as a subset of $\mathbb{R}$. We base our definition of the natural numbers upon induction. Roughly speaking, a natural number is any real number that can be expressed as the sum of a bunch of 1's. We must be more precise.

(1.14)   Definition   A set $I \subset \mathbb{R}$ is called an *inductive set* if (i) $1 \in I$ and (ii) $x \in I$ implies $x + 1 \in I$.

Notice that $\mathbb{R}$ is an inductive set and so is $\{t \in \mathbb{R} : t \geqq 1\}$. Let $\mathcal{G}$ denote the family of all inductive subsets of $\mathbb{R}$ and let $\mathbb{N} = \cap \mathcal{G}$; i.e., $\mathbb{N} = \{x \in \mathbb{R} : x \in I$ for every inductive set $I \subset \mathbb{R}\}$. The elements of $\mathbb{N}$ are called the *natural numbers* [or the *positive integers*]. Notice that if I is any inductive set $\subset \mathbb{R}$, then $\mathbb{N} \subset I$. Note also that $1 \in \mathbb{N}$ and if $x \in \mathbb{N}$, then $x \in I$ for all $I \in \mathcal{G}$ so, by (ii), $x + 1 \in I$ for all $I \in \mathcal{G}$ and hence $x + 1 \in \mathbb{N}$. Thus $\mathbb{N}$ is an inductive set.

We now officially define a few symbols that denote specific natural numbers:

$$2 = 1 + 1, \quad 3 = 2 + 1, \quad 4 = 3 + 1,$$
$$5 = 4 + 1, \quad 6 = 5 + 1, \quad 7 = 6 + 1,$$
$$8 = 7 + 1, \quad 9 = 8 + 1, \quad 10 = 9 + 1.$$

Decimal representation of other natural numbers is discussed later in this chapter.

Note for example that $3 + 2 = 3 + (1 + 1) = (3 + 1) + 1 = 4 + 1 = 5$. We do not choose to dwell on such simple facts.

(1.15) Finite Induction Principle   *Let $S \subset \mathbb{N}$ be such that*
(i) $1 \in S$ *and*
(ii) $x \in S$ *implies* $x + 1 \in S$.
Then $S = \mathbb{N}$

> **Proof**   The hypotheses say that S is an inductive set. Thus, $\mathbb{N} \subset S$. Since $S \subset \mathbb{N}$, we have $S = \mathbb{N}$. □

(1.16) Archimedean Order Property   *If $a, b \in \mathbb{R}$ and $a > 0$, then there exists $n \in \mathbb{N}$ such that $na > b$. In particular, $\mathbb{N}$ is not bounded above in $\mathbb{R}$.*

> **Proof**   Assume that the theorem is false for some $a, b \in \mathbb{R}$, $a > 0$. Then the set $E = \{na : n \in \mathbb{N}\}$ is bounded above by b, so, by the supremum principle, it has a supremum $s \in \mathbb{R}$. Choose $n \in \mathbb{N}$ such that $na > s - a$. Then $(n + 1)a \in E$ and $(n + 1)a > s$, contrary to the choice of s. To see that $\mathbb{N}$ is unbounded take $a = 1$. □

The next theorem exhibits a few more important properties of $\mathbb{N}$.

(1.17) Theorem   *For each $n \in \mathbb{N}$ we have*
(i) $1 \leqq n$,
(ii) $n > 1$ *implies* $(n - 1) \in \mathbb{N}$,
(iii) $x \in \mathbb{R}$, $x > 0$, $x + n \in \mathbb{N}$ *imply* $x \in \mathbb{N}$,
(iv) $m \in \mathbb{N}$, $m > n$ *imply* $(m - n) \in \mathbb{N}$,
(v) $a \in \mathbb{R}$, $n - 1 < a < n$ *imply* $a \notin \mathbb{N}$.

Proof Inequality (i) follows from the fact that $\{x \in \mathbb{R} : x \geqq 1\}$ is an inductive set and hence $\mathbb{N}$ is a subset of it.

To prove (ii) we use induction (1.15). Let $S = \{1\} \cup \{n \in \mathbb{N} : n - 1 \in \mathbb{N}\}$. It suffices to show that $S = \mathbb{N}$. Plainly, $1 \in S$. Suppose that $n \in S$. Then $(n + 1) - 1 = n \in S \subset \mathbb{N}$ and so $(n + 1) \in S$. By induction, $S = \mathbb{N}$.

To prove (iii) we let $T = \{n \in \mathbb{N} : \text{(iii) holds}\}$ and use induction to show that $T = \mathbb{N}$. To see that $1 \in T$, use (ii) with n replaced by $x + 1$. Next let $n \in T$. We must show that $(n + 1) \in T$, so let $x > 0$ be such that $x + (n + 1) \in \mathbb{N}$. Since $x + 1 > 0$ and $(x + 1) + n \in \mathbb{N}$, the fact that $n \in T$ shows that $(x + 1) \in \mathbb{N}$. Using (ii) we see that $x \in \mathbb{N}$. This proves that $(n + 1) \in T$, and so $T = \mathbb{N}$.

Letting $x = m - n$, we see that (iv) follows from (iii).

If (v) were false, we would have $a \in \mathbb{N}$, $n < a + 1$, and $(a + 1) - n < 1$. Combining these statements with (iv) and (i) yields a contradiction, so (v) is established. $\square$

(1.18) Well-Ordering Property *If A is a nonvoid subset of $\mathbb{N}$, then A has a smallest element; i.e., $\inf A \in A$.*

Proof Assume that A has no smallest element. Let $S = \{n \in \mathbb{N} : n < a \text{ for all } a \in A\}$. Surely $1 \in S$ for otherwise (1.17.i) shows that 1 is the smallest element of A. Now suppose that $n \in S$. If we assume that $n + 1 \notin S$, there is an $a \in A$ such that $a \leqq n + 1$. Since there is no natural number between n and $n + 1$ (1.17.v), we see that $a = n + 1$ and a is the smallest member of A. This contradiction shows that $n + 1 \in S$. Thus, by induction, $S = \mathbb{N}$. But $A \neq \varnothing$ and so there exists $a \in A \subset \mathbb{N} = S$. Then $a < a$. This contradiction establishes the theorem. $\square$

Again, using induction, we now show that $\mathbb{N}$ is closed under the operations of addition and multiplication.

(1.19) Theorem *If $m, n \in \mathbb{N}$, then $(m + n) \in \mathbb{N}$ and $mn \in \mathbb{N}$.*

Proof Fix $m \in \mathbb{N}$ and let

$$S = \{n \in \mathbb{N} : (m + n) \in \mathbb{N}\}$$

and

$$T = \{n \in \mathbb{N} : mn \in \mathbb{N}\}.$$

We need only show that $S = \mathbb{N} = T$. Plainly $1 \in S$ and $1 \in T$. If $n \in S$, then $m + (n + 1) = (m + n) + 1 \in \mathbb{N}$ and so $(n + 1) \in S$. Thus $S = \mathbb{N}$. If $n \in T$, then $m(n + 1) = mn + m \in \mathbb{N}$ [since $\mathbb{N}$ is closed under addition] and so $(n + 1) \in T$. Therefore $T = \mathbb{N}$. $\square$

Integers

(1.20) Definition A real number x is called an *integer* if $x = 0$ or $x \in \mathbb{N}$ or $-x \in \mathbb{N}$. The set of all integers is denoted by $\mathbb{Z}$.

The next theorem follows easily from the previous results of this chapter, and we leave its proof as an exercise for the reader.

(1.21) Theorem *If $m, n \in \mathbb{Z}$, then $-n \in \mathbb{Z}$, $m + n \in \mathbb{Z}$, and $mn \in \mathbb{Z}$.*

Next we use the Archimedean and well ordering properties to prove the existence of the *greatest integer function* on $\mathbb{R}$.

(1.22) Theorem *Let $x \in \mathbb{R}$. Then there exists a unique integer n such that*

$$n \leqq x < n + 1, \qquad x - 1 < n \leqq x.$$

This integer n is denoted $[x]$. It is called the *integral part of x*.

Proof The uniqueness of n follows from (1.17) because if m and n are integers and $n < m \leqq x < n + 1$, then $0 < m - n < 1$ and $m - n \in \mathbb{N}$—a contradiction.

Let a be the smallest element of $\mathbb{N}$ that is greater than $|x|$. In case $x \geqq 0$, take $n = a - 1$. In case $x < 0$, take $n = -a + 1$ or $-a$ according as x is an integer or not. □

We now define integral powers of real numbers. We remark that (1.23) and (1.24) apply to any field, thus in particular to the complex numbers, which we define later in this chapter.

(1.23) Definition Let $x \in \mathbb{R}$. Define $x^0 = 1$, $x^1 = x$, and $x^{n+1} = x^n x$ for $n \in \mathbb{N}$. If $x \neq 0$ and $n \in \mathbb{N}$, define $x^{-n} = 1/x^n$.

(1.24) Laws of Exponents *Let $x, y \in \mathbb{R} \backslash \{0\}$ and $m, n \in \mathbb{Z}$. Then*
  (i) $x^m x^n = x^{m+n}$,
  (ii) $x^n y^n = (xy)^n$,
  (iii) $(x^m)^n = x^{mn}$.
If $n > 0$, $x > 0$, $y > 0$, then
  (iv) $x < y$ *if and only if* $x^n < y^n$.

Proof All four conclusions can be proved by fixing x, y, and m and doing inductions on n, considering the cases $n \geqq 0$ and $n < 0$ separately. We leave the details as exercises for the reader. □

Decimal Representation of Natural Numbers

Certainly, all readers are familiar with the usual decimal representation of the natural numbers using only the integers 0 through 9. For example, the symbol 257 is a short way to denote the number $2 \cdot 10^2 + 5 \cdot 10 + 7$. Our next goal is to prove that all natural numbers can be so represented (where the base need not be 10). First we need the following important fact.

(1.25)   Division Algorithm   *If $n, b \in \mathbb{Z}$ with $b > 0$, then there exist unique integers q and r such that*
   (i) *$n = bq + r$ and*
   (ii) *$0 \leqq r < b$.*

   Proof   If (i) and (ii) hold, then $0 \leqq r/b = (n/b) - q < 1$ and so $q \leqq n/b < q + 1$. By (1.22),

$$q = [n/b], \qquad r = n - bq. \tag{1}$$

   This proves the uniqueness of q and r. If, on the other hand, q and r are defined by (1), then (i) and (ii) follow. $\square$

   We next define the notions of a finite sequence and of the sum and product of a finite sequence. The second definition applies equally well to any field.

(1.26)   Definition   Let X be a set and let $a, b \in \mathbb{Z}$ with $a \leqq b$. To say that $(x_j)_{j=a}^b$ is a *finite sequence in X* means that for each $j \in \{j \in \mathbb{Z} : a \leqq j \leqq b\}$ x_j is a single well defined element of X. It may be that $x_j = x_k$ even though $j \neq k$. Another way to say this is that $(x_j)_{j=a}^b$ is a function from $\{j : j \in \mathbb{Z}, a \leqq j \leqq b\}$ into X whose value at j is x_j. For short we write

$$(x_j)_{j=a}^b \subset X.$$

The element x_j is called the jth *term* of the sequence.
   For example, $(j^2 + 1)_{j=0}^3$ is a finite sequence in $\mathbb{N}$. Its terms in order, are

$$1, 2, 5, 10.$$

(1.27)   Definition   Let $(x_j)_{j=a}^b \subset \mathbb{R}$. We define, for $k \in \mathbb{Z}$, $a < k \leqq b$,

$$\sum_{j=a}^a x_j = x_a, \qquad \sum_{j=a}^k x_j = \sum_{j=a}^{k-1} x_j + x_k.$$

The number $\displaystyle\sum_{j=a}^b x_j$ is called the *sum* of the sequence $(x_j)_{j=a}^b$. This sum is also frequently written

$$x_a + x_{a+1} + \ldots + x_b.$$

It is just the sum of all of the terms of the sequence. The capital Greek sigma that appears here is called the *summation symbol*.

For example, the sum of the sequence $(j^2 + 1)_{j=0}^3$ is

$$1 + 2 + 5 + 10 = 8 + 10 = 18.*$$

Similarly, we define

$$\prod_{j=a}^{a} x_j = x_a, \qquad \prod_{j=a}^{k} x_j = \left(\prod_{j=a}^{k-1} x_j \right) x_k.$$

The number $\displaystyle\prod_{j=a}^{b} x_j$ is called the *product* of $(x_j)_{j=a}^b$. This product can also be written

$$x_a x_{a+1} \cdots x_b.$$

The capital Greek pi used here is the *product symbol*.

We shall also need the following theorem, valid in any field.

(1.28)   Geometric Progressions   *Let* $b \in \mathbb{R}$, $b \neq 1$, *and* $p \in \mathbb{N}$. *Then*

(i)
$$\sum_{j=0}^{p} b^j = \frac{b^{p+1} - 1}{b - 1}.$$

Proof   This is an easy induction on p and is left to the reader. $\square$

We now present the main result of this section.

(1.29)   Theorem   *Let* $b \in \mathbb{N}$ *with* $b > 1$. *Then to each* $n \in \mathbb{N}$ *there corresponds a unique integer* $p \geqq 0$ *and a unique finite sequence*

$$(r_j)_{j=0}^{p} \subseteq \{ k : k \in \mathbb{Z}, 0 \leqq k < b \}$$

with $r_p \neq 0$ *such that*

(i)
$$n = \sum_{j=0}^{p} r_j b^j.$$

The symbol $r_p r_{p-1} \cdots r_1 r_0$, obtained by juxtaposing the terms of the sequence, in reverse order, is called the *b-adic expansion* of n and we write

(ii)
$$n = r_p r_{p-1} \cdots r_1 r_0.$$

Of course, the right side of (ii) should *not* be regarded as a product. Also, it is obviously important to understand what the base b is in using such expansions.

---

*We have not yet defined the symbol 18. That is partially the point of Theorem (1.29) where we see that it is a definition that $8 + 10 = 18$.

We shall always use decimal expansions [10-adic, $b = 10$] unless the contrary is specifically stated.

Proof First we prove uniqueness. Suppose that n has two expansions of the form (i):

$$\sum_{j=0}^{p} r_j b^j = n = \sum_{j=0}^{q} s_j b^j.$$

Assume that $p \neq q$: say $p < q$. In view of (1.28) and the conditions on r_j and s_j we obtain

$$n = \sum_{j=0}^{p} r_j b^j \leq (b - 1) \sum_{j=0}^{p} b^j$$

$$= b^{p+1} - 1 < b^q \leq \sum_{j=0}^{q} s_j b^j = n$$

and so $n < n$. This contradiction proves that $p = q$. Next assume that $r_j \neq s_j$ for some j and let a be the largest such j: say $r_a < s_a$. Then, subtracting the two expressions for n, we obtain

$$0 = \sum_{j=0}^{a} (r_j - s_j) b^j \leq \sum_{j=0}^{a-1} (b - 1) b^j - b^{a*} = (b^a - 1) - b^a = -1.$$

This contradiction proves that $r_j = s_j$ for all j.

Now we prove the existence part of the theorem. Using the division algorithm, we have

$$n = r_0 + q_0 b \tag{1}$$

$$\vdots$$

$$q_{j-1} = r_j + q_j b \tag{2}$$

for each $j \in \mathbb{N}$, where q_j and r_j are integers and $0 \leq r_j < b$. Since $n > 0$ and $b > 1$, one can check by induction that $q_j \geq 0$ for each j. Also $q_{j-1} \geq q_j b \geq 2q_j > q_j$ unless $q_j = 0$. Since the set $\{q_j : j \in \mathbb{N} \cup \{0\}\}$ has a smallest member, it follows that $q_j = 0$ for some j. Let p be the smallest element of $\mathbb{N} \cup \{0\}$ such that $q_p = 0$. If $p = 0$, then (1) shows that $n = r_0$ and the proof is complete. Suppose that $p > 0$. Then (2) shows that $r_p = q_{p-1} \neq 0$ [p is minimal]. Using (1) and (2) and induction, one checks that

$$n = r_0 + r_1 b + \ldots + r_j b^j + q_j b^{j+1} \tag{3}$$

for all $j \in \mathbb{N}$. Take $j = p$ in (3) to obtain (i). $\square$

---

*Any sum of the form $\sum_{j=a}^{b} x_j$ where $a > b$ is defined to be 0.

Roots

In this section we use the Supremum Principle to prove the existence of nth roots of nonnegative real numbers. First, we prove a lemma.

(1.30) Lemma   *If $n \in \mathbb{N}$ and $0 < \epsilon < 1$, then*
(i) $(1 + \epsilon)^n < 1 + 3^n\epsilon$.

Proof   Let ϵ be arbitrary, but fixed, with $0 < \epsilon < 1$ and let $S = \{n \in \mathbb{N} : (i)$ obtains$\}$. Since $1 + \epsilon < 1 + 3\epsilon$, we have $1 \in S$. Now suppose that $n \in S$. Then

$$(1 + \epsilon)^{n+1} = (1 + \epsilon)^n(1 + \epsilon)$$
$$< (1 + 3^n\epsilon)(1 + \epsilon)$$
$$= 1 + (3^n\epsilon + 3^n + 1)\epsilon$$
$$< 1 + (3^n + 3^n + 3^n)\epsilon = 1 + 3^{n+1}\epsilon$$

and so $(n + 1) \in S$. By induction $S = \mathbb{N}$. $\square$

(1.31)   Theorem   *Let $a \geq 0$ be a real number and let $n \in \mathbb{N}$. Then there exists a unique real number x such that $x \geq 0$ and $x^n = a$.*
This number x is called the (nonnegative) nth *root* of a and we denote it by $\sqrt[n]{a}$ or by $a^{1/n}$. In case $n = 2$, we write $\sqrt{a}$ instead of $\sqrt[2]{a}$.
   If $a \geq 0$ and $b \geq 0$, then
(i) $\sqrt[n]{ab} = \sqrt[n]{a}\,\sqrt[n]{b}$.

Proof   The uniqueness is clear since if $0 \leq x < y$, then $x^n < y^n$. If $a = 0$, take $x = 0$. Thus suppose that $a > 0$ and let
$$E = \{t \in \mathbb{R} : t > 0, t^n < a\}.$$
To see that $E \neq \varnothing$ let $t_0 = a/(1 + a)$. Then $0 < t_0 < 1$ and so, by induction on n,
$$t_0^n \leq t_0 < a.$$
Also E is bounded above by $1 + a$ because, for $t \in E$, if $t \leq 1$, then $t < 1 + a$ while if $t > 1$, then $t \leq t^n < a < 1 + a$. By the Supremum Principle, E has a supremum $x \in \mathbb{R}$. Plainly $x \geq t_0 > 0$. We assert that $x^n = a$.
    Assume that $x^n < a$. Choose $\epsilon \in \mathbb{R}$ so that $0 < \epsilon < 1$ and $\epsilon < (a - x^n)/(3x)^n$. Then, using (1.30), we obtain
$$x^n(1 + \epsilon)^n < x^n(1 + 3^n\epsilon)$$
$$= x^n + (3x)^n\epsilon < x^n + (a - x^n) = a$$
and so $x < x(1 + \epsilon) \in E$. But $x = \sup E$, and so this contradiction shows

that

$$x^n \geqq a. \tag{1}$$

Assume that $x^n > a$. Now select any $\epsilon \in \mathbb{R}$ such that $0 < \epsilon < 1$ and $\epsilon < (x^n - a)/(3^n a)$. Then $a(1 + 3^n\epsilon) < x^n$ and so, using (1.30), we have

$$a < x^n/(1 + 3^n\epsilon) < \left[x/(1 + \epsilon)\right]^n. \tag{2}$$

Since $x/(1 + \epsilon) < x$ there exists $t \in E$ such that $x/(1 + \epsilon) < t$ and hence

$$\left[x/(1 + \epsilon)\right]^n < t^n < a. \tag{3}$$

Now (2) and (3) are contradictory, and so we conclude, using (1), that $x^n = a$.

To prove (i) simply note that

$$\left(\sqrt[n]{a}\,\sqrt[n]{b}\,\right)^n = \left(\sqrt[n]{a}\,\right)^n\left(\sqrt[n]{b}\,\right)^n = ab$$

so that $\sqrt[n]{a}\,\sqrt[n]{b}$ is the unique nonnegative nth root for ab. $\quad\square$

Rational and Irrational Numbers

(1.32) **Definition** A *rational number* is any real number x that can be expressed in the form $x = a/b$, where a and b are integers and $b \neq 0$. The set of all rational numbers is denoted by $\mathbb{Q}$. The numbers in $\mathbb{R}\backslash\mathbb{Q}$ are called *irrational numbers*.

The reader can easily deduce the following theorem from the properties of the integers.

(1.33) **Theorem** *If x and y are rational numbers, then so are $-x$, $x + y$, xy, and [for $x \neq 0$] x^{-1}. Thus $\mathbb{Q}$ is a subfield of $\mathbb{R}$: it satisfies Axioms I–V.*

The next theorem shows that the rationals are "order dense" in $\mathbb{R}$. Unlike the integers, no rational has a nearest neighbor among the rationals.

(1.34) **Theorem** *If $x, y \in \mathbb{R}$ with $x < y$, then there exists $z \in \mathbb{Q}$ such that $x < z < y$.*

Proof Use the Archimedean property to obtain $b \in \mathbb{N}$ such that $b > (y - x)^{-1}$. Then $b^{-1} < y - x$. Next let $a = [bx] + 1 \in \mathbb{Z}$. Then $a - 1 \leqq bx < a$ and therefore $x < a/b \leqq x + 1/b < y$. Let $z = a/b$. $\quad\square$

Similarly, the irrational numbers are "order dense" in $\mathbb{R}$.

(1.35) **Theorem** *If $x, y \in \mathbb{R}$ with $x < y$, then there exists an irrational number z such that $x < z < y$.*

Proof   According to (1.31) there exists a positive real number s such that $s^2 = 2$. By the footnote on p. 7, we see that s is irrational. Now use (1.34) to get a rational u such that $u \neq 0$ and $x/s < u < y/s$. Let $z = us$. Then $x < z < y$. Also z is irrational since otherwise $s = z/u$ would be rational. $\square$

Notice that the above proof requires only the existence of some positive irrational number s. In the next chapter we give a more detailed proof that such numbers exist.

Complex Numbers

In this section we introduce addition and multiplication into the plane $\mathbb{R} \times \mathbb{R}$ in such a way as to make it a field. This field is called the *complex number field*.

(1.36)   **Definition**   A *complex number* is an ordered pair (a, b), where a and b are real numbers. The set of all complex numbers is denoted by $\mathbb{C}$. For $z = (a, b)$ and $w = (c, d)$ in $\mathbb{C}$, write $z = w$ if and only if $a = c$ and $b = d$. Also define

$$\theta = (0, 0), \qquad u = (1, 0),$$

$$z + w = (a + c, b + d),$$

$$zw = (ac - bd, ad + bc),$$

$$-z = (-a, -b),$$

and, in case $z \neq \theta$,

$$z^{-1} = \left(a/(a^2 + b^2), -b/(a^2 + b^2) \right).$$

(1.37)   **Theorem**   *With $\mathbb{R}$ replaced by $\mathbb{C}$, 0 replaced by θ, and 1 replaced by u, Axioms* I–V *are satisfied. That is, $\mathbb{C}$ is a field.*

Proof   The required verifications are all routine calculations. We prove only that multiplication is associative, leaving the rest as exercises for the reader.
Let $x = (a, b)$, $y = (c, d)$, and $z = (e, f)$ be in $\mathbb{C}$. Then

$$x(yz) = (a, b)(ce - df, cf + de)$$

$$= (ace - adf - bcf - bde, acf + ade + bce - bdf)$$

$$= (ac - bd, ad + bc)(e, f)$$

$$= (xy)z. \quad \square$$

(1.38) **Theorem** *With the replacements mentioned in* (1.37), *Theorems* (1.1) *and* (1.2) *remain true.*

Proof Use (1.37) and the proofs given for (1.1) and (1.2). ☐

(1.39) **Convention** Notice that if $a, b \in \mathbb{R}$, then $(a, 0) + (b, 0) = (a + b, 0)$, $(a, 0)$ $(b, 0) = (ab, 0)$, $-(a, 0) = (-a, 0)$, and, if $a \neq 0$, $(a, 0)^{-1} = (a^{-1}, 0)$. Thus the complex numbers of the form $(a, 0)$ have the same arithmetic properties as the real numbers a. It is customary to call such complex numbers real and to write $(a, 0) = a$ when $a \in \mathbb{R}$. We hereby adopt this convention. Now define i to be the complex number $(0, 1)$. Then, for $b \in \mathbb{R}$ we have

$$bi = (b, 0)(0, 1) = (0, b).$$

Complex numbers of this form are called *purely imaginary*. Now $a, b \in \mathbb{R}$ implies

$$a + bi = (a, 0) + (0, b) = (a, b).$$

Thus each complex number has the form $a + bi$ where $a, b \in \mathbb{R}$. Note also that

$$i^2 = (0, 1)(0, 1) = (-1, 0) = -1.$$

(1.40) **Definitions** If $z = a + bi$ $(a, b \in \mathbb{R})$ is a complex number, define the *real part* of z and the *imaginary part* of z to be the real numbers

$$\mathrm{Re}(z) = a \quad \text{and} \quad \mathrm{Im}(z) = b,$$

respectively. Define the *complex conjugate* of z to be the complex number

$$\bar{z} = a - bi$$

and define the *absolute value* [or *modulus*] of z to be the real number

$$|z| = \sqrt{a^2 + b^2} .$$

(1.41) **Theorem** *For $z, w \in \mathbb{C}$ we have*
   (i) $\mathrm{Re}(z) = (z + \bar{z})/2$,
   (ii) $\mathrm{Im}(z) = (z - \bar{z})/(2i)$,
   (iii) $\overline{z + w} = \bar{z} + \bar{w}$,
   (iv) $\overline{zw} = \bar{z}\bar{w}$,
   (v) $\overline{(1/z)} = 1/\bar{z}$ *if $z \neq 0$*,
   (vi) $z\bar{z} = |z|^2$,
   (vii) $z = \bar{z}$ *if and only if $z \in \mathbb{R}$*,
   (viii) $|zw| = |z||w|$,
   (ix) $|\mathrm{Re}(z)| \leq |z|$, $|\mathrm{Im}(z)| \leq |z|$.

Proof These are simple exercises. ☐

Some Inequalities

One of the chief assets of a good analyst is an ability to estimate the relative sizes of various quantities. In this section we prove a few important inequalities that are frequently useful, or even essential, in making such estimates.

(1.42)   **Triangle Inequalities**   *If $z, w \in \mathbb{C}$, then*
(i) $|z + w| \leq |z| + |w|$ *and*
(ii) $| \, |z| - |w| \, | \leq |z - w|$.

Proof   To prove (i) we use (1.41) to write

$$|z + w|^2 = (z + w)(\bar{z} + \bar{w}) = |z|^2 + z\bar{w} + \overline{z\bar{w}} + |w|^2$$

$$= |z|^2 + 2\,\mathrm{Re}(z\bar{w}) + |w|^2$$

$$\leq |z|^2 + 2|z\bar{w}| + |w|^2$$

$$= (|z| + |w|)^2$$

and then take nonnegative square roots. Finally, we deduce (ii) from (i) in the same way that (1.8.iii) was deduced from (1.8.ii).   □

(1.43)   **Remark**   If we coordinatize the plane with the customary rectangular coordinates of analytic geometry, then the reason for the name "triangle inequality" for (1.42.i) becomes clear. It says that the length of any side of a triangle does not exceed the sum of the lengths of the other two sides. A similar geometrical interpretation can be made for (1.42.ii).

(1.44)   **Cauchy's Inequality***   *If $a_1, \ldots, a_n$ and $b_1, \ldots, b_n$ are complex numbers,* then

(i) $\left| \sum_{j=1}^{n} a_j \bar{b}_j \right|^2 \leq \left(\sum_{j=1}^{n} |a_j|^2 \right)\left(\sum_{j=1}^{n} |b_j|^2 \right).$

Equality obtains in (i) if and only if either $b_j = 0$ for each j or there exists $z \in \mathbb{C}$ such that $a_j = zb_j$ for each j.

Proof   Let $A = \sum_{j=1}^{n} |a_j|^2$, $B = \sum_{j=1}^{n} |b_j|^2$, and $C = \sum_{j=1}^{n} a_j \bar{b}_j$. If $B = 0$, there is nothing to prove, so we suppose that $B > 0$. Using (1.41), we see that, for

---

*This inequality was first proved by the great French analyst A.-L. Cauchy (1789–1857). More general versions of it for integrals were later proved by the Russian, V. Bunyakovskiĭ (1804–1889), and the German, H. A. Schwarz (1843–1921). At present, most mathematicians refer to this inequality and its more general forms as Schwarz's Inequality.

. any complex number z, we have

$$0 \leq \sum_{j=1}^{n} |a_j - zb_j|^2 = \sum_{j=1}^{n} (a_j - zb_j)(\bar{a}_j - \bar{z}\bar{b}_j)$$

$$= A - \bar{z}C - z\bar{C} + |z|^2 B. \tag{1}$$

Letting $z = C/B$ in (1), we obtain $0 \leq A - |C|^2/B - |C|^2/B + |C|^2 B/B^2$ and so

$$|C|^2 \leq AB, \tag{2}$$

which is (i). It is clear that if equality holds in (2), then, for our particular z, equality holds throughout (1) and so $a_j = zb_j$ for all j. If, on the other hand, $a_j = zb_j$ for some z and all j, then it is obvious that equality holds in (i).

$\square$

Using Cauchy's inequality, we are able to deduce the following generalization of (1.42.i).

(1.45) **Minkowski's Inequality*** *If* $a_1, \ldots, a_n$ *and* $b_1, \ldots, b_n$ *are complex numbers, then*

(i) $(\sum_{j=1}^{n} |a_j + b_j|^2)^{1/2} \leq (\sum_{j=1}^{n} |a_j|^2)^{1/2} + (\sum_{j=1}^{n} |b_j|^2)^{1/2}.$

The conditions for equality are the same as those in (1.44) except that $z \in \mathbb{R}$ *and* $z \geq 0$.

Proof Let A, B, C be as in the proof of (1.44). Again suppose $B \neq 0$. Then we have

$$\sum_{j=1}^{n} |a_j + b_j|^2 = A + C + \bar{C} + B$$

$$= A + 2\,\mathrm{Re}\,C + B$$

$$\leq A + 2|C| + B$$

$$\leq A + 2(AB)^{1/2} + B$$

$$= (A^{1/2} + B^{1/2})^2 \tag{1}$$

where the second inequality is Cauchy's Inequality. Taking nonnegative square roots in (1) yields (i).

Suppose that equality obtains in (i). Then equality obtains throughout (1). Since $B \neq 0$, the second inequality in (1) [now an equality] and (1.44) show that $a_j = zb_j$ for some z and all j. Since the first inequality in (1) is now an equality, we see that $|C| = \mathrm{Re}\,C$ so that C is real and $C \geq 0$. But

---

*This is only a special case of the famous inequality by this name.

$C = \sum_{j=1}^{n} a_j \bar{b}_j = z \sum_{j=1}^{n} |b_j|^2 = zB$ and so z is a nonnegative real number. On the other hand, if $a_j = zb_j$ where $z \geq 0$, then both sides of (i) are equal to $(z + 1)B^{1/2}$. $\square$

The next inequality will be useful to us in Chapter 2.

(1.46)   Bernoulli's Inequality   *Let* $x \in \mathbb{R}$ *and let* $n \in \mathbb{N}$. *If* $x > -1$, $x \neq 0$, *and* $n > 1$, *then*

(i) $(1 + x)^n > 1 + nx$.

Proof   We proceed by induction. For $n = 2$ we have

$$(1 + x)^2 = 1 + 2x + x^2 > 1 + 2x$$

and so (i) holds for $n = 2$. Suppose that (i) holds for $n = k \geq 2$. Then, since $(1 + x) > 0$,

$$(1 + x)^{k+1} = (1 + x)(1 + x)^k$$
$$> (1 + x)(1 + kx)$$
$$= 1 + (k + 1)x + kx^2 > 1 + (k + 1)x.$$

and so (i) also holds for $n = k + 1$. $\square$

We next give an inductive proof of another important inequality. Simpler proofs, using logarithms, and related inequalities are taken up as exercises in later chapters. Our induction is done in the following lemma which is a special case of the general inequality.

(1.47)   Lemma   *Let* n *be an integer* > 1 *and let* $b_1, \ldots, b_n$ *be positive real numbers, not all equal, such that the product* $b_1 b_2 \ldots b_n = 1$. *Then*

(i) $b_1 + b_2 + \ldots + b_n > n$.

Proof   If $n = 2$, then, since $b_1 \neq b_2$,

$$0 < \left(\sqrt{b_1} - \sqrt{b_2} \right)^2 = b_1 + b_2 - 2\sqrt{b_1 b_2} = b_1 + b_2 - 2.$$

Thus, the lemma is true if $n = 2$.

Now let k be any integer > 1 such that the lemma is true when $n = k$. We shall prove that the lemma remains true for $n = k + 1$. Thus let $n = k + 1$ in our hypotheses. Since the b_j's do not all have the same value, we may suppose that $b_1 \leq b_j \leq b_{k+1}$ $(j = 1, 2, \ldots, k + 1)$ and $b_1 < b_{k+1}$. Then

$$b_1 < 1 < b_{k+1} \tag{1}$$

because otherwise $b_1 b_2 \ldots b_{k+1} \neq 1$. Since the lemma is true for $n = k$ and

$(b_1 b_{k+1}) b_2 \ldots b_k = 1$ we have

$$b_1 b_{k+1} + b_2 + \ldots + b_k \geqq k \tag{2}$$

where the possible equality results from the fact that the k summands on the left may be all equal and hence all equal to 1. From (1) we obtain

$$(b_{k+1} - 1)(1 - b_1) > 0. \tag{3}$$

Using (2) and (3), we have

$$b_1 + b_2 + \ldots + b_k + b_{k+1} = (b_1 b_{k+1} + b_2 + \ldots + b_k) + 1$$
$$+ (b_{k+1} - 1)(1 - b_1)$$
$$\geqq k + 1 + (b_{k+1} - 1)(1 - b_1)$$
$$> k + 1$$

and so the lemma is true for $n = k + 1$. By induction, the proof is complete. $\square$

(1.48) Geometric Mean–Arithmetic Mean Inequality *Let $a_1, a_2, \ldots, a_n$ be nonnegative real numbers. Then*

(i) $(a_1 a_2 \ldots a_n)^{1/n} \leqq (a_1 + a_2 + \cdots + a_n)/n$,

and equality obtains if and only if $a_1 = a_2 = \ldots = a_n$.

The left side of (i) is called the *geometric mean* and the right side the *arithmetic mean* of the numbers $a_1, a_2, \ldots, a_n$.

Proof The theorem is obvious if $n = 1$ or if $a_j = 0$ for some j or if $a_1 = a_2 = \ldots a_n$. Hence, we suppose that $n > 1$, $a_j > 0$ for all j, and $a_1 < a_n$. Our task then is to prove that

$$(a_1 a_2 \ldots a_n)^{1/n} < (1/n)(a_1 + \ldots + a_n). \tag{1}$$

Let $G = (a_1 a_2 \ldots a_n)^{1/n} > 0$ and write $b_j = a_j/G$ for $j = 1, 2, \ldots, n$. Then $b_1 b_2 \ldots b_n = 1$, $b_j > 0$ for all j, and $b_1 < b_n$. Substituting $a_j = Gb_j$ into (1), we see that (1) is equivalent to inequality (1.47.i). $\square$

Extended Real Numbers

It is convenient for our discussion of limits to adjoin two new objects to $\mathbb{R}$. These objects are denoted ∞ and $-\infty$. They are *not* real (or complex) numbers.

(1.49) Definition Let ∞ and $-\infty$ be two distinct fixed objects neither of which is an element of $\mathbb{R}$. Let $\mathbb{R}^\# = \mathbb{R} \cup \{\infty, -\infty\}$ and call $\mathbb{R}^\#$ the *extended real number system*. Introduce an ordering in $\mathbb{R}^\#$ as follows. For $x, y \in \mathbb{R}$ let $x < y$ have its usual meaning and write $x < \infty$, $x > -\infty$, $-\infty < \infty$. Now the symbols $\leqq, >, \geqq$ have obvious meanings. Also introduce into $\mathbb{R}^\#$ *some* arithmetic operations. For $x, y \in \mathbb{R}$

let $x + y$, $x - y$, and xy have their usual meanings. If $x \in \mathbb{R}$ define

$$x + \infty = \infty + x = x - (-\infty) = \infty,$$
$$x + (-\infty) = -\infty + x = x - \infty = -\infty.$$

If $x > 0$,

$$\infty \cdot x = x \cdot \infty = \infty,$$
$$x \cdot (-\infty) = (-\infty) \cdot x = -\infty.$$

If $x < 0$,

$$\infty \cdot x = x \cdot \infty = -\infty,$$
$$x \cdot (-\infty) = (-\infty) \cdot x = \infty.$$

We also define

$$\infty + \infty = \infty \text{ and } (-\infty) + (-\infty) = -\infty,$$
$$\infty \cdot \infty = \infty,$$
$$\infty \cdot 0 = 0 \cdot \infty = (-\infty) \cdot 0 = 0 \cdot (-\infty) = 0.$$

We *do not* define $\infty + (-\infty)$, $-\infty + \infty$, $\infty \cdot (-\infty)$, $(-\infty) \cdot \infty$, and $(-\infty) \cdot (-\infty)$. Neither do we define division by ∞ or $-\infty$.

The symbol ∞ is called *infinity* and the symbol $-\infty$ is called *minus infinity*.

(1.50)   Definition   Let $E \subset \mathbb{R}^{\#}$. If E is not bounded above by a real number, then we call ∞ the *supremum* of E and write $\sup E = \infty$. If E is not bounded below by a real number, we call $-\infty$ the *infimum* of E and write $\inf E = -\infty$. Notice that $\emptyset$ is bounded both above and below by any real number. We write $\sup \emptyset = -\infty$ and $\inf \emptyset = \infty$. Thus every subset of $\mathbb{R}^{\#}$ has both a supremum and an infimum in $\mathbb{R}^{\#}$.

(1.51)   Definition   For $a \leq b$ in $\mathbb{R}^{\#}$ we define the four *intervals* having *left endpoint a* and *right endpoint b* to be the sets

$$]a,b[= \{x \in \mathbb{R}^{\#} : a < x < b\}$$
$$[a,b[= \{x \in \mathbb{R}^{\#} : a \leq x < b\}$$
$$]a,b] = \{x \in \mathbb{R}^{\#} : a < x \leq b\}$$
$$[a,b] = \{x \in \mathbb{R}^{\#} : a \leq x \leq b\}.$$

We call $]a,b[$ an *open interval* and $[a,b]$ a *closed interval*. Note that if $a = b$, then $[a, b] = \{a\}$ while the other three are void. The *length* of each of these intervals is the number $b - a$.

Finite and Infinite Sets

We begin by defining a term that makes precise the vague notion of "same number of elements."

(1.52) **Definition** Two sets A and B are said to be *equivalent* and we write $A \sim B$ if there exists some one-to-one function from A onto B.

(1.53) **Theorem** *For any sets A, B, and C we have*
(i) $A \sim A$,
(ii) $A \sim B$ *implies* $B \sim A$,
(iii) $A \sim B$ *and* $B \sim C$ *imply* $A \sim C$.
Thus $\sim$ is an equivalence relation on any family of sets.

> **Proof** (i) Define $f(x) = x$ for all $x \in A$ if $A \neq \emptyset$. If $A = \emptyset$, take $f = \emptyset$. In either case f is the required function.
>
> (ii) Suppose $A \sim B$. By (1.52) there exists a one-to-one function f from A onto B. Then f^{-1} is one to one function from B onto A and so $B \sim A$.
>
> (iii) Suppose $A \sim B$ and $B \sim C$. Then there exist one to one functions f and g with dom $f = A$, rng $f = B$, dom $g = B$, and rng $g = C$. Thus the composite function $g \circ f$ is one to one from A onto C and so $A \sim C$. $\qquad\square$

(1.54) **Definitions** A set E is said to be *finite* if either $E = \emptyset$ or else there exists some $n \in \mathbb{N}$ such that $E \sim \{k \in \mathbb{N} : 1 \leq k \leq n\}$. In the latter case, if f is a one to one function from $\{k \in \mathbb{N} : 1 \leq k \leq n\}$ onto E and if $f(k) = x_k$, we can write $E = \{x_k : 1 \leq k \leq n\} = \{x_1, x_2, \ldots, x_n\} = \{x_k\}_{k=1}^n$.

All sets that are not finite are said to be *infinite*. A set E is said to be *denumerable* [or *enumerable*] if $E \sim \mathbb{N}$. If E is denumerable and $f : \mathbb{N} \to E$ is one to one and onto and $x_n = f(n)$, then we write $E = \{x_n : n \in \mathbb{N}\} = \{x_1, x_2, x_3, \ldots\} = \{x_n\}_{n=1}^\infty$.* Such an f is called an *enumeration* of E. A set that is either finite or denumerable is said to be *countable*. Denumerable sets are often called *countably infinite*. Any set that is not countable is called *uncountable*.

The next theorem may seem obvious, but it does require a proof.

(1.55) **Theorem** *Any denumerable set is infinite.*

> **Proof** We need only show that $\mathbb{N}$ is infinite. Assume that $\mathbb{N}$ is finite. Since $\mathbb{N} \neq \emptyset$, there is a smallest $n \in \mathbb{N}$ such that $\mathbb{N} \sim \{k \in \mathbb{N} : 1 \leq k \leq n\}$. Let $f : \{k \in \mathbb{N} : 1 \leq k \leq n\} \to \mathbb{N}$ by one to one and onto. Now define $g : \{k \in \mathbb{N} : 1 \leq k \leq n-1\} \to \mathbb{N}$ by the rule
>
> $$g(k) = \begin{cases} f(k) & \text{if } f(k) < f(n) \\ f(k) - 1 & \text{if } f(k) > f(n). \end{cases}$$

---

*The symbols $\{x_n\}_{n=1}^\infty$ and $\{x_k\}_{k=1}^n$ must not be confused with the symbols $(x_k)_{k=1}^n$ and $(x_n)_{n=1}^\infty$ for sequences. In sequences, repetition is allowed. In enumerations, it is not; i.e., $x_k \neq x_j$ if $k \neq j$.

One checks that g is one to one and onto $\mathbb{N}$. This contradicts the minimality of n. □

(1.56) Example We show that the set $\mathbb{Z}$ of all integers is countable by defining $f : \mathbb{Z} \to \mathbb{N}$ as follows:

$$f(n) = \begin{cases} 2n & \text{if } n > 0 \\ -2n + 1 & \text{if } n \leq 0. \end{cases}$$

It is easy to see that f is one to one and onto $\mathbb{N}$. The function f^{-1} simply enumerates $\mathbb{Z}$ as $\{0, 1, -1, 2, -2, \dots \}$.

To show that $[a, b]$ is uncountable for $a < b$ in $\mathbb{R}$, we use the following important theorem.

(1.57) Nested Interval Principle *Let* $\{ I_n : n \in \mathbb{N} \}$ *be a family of* closed *intervals of* $\mathbb{R}$ *such that*
  (i) $I_{n+1} \subset I_n$ *for each* $n \in \mathbb{N}$,
  (ii) $\epsilon > 0$ *in* $\mathbb{R}$ *implies that the length of* I_n *is* $< \epsilon$ *for some* n.
Then $\cap_{n=1}^{\infty} I_n = \{z\}$ *for some* $z \in \mathbb{R}$.

Proof Write $I_n = [a_n, b_n]$. According to (i), $a_n \leq a_{n+1} \leq b_{n+1} \leq b_n$ for all $n \in \mathbb{R}$. Thus $k \leq n$ in $\mathbb{N}$ implies $a_k \leq a_n \leq b_n$ and $a_n \leq b_n \leq b_k$, and so each element of $A = \{a_n : n \in \mathbb{N}\}$ is $\leq$ each element of $B = \{b_n : n \in \mathbb{N}\}$. In particular,

$$a = \sup A \leq b_n \quad and \quad b = \inf B \geq a_n$$

for every $n \in \mathbb{N}$. Thus, $a \leq b$ and $\{a, b\} \subset I_n$ for all n. Therefore $\cap_{n=1}^{\infty} I_n \neq \varnothing$. If this intersection contains two distinct points $z < w$, then take $\epsilon = w - z$ in (ii) and obtain a contradiction. Accordingly, $\cap_{n=1}^{\infty} I_n = \{a\} = \{b\}$. □

(1.58) Theorem *If* $a < b$ *in* $\mathbb{R}$, *then* $[a, b]$ *is an uncountable set.*

Proof Assume that $[a, b]$ is countable. Since the function f defined by $f(n) = a + (b - a)/n$ is one to one from $\mathbb{N}$ into $[a, b]$, it is plain that $[a, b]$ is infinite.* Thus $[a, b]$ is denumerable. Let $\{x_n\}_{n=1}^{\infty}$ be an enumeration of $[a, b]$. We shall obtain a contradiction by constructing a number $z \in [a, b]$ that does not appear in this list; i.e., $z \neq x_n$ for all n. Divide $[a, b]$ into three equal closed intervals

$$\left[a, a + \frac{b-a}{3} \right], \quad \left[a + \frac{b-a}{3}, b - \frac{b-a}{3} \right], \quad \left[b - \frac{b-a}{3}, b \right]$$

and choose one of them (call it I_1) such that $x_1 \notin I_1$. Once I_n has been

---

*This follows from (1.55) and the fact that any superset of an infinite set is infinite (see Exercise 22).

chosen, divide in into three equal closed intervals and choose one of them (call it I_{n+1}) such that $x_{n+1} \notin I_{n+1}$. By induction, we have defined $\{I_n : n \in \mathbb{N}\}$. This family satisfies the hypotheses of (1.57) [the length of I_n is $(b-a)/3^n < (b-a)/n$ as can be proved by induction]. Therefore $\bigcap_{n=1}^{\infty} I_n = \{z\}$. Plainly $z \neq x_n$ since $z \in I_n$ and $x_n \notin I_n$ and this holds for all $n \in \mathbb{N}$. But $z \in [a, b] = \{x_n\}_{n=1}^{\infty}$—a contradiction. $\square$

We will need a few more facts about countable sets.

(1.59) **Theorem** *Any subset of a countable set is countable.*

Proof Let $E = \{x_n\}_{n=1}^{\infty}$ be a countable set and let $A \subset E$. We have supposed that E is infinite: otherwise, use Exercise 22 to see that A is finite. We may suppose that A is infinite. Let n_1 be the least integer such that $x_{n_1} \in A$. If $n_1, n_2, \ldots, n_k$ have been selected, let n_{k+1} be the least integer such that $x_{n_{k+1}} \in A \setminus \{x_{n_1}, \ldots, x_{n_k}\}$. This defines $n_1 < n_2 < n_3 < \ldots$ and enumerates A as $\{x_{n_k}\}_{k=1}^{\infty}$. $\square$

(1.60) **Theorem** *The set $\mathbb{N} \times \mathbb{N}$ is countable.*

Proof Define $f : \mathbb{N} \times \mathbb{N} \to \mathbb{N}$ by $f(a, b) = 2^{a-1}(2b - 1)$. Given $n \in \mathbb{N}$, let 2^{a-1} be the highest power of 2 that divides n [$a = 1$ if n is odd] so that $n/2^{a-1}$ is odd and therefore has the form $2b - 1$ for some $b \in \mathbb{N}$. Then $f(a, b) = n$ and so f is onto $\mathbb{N}$. It is also clear that (a, b) is uniquely determined by n and so f is one to one. $\square$

(1.61) **Remark** A geometric proof of (1.60) is as follows. Arrange $\mathbb{N} \times \mathbb{N}$ is an infinite matrix

$$(1,1) \quad (1,2) \quad (1,3) \quad \cdots$$
$$(2,1) \quad (2,2) \quad (2,3) \quad \cdots$$
$$(3,1) \quad (3,2) \quad (3,3) \quad \cdots$$
$$\cdots \quad \cdots \quad \cdots \quad \cdots$$

so that (a, b) is in the ath row and bth column. Then count $\mathbb{N} \times \mathbb{N}$ by following successive arrows as indicated. One gets the enumeration

$$(1, 1), (2, 1), (1, 2), (3, 1), (2, 2), \ldots .$$

This correspondence is given by the formula

$$g(a, b) = (a + b - 1)(a + b)/2 - (a - 1),$$

but it is not so easy to check that g is one to one and onto $\mathbb{N}$.

(1.62)   **Theorem**   *Let A and B be nonvoid sets where B is denumerable. Then A is countable if and only if there exists a function f from B onto A.*

> **Proof**   Suppose that A is countable. Enumerate B as $\{b_n\}_{n=1}^{\infty}$. If A is finite, say $A = \{a_1, \ldots, a_p\}$, define $f(b_k) = a_k$ for $1 \leq k \leq p$ and $f(b_k) = a_1$ for $k > p$. If A is denumerable, write $A = \{a_n\}_{n=1}^{\infty}$, and define $f(b_n) = a_n$.
>
> Conversely, suppose that f maps B onto A. For each $a \in A$ choose $b \in B$ such that $f(b) = a$ and define $g(a) = b$; i.e., $g(a) \in f^{-1}(\{a\})$ for each $a \in A$.* Then g is a function from A into B. Since $a_1 \neq a_2$ in A implies $f^{-1}(\{a_1\})$ and $f^{-1}(\{a_2\})$ are disjoint sets, it follows that g is one to one. Thus $A \sim g(A) \subset B$ and so (1.59) shows that A is countable.   □

(1.63)   **Theorem**   *Any countable union of countable sets is a countable set. More precisely, if I is a countable set and A_i is a countable set for each $i \in I$, then the set $A = \bigcup_{i \in I} A_i$ is a countable set.*

> **Proof**   We may suppose $I \neq \emptyset \neq A_i$. According to (1.62) there exist functions g from $\mathbb{N}$ onto I and f_i from $\mathbb{N}$ onto A_i for each $i \in I$. Now define h on $\mathbb{N} \times \mathbb{N}$ by
>
> $$h(m, n) = f_{g(n)}(m).$$
>
> One easily checks that h is onto A: given $a \in A$, choose $i \in I$ such that $a \in A_i$, then choose $m, n \in \mathbb{N}$ such that $g(n) = i$ and $f_i(m) = a$. Apply (1.60) and (1.62) again to see that A is countable.   □

(1.64)   **Theorem**   *The set $\mathbb{Q}$ of all rational numbers is a countable set.*

> **Proof**   Since $\mathbb{Z}$ is countable (1.56) and
>
> $$\mathbb{Q} = \bigcup_{n \in \mathbb{N}} \{m/n : m \in \mathbb{Z}\},$$
>
> we need only invoke (1.63).   □

(1.65)   **Corollary**   *The set S of all irrational numbers is uncountable.*

> **Proof**   Exercise.

(1.66)   **Heine–Borel Theorem**   *Let $-\infty < a < b < \infty$ and let $\mathcal{G}$ be a family of open intervals such that $[a, b] \subset \cup \mathcal{G}$. Then there exists a finite subfamily $\mathcal{F} \subset \mathcal{G}$ such that $[a, b] \subset \cup \mathcal{F}$.*

> **Proof**   Let $T = \{t \in \,]a, b] : [a, t] \subset \cup \mathcal{F}$ for some finite family $\mathcal{F} \subset \mathcal{G}\}$. Choose $I \in \mathcal{G}$ such that $a \in I$ and then choose $t \in I \cap \,]a, b]$. Clearly, $t \in T$

---

*More precisely, let $g(a) = b_{n(a)}$ where $n(a) = \min\{n : f(b_n) = a\}$.

so $T \neq \emptyset$. Also T is bounded so T has a supremum $s \in]a,b]$. Choose $I_0 \in \mathcal{G}$ such that $s \in I_0$. By (1.13.ii), there exists some $x \in I_0 \cap T$. Since $x \in T$ there exists some finite family $\{I_k\}_{k=1}^{n} \subset \mathcal{G}$ such that $[a,x] \subset \bigcup_{k=1}^{n} I_k$. Then

$[a,s] \subset \bigcup_{k=0}^{n} I_k$ and so $s \in T$. To finish the proof we need only show that

$s = b$. Assume $s < b$. Choose $y \in I_0 \cap]s,b]$. Then $[a,y] \subset \bigcup_{k=0}^{n} I_k$ so $y \in T$.

But $s < y$ and s is an upper bound for T. This contradiction proves that $s = b$. $\quad\square$

Newton's Binomial Theorem

(1.67) **Definitions** We inductively define n *factorial*, denoted $n!$, for integers $n \geq 0$ by

$$0! = 1, \qquad (n+1)! = n!\,(n+1).$$

Thus $1! = 1$, $2! = 2$, $3! = 6$, $4! = 24$, etc. If $n \geq 1$ we have $n! = \prod_{k=1}^{n} k$.

For integers n and k with $0 \leq k \leq n$ we define the *binomial coefficient* $\binom{n}{k}$ [read: n by k] by

$$\binom{n}{k} = \frac{n!}{k!\,(n-k)!}\,.$$

(1.68) **Theorem** *If n and k are integers with $1 \leq k \leq n$, then*
(i) $\binom{n}{k-1} + \binom{n}{k} = \binom{n+1}{k}$ *and*
(ii) $\binom{n}{k}$ *is a positive integer.*

Proof For (i), we compute as follows:

$$\binom{n}{k-1} + \binom{n}{k} = \frac{n!}{(k-1)!\,(n-k+1)!} + \frac{n!}{k!\,(n-k)!}$$

$$= \frac{n!\,k}{k!\,(n-k+1)!} + \frac{n!\,(n-k+1)}{k!\,(n-k+1)!}$$

$$= \frac{n!\,(n+1)}{k!\,(n+1-k)!} = \binom{n+1}{k}.$$

The proof of (ii) is an easy induction on n by the use of (i). We leave the details to the reader. $\quad\square$

(1.69) **Binominal Theorem** *If a and b are complex numbers and n is a nonnegative integer, then*

(i) $(a+b)^n = \sum_{k=0}^{n} \binom{n}{k} a^{n-k} b^{k}.$

In particular,

(ii) $2^n = \sum_{k=0}^{n} \binom{n}{k}$.

Proof   Formula (ii) follows from (i) by taking $a = b = 1$.

We prove (i) by induction on n. Let S be the set of all nonnegative integers n for which (i) holds. Plainly $0 \in S$ and $1 \in S$. Suppose that $n \geq 1$ and $n \in S$. We shall show that $(n + 1) \in S$. Indeed, using (1.68.i) and our hypothesis that $n \in S$, we have

$$\sum_{k=0}^{n+1} \binom{n+1}{k} a^{n+1-k} b^k$$

$$= a^{n+1} + b^{n+1} + \sum_{k=1}^{n} \left\{ \binom{n}{k} + \binom{n}{k-1} \right\} a^{n+1-k} b^k$$

$$= \sum_{k=0}^{n} \binom{n}{k} a^{n+1-k} b^k + \sum_{k=1}^{n+1} \binom{n}{k-1} a^{n+1-k} b^k$$

$$= \sum_{k=0}^{n} \binom{n}{k} a^{n+1-k} b^k + \sum_{k=0}^{n} \binom{n}{k} a^{n-k} b^{k+1}$$

$$= (a + b) \sum_{k=0}^{n} \binom{n}{k} a^{n-k} b^k$$

$$= (a + b)(a + b)^n = (a + b)^{n+1}$$

and so $(n + 1) \in S$. □

Our final theorem of this chapter is another useful algebraic identity.

(1.70)   **Theorem**   *Let a and b be complex numbers and let n be a positive integer. Then*

$$a^n - b^n = (a - b) \sum_{k=1}^{n} a^{n-k} b^{k-1}.$$

Proof   Simply compute:

$$(a - b) \sum_{k=1}^{n} a^{n-k} b^{k-1} = \sum_{k=1}^{n} a^{n+1-k} b^{k-1} - \sum_{k=1}^{n} a^{n-k} b^k$$

$$= \sum_{k=0}^{n-1} a^{n-k} b^k - \sum_{k=1}^{n} a^{n-k} b^k$$

$$= a^n - b^n. □$$

We close this chapter with a large number of exercises.

Exercises

1.   Using the definition of $\sum$ [see (1.27)] and induction, prove that if $(x_j)_{j=a}^b$ and $(y_j)_{j=a}^b$ and α and β are all in $\mathbb{C}$, where $a \leqq b$ in $\mathbb{Z}$, then

$$\sum_{j=a}^b (\alpha x_j + \beta y_j) = \alpha \sum_{j=a}^b x_j + \beta \sum_{j=a}^b y_j.$$

Also, show that if x_j and y_j are real and $x_j \leqq y_j$ for all j, then

$$\sum_{j=a}^b x_j \leqq \sum_{j=a}^b y_j.$$

2.   Let $\emptyset \neq S \subset \mathbb{Z}$. Prove that
      (a) if S is bounded below, then S has a smallest member;
      (b) if S is bounded above, then S has a largest member.

3.   Find the b adic expansion of the number n whose decimal expansion is given when
      (a) $b = 2$, $n = 961$;
      (b) $b = 9$, $n = 93$;
      (c) $b = 12$, $n = 212686$ [use the symbols $T = 10$, $E = 11$].

4.   Find the decimal expansion of the number n given in base b [b is given in decimal expansion] when
      (a) $b = 2$, $n = 11010$;
      (b) $b = 5$; $n = 44321$;
      (c) $b = 12$, $n = ETET$ [see Exercise 3(c)].

5.   The complex field cannot be ordered. That is, there is no possible choice of $P \subset \mathbb{C}$ such that Axioms VI and VII hold with $\mathbb{R}$ replaced by $\mathbb{C}$. [Hint: $i \in P$ or $-i \in P$.]

6.   Let a, b, c, d be rational numbers and x an irrational number such that $cx + d \neq 0$. Prove that $(ax + b)/(cx + d)$ is irrational if and only if $ad \neq bc$.

7.   Čebyšev's Inequality   Let $a_1 \leqq a_2 \leqq \ldots \leqq a_n$ and $b_1 \leqq b_2 \leqq \ldots b_n$ be real numbers. Prove that

$$\left(\sum_{j=1}^n a_j \right) \left(\sum_{k=1}^n b_k \right) \leqq n \sum_{j=1}^n a_j b_j$$

and that equality obtains if and only if either $a_1 = a_n$ or $b_1 = b_n$. [Hint: First prove that $0 \leqq \sum_{j=1}^n \sum_{k=1}^n (a_j - a_k)(b_j - b_k)$.]

8.   (a) If $x, y \in \mathbb{R}$, then $2xy \leqq x^2 + y^2$ and $4xy \leqq (x + y)^2$. A necessary and sufficient condition for equality is that $x = y$.
      (b) If $a > 0$, $b > 0$, and $a + b = 1$, then

$$(a + 1/a)^2 + (b + 1/b)^2 \geqq 25/2.$$

When does equality obtain?

9.   (a) If $a_1, a_2, \ldots, a_n$ are all positive, then

$$\left(\sum_{j=1}^{n} a_j \right) \left(\sum_{j=1}^{n} \frac{1}{a_j} \right) \geqq n^2$$

and equality obtains if and only if $a_1 = a_2 = \ldots = a_n$.
(b) If a, b, c are positive and $a + b + c = 1$, then

$$(1/a - 1)(1/b - 1)(1/c - 1) \geqq 8$$

and equality obtains if and only if $a = b = c = 1/3$.

10.   Prove that for all $n \in \mathbb{N}$ we have

$$\frac{1}{2} \cdot \frac{3}{4} \cdot \frac{5}{6} \cdot \ldots \cdot \frac{2n-1}{2n} \leqq \frac{1}{\sqrt{3n+1}}$$

and equality obtains if and only if $n = 1$.

11.   Arithmetic progressions. If $a, d \in \mathbb{C}$ and $n \in \mathbb{N}$, then

$$\sum_{j=0}^{n} (a + jd) = \frac{(2a + nd)(n+1)}{2}.$$

12.   (a) For all $n \in \mathbb{N}$ we have

$$\sqrt{n+1} - \sqrt{n} < \frac{1}{2\sqrt{n}} < \sqrt{n} - \sqrt{n-1}.$$

[Hint: $(\sqrt{n+1} - \sqrt{n})(\sqrt{n+1} + \sqrt{n}) = 1$.]
(b) If $k \in \mathbb{N}$ and $k > 1$, then

$$2\sqrt{k+1} - 2 < \sum_{n=1}^{k} \frac{1}{\sqrt{n}} < 2\sqrt{k} - 1.$$

[Hint: Sum inequalities (a) for $n = 2, 3, \ldots, k$.]

13.   Let $n \in \mathbb{N}$ and $x \in \mathbb{R}$. Then
(a) $-1 < x < 0$ implies $(1 + x)^n \leqq 1 + nx + (n(n-1)/2)x^2$,
(b) $x > 0$ implies $(1 + x)^n \geqq 1 + nx + (n(n-1)/2)x^2$.
Compare Bernoulli's Inequality.

14.   Any positive rational number r can be expressed in exactly one way in the form

$$r = \sum_{j=1}^{n} a_j/j!$$

where $a_1, a_2, \ldots, a_n$ are integers such that $a_1 \geqq 0$, $0 \leqq a_j < j$ for $2 \leqq j \leqq n$, and $a_n \neq 0$.

15.   If $n \in \mathbb{N}$, then $n! \leqq ((n+1)/2)^n$. [Hint: Use (1.48).]

16.   (a) If $a_1, a_2, \ldots, a_n$ are positive real numbers, then

(i)
$$\frac{n}{\dfrac{1}{a_1} + \dfrac{1}{a_2} + \ldots + \dfrac{1}{a_n}} \leqq (a_1 a_2 \ldots a_n)^{1/n}.$$

(b) When does equality hold in (i)?
The left side of (i) is called the *harmonic mean* of $a_1, \ldots, a_n$.

17.   If $a, b \in \mathbb{Z}$ and $a \neq 0$, we say that b is *divisible* by a [or a *divides* b] and write $a \mid b$ if there exists $c \in \mathbb{Z}$ such that $ac = b$. If $n \in \mathbb{N}$, then
     (a) $6 \mid (n^3 + 5n)$,
     (b) $30 \mid (n^5 - n)$,
     (c) $3 \mid (n^3 - n + 1)$ is false,
     (d) $24 \mid (n^3 - n)$ if n is odd.

18.   Find the infimum and the supremum of the set $E = \{2^{-k} + 3^{-m} + 5^{-n} : \{k, m, n\} \subset \mathbb{N}\}$.

19.   If $a, b, c \in \mathbb{C}$ satisfy $|a| = |b| = |c|$ and $a + b + c = 0$, then $|a - b| = |b - c| = |c - a|$. Interpret geometrically. [Hint: Write $|z|^2 = z\bar{z}$.]

20.   If $a, b \in \mathbb{C}$, then $|a + b|^2 + |a - b|^2 = 2|a|^2 + 2|b|^2$. Interpret geometrically.

21.   If $a, b \in \mathbb{C}$ and $\mathrm{Re}(\bar{a}b) = 0$, then $|a - b|^2 = |a|^2 + |b|^2$. Interpret geometrically.

22.   (a) If A is a finite set and $B \subset A$, then B is finite. [Use our definition of "finite set" and induction.]
     (b) If B is an infinite set and A is a set with $B \subset A$, then A is infinite.

23.   If A is an infinite set, then A has a countably infinite subset. [Hint: By induction, one shows that for each $n \in \mathbb{N}$, A has a subset having n elements. Choose one such B_n for each n and let $B = \bigcup \{B_n : n \in \mathbb{N}\}$.]

24.   If E is a set, then E is infinite if and only if E is equivalent to a proper subset of E. [Hint: Use Exercise 23 to get the required one to one function. Conversely, if $f : E \to E$ is one to one and $a \in E \setminus f(E)$, define $a_1 = a$ and $a_{n+1} = f(a_n)$.]

25.   A complex number z is said to be *algebraic* if there exists $n \in \mathbb{N}$ and integers $a_0, a_1, \ldots, a_n$ with $a_n \neq 0$ and
     (i) $\displaystyle\sum_{j=0}^{n} a_j z^j = 0$.
     All other complex numbers are said to be *transcendental*.
     (a) The set of all algebraic numbers is countable. [Hint: For $k \in \mathbb{N}$ there are only finitely many equations (i) for which $n + \displaystyle\sum_{j=0}^{n} |a_j| = k$.]
     (b) Every rational number is algebraic.
     (c) There are uncountably many transcendental numbers in each interval $[\alpha, \beta]$, $\alpha < \beta$.

26.   If $x, y \in \mathbb{R}$ and $n \in \mathbb{N}$, then
     (a) $[x + y] \geq [x] + [y]$,
     (b) $[[x]/n] = [x/n]$,
     (c) $\displaystyle\sum_{k=0}^{n-1} [x + k/n] = [nx]$,
     where $[t]$ is the integer satisfying $[t] \leq t < [t] + 1$.

27.   (a) If a, b, c are positive real numbers, then

$$\left(\frac{1}{2}a + \frac{1}{3}b + \frac{1}{6}c \right)^2 \leq \frac{1}{2}a^2 + \frac{1}{3}b^2 + \frac{1}{6}c^2$$

     with equality if and only if $a = b = c$.

(b) If $a_1, \ldots, a_n$ and $w_1, \ldots, w_n$ are positive real numbers with $\sum_{j=1}^{n} w_j = 1$, then

$$\left(\sum_{j=1}^{n} a_j w_j \right)^2 \leq \sum_{j=1}^{n} a_j^2 w_j$$

where equality obtains if and only if $a_1 = a_2 = \ldots = a_n$.

[Hint: Use Cauchy's Inequality with $x_j = a_j\sqrt{w_j}$ and $y_j = \sqrt{w_j}$.]

28.   Let $n \in \mathbb{N}$ and let $r_p r_{p-1} \ldots r_1 r_0$ be the decimal expansion of n. Then

(a) n is divisible by 9 if and only if $\sum_{j=0}^{p} r_j$ is divisible by 9.

[Hint: $10^j - 1$ is divisible by 9.]

(b) n is divisible by 11 if and only if $\sum_{j=0}^{p} (-1)^j r_j$ is divisible by 11.

[Hint: $10^{2k} - 1$ and $10^{2k+1} + 1$ are divisible by 11.]

29.   If n and k are integers with $0 \leq k \leq n$ and S is a set with exactly n elements, then S has exactly $\binom{n}{k}$ distinct subsets, each having exactly k elements. [Hint: Use (1.68.i) and induction on n.]

30.   If $n \in \mathbb{N}$, then

(a) $\sum_{k=0}^{n} \binom{n}{k}^2 = \binom{2n}{n}$,

(b) $\sum_{k=0}^{2n} (-1)^k \binom{2n}{k}^2 = (-1)^n \binom{2n}{n}$.

[Hint: Use the Binomial Theorem to write the identities $(1 + x)^n (1 + x)^n = (1 + x)^{2n}$ and $(1 - x)^{2n}(1 + x)^{2n} = (1 - x^2)^{2n}$ as identities between polynomials in x and then equate the coefficients of x^n and x^{2n} respectively. How do you know that coefficients of like powers of x must be equal?]

31.   If $m, n \in \mathbb{N}$, then $1 + \sum_{k=1}^{m} \binom{n+k}{k} = \binom{n+m+1}{m}$. [Hint: Write the 1 as $\binom{n+1}{0}$ and use (1.68.i).]

2

SEQUENCES AND SERIES

In this chapter we introduce the notion of convergence, a concept that will occupy us throughout the remainder of this book. The basic definitions and theorems presented here are essential to an understanding of analysis. A more complete treatment of infinite series is given in Chapter 7.

Sequences in $\mathbb{C}$

(2.1)   **Definition**   Let X be a set and $f: \mathbb{N} \to X$. Then f is called a *sequence in X*. If $f(n) = x_n$, then x_n is called the *nth term* of the sequence. It is customary to write such a sequence as $(x_n)_{n=1}^{\infty}$ or as $(x_1, x_2, x_3, \dots)$ or as $x_1, x_2, x_3, \dots$. We also abuse notation and write $(x_n)_{n=1}^{\infty} \subset X$ to indicate that the terms of the sequence are in the set X.

We will have occasion to consider functions whose domains of definition are of the form $\{n \in \mathbb{Z} : n \geqq p\}$ for some fixed $p \in \mathbb{Z}$. Such functions are also called sequences and are denoted as $(x_n)_{n=p}^{\infty}$. It involves only a trivial change of notation to regard such a sequence as being defined on $\mathbb{N}$; $(x_n)_{n=p}^{\infty} = (x_{n+p-1})_{n=1}^{\infty}$, the first term is x_p, etc.

We next define convergence for complex sequences.

(2.2)   **Definition**   A sequence $(x_n)_{n=1}^{\infty} \subset \mathbb{C}$ is said to *converge* if there exists $x \in \mathbb{C}$ such that to each $\epsilon > 0$ corresponds $n_\epsilon \in \mathbb{N}$ for which

$$|x - x_n| < \epsilon \quad \text{for all } n \geqq n_\epsilon.$$

In this case, we call x a *limit* for $(x_n)_{n=1}^{\infty}$ and write

$$\lim_{n \to \infty} x_n = x, \quad \text{or } x_n \to x \text{ as } n \to \infty.$$

A sequence that does not converge is said to *diverge*.

The limit of a convergent sequence is unique as the next theorem shows.

(2.3)   **Theorem**   *If* $(x_n)_{n=1}^\infty \subset \mathbb{C}$ *converges to* $x \in \mathbb{C}$ *and to* $y \in \mathbb{C}$, *then* $x = y$.

Proof   Let $\epsilon > 0$ be given. Choose $n_1 \in \mathbb{N}$ and $n_2 \in \mathbb{N}$ such that

$$n \geq n_1 \quad \text{implies} \quad |x - x_n| < \epsilon$$
$$n \geq n_2 \quad \text{implies} \quad |y - x_n| < \epsilon.$$

Let $n_0 = \max\{n_1, n_2\}$. Then

$$|x - y| \leq |x - x_{n_0}| + |x_{n_0} - y| < 2\epsilon.$$

Thus $x = y$, for otherwise we could have taken $\epsilon = |x - y|/2$ and deduced $\epsilon < \epsilon$. $\square$

(2.4)   **Definition**   A sequence $(z_n)_{n=1}^\infty \subset \mathbb{C}$ is *bounded* if there exists a positive $\beta \in \mathbb{R}$ such that $|z_n| \leq \beta$ for all $n \in \mathbb{N}$.

(2.5)   **Theorem**   *Every convergent sequence in* $\mathbb{C}$ *is bounded.*

Proof   Say $z_n \to z$ in $\mathbb{C}$. Choose n_0 such that

$$n \geq n_0 \quad \text{implies} \quad |z - z_n| < 1.$$

Let $\beta = \max\{|z| + 1, |z_1|, \ldots, |z_{n_0}|\}$. For $1 \leq n \leq n_0$, it is clear that $|z_n| \leq \beta$. For $n > n_0$ we have

$$|z_n| \leq |z_n - z| + |z| < 1 + |z| \leq \beta. \quad \square$$

Our next theorem gives several examples of both convergent and divergent sequences.

(2.6)   **Theorem**   *Let* $z \in \mathbb{C}$ *and* $k = \mathbb{N}$. *Then*
(i) $|z| < 1$ *implies* $\lim\limits_{n \to \infty} z^n = 0$,
(ii) $|z| \geq 1$, $z \neq 1$ *imply* $(z^n)_{n=1}^\infty$ *diverges*,
(iii) $\lim\limits_{n \to \infty} n^{-k} = 0$.

Proof   Suppose $|z| < 1$. Let $\alpha = \inf\{|z^n| : n \in \mathbb{N}\}$. Plainly $\alpha \geq 0$. Assume $\alpha > 0$. Then $z \neq 0$ and $\alpha < \alpha/|z|$. By definition of α, there is an $n \in \mathbb{N}$ such that $|z^n| < \alpha/|z|$ and so $|z^{n+1}| < \alpha$. This contradiction shows that $\alpha = 0$. Let $\epsilon > 0$ and choose n_0 such that $|z^{n_0}| < \epsilon$. Then $n \geq n_0$ implies $|z^n - 0| = |z|^{n - n_0}|z^{n_0}| \leq |z^{n_0}| < \epsilon$. Thus (i) obtains.

Next suppose $|z| > 1$. Let $\beta = \sup\{|z^n| : n \in \mathbb{N}\}$. We claim that $\beta = \infty$. If not, then $\beta/|z| < \beta$ so $\beta/|z| < |z^n|$, $\beta < |z^{n+1}|$, for some $n \in \mathbb{N}$, contrary to the choice of β. Thus $\beta = \infty$ and so $(z^n)_{n=1}^\infty$ is not bounded. By (2.5) this sequence diverges.

Suppose that $|z| = 1$, $z \neq 1$. Assume that $z^n \to z_0 \in \mathbb{C}$ as $n \to \infty$. Let $\epsilon = (1/3)|z - 1| > 0$ and choose n_0 so that $|z_0 - z^n| < \epsilon$ for $n \geq n_0$. Then $3\epsilon = |z - 1| = |z|^{n_0}|z - 1| = |z^{n_0+1} - z^{n_0}| \leq |z^{n_0+1} - z_0| + |z_0 - z^{n_0}| < \epsilon + \epsilon = 2\epsilon$. This contradiction proves that (z^n) diverges. Therefore, (ii) is proven.

Let $k \in \mathbb{N}$ and $\epsilon > 0$ be given. Use the Archimedean property of $\mathbb{R}$ to get $n_0 \in \mathbb{N}$ such that $n_0 > 1/\epsilon$; hence $n_0^{-1} < \epsilon$. Now $n \geq n_0$ implies $|n^{-k} - 0| = n^{-k} \leq n^{-1} \leq n_0^{-1} < \epsilon$ and so (iii) obtains. $\quad\square$

The next theorem gives a relation between real sequences and complex sequences.

(2.7) **Theorem** *Let* $z_n = x_n + iy_n$ *where* $x_n, y_n \in \mathbb{R}$ *for each* $n \in \mathbb{N}$. *Then* (z_n) *converges if and only if both* (x_n) *and* (y_n) *converge. In this case,*

$$\lim z_n = \lim x_n + i \lim y_n$$

and the two limits on the right are real.

Proof If $\lim z_n = z \in \mathbb{C}$ and $z = x + iy$ with $x, y \in \mathbb{R}$, then $|x - x_n| \leq |z - z_n|$ and $|y - y_n| \leq |z - z_n|$. Thus $x_n \to x$ and $y_n \to y$.

If $x_n \to x$ and $y_n \to y$ as $n \to \infty$ and $z = x + iy$, then $|z - z_n| = |(x - x_n) + i(y - y_n)| \leq |x - x_n| + |y - y_n|$ and so $z_n \to z$. $\quad\square$

The next theorem shows how limits team up with the basic arithmetic operations.

(2.8) **Limit Theorems** *Let* $z_n \to z$ *and* $w_n \to w$ *in* $\mathbb{C}$. *Then*
(i) $\lim(z_n + w_n) = z + w$,
(ii) $\lim c z_n = cz$ *for* $c \in \mathbb{C}$,
(iii) $\lim z_n w_n = zw$,
(iv) $\lim(z_n/w_n) = z/w$ *if* $w \neq 0$ *and* $w_n \neq 0$ *for* $n \in \mathbb{N}$.

Proof (i) Given $\epsilon > 0$, choose $n_1, n_2 \in \mathbb{N}$ such that

$$n \geq n_1 \quad \text{implies} \quad |z - z_n| < \epsilon/2$$
$$n \geq n_2 \quad \text{implies} \quad |w - w_n| < \epsilon/2.$$

Then

$$n \geq \max\{n_1, n_2\} \quad \text{implies}$$
$$|(z + w) - (z_n + w_n)| = |(z - z_n) + (w - w_n)|$$
$$\leq |z - z_n| + |w - w_n| < \epsilon/2 + \epsilon/2 = \epsilon.$$

(ii) Given $\epsilon > 0$, choose $n_0 \in \mathbb{N}$ such that $n \geq n_0$ implies $|z - z_n| < \epsilon/(|c| + 1)$. Then

$$n \geq n_0 \quad \text{implies} \quad |cz - cz_n| = |c||z - z_n| < |c|\frac{\epsilon}{|c| + 1} < \epsilon.$$

(iii) Given $\epsilon > 0$, choose $n_1, n_2 \in \mathbb{N}$ such that

$$n \geqq n_1 \quad \text{implies} \quad |z - z_n| < \sqrt{\epsilon},$$
$$n \geqq n_2 \quad \text{implies} \quad |w - w_n| < \sqrt{\epsilon}.$$

Then

$$n \geqq \max\{n_1, n_2\} \quad \text{implies}$$
$$|(z - z_n)(w - w_n) - 0| = |z - z_n||w - w_n| < \epsilon.$$

This proves that

$$\lim(z - z_n)(w - w_n) = 0.$$

Using this fact along with (i) and (ii), we have

$$\lim(z_n w_n - zw) = \lim\left[(z - z_n)(w - w_n) + z(w_n - w) + w(z_n - z)\right]$$
$$= 0 + z \lim(w_n - w) + w \lim(z_n - z)$$
$$= 0 + z \cdot 0 + w \cdot 0 = 0$$

and so (iii) obtains. [Here we have used the obvious fact that $\lim a_n = a$ if and only if $\lim(a_n - a) = 0$.]

(iv) Given $\epsilon > 0$, choose $n_0 \in \mathbb{N}$ such that $n \geqq n_0$ implies $|w - w_n| < |w|/2$. Then

$$n \geqq n_0 \quad \text{implies} \quad |w_n| \geqq |w| - |w - w_n| > |w|/2. \tag{1}$$

Next choose $n_1 > n_0$ such that

$$n \geqq n_1 \quad \text{implies} \quad |w - w_n| < |w|^2 \epsilon / 2. \tag{2}$$

Using (1) and (2) we get

$$n \geqq n_1 \quad \text{implies} \quad \left|\frac{1}{w} - \frac{1}{w_n}\right| = \frac{|w_n - w|}{|w||w_n|} < \frac{2|w_n - w|}{|w|^2} < \epsilon.$$

This proves that

$$\lim(1/w_n) = 1/w.$$

Combine this with (iii) to get (iv). □

(2.9)   Warning The conclusions in (2.8) depend on the hypothesis that the individual sequences (z_n) and (w_n) converge. It can happen that the limits on the left sides in (2.8.i)–(2.8.iv) exist in $\mathbb{C}$ even though the individual limits do not. For example,

$$\lim\left[(-1)^n + (-1)^{n+1}\right] = 0,$$
$$\lim(-1)^n(-1)^n = 1.$$

Sequences in $\mathbb{R}^{\#}$

In this section we make use of the ordering on $\mathbb{R}^{\#}$ to discuss special properties of sequences in $\mathbb{R}^{\#}$. Since $\mathbb{R} \subset \mathbb{C}$, we have already considered convergence in $\mathbb{R}$. We begin with a definition of infinite limits.

(2.10)   **Definition**   Let $(x_n)_{n=1}^{\infty} \subset \mathbb{R}^{\#}$. We say that this sequence has *limit* ∞ [resp. $-\infty$] and write

$$\lim_{n \to \infty} x_n = \infty \left[\text{resp.} \lim_{n \to \infty} x_n = -\infty \right]$$

if to each $\alpha \in \mathbb{R}$ corresponds some $n_0 \in \mathbb{N}$ such that $n \geqq n_0$ implies $x_n > \alpha$ [resp. $x_n < \alpha$]. We *do not* say that (x_n) converges in these cases.

(2.11)   **Definition**   A sequence $(x_n)_{n=1}^{\infty} \subset \mathbb{R}^{\#}$ is said to be

$$\begin{array}{ll} \textit{nondecreasing} & \text{if } x_n \leqq x_{n+1} \\ \textit{nonincreasing} & \text{if } x_n \geqq x_{n+1} \\ \textit{strictly increasing} & \text{if } x_n < x_{n+1} \\ \textit{strictly decreasing} & \text{if } x_n > x_{n+1} \end{array}$$

for all $n \in \mathbb{N}$. All such sequences are said to be *monotone* or, in the latter two cases, *strictly monotone*.

(2.12)   **Theorem**   *Any monotone sequence in $\mathbb{R}^{\#}$ has a limit in $\mathbb{R}^{\#}$. In particular, a monotone sequence of real numbers converges if and only if it is bounded.*

     Proof   Let $x_1 \leqq x_2 \leqq x_3 \ldots$ in $\mathbb{R}^{\#}$ and let $x = \sup\{x_n : n \in \mathbb{N}\}$. We claim that $\lim x_n = x$.

         Case 1. $x = -\infty$. Then $x_n = -\infty$ for all n and so, given $\alpha \in \mathbb{R}$, $x_n < \alpha$ for all n. Thus

$$\lim x_n = -\infty = x.$$

         Case 2. $x = \infty$. Given $\alpha \in \mathbb{R}$, there exists n_0 such that $x_{n_0} > \alpha$. Therefore

$$n \geqq n_0 \quad \text{implies} \quad x_n \geqq x_{n_0} > \alpha$$

and hence

$$\lim x_n = \infty = x.$$

         Case 3. $x \in \mathbb{R}$. Given $\epsilon > 0$, there exists n_0 such that $x_{n_0} > x - \epsilon$. Thus

$$n \geqq n_o \quad \text{implies} \quad x - \epsilon < x_{n_0} \leqq x_n \leqq x < x + \epsilon$$

and so

$$|x - x_n| < \epsilon \quad \text{for all } n \geqq n_o;$$

i.e., $\lim x_n = x$.

The case of nonincreasing sequences is handled similarly. □

Of course a bounded sequence can converge without being monotone. For example, let $x_n = (-1/2)^n$ [see (2.6.i)].

(2.13)   Example   Consider the sequence (x_n) defined recursively by

$$x_1 = \sqrt{2}, \qquad x_{n+1} = \sqrt{2 + x_n} .$$

We prove by induction that (x_n) is strictly increasing. Indeed, $x_1 = \sqrt{2} < \sqrt{2 + \sqrt{2}}$ $= x_2$ and if $x_n < x_{n+1}$, then

$$x_{n+1} = \sqrt{2 + x_n} < \sqrt{2 + x_{n+1}} = x_{n+2}.$$

Also, we prove by induction that (x_n) is bounded above by 2. Indeed, $x_1 = \sqrt{2} < 2$ and if $x_n < 2$, then

$$x_{n+1} = \sqrt{2 + x_n} < \sqrt{2 + 2} = 2.$$

We conclude from (2.12) that (x_n) converges to sup $\{x_n : n \in \mathbb{N}\} = x \leqq 2$. Next, (2.8.iii) yields

$$x^2 = \lim x_{n+1}^2 = \lim(2 + x_n) = 2 + x$$

and so $x = 2$.

Not all sequences in $\mathbb{R}$ are monotone, but sometimes certain subsequences are monotone and this can be useful in testing convergence of the sequence itself. An instance of this type of phenomenon is examined in the next theorem and the example that follows it.

(2.14)   Theorem   *Suppose* $(x_n) \subset \mathbb{R}$,

$$x_{2n} \leqq x_{2n+2} \leqq x_{2n+1} \leqq x_{2n-1}$$

for all $n \in \mathbb{N}$, *and*

$$\lim_{n \to \infty} (x_{2n-1} - x_{2n}) = 0.$$

Then (x_n) *converges to some* $x \in \mathbb{R}$ *and*

$$x_{2n} \leqq x \leqq x_{2n-1}$$

for all $n \in \mathbb{N}$.

Proof We use the Nested Interval Principle (1.57). By hypothesis

$$[x_{2n+2}, x_{2n+1}] \subset [x_{2n}, x_{2n-1}]$$

for all $n \in \mathbb{N}$ and for each $\epsilon > 0$ there exists an n_0 such that

$$x_{2n_0 - 1} - x_{2n_0} < \epsilon. \tag{1}$$

According to (1.57) there exists $x \in \mathbb{R}$ such that

$$\bigcap_{n=1}^{\infty} [x_{2n}, x_{2n-1}] = \{x\}.$$

Plainly $x_{2n} \leqq x \leqq x_{2n-1}$ for all $n \in \mathbb{N}$. To see that (x_n) converges to x, let $\epsilon > 0$ be given and choose n_0 so that (1) holds. Then

$$n \geqq n_0 \quad \text{implies} \quad 0 \leqq x_{2n-1} - x \leqq x_{2n-1} - x_{2n}$$

$$\leqq x_{2n_0 - 1} - x_{2n_0} < \epsilon$$

and similarly

$$n \geqq n_0 \quad \text{implies} \quad 0 \leqq x - x_{2n} < \epsilon.$$

Thus

$$k \geqq 2n_0 \quad \text{implies} \quad |x - x_k| < \epsilon$$

[consider $k = 2n - 1$ and $k = 2n$]. Therefore $\lim_{k \to \infty} x_k = x$. $\quad \square$

(2.15) **Example** Consider the sequence (x_n) defined by $x_1 = 1$, $x_{n+1} = 1/(1 + x_n)$. By induction, $x_n > 0$ for all n. We claim that (x_n) satisfies the hypotheses of (2.14). We show by induction that

$$x_{2n} < x_{2n+2} < x_{2n+1} < x_{2n-1} \tag{1}$$

for all $n \in \mathbb{N}$. We have

$$x_1 = 1, \quad x_2 = 1/2, \quad x_3 = 2/3, \quad x_4 = 3/5$$

and

$$1/2 < 3/5 < 2/3 < 1$$

so (1) holds for $n = 1$. If (1) holds for $n = k \geqq 1$, then, adding 1 to each member of (1) [with $n = k$] and then taking reciprocals, we get

$$x_{2k+1} > x_{2k+3} > x_{2k+2} > x_{2k}. \tag{2}$$

Applying the same process to (2) we get

$$x_{2k+2} < x_{2k+4} < x_{2k+3} < x_{2k+1},$$

which is (1) with $n = k + 1$. Thus (1) holds for all $n \in \mathbb{N}$. To show that $x_{2n-1} - x_{2n} \to 0$ as $n \to \infty$ it clearly suffices to prove that

$$x_{2n-1} - x_{2n} < (5/9)^n \tag{3}$$

for all $n \in \mathbb{N}$ [see (2.6.i)]. We prove (3) by induction. For $n = 1$ we have

$$x_1 - x_2 = 1 - 1/2 < 5/9$$

so (3) holds for $n = 1$. If (3) holds for $n = k \geq 1$, then

$$
\begin{aligned}
x_{2k+1} - x_{2k+2} &= \frac{1}{1 + x_{2k}} - \frac{1}{1 + x_{2k+1}} \\
&= \frac{x_{2k+1} - x_{2k}}{(1 + x_{2k})(1 + x_{2k+1})} \\
&< \frac{x_{2k-1} - x_{2k}}{\left(1 + \frac{1}{2}\right)\left(1 + \frac{1}{2}\right)} < \frac{4}{9}\left(\frac{5}{9}\right)^k < \left(\frac{5}{9}\right)^{k+1},
\end{aligned}
$$

which yields (3) for $n = k + 1$.

By (2.14), our sequence (x_n) converges to some $x \in \mathbb{R}$. Also, $1/2 < x < 1$. Moreover, we have [using (2.8)]

$$
x = \lim x_{n+1} = \lim \frac{1}{1 + x_n} = \frac{1}{1 + x}
$$

and so

$$
x = (1/2)(\sqrt{5} - 1).
$$

Our next theorem gives further nontrivial examples of monotone limits and, at the same time, serves to define a number that is of extreme importance in analysis.

(2.16)   The Number e   *For each $n \in \mathbb{N}$ let*

$$
a_n = (1 + 1/n)^n, \qquad b_n = (1 + 1/n)^{n+1}.
$$

Then

   (i) (a_n) *is strictly increasing;*
   (ii) (b_n) *is strictly decreasing; and*
   (iii) $\lim\limits_{n \to \infty} a_n = \lim\limits_{n \to \infty} b_n$.
The common limit of these two sequences is denoted by e. We have $2 < e < 4$. *

Proof   For $n > 1$ we use Bernoulli's Inequality (1.46) to obtain

$$
\frac{a_n}{b_{n-1}} = \left(\frac{n+1}{n}\right)^n \left(\frac{n}{n-1}\right)^{-n} = \left(\frac{n^2 - 1}{n^2}\right)^n
$$

$$
= \left(1 - \frac{1}{n^2}\right)^n > 1 - \frac{1}{n} = \left(\frac{n}{n-1}\right)^{-1} \tag{1}
$$

and

$$
\frac{b_{n-1}}{a_n} = \left(\frac{n^2}{n^2 - 1}\right)^n = \left(1 + \frac{1}{n^2 - 1}\right)^n > \left(1 + \frac{1}{n^2}\right)^n > 1 + \frac{1}{n}. \tag{2}
$$

---

*Further properties of e and better rational estimates of its value are given in (2.58).

From (1) we have

$$a_n > b_{n-1}(n/(n-1))^{-1} = a_{n-1}$$

and from (2) we have

$$b_{n-1} > a_n(1 + 1/n) = b_n.$$

Thus (i) and (ii) are proven. Plainly, $n > 1$ implies

$$2 = a_1 < a_n < b_n < b_1 = 4$$

and so, using (2.12), (a_n) and (b_n) both converge to limits in $]2, 4[$. Moreover, (2.8.iii) yields

$$\lim_{n\to\infty} b_n = \left[\lim_{n\to\infty} a_n \right]\left[\lim_{n\to\infty} (1 + 1/n) \right] = \lim_{n\to\infty} a_n. \quad \square$$

Even though a sequence in $\mathbb{R}^{\#}$ need not have a limit, there are two important numbers associated with any such sequence. Our next definition makes this precise.

(2.17) **Definition** Let $(x_n)_{n=1}^{\infty} \subset \mathbb{R}^{\#}$. For each $k \in \mathbb{N}$ define

$$y_k = \inf\{ x_n : n \in \mathbb{N}, n \geq k \}$$

and

$$z_k = \sup\{ x_n : n \in \mathbb{N}, n \geq k \}.$$

Plainly,

$$y_1 \leq y_2 \leq y_3 \leq \ldots \quad \text{and} \quad z_1 \geq z_2 \geq z_3 \geq \ldots .$$

Let

$$y = \lim_{k\to\infty} y_k = \sup\{ y_k : k \in \mathbb{N}\}$$

and

$$z = \lim_{k\to\infty} z_k = \inf\{ z_k : k \in \mathbb{N}\}.$$

Then $y, z \in \mathbb{R}^{\#}$. The numbers y and z are called the *limit inferior* of (x_n) and the *limit superior* of (x_n) respectively. We write

$$y = \varliminf_{n\to\infty} x_n = \lim_{n\to\infty} \inf x_n,$$

$$z = \varlimsup_{n\to\infty} x_n = \lim_{n\to\infty} \sup x_n.$$

Since $j \leq k$ in $\mathbb{N}$ implies $y_j \leq y_k \leq z_k \leq z_j$, we have $y \leq z$.

The numbers y and z are characterized in a different, but equivalent, way in the next theorem.

(2.18) **Theorem** *Let $(x_n)_{n=1}^{\infty} \subset \mathbb{R}^{\#}$ and let $a, b \in \mathbb{R}^{\#}$. Then*

(i) $a = \lim_{n\to\infty} x_n$ *if and only if whenever $\alpha < a$ we have $\{n \in \mathbb{N} : x_n < \alpha\}$ is* *finite and whenever* $a < \beta$ *we have* $\{n \in \mathbb{N}: x_n < \beta\}$ *is infinite; and*

(ii) $b = \varlimsup\limits_{n \to \infty} x_n$ *if and only if whenever* $\alpha < b$ *we have* $\{n \in \mathbb{N}: x_n > \alpha\}$ *is infinite and whenever* $b < \beta$ *we have* $\{n \in \mathbb{N}: x_n > \beta\}$ *is finite.*

Proof   We prove only (i). The proof of (ii) is similar.

Suppose that $a = \varliminf x_n$ and let (y_k) and y be defined as in (2.17). Then $y = a$ and $a = \sup\{y_k : k \in \mathbb{N}\}$. Let $\alpha < a$. [If $a = -\infty$, this "α condition" is vacuously satisfied.] Then there is a k_0 such that $y_{k_0} > \alpha$; hence $x_n \geq y_{k_0} > \alpha$ for all $n \geq k_0$. Thus $\{n \in \mathbb{N}: x_n < \alpha\}$ is finite. Now let $a < \beta$. [If $a = \infty$, the "β condition" holds.] Then $y_k < \beta$ for all k, so, by the definition of y_k, for each $k \in \mathbb{N}$ there exists $n_k \geq k$ such that $y_k \leq x_{n_k} < \beta$. Therefore $\{n \in \mathbb{N}: x_n < \beta\} \supset \{n_k : k \in \mathbb{N}\}$ and the latter set is infinite.

Conversely, suppose that a satisfies both the "α condition" and the "β condition" of (i). If $a = -\infty$, then $a \leq y$. If $a > -\infty$, then, given $\alpha < a$, we can choose $k_0 \in \mathbb{N}$ such that $\alpha \leq x_n$ for all $n \geq k_0$. Thus $\alpha \leq y_{k_0}$ and so we obtain

$$\alpha \leq \sup\{y_k : k \in \mathbb{N}\} = y. \tag{1}$$

Since (1) holds for all $\alpha < a$, we have, whether or not $a = -\infty$, that

$$a \leq y. \tag{2}$$

If $a = \infty$, the proof is complete. If $a < \infty$, let $a < \beta$. By the "β condition" of (i), we see that $y_k < \beta$ for each $k \in \mathbb{N}$ and therefore

$$y \leq \beta. \tag{3}$$

Since (3) holds for every $\beta > a$, we have

$$y \leq a. \tag{4}$$

Finally, combine (2) and (4) to get $a = y = \varliminf x_n$. $\square$

The chief value of the lim inf and the lim sup lies in the next theorem.

(2.19)   Theorem   *Let* $(x_n)_{n=1}^{\infty} \subset \mathbb{R}^{\#}$. *Then* $\lim x_n$ *exists in* $\mathbb{R}^{\#}$ *if and only if* $\varliminf x_n = \varlimsup x_n$. *In this case,* $\lim x_n = \varliminf x_n = \varlimsup x_n$.

Proof   Let $a = \varliminf x_n$ and $b = \varlimsup x_n$. Then $a \leq b$. Assume that $\lim x_n = x \in \mathbb{R}^{\#}$ and $a < b$. Choose $\alpha, \beta \in \mathbb{R}$ such that $a < \beta < \alpha < b$. Then (2.18) shows that

$$\{n \in \mathbb{N} : x_n < \beta\} \text{ is infinite} \tag{1}$$

and

$$\{n \in \mathbb{N} : x_n > \alpha\} \text{ is infinite.} \tag{2}$$

If $x \in \mathbb{R}$ and $\epsilon = (\alpha - \beta)/2$, then (1) or (2) shows that there are infinitely many n for which x_n is outside the interval $]x - \epsilon, x + \epsilon[$ contrary to the assumption that $\lim x_n = x$. If $x = -\infty$, then (2) is contrary to $\lim x_n = x$. If $x = \infty$, then (1) is contrary to our assumption. We conclude in every case that our assumption is false and so $\lim x_n$ exists in $\mathbb{R}^{\#}$ only if $a = b$.

Conversely, suppose that $a = b$. If $a \in \mathbb{R}$ and $\epsilon > 0$, then (2.18) shows that $|x_n - a| < \epsilon$ except for finitely many n. Thus $\lim x_n = a = b$. If $a = -\infty$ and $\beta \in \mathbb{R}$, then (2.18.ii) shows that $x_n < \beta$ except for finitely many n. Thus $\lim x_n = b = a$. If $a = \infty$ and $\alpha \in \mathbb{R}$, then (2.18.i) shows that $x_n > \alpha$ except for finitely many n. Thus $\lim x_n = a = b$. $\quad\square$

(2.20) Definitions Let $x \in \mathbb{R}$. By a *neighborhood* of x (in $\mathbb{R}$) we mean any open interval $]a, b[$ containing x. A *neighborhood* of ∞ is any interval of the form $]a, \infty]$ where $a \in \mathbb{R}$. A *neighborhood* of $-\infty$ is any interval of the form $[-\infty, b[$ where $b \in \mathbb{R}$.

In view of these definitions, we see that, for $(x_n)_{n=1}^{\infty} \subset \mathbb{R}^{\#}$ and $x \in \mathbb{R}^{\#}$, $\lim x_n = x$ if and only if each neighborhood of x contains x_n for all but finitely many n.

(2.21) Definition Let $(x_n)_{n=1}^{\infty} \subset \mathbb{R}^{\#}$. A point $x \in \mathbb{R}^{\#}$ is called a *cluster point* of (x_n) if each neighborhood of x contains x_n for infinitely many n.

(2.22) Theorem *Let $(x_n)_{n=1}^{\infty} \subset \mathbb{R}^{\#}$ and let A be the set of all cluster points of this sequence. Then*
  (i) $(\underline{\lim} x_n) \in A$ *and* $(\overline{\lim} x_n) \in A$,
  (ii) $\underline{\lim} x_n \leqq c \leqq \overline{\lim} x_n$ *for all $c \in A$.*

Proof This is an immediate consequence of (2.20), (2.21), and (2.18). $\quad\square$

(2.23) Corollary *A sequence in $\mathbb{R}^{\#}$ has a limit in $\mathbb{R}^{\#}$ if and only if it has just one cluster point.*

Proof Combine (2.22) and (2.19). $\quad\square$

It follows from (2.22) that each sequence in $\mathbb{R}^{\#}$ has at least one cluster point. The next example shows that a sequence can have many cluster points.

(2.24) Example Let $(x_n)_{n=1}^{\infty}$ be an enumeration of the set $\mathbb{Q}$ of all rational numbers. Then each $x \in \mathbb{R}^{\#}$ is a cluster point of (x_n), $\underline{\lim} x_n = -\infty$, and $\overline{\lim} x_n = \infty$.

Our next theorem is often useful in evaluating limits. Its proof shows how we work with $\underline{\lim}$ and $\overline{\lim}$. First, a lemma.

(2.25) Lemma *If c is a positive real number, then*
  (i) $\displaystyle\lim_{n \to \infty} \sqrt[n]{c} = 1$.

Proof Suppose first that $c \geq 1$. Write $\sqrt[n]{c} = 1 + x_n$. Then $x_n \geq 0$, $c = (1 + x_n)^n \geq 1 + nx_n$,

$$0 \leq \sqrt[n]{c} - 1 = x_n \leq (c - 1)/n, \tag{1}$$

and so (i) obtains.

If $0 < c < 1$, we have $\lim_{n \to \infty} \sqrt[n]{c} = \lim_{n \to \infty} 1/\sqrt[n]{1/c} = 1/1 = 1$. Note, for example, that (1) yields $0 < \sqrt[n]{2} - 1 \leq 1/n$. $\square$

(2.26) Theorem *Let $(a_n)_{n=1}^{\infty}$ be a sequence of positive real numbers. Then*

$$\varliminf_{n \to \infty} (a_{n+1}/a_n) \leq \varliminf_{n \to \infty} \sqrt[n]{a_n} \leq \varlimsup_{n \to \infty} \sqrt[n]{a_n} \leq \varlimsup_{n \to \infty} (a_{n+1}/a_n).$$

In particular, if $\lim_{n \to \infty} (a_{n+1}/a_n) = L$, where $0 \leq L \leq \infty$, then $\lim_{n \to \infty} \sqrt[n]{a_n} = L$.

Proof The last assertion follows from the first by using (2.19) twice.

The second inequality is obvious, while the first and third have similar proofs. We prove only the first. Let $\varliminf(a_{n+1}/a_n) = a$. Plainly $a \geq 0$. If $a = 0$, there is nothing to prove, so we suppose that $a > 0$. Let $0 < \alpha < a$. By (2.18.i), there is an integer N such that

$$a_{k+1}/a_k > \alpha \tag{1}$$

for all $k \geq N$. For $n > N$, multiply the inequalities (1) for $k = N, N + 1, \ldots, n - 2, n - 1$ to get

$$a_n/a_N > \alpha^{n-N} \quad \text{and} \quad \sqrt[n]{a_n} > \alpha \sqrt[n]{a_N \alpha^{-N}}. \tag{2}$$

Since (2) holds for all $n > N$, (2.25) shows that

$$\varliminf \sqrt[n]{a_n} \geq \varliminf \alpha \sqrt[n]{a_N \alpha^{-N}} = \alpha.$$

But α was an arbitrary (positive) number $< a$, and so $\varliminf \sqrt[n]{a_n} \geq a = \varliminf(a_{n+1}/a_n)$. $\square$

(2.27) Examples (a) Let $a_n = 1$ for odd n and $a_n = 2^n$ for even n. Then we have

$$\frac{a_{n+1}}{a_n} = \begin{cases} 2^{n+1} & n \text{ odd}, \\ 1/2^n & n \text{ even} \end{cases} \qquad \sqrt[n]{a_n} = \begin{cases} 1 & n \text{ odd} \\ 2 & n \text{ even} \end{cases}$$

and so

$$\varliminf(a_{n+1}/a_n) = 0 < 1 = \varliminf \sqrt[n]{a_n} < \varlimsup \sqrt[n]{a_n} = 2 < \infty = \varlimsup (a_{n+1}/a_n).$$

(b) Let $a_n = n$ for all n. Then $\lim(a_{n+1}/a_n) = \lim(1 + 1/n) = 1$, and so $\lim \sqrt[n]{n} = \lim \sqrt[n]{a_n} = 1$.

(c) Let $a_n = n^n/n!$ for all n. Then $\lim(a_{n+1}/a_n) = \lim(1 + 1/n)^n = e$, and so $\lim(n/\sqrt[n]{n!}) = e$.

Cauchy Sequences

The definition of convergence for a sequence of complex numbers requires knowledge of the limit in order to check that the sequence converges. One can guess at the limit (if any) and then check the candidate against the definition. Other possibilities are to use the limit theorems or to separate into real and imaginary parts in hopes that some theorem about sequences in $\mathbb{R}^{\#}$ might apply; e.g., monotoneness or $\overline{\lim} = \underline{\lim}$. In many cases we do not care to (or cannot) evaluate the actual limit, but we want only to know whether or not the limit exists. In such cases, what is needed is some sort of criterion that can be checked by looking only at the sequence itself, not a proposed limit. One such criterion is monotoneness, but it applies only to certain sequences. One of the many contributions of the great French analyst Augustin Cauchy (1789–1857) was his discovery of an intrinsic criterion that applies to all sequences in $\mathbb{C}$.

(2.28) Definition A sequence $(S_n)_{n=1}^{\infty} \subset \mathbb{C}$ is called a *Cauchy sequence* if for each $\epsilon > 0$ there exists some $n_0 \in \mathbb{N}$ such that, for $m, n \in \mathbb{N}$,

$$m \geqq n_o \text{ and } n \geqq n_o \text{ imply } |S_m - S_n| < \epsilon.$$

If this is so, we write $\lim\limits_{m,\,n \to \infty} |S_m - S_n| = 0$.

(2.29) Cauchy's Criterion *A sequence of complex numbers converges if and only if it is a Cauchy sequence.*

Proof Let $(S_n) \subset \mathbb{C}$. Suppose that $\lim S_n = S \in \mathbb{C}$. Then, given $\epsilon > 0$, there exists n_0 such that

$$n \geqq n_0 \text{ implies } |S - S_n| < \epsilon/2.$$

Thus

$$m, n \geqq n_0 \text{ implies } |S_m - S_n| \leqq |S_m - S| + |S - S_n| < \epsilon/2 + \epsilon/2 = \epsilon.$$

Hence, (S_n) is a Cauchy sequence.

The converse is more difficult. We split its proof into two cases.

Case 1. Suppose that $(x_n) \subset \mathbb{R}$ is a Cauchy sequence. Let $a = \underline{\lim} \, x_n$ and $b = \overline{\lim} \, x_n$. Choose n_0 such that

$$m, n \geqq n_0 \text{ implies } |x_m - x_n| < 1.$$

Let $\alpha = 1 + \max\{|x_1|, |x_2|, \ldots, |x_{n_0}|\}$. Then $|x_n| < \alpha$ for all n, and so

$$-\infty < -\alpha \leqq a \leqq b \leqq \alpha < \infty.$$

Assume that $a < b$. Let $\epsilon = (b - a)/4$. Then $\epsilon > 0$, and so there exists n_1 such that

$$m, n, \geqq n_1 \text{ implies } |x_m - x_n| < \epsilon.$$

Since by (2.22) a and b are cluster points of (x_n), there exist $m, n > n_1$ such

that
$$|x_m - a| < \epsilon \quad \text{and} \quad |x_n - b| < \epsilon.$$
Then
$$4\epsilon = b - a \leq |a - x_m| + |x_m - x_n| + |x_n - b| < \epsilon + \epsilon + \epsilon = 3\epsilon,$$

which implies that $4 < 3$. This contradiction shows that our assumption is false. Thus $a = b$, and so (x_n) converges and $\lim x_n = a = b \in \mathbb{R}$, (2.19).

Case 2. Let $(z_n) \in \mathbb{C}$ be a Cauchy sequence. Write $z_n = x_n + iy_n$ where $x_n, y_n \in \mathbb{R}$. Since
$$|x_m - x_n| \leq |z_m - z_n| \quad \text{and} \quad |y_m - y_n| \leq |z_m - z_n|,$$

the real sequences (x_n) and (y_n) are Cauchy sequences, and so [Case 1] they both converge. Using (2.7), we see that (z_n) is convergent. $\square$

Subsequences

(2.30)   Definition   Let $(x_n)_{n=1}^{\infty}$ be a sequence in some set X. If $n_1 < n_2 < n_3 < \ldots$ are natural numbers, then the sequence $(x_{n_k})_{k=1}^{\infty}$ is called a *subsequence* of $(x_n)_{n=1}^{\infty}$.

(2.31)   Theorem   *If* $(S_k)_{k=1}^{\infty} \subset \mathbb{C}$, $\lim_{k \to \infty} S_k = S \in \mathbb{C}$, *and* $(S_{k_j})_{j=1}^{\infty}$ *is a subsequence of* (S_k), *then* $\lim_{j \to \infty} S_{k_j} = S$.

Proof   Given $\epsilon > 0$, choose k_0 such that
$$k \geq k_o \quad \text{implies} \quad |S - S_k| < \epsilon.$$
Choose j_0 such that $k_{j_0} \geq k_0$ [one can take $j_0 = k_0$ (2.30)]. Then
$$j \geq j_0 \quad \text{implies} \quad k_j \geq k_{j_0} \geq k_0, \text{ which in turn}$$
$$\text{implies} \qquad |S - S_{k_j}| < \epsilon. \qquad \square$$

Even though a sequence need not converge, it may have a convergent subsequence.

(2.32)   Theorem   *Every bounded sequence in* $\mathbb{C}$ *has a convergent subsequence.*

Proof   *Case* 1.   Let (x_n) be a bounded sequence in $\mathbb{R}$. Let c be any cluster point of (x_n); e.g., $c = \overline{\lim} \, x_n$. Since (x_n) is bounded we have $c \in \mathbb{R}$. Choose $n_1 \in \mathbb{N}$ such that $|c - x_{n_1}| < 1$. When $n_1, \ldots, n_k$ have been chosen, choose $n_{k+1} > n_k$ so that $|c - x_{n_{k+1}}| < 1/(k + i)$. Thus we have selected a strictly increasing sequence $(n_k)_{k=1}^{\infty} \subset \mathbb{N}$ such that $|c - x_{n_k}| < 1/k$ for all k. Given

$\epsilon > 0$, choose $k_0 > 1/\epsilon$. Then

$$k \geq k_o \quad \text{implies} \quad |c - x_{n_k}| < 1/k \leq 1/k_0 < \epsilon.$$

Therefore, $\lim_{k \to \infty} x_{n_k} = c$.

    Case 2. Let (z_n) be a bounded sequence in $\mathbb{C}$. Write $z_n = x_n + iy_n$ where $x_n, y_n \in \mathbb{R}$. Then (x_n) and (y_n) are bounded in $\mathbb{R}$: $|x_n| \leq |z_n|$, $|y_n| \leq |z_n|$. By Case 1, there is a subsequence (x_{n_k}) of (x_n) such that $\lim_{k \to \infty} x_{n_k} = x \in \mathbb{R}$. Use Case 1 again to get a subsequence $(y_{n_{k_j}})_{j=1}^{\infty}$ of $(y_{n_k})_{k=1}^{\infty}$ such that $\lim_{j \to \infty} y_{n_{k_j}} = y \in \mathbb{R}$. By (2.31), $\lim_{j \to \infty} x_{n_{k_j}} = x$. Now let $z = x + iy$ and use (2.8) to see that

$$\lim_{j \to \infty} z_{n_{k_j}} = z.$$

Thus $(z_{n_{k_j}})_{j=1}^{\infty}$ is a convergent subsequence of $(z_n)_{n=1}^{\infty}$. $\quad\square$

(2.33) **Examples** (a) A bounded sequence may have many convergent subsequences with distinct limits. For example, let $z_n = (i)^n$ for $n \in \mathbb{N}$. If we successively take $n_k = 4k$, $4k + 1$, $4k + 2$, and $4k + 3$ for $k \in \mathbb{N}$, we have $z_{4k} = 1$, $z_{4k+1} = i$, $z_{4k+2} = -1$, and $z_{4k+3} = -i$, respectively. Thus we have four distinct subsequential limits. Of course, a sequence with more than one subsequential limit cannot converge [see (2.31)].

    (b) No unbounded sequence can converge (2.5), but an unbounded sequence can have a convergent subsequence. For example, take $z_{2k-1} = k$, $z_{2k} = 1/k$ for $k \in \mathbb{N}$. Thus (z_n) is the sequence

$$1, 1, 2, 1/2, 3, 1/3, 4, 1/4, \ldots .$$

We have $z_{2k-1} \to \infty$ and $z_{2k} \to 0$.

Series of Complex Terms

The study of infinite series is one of the most important aspects of analysis. The usefulness of this topic cannot be overemphasized. In this brief section we give only a bare introduction to it. An extensive study of series is carried out in Chapter 7 and elsewhere in this volume. For the moment, we stick to series of complex numbers. Later, we study series of functions.

(2.34) **Definition** Let $(a_n)_{n=1}^{\infty} \subset \mathbb{C}$ and define

$$s_n = \sum_{k=1}^{n} a_k = a_1 + a_2 + \ldots + a_n$$

for each $n \in \mathbb{N}$. The symbol

$$\sum_{n=1}^{\infty} a_n \quad \text{or} \quad a_1 + a_2 + \dots$$

is called an *infinite series* having nth *term* a_n and nth *partial sum* S_n. This series is said to *converge* in case the sequence (S_n) of its partial sums converges (2.2); otherwise, it is said to *diverge*. In case $\lim_{n \to \infty} S_n = S$, we call the number S the *sum of* the series and we write

$$\sum_{n=1}^{\infty} a_n = S.$$

Thus, for convergent series, the symbol $\sum_{n=1}^{\infty} a_n$ plays a dual role: it denotes the series and also its sum. Note carefully that the sum of a convergent series is *not* obtained by performing infinitely many additions, but rather, it is the limit of a sequence.

We will also consider series of the form

$$\sum_{n=p}^{\infty} a_n = a_p + a_{p+1} + \dots$$

where $p \in \mathbb{Z}$ [cf. (2.1)]. Here the partial sums are the numbers

$$S_n = a_p + a_{p+1} + \dots + a_n$$

$n = p, p + 1, \dots$. Note that if $a_n = 0$ for all $n > q \geq p$, then $S_n = \sum_{k=p}^{q} a_k$ for all $n > q$ and so

$$\sum_{n=p}^{\infty} a_n = \lim_{n \to \infty} S_n = \sum_{n=p}^{q} a_n,$$

thus the series is just a finite sum in disguise.

Our first theorem about series gives a *necessary* (but not sufficient) condition that a series converge.

(2.35)   Theorem   *The notation is as in (2.34). If $\displaystyle\sum_{n=1}^{\infty} a_n$ converges, then $\lim_{n \to \infty} a_n = 0$.*

Proof   Let $\displaystyle\sum_{n=1}^{\infty} a_n = S \in \mathbb{C}$. Since $a_n = S_n - S_{n-1}$ for $n > 1$, we have

$$\lim_{n \to \infty} a_n = \lim_{n \to \infty} (S_n - S_{n-1}) = S - S = 0. \quad \square$$

To see that the condition $a_n \to 0$ is *not* sufficient to guarantee the convergence of $\sum a_n$ we refer the reader to Exercise 12 of Chapter 1 or to the next theorem.

(2.36) **Theorem** *The series* $\sum\limits_{n=1}^{\infty} (1/n)$ *diverges. It is called the harmonic series.*

Proof Assume that the series converges. Then its partial sums

$$S_n = 1 + 1/2 + \ldots + 1/n = \sum_{k=1}^{n} (1/k)$$

form a Cauchy sequence. Thus there exists $N \in \mathbb{N}$ such that

$$n \geqq N \quad \text{implies} \quad |S_n - S_N| < 1/3.$$

Therefore,

$$\frac{1}{3} > S_{2N} - S_N$$

$$= \frac{1}{N+1} + \frac{1}{N+2} + \ldots + \frac{1}{2N} > \frac{1}{2N} + \frac{1}{2N} + \ldots + \frac{1}{2N} = \frac{N}{2N}$$

$$= \frac{1}{2}.$$

This contradiction completes the proof. $\square$

Our next theorem treats what is perhaps the most elementary class of convergent series.

(2.37) **Geometric Series** *Let* $a, z \in \mathbb{C}$. *Then the series* $\sum\limits_{n=0}^{\infty} az^n$ *converges and its sum is* $a/(1 - z)$ *if* $|z| < 1$. *If* $a \neq 0$ *and* $|z| \geqq 1$, *then this series diverges.*

Proof By the formula for geometric progressions, we have

$$S_n = \sum_{k=0}^{n} az^k = a \frac{1 - z^{n+1}}{1 - z}.$$

Now, supposing that $|z| < 1$, apply (2.6.i) and (2.8) to obtain

$$\lim_{n \to \infty} S_n = a \frac{1 - 0}{1 - z} = \frac{a}{1 - z}.$$

This proves our first assertion.

If $a \neq 0$ and $|z| \geqq 1$, then $|az^n| \geqq |a| > 0$ for all n and so (2.35) shows that the series cannot converge. $\square$

Geometric series are easy to deal with because the partial sums can be computed explicitly. This can also be done if the terms a_n can be put into the form $a_n = b_n - b_{n+1}$, where the behavior of the sequence (b_n) is known; for then

$$S_n = \sum_{k=1}^{n} (b_k - b_{k+1}) = (b_1 - b_2) + (b_2 - b_3) + \ldots + (b_{n-1} - b_n) + (b_n - b_{n+1})$$

$$= b_1 - b_{n+1}.$$

Such sums are said to be *telescoping*. Of course, this is always possible in theory, for we could take $b_1 = 0$ and $b_{n+1} = -S_n$ for $n > 1$, but we might not know the behavior of (S_n). However, this program does succeed sufficiently often to make it worth mentioning. An instance is the following.

(2.38)   **Examples**   (a) If $a_n = 1/((n+1)n)$ for $n \geq 1$, then $a_n = 1/n - 1/(n+1)$ and

$$S_n = \frac{1}{2 \cdot 1} + \frac{1}{3 \cdot 2} + \frac{1}{4 \cdot 3} + \ldots + \frac{1}{(n+1)n} = 1 - \frac{1}{n+1} \to 1$$

as $n \to \infty$. Thus $\sum_{n=1}^{\infty} 1/n(n+1) = 1$.

(b) We can use (a) to prove that $\sum_{n=1}^{\infty} 1/n^2$ converges to a sum ≤ 2. Let S_n be as in (a). Then the nth partial sum t_n of our present series satisfies

$$t_n = 1 + \frac{1}{2 \cdot 2} + \frac{1}{3 \cdot 3} + \ldots + \frac{1}{n \cdot n}$$

$$< 1 + \frac{1}{2 \cdot 1} + \frac{1}{3 \cdot 2} + \ldots + \frac{1}{n(n-1)} = 1 + S_{n-1} < 2.$$

Since t_n is strictly increasing and bounded above by 2, we have, by (2.12), that (t_n) converges and that

$$\sum_{n=1}^{\infty} \frac{1}{n^2} = \lim_{n \to \infty} t_n \leq 2.$$

This is a good example of a series for which it is not hard to see that it converges, but its sum is quite hard to compute explicitly. This is a feature common to many infinite series. Actually, it is known that

$$\frac{8}{5} < \sum_{n=1}^{\infty} \frac{1}{n^2} = \frac{\pi^2}{6} < \frac{5}{3},$$

but we are hopelessly far at this stage in the book from being able to prove this equality. Indeed, since we have not yet *defined* the number π, the equality has, as yet, no meaning for us.

We can phrase the Cauchy Criterion in terms of convergence of series.

(2.39)   **Cauchy Criterion**   *Let $(a_n)_{n=1}^{\infty} \subset \mathbb{C}$. Then $\sum_{n=1}^{\infty} a_n$ converges if and only if to each $\epsilon > 0$ there corresponds $n_0 \in \mathbb{N}$ such that $q > p \geq n_0$ implies $\left| \sum_{n=p+1}^{q} a_n \right| < \epsilon$.*

Proof   Let $S_p = a_1 + a_2 + \ldots + a_p$. To say that $\sum a_n$ converges means that the sequence $(S_p)_{p=1}^{\infty}$ converges. But this is true if and only if (S_p) is a

Cauchy sequence; i.e., $\left| \sum_{n=p+1}^{q} a_n \right| = |S_q - S_p| < \epsilon$ for all sufficiently large

$p < q$. $\square$

(2.40)   **Corollary**   *If $\sum_{n=0}^{\infty} a_n$ and $\sum_{n=0}^{\infty} b_n$ are series of complex terms and $a_n = b_n$ for all sufficiently large n, then the two series either both converge or both diverge. That is, the convergence or divergence of a series cannot be affected by altering finitely many terms.*

Proof   For some $n_0 \in \mathbb{N}$, $n > n_0$ implies $a_n = b_n$. Thus $q > p > n_0$ implies

$$\sum_{n=p+1}^{q} a_n = \sum_{n=p+1}^{q} b_n. \quad \square$$

To show how Cauchy's Criterion can be used either to prove convergence or to prove divergence, we next present an interesting theorem of the great Norwegian mathematician N. H. Abel (1802–1829).

(2.41)   **Theorem**   *Let $\sum a_n$ be a divergent series of positive terms and let $S_n = a_1 + a_2 + \ldots + a_n$ be its nth partial sum. Then*
   (i) $\sum(a_n/S_n)$ diverges

and

   (ii) $\sum(a_n/S_n^2)$ converges.

Proof   (i) Write $d_n = a_n/S_n$. We shall show that Cauchy's Criterion for convergence of $\sum d_n$ is violated with $\epsilon = 1/2$. In fact, for any given $p \in \mathbb{N}$, no matter how large, we can choose $q \in \mathbb{N}$ such that $S_q > 2S_p$ because $S_n \to \infty$ as $n \to \infty$. Then we have

$$\sum_{n=p+1}^{q} d_n = \frac{a_{p+1}}{S_{p+1}} + \ldots + \frac{a_q}{S_q} \geq \frac{1}{S_q}(a_{p+1} + \ldots + a_q)$$

$$= \frac{1}{S_q}(S_q - S_p) = 1 - S_p/S_q > 1/2 = \epsilon.$$

Thus, there is no n_0 corresponding to $\epsilon = 1/2$ as required in (2.39): hence, $\sum d_n$ diverges.
   (ii) Write $c_n = a_n/S_n^2$. Given $\epsilon > 0$, choose n_0 such that $S_{n_0} > 1/\epsilon$. Then $q > p \geq n_0$ implies

$$\sum_{n=p+1}^{q} c_n < \sum_{n=p+1}^{q} \frac{a_n}{S_n S_{n-1}} = \sum_{n=p+1}^{q} \left(\frac{1}{S_{n-1}} - \frac{1}{S_n} \right)$$

$$= \frac{1}{S_p} - \frac{1}{S_q} < \frac{1}{S_p} \leq \frac{1}{S_{n_0}} < \epsilon.$$

Thus, (2.39) shows that $\sum c_n$ converges. $\square$

Note that if $a_n = 1$ for all $n \geq 1$, then this theorem shows that $\sum(1/n)$ diverges and $\sum(1/n^2)$ converges.

Another simple application of (2.39) is the next theorem.

(2.42) Theorem *Let $(a_n)_{n=0}^{\infty}$ and $(b_n)_{n=0}^{\infty}$ be complex sequences. If $\sum|a_n|$ converges and (b_n) is bounded, then $\sum a_n b_n$ converges.*

> **Proof** Choose $\beta \in \mathbb{R}$ such that $|b_n| \leq \beta$ for all n. Let $\epsilon > 0$ be given. Choose n_0 such that
>
> $$q > p \geq n_0 \quad \text{implies} \quad \sum_{n=p+1}^{q} |a_n| < \frac{\epsilon}{\beta}.$$
>
> Then
>
> $$q > p \geq n_0 \quad \text{implies} \quad |\sum_{n=p+1}^{q} a_n b_n| \leq \sum_{n=p+1}^{q} |a_n b_n| \leq \beta \sum_{n=p+1}^{q} |a_n| < \epsilon.$$
>
> We have used Cauchy's Criterion twice to complete the proof. $\square$

(2.43) Corollary *If $\sum\limits_{n=0}^{\infty} a_n$ is a series of complex terms and $\sum\limits_{n=0}^{\infty} |a_n|$ converges, then $\sum\limits_{n=0}^{\infty} a_n$ converges.*

Proof Let $b_n = 1$ for all n. $\square$

This corollary leads us to define absolute convergence.

(2.44) Definition A series $\sum a_n$ of complex terms is said to *converge absolutely* [or to be *absolutely convergent*] if $\sum|a_n|$ converges.

We note that (2.43) says that every absolutely convergent series is convergent. The converse is false since $\sum\limits_{n=1}^{\infty} (1/n)$ diverges (2.36) while $\sum\limits_{n=1}^{\infty} ((-1)^{n+1}/n)$ converges, as the following theorem shows.

(2.45) Leibnitz's Alternating Series Test *Let $(a_n)_{n=1}^{\infty}$ be a monotone nonincreasing sequence of positive terms such that $\lim\limits_{n\to\infty} a_n = 0$. Then $\sum\limits_{n=1}^{\infty} (-1)^{n+1}a_n$ is convergent. Moreover, if $S_n = \sum\limits_{k=1}^{n} (-1)^{k+1}a_k$ and $S = \sum\limits_{k=1}^{\infty} (-1)^{k+1}a_k$, then $|S - S_n| \leq a_{n+1}$ for all $n \geq 1$.*

Proof For all $n \geq 1$ we have

$$S_{2n+2} - S_{2n} = a_{2n+1} - a_{2n+2} \geq 0,$$
$$S_{2n+2} - S_{2n+1} = -a_{2n+2} \qquad \leq 0,$$
$$S_{2n+1} - S_{2n-1} = -a_{2n} + a_{2n+1} \leq 0,$$

and so

$$S_{2n} \leq S_{2n+2} \leq S_{2n+1} \leq S_{2n-1}.$$

Also,

$$\lim_{n \to \infty} (S_{2n-1} - S_{2n}) = \lim_{n \to \infty} a_{2n} = 0.$$

It follows from (2.14) that (S_n) converges to some $S \in \mathbb{R}$ and $S_{2n} \leq S \leq S_{2n-1}$ for all $n \geq 1$. Accordingly,

$$|S - S_{2n}| \leq |S_{2n+1} - S_{2n}| = a_{2n+1}$$

and

$$|S - S_{2n-1}| \leq |S_{2n} - S_{2n-1}| = a_{2n}$$

for all $n \geq 1$. $\square$

Corresponding to the limit theorems for sequences we have the following facts for series.

(2.46) **Theorem** *If $\sum_{n=1}^{\infty} a_n$ and $\sum_{n=1}^{\infty} b_n$ are both convergent series of complex terms and $c \in \mathbb{C}$, then*

(i) $\sum_{n=1}^{\infty} ca_n = c \sum_{n=1}^{\infty} a_n,$

(ii) $\sum_{n=1}^{\infty} (a_n + b_n) = \sum_{n=1}^{\infty} a_n + \sum_{n=1}^{\infty} b_n.$

In particular, the two series on the left both converge.

Proof This follows at once from (2.8) if we recall the definition (2.34) of convergence for series. The details are left to the reader. $\square$

Series of Nonnegative Terms

In this section we study series with terms in $[0, \infty]$.

(2.46) **Definition** Let $\sum_{n=1}^{\infty} a_n$ be a series with terms $a_n \in [0, \infty]$ and let $S_n = \sum_{k=1}^{n} a_k$

be its nth partial sum [recall that $\infty + \infty = \infty$ and $x + \infty = \infty = \infty + x$ for $x \in \mathbb{R}$ so that $S_n = \infty$ if $a_k = \infty$ for some k with $1 \leq k \leq n$]. Since $(S_n)_{n=1}^{\infty}$ is a nondecreasing sequence it has a limit $s \in \mathbb{R}^{\#}$ (2.12). Thus we write

$$\sum_{n=1}^{\infty} a_n = s.$$

This number s is called the *sum* of the series. If $s < \infty$, this definition agrees with (2.34); if $s < \infty$, the series converges; if $s = \infty$, the series diverges. In any case, the series has a sum and we write

$$\sum_{n=1}^{\infty} a_n < \infty \quad \text{or} \quad \sum_{n=1}^{\infty} a_n = \infty$$

according as the series converges or diverges.

(2.47) Theorem *If $a_n, b_n \in [0, \infty]$ for all $n \in \mathbb{N}$ and $c \in [0, \infty]$, then*

$$\text{(i)} \quad \sum_{n=1}^{\infty} ca_n = c \sum_{n=1}^{\infty} a_n$$

and

$$\text{(ii)} \quad \sum_{n=1}^{\infty} (a_n + b_n) = \sum_{n=1}^{\infty} a_n + \sum_{n=1}^{\infty} b_n.$$

Recall that $x \cdot \infty = \infty \cdot x = \infty$ for $x \in]0, \infty]$ and $0 \cdot \infty = \infty \cdot 0 = 0$.

Proof If both sums on the right of (ii) are finite, apply (2.46). Otherwise, use (2.48). $\square$

Our next theorem, though trivial, is very often useful in testing series for convergence.

(2.48) Comparison Test *Let $0 \leq a_n \leq b_n \leq \infty$ for all $n \in \mathbb{N}$. Then*

$$\text{(i)} \quad \sum_{n=1}^{\infty} a_n \leq \sum_{n=1}^{\infty} b_n.$$

Hence

$$\text{(ii)} \quad \sum_{n=1}^{\infty} a_n \text{ converges if } \sum_{n=1}^{\infty} b_n \text{ converges}$$

and

$$\text{(iii)} \quad \sum_{n=1}^{\infty} b_n \text{ diverges if } \sum_{n=1}^{\infty} a_n \text{ diverges}.$$

If, in addition, $\displaystyle\sum_{n=1}^{\infty} a_n < \infty$ and $a_p < b_p$ for some $p \in \mathbb{N}$, then

$$\text{(iv)} \quad \sum_{n=1}^{\infty} a_n < \sum_{n=1}^{\infty} b_n.$$

Proof We have

$$\sum_{n=1}^{k} a_n \leq \sum_{n=1}^{k} b_n \leq \sup\left\{ \sum_{n=1}^{m} b_n : m \in \mathbb{N} \right\} = \lim_{m \to \infty} \sum_{n=1}^{m} b_n = \sum_{n=1}^{\infty} b_n$$

for all $k \in \mathbb{N}$ and so

$$\sum_{n=1}^{\infty} a_n = \lim_{k \to \infty} \sum_{n=1}^{k} a_n \leqq \sum_{n=1}^{\infty} b_n.$$

This proves (i), from which (ii) and (iii) are obvious.

To prove (iv), note that

$$(b_p - a_p) + \sum_{n=1}^{k} a_n \leqq \sum_{n=1}^{k} b_n \leqq \sum_{n=1}^{\infty} b_n$$

for all $k \geqq p$, and so

$$\sum_{n=1}^{\infty} a_n < (b_p - a_p) + \sum_{n=1}^{\infty} a_n \leqq \sum_{n=1}^{\infty} b_n. \quad \square$$

(2.49) **Examples** (a) $\sum_{n=1}^{\infty} (1/\sqrt{n}) = \infty$ because $\sum_{n=1}^{\infty} (1/n) = \infty$ and $1/\sqrt{n} \geqq 1/n$

for all $n \in \mathbb{N}$.

(b) $\sum_{n=1}^{\infty} \frac{1}{2^n + n} < \sum_{n=1}^{\infty} \frac{1}{2^n} = 1.$

We prove the following rearrangement theorem for later use.

(2.50) **Theorem** *Suppose that $a_{m,n} \in [0, \infty]$ for each $(m, n) \in \mathbb{N} \times \mathbb{N}$ and that φ is any one to one mapping of $\mathbb{N}$ onto $\mathbb{N} \times \mathbb{N}$ [see (1.60)]. Then*

(i) $\displaystyle\sum_{m=1}^{\infty} \left(\sum_{n=1}^{\infty} a_{m,n} \right) = \sum_{k=1}^{\infty} a_{\varphi(k)}$

and

(ii) $\displaystyle\sum_{m=1}^{\infty} \left(\sum_{n=1}^{\infty} a_{m,n} \right) = \sum_{n=1}^{\infty} \left(\sum_{m=1}^{\infty} a_{m,n} \right).$

Proof First let α be any real number $<$ the right hand side of (i). Choose $k_0 \in \mathbb{N}$ such that

$$\sum_{k=1}^{k_0} a_{\varphi(k)} > \alpha.$$

Next choose $m_0, n_0 \in \mathbb{N}$ such that

$$\{\varphi(k) : 1 \leqq k \leqq k_0\} \subset \{(m, n) : 1 \leqq m \leqq m_0, 1 \leqq n \leqq n_0\}.$$

Then we have

$$\alpha < \sum_{k=1}^{k_0} a_{\varphi(k)} \leqq \sum_{m=1}^{m_0} \left(\sum_{n=1}^{n_0} a_{m,n} \right) \leqq \sum_{m=1}^{m_0} \left(\sum_{n=1}^{\infty} a_{m,n} \right) \leqq \sum_{m=1}^{\infty} \left(\sum_{n=1}^{\infty} a_{m,n} \right).$$

Since α was arbitrary, it follows that

$$\sum_{k=1}^{\infty} a_{\varphi(k)} \leqq \sum_{m=1}^{\infty} \left(\sum_{n=1}^{\infty} a_{m,n} \right). \tag{1}$$

Now let β be any real number $<$ the left hand side of (i) and then choose $m_1 \in \mathbb{N}$ such that

$$\beta < \sum_{m=1}^{m_1} \left(\sum_{n=1}^{\infty} a_{m,n} \right).$$

Using (2.47.ii) and induction on m_1, we have

$$\sum_{m=1}^{m_1} \left(\sum_{n=1}^{\infty} a_{m,n} \right) = \sum_{n=1}^{\infty} \left(\sum_{m=1}^{m_1} a_{m,n} \right);$$

hence, we can choose $n_1 \in \mathbb{N}$ such that

$$\beta < \sum_{n=1}^{n_1} \left(\sum_{m=1}^{m_1} a_{m,n} \right). \tag{2}$$

Now choose $k_1 \in \mathbb{N}$ such that

$$\{(m,n) : 1 \leq m \leq m_1, 1 \leq n \leq n_1\} \subset \{\varphi(k) : 1 \leq k \leq k_1\}.$$

Then we have

$$\sum_{m=1}^{m_1} \left(\sum_{n=1}^{n_1} a_{m,n} \right) \leq \sum_{k=1}^{k_1} a_{\varphi(k)} \leq \sum_{k=1}^{\infty} a_{\varphi(k)}. \tag{3}$$

Combining (2) and (3) yields

$$\beta < \sum_{k=1}^{\infty} a_{\varphi(k)}. \tag{4}$$

Finally, (4) and the arbitrariness of β prove the reverse of inequality (1) and so (i) obtains.

Equality (ii) will follow from equality (i) if we can show that

$$\sum_{m=1}^{\infty} \left(\sum_{n=1}^{\infty} a_{n,m} \right) = \sum_{k=1}^{\infty} a_{\varphi(k)} \tag{5}$$

since the left side of (5) is just the right side of (ii) with the roles of m and n interchanged. To this end, write

$$b_{m,n} = a_{n,m} \tag{6}$$

and $\psi(k) = (m,n)$ if $\varphi(k) = (n,m)$ so that

$$b_{\psi(k)} = a_{\varphi(k)}. \tag{7}$$

Then ψ is a one to one mapping of $\mathbb{N}$ onto $\mathbb{N} \times \mathbb{N}$, so applying the first part of our theorem to $b_{m,n}$ and ψ, we have

$$\sum_{m=1}^{\infty} \left(\sum_{n=1}^{\infty} b_{m,n} \right) = \sum_{k=1}^{\infty} b_{\psi(k)}. \tag{8}$$

In view of (6) and (7), (5) is just (8). $\square$

(2.51) **Example** To illustrate the preceding theorem take $\varphi(k) = (m, n)$ when $k = 2^{n-1}(2m - 1)$ and take $a_{m,n} = 2^{-2n}(2m - 1)^{-2}$. Then $a_{\varphi(k)} = 1/(4k^2)$ and so, since $\sum_{n=1}^{\infty} \dfrac{1}{4^n} = \dfrac{1}{3}$, we have

$$\sum_{m=1}^{\infty} \frac{1}{(2m - 1)^2} = 3 \sum_{m=1}^{\infty} \left(\frac{1}{(2m - 1)^2} \sum_{n=1}^{\infty} \frac{1}{4^n} \right)$$

$$= 3 \sum_{m=1}^{\infty} \left(\sum_{n=1}^{\infty} a_{m,n} \right) = 3 \sum_{k=1}^{\infty} a_{\varphi(k)} = \frac{3}{4} \sum_{k=1}^{\infty} \frac{1}{k^2}.$$

This fact can be seen more easily from the equality

$$\sum_{n=1}^{\infty} \frac{1}{(2n)^2} = \frac{1}{4} \sum_{n=1}^{\infty} \frac{1}{n^2}.$$

As an illustration of the use of (2.50.ii) we prove the following inequality.

(2.52) **Carleman's Inequality** *If* $(x_n)_{n=1}^{\infty} \subset [0, \infty[$ *and* $x_n \neq 0$ *for some n, then*

(i) $\displaystyle\sum_{n=1}^{\infty} (x_1 x_2 \ldots x_n)^{1/n} < e \sum_{n=1}^{\infty} x_n$ *if the latter series is convergent.*

Proof* If $(c_n)_{n=1}^{\infty}$ is any sequence of positive real numbers, then it follows from (1.48) that

$$(x_1 x_2 \ldots x_n)^{1/n} \leqq \frac{x_1 c_1 + x_2 c_2 + \ldots + x_n c_n}{n (c_1 c_2 \ldots c_n)^{1/n}} \tag{1}$$

for all $n \in \mathbb{N}$. Let us take

$$c_n = (n + 1)^n / n^{n-1}.$$

Then

$$(c_1 c_2 \ldots c_n)^{1/n} = \left(\frac{2}{1} \cdot \frac{3^2}{2} \cdot \frac{4^3}{3^2} \cdot \ldots \cdot \frac{(n + 1)^n}{n^{n-1}} \right)^{1/n} = n + 1. \tag{2}$$

Now write

$$a_{m,n} = x_m c_m / (n(n + 1)) \quad \text{if } 1 \leqq m \leqq n$$

and $a_{m,n} = 0$ if $m > n$.

---

*This clever proof was kindly communicated to the author by Professor Robert B. Burckel. It is due to Professor G. Pólya.

Invoking (1), (2), and (2.50.ii), we have

$$\sum_{n=1}^{\infty} (x_1 x_2 \dots x_n)^{1/n} \leq \sum_{n=1}^{\infty} \left(\frac{1}{n(n+1)} \sum_{m=1}^{n} x_m c_m \right)$$

$$= \sum_{n=1}^{\infty} \left(\sum_{m=1}^{\infty} a_{m,n} \right) = \sum_{m=1}^{\infty} \left(\sum_{n=1}^{\infty} a_{m,n} \right)$$

$$= \sum_{m=1}^{\infty} \left(\sum_{n=m}^{\infty} \frac{x_m c_m}{n(n+1)} \right)$$

$$= \sum_{m=1}^{\infty} \left(x_m c_m \sum_{n=m}^{\infty} \left\{ \frac{1}{n} - \frac{1}{n+1} \right\} \right)$$

$$= \sum_{m=1}^{\infty} \frac{x_m c_m}{m} . \tag{3}$$

From (2.16) we have

$$c_m / m = (1 + 1/m)^m < e \tag{4}$$

for all $m \in \mathbb{N}$. Now combine (3) and (4) to get (i). $\square$

(2.53) **Caution** Theorem (2.50) can fail without the hypothesis that $a_{m,n} \geq 0$. For example, let $a_{m,n} = 1$ if $n = m + 1$, $a_{m,n} = -1$ if $m = n + 1$, and $a_{m,n} = 0$ otherwise. Then

$$\sum_{m=1}^{\infty} \left(\sum_{n=1}^{\infty} a_{m,n} \right) = 1, \qquad \sum_{n=1}^{\infty} \left(\sum_{m=1}^{\infty} a_{m,n} \right) = -1,$$

and, for any φ as in (2.50), the series

$$\sum_{k=1}^{\infty} a_{\varphi(k)} \text{ diverges}$$

since its terms do not tend to 0. Other conditions for the validity of (i) and (ii) of (2.50) are examined in Chapter 7.

Our next theorem shows that the order in which the terms of a nonnegative termed series are taken cannot affect its sum.

(2.54) **Theorem** *Let S be a countably infinite set and let $\{x_k\}_{k=1}^{\infty}$ be an enumeration of S. Then for any function $f : S \to [0, \infty]$ we have*

(i) $\displaystyle \sum_{k=1}^{\infty} f(x_k) = \sup \left\{ \sum_{x \in F} f(x) : F \text{ is finite, } F \subset S \right\}$.

The symbol $\displaystyle \sum_{x \in F} f(x)$ denotes the sum of the values of f at the points of the

set F. In view of this theorem we denote the right-hand side of (i) by $\displaystyle \sum_{x \in S} f(x)$ or

$\sum \{ f(x) : x \in S \}$ for any nonvoid set S and any $f : S \to [0, \infty]$. If $S = \emptyset$, $\displaystyle \sum_{x \in S} f(x)$

$= 0$ by definition.

Proof   Let $\alpha < \sum_{x \in S} f(x)$. Choose a finite $F \subset S$ such that

$$\alpha < \sum_{x \in F} f(x).$$

Then choose $k_0 \in \mathbb{N}$ such that $F \subset \{x_k : 1 \leq k \leq k_0\}$. It follows that

$$\alpha < \sum_{k=1}^{k_0} f(x_k) \leq \sum_{k=1}^{\infty} f(x_k)$$

and so, since α was arbitrary,

$$\sum_{x \in S} f(x) \leq \sum_{k=1}^{\infty} f(x_k).$$

Plainly

$$\sum_{k=1}^{\infty} f(x_k) = \sup \left\{ \sum_{k=1}^{n} f(x_k) : n \in \mathbb{N} \right\} \leq \sum_{x \in S} f(x). \quad \square$$

Decimal Expansions

Undoubtedly every reader is familiar with such customary decimal notations as 3.1416. Such a symbol is understood to denote a number. Thus, the equality

$$3.1416 = 3 + 1/10 + 4/100 + 1/1000 + 6/10000$$

defines the numerical value of the symbol on its left side. In this section we define and study b-adic expansions of real numbers. Compare this with b-adic expansions of integers given in (1.29).

(2.55)   Theorem   *Let b be an integer > 1 and let $(x_j)_{j=1}^{\infty}$ be a sequence of integers with $0 \leq x_j < b$ for all j. Then the series*

(i) $\displaystyle\sum_{j=1}^{\infty} x_j b^{-j}$

converges and its sum x satisfies $0 \leq x \leq 1$.

Proof   By (2.48) and (2.37) we have $0 \leq \displaystyle\sum_{j=1}^{\infty} x_j b^{-j} \leq \sum_{j=1}^{\infty} (b-1) b^{-j} = 1$.

$\square$

(2.56)   Definition   With the notation of (2.55), we write

(i) $x = \cdot_b x_1 x_2 x_3 \ldots$

where the symbol on the right side of (i) is formed by juxtaposing all of the terms of the sequence $(x_j)_{j=1}^{\infty}$, in order, and preceding them by a period with a subscript b. This symbol is called an *infinite* [or *unending*] *b-adic expansion* for x. In case $x_j = 0$ for all $j > p \geq 1$, we also write

(ii) $x = \cdot_b x_1 x_2 x_3 \ldots x_p$.

This is a *finite* [or *terminating*] *b-adic expansion* for x. In case $b = 10$, it is customary to omit the subscript 10 in (i) and (ii) and then these are *decimal* expansions. A *binary* [or *dyadic*] *expansion* is a 2-adic expansion. A *ternary expansion* is a 3-adic expansion.

Note that some numbers have more than one decimal expansion. For example,

$$1/2 = .5000 \ldots \text{(all 0's)} \ldots$$

and

$$1/2 = .4999 \ldots \text{(all 9's)} \ldots.$$

We do, however, have the following existence and uniqueness theorem.

(2.57)   Theorem   *Let b be an integer > 1 and let $x \in \mathbb{R}$ with $0 < x \le 1$. Then there exists a sequence $(x_j)_{j=1}^{\infty}$ of integers such that $0 \le x_j \le b - 1$ for all j, $x_j \ne 0$ for infinitely many j, and*

(i) $x = \displaystyle\sum_{j=1}^{\infty} x_j b^{-j}.$

Moreover, if $(y_j)_{j=1}^{\infty}$ is any other $[x_j \ne y_j$ for some $j]$ sequence of integers such that $0 \le y_j \le b - 1$ for all j and

(ii) $x = \displaystyle\sum_{j=1}^{\infty} y_j b^{-j},$

then there exists $j_0 \ge 1$ such that
  (iii) $y_j = x_j \quad (1 \le j < j_0),$
  (iv) $y_{j_0} = x_{j_0} + 1,$
  (v) $x_j = b - 1, y_j = 0 \quad (j > j_0).$
Thus, x has a unique b-adic expansion containing infinitely many nonzero digits and x has at most one other b-adic expansion. Indeed, x has two b-adic expansions if and only if $x = a/b^{j_0}$ for some $a, j_0 \in \mathbb{N}$ $(a < b^{j_0})$.

Proof   Let x_1 be the integer such that

$$x_1 < bx \le x_1 + 1 \tag{1}$$

and, once $x_1, \ldots, x_{j-1}$ have been chosen, let x_j be the integer such that

$$x_j < b^j \left(x - \sum_{k=1}^{j-1} x_k b^{-k} \right) \le x_j + 1. \tag{2}$$

It is obvious from (1) and (2) that

$$0 < x - \sum_{k=1}^{j} x_k b^{-k} \le b^{-j} \tag{3}$$

for all $j \ge 1$. Therefore, (i) obtains. Because $0 < x \le 1$, it follows by induction on j, with the use of (1), (2), and (3), that $0 \le x_j < b$ for all j. With (i) and (3), combined, it is clear that $x_k = 0$ for all $k > j$ is impossible, and so $x_k \neq 0$ for infinitely many k. Thus we have the required sequence $(x_j)_{j=1}^{\infty}$.

Next suppose that (ii) obtains for some sequence $(y_j)_{j=1}^{\infty}$ of integers where $0 \leqq y_j < b$ for all j and $y_j \neq x_j$ for some j. Let j_0 be the smallest j such that $x_{j_0} \neq y_{j_0}$. Then (iii) obtains and

$$\sum_{j=j_0}^{\infty} x_j b^{-j} = \sum_{j=j_0}^{\infty} y_j b^{-j}. \tag{4}$$

Since $x_j \neq 0$ for some $j > j_0$, we have

$$x_{j_0} b^{-j_0} < \sum_{j=j_0}^{\infty} x_j b^{-j}. \tag{5}$$

Combining the estimate

$$\sum_{j=j_0}^{\infty} y_j b^{-j} \leqq y_{j_0} b^{-j_0} + \sum_{j=j_0+1}^{\infty} (b-1) b^{-j} = y_{j_0} b^{-j_0} + b^{-j_0}$$

with (4) and (5), we obtain $x_{j_0} < y_{j_0} + 1$ and so, since $x_{j_0} \neq y_{j_0}$,

$$x_{y_0} < y_{j_0}. \tag{6}$$

Now, using (6), we can write

$$y_{j_0} b^{-j_0} \leqq \sum_{j=j_0}^{\infty} y_j b^{-j} = \sum_{j=j_0}^{\infty} x_j b^{-j} \leqq x_{j_0} b^{-j_0} + \sum_{j=j_0+1}^{\infty} (b-1) b^{-j}$$

$$= (x_{j_0} + 1) b^{-j_0} \leqq y_{j_0} b^{-j_0}. \tag{7}$$

Since the equalities must obtain throughout (7), we get (iv) and (v).

If x has a second expansion, then we have just proved that

$$x = \sum_{j=1}^{j_0} y_j b^{-j},$$

$$b^{j_0} x = \sum_{j=1}^{j_0} y_j b^{j_0-j} = a \in \mathbb{N}.$$

On the other hand, if $a, j_0 \in \mathbb{N}$, $a < b^{j_0}$, and $x = a/b^{j_0}$, let $a = \sum_{k=0}^{j_0-1} a_k b^k$ be the base b expansion of a (1.29). Then x has the terminating expansion $\sum_{k=0}^{j_0-1} a_k b^{k-j_0}$. $\square$

The Number e

In (2.16) we defined the number e by

$$e = \lim_{n \to \infty} (1 + 1/n)^n = \lim_{n \to \infty} (1 + 1/n)^{n+1}.$$

Here we obtain an infinite series expression for e, we prove that e is irrational, and we obtain rational approximations for e.

(2.58) Theorem   *We have*

(i) $e = \sum\limits_{k=0}^{\infty} \dfrac{1}{k!}$,

(ii) *e is irrational,*

(iii) $2.666 \ldots < e < 2.722 \ldots$,

and if $S_n = \sum\limits_{k=0}^{n} \dfrac{1}{k!}$, *then*

(iv) $0 < e - S_n < 1/(n!n)$

for all $n \in \mathbb{N}$.

Proof   Write $\sum\limits_{k=0}^{\infty} \dfrac{1}{k!} = S \in \mathbb{R}^{\#}$. Since, in view of (2.48), for $n > 0$ we have

$$\sum_{k=n+1}^{\infty} \frac{1}{k!} = \frac{1}{n!}\left(\frac{1}{n+1} + \frac{1}{(n+1)(n+2)} + \frac{1}{(n+1)(n+2)(n+3)} + \cdots \right)$$

$$< \frac{1}{n!}\left(\frac{1}{n+1} + \frac{1}{(n+1)^2} + \frac{1}{(n+1)^3} + \cdots \right)$$

$$= \frac{1}{n!n},$$

it is clear that $S < 3$ [consider $n = 1$] and

$$0 < S - S_n < 1/(n!n) \tag{1}$$

for all $n \in \mathbb{N}$. As in (2.16), write $a_n = (1 + 1/n)^n$ and then use the Binomial Theorem to see that

$$a_n = 1 + \sum_{k=1}^{n} \frac{n(n-1)(n-2)\ldots(n-k+1)}{k!} \frac{1}{n^k}$$

$$= 1 + \sum_{k=1}^{n} \left(1 - \frac{1}{n}\right)\left(1 - \frac{2}{n}\right)\ldots\left(1 - \frac{k-1}{n}\right)\frac{1}{k!}$$

$$< 1 + \sum_{k=1}^{n} \frac{1}{k!} = S_n < S \tag{2}$$

for all $n > 1$. Thus

$$e = \lim_{n \to \infty} a_n \leqq S. \tag{3}$$

Let $\epsilon > 0$ be given and then fix an integer $p > 1$ such that

$$S - \epsilon < S_p. \tag{4}$$

For arbitrary $n > p$, we estimate as in (2) to obtain

$$e > a_n > 1 + \sum_{k=1}^{p} \left(1 - \frac{1}{n}\right) \cdots \left(1 - \frac{k-1}{n}\right) \frac{1}{k!} .$$

Letting $n \to \infty$ on the right side of this inequality and applying (2.8), we have

$$e \geqq S_p. \tag{5}$$

The combination of (3), (4) and (5) yields

$$S - \epsilon < e \leqq S$$

for all $\epsilon > 0$, and so $S = e$, proving (i). Now (iv) follows from $S = e$ and (1). To prove (iii) take $n = 3$ in (iv).

Finally, assume that e is rational—say $e = m/n$ where $m, n \in \mathbb{N}$. Replace e by m/n in (iv) and then multiply through by $n!$ to get

$$0 < (n-1)!m - n!S_n < 1/n \leqq 1. \tag{6}$$

Clearly, the second member of (6) is an integer between 0 and 1. This contradiction proves (ii). $\square$

(2.59) **Remarks** (a) The number e is actually transcendental. This fact was discovered in 1873 by the French mathematician Charles Hermite (1822–1901). We prove this fact later.

(b) Inequalities (2.58.iv) can be used to calculate decimal approximations to e with a great deal of accuracy. Taking $n = 12$, one obtains

$$2.718281826 < e < 2.718281832.$$

An easy induction [Exercise !], with the use of (2.16), shows that

$$n! > n^n e^{-n} > (n/3)^n$$

for all $n \in \mathbb{N}$, and hence

$$0 < e - S_n < 3^n / n^{n+1}.$$

For example, if $n = 30$, this shows that S_{30} differs from e by $< (1/3) \cdot 10^{-31}$ and therefore yields a decimal approximation accurate to at least 31 digits.

The Root and Ratio Tests

The two tests for absolute convergence given next are of considerable theoretical importance and are sometimes useful in testing particular series. However, they are both rather weak tests in that they show convergence only for series that compare favorably with geometric series and divergence only for series whose terms fail to have limit 0. More delicate tests are studied in Chapter 7.

(2.60) **Root Test** [Cauchy, 1821] *Let $\sum c_n$ be a series of complex terms and let*

$$\rho = \varlimsup_{n\to\infty} \sqrt[n]{|c_n|} \ .$$

(i) *If $\rho < 1$, the series converges absolutely.*
(ii) *If $\rho > 1$, the series diverges.*

Proof Suppose that $\rho < 1$. Fix β such that $\rho < \beta < 1$. By (2.18.ii), there exists $n_0 \in \mathbb{N}$ such that $\sqrt[n]{|c_n|} < \beta$ for all $n \geq n_0$. Thus,

$$\sum_{n=n_0}^{\infty} |c_n| < \sum_{n=n_0}^{\infty} \beta^n = \frac{\beta^{n_0}}{1-\beta} < \infty$$

and (i) follows.

   If $\rho > 1$, then (2.18.ii) shows that $\sqrt[n]{|c_n|} > 1$, hence $|c_n| > 1$, for infinitely many n, and so it is false that $c_n \to 0$. Thus, (ii) follows from (2.35). ☐

(2.61) **Ratio Test** [d'Alembert, 1768] *Let $\sum c_n$ be a series of complex terms with $c_n \neq 0$ for all n.*
   (i) *If $\varlimsup_{n\to\infty} |c_{n+1}/c_n| < 1$, then the series converges absolutely.*
   (ii) *If $\varliminf_{n\to\infty} |c_{n+1}/c_n| > 1$, the series diverges.*

Proof Put $a_n = |c_n|$ and apply (2.26) and (2.60). ☐

(2.62) **Remarks** (a) The strict inequalities are required in the above tests. If $c_n = 1/n$, then $\sum_{n=1}^{\infty} c_n = \infty$ while $\lim \sqrt[n]{c_n} = \lim (c_{n+1}/c_n) = 1$ [cf. (2.27.ii)]. If $c_n = 1/n^2$, then $\sum_{n=1}^{\infty} c_n < \infty$ while $\lim \sqrt[n]{c_n} = \lim (c_{n+1}/c_n) = 1$.
   (b) Our proof of the ratio test shows that any series that can be tested successfully by the ratio test can also be tested successfully by the root test [though the computations involved might be much more difficult]. In fact, the root test is strictly stronger than the ratio test as (c) and (d) show.
   (c) For $n \in \mathbb{N}$ let $c_n = 1/2^n$ if n is even and $c_n = 1/3^n$ if n is odd. Then $\sum c_n < \sum(1/2^n) < \infty$, $\varlimsup \sqrt[n]{|c_n|} = 1/2 < 1$, $\varlimsup |c_{n+1}/c_n| = \lim_{k\to\infty} (c_{2k}/c_{2k-1}) = \lim_{k\to\infty} (3^{2k-1}/2^{2k}) = \infty$. Thus, the root test detects the convergence while the ratio test does not. This also shows that $\varliminf$ cannot be replaced by $\varlimsup$ in (2.61.ii).
   (d) For $n \in \mathbb{N}$ let $c_n = 2^n$ if n is even and $c_n = 2^{-n}$ if n is odd. Then $\sum c_n = \infty$, $\varlimsup \sqrt[n]{c_n} = 2$, $\varliminf(c_{n+1}/c_n) = 0 < 1$. Thus, the root test detects the diver-

gence while the ratio test does not. Also, we have $\underline{\lim} \sqrt[n]{c_n} = 1/2 < 1$, and so $\overline{\lim}$ cannot be replaced by $\underline{\lim}$ in either (2.60.i) or (2.61.i).

Power Series

Among the simplest and most important ways of representing or defining a function as a limit of polynomials is by the use of power series. We use this method to define the exponential and the trigonometric functions.

(2.63) **Definition** A *power series* is a series of the form

$$\sum_{n=0}^{\infty} a_n z^n$$

where (a_n) is a fixed sequence of complex numbers and $z \in \mathbb{C}$. The number a_n is called the *n*th *coefficient* of the series.

(2.64) **Theorem** *Let* $\sum_{n=0}^{\infty} a_n z^n$ *be a power series and let* $\alpha = \overline{\lim_{n \to \infty}} \sqrt[n]{|a_n|}$. *Define R by*

(i)

$$R = \begin{cases} 1/\alpha & \text{if } 0 < \alpha < \infty \\ \infty & \text{if } \alpha = 0 \\ 0 & \text{if } \alpha = \infty. \end{cases}$$

Then the series converges absolutely whenever $|z| < R$ *and diverges whenever* $|z| > R$.

Proof Put $c_n = a_n z^n$. The theorem follows at once from (2.60), the computation

$$\rho = \overline{\lim} \sqrt[n]{|c_n|} = |z| \overline{\lim} \sqrt[n]{|a_n|} = |z| \cdot \alpha,$$

and a consideration of the three cases listed in (i). $\quad\square$

(2.65) **Definitions** With the notation of (2.64), the number R is called the *radius of convergence* of the power series. If $0 < R < \infty$, the circle $\{z \in \mathbb{C} : |z| = R\}$ is called the *circle of convergence* of the series.

(2.66) **Remark** It is clear from (2.64) that, for a given power series, R is the unique number in $[0, \infty]$ for which the conclusion of (2.64) obtains. It is not always necessary to use the *definition* of R in order to compute its value. For instance, if we apply the ratio test to the series

$$\sum_{n=1}^{\infty} \frac{n^n z^n}{n!}$$

we find, writing $c_n = n^n z^n / n!$, that

$$\frac{c_{n+1}}{c_n} = \frac{(n+1)^{n+1} z^{n+1} n!}{n^n z^n (n+1)!} = \left(1 + \frac{1}{n}\right)^n z$$

and so

$$\lim_{n \to \infty} |c_{n+1}/c_n| = e|z|.$$

Thus, the series converges absolutely if $|z| < 1/e$ and it diverges if $|z| > 1/e$; hence, $R = 1/e$. It is more difficult to prove directly that

$$\lim_{n \to \infty} \left(n / \sqrt[n]{n!}\right) = e.$$

     The following examples show that a power series can have quite varied convergence behavior at points on its circle of convergence.

(2.67)   **Examples**   (a) For $\sum_{n=0}^{\infty} z^n$, we have $R = 1$ and the series diverges whenever $|z| = 1$ since the terms do not approach 0.

     (b) For $\sum_{n=1}^{\infty} \frac{z^n}{n(n+1)}$, we have $R = 1$ [ratio test] and the series converges absolutely whenever $|z| = 1$ [comparison test].

     (c) For $\sum_{n=1}^{\infty} \frac{z^n}{n}$, we have $R = 1$ It diverges when $z = 1$ and it fails to converge absolutely when $|z| = 1$ [harmonic series]. It does, however, converge whenever $|z| = 1$ and $z \neq 1$ [cf. (7.38)]. For example, it converges when $z = -1$ [Leibnitz test].

     (d) For $\sum_{n=1}^{\infty} n^n z^n$ we have $R = 0$.

     (e) For $\sum_{n=0}^{\infty} \frac{z^n}{n!}$ we have $R = \infty$ [ratio test].

Multiplication of Series

If $\sum_{n=0}^{\infty} a_n = A$ and $\sum_{n=0}^{\infty} b_n = B$ are two convergent series, then one might hope that, by analogy with multiplication of polynomials, it would be possible to form a series whose terms consist of all possible products $a_m b_n$ $(m \geq 0, n \geq 0)$ in such a way that the new series converges to the sum AB. The immediate problem in doing this is to decide upon a reasonable way of arranging this collection of terms into a single sequence so that partial sums can be meaningfully considered [recall that the *sum* of a series is the limit of its sequence of partial sums if the limit exists]. It might be expected that the arrangement that is chosen can affect the sum (if any). This question is examined in detail in Chapter 7. For the present we consider only the Cauchy product of the two series. The idea behind it is that if we formally multiply the power series $\sum_{k=0}^{\infty} a_k z^k$ and $\sum_{m=0}^{\infty} b_m z^m$ and then collect the coefficients of like powers of z, we find that the coefficient of z^n is $\sum a_k b_m$ where the sum is extended over all ordered pairs (k, m) such that $k + m = n$; i.e., $0 \leq k \leq n$, $m = n - k$.

(2.68) Definition The *Cauchy product* of two series $\sum_{n=0}^{\infty} a_n$ and $\sum_{n=0}^{\infty} b_n$ of complex terms is the series $\sum_{n=0}^{\infty} c_n$ where

$$c_n = \sum_{k=0}^{n} a_k b_{n-k}.$$

(2.69) Example Cauchy himself gave this example of a convergent series whose Cauchy product with itself diverges. Take

$$a_0 = b_0 = 0, \qquad a_n = b_n = \frac{(-1)^{n-1}}{\sqrt{n}} \qquad \text{for } n \geq 1.$$

Then $\sum a_n = \sum b_n$ converges by (2.45), but $c_0 = c_1 = 0$ and, for $n > 1$,

$$c_n = \sum_{k=0}^{n} a_k b_{n-k} = (-1)^n \sum_{k=1}^{n-1} \frac{1}{\sqrt{k} \sqrt{n-k}},$$

$$|c_n| = \sum_{k=1}^{n-1} \frac{1}{\sqrt{k(n-k)}} \geq \sum_{k=1}^{n-1} \frac{1}{\sqrt{(n-1)(n-1)}} = 1;$$

hence, $c_n \not\to 0$ and so $\sum c_n$ diverges.

Despite the negative result of this example our next two theorems give remarkable positive results. First, we prove a lemma which is of considerable independent interest.

(2.70) Lemma Let $(\alpha_n)_{n=0}^{\infty}$ and $(\beta_n)_{n=0}^{\infty}$ be sequences in $\mathbb{C}$ and let α, β, $\in \mathbb{C}$.

(i) If $\sum_{n=0}^{\infty} |\beta_n| < \infty$ and $\alpha_n \to 0$, then

$$\gamma_n = \sum_{k=0}^{n} \alpha_k \beta_{n-k} \to 0.$$

(ii) *If $\alpha_n \to \alpha$ and $\beta_n \to \beta$, then*

$$\frac{1}{n+1} \sum_{k=0}^{n} \alpha_k \beta_{n-k} \to \alpha\beta.$$

(iii) *If $\alpha_n \to \alpha$, then*

$$\frac{1}{n+1} \sum_{k=0}^{n} \alpha_k \to \alpha.$$

Proof (i) Write $b = \sum_{n=0}^{\infty} |\beta_n|$. Let $\epsilon > 0$ and choose $n_0 \in \mathbb{N}$ such that $|\alpha_n|$ $< \epsilon/(2b)$ for all $n > n_0$. Write $a = \sum_{k=0}^{n_0} |\alpha_k|$. Since $\beta_j \to 0$ we can choose $j_0 \in \mathbb{N}$ such that $|\beta_j| < \epsilon/(2a)$ for all $j > j_0$. Then, $n > n_0 + j_0$ implies

$$|\gamma_n| \leq \sum_{k=0}^{n_0} |\alpha_k| |\beta_{n-k}| + \sum_{k=n_0+1}^{n} |\alpha_k| |\beta_{n-k}| < \frac{\epsilon}{2a} \cdot a + \frac{\epsilon}{2b} \cdot b = \epsilon,$$

and so (i) obtains.

(iii) Let $\epsilon > 0$ be given. Choose $n_0 \in \mathbb{N}$ such that $|\alpha - \alpha_k| < \epsilon/2$ for all $k > n_0$. Next choose $n_1 > n_0$ such that $\frac{1}{n+1} \sum_{k=0}^{n_0} |\alpha - \alpha_k| < \epsilon/2$ for all $n > n_1$. Then, $n > n_1$ implies

$$\left| \alpha - \frac{1}{n+1} \sum_{k=0}^{n} \alpha_k \right| = \left| \frac{1}{n+1} \sum_{k=0}^{n} (\alpha - \alpha_k) \right|$$

$$\leq \frac{1}{n+1} \sum_{k=1}^{n_0} |\alpha - \alpha_k| + \frac{1}{n+1} \sum_{k=n_0+1}^{n} |\alpha - \alpha_k|$$

$$< \frac{\epsilon}{2} + \frac{n - n_0}{n+1} \cdot \frac{\epsilon}{2} < \epsilon,$$

and so (iii) obtains.

(ii) Since $\beta_n \to \beta$, the sequence (β_n) is bounded—say $|\beta_n| < B < \infty$ for all n. Next write

$$\frac{1}{n+1} \sum_{k=0}^{n} \alpha_k \beta_{n-k} = \frac{1}{n+1} \sum_{k=0}^{n} (\alpha_k - \alpha) \beta_{n-k} + \frac{\alpha}{n+1} \sum_{k=0}^{n} \beta_k$$

$$= s_n + t_n.$$

From (iii), it is clear that $t_n \to \alpha\beta$. Thus, we need only prove $s_n \to 0$. But $|\alpha_k - \alpha| \to 0$, and so (iii) shows that

$$|s_n| \leq \frac{B}{n+1} \sum_{k=0}^{n} |\alpha_k - \alpha| \to 0. \quad \square$$

Now we return to Cauchy products.

(2.71) **Mertens' Theorem** [1875] *If at least one of the two convergent series* $\sum_{n=0}^{\infty} a_n = A$ *and* $\sum_{n=0}^{\infty} b_n = B$ *of complex terms is absolutely convergent and if* $\sum_{n=0}^{\infty} c_n$ *is their Cauchy product, then* $\sum_{n=0}^{\infty} c_n = AB$. *In particular, the Cauchy product converges.*

Proof Suppose that $\sum_{n=0}^{\infty} |b_n| < \infty$. Write

$$A_n = \sum_{k=0}^{n} a_k, \quad B_n = \sum_{k=0}^{n} b_k, \quad C_n = \sum_{k=0}^{n} c_k,$$

$$\beta_n = b_n, \qquad \alpha_n = A_n - A.$$

Then C_n is the sum of all of the products $a_j b_k$ for which $0 \le j \le n$, $0 \le k \le n$, and $0 \le j + k \le n$. Thus,

$$C_n = A_0 b_n + A_1 b_{n-1} + A_2 b_{n-2} + \ldots + A_n b_0$$

$$= (A + \alpha_0) \beta_n + (A + \alpha_1) \beta_{n-1} + (A + \alpha_2) \beta_{n-2} + \ldots + (A + \alpha_n) \beta_0$$

$$= AB_n + \sum_{k=0}^{n} \alpha_k \beta_{n-k}$$

$$\to AB + 0 = AB$$

by virtue of our hypotheses and (2.70.i). $\square$

(2.72) **Theorem** [Abel, 1826] *If* $\sum_{n-0}^{\infty} a_n = A$ *and* $\sum_{n=0}^{\infty} b_n = B$ *are convergent series of complex terms and if their Cauchy product* $\sum_{n=0}^{\infty} c_n = C$ *also converges, then* $C = AB$.

Proof [Cesàro, 1890] With the notation of the preceding proof, we have

$$C_k = A_0 b_k + A_1 b_{k-1} + \ldots + A_k b_0,$$

and so

$$C_0 + C_1 + \ldots + C_n = A_0 B_n + A_1 B_{n-1} + \ldots + A_n B_0. \tag{1}$$

Now divide both sides of (1) by $n + 1$ and then take the limit as $n \to \infty$, using (2.70.ii) and (2.70.iii), to obtain $C = AB$. $\square$

(2.73) **Corollary** *If the power series* $\sum_{n=0}^{\infty} a_n z^n$ *and* $\sum_{n=0}^{\infty} b_n z^n$ *have radii of convergence* R_1 *and* R_2 *respectively and if* $c_n = \sum_{k=0}^{n} a_k b_{n-k}$ *for* $n \ge 0$, *then* $\sum_{n=0}^{\infty} c_n z^n$
$= \left(\sum_{n=0}^{\infty} a_n z^n \right) \left(\sum_{n=0}^{\infty} b_n z^n \right)$ *whenever* $|z| < \min \{R_1, R_2\}$.

Proof Apply (2.71) and (2.64). $\square$

(2.74) **Examples** (a) Write

$$\text{(i) } \exp(z) = \sum_{n=0}^{\infty} \frac{z^n}{n!}$$

for all $z \in \mathbb{C}$. It follows from the Ratio Test that this series always converges absolutely and so (i) defines a function exp from $\mathbb{C}$ into $\mathbb{C}$. It is called the *exponential function*. We claim that

$$\text{(ii) } \exp(z + w) = \exp(z) \cdot \exp(w)$$

for all $z, w \in \mathbb{C}$. In fact, if we form the Cauchy product of the two series $\exp(z) = \sum_{n=0}^{\infty} \frac{z^n}{n!}$ and $\exp(w) = \sum_{n=0}^{\infty} \frac{w^n}{n!}$ and use the Binomial Theorem, we obtain

$$\exp(z)\exp(w) = \sum_{n=0}^{\infty} \left(\sum_{k=0}^{n} \frac{z^k}{k!} \frac{w^{n-k}}{(n-k)!} \right)$$

$$= \sum_{n=0}^{\infty} \frac{1}{n!} \left(\sum_{k=0}^{n} \frac{n!}{k!(n-k)!} z^k w^{n-k} \right)$$

$$= \sum_{n=0}^{\infty} \frac{1}{n!} (z + w)^n = \exp(z + w),$$

proving (ii).

(b) It may happen that the radius of convergence of the Cauchy product of two power series is strictly larger than that of either of them. For example, let $a_0 = 1$, $a_n = 2$ for $n \geq 1$, and

$$b_n = \frac{1}{5}\left((-1)^n \cdot 8 - \frac{3}{4^n} \right)$$

for $n \geq 0$. Using the Ratio Test, one checks that the two power series $\sum_{n=0}^{\infty} a_n z^n$ and $\sum_{n=0}^{\infty} b_n z^n$ each have radius of convergence equal to 1, while a simple computation, which we leave to the reader, shows that their Cauchy product is the series $\sum_{n=0}^{\infty} (z/4)^n$, which has radius of convergence equal to 4. In particular, taking $z = 1$, we get two divergent series [the terms do not have limit 0] whose Cauchy product is absolutely convergent.

Lebesgue Outer Measure

As an application of some of our work on series and for later use, we next define Lebesgue outer measure and prove a few of its simplest properties. Among other things, we learn that this measure is a function that assigns to each subset of $\mathbb{R}$ a number in $[0, \infty]$ in such a way that each interval is assigned its length.

(2.75) **Definition** For an interval I with endpoints $a \leqq b$ in $\mathbb{R}^{\#}$, let the *length* of I be denoted by

$$|I| = b - a.$$

For $E \subset \mathbb{R}$, we define the *Lebesgue [outer] measure* of E to be the number

$$\lambda(E) = \inf\left\{ \sum_{I \in \mathcal{I}} |I| : \mathcal{I} \text{ is a countable family of open intervals with } E \subset \cup \mathcal{I} \right\}.$$

This defines a function λ on $\mathcal{P}(\mathbb{R})$, the family of all subsets of $\mathbb{R}$, into $[0, \infty]$. This function is called *Lebesgue [outer] measure*. [Recall the definition of $\sum_{I \in \mathcal{I}} |I|$ given in (2.54).]

In the following theorem we give a number of important properties of Lebesgue measure.

(2.76) **Theorem**
    (i) $E \subset \mathbb{R}$ *implies* $0 \leqq \lambda(E) \leqq \infty$.
    (ii) $\lambda(\varnothing) = 0$.
    (iii) $E \subset F \subset \mathbb{R}$ *implies* $\lambda(E) \leqq \lambda(F)$.
    (iv) $(E_n)_{n=1}^{\infty} \subset \mathcal{P}(\mathbb{R})$ *implies* $\lambda(\bigcup_{n=1}^{\infty} E_n) \leqq \sum_{n=1}^{\infty} \lambda(E_n)$.
    (v) *If C is a countable subset of* $\mathbb{R}$, *then* $\lambda(C) = 0$.
    (vi) *If I is an interval of* $\mathbb{R}$, *then* $\lambda(I) = |I|$.
    (vii) *For $E \subset \mathbb{R}$ and $x \in \mathbb{R}$, write $E + x = \{t + x : t \in E\}$.*
Then $\lambda(E + x) = \lambda(E)$.

Proof Statements (i), (ii), and (iii) are obvious. To prove (iv), we suppose that $\lambda(E_n) < \infty$ for all n, since otherwise the conclusion is obvious. Let $\epsilon > 0$ be given. For each $n \in \mathbb{N}$ choose a family $\{I_{n,m}\}_{m=1}^{\infty}$ of open intervals such that $E_n \subset \bigcup_{m=1}^{\infty} I_{n,m}$ and

$$\sum_{m=1}^{\infty} |I_{n,m}| < \lambda(E_n) + \frac{\epsilon}{2^n}.$$

Let φ be a one-to-one mapping of $\mathbb{N}$ onto $\mathbb{N} \times \mathbb{N}$. Then $\bigcup_{k=1}^{\infty} I_{\varphi(k)} \supset \bigcup_{n=1}^{\infty} E_n$ and so, using (2.50.i), we have

$$\lambda\left(\bigcup_{n=1}^{\infty} E_n\right) \leqq \sum_{k=1}^{\infty} |I_{\varphi(k)}| = \sum_{n=1}^{\infty}\left(\sum_{m=1}^{\infty} |I_{n,m}|\right)$$

$$\leqq \sum_{n=1}^{\infty}\left(\lambda(E_n) + \frac{\epsilon}{2^n}\right)$$

$$= \sum_{n=1}^{\infty} \lambda(E_n) + \epsilon.$$

Since $\epsilon > 0$ was arbitrary, (iv) follows.

To prove (v), note first that if $x \in \mathbb{R}$, then $\lambda(\{x\}) \leq |]x - \epsilon, x + \epsilon[|$ $= 2\epsilon$ for all $\epsilon > 0$ and so $\lambda(\{x\}) = 0$. Thus if C is a countable subset of $\mathbb{R}$, we use (iv) [and (ii), if C is finite] to write

$$0 \leq \lambda(C) = \lambda\left(\bigcup_{x \in C} \{x\} \right) \leq \sum_{x \in C} \lambda(\{x\}) = 0,$$

which proves (v).

Suppose $I = [a, b]$ where $-\infty < a < b < \infty$. Plainly $I \subset]a - \epsilon, b + \epsilon[$ and so $\lambda(I) \leq b - a + 2\epsilon$ for all $\epsilon > 0$. Therefore

$$\lambda(I) \leq b - a. \tag{1}$$

To prove the reverse of inequality (1), let $\epsilon > 0$ be given and choose a countable family $\mathcal{G}$ of open intervals such that $I \subset \cup \mathcal{G}$ and

$$\sum_{J \in \mathcal{G}} |J| < \lambda(I) + \epsilon. \tag{2}$$

Use the Heine–Borel Theorem to obtain a finite family $\mathcal{F} \subset \mathcal{G}$ such that $I \subset \cup \mathcal{F}$. Now choose $]a_1, b_1[\in \mathcal{F}$ with $a_1 < a < b_1$. If $b < b_1$, the construction stops. If $n \geq 1$ and $b_n \leq b$ has been selected, choose $]a_{n+1}, b_{n+1}[\in \mathcal{F}$ such that $a_{n+1} < b_n < b_{n+1}$. If $b < b_{n+1}$, the construction stops. Otherwise, it continues inductively. Since $\mathcal{F}$ is finite and the intervals constructed are distinct elements of $\mathcal{F}$ $[b_n < b_{n+1}]$, the process must stop after p intervals have been chosen from $\mathcal{F}$ for some $p \in \mathbb{N}$. Thus, we get

$$\{]a_1, b_1[, \ldots,]a_p, b_p[\} \subset \mathcal{F}$$

with

$$a_1 < a < b_1, \qquad a_p < b < b_p,$$

and, if $p > 1$,

$$b_{p-1} \leq b, \qquad a_n < b_{n-1} < b_n \qquad (2 \leq n \leq p).$$

It follows that $I \subset \bigcup_{n=1}^{p}]a_n, b_n[$ and so, using (2), we have

$$|I| = b - a = (b_1 - a) + \sum_{n=2}^{p-1} (b_n - b_{n-1}) + (b - b_{p-1})$$

$$< \sum_{n=1}^{p} (b_n - a_n) \leq \lambda(I) + \epsilon.$$

Since ϵ was arbitrary, we infer that

$$|I| \leq \lambda(I). \tag{3}$$

Combining (1) and (3) yields (vi) for closed intervals.

If I is any interval with endpoints $a < b$ in $\mathbb{R}$, then we have

$$\lambda(I) \leq \lambda([a, b]) = b - a \leq \lambda(I) + \lambda(\{a, b\}) = \lambda(I),$$

so (vi) obtains for I.

If I is an unbounded interval, then I contains bounded intervals of arbitrarily great length and so (vi) holds for I.

Statement (vii) is true because $|I + x| = |I|$ for every interval I. $\quad\square$

(2.77) **Remarks** (a) Parts (v) and (vi) of (2.76) give another proof that $[0, 1]$ is uncountable. This proof can be easily traced to the completeness of $\mathbb{R}$ via the Heine–Borel Theorem.

(b) There exist disjoint subsets A and B of $\mathbb{R}$ such that

$$\lambda(A \cup B) < \lambda(A) + \lambda(B).$$

This can be proved by using the Axiom of Choice. However, it is true that there is a very large family $\mathfrak{M}_\lambda$ of "λ-measureable" sets having the property that if $\{E_n\}_{n=1}^\infty \subset \mathfrak{M}_\lambda$ is pairwise disjoint, then

$$\lambda\left(\bigcup_{n=1}^\infty E_n\right) = \sum_{n=1}^\infty \lambda(E_n).$$

This family $\mathfrak{M}_\lambda$ contains all intervals and is closed under the formation of countable unions and complements. This matter is examined in detail in Chapter 6.

(c) A set of positive measure need not contain any nonvoid open interval. For example, if E is the set of irrationals in $[0, 1]$ and N is the set of rationals in $[0, 1]$, then

$$1 = \lambda([0, 1]) = \lambda(E \cup N) \leq \lambda(E) + \lambda(N) = \lambda(E) \leq \lambda([0, 1]) = 1$$

so $\lambda(E) = 1$.

(2.78) **Definition** A set $N \subset \mathbb{R}$ is called a λ-*null set* if $\lambda(N) = 0$.

(2.79) **Corollary** (*i*) *Any subset of a λ-null set is a λ-null set.*

(ii) *The union of any countable family of λ-null sets is a λ-null set.*

(iii) *If N is a λ-null set and $E \subset \mathbb{R}$, then*

$$\lambda(E \cup N) = \lambda(E).$$

Proof (i) and (ii) follow from (2.76.iii) and (2.76.iv), respectively. To prove (iii) write

$$\lambda(E) \leq \lambda(E \cup N) \leq \lambda(E) + \lambda(N) = \lambda(E). \quad\square$$

We close this section with a rather weak result which we shall find very useful later. Its proof uses none of (2.76).

(2.80) **Lemma** *Let* $-\infty < a < b < \infty$, *let $\mathcal{I}$ be a nonvoid family of subintervals of* $[a, b]$ (*each of positive length*), *and let $E \subset \cup \mathcal{I}$. Then there exists a nonvoid, finite,* *pairwise disjoint family* $\mathcal{F} \subset \mathcal{I}$ *such that*

$$\sum_{I \in \mathcal{F}} |I| > (1/6)\lambda(E).$$

Proof For each $I \in \mathcal{I}$ let I^* denote the closed interval having the same midpoint as I with $|I^*| = 5|I|$. Note that

$$I, J \in \mathcal{I}, \quad I \cap J \neq \varnothing, \quad |I| < 2|J| \quad \text{imply} \quad I \subset J^*. \tag{1}$$

Write $\delta_0 = \sup\{|I| : I \in \mathcal{I}\}$ [plainly $\delta_0 \leqq b - a < \infty$] and then choose $I_1 \in \mathcal{I}$ such that $|I_1| > \delta_0/2$. Suppose that $n \geqslant 1$ and that pairwise disjoint intervals $I_1, I_2, \ldots, I_n$ have been chosen from $\mathcal{I}$ [write $A_n = I_1 \cup I_2 \cup \ldots \cup I_n$] such that

$$I \in \mathcal{I} \quad \text{implies} \quad I \cap A_n = \varnothing \quad \text{or} \quad I \subset I_1^* \cup \ldots \cup I_n^*. \tag{H_n}$$

It follows from (1) that (H_1) is satisfied.

If $I \cap A_n \neq \varnothing$ for all $I \in \mathcal{I}$, then $E \subset \cup \mathcal{I} \subset I_1^* \cup \ldots \cup I_n^*$ and so $\lambda(E) \leqq \sum_{k=1}^{n} |I_k^*| = 5 \sum_{k=1}^{n} |I_k|$. In this case our construction terminates and we take $\mathcal{F} = \{I_1, \ldots, I_n\}$. Otherwise, we have $\mathcal{I}_n = \{I \in \mathcal{I} : I \cap A_n = \varnothing\} \neq \varnothing$, we let $\delta_n = \sup\{|I| : I \in \mathcal{I}_n\}$, and we choose $I_{n+1} \in \mathcal{I}_n$ such that $|I_{n+1}| > \delta_n/2$. To see that (H_{n+1}) holds for the intervals $I_1, \ldots, I_{n+1}$, suppose that $I \in \mathcal{I}$ and $I \cap A_{n+1} \neq \varnothing$. Then either $I \cap A_n \neq \varnothing$ and (H_{n+1}) follows from (H_n) or else $I \in \mathcal{I}_n$ and $I \cap I_{n+1} \neq \varnothing$, in which case (H_{n+1}) follows from (1). This completes our inductive construction.

Suppose the above construction fails to terminate in finitely many steps. Then $\sum_{n=0}^{\infty} \delta_n \leqq \lim_{p \to \infty} \sum_{n=1}^{p} 2|I_n| \leqq 2(b - a) < \infty$ and so $\lim_{n \to \infty} \delta_n = 0$. Therefore, given $I \in \mathcal{I}$, we can choose n such that $\delta_n < |I|$; hence, $I \notin \mathcal{I}_n$, $I \cap A_n \neq \varnothing$, and by (H_n), $I \subset \bigcup_{k=1}^{\infty} I_k^*$. Thus, $E \subset \cup \mathcal{I} \subset \bigcup_{k=1}^{\infty} I_k^*$ and so $\lambda(E) \leqq \sum_{k=1}^{\infty} |I_k^*| = 5 \sum_{k=1}^{\infty} |I_k|$. Now choose p such that $\sum_{k=1}^{p} |I_k| > (1/6)\lambda(E)$ and let $\mathcal{F} = \{I_1, \ldots, I_p\}$. $\square$

We remark that the constant $1/6$ in this lemma is not the best possible but it suffices for our purposes. A simple reworking of our proof shows that $1/6$ can be replaced by any θ such that $0 < \theta < 1/3$.

Cantor Sets

We close this chapter with the construction of a class of subsets of $\mathbb{R}$ and a study of some of the strange properties of these sets. These sets are important in several constructions in later chapters. Such a set was first constructed by Georg Cantor

(1845–1918), the father of modern set theory, to solve a problem in trigonometric series.

(2.81) **The Construction** Let $\mathbf{a} = (a_n)_{n=0}^\infty$ be a fixed sequence of real numbers such that

$$a_0 = 1 \quad \text{and} \quad 0 < 2a_n < a_{n-1} \tag{1}$$

for all $n \in \mathbb{N}$. [In Cantor's original construction, he took $a_n = 3^{-n}$. There are obviously many sequences that satisfy (1).] Define $(d_n)_{n=1}^\infty$ by

$$d_n = a_{n-1} - 2a_n. \tag{2}$$

Now let $J_{0,1} = [0, 1]$, $J_{1,1} = [0, a_1]$, and $J_{1,2} = [1 - a_1, 1]$. We proceed by induction. When $n \in \mathbb{N}$ and a pairwise disjoint family $\{J_{n-1,k}\}_{k=1}^{2^{n-1}}$ of closed intervals, each having length a_{n-1}, has been designated, let $J_{n,2k-1}$ and $J_{n,2k}$ be the closed intervals of length a_n such that $J_{n,2k-1}$ has the same left endpoint of $J_{n-1,k}$ and $J_{n,2k}$ has the same right endpoint as $J_{n-1,k}$. By induction, this defines

$$\{J_{n,k} : n \geq 0, 1 \leq k \leq 2^n\}.$$

Let $I_{n,k} = J_{n-1,k} \backslash (J_{n,2k-1} \cup J_{n,2k})$ for $n > 0$ and $1 \leq k \leq 2^{n-1}$. Then $I_{n,k}$ is the open interval of length d_n having the same midpoint as $J_{n-1,k}$. For example:

$$J_{2,2} = [a_1 - a_2, a_1]; \qquad J_{3,3} = [a_1 - a_2, a_1 - a_2 + a_3];$$
$$J_{3,4} = [a_1 - a_3, a_1];$$
$$I_{3,2} = \,]a_1 - a_2 + a_3, a_1 - a_3[.$$

Next, write

$$P_n = \bigcup_{k=1}^{2^n} J_{n,k} \tag{3}$$

and

$$P = \bigcap_{n=1}^\infty P_n. \tag{4}$$

This completes the construction of the *Cantor set P corresponding to* $\mathbf{a}$. If $a_n = 3^{-n}$, then P is called *Cantor's ternary set.*

It is clear that the family $\{I_{n,k} : n \in \mathbb{N}, 1 \leq k \leq 2^{n-1}\}$ is pairwise disjoint. Also, we evidently have

$$[0, 1] = P_0 \supset P_1 \supset P_2 \supset \cdots, \tag{5}$$

$$P_{n-1} \backslash P_n = \bigcup_{k=1}^{2^{n-1}} I_{n,k}, \tag{6}$$

$$P_0 \backslash P = \bigcup_{n=1}^\infty (P_{n-1} \backslash P_n). \tag{7}$$

Thus, $P_0 = [0, 1]$ and P_n is obtained from P_{n-1} by removing from the center of $J_{n-1,k}(1 \leq k \leq 2^{n-1})$ the open interval $I_{n,k}$ and P is obtained from $[0, 1]$ by removing all of the open intervals $I_{n,k}(n \geq 1, 1 \leq k \leq 2^{n-1})$. In the case of Cantor's ternary set ($a_n = 3^{-n}$), we have, for example,

$$P_2 = [0, 1/9] \cup [2/9, 1/3] \cup [2/3, 7/9] \cup [8/9, 1],$$

while P_3 is the union of the eight closed intervals obtained by removing the open interval of length $1/27$ from the middle of each of these four.

We next compute the Lebesgue measure of P.

(2.82) Theorem *With the notation of (2.81) we have*
(i) $\lambda(P) = \lim_{n \to \infty} 2^n a_n$;
(ii) *if* $0 \leq \beta < 1$, *then* **a** *can be chosen such that* $\lambda(P) = \beta$;
(iii) *if* P *is Cantor's ternary set, then* $\lambda(P) = 0$.

Proof It is clear from (2.76) that

$$\lambda(P) \leq \lambda(P_n) \leq \sum_{k=1}^{2^n} \lambda(J_{n,k}) = 2^n a_n \tag{1}$$

for all n, and that

$$\lambda(P_0 \backslash P) \leq \sum_{n=1}^{\infty} \left(\sum_{k=1}^{2^{n-1}} \lambda(I_{n,k}) \right) = \sum_{n=1}^{\infty} 2^{n-1} d_n$$

$$= \sum_{n=1}^{\infty} 2^{n-1}(a_{n-1} - 2a_n)$$

$$= \lim_{n \to \infty} \sum_{k=1}^{n} (2^{k-1} a_{k-1} - 2^k a_k) = \lim_{n \to \infty} (1 - 2^n a_n)$$

$$= 1 - \lim_{n \to \infty} 2^n a_n. \tag{2}$$

Since $1 = \lambda(P_0) \leq \lambda(P) + \lambda(P_0 \backslash P)$, (2) yields

$$\lambda(P) \geq 1 - \lambda(P_0 \backslash P) \geq \lim_{n \to \infty} 2^n a_n.$$

Combining this with (1) gives (i).
To prove (ii) take, for example, $a_n = 2^{-n}\beta + 3^{-n}(1 - \beta)$, check that (2.81.1) holds, and apply (i). Taking $\beta = 0$, this also proves (iii). $\square$

Our next theorem gives a precise analytical description of the elements of P in terms of **a**.

(2.83) Theorem *Let* **a** *and* P *be as in (2.81). Define* $r_n = a_{n-1} - a_n$ *for each* $n \in \mathbb{N}$. *Let* S *denote the set of all infinite sequences* $\epsilon = (\epsilon_j)_{j=1}^{\infty} \subset \{0, 1\}$ [$\epsilon_j = 0$ *or* 1] *and write*

$$f(\epsilon) = \sum_{j=1}^{\infty} \epsilon_j r_j \text{ for } \epsilon \in S. \text{ Then } f \text{ is a one-to-one mapping of } S \text{ onto } P. \text{ That is, the}$$

elements of P *are just the numbers expressible in the form* $\sum_{j=1}^{\infty} \epsilon_j r_j$ *where each* $\epsilon_j = 0$ *or*

1. *In particular, if $a_n = 3^{-n}$, then $r_n = 2 \cdot 3^{-n}$, and so Cantor's ternary set consists of just those numbers in $[0, 1]$ having a ternary expansion containing only the digits 0 and 2.*

Proof Notice first that for $n \geq 0$,

$$\sum_{j=n+1}^{\infty} r_j = \lim_{m \to \infty} \sum_{j=n+1}^{m} (a_{j-1} - a_j) = \lim_{m \to \infty} (a_n - a_m) = a_n \qquad (1)$$

because $a_m \leq 2^{-m}$ for all m. Since $r_n - a_n = d_n > 0$, we also have

$$r_n > \sum_{j=n+1}^{\infty} r_j \qquad (2)$$

for all $n \in \mathbb{N}$. For $n \geq 1$ and any finite sequence $(\epsilon_1, \ldots, \epsilon_n)$ of 0's and 1's, define the closed interval

$$J(\epsilon_1, \ldots, \epsilon_n) = \left[\sum_{j=1}^{n} \epsilon_j r_j, a_n + \sum_{j=1}^{n} \epsilon_j r_j \right]. \qquad (3)$$

Clearly,

$$J(0) = [0, a_1] = J_{1,1}$$

and

$$J(1) = [r_1, a_1 + r_1] = [1 - a_1, 1] = J_{1,2}. \qquad (4)$$

We shall show by induction on n that the intervals $J(\epsilon_1, \ldots, \epsilon_n)$ are just the 2^n intervals $J_{n,k}$ $(1 \leq k \leq 2^n)$. For $n = 1$, this is (4). Let $n > 1$ and suppose that the assertion is true for $n - 1$. Given $J_{n-1,k}$ $(1 \leq k \leq 2^{n-1})$, our induction hypothesis says that $J_{n-1,k} = J(\epsilon_1, \ldots, \epsilon_{n-1})$ for one and only one sequence $(\epsilon_1, \ldots, \epsilon_{n-1})$. By our construction (2.81), the two intervals $J_{n,2k-1}$ and $J_{n,2k}$ are, respectively, the initial and final subintervals of $J_{n-1,k}$ having length a_n. But, by (3), $J(\epsilon_1, \ldots, \epsilon_{n-1}, 0)$ and $J(\epsilon_1, \ldots, \epsilon_{n-1}, 1)$ are just these two subintervals of $J(\epsilon_1, \ldots, \epsilon_{n-1}) = J_{n-1,k}$ [recall: $a_n + r_n = a_{n-1}$]. This completes our induction.

Now let $\epsilon \in S$. By (3) and (1), it is obvious that, for each $n \in \mathbb{N}$, $f(\epsilon) \in J(\epsilon_1, \ldots, \epsilon_n)$. Also, our induction shows that for each n there exists $1 \leq k(\epsilon, n) \leq 2^n$ such that $J(\epsilon_1, \ldots, \epsilon_n) = J_{n,k(\epsilon,n)} \subset P_n$. Thus, $f(\epsilon) \in P = \bigcap_{n=1}^{\infty} P_n$ and so f maps S into P.

To show that f is one-to-one, let $\epsilon \neq \epsilon'$ in S. Choose n such that $\epsilon_n \neq \epsilon'_n$. Then, by our induction, $f(\epsilon)$ and $f(\epsilon')$ are in different ones of the 2^n disjoint intervals $J_{n,k}$.

To show that f is onto P, let $x \in P$. Since $x \in P_1$, either $x \in J(0)$ or $x \in J(1)$. Choose ϵ_1 such that $x \in J(\epsilon_1)$. Suppose that $\epsilon_1, \ldots, \epsilon_{n-1}$ have been chosen such that $x \in J(\epsilon_1, \ldots, \epsilon_{n-1})$. Since $x \in P_n \cap J(\epsilon_1, \ldots, \epsilon_{n-1}) = J(\epsilon_1, \ldots, \epsilon_{n-1}, 0) \cup J(\epsilon_1, \ldots, \epsilon_{n-1}, 1)$ we can choose ϵ_n such that $x \in J(\epsilon_1, \ldots, \epsilon_n)$. Inductively, we have chosen $\epsilon \in S$ such that

$$x \in \bigcap_{n=1}^{\infty} J(\epsilon_1, \ldots, \epsilon_n).$$

From (3) and (1), we see that $f(\epsilon)$ is also in this intersection. Since $\lim\limits_{n\to\infty} \lambda(J(\epsilon_1, \ldots, \epsilon_n)) = \lim\limits_{n\to\infty} a_n = 0$, we must have $f(\epsilon) = x$. □

(2.84) **Corollary** *Any Cantor set is uncountable.*

Proof We need only show that S is uncountable. Define $\varphi : S \to [0, 1]$ by

$$\varphi(\epsilon) = \sum_{n=1}^{\infty} \epsilon_n 2^{-n}.$$

By (2.57), φ is onto $[0, 1]$ and, by (1.58) and (1.62), this implies that S is uncountable. □

(2.85) **Remarks** (a) It might seem to the casual reader that a Cantor set P consists of nothing but the endpoints of the closed intervals $J_{n,k}$ that appear in its construction. This notion, however, is seen to be false because P is uncountable and there are only countably many endpoints. For example, the number

$$\frac{1}{4} = \sum_{n=1}^{\infty} \frac{2}{3^{2n}} = {}_{.3}020202 \ldots$$

is in Cantor's ternary set, but it is *not* an endpoint [they all have the form $a/3^n$ where a and n are nonnegative integers].

However, it is true that any $x_0 \in P$ is the limit of a nondecreasing [and a nonincreasing] sequence of endpoints of the $J_{n,k}$. Specifically, for each n there exists a unique $k(n)$ such that $x_0 \in J_{n,k(n)} = [\alpha_n, \beta_n]$. Plainly, (α_n) and (β_n) are monotone and, since $\beta_n - \alpha_n = a_n \to 0$, $x_0 = \lim \alpha_n = \lim \beta_n$.

(b) No Cantor set P contains an open interval $I =]\alpha, \beta[$ with $\alpha < \beta$. Indeed, if $\alpha < \beta$ we can choose n such that $a_n < \beta - \alpha$ and then it is impossible that $I \subset P_n$, much less that $I \subset P$.

Exercises

1. Let $p \in \mathbb{N}$, $a \in [0, \infty[$, and $(a_n)_{n=1}^{\infty} \subset [0, \infty[$. Then $\lim\limits_{n\to\infty} \sqrt[p]{a_n} = \sqrt[p]{a}$ if and only if $\lim\limits_{n\to\infty} a_n = a$.

2. Evaluate
   (a) $\lim\limits_{n\to\infty} \left(\sqrt{n^2 + 2n} - n\right)$,
   (b) $\lim\limits_{n\to\infty} (2^n + n)/(3^n - n)$,
   (c) $\lim\limits_{n\to\infty} (2^{n^2} + 1)/\sqrt{n^4 + n^3}$,
   (d) $\lim\limits_{n\to\infty} \sqrt[n]{n!}$.

3. Evaluate $\lim\limits_{n\to\infty} \prod\limits_{k=0}^{n} (1 + a^{2^k})$ where $a \in \mathbb{C}$. [Hint: $(1 - a)(1 + a)(1 + a^2)(1 + a^4)$
    $= (1 - a^8)$.]

4. Evaluate $\lim\limits_{n\to\infty} \sum\limits_{k=1}^{n} \dfrac{1}{\sqrt{n^2 + k}}$. [Hint: Show that the sum is < 1 and $> n/\sqrt{n^2 + n}$.]

5. Let $(z_n) \subset \mathbb{C}$ and $0 \neq z \in \mathbb{C}$. Then $\lim\limits_{n\to\infty} z_n = z$ if and only if

    $$\lim\limits_{n\to\infty} [(z - z_n)/(z + z_n)] = 0.$$

    [Hint: Consider whether or not (z_n) is bounded.]

6. Let $\alpha = 2 + \sqrt{2}$ and $\beta = 2 - \sqrt{2}$. Then $n \in \mathbb{N}$ implies
    (a) $\alpha^n + \beta^n \in \mathbb{N}$ and $\alpha^n + \beta^n = [\alpha^n] + 1$,
    (b) $\lim\limits_{n\to\infty} \alpha^n - [\alpha^n] = 1$, where $[x]$ is the integer such that $[x] \leq x < [x] + 1$.

7. If $(x_n)_{n=1}^{\infty} \subset \mathbb{R}$, $(y_n)_{n=1}^{\infty} \subset]0, \infty[$ and $(x_n/y_n)_{n=1}^{\infty}$ is monotone, then the sequence $(z_n)_{n=1}^{\infty}$ defined by

    $$z_n = (x_1 + x_2 + \ldots + x_n)/(y_1 + y_2 + \ldots + y_n)$$

    is also monotone. [Hint: If $a/b \leq c/d$, then $a/b \leq (a + c)/(b + d) \leq c/d$.]

8. Let $0 < a < b < \infty$. Define $x_1 = a$, $x_2 = b$, $x_{2n+1} = \sqrt{x_{2n}x_{2n-1}}$, $x_{2n+2} = (1/2)$
    $(x_{2n} + x_{2n-1})$. Then the sequence $(x_n)_{n=1}^{\infty}$ converges. [Hint: For $0 < \alpha < \beta$ we have
    $\alpha < \sqrt{\alpha\beta} < (1/2)(\alpha + \beta) < \beta$ and $(\sqrt{\beta} - \sqrt{\alpha})^2 < \beta - \alpha$.]

9. Let $0 < a < b < \infty$. Define $x_1 = a$, $x_2 = b$, $x_{n+2} = (1/2)(x_n + x_{n+1})$. Does (x_n) converge? If so, to what limit?

10. Let $(x_n)_{n=1}^{\infty} \subset \mathbb{R}$ satisfy $0 < x_n < 1$ and $4x_{n+1}(1 - x_n) \geq 1$ for all $n \in \mathbb{N}$. Then $\lim\limits_{n\to\infty} x_n = 1/2$. [Hint: If $x_n > 1/2$ for some n, then $x_{n+m} \geq 1$ for some m. Sketch a graph of $4y(1 - x) = 1$.]

11. Let $1 < a < \infty$, $x_1 = a$, and

    $$x_{n+1} = a(1 + x_n)/(a + x_n).$$

    Show that $x_n \to \sqrt{a}$.

12. Let (x_n) and (y_n) be bounded real sequences. Then
    (a) $\underline{\lim} x_n + \underline{\lim} y_n \leq \underline{\lim}(x_n + y_n) \leq \overline{\lim}(x_n + y_n) \leq \overline{\lim} x_n + \overline{\lim} y_n$,
    (b) $\underline{\lim} x_n = x \in \mathbb{R}$ implies $\underline{\lim} (x_n + y_n) = x + \underline{\lim} y_n$ and $\overline{\lim} (x_n + y_n)$
    $= x + \overline{\lim} y_n$.
    (c) Give an example to show that the three inequalities in (a) may all be strict.

13. If (x_n) is a bounded real sequence and $c \in \mathbb{R}$, then
    (a) $\underline{\lim} cx_n = c \underline{\lim} x_n$ and $\overline{\lim} cx_n = c \overline{\lim} x_n$ if $c \geq 0$,
    (b) $\underline{\lim} cx_n = c \overline{\lim} x_n$ and $\overline{\lim} cx_n = c \underline{\lim} x_n$ if $c \leq 0$.

14. Define $x_0 = 0$, $x_1 = 1$, and

    $$x_{n+1} = [1/(n + 1)]x_{n-1} + [n/(n + 1)]x_n \qquad (n \geq 1).$$

    Does (x_n) converge? If so, to what limit?

15.   Let $a \in \mathbb{R}$, $a \notin \{0, 1, 2\}$. Define $x_1 = a$ and $x_{n+1} = 2 - 2/x_n$ for $n \in \mathbb{N}$. Find all cluster points of the sequence (x_n). [Hint: Find x_{n+2}, x_{n+3}, and x_{n+4} in terms of x_n.]

16.   For $n \in \mathbb{N}$, write $n = 2^{j-1}(2k - 1)$ where $j, k \in \mathbb{N}$ and write
$$S_n = 1/j + 1/k.$$
Find all cluster points of the sequence (S_n). Evaluate $\underline{\lim}\, S_n$ and $\overline{\lim}\, S_n$.

17.   Let $n \in \mathbb{N}$. Then $(n/e)^n < n!$. If $n > 5$, then $n! < (n/2)^n$. [Hint: Use $2 < (1 + 1/n)^n < e$ to prove that
$$((n + 1)/e)^{n+1}(n/e)^{-n} < (n + 1)!/n! < ((n + 1)/2)^{n+1}(n/2)^{-n}$$
and then use induction.]

18.   Evaluate

   (a) $\lim\limits_{n \to \infty} ((2n)!/(n!)^2)^{1/n}$,

   (b) $\lim\limits_{n \to \infty} (1/n)[(n + 1)(n + 2) \ldots (n + n)]^{1/n}$,

   (c) $\lim\limits_{n \to \infty} \left[(2/1)(3/2)^2(4/3)^3 \ldots ((n + 1)/n)^n \right]^{1/n}$.

   [Hint: Use (2.26).]

19.   Evaluate $\lim\limits_{n \to \infty} (\sqrt[n]{n} - 1)^n$. [Hint: $\sqrt[n]{n} \to 1$.]

20.   If $(x_n)_{n=1}^{\infty} \subset \,]0, \infty[$ and $x_n \to x$ as $n \to \infty$, then $(x_1 x_2 \ldots x_n)^{1/n} \to x$.

21.   (a) Let $S_n = \sum\limits_{k=1}^{n} \dfrac{1}{k}$ for $n \in \mathbb{N}$. Then $\lim\limits_{n \to \infty} |S_{n+p} - S_n| = 0$ for all $p \in \mathbb{N}$, but (S_n) diverges to ∞.

   (b) Find a divergent sequence (x_n) in $\mathbb{R}$ such that $\lim\limits_{n \to \infty} |x_{n^2} - x_n| = 0$.

   (c) Compare the results of (a) and (b) with Cauchy's Criterion.

22.   Let $(S_n)_{n=1}^{\infty} \subset \mathbb{R}$ and $\sigma_n = \dfrac{1}{n} \sum\limits_{k=1}^{n} S_k$ $(n \geq 1)$. Then $\underline{\lim}\limits_{n \to \infty} S_n \leq \underline{\lim}\limits_{n \to \infty} \sigma_n$
$\leq \overline{\lim}\limits_{n \to \infty} \sigma_n \leq \overline{\lim}\limits_{n \to \infty} S_n$.

23.   There exist two divergent series $\sum a_n$ and $\sum b_n$ of positive terms with $a_1 \geq a_2 \geq a_3 \geq \ldots$ and $b_1 \geq b_2 \geq b_3 \geq \ldots$ such that if $c_n = \min\{a_n, b_n\}$, then $\sum c_n$ converges.

24.   Evaluate the sums

   (a) $\sum\limits_{n=1}^{\infty} \dfrac{1}{n(n + 1)(n + 2)}$,

   (b) $\sum\limits_{n=1}^{\infty} \dfrac{(n - 1)!}{(n + p)!}$,

   where $p \in \mathbb{N}$ is fixed. [Hint: $2/(n(n + 1)(n + 2)) = 1/(n(n + 1)) - 1/((n + 1)(n + 2))$.]

25.   If $0 < a < 1$ and $p \in \mathbb{N}$, then

   (a) $\sum\limits_{n=1}^{\infty} n^p a^n < \infty$

   and

   (b) $\lim\limits_{n \to \infty} n^p a^n = 0$.

26. Let $\sum a_n$ be a convergent series of nonnegative terms. Then
(a) $\underline{\lim}\, na_n = 0$,
(b) possibly $\overline{\lim}\, na_n > 0$ [give an example],
(c) if $a_n \geq a_{n+1}$ for all $n > n_0$, then $\lim na_n = 0$.

27. Let $0 < \alpha < \beta < 1$. For odd n let $c_n = \alpha^n$ and for even n let $c_n = \beta^n$. Then
$$\sum_{n=1}^{\infty} c_n < \infty \quad \text{and} \quad \lim_{n\to\infty} \frac{c_{2n}}{c_{2n-1}} = \infty. \text{ Compare this example with the Ratio Test.}$$

28. If $\sum a_n^2$ converges absolutely, then $\sum a_n/n$ converges absolutely. [Hint: Use Cauchy's Inequality.]

29. If $\sum c_n$ converges absolutely, then so do $\sum c_n^2$ and $\sum (c_n/(1 + c_n))$ ($c_n \neq -1$ for all n). [Hint: $|c_n| < 1/2$ for all large n and $|1 + c_n| \geq 1 - |c_n|$.]

30. Find the Cauchy product of
$$3 + \sum_{n=1}^{\infty} 3^n z^n \quad \text{and} \quad -2 + \sum_{n=1}^{\infty} 2^n z^n$$

and then examine this situation at $z = 1$. Also compute the sums of the two given series and then multiply them directly.

31. Prove, using Cauchy products, that
$$\frac{1}{(1 - z)^2} = \sum_{n=1}^{\infty} nz^{n-1}$$

for $|z| < 1$.

32. If S is an uncountable set, $f : S \to [0, \infty]$ and $\sum_{x \in S} f(x) < \infty$, then $\{x \in S : f(x) > 0\}$ is a countable set. [Hint: Let $S_n = \{x \in S : f(x) > 1/n\}$.]

33. If $(c_m)_{m=1}^{\infty} \subset [0, \infty]$ and
$$b_n = \frac{1}{n(n + 1)} \sum_{m=1}^{n} mc_m,$$

then
$$\sum_{n=1}^{\infty} b_n = \sum_{m=1}^{\infty} c_m.$$

[Hint: Put $a_{m,n} = \dfrac{mc_m}{n(n + 1)}$ if $1 \leq m \leq n$ and $= 0$ if $m > n$.]

34. (a) Prove that $\sum_{n=1}^{\infty} \dfrac{1}{n^2} < 2$. [Hint: Use (2.34), (2.44), and $1/n^2 < 1/(n(n - 1))$.]
(b) Prove that
$$\sum_{m=1}^{\infty} \left(\sum_{n=1}^{\infty} \frac{1}{(m + n)^2} \right) = \infty.$$

[Hint: Consider the map φ of $\mathbb{N}$ onto $\mathbb{N} \times \mathbb{N}$ indicated in (1.61) and use (2.50).]

35. If $0 < x \leq 1$, then there exists a unique sequence (p_n) of integers such that

$1 < p_1 \leqq p_2 \leqq p_3 \leqq \ldots$ and

$$x = \sum_{n=1}^{\infty} \frac{1}{p_1 p_2 \cdots p_n} .$$

Moreover, x is rational if and only if there exists some n_0 such that $p_n = p_{n_0}$ for all $n > n_0$. [Hint: If $x = a/b$ and $\pi_k = p_1 \cdot p_2 \cdot \ldots \cdot p_k$, then $0 < a\pi_k - bm_k < b/(p_{k+1} - 1)$ for some $m_k \in \mathbb{N}$ $(k = 1, 2, 3, \ldots)$.]

36. Let b be an integer > 1 and let d be a digit $(0 \leqq d < b)$. Let A denote the set of all $k \in \mathbb{N}$ such that the b-adic expansion of k fails to contain the digit d.

(a) If $a_k = 1/k$ for $k \in A$ and $a_k = 0$ otherwise, then $\sum_{k=1}^{\infty} a_k < \infty$.

(b) For $n \in \mathbb{N}$ let $A(n)$ denote the number of elements of A that are $\leqq n$. Then $\lim_{n \to \infty} (A(n)/n) = 0$.

[Hint: Estimate the number of $k \in A$ such that $b^{m-1} \leqq k < b^m$.]

37. Let $0 < x < 1$. Then x has a terminating decimal expansion if and only if there exist nonnegative integers m and n such that $2^m 5^n x$ is an integer.

38. An unending b-adic expansion $x = ._b x_1 x_2 x_3 \ldots$ is said to be *periodic* of *period* p if there exist integers $p \geqq 1$ and $k \geqq 0$ such that $x_{n+p} = x_n$ for all $n > k$. For example, $3/14 = .2142857\underline{142857} \ldots$ is periodic with $k = 1$ and $p = 6$. The underline indicates that that block of digits recurs indefinitely. Let $0 \leqq x \leqq 1$ and b an integer > 1. Then the b-adic expansions of x are periodic if and only if x is rational.

39. **a**-*adic Expansions.* Let $\mathbf{a} = (a_1, a_2, a_3, \ldots)$ be an arbitrary fixed sequence of integers > 1. Then every $x \in \mathbb{R}$ with $0 \leqq x < 1$ has an expansion of the form

$$x = \sum_{j=1}^{\infty} \frac{x_j}{a_1 a_2 \cdots a_j}$$

where (x_j) is a sequence of integers with $0 \leqq x_j < a_j$ for all j. Find a necessary and sufficient condition that two distinct expansions converge to the same number. Note that if $a_j = b$ for all j, these are b-adic expansions.

40. Evaluate $\lim_{n \to \infty} (n!e - [n!e])$.

41. Which is larger,
(a) 1000^{1000} or 1001^{999},
(b) $(1.000001)^{1000001}$ or 2?
[Hint: Use (2.16).]

42. Find the sum of the series

$$\sum_{n=1}^{\infty} \frac{1}{(n+1)\sqrt{n} + n\sqrt{n+1}} .$$

43. Which of the following series converge?

(a) $\displaystyle\sum_{n=0}^{\infty} \frac{n^{10}}{10^n}$,   (b) $\displaystyle\sum_{n=0}^{\infty} \frac{(n!)^2}{(2n)!}$,

(c) $\displaystyle\sum_{n=1}^{\infty} \left(1 + \frac{1}{n}\right)^{-n^2}$,   (d) $\displaystyle\sum_{n=1}^{\infty} (1 - e^{-1/n})$,

(e) $\sum_{n=1}^{\infty} (1 - e^{-1/n})^2$, (f) $\sum_{n=1}^{\infty} \frac{2^n n!}{n^n}$,

(g) $\sum_{n=1}^{\infty} \frac{3^n n!}{n^n}$, (h) $\sum_{n=1}^{\infty} \frac{n - \sqrt{n}}{n^2 + n}$.

44. Find the radius of convergence of

(a) $\sum_{n=0}^{\infty} n! z^{n!}$, (b) $\sum_{n=0}^{\infty} z^{n!}$,

(c) $\sum_{n=0}^{\infty} \frac{z^{n!}}{n!}$, (d) $\sum_{n=0}^{\infty} \frac{z^{n^2}}{n!}$,

(e) $\sum_{n=0}^{\infty} a^{n^2} z^n$, (f) $\sum_{n=1}^{\infty} \frac{z^{n^2}}{(n!)^n}$.

45. For which $x \in \mathbb{R}$ do the following series converge?

(a) $\sum_{n=1}^{\infty} \frac{1}{n} \frac{x^n}{1 - x^n}$, (b) $\sum_{n=1}^{\infty} \frac{(-1)^n}{x^2 - n^2}$.

What if $x \in \mathbb{C}$? Consider also absolute convergence.

46. Let $a_n > 0$ for each $n \in \mathbb{N}$. Then

(a) $\sum_{n=1}^{\infty} a_n < \infty$ implies $\sum_{n=1}^{\infty} \sqrt{a_n a_{n+1}} < \infty$,

(b) the converse of (a) is false,

(c) $\sum_{n=1}^{\infty} a_n < \infty$ implies $\sum_{n=1}^{\infty} (a_n^{-1} + a_{n+1}^{-1})^{-1} < \infty$,

(d) the converse of (c) is false.

47. Suppose that $d_n > 0$ for all $n \in \mathbb{N}$ and $\sum_{n=1}^{\infty} d_n = \infty$. What can be said of the following series?

(a) $\sum_{n=1}^{\infty} \frac{d_n}{1 + d_n}$, (b) $\sum_{n=1}^{\infty} \frac{d_n}{1 + n d_n}$, (c) $\sum_{n=1}^{\infty} \frac{d_n}{1 + d_n^2}$.

Be sure to consider the cases: $d_n \to 0$; $d_n \to \infty$; $d_n \to d$ $(0 < d < \infty)$; $\underline{\lim} \, d_n < \overline{\lim} \, d_n$.

48. Let $0 < a < b < \infty$ and define $x_1 = a$, $x_2 = b$, and $x_{n+2} = \sqrt{x_n x_{n+1}}$ for $n \in \mathbb{N}$. Find $\lim_{n \to \infty} x_n$.

49. Let $0 < a < b < \infty$ and define $x_1 = a$, $y_1 = b$, $x_{n+1} = 2(x_n^{-1} + y_n^{-1})^{-1}$, and $y_{n+1} = \sqrt{x_n y_n}$. Then $(x_n)_{n=1}^{\infty}$ and $(y_n)_{n=1}^{\infty}$ both converge and have the same limit.

50. The sequence of *Fibonacci numbers* is defined by

$$u_0 = 0, \quad u_1 = 1, \quad u_{n+2} = u_n + u_{n+1}.$$

Writing $x_n = u_{n+1}/u_n$ for $n \in \mathbb{N}$, we have

(a) $x_1 < x_3 < x_5 < \ldots \ldots < x_6 < x_4 < x_2$,

(b) $\lim_{n \to \infty} (x_{2n} - x_{2n-1}) = 0$.

(c) Find $\lim\limits_{n\to\infty} x_n$.

(d) If α and β are the roots of $x^2 = x + 1$ and if $w_n = a\alpha^n + b\beta^n$, then the sequence $(w_n)_{n=0}^{\infty}$ satisfies $w_{n+2} = w_{n+1} + w_n$ for all $n \geq 0$.

(e) Use (d) to find an explicit expression of u_n in terms of n.

(f) Use the result of (e) to solve (c).

51.   If $0 \leq a \leq 2$ and P is Cantor's ternary set, then there exist $x, y \in P$ such that $x + y = a$. For which such a is there only one such set $\{x, y\} \subset P$? [Hint: Consider ternary expansions of $a/2$.]

3

LIMITS AND CONTINUITY

The main distinction between algebra and analysis is that in analysis we take limits. In Chapter 2 we have studied the notion of the limit of a sequence. Roughly speaking, $\lim_{n\to\infty} x_n = x$ means that x_n is as "near" to x as we wish if only n is taken "near enough" to ∞. Of course a sequence is just a function having domain $\mathbb{N}$. More generally, suppose that X and Y are sets, $D \subset X$, and $f : D \to Y$ is a function. For $a \in X$ and $b \in Y$ we want to write

$$\lim_{\substack{x \to a \\ x \in D}} f(x) = b$$

to mean something like this. There are points of D arbitrarily "close to" a and $f(x)$ is as "near" b as we like provided only that $x \in D$ is sufficiently "near" a [but not equal to a]. To make this idea precise, we need a concept of "nearness." As a first step in this direction we give the following definition.

Metric Spaces

(3.1) **Definition** A *metric space* is a nonvoid set X together with a given function ρ from $X \times X$ into $\mathbb{R}$ which satisfies
- (i) $\rho(x, x) = 0$,
- (ii) $\rho(x, y) > 0$ if $x \neq y$,
- (iii) $\rho(y, x) = \rho(x, y)$,
- (iv) $\rho(x, z) \leqq \rho(x, y) + \rho(y, z)$

for all $x, y, z \in X$. The function ρ is called a *metric* [or *distance function*] for X. Inequality (iv) is known as the *triangle inequality* for ρ.

(3.2) **Example** (a) Let X be any nonvoid set and define $\rho(x, y) = 0$ or 1 according as $x = y$ or $x \neq y$. One checks that (i)–(iv) of (3.1) are satisfied. The metric ρ is called the *discrete metric* for X.

91

(b) Let X be any nonvoid subset of $\mathbb{R}$ and define $\rho(x, y) = |x - y|$ for $x, y \in X$. This ρ is the *usual metric* in $\mathbb{R}$.

(c) Let $n \in \mathbb{N}$ and let X be any nonvoid subset of $\mathbb{R}^n = \{(x_1, x_2, \ldots, x_n): x_j \in \mathbb{R}$ for $1 \leq j \leq n\}$. For $x = (x_1, \ldots, x_n)$ and $y = (y_1, \ldots, y_n)$ in X, define

$$\rho(x, y) = \left(\sum_{j=1}^{n} |x_j - y_j|^2 \right)^{1/2}.$$

It is obvious that ρ satisfies (i)–(iii) of (3.1). Inequality (3.1.iv) follows from Minkowski's Inequality by writing $a_j = x_j - y_j$ and $b_j = y_j - z_j$ in (1.45.i). This metric ρ is called the *Euclidean metric* for X. Note that when $n = 1$ this reduces to (b) and that, when $n = 2$ or 3, ρ is just the usual distance function of analytic geometry. For $x \in \mathbb{R}^n$ we define the *Euclidean norm* of x by

$$|x| = \left(\sum_{j=1}^{n} |x_j|^2 \right)^{1/2}.$$

Thus

$$\rho(x, y) = |x - y|$$

where $x - y = (x_1 - y_1, \ldots, x_n - y_n)$.

(d) Let X be any nonvoid subset of $\mathbb{C}$ and define $\rho(z, w) = |z - w|$ for $z, w \in X$. If we write $z = (a, b)$ and $w = (c, d)$ as in (1.36) and (1.39) and then use (1.40), we see that

$$\rho(z, w) = |(a - c, b - d)| = \sqrt{(a - c)^2 + (b - d)^2}.$$

Thus, if $\mathbb{C}$ is identified with $\mathbb{R}^2$, this example is the case $n = 2$ of (c). This is the *usual metric* in $\mathbb{C}$.

(e) Let S be any nonvoid set and let X be any nonvoid set of complex-valued functions defined on S such that, for each $f \in X$, we have

(i) $\|f\|_u = sup\{|f(s)| : s \in S\} < \infty.$

The number $\|f\|_u$ defined by (i) is called the *uniform norm* of f (over S). Functions f for which (i) obtains are said to be *bounded on S*. For $f, g \in X$, define

$$\rho(f, g) = \|f - g\|_u.{}^*$$

We leave it to the reader to check that ρ is a metric for X. It is called the *uniform metric*.

(f) Let X be the set of all sequences $a = (a_j)_{j=1}^{\infty} \subset \mathbb{C}$ such that

(ii) $\sum_{j=1}^{\infty} |a_j|^2 < \infty$

and let $\|a\|_2$ denote the nonnegative square root of the left side of (ii). Using

---

*The function $f - g$ is defined *pointwise* on S by $(f - g)(s) = f(s) - g(s)$. Similarly, we define functions cf $(c \in \mathbb{C})$, $f + g$, fg, $|f|$, $\bar{f}$, $\mathrm{Re}\,f$, $\mathrm{Im}\,f$ etc., by specifying their value at each point of S.

Minkowski's Inequality (1.45), we see that if $a, b \in X$, then

$$\left(\sum_{j=1}^{n} |a_j + b_j|^2 \right)^{1/2} \le \|a\|_2 + \|b\|_2$$

for each $n \in \mathbb{N}$ and so, letting $n \to \infty$, we obtain

$$a + b \in X \text{ and}$$

(iii)
$$\|a + b\|_2 \le \|a\|_2 + \|b\|_2.$$

Using (iii), it is easy to check [as in (c)] that the function ρ defined on $X \times X$ by

$$\rho(x, y) = \|x - y\|_2$$

is a metric. This metric space is usually denoted l_2 and called *complex sequential Hilbert space*. *Real sequential Hilbert space* is

$$l_2^r = \{ a \in l_2 : a_j \in \mathbb{R} \text{ for } j \in \mathbb{N} \}.$$

(g) Let $X = \{ z \in \mathbb{C} : |z| \le 1 \}$ be the unit disk in $\mathbb{C}$, but give it the following unusual metric. For $z, w \in X$, define $\rho(z, w) = |z - w|$ if $w = tz$ for some $t > 0$; i.e., w and z are on the same radius of X. Otherwise, define $\rho(z, w) = |z| + |w|$. We leave it to the reader to check that ρ is a metric. A little reflection shows why this is called the *French railroad space*—Paris is at the center.

We can generalize the notion of a ball in $\mathbb{R}^3$ as follows.

(3.3)   **Definitions**   If X is a metric space with metric ρ, $a \in X$, and r is a positive real number, we define the *ball* of *radius* r with *center* a to be the set

$$B_r(a) = \{ x \in X : \rho(x, a) < r \}.$$

A subset U of X is said to be *open* [relative to ρ] if for each $a \in U$ there exists some $r > 0$ such that $B_r(a) \subset U$ [where r may depend on a].

(3.4)   **Theorem**   *Every ball in a metric space is open.*

> **Proof**   Let $B_r(a)$ be any ball in a metric space X. Let $b \in B_r(a)$. We must find $s > 0$ such that $B_s(b) \subset B_r(a)$. Choose $s = r - \rho(a, b) > 0$. Then $x \in B_s(b)$ implies $\rho(x, a) \le \rho(x, b) + \rho(b, a) < s + \rho(a, b) = r$, and so $x \in B_r(a)$. $\square$

(3.5)   **Examples**   (a) In $\mathbb{R}$, with its usual metric, the balls are just the open intervals $]a - r, a + r[$.

(b) In $\mathbb{C}$, with its usual metric, the balls are just the "inside parts" of ordinary circles.

(c) If $X = [0, 1] \subset \mathbb{R}$ as in (3.2.b), then

$$B_2(a) = X \quad \text{for all } a \in X,$$
$$B_{1/2}(0) = [0, 1/2[.$$

(d) If $X = \mathbb{N} \subset \mathbb{R}$, with the usual metric (3.2.b), then

$$B_1(2) = \{2\}, \qquad B_{3/2}(2) = \{1, 2, 3\}.$$

(e) If $X = \mathbb{Q} \subset \mathbb{R}$, with the usual metric, then

$$B_e(0) = [-e, e] \cap \mathbb{Q} =]-e, e[\cap \mathbb{Q}.$$

These examples are meant to point up that the points in a ball depend on the space X under consideration.

(3.6)   Theorem *Let X be any metric space and let $\mathcal{T}$ be the family of all open subsets of X. Then*

(i) $\emptyset \in \mathcal{T}$ *and* $X \in \mathcal{T}$,

(ii) $\{U_1, U_2, \ldots, U_n\} \subset \mathcal{T}$ *implies* $\bigcap_{j=1}^{n} U_j \in \mathcal{T}$,

(iii) $\mathcal{U} \subset \mathcal{T}$ *implies* $\cup \mathcal{U} \in \mathcal{T}$.

Proof Since $\emptyset$ contains no point a, Definition (3.3) cannot fail for $U = \emptyset$; hence $\emptyset \in \mathcal{T}$. Plainly, $a \in X$ implies $B_r(a) \subset X$ for all $r > 0$ so $X \in \mathcal{T}$.

Next suppose that $\{U_1, U_2, \ldots, U_n\} \subset \mathcal{T}$ and let $a \in \bigcap_{j=1}^{n} U_j$. For each $j \in \{1, 2, \ldots, n\}$, $a \in U_j$ so there is some $r_j > 0$ such that $B_{r_j}(a) \subset U_j$. Letting $r = \min\{r_1, r_2, \ldots, r_n\}$, we see that $B_r(a) \subset U_j$ for each $j = 1$, $2, \ldots, n$, and so $B_r(a) \subset \bigcap_{j=1}^{n} U_j$. Therefore (ii) obtains.

Finally, let $\mathcal{U} \subset \mathcal{T}$ and let $a \in \cup \mathcal{U}$. Then $a \in U$ for some $U \in \mathcal{U}$ and so there exists an $r > 0$ such that

$$B_r(a) \subset U \subset \cup \mathcal{U}. \quad \square$$

We can use the *given* metric in a metric space X to make the following definitions.

(3.7)   Definitions Let X be a metric space and let $\emptyset \neq A \subset X$. Define the *diameter* of A to be the extended real number

$$\operatorname{diam} A = \sup\{\rho(x, y) : x, y \in A\}.$$

If $\operatorname{diam} A < \infty$, we say that A is *bounded* [check that this is consistent with our earlier definitions of boundedness for subsets of $\mathbb{R}$ or $\mathbb{C}$]. For $x \in X$, define the *distance* from x to A to be

$$\operatorname{dist}(x, A) = \inf\{\rho(x, a) : a \in A\}.$$

(3.8)   Theorem *With the notation of (3.7), we have*

(i) $|\operatorname{dist}(x, A) - \operatorname{dist}(y, A)| \leq \rho(x, y)$

for all $x, y \in X$.

Proof For each $a \in A$ we have

$$\text{dist}(x, A) \leqq \rho(x, a) \leqq \rho(x, y) + \rho(y, a).$$

Taking the infimum over all $a \in A$, we obtain

$$\text{dist}(x, A) \leqq \rho(x, y) + \text{dist}(y, A)$$

$$\text{dist}(x, A) - \text{dist}(y, A) \leqq \rho(x, y). \tag{1}$$

Exchanging the roles of x and y, we also obtain

$$\text{dist}(y, A) - \text{dist}(x, A) \leqq \rho(x, y). \tag{2}$$

Now, the conjunction of (1) and (2) is equivalent to (i). $\qquad\square$

Topological Spaces

For the purpose of discussing limits and continuity of functions, it is not necessary to have a metric. All that is needed is the notion of open sets. Thus, we are led to the following definition.

(3.9) **Definition** A *topological space* is a set X together with a specified family $\mathfrak{T}$ of subsets of X such that $\mathfrak{T}$ satisfies (i), (ii), and (iii) of (3.6). Such a family $\mathfrak{T}$ is called a *topology* for the set X.

Notice that Theorem (3.6) tells us that if X is any metric space and $\mathfrak{T}$ is the family of all open subsets of X, as defined in (3.3), then $\mathfrak{T}$ is a topology for X, and so X, together with this topology, is a topological space. In view of this, if X is *any* topological space and $\mathfrak{T}$ is the given topology, then we call the members of $\mathfrak{T}$ *open sets*.

(3.10) **Examples** (a) As just noted, any metric space, with its *metric topology* $\mathfrak{T}$, is a topological space.

(b) Let X be any set and $\mathfrak{T}$ the family of all subsets of X. It is clear that $\mathfrak{T}$ is a topology for X. It is called the *discrete topology* because it is nothing but the metric topology obtained from the discrete metric (3.2.a).

(c) Let X be any set and let $\mathfrak{T} = \{\emptyset, X\}$. This uninteresting topology is called the *indiscrete topology* for X.

(d) Let $X = \mathbb{R}$ and let

$$\mathfrak{T} = \{\,]a, \infty[\ : a \in \mathbb{R}\} \cup \{\emptyset, \mathbb{R}\}.$$

We leave it to the reader to check that $\mathfrak{T}$ is a topology for $\mathbb{R}$ and that it is *not* the metric topology obtained from the usual metric (3.2.b) or, indeed, from *any* metric for $\mathbb{R}$.

(e) If $X \subset \mathbb{R}^n$ $(n \in \mathbb{N})$, then the metric topology on X obtained from the Euclidean metric (3.2.c) is called the *Euclidean* [or *usual*] *topology* on X. This applies, in particular, to $X \subset \mathbb{C} = \mathbb{R}^2$. Whenever we discuss topological notions for subsets of $\mathbb{R}^n$, we shall always be referring to this topology unless we make an explicit statement to the contrary.

(f) Let X be any set and let $\mathcal{S}$ be any collection of subsets of X such that $\cup \mathcal{S} = X$. Next let $\mathcal{B}$ denote the collection of all sets that can be obtained by intersecting a *finite* number of members of $\mathcal{S}$ $[\cap \emptyset = X]$. Finally, let $\mathcal{T}$ denote the collection of all sets that can be obtained as the union of some subfamily of $\mathcal{B}$ $[\cup \emptyset = \emptyset]$. It is left to the reader to check that $\mathcal{T}$ is a topology for X. For instance, if $X = \{1, 2, 3\}$ and $\mathcal{S} = \{\{1, 2\}, \{2, 3\}\}$, then $\mathcal{B} = \mathcal{S} \cup \{X, \{2\}\}$ and $\mathcal{T} = \mathcal{B} \cup \{\emptyset\}$. Any topology $\mathcal{T}$ on any set X can be generated from a collection $\mathcal{S}$, as above, because we can take $\mathcal{S} = \mathcal{T}$. Of course, in most cases, $\mathcal{T}$ will contain many more sets than $\mathcal{S}$ does. Whenever a topology $\mathcal{T}$ is generated by a family $\mathcal{S}$, as above, we call $\mathcal{S}$ a *subbase* for $\mathcal{T}$. Most topologies have many different subbases. To obtain the metric topology for a given metric space, we can start by taking $\mathcal{S}$ to be the family of all balls in the space.

(g) Let $X = \mathbb{R}^{\#}$. Using the scheme in (f), we obtain the *usual topology* on $\mathbb{R}^{\#}$ by letting $\mathcal{S} = \{\,]a, \infty] : a \in \mathbb{R}\} \cup \{[-\infty, b[\,: b \in \mathbb{R}\}$. Then $\mathcal{B} = \mathcal{S} \cup \{\,]a, b[\,: a < b \text{ in } \mathbb{R}\} \cup \{\emptyset, \mathbb{R}^{\#}\}$ and a set U is in $\mathcal{T}$ if and only if for each $x \in U$ there is a $B \in \mathcal{B}$ such that $x \in B \subset U$.

There is a great deal of terminology associated with topological spaces. We collect some of it together here.

(3.11) Definitions Let X be a topological space. [It is understood, of course, that X has a specific topology $\mathcal{T}$ and that the members of $\mathcal{T}$ are the open subsets of X.] A *neighborhood* of a point $x \in X$ is any open set $U \subset X$ such that $x \in U$. If $D \subset X$ and $x \in X$, we call x a *limit point* of D provided that each neighborhood of x contains at least one point of D other than x [x need not be in D]. All points of D that are not limit points of D are called *isolated points* of D. For $x \in D \subset X$, we say that x is an *interior point* of D if x has a neighborhood $U \subset D$. The set of all *interior point* of D is denoted D° and is called the *interior* of D. A set $F \subset X$ is called *closed* if its complement $F' = X \setminus F$ is open. The *closure* of a set $D \subset X$ is the set

$$D^{-} = D \cup \{x \in X : x \text{ is a limit point of } D\}.$$

A set $P \subset X$ is called *perfect* if P is closed and has no isolated points. A set $D \subset X$ is *dense* in X if $D^{-} = X$. The *boundary* of a set $D \subset X$ is the set $\partial D = D^{-} \cap D'^{-}$.

(3.12) Remark For the case $X = \mathbb{R}$, the above definition of neighborhood is an extension of that given in (2.20).

(3.13) Theorem *Let X be a topological space. Then*
(i) *X and $\emptyset$ are closed,*
(ii) *if $F_1, F_2, \ldots, F_n$ are closed subsets of X, then $\displaystyle\bigcup_{k=1}^{n} F_k$ is closed,*
(iii) *if $\mathcal{F}$ is a family of closed subsets of X, then $\cap \mathcal{F}$ is closed.*

Proof Since $\emptyset' = X$ and $X' = \emptyset$, (i) follows from (3.9) and (3.6.i). To prove (ii), notice that

$$\left(\bigcup_{k=1}^{n} F_k \right)' = \bigcap_{k=1}^{n} F_k'$$

and then apply (3.6.ii). Similarly, (iii) follows from (3.6.iii) by noticing that

$$(\cap \mathcal{F})' = \cup \{ F' : F \in \mathcal{F} \}. \quad \square$$

(3.14) **Theorem** *Let X be a topological space and let $D \subset X$. Then*
(i) *D° is open,*
(ii) *D^- is closed,*
(iii) *D is open if and only if $D = D^\circ$,*
(iv) *D is closed if and only if $D = D^-$,*
(v) *$D^\circ = D'^{-\prime}$.*

Proof Plainly $D^\circ \subset D$. If D is open, then D is a neighborhood of each of its points and so each such point is in D°; i.e., $D \subset D^\circ$; $D = D^\circ$. In any case, for each $x \in D^\circ$, we can choose a neighborhood U_x of x such that $U_x \subset D$. Since each point of U_x has U_x as a neighborhood, it follows that each such point is an interior point of D; hence $U_x \subset D^\circ$ for each $x \in D^\circ$. Therefore

$$D^\circ = \cup \{ U_x : x \in D^\circ \}$$

and so (3.6.iii) yields (i) and (iii).

We next prove (ii) by showing that $D^{-\prime}$ is open. Let $x \in D^{-\prime}$. Then $x \notin D$ and so, since x is not a limit point of D, there is a neighborhood U of x such that $U \cap D = \emptyset$. Furthermore, no point of the open set U can be a limit point of D; hence, $U \cap D^- = \emptyset$, $U \subset D^{-\prime}$. This proves that $x \in D^{-\prime\circ}$ for each $x \in D^{-\prime}$; whence, $D^{-\prime}$ is open by (iii).

If $D = D^-$, (ii) proves that D is closed. On the other hand, if D is closed, then D' is a neighborhood of each of its points and so no such point is a limit point of D; i.e., D contains all of its limit points, $D^- \subset D$. Since $D \subset D^-$, our proof of (iv) is complete.

Since $D' \subset D'^-$, we have $D = D'' \supset D'^{-\prime}$. The latter set, being the complement of the closed set D'^- is open and so $D'^{-\prime} \subset D^\circ$. To prove the reverse inclusion, suppose that $x \notin D'^{-\prime}$. Then $x \in D'^-$ and so either $x \in D'$ or x is a limit point of D'. In either case we conclude that $x \notin D^\circ$ because otherwise D° would be a neighborhood of x that is disjoint from D'. $\square$

(3.15) **Corollary** *If X is a topological space and $D \subset X$, then ∂D is closed and $\partial D = D^- \setminus D^\circ$.*

Proof By definition ∂D is the intersection of the two closed sets D^- and D'^- and so it is closed by (3.13.ii). To complete the proof, we apply (3.14.v) to write $\partial D = D^- \cap D'^- = D^- \cap D'^{-\prime\prime} = D^- \cap D^{\circ\prime} = D^- \setminus D^{\circ}$. ☐

(3.16) Corollary   *A subset P of a topological space is perfect if and only if it is equal to its set of limit points.*

Proof Let L be the set of all limit points of P. If P is perfect, then, by definition, P is closed ($P = P^- \supset L$) and has no isolated points ($P \subset L$); hence, $P = L$. Conversely, if $P = L$, then $P \subset L$ (P has no isolated points) and $P = P \cup L = P^-$ (P is closed); hence, P is perfect. ☐

(3.17) Examples We now list a number of subsets of the topological (metric) spaces $\mathbb{R}$ and $\mathbb{C}$ (with their usual metric topologies) and discuss some of their properties.

(a) The set $\mathbb{Q}$ of all rational numbers, as a subset of $\mathbb{R}$, is neither open nor closed because

$$\mathbb{Q}^{\circ} = \varnothing \neq \mathbb{Q} \neq \mathbb{R} = \mathbb{Q}^-$$

[use (1.34) and (1.35)]. Also, this shows that both $\mathbb{Q}$ and $\mathbb{Q}'$ are dense in $\mathbb{R}$ and that $\partial \mathbb{Q} = \mathbb{R}$.

(b) The set $E = \{1/n : n \in \mathbb{N}\}$, as a subset of either $\mathbb{R}$ or $\mathbb{C}$, has only 0 as a limit point (E is not closed) and E contains no nonvoid open interval or disk ($E^{\circ} = \varnothing \neq E$). Every point of E is isolated. What is ∂E?

(c) The *set* $\mathbb{R}$, as a subset of the *space* $\mathbb{C}$, is perfect, not open, and equals its boundary. While the *set* $\mathbb{R}$, as a subset of the *space* $\mathbb{R}$, is open, closed, perfect, dense, and has void boundary. This example shows why it's emphatically necessary to specify the space in which a set is to be considered. Consider also $\mathbb{Q}$ as a subset of $\mathbb{C}$. Is it dense?

(d) Let $a < b$ in $\mathbb{R}$. Then the set $]a, b[$ is open in $\mathbb{R}$ but not in $\mathbb{C}$, and its closure is $[a, b]$ in both $\mathbb{R}$ and $\mathbb{C}$. The set $[a, b]$ is perfect in $\mathbb{R}$ and $\mathbb{C}$.

(e) The set $A = \{z \in \mathbb{C} : |z| < 1, z \notin [0, 1[\}$ is open in $\mathbb{C}$ [prove this]. Also, A^- is the disk $\{z \in \mathbb{C} : |z| \leq 1\}$.

(f) Let P be any Cantor set (2.81). Then P is closed ($P = \bigcap_{n=1}^{\infty} P_n$ and each P_n is a finite union of closed intervals) and has no isolated points (2.85.a); whence, P is perfect. Also, $\partial P = P$. [These statements apply to both of the spaces $\mathbb{R}$ and $\mathbb{C}$.]

We next prove an important structure theorem for the open subsets of $\mathbb{R}$.

(3.18) Theorem   *Let V be any nonvoid open subset of $\mathbb{R}$. Then there exists a unique family $\mathcal{I}$ of open intervals having the following properties:*

(i) $\cup \mathcal{I} = V$;

(ii) *$\mathcal{I}$ is pairwise disjoint;*

(iii) *$]a, b[\in \mathcal{I}$ implies $a < b$, $a \notin V$, and $b \notin V$;*

(iv) $\mathcal{I}$ *is countable.*

The elements of $\mathcal{I}$ are called the *component intervals* of V.

Proof For $x \in V$, there exists a neighborhood $]\alpha, \beta[$ of x such that $]\alpha, \beta[\subset V$. Thus, we define

$$A_x = \{ \alpha \in \mathbb{R}^\# : \alpha < x, \,]\alpha, x] \subset V \},$$

$$B_x = \{ \beta \in \mathbb{R}^\# : x < \beta, \, [x, \beta[\subset V \},$$

$$a_x = \inf A_x, \qquad b_x = \sup B_x.$$

Now let $I_x =]a_x, b_x[$ and

$$\mathcal{I} = \{ I_x : x \in V \}.$$

For each $x \in V$, we have $x \in I_x \subset \cup \mathcal{I}$ and so $V \subset \cup \mathcal{I}$. If $a_x < t \leqq x$, then t is not a lower bound for A_x and so there exists $\alpha \in A_x$ such that $t \in]\alpha, x] \subset V$; whence, $]a_x, x] \subset V$. Similarly, $[x, b_x[\subset V$, and so $I_x \subset V$ for each $x \in V$. Therefore $\cup \mathcal{I} \subset V$ and (i) is established. Incidentally, this shows that $a_x \in A_x$ and $b_x \in B_x$.

Assume that $x \in V$ and $a_x \in V$. Then a_x has a neighborhood $]\alpha, \beta[\subset V$ and it follows that $\alpha \in A_x$, contrary to the definition of a_x. Therefore, $x \in V$ implies $a_x \notin V$. Similarly, $x \in V$ implies $b_x \notin V$; hence, (iii) obtains.

Next, let $x, y \in V$ and $I_x \cap I_y \neq \emptyset$. Choose c such that $a_x < c < b_x$ and $a_y < c < b_y$. If $\alpha \in A_c$, then $a_x \leqq \alpha$, since otherwise $a_x \in]\alpha, c] \subset V$, contrary to (iii). Therefore a_x is a lower bound for A_c and so $a_x \leqq a_c$. Assuming $a_x < a_c$, we would have $a_x < a_c < c < b_x$ so $a_c \in I_x \subset V$, contrary to (iii). Thus $a_x = a_c$. Similar reasoning shows that $a_y = a_c$, $b_x = b_c$, and $b_y = b_c$. Accordingly $I_x = I_c = I_y$. We conclude that any two elements of $\mathcal{I}$ are either identical or disjoint, proving (ii).

For each $I \in \mathcal{I}$ choose just one $r(I) \in I \cap \mathbb{Q}$. In view of (ii), the mapping $I \to r(I)$ is a one-to-one mapping of $\mathcal{I}$ into the countable set $\mathbb{Q}$, and so (iv) obtains.

To prove the uniqueness assertion, suppose that $\mathcal{I}$ and $\mathcal{J}$ are any two families of open intervals satisfying (i)–(iv). If $J =]a, b[\in \mathcal{J}$, then $J \subset V = \cup \mathcal{I}$ [by (i)], $J \subset I$ for some $I \in \mathcal{I}$ [by (ii) and (iii) for $\mathcal{I}$], and $J = I \in \mathcal{I}$ [by (iii) for $\mathcal{J}$ and (i) for $\mathcal{I}$]. Therefore $\mathcal{J} \subset \mathcal{I}$. Similarly, $\mathcal{I} \subset \mathcal{J}$. $\square$

(3.19) Remark There is no such simple structure theorem for the open subsets of $\mathbb{C}$ as the one for $\mathbb{R}$ given in (3.18).

(3.20) Definition A topological space X is called a *Hausdorff space* if each two distinct points of X have disjoint neighborhoods.

(3.21) Theorem *Any metric space [with its metric topology] is a Hausdorff space.*

Proof If $x \neq y$, write $r = (1/2)\rho(x, y) > 0$. Then $B_r(x) \cap B_r(y) = \emptyset$, since otherwise, if z were in the intersection, we would have $2r = \rho(x, y) \leq \rho(x, z) + \rho(z, y) < r + r = 2r$. □

(3.22) **Remark** The spaces given in (c) and (d) of (3.10) are not Hausdorff spaces if X has more than one point.

(3.23) **Theorem** *If X is a Hausdorff space, then any finite subset of X is closed in X.*

Proof Let $x \in X$. For each $y \neq x$ in X choose a neighborhood U_y of y such that $x \notin U_y$. Then the set $\cup \{U_y : y \in \{x\}'\} = \{x\}'$ is open, and so $\{x\}$ is closed. The rest follows from (3.13.ii). □

(3.24) **Theorem** *Let X be a Hausdorff space, $D \subset X$, x a limit point of D, and V a neighborhood of x. Then $D \cap V$ is an infinite set.*

Proof Assume that the set $D \cap V$ is finite. Define $A = \{y \in D \cap V : y \neq x\}$. Then A is closed (3.23), and so $V \cap A'$ is a neighborhood of x. Thus, $D \cap V \cap A'$ contains some $y \neq x$. But then $y \in A \cap A' = \emptyset$. □

We can discuss convergence of sequences in any topological space.

(3.25) **Definition** Let X be a topological space, let $(x_n)_{n=1}^{\infty} \subset X$, and let $x \in X$. We say that the sequence $(x_n)_{n=1}^{\infty}$ *converges* to x and we write $\lim_{n \to \infty} x_n = x$ [or $x_n \to x$ as $n \to \infty$] if for each neighborhood V of x there exists an $n_0 \in \mathbb{N}$ such that $x_n \in V$ for all $n \geq n_0$. All sequences that fail to converge are said to *diverge*.

This definition obviously reduces to (2.2) in case $X = \mathbb{C}$. The following theorem generalizes (2.3).

(3.26) **Theorem** *In a Hausdorff space, a sequence can converge to at most one point.*

Proof Exercise. □

(3.27) **Remark** If a sequence $(x_n)_{n=1}^{\infty}$ converges to two distinct points x and y, then the notation

$$x = \lim_{n \to \infty} x_n = y$$

is dangerous. It might lead one to conclude that $x = y$. In the space given in (3.10.d), the sequence $(1/n)_{n=1}^{\infty}$ converges to *every* nonpositive real number [take $n_0 = 1$]. This is one good reason for preferring Hausdorff spaces, although there are others.

(3.28) **Theorem** *Let X be a topological space and let $A \subset X$. Then*

(i) *if there is a sequence of points of A that converges to some $x \in X$, then* $x \in A^-$;

(ii) *if $x \in A^-$ and X is a metric space, then there exists a sequence in A that converges to x.*

> **Proof** Assertion (i) is trivial. If $x \in A^-$ then $V \cap A \neq \emptyset$ for each neighborhood V of x. Now suppose that $x \in A^-$ and X is a metric space. For $n \in \mathbb{N}$, choose $x_n \in A \cap B_{1/n}(x)$. Then $(x_n)_{n=1}^\infty \subset A$ and if V is any neighborhood of x, we can choose $p \in \mathbb{N}$ such that $B_{1/p}(x) \subset V$, and so $n \geqq p$ implies $x_n \in V$. $\square$

(3.29) **Remark** The condition that X be a metric space in (3.28.ii) cannot be omitted. Indeed, there exist Hausdorff spaces of the form $X = A \cup \{x\}$ such that $x \in A^-$ and the conclusion of (3.28.ii) fails. A remedy is to replace the index set $\mathbb{N}$ by a more general kind of "directed set." Since our chief concern here is metric spaces, we shall not belabor this point.

It is frequently useful to regard subsets of topological spaces as topological spaces in their own right. Let's be precise.

(3.30) **Definition** Let X be a topological space with topology $\mathfrak{T}$ and let $S \subset X$. Define

$$\mathfrak{T}_S = \{V \cap S : V \in \mathfrak{T}\}.$$

One checks that $\mathfrak{T}_S$ is a topology for S. It is called the *relative topology* on S, and S, with this topology, is called a *subspace* of X. The elements of $\mathfrak{T}_S$ are called *relatively open* or are said to be *open in S*.

(3.31) **Example** Let $X = \mathbb{R}$ with its usual topology and let $S = [0, 1]$. Then $[0, 1/2[$ is open *in S* because it is equal to $]-1, 1/2[\cap S$, but it is *not* open *in $\mathbb{R}$*. Note that, for any X and S, all isolated points of S are open *in S*.

(3.32) **Theorem** *If X is a metric space [with its metric topology] and $S \subset X$ is nonvoid, then the relative topology on S is the same as the metric topology on S obtained by restricting the given metric to $S \times S$.*

> **Proof** This follows easily from the observation that, for $x \in S$, the ball in the metric space S with center x and radius $r > 0$ is just
>
> $$\{y \in S : \rho(x, y) < r\} = \{y \in X : \rho(x, y) < r\} \cap S$$
>
> $$= B_r(x) \cap S. \quad \square$$

Compactness

An important class of topological spaces for analysts is the class of compact spaces. We now define compactness.

(3.33)   Definition   Let X be a topological space and let $S \subset X$. A family $\mathfrak{U}$ of open subsets of X is called an *open cover* of S if $S \subset \cup\mathfrak{U}$. A family $\mathfrak{V}$ is called a *subcover* of S if $\mathfrak{V} \subset \mathfrak{U}$ and $\mathfrak{V}$ is a cover of S. The set S is called *compact* if *every* open cover of S has a *finite* subcover. If $S = X$ and S is compact, we call X a compact space.

(3.34)   Examples   (a) Let

$$S = \,]0, 1\,] \subset \mathbb{R} = X.$$

Then S has many finite open covers; e.g., $\{\mathbb{R}\}$. But S is *not* compact because it is possible to find an open cover $\mathfrak{U}$ that has *no* finite subcover. For instance, let

$$\mathfrak{U} = \{\,]x, \infty[\, : x > 0\}.$$

(b) Any finite subset of any space is compact. For any $\mathfrak{U}$, just choose, for each point of the set, one element of $\mathfrak{U}$ that contains that point.

(c) If $a < b$ in $\mathbb{R}$, then $[a, b]$ is a compact subset of $\mathbb{R}$. To see this, let $\mathfrak{U}$ be any open cover of $[a, b]$. Let $\mathcal{I}$ be the collection of all open intervals I such that $I \subset U$ for some U in $\mathfrak{U}$. Check that $\mathcal{I}$ is a cover of $[a, b]$ and then apply the Heine–Borel Theorem (1.66) to obtain a finite family $\mathfrak{F} \subset \mathcal{I}$ such that $[a, b] \subset \cup\mathfrak{F}$. For each $I \in \mathfrak{F}$, choose $U_I \in \mathfrak{U}$ such that $I \subset U_I$ and then let $\mathfrak{V} = \{U_I : I \in \mathfrak{F}\}$. Plainly, $\cup\mathfrak{V} \supset \cup\mathfrak{F} \supset [a, b]$.

(3.35)   Theorem   *Any closed subset of a compact space is compact.*

Proof   If S is closed in a compact space X and $\mathfrak{U}$ is an open cover of S, then $\mathfrak{U} \cup \{S'\}$ is an open cover of X and so has a finite subcover $\{U_1, U_2, \ldots, U_n, S'\}$ $(U_k \in \mathfrak{U})$. Then $\mathfrak{V} = \{U_1, U_2, \ldots, U_n\}$ covers S. □

(3.36)   Theorem   *Any compact subset of a Hausdorff space is a closed set.*

Proof   Let S be a compact subset of a Hausdorff space X and let $x \in S'$. For each $y \in S$ choose disjoint neighborhoods U_y of x and V_y of y. Then $\{V_y : y \in S\}$ is an open cover of S and so there exist $y_1, \ldots, y_n \in S$ such that $S \subset \bigcup_{k=1}^{n} V_{y_k}$. Next let $U = \bigcap_{k=1}^{n} U_{y_k}$. Then U is a neighborhood of x and $U \subset S'$. This proves that S' is open. □

We next want to prove a generalized version of the Heine–Borel Theorem (1.66), but first we need two lemmas. The first is an extension to $\mathbb{R}^n$ of the Nested Interval Principle (1.57).

(3.37) **Definition** If $n \in \mathbb{N}$ and $a = (a_1, \ldots, a_n)$ and $b = (b_1, \ldots, b_n)$ are points of $\mathbb{R}^n$ with $a_j \leq b_j$ $(1 \leq j \leq n)$, then the *closed interval* in $\mathbb{R}^n$ determined by a and b is the set

$$[a, b] = \{ x \in \mathbb{R}^n : a_j \leq x_j \leq b_j;\ (1 \leq j \leq n) \}.$$

(3.38) **Lemma** *Let $(I_k)_{k=1}^\infty$ be a sequence of closed intervals in $\mathbb{R}^n$ such that*
(i) *$I_{k+1} \subset I_k$ for all $k \in \mathbb{N}$,*
(ii) *$\lim\limits_{k \to \infty} \operatorname{diam} I_k = 0$.*

Then $\bigcap\limits_{k=1}^\infty I_k = \{z\}$ for some $z \in \mathbb{R}^n$.

Proof Say $I_k = [a_k, b_k]$ where $a_k = (a_{k,1}, \ldots, a_{k,n})$ and $b_k = (b_{k,1}, \ldots, b_{k,n})$. For $1 \leq j \leq n$, the sequence $([a_{k,j}, b_{k,j}])_{k=1}^\infty$ satisfies the hypotheses of (1.57) and so $\bigcap\limits_{k=1}^\infty [a_{k,j}, b_{k,j}] = \{z_j\}$ for some $z_j \in \mathbb{R}$. Writing $z = (z_1, \ldots, z_n)$, we conclude that $\bigcap\limits_{k=1}^\infty I_k = \{z\}$. $\square$

(3.39) **Lemma** *Any closed interval in $\mathbb{R}^n$ is a compact subset of $\mathbb{R}^n$.*

Proof Assume that this is false and that $I_1 = [a, b]$ is a closed interval in $\mathbb{R}^n$ that is not compact. Let $\mathcal{U}$ be an open cover if I_1 that has no finite subfamily that covers I_1. Now divide I_1 into 2^n congruent closed intervals by means of the midpoints $c_j = (a_j + b_j)/2$ of $[a_j, b_j] \subset \mathbb{R}$; i.e., each of these 2^n intervals has the form $[x, y]$ where, for each j, either $x_j = a_j$ and $y_j = c_j$ or $x_j = c_j$ and $y_j = b_j$. It cannot happen that each of these 2^n intervals can be covered by a finite subfamily of $\mathcal{U}$ since, otherwise, the union of 2^n such subfamilies of $\mathcal{U}$ would constitute a finite cover of I_1. Thus, we can choose I_2 among these 2^n intervals such that I_2 cannot be covered by a finite subfamily of $\mathcal{U}$. Then $I_2 \subset I_1$ and $\operatorname{diam} I_2 = (1/2)\operatorname{diam} I_1$.

Continuing inductively, we obtain a sequence $(I_k)_{k=1}^\infty$ of closed intervals of $\mathbb{R}^n$ such that, for each k, $I_{k+1} \subset I_k$, $\operatorname{diam} I_{k+1} = (1/2)\operatorname{diam} I_k$, and no finite subfamily of $\mathcal{U}$ covers I_k. Applying (3.38), we have $\bigcap\limits_{k=1}^\infty I_k = \{z\}$ for some $z \in \mathbb{R}^n$. Since $z \in I_1$, we can choose $U \in \mathcal{U}$ such that $z \in U$. Since U is open, we can choose $r > 0$ such that $B_r(z) \subset U$. Now choose k so that $(1/2^{k-1})\operatorname{diam} I_1 = \operatorname{diam} I_k < r$. Then $I_k \subset U$ and so

$\{U\} \subset \mathfrak{U}$ covers I_k, contrary to the choice of I_k. This contradiction completes the proof. $\square$

(3.40)   Generalized Heine–Borel Theorem   *A subset of $\mathbb{R}^n$ is compact if and only if it is closed and bounded.*

Proof Suppose that $S \subset \mathbb{R}^n$ is compact. Then (3.36) shows that S is closed in $\mathbb{R}^n$. To see that S is bounded, consider $\mathfrak{U} = \{B_k(0) : k \in \mathbb{N}\}$. Plainly $\mathfrak{U}$ is an open cover of $\mathbb{R}^n$ and, in particular, of S. Thus there is a $k_0 \in \mathbb{N}$ such that $S \subset B_{k_0}(0)$; whence diam $S \leq 2k_0 < \infty$.

Conversely, suppose that $S \subset \mathbb{R}^n$ is closed and bounded. Say, diam $S = \beta < \infty$. Let $p \in S$ [if $S = \emptyset$, there is nothing to prove]. Then $x \in S$ implies

$$|x_j - p_j| \leq \rho(x, p) \leq \beta \qquad (1 \leq j \leq n),$$

and so, writing $a_j = p_j - 2\beta$ and $b_j = p_j + 2\beta$ $(1 \leq j \leq n)$, we see that S is a closed subset of the interval $[a, b]$. By (3.39) and (3.35), it follows that S is compact. $\square$

(3.41)   Warning   Theorem (3.40) applies only to subsets of $\mathbb{R}^n$. The proof of (3.40) shows that any compact subset of any metric space is both closed and bounded. The converse often fails, as the following examples show.

(a) Let X be any infinite set with the discrete metric. Then $\mathfrak{U} = \{B_1(x) : x \in X\}$ is an infinite open cover of X and, since $B_1(x) = \{x\}$, no member of $\mathfrak{U}$ can be removed and still leave a cover. Therefore X is not compact. However, X is closed in X and diam $X = 1 < \infty$.

(b) Let $X = l_2$ (3.2.f). For $k \in \mathbb{N}$, let $e_k = (e_{k,j})_{j=1}^{\infty}$ be defined by $e_{k,j} = 1$ if $j = k$ and $= 0$ if $j \neq k$. Let $S = \{e_k : k \in \mathbb{N}\}$. Then $\|e_k\|_2 = 1$ for all k and $\|e_k - e_m\|_2 = \sqrt{2}$ for $k \neq m$. It follows that $S \subset l_2$, diam $S = \sqrt{2} < \infty$, and S is closed [it has no limit points (3.24)]. However, $\mathfrak{U} = \{B_1(e_k) : k \in \mathbb{N}\}$ is an open cover of S that has no proper subcover, and so S is not compact.

We next give a condition that is equivalent to compactness.

(3.42)   Theorem   *A topological space X is compact if and only if whenever $\mathcal{C}$ is a family of closed subsets of X with $\cap \mathcal{C} = \emptyset$, there is some finite $\mathcal{F} \subset \mathcal{C}$ with $\cap \mathcal{F} = \emptyset$.*

Proof Suppose X is compact and let $\mathcal{C}$ be given as above. Then $X = \emptyset'$ $= (\cap \mathcal{C})' = \cup \{A' : A \in \mathcal{C}\}$ so there exist $A_1, \ldots, A_n \in \mathcal{C}$ such that

$$X = \bigcup_{k=1}^{n} A_k' = \left(\bigcap_{k=1}^{n} A_k\right)' \text{ and so } \bigcap_{k=1}^{n} A_k = \emptyset.$$

Conversely, suppose X has the "$\mathcal{C}$-property." Let $\mathfrak{U}$ be an open cover of X and let $\mathcal{C} = \{U' : U \in \mathfrak{U}\}$. Then $\cap \mathcal{C} = (\cup \mathfrak{U})' = \emptyset$, so we can get a finite $\mathcal{F} = \{U_1', \ldots, U_n'\} \subset \mathcal{C}$ such that $\varnothing = \bigcap_{k=1}^{n} U_k' = (\bigcup_{k=1}^{n} U_k)'$, and so $\{U_1, \ldots, U_n\} \subset \mathcal{U}$ covers X. Hence, X is compact. $\quad\square$

For metric spaces we have the following equivalents to compactness.

(3.43)   **Theorem**   *Let X be a metric space. Then the following three statements are equivalent.*

(i) *X is compact.*

(ii) *Every infinite subset of X has a limit point in X.*

(iii) *Every sequence in X has a subsequence that converges to a point of X.*

Before proving (3.43) we need the following lemma.

(3.44)   **Lemma**   *Any metric space satisfying (3.43.iii) has a countable dense subset.*

Proof   Suppose that X is a metric space satisfying (3.43.iii) and let $\epsilon > 0$ be given. We assert that there exists a finite set $A_\epsilon \subset X$ satisfying

$$\rho(a, b) \geqq \epsilon \quad \text{whenever } a \neq b \text{ in } A_\epsilon, \tag{1}$$

$$B_\epsilon(x) \cap A_\epsilon \neq \varnothing \quad \text{for each } x \in X. \tag{2}$$

Assume that this is false. Let $x_1 \in X$. When $x_1, \ldots, x_n \in X$ have been chosen so that $\rho(x_i, x_j) \geqq \epsilon$ if $i \neq j$, we observe that the set $\{x_1, \ldots, x_n\}$ satisfies (1) so it cannot satisfy (2). Thus, there exists $x_{n+1} \in X$ such that

$$B_\epsilon(x_{n+1}) \cap \{x_1, \ldots, x_n\} = \varnothing;$$

whence $\rho(x_{n+1}, x_i) \geqq \epsilon$ for $1 \leqq i \leqq n$. In this way we inductively choose a sequence $(x_n)_{n=1}^{\infty} \subset X$ such that $\rho(x_i, x_j) \geqq \epsilon$ whenever $i \neq j$ in $\mathbb{N}$. This sequence certainly has no subsequential limit $x \in X$ [otherwise, we could choose suitable $i < j$ such that $\rho(x_{n_i}, x_{n_j}) \leqq \rho(x_{n_i}, x) + \rho(x, x_{n_j}) < \epsilon$]. This contradiction proves the existence of the required set A_ϵ.

Now define $A = \bigcup_{n=1}^{\infty} A_{1/n}$. Then A, being a countable union of countable [in fact, finite] sets, is countable. Also, if $x \in X$ and $\epsilon > 0$, it follows from (2) and $B_\epsilon(x) \cap A \supset B_{1/n}(x) \cap A_{1/n} \neq \varnothing$ where $1/n < \epsilon$, and so either $x \in A$ or x is a limit point of A. Therefore $A^- = X$. $\quad\square$

Proof of (3.43)   Suppose that (iii) holds. We shall prove (i). Apply (3.44) to get a countable dense subset A of X. Let $\mathcal{B} = \{B_r(x) : x \in A, r \in \mathbb{Q} \text{ is positive}\}$. Since both A and $\mathbb{Q}$ are countable, we see that $\mathcal{B}$ is countable. Now let $\mathcal{U}$ be any open cover of X and let $\mathcal{B}_0 = \{B \in \mathcal{B} : B \subset U \text{ for some } U \in \mathcal{U}\}$. For each $B \in \mathcal{B}_0$ choose $U_B \in \mathcal{U}$ such that $B \subset U_B$ and write

$\mathcal{V} = \{ U_B : B \in \mathcal{B}_0 \}$. Then $\mathcal{V}$ is countable because $\mathcal{B}_0$ is. We next show that $\mathcal{V}$ is a cover of X. To this end, let $z \in X$ be arbitrary and choose $U \in \mathcal{U}$ such that $z \in U$. Since U is open we can find $\epsilon > 0$ such that $B_\epsilon(z) \subset U$. Next select $x \in A$ such that $\rho(x, z) < \epsilon/2$ [A is dense] and then find $r \in \mathbb{Q}$ such that $\rho(x, z) < r < \epsilon/2$. Then $z \in B_r(x) \subset B_\epsilon(z) \subset U [y \in B_r(x)$ implies $\rho(y, z) \leq \rho(y, x) + \rho(x, z) < r + \epsilon/2 < \epsilon]$ and so $B_r(x) \in \mathcal{B}_0$; whence $z \in U_{B_r(x)} \in \mathcal{V}$. This proves that $\cup \mathcal{V} = X$. If $\mathcal{V}$ is finite, we have established (i). Otherwise, enumerate $\mathcal{V}$ as $\{ V_k \}_{k=1}^\infty$. For $n \in \mathbb{N}$, write

$$W_n = \bigcup_{k=1}^n V_k.$$ Then $W_n \subset W_{n+1}$ for each n and $\bigcup_{n=1}^\infty W_n = \cup \mathcal{V} = X$. To complete our proof of (i) we need only show that $W_n = X$ for some $n \in \mathbb{N}$. Assume that this is false. Then, for each $n \in \mathbb{N}$, we can choose $x_n \in X \setminus W_n$. According to (iii), the sequence $(x_n)_{n=1}^\infty$ has a subsequence $(x_{n_k})_{k=1}^\infty$ that converges to some $x \in X$. Now choose $N \in \mathbb{N}$ such that $x \in W_N$. Then, on the one hand, $x_n \notin W_N$ for all $n \geq N$, while on the other hand, $x_{n_k} \in W_N$ [a neighborhood of x] for all $k \geq k_0$ for some k_0. Therefore, $k \geq k_0$ implies $n_k < N$. This contradicts the definition of the term subsequence $[n_1 < n_2 < n_3 < \dots]$ and our proof that (iii) implies (i) is complete.

   Next we prove that (i) implies (ii). Assume that (ii) fails. Then there is some infinite subset D of X having no limit point in X. For each $x \in D$ there is some neighborhood V_x of x such that $D \cap V_x = \{x\}$. Since D is a closed set $[D = D^-$ since D has no limit points], the family $\mathcal{U} = \{D'\} \cup \{ V_x : x \in D \}$ is an infinite open cover of X that has no finite subcover [if some V_x is removed, then x is not covered]. Thus (i) fails whenever (ii) does.

   Finally, we show that (ii) implies (iii). Let $(x_n)_{n=1}^\infty$ be any sequence in X. There are two possibilities: either the set $E = \{ x_n : n \in \mathbb{N} \}$ is finite or E is infinite. If E is finite, then we can find a strictly increasing sequence $(n_k)_{k=1}^\infty \subset \mathbb{N}$ such that $x_{n_k} = x_{n_1}$ for all k and then the subsequence $(x_{n_k})_{k=1}^\infty$ converges to x_{n_1}. Thus, we suppose that E is infinite and we apply (ii) to obtain some limit point $x \in X$ of the set E. Choose n_1 so that $x_{n_1} \in E \cap B_1(x)$. When $n_1 < n_2 < \dots n_{k-1}$ have been chosen, use (3.24) to choose $n_k > n_{k-1}$ such that $x_{n_k} \in E \cap B_{1/k}(x)$. This inductively defines a subsequence $(x_{n_k})_{k=1}^\infty$. Given any neighborhood U of x, we can choose $k_0 \in \mathbb{N}$ such that $B_{1/k_0}(x) \subset U$. Then it is clear that $x_{n_k} \in U$ for all $k \geq k_0$ and so $x_{n_k} \to x$ as $k \to \infty$. $\square$

(3.45) Remarks   (a) Our proof that (3.43.i) implies (3.43.ii) applies to any topological space.

   (b) It is an easy exercise to show that (3.43.iii) implies (3.43.ii) in any topological space.

   (c) There exist examples to show that none of the other four implications joining the three statements in (3.43) is true for arbitrary topological spaces.

Connectedness

(3.46) **Definition** A subset S of a topological space X is said to be *connected* if there do not exist two open sets U and V such that $S \cap U \cap V = \emptyset$, $S \subset U \cup V$, and $U \cap S \neq \emptyset \neq V \cap S$. If such U and V do exist, then S is said to be *disconnected*.

(3.47) **Theorem** *A subset of $\mathbb{R}$ is connected if and only if it is an interval.*

Proof Assume that there exists an interval $S \subset \mathbb{R}$ such that S is disconnected. Let U and V be as in (3.46). Choose $a \in U \cap S$ and $b \in V \cap S$. Then $a \neq b$, so we may suppose that $a < b$. Since S is an interval we have $[a, b] \subset S$. Let $c = \sup(U \cap [a, b])$. Then $c \in [a, b] \subset S$, so either $c \in U \cap [a, b]$ or $c \in V \cap [a, b]$.

Assume that $c \in U \cap [a, b]$. Then $c < b$ [otherwise $c \in S \cap U \cap V$]. Since U is open we can choose $c' > c$ such that $[c, c'] \subset U \cap [a, b]$, contrary to our choice of c. Thus $c \notin U \cap [a, b]$.

Assume that $c \in V \cap [a, b]$. Then $c > a$ and there is some $c'' < c$ such that $[c'', c] \subset V \cap [a, b]$. But $U \cap V \cap [a, b] \subset U \cap V \cap S = \emptyset$, and so c'' is an upper bound for $U \cap [a, b]$, contrary to our choice of c. Thus $c \notin V \cap [a, b]$.

This contradiction proves that every interval is connected.

Conversely, suppose that $S \subset \mathbb{R}$ is not an interval. Recall that $\emptyset$ and every singleton are intervals. We can choose $d \in \mathbb{R}$ such that $\inf S < d < \sup S$ and $d \notin S$. Setting $U =]-\infty, d[$ and $V =]d, \infty[$, we see that S is disconnected. $\square$

(3.48) **Remark** Theorem (3.47) solves the problem of characterizing the connected subsets of $\mathbb{R}$. There is no such simple characterization of the connected subsets of $\mathbb{R}^n$ for $n > 1$. For instance, the sets

$$A = \left\{ (x, y) \in \mathbb{R}^2 : x^2 + y^2 < 1 \right\}$$

and

$$B = A \setminus (\{ 1/n : n \in \mathbb{N} \} \cup \{0\})$$

are two connected open subsets of $\mathbb{R}^2$ that are topologically quite different: B has a lot of holes in it; A has none.

Completeness

We can extend the notion of a Cauchy sequence to arbitrary metric spaces. The following definition obviously agrees with (2.28) when X is the complex plane $\mathbb{C}$ with its usual metric.

(3.49)   Definition   Let X be a metric space with metric ρ. A sequence $(x_n)_{n=1}^{\infty} \subset X$ is called a *Cauchy sequence* if for each $\epsilon > o$ there exists $n_0 \in \mathbb{N}$ such that $\rho(x_n, x_m) < \epsilon$ whenever $m, n \geq n_0$. We call X a *complete metric space* if every Cauchy sequence in X converges to a point of X.

(3.50)   Examples   (a) The complex plane $\mathbb{C}$ is complete as (2.29) shows.

(b) The real line $\mathbb{R}$ is complete. This follows from (a) and the fact that the limit of a sequence of real numbers, if it exists, cannot be imaginary.

(c) For $n \in \mathbb{N}$, the space $\mathbb{R}^n$, with its usual metric, is complete. In fact, if $(x^{(k)})_{k=1}^{\infty} \subset \mathbb{R}^n$ is a Cauchy sequence, then, for $1 \leq j \leq n$, the sequence $(x_j^{(k)})_{k=1}^{\infty}$ is a Cauchy sequence in $\mathbb{R}$ $[|x_j^{(k)} - x_j^{(m)}| \leq \rho(x^{(k)}, x^{(m)})]$ and so $x_j^{(k)} \to x_j$ for some $x_j \in \mathbb{R}$ as $k \to \infty$. Putting $x = (x_1, \ldots, x_n) \in \mathbb{R}^n$, one checks that $\rho(x^{(k)}, x) \to 0$ as $k \to \infty$.

(d) The set $\mathbb{Q}$ of rational numbers is not complete, as the following theorem shows $[X = \mathbb{R}, A = \mathbb{Q}]$.

(3.51)   Theorem   *Let X be a complete metric space and $A \subset X$. Then A, with the metric of X, is a complete metric space if and only if A is closed in X.*

Proof   Suppose that A is closed in X and that $(x_n)_{n=1}^{\infty} \subset A$ is a Cauchy sequence. Since X is complete, there is some $x \in X$ such that $x_n \to x$. Now (3.28.i) shows that $x \in A^- = A$. Thus A is complete.

Next, suppose that A is not closed in X. Then there is some $x \in A^- \setminus A$ and (3.28.ii) furnishes a sequence $(x_n)_{n=1}^{\infty} \subset A$ such that $x_n \to x$. Plainly, $(x_n)_{n=1}^{\infty}$ is a Cauchy sequence $[\rho(x_m, x_n) \leq \rho(x_m, x) + \rho(x, x_n)]$ that has no limit in A; hence A is not complete.   □

Another important class of complete metric spaces is given by the following theorem.

(3.52)   Theorem   *Let S be any nonvoid set and let $B(S)$ denote the set of all complex-valued functions f on S such that*

$$\|f\|_u = \sup\{|f(x)| : x \in S\} < \infty.$$

As in (3.2.e), define $\rho(f, g) = \|f - g\|_u$ for $f, g \in B(S)$. Then $B(S)$ is a complete metric space.

Proof   Let $(f_n)_{n=1}^{\infty}$ be a Cauchy sequence in $B(S)$. For fixed $s \in S$, we have

$$|f_m(s) - f_n(s)| \leq \|f_m - f_n\|_u \to 0$$

as $m, n \to \infty$. Since $\mathbb{C}$ is complete, there exists $f(s) \in \mathbb{C}$ such that

$$|f(s) - f_n(s)| \to 0 \tag{1}$$

as $n \to \infty$. This defines a complex-valued function f on S. We shall show that $f \in B(S)$ and that $\|f - f_n\|_u \to 0$ as $n \to \infty$.

Let $\epsilon > 0$ be given. Choose $N \in \mathbb{N}$ such that

$$m, n \geq N \quad \text{implies} \quad \|f_m - f_n\|_u < \epsilon/2. \tag{2}$$

Next use (1) to choose, for each $s \in S$, an $n(s) \in \mathbb{N}$ such that $n(s) > N$ and

$$|f(s) - f_{n(s)}(s)| < \epsilon/4. \tag{3}$$

Then, for $n \geq N$ and any $s \in S$, we have

$$|f(s) - f_n(s)| \leq |f(s) - f_{n(s)}(s)| + |f_{n(s)}(s) - f_n(s)|$$
$$< \epsilon/4 + \|f_{n(s)} - f_n\|_u$$
$$< (3/4)\epsilon,$$
$$|f(s) - f_n(s)| < (3/4)\epsilon. \tag{4}$$

Since (4) holds for all $s \in S$, we may take the supremum of its left side to obtain

$$\|f - f_n\|_u \leq (3/4)\epsilon < \epsilon \tag{5}$$

whenever $n \geq N$. In particular,

$$\|f\|_u \leq \|f - f_N\|_u + \|f_N\|_u < \infty,$$

proving that $f \in B(S)$. Finally, (5) shows that $f_n \to f$ in the metric of $B(S)$.
□

Baire Category

Our next theorem is perhaps the most important theorem about complete metric spaces. First, we make a few definitions.

(3.53) **Definitions** Let X be a topological space. A set $A \subset X$ is said to be *nowhere dense in X* if the closure of A has no interior points: $A^{-\circ} = \emptyset$. A set $B \subset X$ is said to be of *first category in X* if B is the union of some countable family of sets that are nowhere dense in X. All subsets of X that are not of first category in X are said to be of *second category in X*. A set $E \subset X$ is called *residual in X* if E' is of first category in X.

(3.54) **Examples** (a) If $x \in \mathbb{R}$, then $\{x\}$ is nowhere dense in $\mathbb{R}$. Thus $\mathbb{Q}$ is of first category in $\mathbb{R}$ and the set $\mathbb{R} \backslash \mathbb{Q}$ of irrational numbers is residual in $\mathbb{R}$.

(b) Any Cantor set is nowhere dense in $[0, 1]$ since it is closed and contains no nonvoid open interval (2.85.b). If, for each $n \in \mathbb{N}$, we choose a Cantor set $A_n \subset [0, 1]$ such that $\lambda(A_n) = 1 - 1/n$ (2.82.ii) and then write $B = \bigcup_{n=1}^{\infty} A_n$, then B is of first category in $[0, 1]$ and

$$1 - 1/n = \lambda(A_n) \leq \lambda(B) \leq \lambda([0, 1]) = 1$$

for all $n \in \mathbb{N}$; hence $\lambda(B) = 1$.

(3.55)   The Baire Category Theorem (1899)   *If X is a complete metric space and E is residual in X, then E is dense in X.*

Proof   Since E is residual, we can write $E' = \bigcup_{n=1}^{\infty} A_n$ where each A_n is no-
where dense in X. To show that E is dense, we must show that E contains points in every nonvoid open set. Thus let V be any given nonvoid open set. We shall construct a Cauchy sequence whose limit is in $V \cap E$.
        Since $A_1^{-\circ} = \emptyset$, we can choose x_1 in the open set $V \backslash A_1^{-}$ and then we can choose $0 < r_1 < 1$ such that $B_{r_1}(x_1)^{-} \subset V \backslash A_1^{-}$ [check that $B_r(x)^{-} \subset B_{2r}(x)$]. If $n > 1$ and $x_1, \ldots, x_{n-1}$ and $r_1, \ldots, r_{n-1}$ have been chosen, then, since $A_n^{-\circ} = \emptyset$, the open set $B_{r_{n-1}}(x_{n-1}) \backslash A_n^{-}$ contains some point x_n. Next choose $0 < r_n < 1/n$ such that

$$B_{r_n}(x_n)^{-} \subset B_{r_{n-1}}(x_{n-1}) \backslash A_n^{-}. \tag{1}$$

The sequence $(x_n)_{n=1}^{\infty}$, so chosen, is a Cauchy sequence because

$$\rho(x_m, x_n) \leq \rho(x_m, x_N) + \rho(x_N, x_n) < r_N + r_N < 2/N$$

for $m, n \geq N$. Therefore $x_n \to x$ for some $x \in X$. Since $(x_n)_{n=N}^{\infty} \subset B_{r_N}(x_N)^{-}$ it follows from (3.28.i) that $x \in B_{r_N}(x_N)^{-}$ for every $N \in \mathbb{N}$. Therefore, with the use of (1),

$$x \in \bigcap_{n=1}^{\infty} B_{r_n}(x_n)^{-} \subset \bigcap_{n=1}^{\infty} A_n^{-\prime} \cap V$$

$$\subset V \cap \bigcap_{n=1}^{\infty} A_n' = V \cap \left(\bigcup_{n=1}^{\infty} A_n \right)' = V \cap E;$$

whence $V \cap E \neq \emptyset$.   □

(3.56)   Definition   A subset E of a topological space is called a G_δ if $E = \bigcap_{n=1}^{\infty} U_n$ for some sequence $(U_n)_{n=1}^{\infty}$ of open sets. A subset F of a topological space is called an F_σ if $F = \bigcup_{n=1}^{\infty} F_n$ for some sequence $(F_n)_{n=1}^{\infty}$ of closed sets; i.e., if F' is a G_δ.

(3.57)   Corollary   *If $(U_n)_{n=1}^{\infty}$ is any sequence of dense open subsets of a complete metric space X, then the G_δ set $E = \bigcap_{n=1}^{\infty} U_n$ is dense in X. Moreover, if V is any open subset of X having no isolated points, then $V \cap E$ is uncountable.*

Proof   For each n, the closed set U_n' is nowhere dense in X because $U_n^{-} = X$. Thus, $E' = \bigcup_{n=1}^{\infty} U_n'$ is of first category in X. This means that E is residual in X, and so (3.55) shows that E is dense in X.
        Assume that there exists a nonvoid open set V without isolated

points such that $V \cap E$ is countable. Then, for each $x \in V \cap E$, the closed set $\{x\}$ is nowhere dense in X, and so $V \cap E$, being the countable union of such singletons, is of first category in X. Thus, $E' \cup (V \cap E)$ is of first category in X, and so its complement $E \cap (V' \cup E') = E \cap V'$ is residual and therefore dense in X. But this is impossible because $E \cap V'$ contains no points of the nonvoid open set V. $\square$

(3.58) **Corollary** *If X is a complete metric space, then any dense G_δ in X is residual in X.*

(3.59) **Corollary** *If X is a (nonvoid) complete metric space having no isolated points and E is a countable dense subset of X, then E is not a G_δ. In particular, $\mathbb{Q}$ is not a G_δ in $\mathbb{R}$.*

> **Proof** If E were a G_δ, it would have the form given in (3.57) and we could take $V = X$ to obtain a contradiction. $\square$

(3.60) **Theorem** *Let X be a (nonvoid) complete metric space and E a residual set in X. Then E is of second category in X. In particular, X is of second category in itself.*

> **Proof** Assume that E is of first category in X. Then $X = E \cup E'$, being a countable union of nowhere dense sets, is of first category in X, and so, using (3.55), we deduce that $\emptyset = X'$ is dense in X—an obvious absurdity. $\square$

(3.61) **Remark** The converse of (3.60) is false. If $X = \mathbb{R}$ and $E = [0, 1]$, then E is of second category in $\mathbb{R}$ [why?] but E is not residual in $\mathbb{R}$ $[E^- = E \neq \mathbb{R}]$.

Exercises

1. Give an example of a sequence of open intervals of $\mathbb{R}$ whose intersection is not open in $\mathbb{R}$.

2. If A and B are subsets of a topological space X, then $A^- \cup B^- = (A \cup B)^-$ and $A^\circ \cap B^\circ = (A \cap B)^\circ$, but the equalities $A^- \cap B^- = (A \cap B)^-$ and $A^\circ \cup B^\circ = (A \cup B)^\circ$ may fail. Prove the first two equalities and give an example of two subsets of $\mathbb{R}$ for which both of the latter two equalities fail.

3. Let ρ be the usual metric on $\mathbb{R}^n$ defined in (3.2.c) and define σ by $\sigma(x, y) = \max\{|x_j - y_j| : 1 \leq j \leq n\}$. Then σ is a metric on $\mathbb{R}^n$ and the topology on $\mathbb{R}^n$ generated by σ is the same as that generated by ρ. That is, a set $U \subset \mathbb{R}^n$ is open relative to ρ if and only if it is open relative to σ. The σ-balls are open intervals in $\mathbb{R}^n$.

4. Let X be a metric space and let A and B be nonvoid subsets of X. Define
$$\text{dist}(A, B) = \inf\{\rho(x, y) : x \in A, y \in B\}.$$

Then

(a) If A and B are compact, then there exist $a \in A$ and $b \in B$ such that $\text{dist}(A, B)$ $= \rho(a, b)$ [we say that the distance from A to B is attained].

(b) There exist disjoint nonvoid closed subsets A and B of $\mathbb{R}$ for which $\text{dist}(A, B)$ $= 0$.

5. Let X be a metric space and let A be a nonvoid subset of X. Define $\text{dist}(x, A)$ as in (3.7).

(a) For $x \in X$; $x \in A^-$ if and only if $\text{dist}(x, A) = 0$.

(b) If A is compact and $x \in X$, then there exists an $a \in A$ such that $\text{dist}(x, A)$ $= \rho(x, a)$. Is a unique?

(c) If $X = \mathbb{R}^n$ and A is closed, then the conclusion of (b) holds.

(d) It can happen that $x \in X$, A is closed, and $\text{dist}(x, A) < \rho(x, a)$ for all $a \in A$.

6. Let X be a metric space and let

$$B_r(x) = \{ y \in X : \rho(x, y) < r \}$$

be any ball in X. The set

$$A_r(x) = \{ y \in X : \rho(x, y) \leq r \}$$

is a *closed ball*. It is also a closed set and $B_r(x)^- \subset A_r(x)$. There exist examples for which $B_r(x)^- \neq A_r(x)$.

7. A topological space X is said to be *second countable* if there exists some countable family $\mathscr{B}$ of open sets such that each open set $V \subset X$ can be expressed as the union of some subfamily of $\mathscr{B}$. Prove that $\mathbb{R}^n$ is second countable. Give an example of a metric space that is not second countable. [See Exercise 8.]

8. If X is a metric space, then X is second countable if and only if X has a countable dense subset. [See Exercise 7.]

9. Any uncountable subset of a second countable space has at least one limit point. [See Exercise 7.]

10. If X is a second countable space and A is a closed subset of X, then there exist a perfect set P and a countable set C such that $A = C \cup P$. [Hint: Let P be the set of all $x \in X$ such that, whenever V is a neighborhood of x, the set $V \cap A$ is uncountable. Use Exercise 9.]

11. Any subset of a second countable space can have only countably many isolated points. [See Exercise 7.]

12. (a) If A is a perfect nonvoid subset of a complete metric space, then A is uncountable. [Hint: Use (3.51) and (3.59).]

(b) Any countable and infinite closed subset of a complete metric space has infinitely many isolated points.

(c) There exists a countable closed subset of $\mathbb{R}$ having infinitely many limit points.

13. It is impossible to express $[0, 1]$ as a union of disjoint closed intervals of positive length < 1. [Hint: Assuming such an expression possible, consider the set of all endpoints of the intervals involved. Use Exercise 12(a).] Can $[0, 1]$ be expressed as the union of a countable infinite number of disjoint closed sets?

14. Let A be any given bounded subset of $\mathbb{R}$ that is not closed. Construct explicitly an open cover of A that has no finite subcover.

15. A metric space X is said to be *totally bounded* if for each $r > 0$ there exists a finite set $F \subset X$ such that $\text{dist}(x, F) < r$ for each $x \in X$; i.e., $X = \cup \{ B_r(y) : y \in F \}$. A metric space is compact if and only if it is both complete and totally bounded. [Hint: Use (3.43) to prove compactness. Given $(x_n)_{n=1}^\infty \subset X$, use total boundedness to choose an infinite set $A_1 \subset \mathbb{N}$ such that $\rho(x_m, x_n) < 1$ for all $m, n \in A_1$. When an infinite set A_{k-1} has been chosen, use total boundedness to choose an infinite set $A_k \subset A_{k-1}$ such that $\rho(x_m, x_n) < 1/k$ for all $m, n \in A_k$. Next select $n_1 < n_2 \ldots$ such that $n_k \in A_k$ for all k and prove that $(x_{n_k})_{k=1}^\infty$ is a Cauchy sequence.]

16. [*Lebesgue's Covering Theorem*]. If X is a compact metric space and $\mathfrak{U}$ is an open cover of X, then there exists some $r > 0$ such that to each $x \in X$ corresponds some $V_x \in \mathfrak{U}$ such that $B_r(x) \subset V_x$. [Hints: For $x \in X$, select $U_x \in \mathfrak{U}$ such that $x \in U_x$ and then choose $r(x) > 0$ such that $B_{2r(x)}(x) \subset U_x$. If $X = \bigcup_{j=1}^n B_{r(x_j)}(x_j)$, take $r = \min\{r(x_1), \ldots, r(x_n)\}$.]

17. Let λ be Lebesgue measure on $\mathbb{R}$.
    (a) If V is a nonvoid open subset of $\mathbb{R}$ and $\alpha > 0$, then there exists a compact nowhere dense set $A \subset V$ such that $0 < \lambda(A) < \alpha$.
    (b) There exists an F_σ-set $B \subset \mathbb{R}$ that is of first category in $\mathbb{R}$ which has the further property that $\lambda(U \cap B) > 0$ and $\lambda(U \cap B') > 0$ for every nonvoid open subset U of $\mathbb{R}$. [Hints: Fill in the many missing details in the following. Let $\{I_n\}_{n=1}^\infty$ be an enumeration of the set of all nonvoid open intervals having rational endpoints. Write $r_1 = \lambda(I_1)$ and $B_0 = \emptyset$. When a compact nowhere dense set B_{n-1} and positive numbers $r_1, \ldots, r_n$ have been selected for some $n \in \mathbb{N}$, choose a compact nowhere dense set $A_n \subset I_n \cap B'_{n-1}$ such that $0 < \lambda(A_n) < 2^{-n}\min\{r_1, \ldots, r_n\}$. Write

    $$B_n = B_{n-1} \cup A_n \quad \text{and} \quad r_{n+1} = \lambda(I_{n+1}) - \lambda(I_{n+1} \cap B_n).$$

    Finally, write $B = \bigcup_{n=1}^\infty B_n$ and note that $\lambda(I_p \cap B) \geq \lambda(A_p) > 0$ and that

    $$\lambda(I_p \cap B) \leq \lambda(I_p \cap B_{p-1}) + \sum_{n=p}^\infty \lambda(A_n)$$
    $$< \lambda(I_p \cap B_{p-1}) + \sum_{n=p}^\infty \frac{r_p}{2^n}$$
    $$\leq \lambda(I_p \cap B_{p-1}) + r_p = \lambda(I_p)$$

    so that $\lambda(I_p \cap B') > 0$.]

18. A subset A of a topological space X is called *totally disconnected* if the only connected subsets of A are $\emptyset$ and the sets $\{x\}$ for $x \in A$.
    (a) Show that $\mathbb{Q}$ is a totally disconnected subset of $\mathbb{R}$.
    (b) Show that any Cantor set $P \subset \mathbb{R}$ is totally disconnected.

19. [*Banach's Fixed Point Theorem*] Let X be a complete metric space with metric ρ and let $f: X \to X$ satisfy

    $$\rho(f(x), f(y)) \leq \alpha\rho(x, y)$$

    for all $x, y \in X$, where α is some constant independent of x and y with $0 < \alpha < 1$. Then there is a unique $p \in X$ such that $f(p) = p$. [Hint: obtain p as the limit of a Cauchy sequence. Let $x_1 \in X$ and define $x_{n+1} = f(x_n)$.]

20. Let S be any metric space with metric ρ and let $B(S)$ be as in (3.52). Let $p \in S$ be fixed. For $s \in S$ define f_s on S by

    $$f_s(x) = [\rho(x, s) - \rho(x, p)].$$

Then the mapping $s \to f_s$ is an *isometry* [distance preserving] of S into $B(S)$; i.e.,

$$\| f_s - f_t \|_u = \rho(s, t)$$

for all $s, t \in S$. Thus any metric space is isometric to a subspace of some complete metric space.

Limits of Functions
at a Point

We now return to the problem of giving a rather general definition of the term limit.

(3.62)   Definition   Let X and Y be topological spaces, let $S \subset X$, and let $f : S \to Y$ be a function from S into Y. Suppose that $a \in X$ is a limit point of S and that $b \in Y$. We say that *the limit of f at a is b* and we write

(i) $\lim\limits_{\substack{x \to a \\ x \in S}} f(x) = b$

or

$$f(x) \to b \quad \text{as } x \to a \quad (x \in S)$$

if for each neighborhood V of b $[V \subset Y]$ there exists a neighborhood U of a $[U \subset X]$ such that $f(x) \in V$ whenever $x \in U \cap S$ and $x \neq a$. Note that the point a need not be in S and, even if $a \in S$, we need not have $f(a) = b$.

Frequently the domain S under consideration is clear from the context, and so we often omit the $x \in S$ part of the above notation. In particular, if $X = \mathbb{R}$ and S is some interval having a as a left [resp. right] endpoint we often write

$$\lim_{x \downarrow a} f(x) = b \quad \left[\text{resp. } \lim_{x \uparrow a} f(x) = b \right]$$

in place of (i). In this case we call b a *right hand* [resp. *left hand*] *limit* of f at a.

We emphasize that the value of f at a [if defined] need not be related to a limit b at a. The condition in this definition explicitly states $x \neq a$.

(3.63)   Examples   (a) If $X = Y = \mathbb{R}^{\#}$ and $S = \mathbb{R}$, then

$$\lim_{x \to \infty} \left[(x + 1)/(x - 1) \right] = 1$$

because

$$|(x + 1)/(x - 1) - 1| < \epsilon \quad \text{if } 1 + 2/\epsilon < x < \infty.$$

That is,

$(x + 1)/(x - 1) \in V = B_\epsilon(1)$ whenever $x \in U = \,]1 + 2/\epsilon, \infty]$ and $x \neq \infty$.

(b) Let $X = [0, 1]$, $S = \,]0, 1]$,

$$Y = \mathbb{R}^{\#} \quad \text{and} \quad f(x) = 1/x.$$

Then $\lim\limits_{x \downarrow 0}(1/x) = \infty$.

(c) Let $X = Y = \mathbb{R}^\#$, $S = \mathbb{R}\backslash\{0\}$ and $f(x) = 1/x$. Then $\lim\limits_{x \to 0}(1/x)$ does not exist, but $\lim\limits_{x \to -\infty}(1/x) = \lim\limits_{x \to \infty}(1/x) = 0$.

(d) If $X = \mathbb{N} \cup \{\infty\}$, $S = \mathbb{N}$, and $a = \infty$, $Y = \mathbb{C}$ and if we topologize X as a subspace of $\mathbb{R}^\#$, then (3.62) is just the definition of the limit of a complex sequence as given in Chapter 2.

In general, limits need not be unique, but we do have the following simple theorem.

(3.64) **Theorem** *With the notation of (3.62), if Y is a Hausdorff space and if*

$$f(x) \to b_1 \in Y, \quad f(x) \to b_2 \in Y \quad \text{as } x \to a \quad (x \in S),$$

then $b_1 = b_2$.

> **Proof** Assume that $b_1 \neq b_2$. Choose disjoint neighborhoods V_1 and V_2 of b_1 and b_2 respectively. By (3.62) there exist neighborhoods U_1 and U_2 of a such that $f(x) \in V_j$ whenever $x \in U_j \cap S$ and $x \neq a$ for $j = 1, 2$. But $U_1 \cap U_2$ is a neighborhood of a and a is a limit point of S, so we can choose $x_0 \in U_1 \cap U_2 \cap S$, $x_0 \neq a$. Then $f(x_0) \in V_1 \cap V_2 = \emptyset$, a contradiction. $\square$

If X is a metric space, the following theorem gives a very useful equivalent to our definition of limit.

(3.65) **Theorem** *If X is a metric space and the notation is as in (3.62), then*

(i) $\lim\limits_{\substack{x \to a \\ x \in S}} f(x) = b$

if and only if

(ii) $\lim\limits_{n \to \infty} f(x_n) = b$ *whenever*

$$(x_n)_{n=1}^\infty \subset S\backslash\{a\} \quad \text{and} \quad \lim\limits_{n \to \infty} x_n = a.$$

> **Proof** Suppose that (i) obtains and let $x_n \to a$ as in (ii). If V is any neighborhood of b, then choose a neighborhood U of a as in (3.62). Next choose $N \in \mathbb{N}$ such that $x_n \in U$ whenever $n \geq N$. Then we have $f(x_n) \in V$ whenever $n \geq N$, and so $f(x_n) \to b$.
>
> Next suppose that (i) fails. Then there is some neighborhood V of b such that no neighborhood U of a satisfies the condition of (3.62). Fixing such a V, it follows that for each $n \in \mathbb{N}$ there is some $x_n \in B_{1/n}(a) \cap S$ with $x_n \neq a$ and $f(x_n) \notin V$. But then $x_n \to a$ and $f(x_n) \nrightarrow b$, and so (ii) fails. $\square$

(3.66) **Continuity Definition** Let X and Y be topological spaces and let $f: X \to Y$ [f is a function from X into Y]. For $p \in X$, we say that f is *continuous at p* if for each neighborhood V of $f(p)$ $[V \subset Y]$ there exists some neighborhood U of p $[U \subset X]$ such that $f(U) \subset V$; i.e., $f(x) \in V$ for every $x \in U$. We say that f is *continuous on X* if f is continuous at every $p \in X$.

(3.67) Warning In the statement that a function f is continuous at a point p, it is crucial to understand just which space X is the domain of f. The above definition of "f is continuous on X" entails that X is the domain of f. If $S \subset X$, we will use the clause "f is continuous on S" to mean that f, with its domain restricted to the subspace S [see (3.30)], is a continuous function from the topological space S into Y. That is, for each $p \in S$ and neighborhood V of $f(p)$ there is a neighborhood $U \cap S$ of p in the relative topology on S [U open in X] such that $f(U \cap S) \subset V$. It is clear that if f, with domain X, is continuous at every point of S [$f(U) \subset V$], then f is continuous on S. However, the converse can fail. A couple of examples may illuminate this point.

 Let $X = Y = \mathbb{R}$, let $S = [0, 1]$, and define $f(x) = 1$ for $x \in S$ and $f(x) = 0$ for $x \in S'$. Then f is not continuous at either 0 or 1 [take $V =]0, 2[$], but f is continuous on S because its restriction to S is a constant [given V, take $U = S$ $= S \cap\,] - 1, 2[$]. Similarly, if $g(x) = 1$ for $x \in \mathbb{Q}$ and $g(x) = 0$ for $x \in \mathbb{Q}'$, then g is not continuous at any point of $\mathbb{R}$, yet g is continuous on $\mathbb{Q}$.

 The relation between limits and continuity is as follows.

(3.68) Theorem *Let the notation be as in (3.66). Then f is continuous at p if and only if either p is an isolated point of X $[\{p\}$ is open in $X]$ or p is a limit point of X and $\lim_{x \to p} f(x) = f(p)$.*

 Proof If p is a limit point of X, then this is obvious when we compare (3.66) and (3.62) [with $S = X$, $a = p$, and $b = f(p)$]. If p is isolated, we can always take $U = \{p\}$ in (3.66) to prove continuity at p. □

(3.69) Theorem *With the notation of (3.66), the following three assertions are equivalent.*
 (i) *f is continuous on X.*
 (ii) *$f^{-1}(V)$ is open in X whenever V is open in Y.*
 (iii) *$f^{-1}(B)$ is closed in X whenever B is closed in Y.*
Recall that $f^{-1}(C) = \{x \in X : f(x) \in C\}$.

 Proof Suppose that (i) obtains and let V be open in Y. For each p $\in f^{-1}(V)$ we have $f(p) \in V$, and so, since f is continuous at p, there exists a neighborhood U_p of p such that $f(U_p) \subset V$. Thus, to each $p \in f^{-1}(V)$ corresponds an open set U_p such that $p \in U_p \subset f^{-1}(V)$. Plainly, $f^{-1}(V)$ $= \cup\{U_p : p \in f^{-1}(V)\}$, being a union of open sets, is open. Therefore (i) implies (ii).
 That (ii) implies (iii) is a consequence of the fact that

$$\left[f^{-1}(B) \right]' = f^{-1}(B') \tag{1}$$

for any set $B \subset Y$, because if B is closed, then B' is open and, by (ii) and (1), $f^{-1}(B)$ is the complement of an open set, so is closed.

Suppose that (iii) obtains. To show that (i) follows, let $p \in X$ be arbitrary and let V be any neighborhood of $f(p)$. Let $U = f^{-1}(V)$. Then $f(U) \subset V$ and $p \in U$. Moreover, using (iii) and (1) [with $B = V'$], we see that

$$U = f^{-1}(V) = \left[f^{-1}(V') \right]'$$

is the complement of a closed set, so is open. $\square$

For metric spaces, the next theorem gives two useful equivalents to the definition of continuity at a point.

(3.70) **Theorem** *Let X and Y be metric spaces with metrics ρ and σ, respectively. Let $f : X \to Y$ and let $p \in X$. Then the following three statements are equivalent.*
(i) *f is continuous at p.*
(ii) *To each $\epsilon > 0$ corresponds a $\delta > 0$ such that*

$$x \in X \quad \text{and} \quad \rho(x, p) < \delta \quad \text{imply} \quad \sigma(f(x), f(p)) < \epsilon.$$

(iii) *Whenever $(x_n)_{n=1}^{\infty} \subset X$ and $\lim_{n \to \infty} x_n = p$, we have $\lim_{n \to \infty} f(x_n) = f(p)$.*

Proof This follows at once from (3.68), (3.65), and the definition of the metric topology. $\square$

(3.71) **Corollary** *The functions*

$$(z, w) \to z + w, \qquad (z, w) \to zw$$

from $\mathbb{C} \times \mathbb{C}$ into $\mathbb{C}$ are continuous. Also, the function

$$(z, w) \to z/w$$

from $\mathbb{C} \times (\mathbb{C} \setminus \{0\})$ into $\mathbb{C}$ is continuous.

Proof Apply (2.8) and (3.70). $\square$

(3.72) **Theorem** *Let X, Y, and Z be topological spaces. Suppose that $p \in X$, $f : X \to Y$ is continuous at p, and $g : Y \to Z$ is continuous at $f(p)$. The composite function $g \circ f : X \to Z$ defined by $g \circ f(x) = g(f(x))$ is continuous at p. In particular, if f is continuous on X and g is continuous on Y, then $g \circ f$ is continuous on X.*

Proof Let W be any neighborhood of $g \circ f(p)$. Then there is a neighborhood V of $f(p)$ such that $g(V) \subset W$. Next we can choose a neighborhood U of p such that $f(U) \subset V$. We have $g \circ f(U) \subset g(V) \subset W$. $\square$

(3.73)   Theorem *Let X be a topological space, let $n \in \mathbb{N}$, and let $(\phi_k)_{k=1}^n$ be complex-valued functions on X. Define Φ from X into $\mathbb{C}^n$ by*

$$\Phi(x) = (\phi_1(x), \ldots, \phi_n(x)).$$

Then, for $p \in X$, Φ is continuous at p if and only if each ϕ_k is continuous at p.

> **Proof** Suppose that Φ is continuous at p. Then $\phi_k = \pi_k \circ \Phi$ where $\pi_k : \mathbb{C}^n \to \mathbb{C}$ is the projection defined by $\pi_k(z_1, \ldots, z_n) = z_k$. But π_k is continuous on $\mathbb{C}^n$ because
>
> $$|z_k - w_k| \leq \left(\sum_{j=1}^n |z_j - w_j|^2 \right)^{1/2}.$$
>
> We conclude, using (3.72), that ϕ_k is continuous at p.
>   Conversely, suppose that each ϕ_k is continuous at p. Given $\epsilon > 0$ and $1 \leq k \leq n$, choose a neighborhood U_k of p such that
>
> $$|\phi_k(x) - \phi_k(p)| < \epsilon/n \quad \text{for} \quad x \in U_k.$$
>
> Let $U = \bigcap_{k=1}^n U_k$. Then U is a neighborhood of p and $x \in U$ implies
>
> $$|\Phi(x) - \Phi(p)| = \left(\sum_{k=1}^n |\phi_k(x) - \phi_k(p)|^2 \right)^{1/2} < \left(n\epsilon^2/n^2 \right)^{1/2} \leq \epsilon.$$
>
> Therefore Φ is continuous at p. $\quad\square$

The preceding three results allow us to construct new continuous functions from old ones.

(3.74)   Theorem *Let X be a topological space, let $p \in X$, and let ϕ and ψ be complex-valued functions on X that are continuous at p. Then the functions $\phi + \psi$, $\phi\psi$, $|\phi|$, $\bar{\phi}$, $\operatorname{Re}\phi$, $\operatorname{Im}\phi$ (defined pointwise on X) are all continuous at p. If $\psi(x) \neq 0$ for all $x \in X$, then ϕ/ψ is also continuous at p.*

> **Proof** These are all composites of continuous functions. For example,
>
> $$(\phi/\psi)(x) = \phi(x)/\psi(x) = g \circ f(x)$$
>
> where
>
> $$f(x) = (\phi(x), \psi(x)) \in \mathbb{C} \times (\mathbb{C} \backslash \{0\})$$
>
> and $g(z, w) = z/w$. We leave the details to the reader. $\quad\square$

(3.75)   Corollary *Any complex polynomial*

$$P(z) = \sum_{k=0}^n c_k z^k \qquad (z \in \mathbb{C})$$

$[(c_k)_{k=0}^n \subset \mathbb{C}]$ is continuous on $\mathbb{C}$.

Proof Since any constant function is continuous and the function $z \to z$ is continuous, this follows from (3.74) by induction. $\square$

(3.76) **Corollary** *Any rational function $P(z)/Q(z)$, where $P(z)$ and $Q(z)$ are complex polynomials, is continuous at every $p \in \mathbb{C}$ such that $Q(p) \neq 0$.*

(3.77) **Theorem** *Let X be a topological space, let $S \subset X$, and let $a \in X$ be a limit point of S. Suppose that ϕ and ψ are complex-valued functions defined on S, and that*

$$\lim_{\substack{x \to a \\ x \in S}} \phi(x) = b, \qquad \lim_{\substack{x \to a \\ x \in S}} \psi(x) = c$$

where $b, c \in \mathbb{C}$. Then

$$\lim_{x \to a} \left[\phi(x) + \psi(x) \right] = b + c,$$

$$\lim_{x \to a} \left[\phi(x)\psi(x) \right] = bc,$$

$$\lim_{x \to a} \left[\phi(x)/\psi(x) \right] = b/c \quad \text{if } c \neq 0.$$

Proof We may suppose that $a \notin S$ and that $X = S \cup \{a\}$. Define $\phi(a) = b$ and $\psi(a) = c$. Using (3.68) we see that ϕ and ψ are both continuous at a. Now apply (3.74). [In the case $c \neq 0$, we may suppose $\psi(x) \neq 0$ for all $x \in X$ since this is true on some neighborhood of a.] $\square$

Exercises

1. Define f_n on $\mathbb{R}^2$ for $1 \leq n \leq 4$ by $f_n(0,0) = 0$ for all n and, for $(x, y) \neq (0,0)$,

$$f_1(x, y) = \frac{xy}{x^2 + y^2}, \quad f_2(x, y) = \frac{x^2 - y^2}{x^2 + y^2}, \quad f_3(x, y) = \frac{x^2 y}{x^4 + y^2},$$

$$f_4(x, y) = \frac{x^2 y^2}{(x - y)^2 + x^2 y^2}.$$

Let $a \in \mathbb{R}$ and define

$$S_1 = \{(x, y) \in \mathbb{R}^2 : y = ax\}, \qquad S_2 = \{(x, y) \in \mathbb{R}^2 : y = ax^2\},$$

$$S_3\{(x, y) \in \mathbb{R}^2 : x^2 = y^3\}, \qquad S_4 = \mathbb{R}^2.$$

Evaluate each of the sixteen limits [if they exist]

$$\lim_{\substack{(x,y) \to (0, 0) \\ (x, y) \in S_m}} f_n(x, y).$$

Also, evaluate the eight iterated limits [if they exist]

$$\lim_{x \to 0} \left(\lim_{y \to 0} f_n(x, y) \right), \qquad \lim_{y \to 0} \left(\lim_{x \to 0} f_n(x, y) \right).$$

2.   If X is a metric space and A and B are nonvoid disjoint closed subsets of X, then there is a continuous $f: X \to [0, 1]$ such that $f(x) = 0$ for $x \in A$, $f(x) = 1$ for $x \in B$, and $0 < f(x) < 1$ otherwise. [Hint: Let

$$f(x) = \frac{\operatorname{dist}(x, A)}{\operatorname{dist}(x, A) + \operatorname{dist}(x, B)}$$

and use (3.8) and (3.74).]

3.   Let X be a topological space, let Y be a metric space with metric ρ, and let f be a function from X into Y. Define the *oscillation function* ω of f on X by

$$\omega(x) = \inf\{\operatorname{diam} f(U) : U \text{ is a neighborhood of } x\}$$

[see (3.7)]. Then
  (a) f is continuous at $p \in X$, if and only if $\omega(p) = 0$;
  (b) for each $\beta \leq \infty$, the set $\{x \in X : \omega(x) < \beta\}$ is open in X;
  (c) the set $\{p \in X : f \text{ is continuous at } p\}$ is a G_δ set in X [see (3.56)].
      Suppose also that there exists a set $D \subset X$ such that D and $D' = X \setminus D$ are both dense in X [for example, $X = \mathbb{R}$, $D = \mathbb{Q}$].
  (d) If G is any G_δ subset of X, then there exists a function $f: X \to \mathbb{R}$ such that

$$\{p \in X : f \text{ is continuous at } p\} = G. \quad [\text{Hints: Write } G = \bigcap_{n=1}^{\infty} G_n \text{ where } X = G_1$$

$\supset G_2 \supset \dots$ and each G_n is open in X. Define f by $f(p) = 0$ if $p \in G$, $f(p) = 1/n$ if $p \in D \cap (G_n \setminus G_{n+1})$, $f(p) = -1/n$ if $p \in D' \cap (G_n \setminus G_{n+1})$.]

4.   (a) Define $f: \mathbb{R} \to \mathbb{R}$ as follows. If $x = m/n$ where $m \in \mathbb{Z}$, $n \in \mathbb{N}$, and m and n have no common divisor > 1, define $f(x) = 1/n$. If x is irrational, define $f(x) = 0$. Then f is continuous at every irrational x and discontinuous at every rational x.
  (b) There exists no function $f: \mathbb{R} \to \mathbb{R}$ such that f is continuous at every rational x and discontinuous at every irrational x. [Hint: Use the preceding exercise and (3.59).]

5.   Define $f: \mathbb{R}^2 \to \mathbb{R}$ by $f(0, 0) = 0$, $f(x, y) = x^2 y / (x^6 + y^2)$ if $(x, y) \neq (0, 0)$. Then f is bounded on no neighborhood of $(0, 0)$, but the restriction of f to any line $L \subset \mathbb{R}^2$ is continuous on L.

6.   Let X be a topological space, let Y be a metric space with metric ρ, and let f and g be continuous functions from X into Y. Define h on X by

$$h(x) = \rho(f(x), g(x)).$$

Then $h: X \to \mathbb{R}$ is continuous on X. [Hint: If $p \in X$ and $\epsilon > 0$, choose a neighborhood U of p such that $\rho(f(x), f(p)) < \epsilon/2$ and $\rho(g(x), g(p)) < \epsilon/2$ for $x \in U$. Then $|h(x) - h(p)| < \epsilon$ for $x \in U$.]

7.   *Osgood's Theorem.*
  (a) Let X be a topological space and let Y be a metric space with metric ρ. Suppose that $(f_n)_{n=1}^{\infty}$ is a sequence of continuous functions from X into Y and that $f: X \to Y$ is some function such that $\lim_{n \to \infty} \rho(f_n(x), f(x)) = 0$ for all $x \in X$ [i.e., $f_n(x) \to f(x)$ for each $x \in X$]. Then there exists a set $E \subset X$ that is of first category in X such that f is continuous at each $p \in X \setminus E$. In particular, if X is a complete metric space, then f is continuous at every point of a dense subset of X. [Hints: Fill in the details in the following outline. Let $A_{k,m} = \{x \in X : \rho(f_m(x), f_n(x)) \leq 1/k \text{ for all } n \geq m\}$. Then each $A_{k,m}$ is closed in X (see

preceding exercise), and so $B_{k,m} = A_{k,m} \setminus A_{k,m}^{\circ}$ is nowhere dense in X. Write

$$W_k = \bigcup_{m=1}^{\infty} A_{k,m}^{\circ}, \quad A_k = \bigcup_{m=1}^{\infty} A_{k,m}, \quad G = \bigcap_{k=1}^{\infty} W_k, \quad \text{and} \quad E = \bigcup_{k=1}^{\infty} \bigcup_{m=1}^{\infty} B_{k,m}. \text{ Note that}$$

$A_k = X$ for all k, and so $X \setminus G = \bigcup_{k=1}^{\infty} (X \setminus W_k) = \bigcup_{k=1}^{\infty} (A_k \setminus W_k) \subset E$. To complete the proof we show that f is continuous at each $p \in G$. Given $\epsilon > 0$, choose $k > 5/\epsilon$. Since $p \in G \subset W_k$, there is some m with $p \in A_{k,m}^{\circ}$. Since f_m is continuous at p, there is a neighborhood U of p with $U \subset A_{k,m}^{\circ}$ such that

$$\rho(f_m(x), f_m(p)) < \epsilon/5 \quad \text{for all } x \in U.$$

Now, given $x \in U$, choose $n > m$ so that $\rho(f(x), f_n(x)) < \epsilon/5$ and $\rho(f(p), f_n(p)) < \epsilon/5$. Then

$$\rho(f(x), f(p)) \leq \rho(f(x), f_n(x))$$
$$+ \rho(f_n(x), f_m(x)) + \rho(f_m(x), f_m(p))$$
$$+ \rho(f_m(p), f_n(p)) + \rho(f_n(p), f(p))$$
$$< \epsilon/5 + 1/k + \epsilon/5 + 1/k + \epsilon/5 < \epsilon.$$

Therefore $f(U) \subset B_{\epsilon}(f(p))$.]

(b) If f is as in (a), then $f^{-1}(V)$ is an F_σ set in X [see (3.56)] for every open $V \subset Y$.

[Hint: $f^{-1}(V) = \bigcup_{k=1}^{\infty} \bigcup_{m=1}^{\infty} \{x \in X : \text{dist}(f_n(x), V') \geq 1/k \text{ for all } n \geq m\}$ if $V \neq Y$.]

(c) There exists no sequence $(f_n)_{n=1}^{\infty}$ of continuous real-valued functions on $\mathbb{R}$ such that $f_n(x) \to 1$ for $x \in \mathbb{Q}$ and $f_n(x) \to 0$ for $x \in \mathbb{R} \setminus \mathbb{Q}$.

(d) Define $\phi : \mathbb{R} \to \mathbb{R}$ by $\phi(t) = |2t - 2k - 1|$ for $k \in \mathbb{Z}$ and $k \leq t \leq k + 1$. Then ϕ is continuous on $\mathbb{R}$ and

$$\lim_{m \to \infty} \left[\lim_{n \to \infty} \phi(m! x)^n \right] = \xi_{\mathbb{Q}}(x) \quad \text{for all } x \in \mathbb{R},$$

where $\xi_{\mathbb{Q}}$ is 1 on $\mathbb{Q}$ and 0 on $\mathbb{R} \setminus \mathbb{Q}$.

8. Suppose that f and g are continuous functions from a topological space X into a Hausdorff space Y and that $f(d) = g(d)$ for all $d \in D$, where D is a dense subset of X. Then $f(x) = g(x)$ for all $x \in X$.

9. Suppose that $f : \mathbb{R} \to \mathbb{R}$ satisfies $f(x + y) = f(x) + f(y)$ for all $x, y \in \mathbb{R}$. Then
   (a) $f(rx) = rf(x)$ for $r \in \mathbb{Q}, x \in \mathbb{R}$ [Hint: Fix x and consider successively $r = 0$, $r \in \mathbb{N}$, $-r \in \mathbb{N}$, $r \in \mathbb{Q}$.];
   (b) if $f(I)$ is bounded for some nonvoid open interval I, then f is continuous at 0 [Hint: First show that we may suppose $0 \in I$ and then use (a)];
   (c) if f is continuous at 0, then f is continuous on $\mathbb{R}$;
   (d) if f is continuous on $\mathbb{R}$, then $f(x) = ax$ for all $x \in \mathbb{R}$, where $a = f(1)$. [Hint: Use the preceding exercise and (a).]

10. Let S be a nonvoid closed subset of $\mathbb{R}$ and let $f : S \to \mathbb{C}$ be continuous [S has the relative topology inherited from $\mathbb{R}$]. Then there exists a continuous $g : \mathbb{R} \to \mathbb{C}$ such that $g(x) = f(x)$ for all $x \in S$ and $\sup_{x \in \mathbb{R}} |g(x)| = \sup_{x \in S} |f(x)|$. This is false for every nonclosed $S \subset \mathbb{R}$.

Compactness, Connectedness, and Continuity

We now show that continuous functions preserve both compactness and connectedness.

(3.78) Theorem *Let X and Y be topological spaces, let $f: X \to Y$ be continuous, and let $A \subset X$. Then*

(i) *A is compact implies $f(A)$ is compact,*

(ii) *A is connected implies $f(A)$ is connected.*

Recall that $f(A) = \{ f(x) : x \in A \}$.

Proof Suppose that A is compact. Let $\mathcal{V}$ be an arbitrary open cover of $f(A)$. Since f is continuous the family

$$\mathcal{U} = \{ f^{-1}(V) : V \in \mathcal{V} \}$$

is an open cover of A. But A is compact so $\mathcal{U}$ has a finite subcover. This means that there exist $V_1, \ldots, V_n \in \mathcal{V}$ such that

$$A \subset \bigcup_{k=1}^{n} f^{-1}(V_k).$$

This implies immediately that

$$f(A) \subset \bigcup_{k=1}^{n} V_k.$$

That is, the subfamily $\{ V_1, \ldots, V_n \}$ of $\mathcal{V}$ covers $f(A)$. Therefore $f(A)$ is compact and (i) obtains.

To prove (ii), suppose that $f(A)$ is disconnected. Then, according to (3.46), there exist open subsets U and V of Y such that $U \cap V \cap f(A) = \varnothing$, $f(A) \subset U \cup V$, and $U \cap f(A) \neq \varnothing \neq V \cap f(A)$. Let $U_0 = f^{-1}(U)$ and $V_0 = f^{-1}(V)$. One checks at once that the sets U_0 and V_0 suffice to show that A is disconnected. $\square$

(3.79) Corollary *Let X be a compact space and let $f: X \to \mathbb{R}$ be continuous. Then there exist $u, v \in X$ such that $f(u) \leq f(x) \leq f(v)$ for all $x \in X$.*

Proof By (3.78.i), $f(X)$ is a compact subset of $\mathbb{R}$, and so (3.40) shows that it is closed and bounded. Write $a = \inf f(X)$ and $b = \sup f(X)$. Since $f(X)$ is bounded, we have $a, b \in \mathbb{R}$. Since $f(X)$ is closed, we have $a, b \in f(X)$. Thus $a = f(u)$ and $b = f(v)$ for some $u, v \in X$. Now $x \in X$ implies $f(x) \in f(X)$, and so $a \leq f(x) \leq b$. $\square$

(3.80) Corollary *Let X be a connected space and let $f: X \to \mathbb{R}$ be continuous. Then $f(X)$ is an interval.*

Proof This follows from (3.78.ii) and (3.47). $\square$

(3.81) **Corollary** *Let $[a, b] \subset \mathbb{R}$ and let $f:[a, b] \to \mathbb{R}$ be continuous. Then $f([a, b])$ $= [c, d]$ for some $c \leq d$ in $\mathbb{R}$.*

> **Proof** Since $[a, b]$ is compact (3.34.c) and connected (3.47), it follows from (3.78.i) and (3.80) that $f([a, b])$ is a compact interval. $\square$

(3.82) **Intermediate Value Theorem** *Let $[a, b] \subset \mathbb{R}$ and let $f:[a, b] \to \mathbb{R}$ be continuous. If y_0 is between $f(a)$ and $f(b)$, then $y_0 = f(x_0)$ for some $x_0 \in [a, b]$.*

> **Proof** Let $f([a, b]) = [c, d]$ as in (3.81). Then $f(a)$ and $f(b)$ are in $[c, d]$, so $y_0 \in [c, d]$. $\square$

Next, we have a useful theorem on the continuity of inverse functions.

(3.83) **Theorem** *Let X be a compact space, let Y be a Hausdorff space, and let $f: X \to Y$ be continuous and one-to-one. Then the inverse function f^{-1} defined on $f(X)$ by $f^{-1}(f(x)) = x$ is continuous from $f(X)$ onto X.*

> **Proof** Let $A \subset X$ be closed. Then A is compact (3.35), and so $(f^{-1})^{-1}(A)$ $= f(A)$ is compact (3.78.i) and therefore closed in $f(X)$ (3.36). It follows from (3.69) that f^{-1} is continuous on $f(X)$. $\square$

A frequently useful property of functions between *metric* spaces is that of uniform continuity.

(3.84) **Definition** Let X be a metric space with metric ρ and Y a metric space with metric σ. A function $f: X \to Y$ is said to be *uniformly continuous on X* if to each $\epsilon > 0$ there corresponds a $\delta > 0$ such that $\sigma(f(p), f(q)) < \epsilon$ whenever $p, q \in X$ and $\rho(p, q) < \delta$.

It follows at once from (3.70) that if f is uniformly continuous on X, then f is continuous at each $p \in X$. The converse is definitely false in general. The difference is that in (3.70) the δ may depend on both ϵ and p, while in (3.84) the δ must depend only on ϵ [the same δ must work for all choices of p and q]. We illustrate this with an example.

(3.85) **Example** Let $X = Y = \mathbb{R}$ and let $f(x) = x^2$ for all $x \in \mathbb{R}$. Then f is certainly continuous on $\mathbb{R}$. However, we reprove this fact.

Let $p \in \mathbb{R}$ and $\epsilon > 0$ be arbitrary. Choose $0 < \delta < \min\{1, \epsilon/(2|p| + 1)\}$. If $|x - p| < \delta$, then $|x| < |p| + 1$ and $|f(x) - f(p)| = |x + p| \cdot |x - p| \leq \delta \cdot (|x| + |p|)$ $< \delta \cdot (2|p| + 1) < \epsilon$. Thus f is continuous at p.

However, f is not uniformly continuous on $\mathbb{R}$ because there is no δ corresponding to $\epsilon = 1$ as in (3.84). In fact, try a $\delta > 0$. Choose $p > 1/\delta$ and $q = p + \delta/2$. Then $|p - q| < \delta$, but $|f(p) - f(q)| = |p - q| \cdot |p + q| > (\delta/2)(2p)$ > 1. Therefore this (arbitrary) δ does not work for all $p, q \in \mathbb{R}$.

We do, however, have the following partial converse.

(3.86)   Theorem   *Let X and Y be as in (3.84). If X is compact and $f: X \to Y$ is continuous on X, then f is uniformly continuous on X.*

Proof   Assume that f is not uniformly continuous on X. Then there exists some $\epsilon > 0$ so that no $\delta > 0$ satisfies the requirement of (3.84). Thus, for each $n \in \mathbb{N}$ there exists $p_n, q_n \in X$ such that $\rho(p_n, q_n) < 1/n$ and $\sigma(f(p_n), f(q_n)) \geqq \epsilon$. By (3.43), the sequence $(p_n)_{n=1}^{\infty}$ has a subsequence $(p_{n_k})_{k=1}^{\infty}$ convergent to some $p \in X$. But then

$$\rho(q_{n_k}, p) \leqq \rho(q_{n_k}, p_{n_k}) + \rho(p_{n_k}, p) \to 0 \quad \text{as } k \to \infty,$$

and so $(q_{n_k})_{k=1}^{\infty}$ also converges to p. It follows from (3.70) that since f is continuous at p we have

$$\epsilon \leqq \sigma(f(p_{n_k}), f(q_{n_k}))$$
$$\leqq \sigma(f(p_{n_k}), f(p)) + \sigma(f(p), f(q_{n_k})) < \epsilon/2$$

for all sufficiently large k. This contradiction completes the proof.   ☐

Exercises

1. If X is a compact metric space and $A \subset X$ is nonvoid, then there is some $u \in X$ such that $\text{dist}(u, A) \geqq \text{dist}(x, A)$ for all $x \in X$.

2. The function $f: [0, 1] \to [0, 1]$ defined by $f(x) = x$ for x rational and $f(x) = 1 - x$ for x irrational satisfies $f([0, 1]) = [0, 1]$, but f is continuous only at $p = 1/2$. Compare this with the Intermediate Value Theorem.

3. Let f and g be continuous real-valued functions on $[a, b] \subset \mathbb{R}$ such that $f(a) < g(a)$ and $f(b) > g(b)$. Then $f(x_0) = g(x_0)$ for some $x_0 \in [a, b]$.

4. Let $P(x)$ be a polynomial of odd degree with real coefficients. Then $P(x_0) = 0$ for some $x_0 \in \mathbb{R}$.

5. If $f: [0, 1] \to [0, 1]$ is continuous, then $f(x_0) = x_0$ for some $x_0 \in [0, 1]$.

6. Let $f: [a, b] \to \mathbb{R}$ be continuous. Define

$$g(x) = \sup f([a, x])$$

for $x \in [a, b]$. Then g is continuous on $[a, b]$.

7. Construct an example of a two-to-one function $f: [0, 1] \to \mathbb{R}$ [for each $y \in \mathbb{R}$, the set $\{x : f(x) = y\}$ is either empty or contains exactly two elements]. Prove that no such f can be continuous on $[0, 1]$.

8. Let $X \subset \mathbb{R}$ be noncompact.
    (a) There exists a continuous $f: X \to \mathbb{R}$ such that $f(X)$ is not bounded.
    (b) There exists a continuous $g: X \to \mathbb{R}$ such that $g(X)$ is bounded but $g(X)$ has no largest element.

9. There exists a bounded continuous $f: \mathbb{R} \to \mathbb{R}$ such that f is not uniformly continuous on $\mathbb{R}$. [Hint: Define $f(n) = 1$ and $f(n + 1/(2n)) = 0$ for $n \in \mathbb{N}$.]

10. Let X be a metric space. Then there exists a continuous $f : X \to \mathbb{R}$ that is not uniformly continuous on X if and only if there exist two nonvoid disjoint closed subsets A and B of X such that

$$\text{dist}(A, B) = \inf\{\rho(a, b) : a \in A, b \in B\} = 0.$$

[Hint: Use Exercise 2 of the preceding section and examine our proof of (3.86).]

11. Let $n \in \mathbb{N}$ and define g on $[0, \infty[$ by $g(x) = \sqrt[n]{x}$. Then g is continuous on $[0, \infty[$. [Hint: Note that $g = f^{-1}$ where $f(y) = y^n$ or give a simple direct ϵ-δ proof or use (1.70) with $a = \sqrt[n]{x}$ and $b = \sqrt[n]{p}$.] Is g uniformly continuous on $[0, \infty[$?

12. Let $S = \{z \in \mathbb{C} : |z - z_0| = r\}$ be a circle ($z_0 \in \mathbb{C}, r > 0$). Then
    (a) S is connected. [Hint: Suppose $z_0 = 0$, $r = 1$. The function g defined by

    $$g(t) = (1 + t) + i\sqrt{1 - (1 + t)^2} \quad \text{for} \quad -2 \leqq t \leqq 0 \quad \text{and} \quad g(t) = (1 - t) -$$
    $$i\sqrt{1 - (1 - t)^2} \quad \text{for } 0 \leqq t \leqq 2 \text{ maps } [-2, 2] \text{ onto } S. \text{ Use Exercise 11.}]$$

    (b) if $f : S \to \mathbb{R}$ is continuous, then it is not one-to-one. [Hint: Choose $u, v \in S$ such that $f(u) \leqq f(z) \leqq f(v)$ for all $z \in S$. Consider the two arcs of S determined by u and v.]
    (c) if $h : \mathbb{C} \to \mathbb{C}$ is continuous and $z_0 \in \mathbb{C}$, then in any neighborhood of z_0 there exist $z \neq z'$ such that $|h(z)| = |h(z')|$. [Hint: Let $f = |h|$.]
    (d) let $g : [-2, 2[\to S$ be defined as in (a). Then g is continuous and one-to-one but g^{-1} is not continuous on S. Compare (3.83).

13. (a) A topological space X is connected if and only if every continuous function from X into the discrete two element space $\{0, 1\}$ is a constant.
    (b) Let X be a topological space, let $A \subset X$ be connected, and let $A \subset B \subset A^-$. Then B is connected.
    (c) Let $\mathcal{B}$ be a nonvoid family of connected subsets of a topological space X such that $\cap \mathcal{B} \neq \emptyset$. Then $\cup \mathcal{B}$ is connected.

14. If X is a connected metric space with at least two points, then X is uncountable. [Hint: Use Exercise 2 of the preceding section.]

15. Suppose that $f : [a, b] \to \mathbb{R}$ satisfies $f^{-1}(\{y\})$ is closed for all $y \in \mathbb{R}$ and $f([c, d])$ is connected for all $[c, d] \subset [a, b]$. Then f is continuous on $[a, b]$. [Hint: If $f(p) > y$ and $x_n \to p$ with $f(x_n) \leqq y$, then one can find $t_n \to p$ with $f(t_n) = y$ for all n.]

16. Let $f : \mathbb{R} \to \mathbb{R}$ be continuous and satisfy $|f(x) - f(t)| \geqq c|x - t|$ for all $x, t \in \mathbb{R}$, where $c > 0$ does not depend on x and t. Then $f(\mathbb{R}) = \mathbb{R}$. [Hint: $f(\mathbb{R})$ is closed and f is one-to-one.]

17. Let X be a metric space and let f and g be complex-valued functions on X that are uniformly continuous on X. Then
    (a) $f + g$ is uniformly continuous on X;
    (b) fg need not be uniformly continuous on X;
    (c) if f and g are bounded, then fg is uniformly continuous on X.

18. (a) Let X and Y be metric spaces where Y is complete, let $D \subset X$ be dense in X, and let $f : D \to Y$ be uniformly continuous on D. Then there exists $g : X \to Y$ that is uniformly continuous on X such that $g(d) = f(d)$ for all $d \in D$. [Hint: f maps Cauchy sequences to Cauchy sequences. If $d_n \to x \in X$, let $g(x) = \lim f(d_n)$.]

(b) Assertion (a) can fail if the word "uniformly" is omitted even if f is bounded. [Hint: Try some f that takes values 0 and 1 only.]

19. We say that two points p and q of a topological space X can be *joined by an arc* if for some $[a, b] \subset \mathbb{R}$, there is a continuous function $f : [a, b] \to X$ such that $f(a) = p$ and $f(b) = q$. The space X is called *arcwise connected* if each two points of X can be joined by an arc.

(a) Every arcwise connected space is connected.

(b) Every open connected set $X \subset \mathbb{R}^n$ is arcwise connected. [Hint: For $p \in X$, the set $U_p = \{q \in X : p \text{ and } q \text{ can be joined by an arc}\}$ is open in X because every ball is arcwise connected.]

20. This exercise outlines elementary proofs of several well known theorems. The technique used is essentially that of M.K. Fort, Jr. For $r \geq 0$, write

$$S_r = \{z \in \mathbb{C} : |z| = r\}, \qquad D_r = \{z \in \mathbb{C} : |z| \leq r\}.$$

If X is a topological space and $f : X \to \mathbb{C}$ is continuous, then by a *continuous square root* of f we mean a continuous $\phi : X \to \mathbb{C}$ such that $\phi(z)^2 = f(z)$ for all $z \in X$.

(a) If $c \in \mathbb{C}$, $c \neq 0$, then the equation $z^2 = c$ has exactly two solutions $z \in \mathbb{C}$. [Do not use trigonometric functions.]

(b) If $X = S_1 \setminus \{-1\}$, then the identity function $f(z) = z$ $(z \in X)$ has a continuous square root. [Hint: Take $\phi(z) = (x + 1 + iy)/\sqrt{2x + 2}$ where $z = x + iy$, $x, y \in \mathbb{R}$.]

(c) If $f : S_1 \to \mathbb{C}$ is continuous and $f(S_1)$ is a finite set, then f is a constant. [Hint: S_1 is connected (see Exercise 12).]

(d) Let X be any topological space and let f and g be continuous on X into S_1. Suppose that $f(z) + g(z) \neq 0$ for all $z \in X$. Then f has a continuous square root if and only if g does. [Hint: The function f/g maps X into $S_1 \setminus \{-1\}$, so if ϕ is as in (b), then $\psi = \phi \circ (f/g)$ is a continuous square root of f/g.]

(e) Let $h : D_r \to S_1$ be continuous. Then h has a continuous square root. [Hints: Choose $\delta > 0$ such that $|h(u) - h(v)| < 2$ if $u, v \in D_r$ and $|u - v| < \delta$. Choose $n > r/\delta$ and let $h_j(z) = h(jz/n)$ for $z \in D_r$ and $0 \leq j \leq n$. Check that $|h_j(z) - h_{j-1}(z)| < 2$ so $h_j(z) + h_{j-1}(z) \neq 0$ on D_r. Use (a) to see that h_0 has a continuous square root. Finally, invoke (d).]

(f) If $f : D_r \to \mathbb{C}$ is continuous and never zero, then f has a continuous square root. [Hint: Apply (e) to $h = f/|f|$.]

(g) If $f : S_1 \to \mathbb{C}$ is continuous, never zero, and $f(-z) = -f(z)$ for all $z \in S_1$, then f has no continuous square root. [Hints: Assume ϕ is a continuous square root and write $g(z) = \phi(-z)/\phi(z)$. Then $g(z)^2 = -1$, so $g(z) = i$ or $-i$ for each z. By (c), g is a constant on S_1. Thus $-1 = g(z)^2 = g(z)g(-z) = 1$.]

(h) If $f : S_r \to \mathbb{C}$ is continuous and never zero, then f has at most two continuous square roots. [Hint: If ϕ and ψ are continuous square roots and $g = \phi/\psi$, then $g^2 = 1$, so by (c), $g = 1$ or $g = -1$.]

(i) Define f on S_1 by $f(z) = |\text{Re } z|$. Then f has four continuous square roots.

(j) Let $f : D_r \to \mathbb{C}$ be continuous and satisfy $f(-z) = -f(z)$ for all $z \in S_r$. Then $f(z_0) = 0$ for some $z_0 \in D_r$. [Hint: If false, then (f) furnishes a continuous square root ϕ for f. Now obtain a contradiction to (g) by defining $g(z) = f(rz)$ and $\psi(z) = \phi(rz)$ for $z \in S_1$.]

(k) Let $f : D_r \to \mathbb{C}$ be continuous such that, for some integer $n \neq 0$, $f(z) = z^n$ for all

$z \in S_r$. Then $f(z_0) = 0$ for some $z_0 \in D_r$. [Hints: If n is odd, apply (j). Assume the result is false and let n be an integer nearest to 0 for which it fails for some f. Then $n = 2m$ is even. Use (f) to obtain a continuous square root ϕ for f. Use (h) to show that either $\phi(z) = z^m$ for all $z \in S_r$ or $\phi(z) = -z^m$ for all $z \in S_r$. We may suppose the former (otherwise replace ϕ by $-\phi$). The result is true for m, and so $\phi(z_0) = 0$ for some $z_0 \in D_r$. But then $f(z_0) = 0$.]

(l) *Fundamental Theorem of Algebra.* If $P(z) = \sum\limits_{k=0}^{n} c_k z^k$ is a polynomial of degree $n > 0$ ($n \in \mathbb{N}$, $c_k \in \mathbb{C}$, $c_n \neq 0$), then $P(z_0) = 0$ for some $z_0 \in \mathbb{C}$. In fact, we can take $|z_0| \leq \sum\limits_{k=0}^{n} |c_k/c_n|$. [Hints: We may suppose $c_n = 1$ by dividing out. Let $r = 1 + \sum\limits_{k=0}^{n-1} |c_k| > 1$. Then $|P(z) - z^n| < |z|^n$ if $|z| \geq r$. Put $A_r = \{z \in \mathbb{C} : r \leq |z| \leq 2r\}$. Define functions a, b, g on A_r by $a(z) = |z|/r - 1$, $b(z) = 2 - |z|/r$, $g(z) = a(z)z^n + b(z)P(z)$. Then $|z| > r$ implies $|g(z)| = |z^n - b(z)[z^n - P(z)]| > |z|^n - b(z)|z|^n = a(z)|z|^n > 0$. Define f on D_{2r} by $f = P$ on D_r, $f = g$ on A_r. Then f is well-defined and continuous and $f(z) = z^n$ if $z \in S_{2r}$. By (k), $f(z_0) = 0$ for some $z_0 \in D_{2r}$. But f is never 0 on A_r, so $z_0 \in D_r$ and $P(z_0) = f(z_0) = 0$.]

(m) There exists no continuous $f : D_1 \to S_1$ such that $f(z) = z$ for all $z \in S_1$. [Hint: Such an f would contradict (j).]

(n) *Brouwer's Fixed Point Theorem.* If $g : D_1 \to D_1$ is continuous, then $g(z_0) = z_0$ for some $z_0 \in D_1$. [Hints: Assume false. The idea is to obtain a contradiction by constructing an f as in (m). Briefly, given a $z \in D_1$, we let $f(z)$ be the point on S_1 that lies on the line through z and $g(z)$ and has z between $f(z)$ and $g(z)$. Specifically, let $f(z) = g(z) + t(z)[z - g(z)]$, where $t(z)$ is the unique positive number such that $|f(z)|^2 = f(z)\overline{f(z)} = 1$. That is, $t(z)$ is the positive root of the quadratic equation

$$|z - g(z)|^2 t^2 + 2\,\mathrm{Re}\left[(z - g(z))\,\overline{g(z)}\,\right]t + \left[|g(z)|^2 - 1\right] = 0.$$

Note that this quadratic polynomial has nonpositive values at $t = 0$ and at $t = 1$ because $|g(z)| \leq 1$ and $|z| \leq 1$. Solving explicitly for $t(z)$ shows that $t(z)$ is a continuous function of z. Thus f is continuous on D_1. If $|z| = 1$, then $t = 1$ satisfies the equation, so $t(z) = 1$ and $f(z) = z$.]

21. Let X be any (nonvoid) compact metric space and let P be any Cantor set as in (2 81). Then there exists a continuous mapping ϕ from P onto X. [Hints: Show that there exists a family $\{A_{n,k} : n \in \mathbb{N},\ 1 \leq k \leq 2^n\}$ of nonvoid closed sets such that $\lim\limits_{n \to \infty} \mathrm{diam}\, A_{n,k} = 0$ and, for each n,

$$X = \bigcup_{k=1}^{2^n} A_{n,k} \quad \text{and} \quad A_{n,k} = A_{n+1,2k-1} \cup A_{n+1,2k}.$$

To do this, first show that for some $p \in \mathbb{N}$ there exists a cover $\{B_1, B_2, \ldots, B_{2^p}\}$ of X by nonvoid closed sets such that $\mathrm{diam}\, B_j \leq (1/2)\mathrm{diam}\, X$ for all j, then form the A's for $n \leq p$ from these B's. Next, repeat the operation on each compact metric space $A_{p,k}$. Given $x \in P$, there exists a sequence $(k_n(x))_{n=1}^{\infty}$ such that $\{x\}$ $= \bigcap\limits_{n=1}^{\infty} J_{n,k_n(x)}$.]

Simple Discontinuities
and Monotone Functions

(3.87)   Definition   Let $a < b$ be in $\mathbb{R}^{\#}$ and let f be a real-valued function on $]a, b[$. For $c \in [a, b[$, define

$$f(c+) = \lim_{x \downarrow c} f(x)$$

and, for $c \in]a, b]$, define

$$f(c-) = \lim_{x \uparrow c} f(x)$$

whenever these limits exist in $\mathbb{R}^{\#}$. This notation is merely a convenience. It is not meant to imply that $c+$ or $c-$ are in the domain of f. For $a < c < b$, it is clear [exercise] that f is continuous at c if and only if $f(c-) = f(c) = f(c+)$. If $f(c-)$ and $f(c+)$ are both finite and f is discontinuous at c, then we say that f has a *simple discontinuity* at c [that is, $f(c-), f(c+) \in \mathbb{R}$ and either $f(c-) \neq f(c+)$ or $f(c-) = f(c+) \neq f(c)$].

   The function f is said to be *nondecreasing* [resp. *nonincreasing*] on $]a, b[$ if $a < x < y < b$ implies $f(x) \leqq f(y)$ [resp. $f(x) \geqq f(y)$]. We say that f is *strictly increasing* [resp. *strictly decreasing*] on $]a, b[$ if $a < x < y < b$ implies $f(x) < f(y)$ [resp. $f(x) > f(y)$].

   If f has any of these properties, it is said to be *monotone* [or *monotonic*] on $]a, b[$. Similarly, we define these notions on nonopen intervals.

(3.88)   Example   Define the *signum function* on $\mathbb{C}$ by $\operatorname{sgn} 0 = 0$ and $\operatorname{sgn} z = z/|z|$ if $z \neq 0$. For $t \in \mathbb{R}$, write $[t]$ for the integer such that $[t] \leqq t < [t] + 1$. Now define f on $\mathbb{R}$ by $f(0) = 0$ and

$$f(x) = (-1)^{[1/x]} x^{\operatorname{sgn} x}$$

for $x \neq 0$. Then
   (a) $f(x-) = f(x)$ for all $x \neq 0$,
   (b) $f(0+) = f(0)$ and $f(0-)$ does not exist,
   (c)   if $x = 1/n$ for an integer $n \neq 0$, then $f(x+) \neq f(x)$ and f has a simple discontinuity at x,
   (d) f is continuous at every x not covered by (b) and (c),
   (e) $\lim_{x \to \infty} f(x) = \infty$ and $\lim_{x \to -\infty} f(x) = 0$.
The reader should sketch a graph of f.

   We next see that every discontinuity of a monotone function is simple.

(3.89)   Theorem   *Let $f :]a, b[\to \mathbb{R}$ be nondecreasing. Suppose that $a < c < b$. Then*
   (i) $f(c-)$ *and* $f(c+)$ *both exist,*
   (ii) $f(c-) = \sup\{ f(x) : a < x < c\}$
   (iii) $f(c+) = \inf\{ f(x) : c < x < b\}$
   (iv) $-\infty < f(c-) \leqq f(c) \leqq f(c+) < \infty$.

Also,

(v) $a < c < d < b$ *implies* $f(c+) \leq f(d-)$.

A similar theorem obtains for nonincreasing functions.

Proof   Let $\alpha = \sup\{f(x) : a < x < c\}$ and $\beta = \inf\{f(x) : c < x < b\}$. Since $a < x < c$ implies $-\infty < f(x) \leq f(c)$, it is clear that $-\infty < \alpha \leq f(c)$. Similarly $f(c) \leq \beta < \infty$. Thus (i)–(iv) will follow if we can prove $\lim_{x \uparrow c} f(x) = \alpha$ and $\lim_{x \downarrow c} f(x) = \beta$. Let $\epsilon > 0$ be given and choose x_1 and x_2 such that $a < x_1 < c < x_2 < b$ and $f(x_1) > \alpha - \epsilon$, $f(x_2) < \beta + \epsilon$. Let $U =]x_1, x_2[$. Then $x \in]a, c[\cap U$ implies

$$\alpha - \epsilon < f(x_1) \leq f(x) \leq \alpha < \alpha + \epsilon$$

so $\lim_{x \uparrow c} f(x) = \alpha$. Also, $x \in]c, b[\cap U$ implies

$$\beta - \epsilon < \beta \leq f(x) \leq f(x_2) < \beta + \epsilon$$

so $\lim_{x \downarrow c} f(x) = \beta$.

To prove (v), choose x such that $c < x < d$. Then

$$f(c+) \leq f(x) \leq f(d-). \quad \square$$

(3.90)   Theorem   *Let $f :]a, b[\to \mathbb{R}$ be monotone. Then f is continuous except at a countable set of points and each discontinuity of f is simple.*

Proof   Suppose f is nondecreasing. Let $D = \{c : a < c < b, f(c-) < f(c+)\}$. Then (3.89) shows that we need only prove that D is countable. For $c \in D$, let $I_c =]f(c-), f(c+)[$ and choose $r_c \in I_c \cap \mathbb{Q}$. By (3.89.v), $c < d$ in D implies $r_c < r_d$. Thus the mapping $c \to r_c$ of D into $\mathbb{Q}$ is one-to-one. Since $\mathbb{Q}$ is countable, the proof is complete. $\square$

(3.91)   Example   Let D be any countable infinite subset of $\mathbb{R}$ and enumerate it as $D = \{r_n\}_{n=1}^{\infty}$. Define $f : \mathbb{R} \to \mathbb{R}$ by the rule

$$f(x) = \sum_{r_n < x} 1/2^n$$

[the sum is taken over all n such that $r_n < x$ and is 0 if there are no such n]. Then f is nondecreasing, $f(r_k -) = f(r_k) = f(r_k +) - 1/2^k$ for all k, and f is continuous at each $x \in \mathbb{R} \setminus D$. If D is dense in $\mathbb{R}$[e.g., $D = \mathbb{Q}$], then f is strictly increasing: $x < r_k < y$ implies $f(x) + 1/2^k \leq f(y)$. This shows that monotone functions, though special, are not necessarily so simple. Even continuous monotone functions might not be easy to graph, as our next example shows.

(3.92)   Example   Let P be any Cantor set as in (2.81). With the notation of (2.83), S is the set of all sequences $\epsilon = (\epsilon_j)_{j=1}^{\infty}$ where each ϵ_j is 0 or 1. By (2.83) the function

$f: S \to P$ defined by $f(\epsilon) = \sum_{j=1}^{\infty} \epsilon_j r_j$ is one-to-one and onto P. Let us define ψ on P by

$$\psi(f(\epsilon)) = \sum_{j=1}^{\infty} \epsilon_j 2^{-j}.$$

According to (2.84), ψ maps P onto $[0, 1]$ $[\psi = \phi \circ f^{-1}]$. Note also that if $\epsilon \neq \epsilon'$ in S: say $\epsilon_n = 0$, $\epsilon'_n = 1$, and $\epsilon_j = \epsilon'_j$ for $1 \leq j < n$, then

$$f(\epsilon') - f(\epsilon) = r_n + \sum_{j=n+1}^{\infty} (\epsilon'_j - \epsilon_j) r_j$$

$$\geq r_n - \sum_{j=n+1}^{\infty} r_j = d_n > 0$$

and

$$\psi(f(\epsilon')) - \psi(f(\epsilon)) = 2^{-n} + \sum_{j=n+1}^{\infty} (\epsilon'_j - \epsilon_j) 2^{-j}$$

$$\geq 2^{-n} - \sum_{j=n+1}^{\infty} 2^{-j} = 0.$$

This proves that

$$x < y \text{ in } P \quad \text{implies } \psi(x) \leq \psi(y)$$

and that equality obtains if and only if $]x, y[$ is a component interval of $[0, 1] \backslash P$.

$$\left[x = \sum_{j=1}^{n-1} \epsilon_j r_j + \sum_{j=n+1}^{\infty} r_j, \quad y = \sum_{j=1}^{n-1} \epsilon_j r_j + r_n. \right]$$

In this case, define $\psi(t) = \psi(x) = \psi(y)$ for $x < t < y$. Also define $\psi(t) = 0$ if $t < 0$, $\psi(t) = 1$ if $t > 1$. Now ψ is a nondecreasing continuous map of $\mathbb{R}$ onto $[0, 1]$ and ψ is constant on each component interval of $\mathbb{R} \backslash P$.

   Perhaps the above description of ψ is a bit too analytical to digest easily. Here is the idea geometrically. Let $\{I_{n,k} : n \in \mathbb{N}, 1 \leq k \leq 2^{n-1}\}$ be the open intervals removed from $[0, 1]$ in forming P [see (2.81)]. Define ψ to be a constant on each $I_{n,k} : \psi = 1/2$ on $I_{1,1}$; $\psi = 1/4$ on $I_{2,1}$; $\psi = 3/4$ on $I_{2,2}$; $\ldots$; $\psi = (2k - 1)/2^n$ on $I_{n,k}$; $\ldots$. Draw a picture! Next define $\psi = 0$ on $]-\infty, 0[$ and $\psi = 1$ on $]1, \infty[$. For $x \in P$, define

$$\psi(x) = \sup\{\psi(t) : t \in \mathbb{R} \backslash P, t < x\}.$$

Clearly ψ is nondecreasing into $[0, 1]$. It is also continuous since otherwise there would exist some $x \in [0, 1]$ such that $\psi(x -) < \psi(x +)$, and so $\psi([0, 1])$ misses the open interval $]\psi(x -), \psi(x +)[$. But this is impossible because $\psi([0, 1])$ contains every dyadic rational $(2k - 1)/2^n$. Also, $\psi([0, 1]) = [0, 1]$ by (3.78.i).

   If P is Cantor's ternary set, this function ψ is called *Lebesgue's singular function*.

   The question of continuous inverses for monotone functions is extremely simple, as we now show.

(3.93) **Theorem** *Let $I \subset \mathbb{R}$ be an open interval and let $f : I \to \mathbb{R}$ be continuous and strictly monotone. Then*

(i) *f is one-to-one,*

(ii) *$f(I)$ is an open interval,*

(iii) *$f^{-1} : f(I) \to I$ is continuous and strictly monotone [in the same direction as f].*

Proof We suppose that f is strictly increasing [otherwise consider $-f$]. Assertion (i) is trivial, for, by definition, $x < y$ in I implies $f(x) < f(y)$. Let $a = \inf f(I)$ and $b = \sup f(I)$. Assuming that $a \in f(I)$, choose $u \in I$ such that $f(u) = a$. Since I is open there exists $x \in I$ such that $x < u$; hence $f(x) \in f(I)$ and $f(x) < f(u) = a$, contrary to our choice of a. Therefore, $a \notin f(I)$. Similarly $b \notin f(I)$. By (3.80), $f(I)$ is an interval and, since it contains points in every neighborhood of a and of b, we conclude that $f(I) = {]}a, b{[}$, proving (ii).

To see that f^{-1} is strictly increasing on $f(I)$, let $y < y'$ in $f(I)$. If we had $f^{-1}(y) \geq f^{-1}(y')$, we would have $y = ff^{-1}(y) \geq ff^{-1}(y') = y'$. Therefore $f^{-1}(y) < f^{-1}(y')$. To see that f^{-1} is continuous on $f(I)$, let $y \in f(I)$ and let V be any neighborhood of $f^{-1}(y)$. Choose $c, d \in I$ such that $c < f^{-1}(y) < d$ and ${]}c, d{[} \subset V$. Then $f(c) < y < f(d)$ so $U = {]}f(c), f(d){[}$ is a neighborhood of y and, since f^{-1} is increasing, $f^{-1}(U) \subset V$. Thus f^{-1} is continuous at y. But $y \in f(I)$ was arbitrary, so (iii) obtains. $\square$

Exercises

1. If $I \subset \mathbb{R}$ is an open interval and $f : I \to \mathbb{R}$ is continuous and one-to-one, then f is strictly monotone on I.

2. Let $I \subset \mathbb{R}$ be an open interval and let $f : I \to \mathbb{R}$ be any function. For $x_0 \in I$, we say that f is *increasing at* x_0 if $f(x_0) < f(x)$ for all x in some open interval with left endpoint x_0 and $f(x) < f(x_0)$ for all x in some open interval with right endpoint x_0.
   (a) If f is increasing at every x_0 in I, then f is strictly increasing on I. [Hint: Let $a < b$ in I be arbitrary. Define $c = \sup S$ where $S = \{ t \in I : t > a, f(a) < f(x)$ whenever $a < x \leq t \}$. One needs only to show that c is the right endpoint of I. Assuming this false, contradict the choice of c.]
   (b) There exists a dense set $D \subset {]}0, 1{[}$ and a continuous $f : [0, 1] \to \mathbb{R}$ such that f is increasing at each $x_0 \in D$ but $f(0) > f(1)$. [Hint: Use Lebesgue's singular function to construct f.]

3. Let $f : \mathbb{R} \to \mathbb{R}$ be arbitrary. Show that the set E of $x \in \mathbb{R}$ such that f has a *simple discontinuity* at x is a countable set. Of course f may have many other discontinuities. [Hint: Let $E_1 = \{ x \in E : f(x-) = f(x+) < f(x) \}$, $E_2 = \{ x \in E : f(x-) = f(x+) > f(x) \}$, $E_3 = \{ x \in E : f(x-) < f(x+) \}$, $E_4 = \{ x \in E : f(x-) > f(x+) \}$. For each E_j define a one-to-one map of E_j into the countable set $\mathbb{Q}^3$. For example, for $x \in E_1$, choose $a(x) \in \mathbb{Q}$ such that $f(x-) < a(x) < f(x)$ and then choose $b(x) < x < c(x)$ such that $b(x) < t < c(x)$ $(t \neq x)$ implies $f(t) < a(x)$.

Then consider the mapping

$$x \to (a(x), b(x), c(x)) \in \mathbb{Q}^3.$$

For $x \in E_4$, take $f(x-) > a(x) > f(x+)$ and choose $b(x) < x < c(x)$ such that $f(t) > a(x)$ for $b(x) < t < x$ and $f(t) < a(x)$ for $x < t < c(x)$.]

4. *Semicontinuity and Tietze's Theorem.* Let X be a topological space. A function $f : X \to \mathbb{R}^\#$ is said to be *lower* (resp. *upper*) *semicontinuous* at a point $a \in X$ if for each real α with $\alpha < f(a)$ (resp., $\alpha > f(a)$) there exists a neighborhood V of a such that $f(x) > \alpha$ (resp., $f(x) < \alpha$) for all $x \in V$. We say that f is *lower* (resp., *upper*) *semicontinuous* on X if f has that property at each $a \in X$.

(a) f is lower semicontinuous at a (resp., on X) if and only if $-f$ is upper semicontinuous at a (resp., on X). In view of (a) we restrict most of our attention to lower semicontinuity.

(b) f is lower semicontinuous on X if and only if $\{x \in X : f(x) > \alpha\}$ is open in X for each $\alpha \in \mathbb{R}$.

(c) If f and g are $\mathbb{R}^\#$-valued functions on X, we define $f \wedge g$ and $f \vee g$ on X by $(f \wedge g)(x) = \min\{f(x), g(x)\}$, $(f \vee g)(x) = \max\{f(x), g(x)\}$. If f and g are both lower semicontinuous at $a \in X$, then so are $f \wedge g$, $f \vee g$, and (provided f and g are never $-\infty$) $f + g$.

(d) If $\{f_i : i \in I\}$ is a family of $\mathbb{R}^\#$-valued functions defined on X, we define the *lower envelope* and the *upper envelope* of this family to be the functions on X defined by

$$\left(\bigwedge_{i \in I} f_i\right)(x) = \inf\{f_i(x) : i \in I\},$$

and

$$\left(\bigvee_{i \in I} f_i\right)(x) = \sup\{f_i(x) : i \in I\},$$

respectively. If each f_i is lower semicontinuous at $a \in X$, then the upper envelope is also, but the lower envelope need not be. [Hint: If $X = \mathbb{R}$, the function f that equals 0 at every rational and equals 1 at every irrational is the lower envelope of such a (countable) family.]

(e) The upper (resp., lower) envelope of any family of continuous real-valued functions on X is lower (resp., upper) semicontinuous on X.

(f) Suppose that I is a nonvoid interval of $\mathbb{R}^\#$ and that $\phi : I \to \mathbb{R}^\#$ is a monotone nondecreasing function that is left continuous on I (i.e., if $x \in I$ is not the left endpoint of I, then $\phi(x) = \sup\{\phi(t) : t \in I, t < x\}$). If f is lower semicontinuous on X and $f(X) \subset I$, then the composite function $\phi \circ f$ is also lower semicontinuous on X.

The hints for the remaining parts of this exercise are due to Felix Hausdorff.

(g) [*R. Baire*] Let X be a metric space and let $f : X \to \mathbb{R}^\#$ be lower semicontinuous on X and satisfy $f(x) > -\infty$ for all $x \in X$. Then there exists a sequence of real-valued continuous functions $f_1 \leqq f_2 \leqq f_3 \leqq \cdots$ on X such that $\lim_{n \to \infty} f_n(x) = f(x)$ for all $x \in X$. [Hints: Write $\phi(t) = t/(1 + |t|)$ for $t \in \mathbb{R}$, $\phi(-\infty) = -1$, and $\phi(\infty) = 1$. Define $g = \phi \circ f$. Then g is lower semicontinuous on X and $-1 < g \leqq 1$. For $n \in \mathbb{N}$ and $x \in X$, define

$$g_n(x) = \inf\{g(z) + nd(x, z) : z \in X\},$$

where d is the metric for X. Then $-1 \leq g_1 \leq g_2 \leq \ldots \leq g$ and, since $g_n(x) \leq g(z) + nd(y, z) + nd(x, y)$, $g_n(x) \leq g_n(y) + nd(x, y)$, $|g_n(x) - g_n(y)| \leq nd(x, y)$ for all $x, y \in X$, g_n is (uniformly) continuous on X. Given $x \in X$, choose z_n such that

$$g(z_n) + nd(x, z_n) < g_n(x) + 1/n \leq g(x) + 1/n$$

and conclude that

$$nd(x, z_n) < g(x) + 1/n + 1, \qquad z_n \to x,$$

and so

$$g(x) \leq \lim_{n \to \infty} g(z_n) \leq \lim_{n \to \infty} g_n(x) \leq g(x).$$

The functions h_n defined by

$$h_n = g_n + \tfrac{1}{2}(g_{n+1} - g_n) + \tfrac{1}{4}(g_{n+2} - g_{n+1}) + \ldots + \frac{1}{2^j}(g_{n+j} - g_{n+j-1}) + \ldots$$

are continuous on X and satisfy

$$g_n \leq h_n \leq g_n + (g_{n+1} - g_n) + (g_{n+2} - g_{n+1}) + \ldots = g \leq 1$$

and

$$h_{n+1} = g_n + (g_{n+1} - g_n) + (1/2)(g_{n+2} - g_{n+1}) + \ldots \geq h_n.$$

Also, $h_n > -1$ because either $h_n > g_n \geq -1$ or $g_n = g_{n+1} = g_{n+2} = \ldots$ and $h_n = g > -1$. The functions $k_n = (nh_n - 1)/(n + 1)$ satisfy $-1 < k_n \leq k_{n+1} < 1$ and $k_n(x) \to g(x)$ for all x. Finally, the required functions are $f_n = \phi^{-1} \circ k_n = k_n/(1 - |k_n|)$.

(h) State and prove the analog of (g) for upper semicontinuous functions.

(i) [*H. Hahn*] If X is a metric space and f and g are extended real-valued functions such that f is lower semicontinuous on X, g is upper semicontinuous on X, and, for each $x \in X$,

$$f(x) > -\infty, \quad g(x) < \infty, \quad g(x) \leq f(x),$$

then there exists a continuous real-valued function h on X such that $g(x) \leq h(x) \leq f(x)$ for all $x \in X$. [Hints: Choose a nondecreasing sequence $(f_n)_{n=1}^{\infty}$ and a nonincreasing sequence $(g_n)_{n=1}^{\infty}$ of continuous real-valued functions on X such that $f_n(x) \to f(x)$ and $g_n(x) \to g(x)$ for all $x \in X$. For any extended real-valued function ϕ on X, define ϕ^+ on X by $\phi^+(x) = \max\{\phi(x), 0\}$. Then

$$(g_1 - f_1)^+ \geq (g_1 - f_2)^+ \geq (g_2 - f_2)^+ \geq \ldots \geq (g_n - f_n)^+$$

$$\geq (g_n - f_{n+1})^+ \geq (g_{n+1} - f_{n+1})^+ \geq \ldots$$

and this sequence has limit 0 at every $x \in X$. Therefore the alternating series $s = (g_1 - f_1)^+ - (g_1 - f_2)^+ + (g_2 - f_2)^+ - \ldots$ formed from this sequence converges at every $x \in X$. The partial sums s_n of this series are continuous and satisfy $s_1 \geq s_3 \geq s_5 \geq \ldots$ and $s_2 \leq s_4 \leq s_6 \leq \ldots$, so s is both upper and lower semicontinuous and thus continues on X. Write $h = f_1 + s$. If $f(x) = g(x)$, then $f_j(x) \leq g_k(x)$ for all j and k, so

$$h(x) = \lim_{n \to \infty} (f_1(x) + s_{2n-1}(x)) = \lim_{n \to \infty} g_n(x) = g(x)$$

and

$$h(x) = \lim_{n \to \infty} (f_1(x) + s_{2n}(x)) = \lim_{n \to \infty} f_{n+1}(x) = f(x).$$

Suppose $g(x) < f(x)$. If $(g_n - f_n)^+(x)$ is the first 0 term in the series for $s(x)$, then $h(x) = f_n(x) \leqq f(x)$ and $h(x) = f_n(x) \geqq g_n(x) \geqq g(x)$. If the first 0 term is $(g_n - f_{n+1})^+(x)$, then $f(x) \geqq f_{n+1}(x) \geqq g_n(x) = h(x) \geqq g(x)$.]

(j) Show by an example that (i) is false if "upper" and "lower" are interchanged.

(k) [*H. Tietze*] Let X be a metric space and A a closed subset of X. Suppose that $\phi : A \to \mathbb{R}$ is continuous on A. Write $a = \inf \phi(A)$ and $b = \sup \phi(A)$. Then there exists a continuous $h : X \to \mathbb{R}$ such that $h(x) = \phi(x)$ for all $x \in A$ and $a \leqq h(x) \leqq b$ for all $x \in X$. [Hint: Apply part (i) to the functions f and g defined by $f = g = \phi$ on A, $f = b$ and $g = a$ on $X \backslash A$.]

Exp and Log

At this point we are able to study some of the most important properties of the real exponential and the real logarithm. We postpone introduction of the trigonometric functions until Chapter 5 after we have studied differentiation.

Recall that we introduced the exponential function in (2.74.a). It is the function $\exp : \mathbb{C} \to \mathbb{C}$ defined by the series

$$\exp(z) = \sum_{n=0}^{\infty} \frac{z^n}{n!} = 1 + \sum_{n=1}^{\infty} \frac{z^n}{n!},$$

which converges absolutely for every $z \in \mathbb{C}$. We next list some important properties of this function.

(3.94)   **Theorem**   *We have*

(i) $\exp(z)\exp(w) = \exp(z + w)$ *for all* $z, w \in \mathbb{C}$;

(ii) $\exp(0) = 1$ *and* $\exp(1) = e$;

(iii) $\exp(z) \neq 0$ *for all* $z \in \mathbb{C}$;

(iv) $\exp(-z) = 1/\exp(z)$ *for all* $z \in \mathbb{C}$;

(v) $\exp$ *is continuous on* $\mathbb{C}$.

Proof   We proved (i) in (2.74.a) by using Cauchy products. Part (ii) follows by substituting $z = 0$ and $z = 1$ in the defining series and invoking (2.58.i). From (i) and (ii) we obtain

$$\exp(z)\exp(-z) = 1 \quad \text{for all } z \in \mathbb{C}$$

and both (iii) and (iv) then follow. To prove (v), let $z \in \mathbb{C}$ and $\epsilon > 0$ be given. Letting $\delta = \min\{1, \epsilon/(2|\exp(z)|)\}$, we see that $h \in \mathbb{C}$ and $|h| < \delta$ implies

$$|\exp(z + h) - \exp(z)| = |\exp(z)| \cdot |\exp(h) - 1|$$

$$\leqq \frac{\epsilon}{2\delta} \cdot |\sum_{n=1}^{\infty} \frac{h^n}{n!}| \leqq \frac{\epsilon}{2\delta} |h| \cdot \sum_{n=1}^{\infty} \frac{|h|^{n-1}}{n!}$$

$$< \frac{\epsilon}{2} \cdot \sum_{n=1}^{\infty} \frac{1}{n!} = \frac{\epsilon}{2}(e - 1) < \epsilon.$$

Thus exp is continuous at z. $\square$

When we restrict exp to $\mathbb{R}$, we have more important properties.

(3.95) **Theorem** *We have*

(i) $\exp(x) > 0$ *for all* $x \in \mathbb{R}$;

(ii) $\exp$ *is strictly increasing on* $\mathbb{R}$;

(iii) $\lim_{x \to \infty} \exp(x) = \infty$ $(x \in \mathbb{R})$;

(iv) $\lim_{x \to -\infty} \exp(x) = 0$ $(x \in \mathbb{R})$;

(v) $\exp(\mathbb{R}) = \,]0, \infty[$;

(vi) $\lim_{x \to \infty} x^{-n}\exp(x) = \infty$ *for every integer* $n \geq 0$ $(x \in \mathbb{R})$.

Proof It is clear from the series definition that $\exp(\mathbb{R}) \subset \mathbb{R}$ and

$$\exp(x) > 1 + x \quad \text{if } x > 0. \tag{1}$$

Thus $x \geq 0$ implies $\exp(x) > 0$ and $\exp(-x) = 1/\exp(x) > 0$, so we have (i). To prove (ii), let $x \in \mathbb{R}$ and $h > 0$ be arbitrary. Then $\exp(x + h) = \exp(x) \cdot \exp(h) > \exp(x)$, as (1) and (i) show. Also, (iii) follows from (1) and then (iv) follows from (iii) and (3.94.iv). To prove (vi), again consult the series to see that $x > 0$ implies that

$$x^{-n} \cdot \exp(x) > x^{-n} \cdot x^{n+1}/(n+1)!$$
$$= x/(n+1)!.$$

Thus $x^{-n} \cdot \exp(x) > \beta > 0$ if $x > (n+1)! \cdot \beta$. Finally, (v) follows from (i), (iii), (iv), and (3.93). $\square$

Since exp is one-to-one on $\mathbb{R}$ (3.95.ii), we can make the following definition.

(3.96) **Definition** The *real logarithm* is the inverse of the restriction of the exponential function to $\mathbb{R}$:

$$\log = (\exp|_{\mathbb{R}})^{-1}.$$

Thus (3.95.v) shows that the domain of log is $]0, \infty[$ and its range is $\mathbb{R}$. For $y > 0$, $\log(y) = x$ means $x \in \mathbb{R}$ and $\exp(x) = y$.

(3.97) **Theorem** *We have*

(i) $\log(]0, \infty[) = \mathbb{R}$;

(ii) $\log$ *is strictly increasing on* $]0, \infty[$;

(iii) $\log$ *is continuous on* $]0, \infty[$;

(iv) $\lim_{x \to \infty} \log(x) = \infty$;

(v) $\lim_{x \downarrow 0} \log(x) = -\infty$;

(vi) $\log(1) = 0$ *and* $\log(e) = 1$;

(vii) $\log(ab) = \log(a) + \log(b)$ *for all* $a > 0$, $b > 0$;

(viii) $\log(a^n) = n \cdot \log(a)$ *for all* $a > 0$ *and integers* n;

(ix) $\log(a^{1/n}) = (1/n) \cdot \log(a)$ *for all* $a > 0$ *and* $n \in \mathbb{N}$;

(x) $\lim_{x \to \infty} (\log(x)/\sqrt[n]{x}) = 0$ *for all* $n \in \mathbb{N}$.

Proof Items (i), (ii), and (iii) follow from (3.96), (3.95), and (3.93). Items (iv) and (v) follow from (i) and (ii). Item (vi) is a consequence of (3.96) and (3.94.ii). For $a > 0$, $b > 0$, we have

$$\exp\big[\log(a) + \log(b)\big] = \exp\big[\log(a)\big] \cdot \exp\big[\log(b)\big] = ab$$

and so (vii) obtains [see the last sentence of (3.96)].

Let $a > 0$. We want to prove (viii). Recall the definition of a^n (1.23). Clearly (viii) is true when $n = 0$ or 1. If (viii) holds from some $n \geq 0$, then $\log (a^{n+1}) = \log(a^n a) = \log(a^n) + \log(a) = n \cdot \log(a) + \log(a) = (n + 1) \cdot \log(a)$. By induction, we have proven (viii) for all $n \geq 0$. For $n < 0$, we have

$$-n \log(a) + \log(a^n) = \log(a^{-n}) + \log(1/a^{-n}) = \log 1 = 0,$$

and so (viii) also holds for $n < 0$. Using (viii), we have $n \cdot \log(a^{1/n}) = \log[(a^{1/n})^n] = \log(a)$ for $n \in \mathbb{N}$, and so (ix) obtains.

To prove (x), let $\epsilon > 0$ and $n \in \mathbb{N}$ be given. Using (3.95.vi), we can choose $\alpha > 1$ such that

$$y^{-n} \cdot \exp(y) > \epsilon^{-n} \quad \text{if } y > \alpha \tag{1}$$

and, using (iv), we can choose $\beta > 1$ such that

$$\log(x) > \alpha \quad \text{if } x > \beta. \tag{2}$$

Combining (1) and (2), we see that $x > \beta$ implies

$$\big[\log(x)\big]^{-n} \cdot x > \epsilon^{-n},$$

$$0 < \big[\log(x)\big]^n \cdot x^{-1} < \epsilon^n,$$

$$0 < \log x / \sqrt[n]{x} < \epsilon.$$

Therefore (x) obtains. ☐

Powers

To this point in the book, we have only defined the power a^b when b is an integer and a is a complex number [$b \geq 0$ if $a = 0$] and when $1/b$ is a natural number and a is a nonnegative real number. If $a > 0$, it follows from (viii) and (ix) of (3.97) that

$$a^b = \exp(b \log(a))$$

whenever b is an integer or $1/b$ is a natural number. We use this formula to extend our notion of powers. We must stick to $a > 0$ at present since we do not discuss logarithms of other complex numbers until Chapter 5.

(3.98)   Definition Let $a > 0$ be a real number and let b be any complex number. We define the *power* a^b with *base* a and *exponent* b to be the complex number given by

(i) $$a^b = \exp(b \cdot \log(a)).$$

We also define $0^b = 0$ if $b \neq 0$ and, as usual, $0^0 = 1$. The above remarks show that this definition is consistent with our earlier definitions of powers. In particular, since $\log(e) = 1$, we now have

$$e^z = \exp(z) \quad \text{for all } z \in \mathbb{C}.$$

If $b \in \mathbb{R}$, we have $a^b > 0$ and $\log(a^b) = b \log(a)$.

(3.99) **Laws of Exponents** *Let* $a, b, c \in \mathbb{C}$. *Then*
  (i) $a^{b+c} = a^b a^c$ *if* $a > 0$;
  (ii) $(ab)^c = a^c b^c$ *if* $a > 0$, $b > 0$;
  (iii) $(a^b)^c = a^{bc}$ *if* $a > 0$, $b \in \mathbb{R}$.

Proof Easy exercises. $\square$

We can use powers to define certain functions as the next two theorems illustrate.

(3.100) **Theorem** *Let* $b \in \mathbb{C}$. *Define* $f:]0, \infty[\to \mathbb{C}$ *by* $f(x) = x^b$. *Then*
  (i) f *is continuous on* $]0, \infty[$;
  (ii) *if* $b \in \mathbb{R}$, $b > 0$, *then*

$$\lim_{x \to \infty} f(x) = \infty, \qquad \lim_{x \downarrow 0} f(x) = 0,$$

f is strictly increasing on $]0, \infty[$ *and* $f(]0, \infty[) =]0, \infty[$;
  (iii) *if* $b \in \mathbb{R}$, $b < 0$, *then*

$$\lim_{x \to \infty} f(x) = 0, \qquad \lim_{x \downarrow 0} f(x) = \infty,$$

f is strictly decreasing on $]0, \infty[$, *and* $f(]0, \infty[) =]0, \infty[$.

Proof Exercise. $\square$

(3.101) **Theorem** *Let* $a \in \mathbb{R}$ *be positive. Define* $g: \mathbb{C} \to \mathbb{C}$ *by*
$$g(z) = a^z.$$

Then
  (i) g *is continuous on* $\mathbb{C}$:
  (ii) *if* $a > 1$, *then*

$$\lim_{x \to \infty} g(x) = \infty, \qquad \lim_{x \to -\infty} g(x) = 0,$$

g is strictly increasing on $\mathbb{R}$, *and* $g(\mathbb{R}) =]0, \infty[$;
  (iii) *if* $0 < a < 1$, *then*

$$\lim_{x \to \infty} g(x) = 0, \qquad \lim_{x \to -\infty} g(x) = \infty,$$

g is strictly decreasing on $\mathbb{R}$, *and* $g(\mathbb{R}) =]0, \infty[$.

Proof Another exercise. $\square$

Exercises

1.   For all $x \in \mathbb{R}$ we have
$$e^x \geq 1 + x$$
and equality obtains if and only if $x = 0$. [Hint: For $-1 < x < 0$, use Leibnitz's Test (2.45).]

2.   If $a > 0$ is a real number, then
(a) $\lim\limits_{x \to \infty} x^{-a} e^x = \infty$,

(b) $\lim\limits_{x \to \infty} x^{-a} \log(x) = 0$,

(c) $\lim\limits_{x \downarrow 0} x^a \log x = 0$ [Hint: Put $x = 1/t$ in (b) to obtain (c)].

3.   (a) $\lim\limits_{z \to 0} z^{-1}[e^z - 1] = 1$ $(z \in \mathbb{C})$ [Hint: Use power series to estimate as follows. For $0 < |z| < 1$,
$$|z^{-1}[e^z - 1] - 1| = \left| z \cdot \sum_{n=2}^{\infty} \frac{z^{n-2}}{n!} \right| < |z|(e - 2) < |z|],$$

(b) $\lim\limits_{t \to 0} t^{-1} \log(1 + t) = 1$ $(t \in \mathbb{R})$ [Hint: Put $z = \log(1 + t)$ and use (a)].

4.   If $a, b \in \mathbb{R}$ and $a > 0$, then
(a) $\lim\limits_{z \to 0} ((a^z - 1)/z) = \log a$,

(b) $\lim\limits_{n \to \infty} n(\sqrt[n]{a} - 1) = \log a$,

(c) $\lim\limits_{h \to 0} (1 + bh)^{1/h} = e^b$,

(d) $\lim\limits_{n \to \infty} (1 + b/n)^n = e^b$.

Here $z \in \mathbb{C}, h \in \mathbb{R}, n \in \mathbb{N}$. [Hint: Use the preceding exercise.]

5.   Let $(x_n)_{n=1}^{\infty} \subset \mathbb{R}$ and suppose that $\lim\limits_{n \to \infty} n x_n = x \in \mathbb{R}^{\#}$.
(a) If $x \in \mathbb{R}$, then $\lim\limits_{n \to \infty} (1 + x_n)^n = e^x$.

(b) If $x = \infty$, then $\lim\limits_{n \to \infty} (1 + x_n)^n = \infty$.

(c) If $x = -\infty$ and $x_n > -1$ for all n, then $\lim\limits_{n \to \infty} (1 + x_n)^n = 0$.

[Hint: Write $n \cdot \log(1 + x_n) = n x_n \cdot x_n^{-1} \log(1 + x_n)$ and use Exercise 3(b).]

6.   Let $a \in \mathbb{R}, a > 0, a \neq 1$. The function $\log_a$ is defined on $]0, \infty[$ to be the inverse of the function $x \to a^x$. That is, $\log_a(y) = x$ means that $a^x = y$. In particular, we have $\log(y) = \log_e(y)$ for all $y > 0$.
(a) The function $\log_a$ is well-defined and continuous on $]0, \infty[$.
(b) If $b \in \mathbb{R}, b > 0, b \neq 1$, then
$$\log_b(y) = \log_a(y)/\log_a(b) = \log_b(a) \cdot \log_a(y)$$
for all $y > 0$.

7.   Evaluate
(a) $\lim\limits_{x \downarrow 0} x^x$,   (b) $\lim\limits_{x \downarrow 0} x^{\log x}$,

(c) $\lim\limits_{x \downarrow 1} (\log x)^x$,   (d) $\lim\limits_{x \downarrow 1} (\log x)^{\log x}$.

[Hint: See Exercise 2(c).]

8.   If $a \in \mathbb{R}$, then there exists exactly one $x > 0$ such that $x + \log x = a$. If we write $f(a) = x$, the function f so defined is continuous and strictly increasing on $\mathbb{R}$.

9.   (a) Let $g : \mathbb{R} \to \mathbb{R}$ be continuous, not identically zero, and satisfy

$$g(x + y) = g(x) g(y)$$

for all $x, y \in \mathbb{R}$. Then $g(x) = a^x$ for all $x \in \mathbb{R}$ where $a = g(1)$. [Hint: Show that $g(x) > 0$ for all x, define $f(x) = \log g(x)$, and use Exercise 9 on p. 121.]
   (b) Let $h :]0, \infty[\to \mathbb{R}$ be continuous and satisfy

$$h(xy) = h(x) + h(y)$$

for all $x > 0$ and $y > 0$. Then $h(x) = b \log x$ for all $x > 0$ where $b = h(e)$.
   (c) Let $k :]0, \infty[\to \mathbb{R}$ be continuous, not identically zero, and satisfy

$$k(xy) = k(x)k(y)$$

for all $x > 0$ and $y > 0$. Then $k(x) = x^b$ for all $x > 0$ where $b = \log k(e)$.

10. Define the *hyperbolic sine* and the *hyperbolic cosine* on $\mathbb{R}$ by

$$\sinh(x) = \frac{e^x - e^{-x}}{2} \quad \text{and} \quad \cosh(x) = \frac{e^x + e^{-x}}{2}, \quad \text{respectively.}$$

   (a) The function sinh is strictly increasing from $\mathbb{R}$ onto $\mathbb{R}$.
   (b) The function cosh is strictly increasing from $[0, \infty[$ onto $[1, \infty[$ and is strictly decreasing from $] - \infty, 0]$ onto $[1, \infty[$.
   (c) We have

$$\cosh^2(x) - \sinh^2(x) = 1,$$

$$\cosh(x + y) = \cosh(x)\cosh(y) + \sinh(x)\sinh(y),$$

$$\sinh(x + y) = \sinh(x)\cosh(y) + \cosh(x)\sinh(y)$$

   for all $x, y \in \mathbb{R}$.
   (d) Find formulas for $\cosh(x)$ and for $\sinh(x)$ in terms of $\cosh(2x)$ and indicate the sets of x's for which they are valid.
   (e) The function $\tanh = \sinh/\cosh$ is strictly increasing from $\mathbb{R}$ onto $] - 1, 1[$.
   (f) Let $s, c, t \in \mathbb{R}$. Solve the equations

$$\sinh(x) = s, \quad \cosh(x) = c, \quad \tanh(x) = t$$

   for $x \in \mathbb{R}$ in terms of logarithms whenever possible.

Uniform Convergence

An important way to construct nontrivial functions is to obtain them as limits of sequences or as sums of series of given functions. For example, the exponential function is defined to be the sum of a certain power series. Exercise 4(b) of the preceding section shows that the real logarithm is obtainable as the limit of a certain sequence of functions:

$$\log x = \lim_{n \to \infty} n(\sqrt[n]{x} - 1) \quad \text{for } x > 0.$$

If $(f_n)_{n=1}^{\infty}$ is a sequence of functions that each have a property P, does it follow that a limit $f = \lim_{n \to \infty} f_n$ or a sum $s = \sum_{n=1}^{\infty} f_n$ must also have the property P?

This very general question is a central one in analysis. Its answer depends not only on the property P under consideration, but also on the *kind* of convergence at hand. In this section, we examine this question when P is the property of being continuous at a point or on a space. The following examples show that the answer can be no.

(3.102) Examples (a) For $x \in \mathbb{R}$ and $n \in \mathbb{N}$, write $f_n(x) = x^{2n}(1 + x^{2n})^{-1}$ and $f(x) = \lim_{n \to \infty} f_n(x)$. Then $f(x) = 0$ if $|x| < 1$, $f(1) = f(-1) = 1/2$, and $f(x) = 1$ if $|x| > 1$. Thus each f_n is continuous on $\mathbb{R}$ while f is discontinuous at both $x = 1$ and $x = -1$.

(b) Define $(f_n)_{n=0}^{\infty}$ on $\mathbb{C}$ by $f_n(z) = z(1 + |z|)^{-n}$ and write $s(z) = \sum_{n=0}^{\infty} f_n(z)$ for each $z \in \mathbb{C}$. Then, summing a geometric series when $z \neq 0$, we have $s(z) = (1 + |z|) \operatorname{sgn}(z)$ for all $z \in \mathbb{C}$ where sgn is as in (3.88). Thus each f_n is continous on $\mathbb{C}$ but s is discontinuous at $z = 0$.

Let us clearly define two different kinds of convergence.

(3.103) Definitions Let X be a set and let $(f_n)_{n=1}^{\infty}$ be a sequence of complex-valued functions on X. The sequence $(f_n)_{n=1}^{\infty}$ is said to be *pointwise convergent on X* if there exists a complex-valued function f on X such that

$$f(x) = \lim_{n \to \infty} f_n(x) \quad \text{for all } x \in X.$$

In this case, f is called the *pointwise limit on X* of the sequence $(f_n)_{n=1}^{\infty}$.

The sequence $(f_n)_{n=1}^{\infty}$ is said to be *uniformly convergent on X* if there exists a complex-valued function f on X such that for every $\epsilon > 0$ there exists an $N \in \mathbb{N}$ such that

$$|f(x) - f_n(x)| < \epsilon \quad \text{whenever } n \geq N \text{ and } x \in X.$$

The number N depends on ϵ but *not* on x. In this case, we call f the *uniform limit on X* of the sequence $(f_n)_{n=1}^{\infty}$.

Let $s_n(x) = \sum_{k=1}^{n} f_k(x)$ for all $x \in X$ and $n \in \mathbb{N}$. If $(s_n)_{n=1}^{\infty}$ is pointwise convergent on X to the pointwise limit s, then we say that the series $\sum_{k=1}^{\infty} f_k$ is *pointwise convergent on X* and that s is its *pointwise sum on X*. If $(s_n)_{n=1}^{\infty}$ is uniformly convergent on X to the uniform limit s, then we say that the series $\sum_{k=1}^{\infty} f_k$ is *uniformly convergent on X* and that s is its *uniform sum on X*.

(3.104) Remarks (a) To say that $f_n \to f$ pointwise on X means that $f_n(x) \to f(x)$ for each $x \in X$. That is, given $\epsilon > 0$ and $x \in X$, there exists some $N \in \mathbb{N}$ such that

$$|f(x) - f_n(x)| < \epsilon \quad \text{whenever } n \geq N.$$

The number N may depend upon x as well as ϵ. In the case of uniform convergence, the same N must work for every $x \in X$. To illustrate the difference concretely, let $(f_n)_{n=1}^{\infty}$ and f be as in (3.102.a). Then $f_n \to f$ pointwise on $\mathbb{R}$, but the convergence is *not* uniform on $\mathbb{R}$. Indeed, there is no N corresponding to $\epsilon = 1/4$. In fact, let us try a fixed N. If we consider x such that $3^{-1/(2N)} < x < 1$, then we have

$$|f(x) - f_N(x)| = x^{2N}/(1 + x^{2N}) = 1/(x^{-2N} + 1) > 1/(3 + 1) = 1/4.$$

(b) It is clear that uniform convergence implies pointwise convergence, but, by (a), the converse is false. It is also clear that a sequence of functions can have at most one pointwise limit.

(c) Plainly, $f_n \to f$ uniformly on X if and only if $\|f - f_n\|_u \to 0$ where

$$\|f - f_n\|_u = \sup_{x \in X} |f(x) - f_n(x)|.$$

The subscript u on this norm symbol stands for "uniform." Compare (3.52).

(d) A sequence $(f_n)_{n=1}^{\infty}$ is pointwise convergent on X if and only if it is a pointwise Cauchy sequence [this means $(f_n(x))_{n=1}^{\infty} \subset \mathbb{C}$ is a Cauchy sequence for each $x \in X$.] Our next theorem shows that this obvious assertion is still true if we replace "pointwise" by "uniform."

(3.105) Theorem *Let X be a set and let $(f_n)_{n=1}^{\infty}$ be a sequence of complex-valued functions on X. Then the following two assertions are equivalent.*

(i) *The sequence $(f_n)_{n=1}^{\infty}$ is uniformly convergent on X.*

(ii) *For every $\epsilon > 0$ there exists some $N \in \mathbb{N}$ such that*

$$|f_n(x) - f_m(x)| < \epsilon \quad \text{whenever } m, n \geq N \text{ and } \quad x \in X.$$

If (ii) obtains, we say that $(f_n)_{n=1}^{\infty}$ is *uniformly Cauchy* on X.

Proof Suppose that (i) holds and let f be the uniform limit. Given $\epsilon > 0$, choose N such that

$$|f(x) - f_n(x)| < \epsilon/2 \quad \text{whenever } n \geq N \text{ and } x \in X.$$

Then $m, n \geq N$ and $x \in X$ imply

$$|f_m(x) - f_n(x)| \leq |f_m(x) - f(x)| + |f(x) - f_n(x)|$$
$$< \epsilon/2 + \epsilon/2 = \epsilon,$$

and so (ii) holds.

The proof that (ii) implies (i) is very like that of (3.52). The boundedness condition there is not important here. $\square$

As a corollary, we obtain the following important test for uniform convergence of series.

(3.106) Weierstrass M-test *Let X be a set and let $(f_n)_{n=1}^{\infty}$ be a sequence of complex-valued functions on X. Suppose that there exists a sequence $(M_n)_{n=1}^{\infty}$ of nonnegative* *real numbers such that* $\sum\limits_{n=1}^{\infty} M_n < \infty$ *and* $|f_n(x)| \leq M_n$ *for all* $x \in X$ *and* $n \in \mathbb{N}$. *Then*

$\sum\limits_{n=1}^{\infty} f_n$ *is uniformly convergent on* X.

Proof Write $s_n = \sum\limits_{k=1}^{n} f_k$. Given $\epsilon > 0$, choose N such that $\sum\limits_{k=m+1}^{n} M_k < \epsilon$ whenever $n > m \geq N$. Then $n > m \geq N$ and $x \in X$ imply

$$|s_n(x) - s_m(x)| = |\sum_{k=m+1}^{n} f_k(x)|$$

$$\leq \sum_{k=m+1}^{n} |f_k(x)| \leq \sum_{k=m+1}^{n} M_k < \epsilon.$$

That is, $(s_n)_{n=1}^{\infty}$ is uniformly Cauchy on X, and so we invoke (3.105) to complete the proof. □

One immediate advantage of uniform convergence over pointwise convergence is that uniform convergence preserves continuity.

(3.107) Theorem *Let* X *be a topological space, let* $p \in X$, *and let* $(f_n)_{n=1}^{\infty}$ *be a sequence of complex-valued functions on* X *that are each continuous at* p.

(i) *If* $f = \lim\limits_{n \to \infty} f_n$ *uniformly on* X, *then* f *is continuous at* p.

(ii) *If* $s = \sum\limits_{n=1}^{\infty} f_n$ *uniformly on* X, *then* s *is continuous at* p.

Proof We first prove (i). Let $\epsilon > 0$ be given. Choose N such that
$$|f(x) - f_N(x)| < \epsilon/3 \quad \text{for all} \quad x \in X.$$
Since f_N is continuous at p, there is a neighborhood U of p such that
$$|f_N(x) - f_N(p)| < \epsilon/3 \quad \text{for all} \quad x \in U.$$
Then $x \in U$ implies
$$|f(x) - f(p)| \leq |f(x) - f_N(x)| + |f_N(x) - f_N(p)| + |f_N(p) - f(p)|$$
$$< \epsilon/3 + \epsilon/3 + \epsilon/3 = \epsilon.$$
Therefore f is continuous at p.

Assertion (ii) follows from (i) since, by (3.74) and induction, the partial sums s_n are each continuous at p. □

(3.108) Remark Theorem (3.107) is often useful in proving that certain limits are not uniform. For instance, the convergence in (3.102.a) cannot be uniform on any interval X containing 1 since each f_n is continuous at 1 but f is not. However, if $0 < \alpha < 1 < \beta < \infty$, then

$$|f(x) - f_n(x)| = x^{2n}/(1 + x^{2n}) \leq 1/(\alpha^{-2n} + 1)$$

for $-\alpha \leqq x \leqq \alpha$ and

$$|f(x) - f_n(x)| = 1/(1 + x^{2n}) \leqq 1/(1 + \beta^{2n})$$

for $|x| \geqq \beta$. Thus $f_n \to f$ uniformly on $Y = \,]-\infty, -\beta] \cup [-\alpha, \alpha] \cup [\beta, \infty[$ because the right sides of these two inequalities do not depend on x and they have limit 0 as $n \to \infty$. More precisely, given $1 > \epsilon > 0$, choose N such that $1/\epsilon - 1 < \min\{\alpha^{-2N}, \beta^{2N}\}$. Then

$$|f(x) - f_n(x)| < \epsilon \quad \text{if } n \geqq N \quad \text{and} \quad x \in Y.$$

(3.109) Theorem   *Let $(c_n)_{n=0}^{\infty} \subset \mathbb{C}$ and suppose that the power series $\sum\limits_{n=0}^{\infty} c_n z^n$ has radius of convergence $R > 0$. If $0 < r < R$, then the given series converges uniformly on $D_r = \{z \in \mathbb{C} : |z| \leqq r\}$. The function s defined by*

$$s(z) = \sum_{n=0}^{\infty} c_n z^n, \qquad |z| < R,$$

is continuous on the open disk $\{z \in \mathbb{C} : |z| < R\}$.

Proof   Given $0 < r < R$, take $M_n = |c_n| r^n$. Using (2.64), we have $\sum\limits_{n=0}^{\infty} M_n < \infty$. Also, $|c_n z^n| \leqq M_n$ for all $z \in D_r$. It follows from (3.106) that the series converges uniformly on D_r. Let $p \in \mathbb{C}$ with $|p| < R$ be given. Choose r such that $|p| < r < R$. Since the series converges uniformly on D_r, and each term is continuous at p, it follows from (3.107.ii) that s is continuous at p. $\square$

The next result is a partial converse of (3.107.i).

(3.110) Dini's Theorem   *Let X be a compact space and let $(f_n)_{n=1}^{\infty}$ be a sequence of real valued functions that are each continuous on X. Suppose that $(f_n)_{n=1}^{\infty}$ is nondecreasing on X [this means that $f_1(x) \leqq f_2(x) \leqq f_3(x) \leqq \ldots$ for each $x \in X$] and converges pointwise on X to a real valued limit f that is continuous on X. Then $f_n \to f$ uniformly on X.*

Proof   Let $\epsilon > 0$ be given. For each $n \in \mathbb{N}$, write

$$A_n = \{x \in X : |f(x) - f_n(x)| \geqq \epsilon\}.$$

Since all of our functions are continuous, each A_n is closed in X. Since the sequence is monotone, we have

$$A_1 \supset A_2 \supset A_3 \ldots . \tag{1}$$

Since $f_n \to f$ pointwise on X we have $\bigcap\limits_{n=1}^{\infty} A_n = \varnothing$. Using (1) and (3.42), we see that there is some $N \in \mathbb{N}$ such that $A_n = \varnothing$ for all $n \geqq N$. Thus $n \geqq N$

and $x \in X$ imply

$$|f(x) - f_n(x)| < \epsilon. \qquad \square$$

(3.111) Examples Dini's Theorem fails if any of its hypotheses are omitted. The following are examples in which all but one hypothesis is satisfied and the conclusion fails.

(a) Let $X = [0, 1]$ and $f_n(x) = 1 - x^n$.

(b) Let $X = \mathbb{R}$ and $f_n(x) = -x^2/n$.

(c) Let $X = [0, 1]$ and $f_n(x) = 0$ if $0 < x < 1/n$, $f_n(x) = 1$ if $x = 0$ or $1/n \leqq x \leqq 1$.

(d) Let $X = [0, 2]$ and $f_n(x) = 0$ if $2/n < x \leqq 2$, $f_n(x) = n^2x^2 - 2nx$ if $0 \leqq x \leqq 2/n$. Note that $f_n(1/n) = -1$ for all $n \in \mathbb{N}$.

Exercises

1. Consider the series

$$f(x) = \sum_{n=1}^{\infty} f_n(x) \qquad (x \in \mathbb{R})$$

where

(a) $f_n(x) = x^n$,

(b) $f_n(x) = 1/(1 - x^n)$,

(c) $f_n(x) = x^n/(1 - x^n)$,

(d) $f_n(x) = (-1)^n/(x + n)$,

(e) $f_n(x) = x/[n^\alpha(1 + nx^2)]$ $(\alpha > 0)$,

(f) $f_n(x) = \{nx/(1 + n^2x^2) - (n + 1)x/[1 + (n + 1)^2x^2]\}$,

(g) $f_n(x) = (-1)^n(x^2 + n)/n^2$.

In each case, find all $x \in \mathbb{R}$ at which the series (1) converges, (2) converges absolutely, and all intervals of $\mathbb{R}$ on which (3) the series converges uniformly, (4) f is continuous.

2. Let $f_n(x) = x + 1/n$, $f(x) = x$ for $n \in \mathbb{N}$, $x \in \mathbb{R}$. Then $f_n \to f$ uniformly on $\mathbb{R}$, but it is false that $f_n^2 \to f^2$ uniformly on $\mathbb{R}$. Of course $f_n^2 \to f^2$ pointwise on $\mathbb{R}$.

3. Let $f_n : X \to \mathbb{C}$ where X is some nonvoid set and $n \in \mathbb{N}$ satisfy $|f_n(x)| \leqq \beta < \infty$ for all $x \in X$, $n \in \mathbb{N}$. Suppose $f_n \to f$ uniformly on X. Let $D = \{z \in \mathbb{C} : |z| \leqq \beta\}$ and suppose that $g : D \to \mathbb{C}$ is continuous on D. Then $g \circ f_n \to g \circ f$ uniformly on X. [Hint: g is uniformly continuous on D.]

4. Define $f_n(x) = nxe^{-nx^2}$. Then $f_n \to 0$ pointwise on $\mathbb{R}$ but not uniformly on any open interval containing 0. In fact, $\|f_n\|_u \to \infty$ on such intervals. On which intervals is the convergence uniform? [Try $x = 1/\sqrt{n}$.]

5. Let X be a topological space, let $D \subset X$ be dense in X, let $f_n : X \to \mathbb{C}$ be continuous on X for each $n \in \mathbb{N}$, and let $f : X \to \mathbb{C}$. Suppose that $f_n \to f$ pointwise on X and uniformly

on D. Then f is continuous on X. [Hint: If $|f_n(x) - f_m(x)| \leqq \epsilon$ for all $x \in D$, then $|f_n(x) - f_m(x)| \leqq \epsilon$ for all $x \in X$.]

6. For $n \in \mathbb{N}$ and $x \geqq 0$ define $f_1(x) = \sqrt{x}$, $f_{n+1}(x) = \sqrt{x + f_n(x)}$. Then (a) $0 = f_n(0) < f_n(x) < f_{n+1}(x) < 1 + x$ for all $n \in \mathbb{N}$ and $x > 0$ and (b) $(f_n)_{n=1}^{\infty}$ converges uniformly on $[a, b]$ whenever $0 < a < b < \infty$ but not on $[0, 1]$. [Hint: Use induction and Dini's Theorem.]

7. A Space Filling Curve. Let $n \in \mathbb{N}$ be fixed, let

$$I = \left\{ x \in \mathbb{R}^n : x = (x_k)_{k=1}^n, x_k \in [0, 1] \text{ for } 1 \leq k \leq n \right\},$$

let P be Cantor's ternary set:

$$P = \left\{ \sum_{j=1}^{\infty} \frac{2\epsilon_j}{3^j} : (\epsilon_j)_{j=1}^{\infty} \subset \{0, 1\} \right\},$$

and let ϕ be any continuous function from $[0, \infty[$ into $[0, 1]$ such that for each integer $m \geqq 0$ we have

$$\phi(t) = 0 \quad \text{if } t \in [2m, 2m + 1/3]$$

$$\phi(t) = 1 \quad \text{if } t \in [2m + 2/3, 2m + 1].$$

[Give an example of such a ϕ.] Define $f_k : [0, \infty[\to [0, 1]$ $(1 \leqq k \leqq n)$ and $f : [0, \infty[\to I$ by

$$f_k(t) = \sum_{j=0}^{\infty} \left[\phi(3^{k-1+jn}t)/2^{j+1} \right],$$

$$f(t) = (f_1(t), f_2(t), \ldots, f_n(t)).$$

Prove the following.

(a) If $t = \sum_{j=1}^{\infty} [2\epsilon_j/3^j] \in P$ and $a \geqq 0$ is an integer, then $\phi(3^a t) = \epsilon_{a+1}$ (each ϵ_j is 0 or

   1). [Hint: $2m + 2\epsilon_{a+1} \cdot 3^{-1} \leqq 3^a t \leqq 2m + 2\epsilon_{a+1} \cdot 3^{-1} + \sum_{j=2}^{\infty} (2/3^j)$ for some

   $m \geqq 0$.]
(b) The functions f_k and f are continuous on $[0, \infty[$.
(c) We have $f(P) = f([0, 1]) = f([0, \infty[) = I$. [Hint: For $x = (x_k)_{k=1}^n \in I$, write

$$x_k = \sum_{j=0}^{\infty} (x_{k, j}/2^{j+1}) \text{ where each } x_{k, j} \text{ is 0 or 1. Then define}$$

$$t = \sum_{j=0}^{\infty} \sum_{k=1}^{n} \left(2x_{k, j}/3^{jn+k} \right)$$

   to get $t \in P$ and $f(t) = x$.]
(d) If $n = 1$, then f agrees on P with Lebesgue's singular function. Remark: Since $f([0, 1]) = I$, the equations

$$x_k = f_k(t) \qquad (1 \leqq k \leqq n)$$

may be regarded as parametric equations of a curve in $\mathbb{R}^n$ that passes through every point of the unit n cube I at least once. For this reason f is called a *space filling curve*.

Stone–Weierstrass Theorems

In the preceding section we saw that any uniform limit of continuous functions is itself continuous. Frequently one can obtain quite arbitrary continuous functions as uniform limits of very special functions. In 1885, the German analyst Karl Weierstrass proved that any continuous real-valued function on a closed interval of $\mathbb{R}$ is the uniform limit on that interval of some sequence of polynomials with real coefficients. Since then, many different proofs of the Weierstrass Approximation Theorem have been given. In this section we present several versions of a vast generalization of this theorem that is due to the contemporary American analyst M.H. Stone. First, we need some definitions.

(3.112) Definitions Let X be a nonvoid set. A nonvoid set A of real-valued [resp., complex-valued] functions on X is called a *real* [resp., *complex*] *function algebra* if whenever $f, g \in A$ and $\alpha \in \mathbb{R}$ [resp., $\alpha \in \mathbb{C}$] the three functions αf, $f + g$, and fg are also in A. If X is a topological space, we denote by $C'(X)$ [resp., $C(X)$] the set of all bounded real-valued [resp., complex-valued] continuous functions on X. Thus $f \in C'(X)$ [resp., $f \in C(X)$] means that $f : X \to \mathbb{R}$ [resp., $f : X \to \mathbb{C}$] is continuous and

$$\|f\|_u = \sup\{|f(x)| : x \in X\} < \infty.$$

A nonvoid set $A \subset C'(X)$ [resp., $C(X)$] is called a *subalgebra of* $C'(X)$ [resp., $C(X)$] if A is a real [resp., complex] function algebra. We make $C'(X)$ an $C(X)$ into metric spaces by defining

$$\rho(f, g) = \|f - g\|_u.$$

The *uniform closure* of a subset A of $C'(X)$ or $C(X)$ is just the closure A^- with respect to the topology generated by this metric. Thus $f \in A^-$ means that $f_n \to f$ uniformly on X for some sequence $(f_n)_{n=1}^\infty \subset A$ [cf. (3.104.c)]. It is clear that $C'(X)$ and $C(X)$ are complete metric spaces [cf. (3.52), (3.105), and (3.107.i)]

If f and g are real-valued functions on X, we define functions $f \vee g$ and $f \wedge g$ on X by

$$(f \vee g)(x) = \max\{f(x), g(x)\},$$
$$(f \wedge g)(x) = \min\{f(x), g(x)\}.$$

It is easy to check that

$$f \vee g = (f + g + |f - g|)/2,$$
$$f \wedge g = (f + g - |f - g|)/2$$

pointwise on X. Thus, if f and g are continuous, so are $f \vee g$ and $f \wedge g$.

(3.113) Theorem *Let X be a topological space and let A be a subalgebra of $C'(X)$. Then the uniform closure A^- of A is also a subalgebra of $C'(X)$. Moreover, if f and g are in A^-, then $|f|$, $f \vee g$, and $f \wedge g$ are also in A^-.*

We need a lemma.

(3.114) Lemma *Define a sequence* $(P_n)_{n=0}^{\infty}$ *of polynomials recursively by the relations* $P_0(t) = 0$ *and*

(i) $P_{n+1}(t) = P_n(t) + (1/2)[t^2 - P_n(t)^2]$. *Then* $-1 \leq t \leq 1$ *implies.*

(ii) $0 \leq |t| - P_n(t) \leq 2|t|/(2 + n|t|) \leq 2/(2 + n)$. *Thus* $P_n(t) \to |t|$ *uniformly on* $[-1, 1]$.

Proof   The last inequality in (ii) is obvious on cross multiplication. Notice that for all $n \geq 0$ and all $t \in \mathbb{R}$ we have

$$|t| - P_{n+1}(t) = \big[|t| - P_n(t)\big]\big[1 - (1/2)(|t| + P_n(t))\big]. \tag{1}$$

Using (1) and (i), we show by an easy induction that

$$0 \leq P_n(t) \leq |t| \tag{2}$$

for $n \geq 0$ and $|t| \leq 1$, and so the first inequality in (ii) obtains. Next we have

$$\big[2 + (n + 1)|t|\big]\big[1 - (1/2)(|t| + P_n(t))\big]$$
$$\leq 2 + n|t| + \{|t| - 2 \cdot (1/2)(|t| + P_n(t))\}$$
$$\leq 2 + n|t|,$$
$$\frac{2|t|}{2 + n|t|}\left[1 - \frac{1}{2}(|t| + P_n(t))\right] \leq \frac{2|t|}{2 + (n + 1)|t|} \tag{3}$$

for $n \geq 0$ and $|t| \leq 1$. The second inequality in (ii) follows from (1), (2), and (3) by induction. $\square$

Proof of (3.113)   Let $f, g \in A^-$. Then we can choose sequences $(f_n)_{n=1}^{\infty}$, $(g_n)_{n=1}^{\infty} \subset A$ such that

$$\|f - f_n\|_u \to 0, \qquad \|g - g_n\|_u \to 0.$$

It follows that, for $\alpha \in \mathbb{R}$,

$$\|(\alpha f) - (\alpha f_n)\|_u = |\alpha| \cdot \|f - f_n\|_u \to 0,$$
$$\|(f + g) - (f_n + g_n)\|_u$$
$$\leq \|f - f_n\|_u + \|g - g_n\|_u \to 0,$$
$$\|fg - f_n g_n\|_u = \|f(g - g_n) + g_n(f - f_n)\|_u$$
$$\leq \|f\|_u\|g - g_n\|_u + \|g_n\|_u\|f - f_n\|_u$$
$$\to \|f\|_u \cdot 0 + \|g\|_u \cdot 0 = 0;$$

hence $\alpha f, f + g$, and fg are all in A^-. Thus A^- is an algebra.

To prove that $|f| \in A^-$, we may suppose that $\beta = \|f\|_u > 0$. Letting P_n be as in (3.114), we have $\beta P_n(\beta^{-1}f) \in A^-$ for all $n \geq 0$ and, since $-1 \leq \beta^{-1}f(x) \leq 1$ for all $x \in X$, (3.114.ii) yields

$$0 \leq |f(x)| - \beta P_n(\beta^{-1}f(x)) \leq 2\beta/(2 + n)$$

for all $x \in X$. Therefore

$$\||f| - \beta P_n(\beta^{-1}f)\|_u \leq 2\beta/(2 + n) \to 0$$

as $n \to \infty$, and so $|f| \in A^-$. The rest follows from

$$f \vee g = (f + g + |f - g|)/2,$$
$$f \wedge g = (f + g - |f - g|)/2. \quad \square$$

(3.115) Definition   Let X be a nonvoid set and let A be a set of complex-valued functions on X. We say that A *separates the points of* X if, whenever $x \neq y$ in X, there exists some $f \in A$ such that $f(x) \neq f(y)$. We say that A *vanishes nowhere on* X if for each $x \in X$ there is some $f \in A$ such that $f(x) \neq 0$.

(3.116) Theorem   *Let X be a nonvoid set and let A be a real function algebra on X that separates the points of X and vanishes nowhere on X. If $x \neq y$ in X and a and b are real numbers, then there is a function $h \in A$ such that $h(x) = a$ and $h(y) = b$.*
The same is true if "real" is replaced by "complex."

> **Proof**   Let $x \neq y$, a, and b be given. Choose $f_1, f_2, f_3 \in A$ such that $f_1(x) \neq f_1(y)$, $f_2(x) \neq 0$, $f_3(y) \neq 0$.
>   Suppose that $f_1(x) = 0$. Then $f_1(y) \neq 0$ and we can take $h = \alpha f_1 + \beta f_2$ where α and β are the solutions of the system
>
> $$\alpha f_1(x) + \beta f_2(x) = a,$$
> $$\alpha f_1(y) + \beta f_2(y) = b.$$
>
> These solutions exist because the determinant of coefficients is $f_1(x)f_2(y) - f_1(y)f_2(x) = -f_1(y)f_2(x) \neq 0$. Similarly, if $f_1(y) = 0$, we can take $h = \alpha f_1 + \beta f_3$ for appropriate $\alpha, \beta \in \mathbb{R}$.
>   Suppose finally that $f_1(x) \neq 0$ and $f_1(y) \neq 0$. Then we take $h = \alpha f_1 + \beta f_1^2$ where α and β are the solutions of the system
>
> $$\alpha f_1(x) + \beta f_1(x)^2 = a$$
> $$\alpha f_1(y) + \beta f_1(y)^2 = b.$$
>
> These solutions exist because the determinant of coefficients is $f_1(x)f_1(y)$ $[f_1(y) - f_1(x)] \neq 0$. $\quad \square$

We can now give our first version of the main theorem of this section.

(3.117) Stone–Weierstrass Theorem [compact-real]   *Let X be a compact space and let A be a subalgebra of $C'(X)$ that separates the points of X. Then the uniform closure A^- of A satisfies either*
  (i) $$A^- = C'(X)$$

or

  (ii) *there is some $p \in X$ such that*
$$A^- = \{f \in C'(X) : f(p) = 0\}.$$

> **Proof**   *Case 1:* Suppose that A vanishes nowhere on X. Let $f \in C'(X)$ and $\epsilon > 0$ be given. To show that $f \in A^-$, we need only find $g \in A^-$ such that

$\|f - g\|_u < \epsilon$ [this would show that no ball centered at f misses A^-]. For every pair $x \neq y$ in X we can use (3.116) to choose $h_{x,\,y} \in A$ such that $h_{x,\,y}(x) = f(x)$ and $h_{x,\,y}(y) = f(y)$. If $y = x$, let $h_{x,\,x} = (f(x)/h(x))h$ where $h \in A$ and $h(x) \neq 0$. Now define

$$V_{x,\,y} = \{t \in X : h_{x,\,y}(t) < f(t) + \epsilon\}.$$

Obviously $x \in V_{x,\,y}$ and $y \in V_{x,\,y}$. Also, since $h_{x,\,y} - f$ is continuous on X, the set $V_{x,\,y}$ is open. For fixed $x \in X$, the family $\{V_{x,\,y} : y \in X\}$ is an open cover of X so there exists a finite set $\{y_1, \ldots, y_n\} \subset X$ such that $X = V_{x,\,y_1} \cup \ldots \cup V_{x,\,y_n}$ and then we define

$$g_x = h_{x,\,y_1} \wedge h_{x,\,y_2} \wedge \ldots \wedge h_{x,\,y_n}.$$

According to (3.113), each g_x is in A^-. It is also clear that

$$g_x(t) < f(t) + \epsilon \quad \text{for all } t \in X,$$

$$g_x(x) = f(x).$$

Writing $W_x = \{t \in X : g_x(t) > f(t) - \epsilon\}$, we see that $x \in W_x$ and W_x is open. Thus the open cover $\{W_x : x \in X\}$ has a finite subcover: $X = W_{x_1} \cup W_{x_2} \cup \ldots \cup W_{x_m}$. Applying (3.113) again, we find that the function

$$g = g_{x_1} \vee g_{x_2} \vee \ldots \vee g_{x_m}$$

is in A^-, and it is also clear that

$$f(t) - \epsilon < g(t) < f(t) + \epsilon \quad \text{for all } t \in X.$$

Therefore $\|f - g\|_u < \epsilon$, and so (i) obtains.

  Case 2: Suppose that there exists a $p \in X$ such that $h(p) = 0$ for every $h \in A$. Let $B = \{h + c : h \in A, c \in \mathbb{R}\}$, that is, the elements of B are obtained by adding constant functions to the elements of A. Plainly, B is a subalgebra of $C'(X)$ that separates the points of X and vanishes nowhere on X. It follows from Case 1 that if $f \in C'(X)$ with $f(p) = 0$ and $\epsilon > 0$ are given, then there is some $h + c \in B$ such that

$$\|f - (h + c)\|_u < \epsilon/2.$$

Evaluating at p, we find that

$$|c| = |f(p) - \{h(p) + c\}| < \epsilon/2.$$

Therefore the function $h \in A$ satisfies

$$\|f - h\|_u \leq \|f - (h + c)\|_u + \|c\|_u$$
$$= \|f - (h + c)\|_u + |c| < \epsilon.$$

This proves that (ii) obtains. □

(3.118) Weierstrass Approximation Theorem *Let $a < b \in \mathbb{R}$, let $f:[a,b] \to \mathbb{R}$ be continuous, and let $\epsilon > 0$ be given. Then there exists a polynomial*

$$P(x) = \sum_{k=0}^{n} a_k x^k \quad (n \in \mathbb{N}, a_k \in \mathbb{R})$$

such that

$$|f(x) - P(x)| < \epsilon \qquad \text{for all } x \in [a, b].$$

Proof Let A be the set of all polynomials $P(x)$ of the stated form with domains restricted to $[a, b]$. Since $[a, b]$ is compact and A is a subalgebra of $C'([a, b])$ that separates points $[P(x) = x]$ and vanishes nowhere $[P(x) = 1]$, it follows from (3.117) that there is some $P \in A$ such that $\| f - P \|_u < \epsilon$. $\square$

To show in (3.120) that the exact analog of (3.117) may fail for the complex case [A a subalgebra of $C(X)$], we use the following lemma. Its proof requires a solution u of an equation of the form $u^m = a$ where $m \in \mathbb{N}$ and $|a| = 1$, $a \in \mathbb{C}$. Those readers who have proven the Fundamental Theorem of Algebra (p. 127) may invoke it. Other readers may wish to postpone (3.119) and (3.120) until Chapter 5, where we learn that such an a has the form $e^{i\theta}$ and we can take $u = e^{i\theta/m}$.

(3.119) Lemma   *Let*

$$P(z) = \sum_{k=0}^{n} c_k z^k \qquad (n \in \mathbb{N}, c_k \in \mathbb{C}, c_n \neq 0)$$

be a nonconstant complex polynomial and let $W \subset \mathbb{C}$ be a nonvoid open set. If $z_0 \in W$ and $P(z_0) \neq 0$, then there exist $z_1, z_2 \in W$ such that

$$|P(z_1)| < |P(z_0)| < |P(z_2)|.$$

Therefore, writing

$$\alpha = \inf\{|P(z)| : z \in W\},$$

$$\beta = \sup\{|P(z)| : z \in W\},$$

we have

(i) $|P(z_0)| < \beta$ *for all $z_0 \in W$,*
(ii) *either $P(z_0) = 0$ for some $z_0 \in W$ or $\alpha < |P(z_0)|$ for all $z_0 \in W$.*

Proof Since P is not identically zero, there are only finitely many $z \in \mathbb{C}$ such that $P(z) = 0$ [Why?]. Thus $P(z_0) \neq 0$ for some $z_0 \in W$. Let such a z_0 be given. There exists a smallest m $(1 \leq m \leq n)$ such that

$$Q(z) = P(z + z_0)/P(z_0) = 1 + b_m z^m + \ldots + b_n z^n$$

and $b_m \neq 0$. Fix any $t > 0$ such that

$$|b_m| - \sum_{k=m+1}^{n} |b_k| t^{k-m} > 0, \qquad 1 - |b_m| t^m > 0,$$

and

$$z_0 + z \in W \quad \text{whenever } |z| \leq t.$$

Next choose u and v in $\mathbb{C}$ such that

$$u^m = -|b_m|/b_m, \qquad v^m = |b_m|/b_m.$$

Then, $|u|^m = |v|^m = 1$, $|u| = |v| = 1$,

$$z_0 + tu \in W, \quad \text{and} \quad z_0 + tv \in W.$$

We have

$$Q(tu) = 1 - |b_m|t^m + \sum_{k=m+1}^{n} b_k u^k t^k,$$

$$|Q(tu)| \leqq 1 - |b_m|t^m + \sum_{k=m+1}^{n} |b_k|t^k$$

$$= 1 - t^m \left(|b_m| - \sum_{k=m+1}^{n} |b_k|t^{k-m} \right) < 1,$$

$$Q(tv) = 1 + |b_m|t^m + \sum_{k=m+1}^{n} b_k t^k v^k,$$

$$|Q(tv)| \geqq 1 + |b_m|t^m - \sum_{k=m+1}^{n} |b_k|t^k$$

$$= 1 + t^m \left(|b_m| - \sum_{k=m+1}^{n} |b_k|t^{k-m} \right) > 1.$$

Choosing $z_1 = z_0 + tu$ and $z_2 = z_0 + tv$, we therefore have $z_1, z_2 \in W$ and

$$|P(z_1)| = |P(z_0)| \cdot |Q(tu)| < |P(z_0)|$$
$$< |P(z_0)| \cdot |Q(tv)| = |P(z_2)|. \quad \square$$

(3.120) Example Let $X = \{z \in \mathbb{C} : |z| = 1\}$ and let A be the set of all complex polynomials

$$h(z) = \sum_{k=0}^{n} c_k z^k \qquad (n \geqq 0, c_k \in \mathbb{C})$$

with their domains restricted to X. Then X is compact and A is a subalgebra of $C(X)$ that separates the points of X [$h(z) = z$] and vanishes nowhere on X [$h(z) = 1$]. However, $A^- \neq C(X)$. In fact, let $f(z) = \bar{z}$ for $z \in X$. Then $f \in C(X)$, but if $h \in A$, then

$$|f(z) - h(z)| = |z| \cdot |\bar{z} - h(z)| = |1 - zh(z)| \tag{1}$$

for all $z \in X$. Let $P(z) = 1 - zh(z)$ and $W = \{z \in \mathbb{C} : |z| < 1\}$. Since the disk $W \cup X$ is compact, we can find some z_0 in $W \cup X$ such that

$$|P(z_0)| = \sup\{|P(z)| : z \in W \cup X\} = \sup\{|P(z)| : z \in W\} = \beta$$

If $h \neq 0$, then P is not a constant, and so (3.119.i) shows that $z_0 \in X$; hence, using

(1) and (3.119, i), we have

$$\|f - h\|_u = \sup\{|P(z)| : z \in X\} = |P(z_0)| = \beta > |P(0)| = 1.$$

If $h = 0$, then $\|f - h\|_u = \|f\|_u = 1$. It follows that $\mathrm{dist}(f, A) = 1$, and so $f \notin A^-$.

If we require A to be closed under complex conjugation, then this difficulty dissolves.

(3.121) Definition   We say that a complex function algebra A is *self-adjoint* if $\bar{f} \in A$ whenever $f \in A$.

(3.122) Stone–Weierstrass Theorem [compact-complex]   *Let X be a compact space and let A be a self-adjoint subalgebra of $C(X)$ that separates the points of X. Then the uniform closure A^- of A satisfies either*

(i) $A^- = C(X)$

or

(ii) *there is some $p \in X$ such that $A^- = \{f \in C(X) : f(p) = 0\}$.*

Proof   Let $A' = \{f \in A : f(X) \subset \mathbb{R}\}$. Then $f \in A$ implies

$$\mathrm{Re}\, f = (f + \bar{f})/2 \in A', \qquad \mathrm{Im}\, f = (f - \bar{f})/(2i) \in A'.$$

Also, for $x \neq y$ in X, we can choose $f \in A$ such that $f(x) \neq f(y)$, hence

$$\mathrm{Re}\, f(x) \neq \mathrm{Re}\, f(y) \quad \text{or} \quad \mathrm{Im}\, f(x) \neq \mathrm{Im}\, f(y),$$

and so A' separates the points of X. Obviously A' is a subalgebra of $C'(X)$. Therefore, either (3.117.i) or (3.117.ii) obtains for A'^-. Thus, given $f \in C(X)$ and $\epsilon > 0$ [if (3.117.ii) obtains for A'^-, we suppose $f(p) = 0$], we can choose $h_1, h_2 \in A'$ such that

$$\|\mathrm{Re}\, f - h_1\|_u < \epsilon/2, \qquad \|\mathrm{Im}\, f - h_2\|_u < \epsilon/2.$$

Writing $h = h_1 + ih_2$, we have $h \in A$ and $\|f - h\|_u < \epsilon$.   □

(3.123) Example   This example shows that compactness is important in (3.117) and (3.122). Let X be any noncompact metric space and let $(x_n)_{n=1}^\infty$ be a sequence of distinct points of X having no convergent subsequence. Define $A = \{f \in C'(X) : \lim_{n \to \infty} f(x_n) \text{ exists}\}$. Plainly, A is a subalgebra of $C'(X)$. Let $a \neq b$ in X. Choose $N \in \mathbb{N}$ such that $a \neq x_n$ for all $n > N$. Write $E = \{b\} \cup \{x_n : n > N\}$. The function f defined on X by

$$f(x) = \frac{\mathrm{dist}(x, E)}{\mathrm{dist}(x, \{a\}) + \mathrm{dist}(x, E)}$$

satisfies $f(a) = 1$, $f(b) = 0$, $f \in C'(X)$, and $\lim_{n \to \infty} f(x_n) = 0$; hence, $f \in A$ and so A separates the points of X. Since A contains 1, A vanishes nowhere on X. However,

$A^- \neq C^r(X)$ [actually A is closed already]. To see this, let

$$S = \{x_{2k} : k \in \mathbb{N}\}, \qquad T = \{x_{2k-1} : k \in \mathbb{N}\},$$

$$g(x) = \frac{\mathrm{dist}(x, T)}{\mathrm{dist}(x, T) + \mathrm{dist}(x, S)}$$

for $x \in X$. Then $g \in C^r(X)$ and $\|g - f\|_u \geq 1/2$ for all $f \in A$.

To give our final version of the main theorem we need more definitions.

(3.124) Definitions A *locally compact space* is a topological space X such that each $x \in X$ has some neighborhood V such that V^- is compact. If X is such a space, we denote by $C_0(X)$ the set of all $f \in C(X)$ such that for each $\epsilon > 0$ the set $\{x \in X : |f(x)| \geq \epsilon\}$ [obviously closed] is compact. The elements of $C_0(X)$ are said to *vanish at infinity on* X. We write $C_0^r(X)$ for the set of real-valued functions in $C_0(X)$.

It is a simple exercise to check that $C_0^r(X)$ and $C_0(X)$ are closed subalgebras of $C^r(X)$ and $C(X)$, respectively. If X is compact, then $C_0^r(X) = C^r(X)$ and $C_0(X) = C(X)$.

Before giving examples of locally compact spaces, we prove an important fact about such spaces.

(3.125) Theorem *Let X be a locally compact Hausdorff space and let $S \subset X$ be an open set. Then, for each $x \in S$ there exists a neighborhood V of x such that $V^- \subset S$ and V^- is compact.*

> **Proof** Let $x \in S$ be given. Choose a neighborhood W of x such that W^- is compact. Then $S_0 = W \cap S$ is a neighborhood of x, and according to (3.35), the closed subsets S_0^- and $T = S_0^- \setminus S_0$ of W^- are compact. If $T = \emptyset$, then $S_0^- = S_0 \subset S$, and so we can take $V = S_0$. Suppose $T \neq \emptyset$. For each $t \in T$, choose disjoint neighborhoods U_t and V_t of t and x, respectively. Then $\{U_t : t \in T\}$ is an open cover of T, and so there exist $t_1, \ldots, t_n$ in T such that $T \subset \bigcup_{j=1}^{n} U_{t_j} = U$. Let $V = \bigcap_{j=1}^{n} V_{t_j} \cap S_0$. Then V is a neighborhood of x and $V \subset S_0 \setminus U \subset S_0^- \setminus U$. But $S_0^- \setminus U$ is closed so $V^- \subset S_0^- \setminus U \subset S_0^- \setminus T = S_0 \subset S$. Since S_0^- is compact, its closed subset V^- is also compact. □

(3.126) Examples (a) Any compact space X is locally compact because we can take $V = X$ for any $x \in X$.

(b) If $n \in \mathbb{N}$, then $\mathbb{R}^n$ is locally compact by (3.40) since we can take $V = B_r(x)$ for any $x \in \mathbb{R}^n$ and any $r > 0$.

(c) Let X be any locally compact space and let $S \subset X$ be closed. Then S, with its relative topology [see (3.30)], is locally compact because if $x \in S$ and V is a neighborhood of x such that V^- is compact, then $V \cap S$ is a relative neighborhood of x, $(V \cap S)^- \subset V^- \cap S$, and $V^- \cap S$ is compact so $(V \cap S)^-$ is compact.

(d) Let X be a locally compact Hausdorff space and let $S \subset X$ be open. Then S, with its relative topology, is locally compact. This is because if $x \in S$ and V are as in (3.125), then $V = V \cap S$ is open in S and $(V \cap S)^- = V^- = V^- \cap S$ is the closure in S of $V \cap S$.

(e) The subspace $\mathbb{Q}$ of $\mathbb{R}$ is *not* locally compact. Indeed, let $x \in \mathbb{Q}$ be arbitrary. If $V = U \cap \mathbb{Q}$ is any neighborhood of x [U is open in $\mathbb{R}$], we can choose $y \in U \setminus \mathbb{Q}$ and $(x_n)_{n=1}^\infty \subset V$ such that $x_n \to y$. Since no subsequence of $(x_n)_{n=1}^\infty$ converges to a point of $V^- \subset U^- \cap \mathbb{Q}$, we see that V^- is not compact.

We can now present our final version of the main theorem. It contains (3.117) and (3.122) as special cases.

(3.127) Stone–Weierstrass Theorem   *Let X be a locally compact space and let A be a self-adjoint subalgebra of $C_0(X)$ [or a subalgebra of $C_0^r(X)$] that separates the points of X and vanishes nowhere on X. Then the uniform closure A^- of A satisfies*

$$A^- = C_0(X) \quad \left[\text{or } A^- = C_0^r(X) \right].$$

Proof   We define a new space Y as follows. Let p be any point *not* in X. Let $Y = X \cup \{p\}$ and let $\mathfrak{U}$ be the family of all subsets U of Y such that either $U \subset X$ and U is open in X or $p \in U$ and $X \setminus U$ is compact and closed in X.* One checks that $\mathfrak{U}$ is a compact Hausdorff topology for Y whose relativization to X is the original topology there.

For $f \in C_0(X)$, define f_0 on Y by $f_0(p) = 0$ and $f_0(x) = f(x)$ for $x \in X$. It is easy to see from (3.124) that $f_0 \in C(Y)$ for each $f \in C_0(X)$. Let $A_0 = \{h_0 : h \in A\}$. Then A_0 is a self adjoint subalgebra of $C(Y)$ [or $C'(Y)$] that separates the points of Y [$x \in X, h(x) \neq 0$ imply $h_0(x) \neq h_0(p)$]. Applying (3.122) [or (3.117)], we have $A_0^- = \{f \in C(Y) : f(p) = 0\}$. Thus, given $f \in C_0(X)$ [or $C_0^r(X)$] and $\epsilon > 0$, we can choose h_0 in A_0 such that $|f_0(x) - h_0(x)| < \epsilon$ for all $x \in Y$. Then $h \in A$ and $\|f - h\|_u < \epsilon$. $\square$

As a nice application of (3.122), we prove a famous extension theorem. Those readers who know Urysohn's Lemma will see that our "metric" hypothesis can be replaced by "Hausdorff."

(3.128) Tietze's Extension Theorem   *Let Y be a locally compact metric space and let $X \subset U \subset Y$ satisfy X is compact and U is open. Then every $f \in C(X)$ has an extension $f^\dagger \in C(Y)$ [$f^\dagger(x) = f(x)$ for $x \in X$] such that*
(i) $f^\dagger(y) = 0$ for all $y \in Y \setminus U = U'$,
(ii) $\|f^\dagger\|_u = \sup\{|f^\dagger(y)| : y \in Y\}$
$= \sup\{|f(x)| : x \in X\} = \|f\|_u.$

---

*In the present setting the word "closed" is not needed because the fact that A separates the points of X entails that X is a Hausdorff space.

Proof Let A be the set of all $f \in C(X)$ that have an extension $f^\dagger \in C(Y)$ that satisfies (i). Then A is a self-adjoint subalgebra of $C(X)$ [if f, $g \in A$, we can choose $(f + g)^\dagger = f^\dagger + g^\dagger$, etc.]. To see that A separates the points of X, X, let $a \neq b$ in X. Define g on Y by

$$g(y) = \frac{\text{dist}(y, \{b\} \cup U')}{\text{dist}(y, \{a\}) + \text{dist}(y, \{b\} \cup U')}$$

and let $f = g|_X$ be the restriction of g to X. Then $f(b) = 0$, $f(a) = 1$, $g(U') \subset \{0\}$, and we can choose $f^\dagger = g$ to see that $f \in A$. This also shows that A vanishes nowhere on X. By (3.122), $A^- = C(X)$.

We next show that any $f \in A$ satisfies (ii) for some $f^\dagger$. Let $f \in A$ and let $g \in C(Y)$ be any extension of f that vanishes on U'. Let $D = \{z \in \mathbb{C} : |z| \leq \|f\|_u\}$ and define $\phi : \mathbb{C}' \to D$ by $\phi(z) = z$ for $z \in D$ and $\phi(z) = \|f\|_u \cdot (z/|z|) = \|f\|_u \cdot \text{sgn}(z)$ for $z \in D' = \mathbb{C} \backslash D$. It is easy to see that ϕ is continuous on $\mathbb{C}$ [sgn is continuous on $\mathbb{C} \backslash \{0\}$] and so $f^\dagger = \phi \circ g \in C(Y)$. Moreover, $f^\dagger$ is an extension of f and satisfies both (i) and (ii).

Finally, let $f \in C(X)$ be arbitrary. Choose $f_1 \in A$ such that $\|f - f_1\|_u < 1/2$. If $n \in \mathbb{N}$ and $f_1, f_2, \ldots, f_n$ have been chosen from A such that

$$\|f - (f_1 + \cdots + f_n)\|_u < 1/2^n, \tag{1}$$

then choose $f_{n+1} \in A$ such that

$$\|f - (f_1 + \cdots + f_n + f_{n+1})\|_u$$
$$= \|[f - (f_1 + \cdots + f_n)] - f_{n+1}\|_u < 1/2^{n+1}. \tag{2}$$

Thus, we inductively choose a sequence $(f_n)_{n=1}^\infty \subset A$ such that

$$f = \sum_{n=1}^\infty f_n, \tag{3}$$

where the series converges uniformly on X. Adding inequalities (1) and (2), we obtain

$$\|f_{n+1}\|_u < 1/2^n + 1/2^{n+1} = 3/2^{n+1} = M_{n+1} \tag{4}$$

for all $n \in \mathbb{N}$. Write $M_1 = \|f_1\|_u$. Now choose $f_n^\dagger \in C(Y)$ that extends f_n and satisfies (i) and (ii). From (ii), (4) and the Weierstrass M-test it follows that the series

$$\sum_{n=1}^\infty f_n^\dagger$$

converges uniformly on Y and, using (3.107), we see that its sum $g \in C(Y)$. By (3), g extends f. Since each $f_n^\dagger$ vanishes on U', so does g. Now use the preceding paragraph to replace g by $f^\dagger = \phi \circ g \in C(Y)$. $\square$

A simple corollary of (3.122), important in Fourier analysis, is the following.

(3.129) Theorem   *Let $X = \{z \in \mathbb{C} : |z| = 1\}$, let $f \in C(X)$ and let $\epsilon > 0$ be given. Then there exist $N \in \mathbb{N}$ and $(c_n)_{n=-N}^{N} \subset \mathbb{C}$ such that the function*

$$P(z) = \sum_{n=-N}^{N} c_n z^n$$

satisfies

$$|f(z) - P(z)| < \epsilon \quad \text{for all } z \in X.$$

Proof   Since $\bar{z} = z^{-1}$ for $z \in X$, the set A of all functions of the form P is a self-adjoint subalgebra of $C(X)$ that separates the points of X [$z \in A$] and vanishes nowhere on X [$1 \in A$]. Just apply (3.122).   □

Exercises

1.   Let $a < b$ in $\mathbb{R}$, $f \in C([a,b])$, $g \in C'([a,b])$, and $\epsilon > 0$ be given. Suppose that g is strictly monotone on $[a,b]$. Then there exist complex numbers $c_0, c_1, \ldots, c_n$ such that

$$|f(x) - \sum_{k=0}^{n} c_k g^k(x)| < \epsilon$$

for all $x \in [a,b]$.

2.   If $x = (x_1, x_2, \ldots, x_n) \in \mathbb{R}^n$ and $\alpha = (\alpha_1, \alpha_2, \ldots, \alpha_n)$ is an n-tuple of nonnegative integers, define

$$x^\alpha = \prod_{j=1}^{n} x_j^{\alpha_j}.$$

A function of the form

$$P(x) = \sum_{\alpha \in F} c_\alpha x^\alpha \qquad (x \in \mathbb{R}^n)$$

where F is a finite set of n-tuples of nonnegative integers and $\{c_\alpha : \alpha \in F\} \subset \mathbb{R}$ (or $\mathbb{C}$) is called a *real* (or *complex*) *polynomial in n real variables*. If $X \subset \mathbb{R}^n$ is compact, $f \in C'(X)$, and $\epsilon > 0$, then there is some P, as above such that

$$|f(x) - P(x)| < \epsilon \quad \text{for all} x \in X.$$

3.   If X is a topological space and $f : X \to \mathbb{C}$ is a function, we define the *support of f* to be the set $\{x \in X : f(x) \neq 0\}^-$. Write

$$C_{00}(X) = \{f \in C(X) : f \text{ has compact support}\}.$$

If X is a locally compact metric* space, then $C_{00}(X)$ is dense in $C_0(X)$.

---

*See the remark preceding (3.128).

4. If $f \in C_0^r(\mathbb{R})$ and $\epsilon > 0$ are given, then there exists a function g of the form

$$g(x) = \sum_{k=1}^{n} c_k e^{-a_k x^2 - b_k x},$$

where $n, a_k, b_k \in \mathbb{N}$, $c_k \in \mathbb{R}$ $(1 \leq k \leq n)$, such that $|f(x) - g(x)| < \epsilon$ for all $x \in \mathbb{R}$.

5. (a) Let p be an integer and let $p < a < b < p + 1$. Suppose that $f : [a, b] \to \mathbb{R}$ is continuous and $\epsilon > 0$ are given. Then there exists a polynomial

$$P(x) = \sum_{k=0}^{n} a_k x^k \quad (n \geq 0, a_k \in \mathbb{Z})$$

with *integral* coefficients such that

$$|P(x) - f(x)| < \epsilon \quad \text{for all } x \in [a, b].$$

[Hints: Let $A \subset C'[a, b]$) be the set of all polynomials with integral coefficients. Then $f, g \in A^-$ implies $f + g$ and $fg \in A^-$. Show that $1/2 \in A^-$ by noting (with S. Saeki) that

$$\frac{1}{2} = \frac{x - p}{1 - (1 - 2(x - p))} = \sum_{k=0}^{\infty} (x - p)(1 - 2(x - p))^k$$

if $p < x < p + 1$. Deduce that $\alpha \in A^-$ for all $\alpha \in \mathbb{R}$ so that A^- is a real function algebra.]
    (b) Show that (a) fails if $[a, b]$ contains an integer.

6. Define polynomials Q_n $(n \geq 0)$ by $Q_0(x) = 1$, $Q_{n+1}(x) = (1/2)[Q_n(x)^2 + (1 - x^2)]$ Then we have
    (a) $Q_n(x) \geq Q_{n+1}(x) \geq 0$ for $n \geq 0$ and $x \in [-1, 1]$ [Hint: induction],
    (b) $Q_n(x) \to 1 - |x|$ for each $x \in [-1, 1]$,
    (c) $Q_n(x) \to 1 - |x|$ uniformly on $[-1, 1]$ [Hint: Dini's Theorem].

7. Let X be a compact space and let $L \subset C'(X)$ be a lattice [$f, g \in L$ implies $f \wedge g$, $f \vee g \in L$].
    (a) If $f \in C'(X)$ and if for each $x, y \in X$ and $\epsilon > 0$ there is some $h \in L$ such that $|f(z) - h(z)| < \epsilon$ for $z = x$ and $z = y$, then there is an $h \in L$ such that $|f(z) - h(z)| < \epsilon$ for all $z \in X$.
    (b) If L is a linear space [$f, g \in L$ and $\alpha \in \mathbb{R}$ imply αf, $f + g \in L$], L separates the points of X, and $1 \in L$, then $L^- = C'(X)$. [Hint: $h \in L$ implies $\alpha h + \beta \in L$ for all $\alpha, \beta \in \mathbb{R}$.]

8. This exercise outlines an elementary proof of the Weierstrass Approximation Theorem given by the Russian analyst S. N. Bernstein. Let $f \in C'([0, 1])$. For $n \in \mathbb{N}$, the nth *Bernstein polynomial* for f is defined by

$$B_n(f, x) = \sum_{k=0}^{n} f(k/n) \binom{n}{k} x^k (1 - x)^{n-k}.$$

Prove the following.

    (a) $1 = \sum_{k=0}^{n} \binom{n}{k} x^k (1 - x)^{n-k}$ for $x \in \mathbb{R}, n \in \mathbb{N}$.

(b) $x = \sum_{k=0}^{n} (k/n) \binom{n}{k} x^k (1-x)^{n-k}$ for $x \in \mathbb{R}, n \in \mathbb{N}$. [Hint: Replace n by $n-1$ in

(a), multiply by x, replace k by $j-1$, and then use

$$\binom{n-1}{j-1} = \frac{j}{n} \binom{n}{j}.\Big]$$

(c) $(n^2 - n)x^2 = \sum_{k=0}^{n} (k^2 - k)\binom{n}{k} x^k (1-x)^{n-k}$ for $x \in \mathbb{R}, n \in \mathbb{N}$. [Hint: Replace n

by $n-2$ in (a), multiply by x^2, replace k by $j-2$, and use

$$\binom{n-2}{j-2} = \frac{j^2 - j}{n^2 - n} \binom{n}{j}.\Big]$$

(d) $\sum_{k=0}^{n} \left(\frac{nx-k}{n}\right)^2 \binom{n}{k} x^k (1-x)^{n-k} = \frac{x(1-x)}{n}$ for $n \in \mathbb{N}$ and $x \in \mathbb{R}$. [Hint:

Divide (c) by n^2, multiply (a) by x^2, multiply (b) by $1/n - 2x$ and then add the
resulting three equations.]

(e) $x(1-x) \leq 1/4$ for $x \in \mathbb{R}$.

(f) $|f(x) - B_n(f,x)| \leq \sum_{k=0}^{n} \left|f(x) - f\left(\frac{k}{n}\right)\right|\binom{n}{k} x^k (1-x)^{n-k}$ for $n \in \mathbb{N}$, $x \in [0,1]$.

[Hint: Multiply (a) by $f(x)$.]

(g) Given $\epsilon > 0$, we have

$$|f(x) - B_n(f,x)| < \epsilon \quad \text{for } x \in [0,1]$$

provided $n \geq \max\{\epsilon^{-2}\|f\|_u^2, \delta^{-4}\}$ where $\delta > 0$ satisfies

$$|f(x) - f(t)| < \epsilon/2 \quad \text{whenever } |x - t| < \delta \quad \text{and } x, t \in [0,1].$$

[Hints: Write the sum in (f) as $\Sigma_1 + \Sigma_2$ where Σ_1 is the sum over those k's for
which $|x - k/n| < n^{-1/4}$ and Σ_2 is the sum over the other k's. Use the choice
of δ and (a) to show that $\Sigma_1 < \epsilon/2$. In Σ_2 use $|f(x) - f(k/n)| \leq 2n^{1/2}(x - k/n)^2\|f\|_u$, (d), and (e) to show that $\Sigma_2 \leq \epsilon/2$.]

(h) If $g \in C'([a,b])$ $(a < b$ in $\mathbb{R})$, then there exists a sequence $(P_n)_{n=1}^{\infty}$ of real
polynomials such that $P_n \to g$ uniformly on $[a,b]$. [Hint: Define $\phi:[0,1] \to [a,b]$
by $\phi(x) = a + (b - a)x$ and let

$$P_n(t) = B_n\Big(g \circ \phi, \phi^{-1}(t)\Big).\Big]$$

9.   Let $b > 0$, let $B = \{f \in C'([-b,b]): f(x) = f(-x)$ for $0 \leq x \leq b\}$, and let A be the
set of all polynomials that contain only terms of even degree (with domains
restricted to $[-b,b]$). Then the uniform closure of A is B. [Hint: Let $g(t) = f(\sqrt{t})$
and approximate g on $[0, b^2]$.]

10. This exercise generalizes the preceding one. Let X be a compact space and let A be
a subalgebra of $C'(X)$ that vanishes nowhere on X. Let $B = \{f \in C'(X): f(x)$
$= f(y)$ whenever $x, y \in X$ and $h(x) = h(y)$ for all $h \in A\}$. Then the uniform
closure of A is B. [Hints: The trick in Exercise 9 was to "identify" x and $-x$ by the
"quotient map" $x \to x^2$. We can imitate that method. Write $x \sim y$ if $h(x) = h(y)$ for
all $h \in A$ and let $\mathcal{E}$ be the family of equivalence classes. That is, $E \in \mathcal{E}$ means that
$\phi \neq E \subset X$, each $h \in A$ is constant on E, and if $E \subsetneq F \subset X$, then some $h \in A$ is not
constant on F. Then $\mathcal{E}$ is pairwise disjoint and $\cup \mathcal{E} = X$. Define $\phi: X \to \mathcal{E}$ by
$\phi(x) = \{y \in X: y \sim x\}$. Call $\mathcal{V} \subset \mathcal{E}$ open if and only if $\phi^{-1}(\mathcal{V}) = \cup \mathcal{V}$ is open in X.

This makes $\mathscr{E}$ a topological space, ϕ is continuous and $\mathscr{E}$ is compact. For each $f \in B$, there is a well-defined $f_0 \in C'(\mathscr{E})$ such that $f_0 \circ \phi = f$. The collection A_0 of all h_0 for $h \in A$ is dense in $C'(\mathscr{E})$. Also $\|f - h\|_u = \|f_0 - h_0\|_u$ whenever $f \in B$ and $h \in A$.]

Total Variation

(3.130) Definition Let $[a,b] \subset \mathbb{R}$ with $a < b$. A *subdivision of* $[a,b]$ is a finite ordered set

$$P = \{ a = x_0 < x_1 < \ldots < x_{n-1} < x_n = b \}.$$

If $f : [a,b] \to \mathbb{C}$ is a function and P is as above, we define

$$V(P, f) = \sum_{k=1}^{n} |f(x_k) - f(x_{k-1})|.$$

We define the *total variation of f on* $[a,b]$ to be the extended real number

$$V_a^b f = \sup \{ V(P, f) : P \text{ is a subdivision of } [a,b] \}.$$

Also define $V_a^a f = 0$. If $V_a^b f < \infty$, we say that f is *of finite variation on* $[a,b]$ [or *of bounded variation on* $[a,b]$] and we write $f \in BV([a, b])$.

(3.131) Theorem *Let* $[a,b] \subset \mathbb{R}$, *let* $c \in [a,b]$, *and let* f *and* g *be complex-valued functions on* $[a,b]$. *Then*
(i) $$V_a^b(f + g) \leq V_a^b f + V_a^b g,$$
(ii) $$V_a^b f = V_a^c f + V_c^b f.$$

Proof Let P as in (3.130) be arbitrary. From the triangle inequality

$$|[f(x_k) + g(x_k)] - [f(x_{k-1}) + g(x_{k-1})]|$$
$$\leq |f(x_k) - f(x_{k-1})| + |g(x_k) - g(x_{k-1})|.$$

Summing for $k = 1, 2, \ldots, n$ shows that

$$V(P, f + g) \leq V(P, f) + V(P, g) \leq V_a^b f + V_a^b g.$$

Taking the supremum over all P's, we get (i).

Again let P as in (3.130) be an arbitrary subdivision of $[a,b]$. Choose m $(1 \leq m \leq n)$ such that $x_{m-1} \leq c \leq x_m$ and write

$$P_1 = \{ a = x_0 \leq \ldots \leq x_{m-1} \leq c \},$$
$$P_2 = \{ c \leq x_m \leq \ldots \leq x_n = b \}.$$

Since

$$|f(x_m) - f(x_{m-1})| \leq |f(c) - f(x_{m-1})| + |f(x_m) - f(c)|,$$

we have

$$V(P, f) \leq V(P_1, f) + V(P_2, f) \leq V_a^c f + V_c^b f.$$

Taking the supremum over all P's yields

$$V_a^b f \leq V_a^c f + V_c^b f. \tag{1}$$

Now let $\alpha < V_a^c f$ and $\beta < V_c^b f$ be arbitrary. Choose subdivisions P_α and P_β of $[a, c]$ and $[c, b]$, respectively, such that

$$V(P_\alpha, f) > \alpha, \qquad V(P_\beta, f) > \beta.$$

Then

$$\alpha + \beta < V(P_\alpha, f) + V(P_\beta, f) = V(P_\alpha \cup P_\beta, f) \leq V_a^b f$$

Letting $\alpha \uparrow V_a^c f$ and $\beta \uparrow V_c^b f$, we obtain

$$V_a^c f + V_c^b f \leq V_a^b f. \tag{2}$$

Now combine (1) and (2) to obtain (ii). $\square$

(3.132) Examples (a) If $f : [a, b] \to \mathbb{R}$ is monotone, then for all subdivisions P of $[a, b]$, we have $V(P, f) = |f(b) - f(a)|$, and so $V_a^b f = |f(b) - f(a)|$. This is because all differences $f(x_k) - f(x_{k-1})$ have the same sign.

(b) A function $f : [a, b] \to \mathbb{C}$ is said to satisfy a *Lipschitz condition* [of degree 1] *on* $[a, b]$ if there exists $M \in \mathbb{R}$ such that

$$|f(x) - f(y)| \leq M |x - y|$$

for all $x, y \in [a, b]$. Plainly $V(P, f) \leq M(b - a)$ for all P, and so $V_a^b f \leq M(b - a)$.

(c) Let $f : [0, 1] \to [0, 1]$ be defined by $f(0) = 0$, $f(1/n) = (-1)^n/n$ for $n \in \mathbb{N}$, and f is linear on each interval $[1/(n + 1), 1/n]$. [Draw (part of) the graph of f.] Then f is continuous on $[0, 1]$ $[|f(x) - f(0)| \leq |x - 0|]$. If $P_n = \{0 < 1/n < 1/(n - 1) < \ldots < 1/2 < 1\}$, then

$$V(P_n, f) = \frac{1}{n} + \sum_{k=2}^{n} \left| \frac{(-1)^k}{k} - \frac{(-1)^{k-1}}{k-1} \right| = \frac{1}{n} + \sum_{k=2}^{n} \left(\frac{1}{k} + \frac{1}{k-1} \right) > \sum_{k=1}^{n} \frac{1}{k}$$
$$\to \infty \quad \text{as } n \to \infty$$

and so $V_0^1 f = \infty$.

Functions of finite variation have a remarkably simple representation in terms of montone functions as the next result shows.

(3.133) Jordan Decomposition Theorem *Let* $f : [a, b] \to \mathbb{C}$ *be a given function. A necessary and sufficient condition that* $V_a^b f < \infty$ *is that there exist four nondecreasing functions* $f_j : [a, b] \to \mathbb{R}$ $(1 \leq j \leq 4)$ *such that*

(i) $\qquad\qquad f = (f_1 - f_2) + i(f_3 - f_4)$

on $[a, b]$.

The expression (i) is called a *Jordan decomposition* of f.

Proof Since $V_a^b \alpha f_j = |\alpha| V_a^b f_j$ for any $\alpha \in \mathbb{C}$ and any f_j, the sufficiency of the condition follows from (3.131.i) and (3.132.a).

Now suppose that $V_a^b f < \infty$ and f is real-valued. Define $g(x) = V_a^x f$ for $x \in [a, b]$. Then $g(b) < \infty$ and, using (3.131.ii), $x < y$ in $[a, b]$ implies

$$0 \leq g(x) \leq g(x) + V_x^y f = g(y) \leq g(b). \tag{1}$$

Thus g is real-valued and nondecreasing on $[a, b]$. Define $h = g - f$. Then $f = g - h$ and, with the use of (1), $x < y$ in $[a, b]$ implies

$$h(y) - h(x) = g(y) - g(x) - \big[f(y) - f(x) \big]$$
$$\geq V_x^y f - |f(y) - f(x)| \geq 0.$$

Therefore h is also nondecreasing. Take $f_1 = g$, $f_2 = h$, and $f_3 = f_4 = 0$.

It follows from $|\mathrm{Re}\ c| \leq |c|$ and $|\mathrm{Im}\ c| \leq |c|$ $(c \in \mathbb{C})$ and (3.130) that if $f : [a, b] \to \mathbb{C}$, then $V_a^b\ \mathrm{Re}\ f \leq V_a^b f$ and $V_a^b\ \mathrm{Im}\ f \leq V_a^b f$. Thus, if $V_a^b f < \infty$, we can apply the preceding paragraph to $\mathrm{Re}\ f$ and to $\mathrm{Im}\ f$ to write

$$\mathrm{Re}\ f = f_1 - f_2, \ \mathrm{Im}\ f = f_3 - f_4$$

where f_j is nondecreasing for $1 \leq j \leq 4$ and thereby obtain (i). $\quad \square$

(3.134) Corollary *Let $f : [a, b] \to \mathbb{C}$ be of finite variation on $[a, b]$. Then every point of discontinuity of f is a simple discontinuity $[f(x +)$ and $f(x -)$ exist in $\mathbb{C}]$ and the set of such discontinuities is countable.*

Proof This follows from (3.133) and (3.90). $\quad \square$

We next prove a technical lemma for use in the next chapter. It is geometrically obvious when one realizes that the total variation of a real-valued function is the sum of all of its ups and downs.

(3.135) Lemma *Let $x < y$ in $\mathbb{R}$ be given. Suppose that*

$$f : [x, y] \to \mathbb{R}, \quad \epsilon > 0, \quad \{]\alpha_j, \beta_j] \}_{j=1}^n$$

is a finite pairwise disjoint family of subintervals of $[x, y]$, and either

  (i) $f(x) \leq f(y)$ and $f(\beta_j) - f(\alpha_j) < -\epsilon(\beta_j - \alpha_j)$ for $1 \leq j \leq n$

or

  (ii) $f(x) > f(y)$ and $f(\beta_j) - f(\alpha_j) > \epsilon(\beta_j - \alpha_j)$ for $1 \leq j \leq n$.

Then

  (iii) $V_x^y f > |f(x) - f(y)| + 2\epsilon \cdot \displaystyle\sum_{j=1}^n (\beta_j - \alpha_j)$.

Proof First, suppose that (i) obtains. Changing the labeling if need be, we suppose that $\beta_j \leq \alpha_{j+1}$ for $1 \leq j < n$ and we write $\beta_0 = x$, $\alpha_{n+1} = y$. By (i), we have

$$|f(\beta_j) - f(\alpha_j)| = -(f(\beta_j) - f(\alpha_j))$$
$$> (f(\beta_j) - f(\alpha_j)) + 2\epsilon(\beta_j - \alpha_j) \tag{1}$$

for $1 \leqq j \leqq n$. Obviously

$$|f(\alpha_{j+1}) - f(\beta_j)| \geqq (f(\alpha_{j+1}) - f(\beta_j)) \tag{2}$$

for $0 \leqq j \leqq n$. Thus, considering the subdivision

$$P = \{x = \beta_0 \leqq \alpha_1 < \beta_1 \leqq \ldots < \beta_n \leqq \alpha_{n+1} = y\},$$

we add together the $2n + 1$ inequalities (1) and (2) to obtain

$$V(P, f) > f(y) - f(x) + 2\epsilon \sum_{j=1}^{n} (\beta_j - \alpha_j).$$

Since $f(x) \leqq f(y)$ and $V(P, f) \leqq V_x^y f$, this implies (iii).

In case (ii) obtains, we note that (i) obtains for $-f$ and that $V_x^y(-f) = V_x^y f$ so that we need only apply the above paragraph to $-f$ to conclude that (iii) obtains for f. $\quad\square$

Absolute Continuity

There is a type of continuity more restrictive than uniform continuity that is needed for understanding the calculus of the Lebesgue integral.

(3.136) Definition Let $a < b$ in $\mathbb{R}$. A function $f : [a, b] \to \mathbb{C}$ is said to be *absolutely continuous on* $[a, b]$ if for every $\epsilon > 0$ there exists a $\delta > 0$ such that if $\{]a_j, b_j[\}_{j=1}^n$ is a finite *pairwise disjoint* family of subintervals of $[a, b]$ satisfying

(i) $$\sum_{j=1}^{n} (b_j - a_j) < \delta, \quad \text{then} \quad \sum_{j=1}^{n} |f(b_j) - f(a_j)| < \epsilon.$$

Notice that if we were to only insist on (i) for $n = 1$, this would be a definition of uniform continuity on $[a, b]$. Clearly, then, every absolutely continuous function is uniformly continuous. The following theorem, together with (3.132.c) and (3.86), shows that the converse is false.

(3.137) Theorem *If $a < b$ in $\mathbb{R}$ and if $f : [a, b] \to \mathbb{C}$ is absolutely continuous on $[a, b]$, then $V_a^b f < \infty$.*

Proof Let $\delta > 0$ correspond to $\epsilon = 1$ as in (3.136). Let $n \in \mathbb{N}$ be chosen so that $n > (b - a)/\delta$. Subdivide $[a, b]$ as $a = x_0 < x_1 < \ldots < x_n = b$ where $x_k - x_{k-1} = (b - a)/n < \delta$ for $1 \leqq k \leqq n$. From (3.136), it follows that $V_{x_{k-1}}^{x_k} f \leqq 1$ since $V(P, f) < \epsilon = 1$ for every subdivision P of $[x_{k-1}, x_k]$. Thus

$$V_a^b f = \sum_{k-1}^{n} V_{x_{k-1}}^{x_k} f \leqq n < \infty. \quad\square$$

The converse of (3.137) is false even for continuous monotone functions.

(3.138) Example Let ψ be Lebesgue's singular function constructed in terms of Cantor's ternary set P as in (3.92). Then ψ is continuous and nondecreasing, but *not* absolutely continuous on $[0, 1]$. In fact, there is no $\delta > 0$ corresponding to $\epsilon = 1/2$ as in (3.136). Indeed, given $\delta > 0$, fix $n \in \mathbb{N}$ such that $(2/3)^n < \delta$. Let $(J_{n,k})_{k=1}^{2^n}$ be as in (2.81) [here $a_n = 3^{-n}$] and write $J_{n,k} = [c_k, d_k]$. Then

$$\sum_{k=1}^{2^n} (d_k - c_k) = (2/3)^n < \delta$$

while

$$\sum_{k=1}^{2^n} |\psi(d_k) - \psi(c_k)| = 2^n \cdot 2^{-n} = 1 > \frac{1}{2} = \epsilon.$$

There are many absolutely continuous functions, as we shall see later. The following examples give some.

(3.139) Examples (a) If $f : [a, b] \to \mathbb{C}$ satisfies a Lipschitz condition

$$|f(x) - f(y)| \leq M|x - y|$$

for $x, y \in [a, b]$, then it is clear that f is absolutely continuous on $[a, b]$ since we need only take $\delta = \epsilon/M$ in order to fulfill (3.136).

(b) The function $f(x) = \sqrt{x}$ satisfies no such Lipschitz condition on $[0, 1]$ because, given M, we have

$$|f(1/n^2) - f(0)| = 1/n > M|1/n^2 - 0|$$

if $n > M$. However, f is absolutely continuous on $[0, 1]$. [See (4.50) below.]

Exercises

1. Let $[a, b] \subset \mathbb{R}$ and let $BV([a, b])$ be the set of all complex-valued functions f on $[a, b]$ such that $V_a^b f < \infty$. For $f \in BV([a, b])$ define the *variation norm* of f to be the number

$$\|f\|_v = |f(a)| + V_a^b f$$

and let

$$\|f\|_u = \sup\{|f(x)| : x \in [a, b]\}$$

denote, as usual, the uniform norm of f. If $f, g \in BV([a, b])$ and $\alpha \in \mathbb{C}$, then
   (a) $f + g \in BV([a, b])$ and $\|f + g\|_v \leq \|f\|_v + \|g\|_v$,
   (b) $\alpha f \in BV([a, b])$ and $\|\alpha f\|_v = |\alpha| \cdot \|f\|_v$,
   (c) $\|f\|_u \leq \|f\|_v$,
   (d) $\|f\|_v = 0$ implies $f = 0$,
   (e) $fg \in BV([a, b])$ and $\|fg\|_v \leq \|f\|_u \|g\|_v + \|f\|_v \|g\|_u$,
   (f) $1/g \in BV([a, b])$ if and only if there exists some $\delta > 0$ such that $|g(x)| \geq \delta$ for all $x \in [a, b]$ [if so, then $\|1/g\|_v \leq \delta^{-2} \|g\|_v$].
   (g) Defining $\rho(f, g) = \|f - g\|_v$ makes $BV([a, b])$ a *complete* metric space. [Hint: If

a sequence is Cauchy in this metric, then it converges uniformly. The uniform limit is in this space and the sequence converges to it in this metric.]

(h) Find a sequence $(f_n)_{n=1}^{\infty} \subset BV([0, 1])$ such that $\|f_n\|_u \to 0$ and $\|f\|_v \to \infty$.

2. Let $AC([a, b])$ be the set of all complex-valued functions on $[a, b]$ that are absolutely continuous on $[a, b]$. Then $AC([a, b])$ is a closed subalgebra of $BV([a, b])$. Here $BV([a, b])$ is as in Exercise 1 and it has the topology induced by the metric of Exercise 1(g). [Hint: $|f(b_j) - f(a_j)| \leq |(f - f_n)(b_j) - (f - f_n)(a_j)| + |f_n(b_j) - f_n(a_j)|$.]

3. Suppose that a function $f : [a, b] \to \mathbb{C}$ satisfies the definition (3.136) of absolute continuity with "pairwise disjoint" omitted. Then f satisfies a Lipschitz condition as in (3.132.b). [Hints: Let $\delta > 0$ correspond to $\epsilon = 1$ and let $M = 2/\delta$. Assume $|f(x) - f(y)| > M|x - y|$ for some $x < y$ in $[a, b]$. Choose $m, n \in \mathbb{N}$ such that $\delta/2 < m|x - y|/n < \delta$ and write $x = x_0 < \ldots < x_n = y$ where $x_k - x_{k-1} = |x - y|/n$ for all k. Then $n < mM|x - y| < \sum_k m|f(x_k) - f(x_{k-1})| < n$ because $m|x_k - x_{k-1}| < \delta$ for all k.]

4. Let $f : [a, b] \to \mathbb{R}$ be of finite variation on $[a, b]$. Define g on $[a, b]$ by $g(x) = V_a^x f$. Then f is continuous at $x \in [a, b]$ if and only if g is continuous at x.

5. If $f : [a, b] \to \mathbb{C}$ is of finite variation on $[a, b]$, then a Jordan decomposition of f can be chosen such that each f_j is continuous at all points of continuity of f. [Hint: Use the preceding exercise and the proof of (3.133).]

6. If f is absolutely continuous on $[a, b]$, then a Jordan decomposition of f can be chosen so that each f_j is absolutely continuous on $[a, b]$. [Hint: If f is real-valued, show that $g(x) = V_a^x f$ is absolutely continuous.]

7. If $f : [a, b] \to \mathbb{C}$ is continuous and of finite variation on $[a, b]$ and if f is absolutely continuous on $[c, d]$ whenever $a < c < d < b$, then f is absolutely continuous on $[a, b]$. [Compare (3.139.b).]

Equicontinuity

It is often the case in analysis that in order to obtain functions that are solutions to certain differential or integral equations (or for other purposes), we wish to know conditions under which certain sequences of "approximating functions" possess subsequences that converge in some sense or other. We are very often desirous of obtaining uniform convergence or, at least, uniform convergence on compact subsets of the domain. One such condition that is useful is equicontinuity. We proceed to some definitions.

Throughout this short section X denotes an arbitrary topological space, Y denotes a metric space with metric ρ, and $C(X, Y)$ denotes the set of all functions from X into Y that are continuous on X.

(3.140) Definition Let K be a nonvoid subset of X. A sequence $(f_n)_{n=1}^{\infty} \subset C(X, Y)$ is said to be *uniformly convergent on K* to the function $f \in C(X, Y)$ if to each $\epsilon > 0$ corresponds some $N \in \mathbb{N}$ such that $\rho(f(x), f_n(x)) < \epsilon$ for all $x \in K$ whenever $n \geq N$.

That is

$$\lim_{n\to\infty}\left(\sup_{x\in K}\rho(f(x),f_n(x))\right)=0.$$

It is clear that this definition of uniform convergence reduces to that given in (3.103) for the case that $K=X$ and $Y=\mathbb{C}$.

(3.141) Definition Let F be a nonvoid subset of $C(X,Y)$. For a given $x\in X$, the set F is said to be *equicontinuous at* x if to each $\epsilon>0$ corresponds a neighborhood U of x such that

$$\rho(f(t),f(x))<\epsilon$$

whenever $t\in U$ and $f\in F$. [This requires that the same U works for all f.] We say that F is *equicontinuous on* X if F is equicontinuous at each $x\in X$.

(3.142) Example Let $X=[0,1]$, $Y=\mathbb{R}$, and $F=\{f_n\}_{n=1}^\infty$ where $f_n(x)=x^n$. The set F is *not* equicontinuous at $x=1$ because, given any t with $0<t<1$, we have $|f_n(1)-f_n(t)|=1-t^n>1/2$ for all sufficiently large n, and so the condition for equicontinuity of F at 1 fails when $\epsilon=1/2$. However, F is equicontinuous at all other $x\in X$. In fact, given $0\le x<1$ and $\epsilon>0$, choose a with $x<a<1$ and notice that since the series $\sum_{n=1}^\infty na^{n-1}$ converges [Ratio Test], there exists $b\in\mathbb{R}$ with $na^{n-1}\le b$ for all $n\in\mathbb{N}$. Choosing $\delta=\min\{a-x,\epsilon/b\}$, we see that for $0\le t<x+\delta\le a$ and for all $n\in\mathbb{N}$ we have

$$|f_n(t)-f_n(x)|=|t^n-x^n|=\left|(t-x)\cdot\sum_{k=1}^n t^{n-k}x^{k-1}\right|\le|t-x|na^{n-1}<(\epsilon/b)\cdot b=\epsilon.$$

Plainly, the sequence $(f_n)_{n=1}^\infty$ converges uniformly on $[0,a]$ whenever $0<a<1$, but it does *not* converge uniformly on $[0,1]$.

Our next theorem gives sufficient conditions to deduce uniform convergence on all compact subsets from mere pointwise convergence on a dense subset.

(3.143) Theorem *Suppose that the metric space Y is complete and that $(f_n)_{n=1}^\infty$ is an equicontinuous sequence in $C(X,Y)$ that converges at each point of a dense subset D of the topological space X. Then there is a function f in $C(X,Y)$ such that $(f_n)_{n=1}^\infty$ converges to f uniformly on each compact subset K of X.*

Proof Let $x\in X$ and $\epsilon>0$ be given. By equicontinuity there is a neighborhood U_x of x for which

$$\rho(f_n(t),f_n(x))<\epsilon/3\quad\text{for all }t\in U_x\quad\text{and all }n\in\mathbb{N}\qquad(1)$$

Since D is dense, we can find a $t\in U_x\cap D$. Since the sequence $(f_n(t))_{n=1}^\infty$ converges, it is a Cauchy sequence; hence, there is an $N\in\mathbb{N}$ such that

$$\rho(f_m(t),f_n(t))<\epsilon/3\quad\text{for all }\quad m,n\ge N.\qquad(2)$$

Now (1) and (2) show that

$$\rho(f_m(x), f_n(x)) \leqq \rho(f_m(x), f_m(t)) + \rho(f_m(t), f_n(t))$$
$$+ \rho(f_n(t), f_n(x)) < \epsilon$$

for all $m, n \geqq N$. This proves that the sequence $(f_n(x))_{n=1}^{\infty} \subset Y$ is a Cauchy sequence. But Y is complete, so there is an element $f(x) \in Y$ such that $\lim_{n\to\infty} \rho(f(x), f_n(x)) = 0$. Thus, we obtain a function $f: X \to Y$ such that $(f_n)_{n=1}^{\infty}$ converges to f pointwise on X.

To see that f is continuous on X, let $x \in X$ and $\epsilon > 0$ be given. Choose U_x as above so that (1) obtains. For any given $t \in U_x$, choose n so large that both $\rho(f(t), f_n(t)) < \epsilon/3$ and $\rho(f_n(x), f(x)) < \epsilon/3$ and combine these two with (1) to obtain $\rho(f(t), f(x)) < \epsilon$. Thus we have

$$\rho(f(t), f(x)) < \epsilon \quad \text{for all } t \in U_x, \tag{3}$$

and so f is continuous at x.

Finally, for the *uniform* convergence assertion, let K be any compact subset of X and let $\epsilon > 0$ be given. For each $x \in K$ choose a neighborhood U_x of x such that (1) obtains and then reason as in the preceding paragraph to show that (3) also obtains. Since K is compact there exists a finite set $\{x_1, \ldots, x_p\} \subset K$ such that $\{U_{x_1}, \ldots, U_{x_p}\}$ covers K. Now choose N so large that

$$\rho(f(x_j), f_n(x_j)) < \epsilon/2 \quad \text{whenever } n \geqq N, \quad 1 \leqq j \leqq p. \tag{4}$$

Now, given $t \in K$, choose j such that $t \in U_{x_j}$ and infer from (3), (4), and (1) that for all $n \geqq N$ we have

$$\rho(f(t), f_n(t)) \leqq \rho(f(t), f(x_j)) + \rho(f(x_j), f_n(x_j))$$

$$+ \rho(f_n(x_j), f_n(t)) < \epsilon + \epsilon/2 + \epsilon/3 < 2\epsilon.$$

Thus $\rho(f(t), f_n(t)) < 2\epsilon$ for all $t \in K$ whenever $n \geqq N$. This proves that $(f_n)_{n=1}^{\infty}$ converges to f uniformly on K. $\square$

The proof of our next theorem is the classical *diagonal sequence argument*. This theorem is useful in many contexts.

(3.144) Theorem *Let D be a nonvoid countable set and let $(f_n)_{n=1}^{\infty}$ be a sequence of functions from D into the metric space Y having the property that for each $x \in D$ the closure (in Y) of the set $\{f_n(x) : n \in \mathbb{N}\}$ is compact. Then there exists a subsequence $(f_{n_k})_{k=1}^{\infty}$ which converges at each point of D to some point of Y.*

Proof Since D is countable, there is a function $k \to x_k$ from $\mathbb{N}$ onto D. That is, each point of D appears at least once as a term of the sequence $(x_k)_{k=1}^{\infty}$.

Since each sequence in a compact metric space has a convergent subsequence (3.43) and the metric space $\{f_n(x_1) : n \in \mathbb{N}\}^-$ is compact, there are an infinite set $J_1 \subset \mathbb{N}$ and a point $y_1 \in Y$ such that

$$\lim_{\substack{n \to \infty \\ n \in J_1}} \rho(y_1, f_n(x_1)) = 0.$$

After an infinite set $J_{k-1} \subset \mathbb{N}$ has been chosen for some $k > 1$, we observe that the sequence $(f_n(x_k))_{n \in J_{k-1}}$ has a convergent subsequence. That is, there are an infinite set $J_k \subset J_{k-1}$ and a point $y_k \in Y$ such that

$$\lim_{\substack{n \to \infty \\ n \in J_k}} \rho(y_k, f_n(x_k)) = 0. \tag{1}$$

Thus, we inductively obtain a decreasing sequence $(J_k)_{k-1}^{\infty}$ of infinite subsets of $\mathbb{N}$ and a sequence $(y_k)_{k=1}^{\infty} \subset Y$ such that (1) holds good for each $k \in \mathbb{N}$. Now choose any sequence $(n_k)_{k=1}^{\infty}$ such that $n_k < n_{k+1}$ and $n_k \in J_k$ for all $k \in \mathbb{N}$. [For instance, one can list J_k in order as $n_{k.1} < n_{k.2} < n_{k.3} < \cdots$ and choose the "diagonal" $n_k = n_{k.k}$.] Then $(f_{n_k})_{k=1}^{\infty}$ is a subsequence of $(f_n)_{n=1}^{\infty}$. Given any $x_j \in D$, we observe that $\{n_k : k \geq j\} \subset J_j$, so the sequence $(f_{n_k}(x_j))_{k=j}^{\infty}$ is a subsequence of $(f_n(x_j))_{n \in J_j}$, and hence it follows from (1) that $\lim_{k \to \infty} \rho(y_j, f_{n_k}(x_j)) = 0$. Thus, the subsequence $(f_{n_k})_{k=1}^{\infty}$ converges at each point of D. ☐

We are now ready to prove the main theorem of this section.

(3.145) Arzelà-Ascoli Theorem.   *Suppose that the topological space X is separable* and the metric space Y is complete. Let F be a nonvoid equicontinuous subset of $C(X, Y)$ having the property that for each $x \in X$, the closure of the set $\{f(x) : f \in F\}$ is a compact subset of Y. Then each sequence $(f_n)_{n=1}^{\infty} \subset F$ has a subsequence that converges pointwise on X to a function $f \in C(X, Y)$. Moreover, this convergence is uniform on each compact subset of X.*

> **Proof**   Let D be a countable dense subset of X. Given a sequence $(f_n)_{n=1}^{\infty} \subset F$, first apply (3.144) to obtain a subsequence $(f_{n_k})_{k=1}^{\infty}$ that converges at each point of D. Now apply (3.143) to this subsequence to obtain the desired f. ☐

Recall that $C(X)$ denotes the set of all complex-valued bounded continuous functions on X and that $C(X)$ is made a complete metric space via the uniform metric $\| f - g \|_u = \sup\{|f(x) - g(x)| : x \in X\}$.

(3.146) Corollary   *Suppose that X is a compact metric space and that F is a nonvoid subset of $C(X)$. If*
  (i) *F is equicontinuous on X and*
  (ii) *$\sup\{|f(x)| : f \in F\} < \infty$ for each $x \in X$,*
then the closure F^- of F is a compact subset of $C(X)$. In particular, each sequence in F^- has a subsequence that converges uniformly on X to a function in F^-.

---

*X is called *separable* if X has a countable dense subset.

Proof   We first show that F^- is equicontinuous on X. Given $x \in X$ and $\epsilon > 0$, choose a neighborhood U of x such that $|f(t) - f(x)| < \epsilon/3$ for all $f \in F$ and all $t \in U$. For given $g \in F^-$, choose $f \in F$ such that $|f(t) - g(t)| < \epsilon/3$ for all $t \in X$. Then $t \in U$ implies

$$|g(t) - g(x)| \leq |g(t) - f(t)| + |f(t) - f(x)| + |f(x) - g(x)| < \epsilon.$$

Thus F^- is equicontinuous on X. Plainly (ii) holds with F^- in place of F.

Since X is separable [see (3.44) and (3.43)], we can apply (3.145) with $Y = \mathbb{C}$ and F replaced by F^- to see that each sequence in F^- has a subsequence that converges uniformly on X to some $f \in C(X)$. Since F^- is closed, $f \in F^-$. By (3.43), we conclude that F^- is compact. $\square$

The converse of (3.146) is also true, but we relegate its proof to the exercises.

Exercises

1.   Define ϕ on $\mathbb{R}$ by $\phi(x) = \text{dist}(x, \mathbb{Z})$. That is, $\phi(x)$ is the distance from x to the nearest integer to x. Define f_n on $\mathbb{R}$ by $f_n(x) = \phi(2^n x)$. Then each f_n is continuous on $\mathbb{R}$, $\|f_n\|_u = 1/2$ for all n, and the sequence $(f_n)_{n=1}^{\infty}$ has *no* subsequence that converges pointwise on $[0, 1]$. [Hint: Given $n_1 < n_2 < n_3 < \ldots$ in $\mathbb{N}$, define $x \in [0,$

$1]$ by $x = \sum_{j=1}^{\infty} 2^{-j} x_j$ where $x_j = 1$ if $j = n_k + 1$ and k is divisible by 3 and $x_j = 0$ for

all other j. Check that $f_{n_k}(x) \geq 3/7$ if k is divisible by 3 while $f_{n_k}(x) \leq 1/7$ if $k - 1$ is divisible by 3.]

2.   Let α, β, and γ be positive real numbers and let $a < b$ be real numbers. Denote by F the set of all complex-valued functions on $[a, b]$ such that $|f(a)| \leq \gamma$ and $|f(x) - f(t)| \leq \beta |x - t|^{\alpha}$ for all $x, t \in [a, b]$. Then F is a compact subset of $C([a, b])$.

3.   For $n \in \mathbb{N}$, define f_n on $\mathbb{R}$ by $f_n(x) = x - n + 1$ if $n - 1 \leq x \leq n$, $f_n(x) = n + 1 - x$ if $n \leq x \leq n + 1$, and $f_n(x) = 0$ for all other x. Then the set $F = \{f_n : n \in \mathbb{N}\}$ is a uniformly bounded equicontinuous subset of $C(\mathbb{R})$ and the sequence $(f_n)_{n=1}^{\infty} \subset F$ converges to 0 at every $x \in \mathbb{R}$, but it has no subsequence that converges uniformly on $\mathbb{R}$. This shows that we cannot delete "each compact subset of" from (3.145) and have it remain true.

4.   Let X and Y be metric spaces with metrics d and ρ, respectively. A set $F \subset C(X, Y)$ is said to be *uniformly equicontinuous on X* if for each $\epsilon > 0$ there is some $\delta > 0$ such that $\rho(f(x), f(y)) < \epsilon$ whenever $f \in F, x, y \in X$, and $d(x, y) < \delta$.
   (a) If X is compact and the set $F \subset C(X, Y)$ is equicontinuous on X, then F is uniformly equicontinuous on X.
   (b) If X is compact and F is a compact subset of the metric space $C(X)$, then F is uniformly bounded and uniformly equicontinuous on X. [Hint: Given $\epsilon > 0$, show that there is a finite set $\{f_1, \ldots, f_n\} \subset F$ such that each $f \in F$ satisfies $\|f - f_j\|_u < \epsilon/3$ for some j.]
   (c) For $f, g \in C(X, Y)$, define $\sigma(f, g) = \sup\{\rho(f(x), g(x)) : x \in X\}$. Supposing X to

be compact, show that σ is a metric on $C(X, Y)$ and then generalize (b) to subsets F of this space.

(d) Fine a uniformly bounded set $F \subset C(\mathbb{R})$ that is equicontinuous on $\mathbb{R}$, but *not* uniformly equicontinuous on $\mathbb{R}$.

5. Let X be the unit square $[0, 1] \times [0, 1]$ and let $f \in C(X)$. For $0 \leq y \leq 1$ define f_y on $[0, 1]$ by $f_y(x) = f(x, y)$. Then the set $F = \{ f_y : 0 \leq y \leq 1 \}$ is equicontinuous on $[0, 1]$.

4

DIFFERENTIATION

Dini Derivates

A great many elementary facts about derivatives can be proved without the notion of Dini derivates and we shall do so. However, more sophisticated facts, such as Lebesgue's Differentiation Theorem (also proved in this chapter), require some such notion. Thus, we find it convenient to begin with these derivates in order to avoid needless repetition in definition. First, another definition.

(4.1)   Definition Let $]a, b[\subset \mathbb{R}$ and $\phi :]a, b[\to \mathbb{R}$ be given. If $a \leqq c < b$, we define the *right limit inferior* and the *right limit superior* of ϕ at c to be the respective extended real numbers

$$\varliminf_{h \downarrow c} \phi(h) = \lim_{t \downarrow c} (\inf\{\phi(h) : c < h < t\}),$$

$$\varlimsup_{h \downarrow c} \phi(h) = \lim_{t \downarrow c} (\sup\{\phi(h) : c < h < t\}).$$

[Note that these limits exist since they are limits of monotone functions of $t \in]c, b]$.] Similarly, if $a < c \leqq b$, we define the *left limit inferior* and the *left limit superior* of ϕ at c to be

$$\varliminf_{h \uparrow c} \phi(h) = \lim_{t \uparrow c} (\inf\{\phi(h) : t < h < c\}),$$

$$\varlimsup_{h \uparrow c} \phi(h) = \lim_{t \uparrow c} (\sup\{\phi(h) : t < h < c\}).$$

(4.2)   Remarks (a) It is obvious that

$$\varliminf_{h \downarrow c} \phi(h) \leqq \varlimsup_{h \downarrow c} \phi(h)$$

170

and a moment's thought shows that equality obtains if and only if $\phi(c^+)$ exists [cf. (3.87)], in which case $\phi(c^+)$ is the common value. A similar remark applies to $h\uparrow c$ and $\phi(c-)$.

(b) Among all limits of the form $\lim_{n\to\infty}\phi(h_n)$, where $(h_n)_{n=1}^\infty\subset\,]c,b[$, the largest is $\overline{\lim}_{h\downarrow c}\phi(h)$ and the smallest is $\underline{\lim}_{h\downarrow c}\phi(h)$. A similar statement obtains for approaches to c from the left. Compare (2.22).

(4.3) **Example** Define ϕ on $\mathbb{R}$ by $\phi(h) = 1/h$ if h is irrational and $\phi(h) = \text{sgn}(h)$ if h is rational. Then $\phi(0) = 0$,

$$\underline{\lim_{h\downarrow 0}}\,\phi(h) = 1, \qquad \overline{\lim_{h\downarrow 0}}\,\phi(h) = \infty,$$

$$\underline{\lim_{h\uparrow 0}}\,\phi(h) = -\infty, \qquad \overline{\lim_{h\uparrow 0}}\,\phi(h) = -1.$$

(4.4) **Definitions** Let $[a, b] \subset \mathbb{R}$ and $f:[a, b] \to \mathbb{R}$ be given. If $a \leq x < b$, write

$$D_+ f(x) = \underline{\lim_{h\downarrow 0}}\,\frac{f(x+h) - f(x)}{h},$$

$$D^+ f(x) = \overline{\lim_{h\downarrow 0}}\,\frac{f(x+h) - f(x)}{h}.$$

These numbers are called the *lower right derivate* and *upper right derivate* of f at x, respectively. Similarly, for $a < x \leq b$, we define the *lower left derivate* and the *upper left derivate* of f at x to be the respective numbers

$$D_- f(x) = \underline{\lim_{h\uparrow 0}}\,\frac{f(x+h) - f(x)}{h},$$

$$D^- f(x) = \overline{\lim_{h\uparrow 0}}\,\frac{f(x+h) - f(x)}{h}.$$

These four derivates are known as the *Dini derivates* of f at x.
In case $D_+ f(x) = D^+ f(x)$, we write

$$f'_+(x) = D_+ f(x) = D^+ f(x)$$

and call this common value the *right derivative* of f at x. Similarly, we define the *left derivative* of f at x to be

$$f'_-(x) = D_- f(x) = D^- f(x)$$

if the second equality obtains. Thus

$$f'_+ (x) = \lim_{h \downarrow 0} \frac{f(x + h) - f(x)}{h} ,$$

$$f'_- (x) = \lim_{h \uparrow 0} \frac{f(x + h) - f(x)}{h}$$

if these limits exist.

In case $f'_+ (x) = f'_- (x)$ [both extant in $\mathbb{R}^{\#}$], we call the common value the *derivative* of f at x and denote it by $f'(x)$. Thus

$$f'(x) = \lim_{h \to 0} \frac{f(x + h) - f(x)}{h}$$

if the limit exists.

We say that f is *differentiable* at x if $f'(x)$ exists *and is finite*.

We see from the above that, given $f : [a, b] \to \mathbb{R}$, there are several $\mathbb{R}^{\#}$-valued functions associated with f. First, the four Dini derivates: the common domain of $D_+ f$ and $D^+ f$ is $[a, b[$ and the common domain of $D_- f$ and $D^- f$ is $]a, b]$. Second, there are the one-sided derivatives: the domain of f'_+ is

$$\{x : a \leqq x < b, D_+ f(x) = D^+ f(x)\}$$

and that of f'_- is $\{x : a < x \leqq b, D_- f(x) = D^- f(x)\}$. [It might be that these sets are empty.] Third, there is the derivative of f: the domain of f' is the set of points in $]a, b[$ at which all four Dini derivates are equal to one another. If we say that f is *differentiable on* a set $E \subset [a, b]$, we mean that f' exists and is finite at each point of $E \cap]a, b[$ and that the relevant one-sided derivatives exist and are finite at points of $E \cap \{a, b\}$.

The following examples may help clarify these many definitions. Of course these notions apply to any interval of positive length, closed or not.

(4.5)   Examples   (a) Any constant function is differentiable everywhere (in its domain) with derivative 0.

(b) If $f(x) = \text{sgn}(x)$ for $x \in \mathbb{R}$, then $f'(x) = 0$ for $x \neq 0$ and

$$\frac{f(0 + h) - f(0)}{h} = \frac{h/|h| - 0}{h} = 1/|h| \to \infty \quad \text{as } h \to 0,$$

so $f'(0) = \infty$. Thus f is differentiable on $\mathbb{R} \setminus \{0\}$ but not at 0.

(c) If $f(x) = |x|$ for $x \in \mathbb{R}$, then $f'(x) = 1 = f'_+ (0)$ for $x > 0$ and $f'(x) = -1 = f'_- (0)$ for $x < 0$. Thus, f is differentiable on $[0, \infty[$ and on $]-\infty, 0]$, but not on $\mathbb{R}$.

(d) Let $f(x) = |x|$ if x is rational and $f(x) = |2x|$ if x is irrational. Then $D_+ f(0) = 1$, $D^+ f(0) = 2$, $D_- f(0) = -2$, $D^- f(0) = -1$. Thus at a single point the four derivates take four distinct finite values. At all other points one derivate is ∞, one is $-\infty$, and the other two are finite and equal.

(e) Define f on $\mathbb{R}$ by $f(x) = e^x$. For $x, h \in \mathbb{R}$ with $h \neq 0$, we have

$$\frac{f(x+h) - f(x)}{h} = h^{-1}[e^{x+h} - e^x] = e^x h^{-1}[e^h - 1]$$

$$= e^x \sum_{n=1}^{\infty} \frac{h^{n-1}}{n!}$$

$$= e^x \cdot g(h),$$

where g is defined on $\mathbb{R}$ by the power series. Since g is continuous at $h = 0$ by (3.109), we have $g(h) \to g(0) = 1$ as $h \to 0$. Consequently, $f'(x) = e^x$ for all $x \in \mathbb{R}$; i.e., $\exp' = \exp$ on $\mathbb{R}$.

(f) Consider the function log defined on $]0, \infty[$. Fix any $x > 0$. Since $\log t \to \log x$ as $t \to x$ by (3.97.iii), it follows from (e) that

$$\lim_{t \to x} \frac{t - x}{\log t - \log x} = \lim_{t \to x} \frac{e^{\log t} - e^{\log x}}{\log t - \log x} = e^{\log x} = x.$$

This limit is nonzero, and so

$$\lim_{t \to x} \frac{\log t - \log x}{t - x} = \frac{1}{x}$$

[take $h = t - x$ if necessary]. Therefore, $\log' x = 1/x$ for all $x > 0$.

Examples (c) and (d) show that a point of continuity need not be a point of differentiability. However, continuity is necessary for differentiability.

(4.6) **Theorem** *Let $f : [a, b] \to \mathbb{R}$ be given. If $x \in [a, b]$ and f is differentiable at x, then f is continuous at x.*

Proof For $x \neq t = x + h \in [a, b]$, we have

$$f(t) - f(x) = \frac{f(t) - f(x)}{t - x} \cdot (t - x) \to f'(x) \cdot 0 = 0 \qquad (1)$$

as $t \to x$. Thus

$$\lim_{t \to x} f(t) = f(x). \qquad \square$$

(4.7) **Remark** It can happen that $f'(x)$ exists [$= \infty$ or $-\infty$] and yet f is discontinuous at x [see (4.5.b)]. In such cases, the proof of (4.6) breaks down because the theorem that

$$\lim(u \cdot v) = (\lim u)(\lim v)$$

is only assured if the two limits on the right are *finite*; hence, it is the arrow that is incorrect in (4.6.1) and *not* the last equality.

A Nowhere Differentiable,
Everywhere Continuous, Function

The reader may have the impression that a continuous function must be differentiable except at isolated "sharp corners." This is emphatically not the case. We next construct an example of a function $f: \mathbb{R} \to \mathbb{R}$ that is continuous at *all* points of $\mathbb{R}$ and is differentiable at *no* point of $\mathbb{R}$. Our construction is essentially that given by van derWaerden (1930). The first publication of such an example was by Karl Weierstrass (1872), although there seems to be good evidence to believe that examples were known to Bolzano as early as 1830.

(4.8)   **Example**   Let ϕ be the function on $\mathbb{R}$ such that

$$\phi(x) = |x| \qquad \text{if } |x| \leq 2,$$

$$\phi(x + 4p) = \phi(x) \quad \text{if } x \in \mathbb{R} \text{ and } p \in \mathbb{Z}. \tag{1}$$

It is obvious that ϕ is continuous on $\mathbb{R}$: in fact, $\phi(x) = \text{dist}(x, A)$ where $A = \{4m : m \in \mathbb{Z}\}$. It is also obvious that

$$|\phi(s) - \phi(t)| = |s - t| \tag{2}$$

whenever $s, t \in \mathbb{R}$ are such that there is no even integer in the open interval having endpoints s and t. For $n \in \mathbb{N}$ and $x \in \mathbb{R}$, define $f_n(x) = 4^{-n}\phi(4^n x)$ and

$$f(x) = \sum_{n=1}^{\infty} f_n(x).$$

Since each f_n is continuous on $\mathbb{R}$ and $0 \leq f_n(x) \leq 2/4^n$ for all x, it follows from the Weierstrass M-test that f is continuous at each point of $\mathbb{R}$. To see that f fails to have a finite derivative anywhere on $\mathbb{R}$, let $a \in \mathbb{R}$ be arbitrary and fixed. For each $k \in \mathbb{N}$, put $\epsilon_k = 1$ or $\epsilon_k = -1$ [depending on k] so that there is no even integer in the open interval between $4^k a$ and $4^k a + \epsilon_k$ [in the ambiguous case that $4^k a$ is an integer, either choice will do]. If $1 \leq n \leq k$ are integers, then it is clear that there is no even integer between $4^n a$ and $4^n a + 4^{n-k}\epsilon_k$ [if $2p$ were between them, then $4^{k-n} \cdot 2p$ would be between $4^k a$ and $4^k a + \epsilon_k$], and so (2) implies that

$$\left| f_n(a + 4^{-k}\epsilon_k) - f_n(a) \right|$$

$$= 4^{-n}\left| \phi(4^n a + 4^{n-k}\epsilon_k) - \phi(4^n a) \right| = 4^{-k} \qquad (1 \leq n \leq k). \tag{3}$$

On the other hand, it is clear from (1) and the first equality in (3) that

$$f_n(a + 4^{-k}\epsilon_k) = f_n(a) \qquad (1 \leq k < n). \tag{4}$$

Writing $h_k = 4^{-k}\epsilon_k$ and applying (3) and (4), we have that

$$\frac{f(a + h_k) - f(a)}{h_k} = \sum_{n=1}^{k} \frac{f_n(a + h_k) - f_n(a)}{h_k} = \sum_{n=1}^{k} (\pm 1) \tag{5}$$

is an integer which is even if k is even and odd if k is odd. Thus the left side of (5) cannot have a finite limit as $k \to \infty$. Since $h_k \to 0$, this proves that f is not differentiable at a.

Some Elementary Formulas

Now we present a collection of formulas for differentiation that are familiar to every student of elementary calculus. Perhaps their proofs are not as familiar.

(4.9) **Theorem** *Let f and g be real-valued functions on an interval $I \subset \mathbb{R}$ that are differentiable at a point $c \in I$. Let $\alpha \in \mathbb{R}$. Then the functions αf, $f + g$, fg, and (provided $g(c) \neq 0$) f/g are all differentiable at c. Moreover,*

- (i) $(\alpha f)'(c) = \alpha f'(c)$,
- (ii) $(f + g)'(c) = f'(c) + g'(c)$,
- (iii) $(fg)'(c) = f(c)g'(c) + g(c)f'(c)$,
- (iv) $\left(\dfrac{f}{g}\right)'(c) = \dfrac{g(c)f'(c) - f(c)g'(c)}{[g(c)]^2}$ *if $g(c) \neq 0$.*

Proof This all follows easily from (3.77), (4.6), and the identities

$$\frac{(\alpha f)(c + h) - (\alpha f)(c)}{h} = \alpha \cdot \frac{f(c + h) - f(c)}{h},$$

$$\frac{(f + g)(c + h) - (f + g)(c)}{h} = \frac{f(c + h) - f(c)}{h} + \frac{g(c + h) - g(c)}{h},$$

$$\frac{(fg)(c + h) - (fg)(c)}{h} = f(c + h) \cdot \frac{g(c + h) - g(c)}{h} + g(c)\frac{f(c + h) - f(c)}{h},$$

$$\frac{(f/g)(c + h) - (f/g)(c)}{h}$$

$$= \frac{1}{g(c + h)g(c)} \left\{ g(c)\left(\frac{f(c + h) - f(c)}{h} \right) - f(c)\left(\frac{g(c + h) - g(c)}{h} \right) \right\},$$

which are valid for all $h \neq 0$ such that $c + h \in I$. [For the last one, we have $\lim_{h \to 0} g(c + h) = g(c) \neq 0$ so we may suppose I is so small that $g(c + h) \neq 0$ if $c + h \in I$.] □

(4.10) **Chain Rule** *Let I and J be intervals of $\mathbb{R}$ having positive length. Suppose that $f : I \to J$ is differentiable at some $c \in I$ and that $g : J \to \mathbb{R}$ is differentiable at $f(c)$. Then the composite function $g \circ f : I \to \mathbb{R}$ defined by $g \circ f(x) = g(f(x))$ is differentiable at c and*

- (i) $(g \circ f)'(c) = g'(f(c))f'(c)$.

Proof Write $d = f(c)$. Define functions $\alpha : I \to \mathbb{R}$ and $\beta : J \to \mathbb{R}$ by requiring that $\alpha(c) = 0$, $\beta(d) = 0$,

$$f(x) - f(c) = (x - c)[f'(c) + \alpha(x)],$$
$$g(y) - g(d) = (y - d)[g'(d) + \beta(y)]$$

for $x \in I$, $x \neq c$, $y \in J$, $y \neq d$. By the definition of derivative, $\alpha(x) \to 0$ as $x \to c$ and $\beta(y) \to 0$ as $y \to d$. Making the substitution $y = f(x)$, we have

$$g \circ f(x) - g \circ f(c) = g(y) - g(d)$$
$$= \{f(x) - f(c)\}[g'\{f(c)\} + \beta\{f(x)\}]$$
$$= (x - c)[f'(c) + \alpha(x)][g'\{f(c)\} + \beta\{f(x)\}].$$

Thus $x \in I \setminus \{c\}$ implies

$$\qquad \frac{g \circ f(x) - g \circ f(c)}{x - c} = [g'(f(c)) + \beta(f(x))][f'(c) + \alpha(x)]. \qquad (1)$$

Letting $x \to c$, we have $\alpha(x) \to 0$, $f(x) \to f(c) = d$ (4.6), and $\beta(f(x)) \to 0$; hence, (i) follows from (1). □

(4.11) Examples (a) For $n \in \mathbb{N}$, define f_n on $\mathbb{R}$ by $f_n(x) = x^n$. We claim that $f_n'(x) = nx^{n-1}$ for all $x \in \mathbb{R}$ and $n \in \mathbb{N}$. This is obvious for $n = 1$ because $[f_1(x + h) - f_1(x)]/h = 1 = 1 \cdot x^0$ for all $h \neq 0$ and all x. If the claim obtains for some $n \geq 1$, then we use (4.9.iii) to write

$$f_{n+1}'(x) = (f_n f_1)'(x) = x^n \cdot 1 + x \cdot nx^{n-1} = (n + 1)x^n$$

for all x. Thus, induction proves the claim.

(b) For an integer $n < 0$, define g_n on $\mathbb{R} \setminus \{0\}$ by $g_n(x) = x^n$. With f_{-n} as in (a), we can use (a) and (4.9.iv) to write

$$g_n'(x) = (1/f_{-n})'(x) = \frac{x^{-n} \cdot 0 - 1 \cdot (-nx^{-n-1})}{(x^{-n})^2} = nx^{n-1}$$

for all $x \in \mathbb{R} \setminus \{0\}$.

(c) Let $b \in \mathbb{R}$ and define f on $]0, \infty[$ by $f(x) = x^b = \exp(b \log x)$. Using (4.10), (4.9.i), (4.5.e), and (4.5.f), we have

$$f'(x) = \exp(b \log x) \cdot b/x$$
$$= b \cdot \exp((b - 1)\log x) = bx^{b-1}$$

for all $x > 0$.

(d) Let $a > 0$ be a real number and define f on $\mathbb{R}$ by $f(x) = a^x = \exp(x \cdot \log a)$. Using (4.5.e), (4.9.i), and (a), we have

$$f'(x) = \exp(x \log a) \cdot \log a = a^x \log a$$

for all $x \in \mathbb{R}$.

(e) For $x > 0$, write $f(x) = x^x$. Then $f(x) = \exp(x \cdot \log x)$, so

$$f'(x) = \exp(x \log x)\big[x \cdot 1/x + (\log x) \cdot 1 \big] = x^x\big[1 + \log x \big]$$

for all $x > 0$.

(f) Let f be any rational function over $\mathbb{R}$: i.e., $f(x) = P(x)/Q(x)$ for $x \in \mathbb{R}$ such that $Q(x) \neq 0$, where P and Q are polynomials with real coefficients and $Q(x) \neq 0$ for some x. It follows from (a) and (4.9) that f is differentiable at each $x \in \mathbb{R}$ such that $Q(x) \neq 0$.

Local Extrema

(4.12) **Definition** Let X be a topological space, let $c \in X$, and let f be a real-valued function on X. We say that f has a *local maximum* [resp., *local minimum*] at c if there exists a neighborhood U of c such that $f(x) \leq f(c)$ [resp., $f(x) \geq f(c)$] for all $x \in U$. If f has either a local maximum or a local minimum at c, we say that f has a *local extremum at c*.

The next theorem gives a necessary, but not sufficient, condition that a local extremum exists at a point.

(4.13) **Theorem** *Let $a < c < b$ in $\mathbb{R}$ and $f :]a,b[\to \mathbb{R}$ be given. If f has a local extremum at c and $f'(c)$ exists in $\mathbb{R}^{\#}$, then $f'(c) = 0$.*

Proof We suppose that f has a local maximum at c [otherwise consider $-f$]. By (4.12), there exists a $\delta > 0$ such that $a \leq c - \delta < c + \delta \leq b$ and $f(x) \leq f(c)$ whenever $|x - c| < \delta$. Thus, the difference quotient

$$q(x) = (f(x) - f(c))/(x - c)$$

satisfies

$$q(x) \geq 0 \quad \text{if } c - \delta < x < c,$$

$$q(x) \leq 0 \quad \text{if } c < x < c + \delta.$$

Since $q(x)$ has the limit $f'(c)$ as $x \to c$, it follows that $f'(c) = 0$. $\square$

(4.14) **Remarks** (a) It is important that c is *not* an endpoint of $[a,b]$. For instance, the function defined by $f(x) = \sqrt{x}$ on $[0,1]$ has a local minimum at 0, a local maximum at 1, $f'_+(0) = \infty$, and $f'_-(1) = 1/2$—neither is 0.

(b) The function given by $f(x) = x^3$ on $]-1,1[$ satisfies $f'(0) = 0$ but does not have a local extremum at 0.

(c) Theorem (4.13) assures that if we are seeking all local extrema of a *differentiable* function on a *open* interval, then we need only consider, as *candidates*, those c for which $f'(c) = 0$.

(d) If $f(x) = |x|$ for $x \in \mathbb{R}$, then f has a local minimum at $c = 0$, but $f'(0)$ does not exist.

Mean Value Theorems

The theorems in this section are extremely important. They lie at the heart of unbelievably many analytical proofs, and so are essential tools for every analyst. The first one is "geometrically obvious," but it is the key to untold "nonobvious" corollaries. It is interesting to trace its proof back to the axioms of Chapter 1.

(4.15) **Rolle's Theorem** *Let $a < b$ in $\mathbb{R}$ and $f:[a,b] \to \mathbb{R}$ be given. Suppose that f is continuous on $[a,b]$, f is differentiable on $]a,b[$, and $f(b) = f(a)$. Then there exists a real number ξ such that $a < \xi < b$ and $f'(\xi) = 0$.*

> **Proof** If $f(x) = f(a)$ for all $x \in [a,b]$, then any $\xi \in]a,b[$ will suffice. Thus, we suppose that $f(x_0) > f(a)$ for some $x_0 \in [a,b]$ [if $f(x) \leq f(a)$ for all $x \in [a,b]$, consider $-f$ in place of f]. According to (3.81), there is some $\xi \in [a,b]$ such that $f(x) \leq f(\xi)$ for all $x \in [a,b]$. Then $f(\xi) \geq f(x_0) > f(a) = f(b)$; whence $a < \xi < b$. Since f has a local maximum on $]a,b[$ at ξ and $f'(\xi)$ exists, it follows from (4.13) that $f'(\xi) = 0$. $\square$

(4.16) **Generalized Mean Value Theorem** [Cauchy] *Let $a < b$ be in $\mathbb{R}$ and let f and g be continuous real-valued functions on $[a,b]$ that are both differentiable on $]a,b[$. Then there exists $\xi \in]a,b[$ such that*

(i) $$\left[f(b) - f(a) \right] g'(\xi) = \left[g(b) - g(a) \right] f'(\xi).$$

> **Proof** Define a function ϕ on $[a,b]$ by
>
> $$\phi(x) = \left[f(b) - f(a) \right] g(x) - \left[g(b) - g(a) \right] f(x). \tag{1}$$
>
> One checks that ϕ satisfies the hypotheses of (4.15), and so $\phi'(\xi) = 0$ for some $a < \xi < b$. Now differentiate (1) and then substitute $x = \xi$ to obtain (i). $\square$

(4.17) **Mean Value Theorem** [Lagrange] *Let $a < b$ be in $\mathbb{R}$ and let $f:[a,b] \to \mathbb{R}$ be continuous on $[a,b]$ and differentiable on $]a,b[$. Then there exists some $\xi \in]a,b[$ such that*

(i) $$f(b) - f(a) = (b - a) f'(\xi).$$

> **Proof** Write $g(x) = x$ for all $x \in [a,b]$ and apply (4.16). $\square$

As an immediate application of (4.17), we deduce the following important facts.

(4.18) **Corollary** *Let f be as in (4.17). Then*
(i) *$f'(x) = 0$ whenever $a < x < b$ implies f is a constant on $[a,b]$;*
(ii) *$f'(x) \geq 0$ whenever $a < x < b$ implies f is nondecreasing on $[a,b]$;*

(iii) $f'(x) \leq 0$ *whenever* $a < x < b$ *implies* f *is nonincreasing on* $[a,b]$;

(iv) $f'(x) > 0$ *whenever* $a < x < b$ *implies* f *is strictly increasing on* $[a,b]$;

(v) $f'(x) < 0$ *whenever* $a < x < b$ *implies* f *is strictly decreasing on* $[a,b]$.

Proof For arbitrary $c < d$ in $[a,b]$, simply apply (4.17) to the restriction of f to $[c,d]$. For example, if $f'(x) \geq 0$ for all $x \in {]a,b[}$ and if $c < d$ are given, then

$$f(d) - f(c) = (d - c)f'(\xi) \geq 0$$

for some $\xi \in {]c,d[}$ so $f(c) \leq f(d)$, which proves (ii). □

(4.19) **Remark** Let f be as in (4.17). If we define *the tangent line to f at* $(\xi, f(\xi))$ to be the line through this point having slope $f'(\xi)$, then all that (4.17) says is that there is some $\xi \in {]a,b[}$ such that this tangent line is parallel to the line through $(a, f(a))$ and $(b, f(b))$. A geometrical interpretation of (4.16) seems remote.

(4.20) **Example** The results in (4.18) are frequently useful in proving inequalities whose truth might otherwise be difficult to establish. As an example, we prove the following generalizations of *Bernoulli's Inequality* (1.46). Let $\alpha \in \mathbb{R}$, $0 \neq \alpha \neq 1$, be fixed. Then $x \in \mathbb{R}, x > -1, x \neq 0$ implies

(a) $\qquad\qquad (1 + x)^{\alpha} > 1 + \alpha x \quad$ if $\alpha < 0$ or $\alpha > 1$,

(b) $\qquad\qquad (1 + x)^{\alpha} < 1 + \alpha x \quad$ if $0 < \alpha < 1$.

In fact, write $f(x) = (1 + x)^{\alpha} - (1 + \alpha x)$ for $x > -1$. Using (4.11.c), we have

$$f'(x) = \alpha(1 + x)^{\alpha - 1} - \alpha, \qquad\qquad (1)$$

$$f''(x) = \alpha(\alpha - 1)(1 + x)^{\alpha - 2} \qquad\qquad (2)$$

where f'' is the derivative of f'. Recall that $a^b = \exp(b \log a) > 0$ for all real a and b with $a > 0$. From (4.18) and (2), we infer that f' is strictly monotone on $]-1, \infty[$. Note also that $f'(0) = 0$. Suppose $\alpha < 0$ or $\alpha > 1$. Then f' is increasing: $f' < 0$ on $]-1, 0[$; $f' > 0$ on $]0, \infty[$. By (4.18), f is decreasing on $]-1, 0[$ and increasing on $]0, \infty[$. Since $f(0) = 0$, it follows that $f(x) > 0$ if $-1 < x \neq 0$, and so (a) obtains. In case $0 < \alpha < 1$, similar reasoning yields (b).

L' Hospital's Rule

Our main purpose in this section is to prove a theorem which is frequently useful in evaluating limits which "lead to" one of the "indeterminate forms"

$$0/0, \quad \infty/\infty, \quad \infty - \infty, \quad 1^{\infty}, \quad \infty^0, \quad 0^0, \quad 0 \cdot \infty.$$

The first five of these symbols have not been [and will not be] assigned any meaning. The last two have been defined $[0^0 = 1, 0 \cdot \infty = 0]$, but incorrect answers can result from combining these perfectly legitimate definitions with false "rules."

In the case of 0^0, this is because the function $f(x, y) = x^y$ $(x > 0 : y \in \mathbb{R})$ does *not* have limit 1 as $(x, y) \rightarrow (0, 0)$. In the case of $0 \cdot \infty$, the reason is that the theorem (3.77) that the limit of a product equals the product of the limits requires that the limits of the two factors be complex numbers [not ∞ or $-\infty$]. We illustrate with some examples.

(4.21)   Example   (a) Let $a \in \mathbb{R}$ and write $f(x) = x^{a/\log x}$ for $x > 0$. If one takes limits as $x \downarrow 0$ of base and exponent separately, one obtains the "answer"

$$\lim_{x \downarrow 0} f(x) = 0^0 = 1, \tag{1}$$

which is incorrect unless $a = 0$. Writing $f(x) = \exp[(a/\log x)\log x] = e^a$, we see that the correct value of this limit is e^a. It is the first equality in (1) which is false, not the second.

   (b) If we apply the erroneous "rule" on the limit of a product mentioned above, we obtain, for $a > 0$,

$$\lim_{x \downarrow 0} (e^x - 1) \cdot a/(x^2 + x) = 0 \cdot \infty = 0. \tag{2}$$

The first equality in (2) is false, the second is not. In fact, by (4.5.e) and (3.77), we have

$$\lim_{x \downarrow 0} (e^x - 1) \cdot \frac{a}{x^2 + x} = \left\{ \lim_{x \downarrow 0} \frac{e^x - 1}{x} \right\} \left\{ \lim_{x \downarrow 0} \frac{a}{x + 1} \right\} = 1 \cdot a = a.$$

   (c) By (4.5.f) and the definition of derivative, we have

$$\lim_{h \to 0} h^{-1} \cdot \log(1 + h) = \log' 1 = 1.$$

Thus, for $a \in \mathbb{R}$, $a \neq 0$, we can write $h = a/x$ and use the continuity of exp to obtain

$$\lim_{x \to \infty} (1 + a/x)^x = \lim_{h \to 0} \exp \left[ah^{-1} \log(1 + h) \right] = e^a.$$

A specious attempt to evaluate this limit by finding the limit of base and exponent separately leads to the meaningless expression 1^∞.

   Examples (b) and (c) show that derivatives can often be useful in evaluating limits. This vague idea is clarified by the following theorem which, in special cases, dates back to a book published by Guillaume l'Hospital in 1696.

(4.22)   L'Hospital's Rule   *Let $a < b$ be in $\mathbb{R}^\#$ and let f and g be real-valued functions that are differentiable on $]a, b[$ such that $g'(x) \neq 0$ for all $x \in]a, b[$. Suppose that either*

   (i) $$\lim_{x \downarrow a} f(x) = \lim_{x \downarrow a} g(x) = 0$$

or

   (ii) $$\lim_{x \downarrow a} g(x) = \infty.$$

Suppose also that

(iii)
$$\lim_{x\downarrow a}\left[\,f'(x)/g'(x)\right] = L \in \mathbb{R}^{\#}.$$

Then

(iv)
$$\lim_{x\downarrow a}\left[\,f(x)/g(x)\right] = L.$$

The same statement is true if "$x\downarrow a$" is replaced by "$x\uparrow b$" and/or "∞" is replaced by "$-\infty$".

Proof Since g' is never 0 on $]a,b[$, the Intermediate Value Theorem and Rolle's Theorem imply that

$$g \text{ is strictly monotone on }]a,b[. \tag{1}$$

Suppose that $L > -\infty$. Let $\alpha < L$ be arbitrary and choose any β with $\alpha < \beta < L$. By (iii) and (1), there exists some c with $a < c < b$ such that

$$f'(t)/g'(t) > \beta, \qquad g(t) \neq 0 \quad \text{if } a < t < c. \tag{2}$$

Given $a < x < y < c$, (4.16) provides some ξ with $x < \xi < y$ such that $[f(y) - f(x)]g'(\xi) = [g(y) - g(x)]f'(\xi)$. Since $g'(\xi) \neq 0$ and, by (1), $g(x) \neq g(y)$, we can divide and then use (2) with $t = \xi$ to get

$$\frac{f(y) - f(x)}{g(y) - g(x)} > \beta \quad \text{if } a < x < y < c. \tag{3}$$

If (i) obtains, then, since $g(y) \neq 0$ by (2), we can let $x\downarrow a$ in (3) and conclude that

$$f(y)/g(y) \geq \beta > \alpha \quad \text{if } a < y < c;$$

whence follows

$$\lim_{x\downarrow a}\left[\,f(x)/g(x)\right] \geq \beta > \alpha. \tag{4}$$

Suppose, on the other hand, that (ii) obtains. Then (1) and (2) show that $a < x < y < c$ implies that $g(x) > g(y) > 0$. If we multiply (3) by $[g(x) - g(y)]/g(x)$ and rearrange, we obtain

$$f(x)/g(x) > \beta\{1 - g(y)/g(x)\} + f(y)/g(x). \tag{5}$$

Fixing y and letting $x\downarrow a$ in (5), we find that (4) holds in this case also. Since α was arbitrary, it follows that, in either case,

$$\lim_{x\downarrow a}\left[\,f(x)/g(x)\right] \geq L. \tag{6}$$

If $L = \infty$, this completes the proof.

Supposing $L < \infty$, an argument similar to the above [let $L < \beta < \alpha$ etc.] shows that

$$\overline{\lim_{x\downarrow a}}\left[\,f(x)/g(x)\right] \leq L. \tag{7}$$

If $L = -\infty$, (iv) follows from (7). If L is finite, (iv) is obtained by combining (6) and (7). $\square$

(4.23) **Examples** (a) To evaluate the limit

$$\lim_{x \to \infty} (e^x + x)^{1/x}, \tag{1}$$

write $(e^x + x)^{1/x} = \exp[(1/x)\log(e^x + x)]$, $f(x) = \log(e^x + x)$, and $g(x) = x$. Then $g(x) \to \infty$ and

$$\frac{f'(x)}{g'(x)} = \frac{e^x + 1}{e^x + x} \to 1 \quad \text{as } x \to \infty,$$

so (4.22) and the continuity of exp at 1 show that the limit in (1) is e. Taking limits in base and exponent separately in (1) yields ∞^0, which is meaningless.

(b) Consider

$$\lim_{x \to 1} \left[x/(x-1) - 1/(\log x) \right]. \tag{2}$$

If we take limits separately, we are led to $\infty - \infty$, which means nothing. If we combine fractions first, we obtain

$$\left[x \log x - x + 1 \right] / \left[(x-1)\log x \right]$$

which is a $0/0$ form at $x = 1$. Differentiating above and below, we get

$$\log x / \left[1 - x^{-1} + \log x \right],$$

another $0/0$ form. Another differentiation above and below gives

$$x^{-1} / \left[x^{-2} + x^{-1} \right]$$

which has limit $1/2$ as $x \to 1$. Thus, two applications of (4.22) show that the limit in (2) is $1/2$.

(c) If one attempts to evaluate

$$\lim_{x \downarrow 0} \left[(\log x)/x \right] \tag{3}$$

by using L'Hospital's Rule, one might write

$$\lim_{x \downarrow 0} \frac{\log x}{x} = \lim_{x \downarrow 0} \frac{1/x}{1} = \infty.$$

This is far from being correct, since the actual value of the limit in (3) is $-\infty$. *Moral*: Don't apply the conclusion of a theorem unless its hypotheses are satisfied.

Exercises

1.   If $I \subset \mathbb{R}$ is a nonvoid open interval and if $f : I \to \mathbb{R}$ is continuous on I and has a local extremum at *no* point of I, then f is strictly monotone on I.

2.   (a) If $f : \mathbb{R} \to \mathbb{R}$ has a local maximum at each point of $\mathbb{R}$, then the set $f(\mathbb{R})$ is countable. [Hint: $f(\mathbb{R}) \subset \{\sup f(I) : I$ is an interval having rational endpoints$\}$.]
       (b) Give an example of an f as in (a) such that there is no open interval containing 0 on which f is either bounded or monotone.

3. There exist real polynomials f and g such that $g(0) \neq g(1)$ and

$$\frac{f(1) - f(0)}{g(1) - g(0)} \neq \frac{f'(\xi)}{g'(\xi)}$$

for every $\xi \in]0, 1[$.

4. Suppose that f satisfies the hypotheses of the Mean Value Theorem on $[a, b]$. If $\lim_{x \downarrow a} f'(x) = A$, then $f'_+(a)$ exists and is equal to A. [A may be finite or infinite.]

5. Let $a \in \mathbb{R}$ with $0 < a \neq 1$. Define f on $\mathbb{R}$ by $f(0) = \log a$ and $f(x) = (a^x - 1)/x$ if $x \neq 0$. Then f is differentiable and strictly increasing on $\mathbb{R}$.

6. (a) For $0 < x \neq 1$, we have

$$(x - 1)/x < \log x < x - 1.$$

   (b) For $j \in \mathbb{N}, j > 1$, we have

$$\log[(j + 1)/j] < 1/j < \log[j/(j - 1)].$$

   (c) For $n, k \in \mathbb{N}, n > 1$, we have

$$\log(k + 1/n) < \sum_{j=n}^{kn} \frac{1}{j} < \log\left(k + \frac{k}{n-1}\right)$$

   and

$$\lim_{n \to \infty} \sum_{j=n}^{kn} \frac{1}{j} = \log k.$$

   (d) Use (c) to prove that

$$\sum_{j=1}^{\infty} \frac{(-1)^{j+1}}{j} = \log 2.$$

   [Hint:

$$\sum_{j=1}^{2n} \frac{(-1)^{j+1}}{j} = \sum_{j=1}^{2n} \frac{1}{j} - 2 \sum_{k=1}^{n} \frac{1}{2k} .]$$

7. (a) If $0 < \alpha < 1$ and $0 \leq s \leq t < \infty$, then

$$s^\alpha t^{1-\alpha} \leq \alpha s + (1 - \alpha)t$$

   and equality obtains if and only if $s = t$. [Hint: Minimize $f(x) = \alpha x - x^\alpha$ to prove that $f(x) > f(1)$ if $0 < x \neq 1$, then put $x = s/t$.]

   (b) Let $(\alpha_j)_{j=1}^n$ and $(x_j)_{j=1}^n$ be finite sequences of positive real numbers such that $\sum_{n=1}^{n} \alpha_j = 1$. Then

   (i)

$$\prod_{j=1}^{n} x_j^{\alpha_j} \leq \sum_{j=1}^{n} \alpha_j x_j$$

   and equality obtains if and only if $x_1 = x_2 = \cdots = x_n$. [Hint: Prove that $\log x < x - 1$ for $0 < x \neq 1$, write A for the right side of (i), replace x by x_j/A, multiply by α_j, and sum.]

   (c) What does inequality (i) say if $\alpha_j = 1/n$ for all j?

8. For $a \in \mathbb{R}, a \neq 0$, the function f defined on $]|a|, \infty[$ by

$$f(x) = (1 - a/x)^x$$

is strictly increasing and $\lim_{x \to \infty} f(x) = e^{-a}$.

9. If $\alpha > 0$, then

$$n^{\alpha+1}/(\alpha + 1) < 1^\alpha + 2^\alpha + 3^\alpha + \ldots + n^\alpha < (n + 1)^{\alpha+1}/(\alpha + 1).$$

[Hint: Use Bernoulli's Inequality to write

$$(1 + 1/k)^{1+\alpha} > 1 + (1 + \alpha)/k \quad \text{and} \quad (1 - 1/k)^{1+\alpha} > 1 - (1 + \alpha)/k.$$

Multiply by $k^{1+\alpha}$, solve each for k^α, then sum for $k = 1, 2, \ldots, n$.]

10. If $a_1, a_2, \ldots, a_n$ are positive real numbers and p is any nonzero real number, then the *mean of order p* is defined as

$$M_p = \left\{ \frac{a_1^p + a_2^p + \ldots + a_n^p}{n} \right\}^{1/p}.$$

Also define

$$M_0 = (a_1 a_2 \ldots a_n)^{1/n}.$$

Some of these means have special names: M_1 is the *arithmetic mean*, M_2 is the *quadratic mean*, M_0 is the *geometric mean*, and M_{-1} is the *harmonic mean*.

(a) If $p < q$ are real numbers, then $M_p \leq M_q$ and equality obtains if and only if $a_1 = a_2 = \cdots = a_n$. [Hints: For $p \leq 0 \leq q$, use (1.48) to prove that $M_p \leq M_0 \leq M_q$. Otherwise, let $q/p = \alpha > 0$ and $b_j = (a_j/M_p)^p$. Check that $b_1 + b_2 + \ldots + b_n = n$ and $(M_q/M_p)^q = (b_1^\alpha + b_2^\alpha + \cdots + b_n^\alpha)/n$. Let $b_j = 1 + x_j$ and then obtain $b_j^\alpha \geq 1 + \alpha x_j$ if $\alpha > 1$ or $b_j^\alpha \leq 1 + \alpha x_j$ if $\alpha < 1$. Summing on j gives that $b_1^\alpha + b_2^\alpha + \ldots + b_n^\alpha$ is $\geq n$ or $\leq n$ according as $\alpha > 1$ or $\alpha < 1$.]

(b) $$\lim_{p \to -\infty} M_p = \min\{a_1, a_2, \ldots, a_n\}.$$

[Hint: With $a_1 \leq a_2 \leq \ldots \leq a_n$ and $p < 0$, we have $na_1^p \geq a_1^p + a_2^p + \ldots + a_n^p > a_1^p$.]

(c) $\lim_{q \to \infty} M_q = \max\{a_1, a_2, \ldots, a_n\}$.

11. (a) If $1 < x \neq e$, then there is a unique number $f(x) > 0$ such that $f(x) \neq x$ and $x^{f(x)} = [f(x)]^x$. [Hint: $x^y = y^x$ if and only if $(\log x)/x = (\log y)/y$.]

(b) Sketch a graph of f.

(c) If m and n are two natural numbers such that $m < n$ and $m^n = n^m$, then $m = 2$ and $n = 4$.

12. Let $a > 0$ and $x > 0$. Then

(a) $(ex/a)^a \leq e^x$
with equality if and only if $x = a$,

(b) $e \log x \leq ax^{1/a}$
with equality if and only if $x = e^a$,

(c) $-ae \log x \leq x^{-a}$
with equality if and only if $x = e^{-1/a}$.

[Hint: Find minima for certain functions of x.]

13.   For $n \in \mathbb{N}$ and $0 < x < n + 1$ we have

$$\sum_{k=0}^{n} \frac{x^k}{k!} < e^x < \sum_{k=0}^{n} \frac{x^k}{k!} + \frac{x^{n+1}}{n!(n+1-x)} .$$

[Hint: Multiply by e^{-x} and differentiate.]

14.   (a) If $x > 0$, then

$$2/(2x + 1) < \log(1 + 1/x) < 1/\sqrt{x^2 + x} .$$

(b) If $x > 0, x \neq 1$, then

$$\log x/(x - 1) < 1/\sqrt{x} .$$

(c) If $x > a > 0$, then

$$(x + a)/2 > (x - a)/(\log x - \log a).$$

[Hint: In (a) replace $1 + 1/x$ by x/a.]

15.   If $a > 0$ and $b > 0$, then $a^b + b^a > 1$.

16.   Let $0 < a < \infty$ and define $a_1 = a$ and $a_{n+1} = a^{a_n}$ for $n \in \mathbb{N}$. Then the sequence $(a_n)_{n=1}^{\infty}$ converges if and only if $e^{-e} \leq a \leq e^{1/e}$. The following steps may be helpful in establishing the preceding proposition.
Write $b = \log a$ and $f(x) = e^{bx}$ for $x \in \mathbb{R}$. Then
(a) $f(1) = a_1$ and $f(a_n) = a_{n+1}$,
(b) $a_n \to \alpha$ implies $f(\alpha) = \alpha$.
For $b > 0$ $(a > 1)$, we have
(c) f is strictly increasing,
(d) $(a_n)_{n=1}^{\infty}$ is strictly increasing [f preserves inequalities, and so: $0 < 1, 1 < a_1, a_1 < a_2, a_2 < a_3, \ldots$],
(e) the function $x \to f(x)/x$ $(x > 0)$ has an absolute minimum value eb at $x = 1/b$.
For $b > 1/e$, we have
(f) the equation $f(x) = x$ has no positive real root [use (e)],
(g) $a_n \to \infty$.
For $0 < b \leq 1/e$,
(h) the equation $f(x) = x$ has just one root α in the interval $]1, e]$ [use (e) and the Intermediate Value Theorem],
(i) $1 < a_n < \alpha$ for all $n \in \mathbb{N}$ [$1 < \alpha$, $a_1 = f(1) < f(\alpha) = \alpha, \ldots$],
(j) $a_n \to \alpha$.
In all remaining parts suppose that $b < 0$ $(0 < a < 1)$. Then
(k) the function f is strictly decreasing,
(l) the equation $f(x) = x$ has just one root α in the interval $]0, 1[$,
(m) $0 < a_{2n-1} < a_{2n+1} < \alpha < a_{2n+2} < a_{2n} < 1$ for all $n \in \mathbb{N}$ [since f reverses inequalities, we have: $0 < \alpha < 1$, $1 = f(0) > f(\alpha) = \alpha > f(1) = a_1 > 0$, $a_1 < \alpha < a_2 < 1$, $a_2 > \alpha > a_3 > a_1$, $a_3 < \alpha < a_4 < a_2$; hence, $a_1 < a_3 < \alpha < a_4 < a_2$ and we can use induction],
(n) $\lim_{n \to \infty} a_{2n-1} = \beta \leq \alpha \leq \lim_{n \to \infty} a_{2n} = f(\beta) = \gamma$ for some β and γ,
(o) $(a_n)_{n=1}^{\infty}$ converges if and only if $\beta = \alpha$. Writing $\phi(x) = f(x) - f^{-1}(x) = e^{bx} - (1/b)\log x$ for $x > 0$, we have

(p) $\phi(\alpha) = 0$ and $\phi(\beta) = 0$,

(q) $\phi'(x) > 0$ if and only if $bxe^{bx} > 1/b$,

(r) $bxe^{bx} \geqq -1/e$ for all $x > 0$, and equality obtains if and only if $x = -1/b$,

(s) if $-e \leqq b < 0$, then ϕ is strictly increasing, $\beta = \alpha$, and $a_n \to \alpha$.

Finally, we suppose that $b < -e$. Then

(t) $-1/b < \alpha$ [recall that $f(x) \leqq x$ for $x \geqq \alpha$],

(u) $\lim_{x \downarrow 0} \phi(x) = -\infty$,

(v) $\phi(-1/b) > 0$ [in fact $x \log x > -1/e$ unless $x = 1/e$ $(x > 0)$],

(w) $\phi(\delta) = 0$ for some δ satisfying $0 < \delta < -1/b$,

(x) with $\epsilon = f(\delta)$, we have $f(\epsilon) = \delta$ and $a_{2n-1} < \delta$ for all $n \in \mathbb{N}$ [as in (m), we have:

$$\epsilon < 1, \quad \delta > a_1, \quad \epsilon < a_2, \quad \delta > a_3, \quad \dots],$$

(y) $\beta = \lim_{n \to \infty} a_{2n-1} \leqq \delta < \alpha$,

(z) if $b < -e$, then $(a_n)_{n=1}^\infty$ does not converge.

17. *Darboux's Theorem.* Let $f: [a,b] \to \mathbb{R}$ be differentiable on $[a,b]$ (f' need not be continuous) and suppose that γ is a number strictly between $f'_+(a)$ and $f'_-(b)$. Then there exists $\xi \in \,]a,b[$ such that $f'(\xi) = \gamma$. [Hint: Letting $g(x) = f(x) - \gamma x$, we see that $g'_+(a)$ and $g'_-(b)$ have strictly opposite signs and so g is not monotone on $[a,b]$. If $a \leqq \alpha < \beta \leqq b$ and $g(\alpha) = g(\beta)$, then Rolle's Theorem provides ξ.]

18. To illustrate Exercise 17, write $P(x) = (x^2 - 1)^2$ and define f on $[0,1]$ by $f(0) = 0$ and $f(x) = n^{-3/2}P(2n(n+1)x - 2n - 1)$ if $n \in \mathbb{N}$ and $(n+1)^{-1} \leqq x \leqq n^{-1}$. Then f is differentiable on $[0,1]$ and f' is not continuous or even bounded on $[0,1]$. In fact, $f'_+(0) = 0$ and if $b_n = (4n+1)/[4n(n+1)]$, then $b_n \to 0$ and $f'(b_n) \to \infty$. Thus f' attains every positive value γ on every interval $[0, b_n]$.

19. If $f:[a,b] \to \mathbb{R}$ is differentiable on $[a,b]$ and if for each $y \in \mathbb{R}$ the set $E_y = \{x \in [a,b]: f'(x) = y\}$ is a closed set, then the function f' is continuous on $[a,b]$. [Hint: For fixed $y \in \mathbb{R}$, $\{x \in [a,b]: f'(x) \geqq y\}$ is closed since otherwise there exist x and x_n in $[a,b]$ such that

$$f'(x_n) \geqq y, \quad f'(x) < y, \quad \text{and} \quad x_n \to x.$$

In this case, Exercise 17 supplies ξ_n between x and x_n such that $f'(\xi_n) = y$. Then $\xi_n \in E_y$, $x \notin E_y$, and $\xi_n \to x$.]

20. Suppose that $a \in \mathbb{R}$ and f is a real-valued function that is differentiable on $[a, \infty[$.

(a) If $\inf\{f'(x): x > a\} > 0$, then $\lim_{x \to \infty} f(x) = \infty$.

(b) If $\lim_{x \to \infty}\{f(x) + f'(x)\} = L \in \mathbb{R}^\#$, then $\lim_{x \to \infty} f(x) = L$. [Hint: Let $h(x) = e^x f(x)$ and apply (4.22) to $h(x)/e^x$. This is a slick trick.]

21. Evaluate each of the following limits where $a, b \in \mathbb{R}$:

(a) $\lim_{x \to \infty} (\log x - x^a)$,

(b) $\lim_{x \to \infty} x^a b^{-x}$ $(b > 0)$,

(c) $\lim_{x \to 0} (b^x - 1)/x$ $(b > 0)$,

(d) $\lim_{x \to 0} (1 + ax)^{1/x}$,

(e) $\lim_{x \to \infty} x^a \log(1 + b/x)$,

(f) $\lim_{x \to \infty} x^a (e^b - (1 + b/x)^x)$

(g) $\lim_{x \to \infty} (b^x + x)^{1/x}$ $(b > 0)$,

(h) $\lim_{x \downarrow 0} x^{a \log x}$,

(i) $\lim_{x \downarrow 0} x^a \log x$,

(j) $\lim_{x \to \infty} x(b^{1/x} - 1)$ $(b > 0)$,

(k) $\lim_{x \to \infty} [x(x^{1/x} - 1)/(\log x)]$.

22. Prove that

$$\lim_{n \to \infty} \sum_{k=1}^{n} \left(\frac{k}{n} \right)^n = \frac{e}{e-1}.$$

[Hint:

$$\sum_{j=0}^{n-1} \left(1 - \frac{j}{n} \right)^n > \sum_{j=0}^{n_0} \left(1 - \frac{j}{n} \right)^n \to \sum_{j=0}^{n_0} e^{-j}$$

for fixed n_0 as $n \to \infty$.]

23. (a) If f is a real-valued function on $[0, \delta]$ $(\delta > 0)$ such that $f(0) = 0$ and $\lim_{x \downarrow 0} (f(x)/x)$ exists in $\mathbb{R}$, then $\lim_{x \downarrow 0} x^{f(x)} = 1$.

(b) Define $f(x) = (\log(1/x))^{-1}$ for $x > 0$ and $f(0) = 0$ and then evaluate $\lim_{x \downarrow 0} x^{f(x)}$.

24. A complex number α is said to be *algebraic* if there exists a polynomial

$$P(x) = \sum_{k=0}^{n} c_k x^k, \quad c_n \neq 0, \quad (c_k)_{k=0}^{n} \subset \mathbb{Z}, \ n \in \mathbb{N}$$

such that $P(\alpha) = 0$. If α is algebraic and n is the least natural number for which such a $P(x)$ exists, then α is said to have *degree n*. All complex numbers that are not algebraic are called *transcendental numbers*.

(a) If $\alpha \in \mathbb{C} \backslash \mathbb{Q}$ is algebraic of degree n and $P(x)$ is as above, then $n > 1$ and $P(r) \neq 0$ for all $r \in \mathbb{Q}$. [Hint: If $P(r) = 0$, write $P(x) = (x - r) \cdot Q(x)$ and contradict the minimality of n.]

(b) If $\alpha \in \mathbb{R} \backslash \mathbb{Q}$ is algebraic of degree n, then there exists $M \in \mathbb{N}$ such that

$$|\alpha - p/q| > 1/(Mq^n)$$

for all $p \in \mathbb{Z}$ and $q \in \mathbb{N}$. [Hints: Choose $M > |P'(x)|$ for $|x - \alpha| \leq 1$. Then $|\alpha - p/q| \leq 1$ implies $1 \leq q^n |P(p/q)| = q^n |P(\alpha) - P(p/q)| < q^n M |\alpha - p/q|$ because $q^n P(p/q)$ is a nonzero integer.]

The set L of *Liouville numbers* is the set of all $\alpha \in \mathbb{R} \backslash \mathbb{Q}$ such that for each $m \in \mathbb{N}$ there exist integers p and q with $q > 1$ such that $|\alpha - p/q| < q^{-m}$.

(c) If $(a_k)_{k=1}^{\infty} \subset \mathbb{N}$ with $1 \leq a_k \leq 9$ for all k and if $\alpha = \sum_{k=1}^{\infty} (a_k/10^{k!})$, then $\alpha \in L$. [Hint: First $\alpha \notin \mathbb{Q}$. Given m, choose $q = 10^{m!}$.]

(d) If $\alpha \in L$, then α is transcendental. [Hint: Assuming α is as in (b), choose m such that $2^m > M \cdot 2^n$.]

(e) $\mathbb{R} \backslash L$ is of first Baire category in $\mathbb{R}$ (see (3.53)). [Hint: For $m \in \mathbb{N}$, write

$$A_m = \mathbb{R} \backslash \left(\bigcup_{q=2}^{\infty} \bigcup_{p=-\infty}^{\infty} \,]p/q - q^{-m}, p/q + q^{-m}[\right)$$

and check that $\mathbb{R} \backslash L = \mathbb{Q} \cup \bigcup_{m=1}^{\infty} A_m$.]

(f) If I is a nonvoid open interval of $\mathbb{R}$, then $I \cap L$ is uncountable.

(g) If λ denotes Lebesgue measure, then $\lambda(L) = 0$. [Hint: For any fixed $k \in \mathbb{N}$ and any $m > 2$ we have

$$L \cap]-k, k[\subset \bigcup_{q_1 = 2}^{\infty} \bigcup_{p = -kq}^{kq} \;] p/q - 1/q^m, p/q + 1/q^m [$$

and the sum of the lengths of these intervals does not exceed $(4k + 1) \sum_{q=2}^{\infty} (1/q^{m-1})$. Also, $\sum_{q=2}^{\infty} (1/q^n) < \sum_{j=1}^{\infty} (2^j/(2^j)^n) = 1/(2^{n-1} - 1)$ for $n > 1$.]

25.   The hypothesis that $g' \neq 0$ in (4.22) is important. In a careless application of L'Hospital's Rule in which zeros of g' are "canceled" by zeros of f', erroneous results can be obtained. This remark is strikingly illustrated by the following example.* Define f and g on $\mathbb{R}$ by

$$f(x) = x + \cos x \sin x, \qquad g(x) = e^{\sin x}(x + \cos x \sin x).$$

Then we have
(a) $f(x) > x - 1$ and $g(x) > e^{-1}(x - 1)$ if $x > 1$,
(b) $\lim_{x \to \infty} f(x) = \lim_{x \to \infty} g(x) = \infty$,
(c) $f'(x) = 2(\cos x)^2$,
(d) $g'(x) = [e^{\sin x} \cos x][2 \cos x + f(x)]$,
(e) $f'(x)/g'(x) = (2e^{-\sin x} \cos x)/(2 \cos x + f(x))$ if $\cos x \neq 0$,
(f) the right side of equality (e) has limit 0 as $x \to \infty$,
(g) $\lim_{x \to \infty} [f(x)/g(x)]$ does *not* exist.

Higher Order Derivatives

(4.24)   Definition   Let $I \subset \mathbb{R}$ be an interval of positive length, let $c \in I$, and let $f: I \to \mathbb{R}$ be differentiable in some neighborhood of c [a one-sided neighborhood if c is an endpoint of I]. Then f', the derivative of f, is defined in this neighborhood. If f' has a derivative at c, then this derivative is called *the second derivative of f at c* and it is denoted by $f''(c)$. Thus, $f''(c) = (f')'(c)$. We also write $f^{(0)} = f$, $f^{(1)} = f'$, and $f^{(2)} = f''$. If, for some $n \in \mathbb{N}$, $f^{(n-1)}$ is defined and real-valued in some neighborhood of c and if $f^{(n-1)}$ has a derivative at c, then we call this derivative *the nth derivative of f at c* and we denote it by $f^{(n)}(c)$. Thus, $f^{(n)}(c) = (f^{(n-1)})'(c)$. We also call $f^{(n)}(c)$ *the derivative of order n of f at c*. If f has a finite derivative of order n at each point of I, then we say that *f is n times differentiable on I*. If f has finite derivatives of all orders $n \geq 0$ at c, then we say that *f is infinitely differentiable at c*. If f is infinitely differentiable at every point of I, then we say that *f is infinitely differentiable on I* or that *f is a C^∞-function on I* and we write $f \in C^\infty(I)$. If $f^{(n)}$ is continuous on I for some $n \geq 0$, then we write $f \in C^{(n)}(I)$.

Notice that a necessary [but not sufficient] condition that $f^{(n)}(c)$ should

---

*This example was called to the author's attention by Professor R. B. Burckel. The trigonometric functions are studied in Chapter 5.

exist is that $f^{(n-2)}$ be differentiable on a neighborhood of c. Plainly, $C^{\infty}(I)$ $\subset C^{(n)}(I) \subset C^{(n-1)}(I)$ for all $n \in \mathbb{N}$.

(4.25) **Examples** (a) Clearly $C^{\infty}(\mathbb{R})$ contains every real polynomial.
(b) Define f on $\mathbb{R}$ by $f(x) = |x|^3$. Then $f'(x) = 3x^2$ if $x \geq 0$, $f'(x) = -3x^2$ if $x \leq 0$, $f''(x) = 6|x|$ for all x, $f^{(3)}(x) = 6$ if $x > 0$, $f^{(3)}(x) = -6$ if $x < 0$, and $f^{(3)}(0)$ does not exist. Therefore,

$$f \in C^{(2)}(\mathbb{R}) \backslash C^{(3)}(\mathbb{R}).$$

(4.26) **Theorem** *Let f and g be real-valued functions defined on some interval $I \subset \mathbb{R}$ which are both n times differentiable at some $c \in I$ for some $n \in \mathbb{N}$. Let $\alpha \in \mathbb{R}$. Then αf, $f + g$, and fg are all n times differentiable at c and we have*

(i) $(\alpha f)^{(n)}(c) = \alpha \cdot f^{(n)}(c),$

(ii) $(f + g)^{(n)}(c) = f^{(n)}(c) + g^{(n)}(c),$

(iii) $(fg)^{(n)}(c) = \sum_{k=0}^{n} \binom{n}{k} f^{(n-k)}(c) g^{(k)}(c).$

Formula (iii) is known as *Leibnitz's Formula*.

Proof This all follows very easily by using (4.9) and induction on n. For (iii) it is necessary to note that

$$\binom{n}{k} + \binom{n}{k-1} = \binom{n+1}{k}.$$

We leave the details to the reader. [Compare the Binomial Theorem.] $\square$

Next, we consider composite functions.

(4.27) **Theorem** *Let I and J be intervals of $\mathbb{R}$, let $c \in I$, and let $f: I \to J$ and $g: J \to \mathbb{R}$ be given. Suppose that for some $n \in \mathbb{N}$, f is n times differentiable at c and g is n times differentiable at $f(c)$. Then $g \circ f$ is n times differentiable at c.*

Proof If $n = 1$, this follows from (4.10). Suppose, inductively, that the theorem is true when n is replaced by $n - 1$ for some fixed $n > 1$. By (4.10) we have

$$(g \circ f)^{(n)} = \left[(g \circ f)' \right]^{(n-1)} = \left[(g' \circ f) \cdot f' \right]^{(n-1)}. \tag{1}$$

Since f' and, by our inductive hypothesis, $g' \circ f$ are both $n - 1$ times differentiable at c, it follows from (4.26) that, at c, the right side of (1), and so also its left, exists and is finite. By induction, the proof is complete. $\square$

(4.28) **Definition** The notation is as in (4.24). If f is infinitely differentiable at c, then we define the *Taylor series of f centered at c* to be the power series

(i) $\sum_{n=0}^{\infty} \frac{f^{(n)}(c)}{n!} (x - c)^n.$

In case $c = 0$, this is also called the *Maclaurin series of f*. The reason for this definition will become clear later in this chapter.

(4.29)   Remark   Notice that we make no claim that the sum of the series (4.28.i), when it converges, is equal to $f(x)$ except when $x = c$. In fact, it can happen that $f \in C^\infty(\mathbb{R})$ and the Maclaurin series of f converges everywhere on $\mathbb{R}$, but its sum is $f(x)$ only when $x = 0$. This is shown by the next example. The question of sufficient conditions that (4.28.i) converge to $f(x)$ is taken up in the next section.

(4.30)   Example   Define f on $\mathbb{R}$ by $f(x) = \exp(-x^{-2})$ for $x \neq 0$ and $f(0) = 0$. We claim that $f \in C^\infty(\mathbb{R})$ and $f^{(n)}(0) = 0$ for all $n \geq 0$. First, we shall show that, for all $n \geq 0$ and all $x \neq 0$, we have

$$f^{(n)}(x) = P_n(x^{-1})\exp(-x^{-2}) \tag{1}$$

where P_n is a real polynomial of degree $3n$. For $n = 0$, this is obvious with $P_0 \equiv 1$. If (1) holds for $n = k \geq 0$, then

$$f^{(k+1)}(x) = P_k(x^{-1})\exp(-x^{-2})(2x^{-3}) + \exp(-x^{-2})P_k'(x^{-1})(-x^{-2})$$
$$= \left(2x^{-3}P_k(x^{-1}) - x^{-2}P_k'(x^{-1})\right)\exp(-x^{-2}),$$

and so (1) also holds for $n = k + 1$ with $P_{k+1}(t) = 2t^3 P_k(t) - t^2 P_k'(t)$. By induction, (1) has been established for all $n \geq 0$. We next show that

$$f^{(n)}(0) = 0 \tag{2}$$

for all $n \geq 0$. This is true by definition if $n = 0$. Suppose (2) holds for $n = k \geq 0$. Then, since $t^r \exp(-t^2) \to 0$ as $|t| \to \infty$ for every integer r (3.95.vi), we have by (1)

$$\lim_{x \to 0} \frac{f^{(k)}(x) - f^{(k)}(0)}{x - 0} = \lim_{x \to 0} x^{-1} P_k(x^{-1})\exp(-x^{-2})$$

$$= \lim_{|t| \to \infty} t P_k(t)\exp(-t^2) = 0.$$

Thus, by definition, $f^{(k+1)}(0) = 0$, and so (2) follows by induction. Our claim is now established.

(4.31)   Theorem   *There exists a function $\phi \in C^\infty(\mathbb{R})$ such that*

$$\phi(x) = 1 \quad \text{if } |x| \leq 1/2,$$

$$0 < \phi(x) < 1 \quad \text{if } 1/2 < |x| < 1,$$

$$\phi(x) = 0 \quad \text{if } |x| \geq 1.$$

In particular, $\phi^{(n)}(0) = 0$ for all $n > 0$, so the Maclaurin series of ϕ converges everywhere to 1, but its sum is ϕ only on $[-1/2, 1/2]$.

Proof Let f be as in (4.30). It is clear from (4.30) that the functions g and h defined on $\mathbb{R}$ by $g(x) = 0$ if $x \geq 1$, $g(x) = ef(x - 1)$ if $x \leq 1$,

$$h(x) = e^4f(x - 1/2) \quad \text{if } x \geq 1/2,$$

$$h(x) = 0 \qquad\qquad\quad \text{if } |x| \leq 1/2,$$

$$h(x) = e^4f(x + 1/2) \quad \text{if } x \leq -1/2$$

are in $C^\infty(\mathbb{R})$. Moreover, g is strictly decreasing on $]-\infty, 1]$, $g(0) = 1$, h is strictly monotone on $]-\infty, -1/2]$ and on $[1/2, \infty[$, and $h(-1) = h(1) = 1$ $[f'(t) = 2t^{-3}e^{-t^{-2}}]$. These facts and (4.27) show that the composite function $\phi = g \circ h$ has the required properties. $\square$

The following remarkable theorem is due to E. Borel.

(4.32) **Theorem** *Let $(a_n)_{n=0}^\infty$ be an arbitrary sequence of real numbers. Then there exists a function $f \in C^\infty(\mathbb{R})$ such that $f^{(n)}(0) = a_n$ for all $n \geq 0$. In particular, if $a_n = (n!)^2$ for all n, then the Maclaurin series of f has radius of convergence $R = 0$.*

Proof [Gårding] Write $c_n = n + |a_n|$. Let ϕ be as in (4.31) and define f on $\mathbb{R}$ by

$$f(x) = \sum_{k=0}^\infty \frac{a_k}{k!} x^k \phi(c_k x). \tag{1}$$

Plainly, $f^{(0)}(0) = f(0) = a_0$. Notice that if $|x| > \delta > 0$ and $k > 1/\delta$, then $|c_k x| > k\delta > 1$, and so $\phi(c_k x) = 0$. Thus, for each $\delta > 0$ only a finite number of summands of the series (1) fail to vanish identically on $\mathbb{R} \setminus [-\delta, \delta]$. It follows that f is infinitely differentiable at every point of $\mathbb{R} \setminus \{0\}$. For $n \geq 0$, define f_n on $\mathbb{R}$ by

$$f_n(x) = \sum_{k=0}^\infty \left\{ \sum_{j=0}^n \binom{n}{j} \frac{a_k}{k!} k(k-1) \ldots (k-j+1) x^{k-j} \cdot \phi^{(n-j)}(c_k x) c_k^{n-j} \right\}, \tag{2}$$

where $k(k-1) \ldots (k-j+1)x^{k-j}$ denotes 0 if $k < j$ and $k(k-1) \ldots (k-j+1)$ denotes $k!/(k-j)!$ if $k \geq j$. Clearly $f_0 = f$. By Leibnitz's Formula (4.26.iii), it is clear that $f_n(x) = f^{(n)}(x)$ for all $x \neq 0$ and all $n > 0$. In view of the notational convention following (2) and because $\phi^{(n-j)}(0) = 0$ unless $j = n$, it follows that, for $x = 0$, the only nonzero summand in (2) is the one for which $j = k = n$. Therefore, $f_n(0) = a_n$ for all n. It remains only to show that $f^{(n)}(0)$ exists and is equal to $f_n(0)$ for all n. We first show that each f_n is continuous on $\mathbb{R}$ by using the Weierstrass M-test.

Fix any $n \geq 0$. For each $k > n$, define

$$M_k = \frac{2^n}{(k+n)!} \cdot \sup\{|\phi^{(n-j)}(x)| : 0 \leq j \leq n, x \in \mathbb{R}\}.$$

Plainly, $\sum\limits_{k=n+1}^{\infty} M_k < \infty$. We shall show that the kth term of the series (2) is dominated by M_k if $k > n$. If $|c_k x| > 1$, this term vanishes. If $k > n$ is fixed, $|c_k x| \leq 1$ and $0 \leq j \leq n$, then

$$\left| a_k x^{k-j} c_k^{n-j} \right| = |a_k| \cdot |c_k x|^{k-j} \cdot c_k^{n-k}$$

$$\leq |a_k| \cdot c_k^{n-k} < c_k^{n+1-k} \leq 1.$$

Therefore, $x \in \mathbb{R}$ and $k > n$ imply

$$\left| \sum_{j=0}^{n} \binom{n}{j} \frac{a_k}{k!} k(k-1) \ldots (k-j+1) x^{k-j} \phi^{(n-j)}(c_k x) c_k^{n-j} \right|$$

$$\leq \sum_{j=0}^{n} \binom{n}{j} \frac{|\phi^{(n-j)}(c_k x)|}{(k-j)!}$$

$$\leq \frac{M_k}{2^n} \sum_{j=0}^{n} \binom{n}{j} = M_k.$$

We infer that the series (2) converges uniformly on $\mathbb{R}$, and so f_n is continuous on $\mathbb{R}$. For each $x \neq 0$, we can now apply the Mean Value Theorem (4.17) to obtain ξ_x in the open interval between 0 and x such that $f_n(x) - f_n(0) = x f_n'(\xi_x) = x f_{n+1}(\xi_x)$. Since f_{n+1} is continuous at 0, this implies that

$$\lim_{x \to 0} \frac{f_n(x) - f_n(0)}{x - 0} = \lim_{x \to 0} f_{n+1}(\xi_x) = f_{n+1}(0).$$

Therefore, for all $n \geq 0$, f_n is differentiable at 0 and

$$f_n'(0) = f_{n+1}(0). \tag{3}$$

Finally, we show by induction that $f^{(n)}(x)$ exists and

$$f^{(n)}(x) = f_n(x) \quad \text{for all } x \in \mathbb{R}. \tag{4}$$

We already know this for $x \neq 0$. For $n = 0$, (4) holds because $f^{(0)} = f = f_0$. Supposing (4) for n, we apply (3) to write $f^{(n+1)}(0) = f^{(n)'}(0) = f_n'(0) = f_{n+1}(0)$. $\square$

Taylor Polynomials

In this section, we examine the partial sums of a Taylor series for a function. We use them to obtain a very general test for local extrema and we find estimates of their degree of approximation to the given function.

(4.33)   **Definition**   Let $\alpha < \beta$ be real numbers, let $n \geq 0$ be an integer, and let $a \in [\alpha, \beta]$. Suppose that $f : [\alpha, \beta] \to \mathbb{R}$ is n times differentiable at a. The polynomial

$$P_n(h) = \sum_{k=0}^{n} \frac{f^{(k)}(a)}{k!} h^k$$

is called the *Taylor polynomial of order n for f at a*. [If $a = \alpha$ or β, all derivatives are one-sided.] If we write $h = x - a$, this polynomial is, of course, the nth partial sum of the Taylor series of f centered at a.

Here is a kind of approximation theorem.

(4.34)   **Theorem**   *The notation is as in (4.33) with $n > 0$. Then*

(i) $$\lim_{h \to 0} \left[(f(a + h) - P_n(h))/h^n \right] = 0$$

where the limit is one-sided if $a = \alpha$ or β.

Proof   A simple calculation shows that the function g defined by $g(h) = f(a + h) - P_n(h)$ satisfies

$$g^{(j)}(0) = 0 \qquad (0 \leq j \leq n).$$

If we apply L'Hospital's Rule $n - 1$ times, we obtain

$$n! \lim_{h \to 0} \frac{g(h)}{h^n} = \lim_{h \to 0} \frac{g^{(n-1)}(h)}{h}$$

$$= \lim_{h \to 0} \frac{f^{(n-1)}(a + h) - f^{(n-1)}(a) - f^{(n)}(a) \cdot h}{h} = 0$$

by the definition of $f^{(n)}(a)$. [Notice that we could not apply L'Hospital n times since $f^{(n)}(a + h)$ need not exist for $h \neq 0$ and, even if it does, $f^{(n)}$ need not be continuous at a.] $\square$

(4.35)   **Corollary**   *The notation is as in (4.33) with $n > 1$, $\alpha < a < \beta$, and*

$$f'(a) = \ldots = f^{(n-1)}(a) = 0, \qquad f^{(n)}(a) \neq 0.$$

(i) *If n is even, then f has a local minimum [resp., local maximum] at a if $f^{(n)}(a) > 0$ [resp., $f^{(n)}(a) < 0$].*
(ii) *If n is odd, then f has neither a local maximum nor a local minimum at a.*

Proof   With P_n as in (4.33), we have

$$\frac{f(a + h) - P_n(h)}{h^n} = \frac{f(a + h) - f(a)}{h^n} - \frac{f^{(n)}(a)}{n!}.$$

Hence, (4.34) shows that

$$\lim_{h \to 0} \left[(f(a + h) - f(a))/h^n \right] = (f^{(n)}(a))/n!,$$

and so there is a $\delta > 0$ such that $0 < |h| < \delta$ implies

$$(f(a + h) - f(a))/h^n \text{ has the same sign as } f^{(n)}(a). \tag{1}$$

Suppose n is even. If $f^{(n)}(a) > 0$, then, since $h^n > 0$, it follows from (1) that $f(a + h) > f(a)$ whenever $0 < |h| < \delta$, and so f has a local minimum at a. The case $f^{(n)}(a) < 0$ is similar. Thus (i) obtains.

Suppose n is odd. If $f^{(n)}(a) > 0$, then, using (1), we have

$$-\delta < h < 0 \quad \text{implies} \quad f(a + h) < f(a) \tag{2}$$

and

$$0 < h < \delta \quad \text{implies} \quad f(a + h) > f(a) \tag{3}$$

and so (ii) obtains. If $f^{(n)}(a) < 0$, then (2) and (3) still prevail after the inequalities in their consequents have been reversed, and so (ii) obtains in this case also. $\square$

(4.36) Remark Though (4.35) is a very good test for local extrema, Example (4.30) shows that it may fail to detect such extrema since there may be no n that satisfies the hypotheses. Of course this is banal if f is constant in a neighborhood of a.

We now turn to a consideration of the difference between $f(a + h)$ and $P_n(h)$.

(4.37) Taylor's Theorem *Let* α, β, n, a, f, *and* P_n *be as in* (4.33). *Suppose, in addition, that* $f^{(k)}$ *is continuous on* $[\alpha, \beta]$ *for* $0 \leq k \leq n$ *and* $f^{(n+1)}(x)$ *exists and is finite for all* $\alpha < x < \beta$. *Write*

(i) $R_n(h) = f(a + h) - P_n(h)$

for $\alpha - a \leq h \leq \beta - a$. *Then, for each fixed* $h \neq 0$ $(\alpha - a \leq h \leq \beta - a)$ *and each integer* $u < n + 1$, *there exists* $0 < \theta < 1$ [θ *depends on* n, h, *and* u] *such that*

(ii) $$R_n(h) = \frac{h^{n+1}(1 - \theta)^u}{(n + 1 - u) \cdot n!} f^{(n+1)}(a + \theta h).$$

Obviously, (ii) *obtains when* $h = 0$ *and* $\alpha < a < \beta$.

Proof Fix u and h and define $A \in \mathbb{R}$ by the relation

$$h^{n+1-u}A = f(a + h) - \sum_{k=0}^{n} \frac{h^k}{k!} f^{(k)}(a) = R_n(h). \tag{1}$$

Next define a function g on the closed interval having endpoints a and $a + h$ by the rule

$$g(t) = -(a + h - t)^{n+1-u}A + f(a + h) - \sum_{k=0}^{n} \frac{(a + h - t)^k}{k!} f^{(k)}(t). \tag{2}$$

We wish to apply Rolle's Theorem to g on this closed interval. Plainly the required continuity and differentiability conditions are met. Also, $g(a + h) = g(a) = 0$. Thus there exists $0 < \theta < 1$ such that

$$g'(a + \theta h) = 0. \tag{3}$$

From (2), we see by direct computation that

$$g'(t) = (n + 1 - u)(a + h - t)^{n-u}A - f'(t)$$

$$- \sum_{k=1}^{n} \frac{(a + h - t)^k}{k!} f^{(k+1)}(t) + \sum_{k=1}^{n} \frac{(a + h - t)^{k-1}}{(k-1)!} f^{(k)}(t)$$

$$= (n + 1 - u)(a + h - t)^{n-u}A - \frac{(a + h - t)^n}{n!} f^{(n+1)}(t). \tag{4}$$

[In case $n = 0$, the two summations in (4) do not appear.]

Combining (3) and (4) gives

$$0 = g'(a + \theta h) = (n + 1 - u)(h - \theta h)^{n-u}A - \frac{(h - \theta h)^n}{n!} f^{(n+1)}(a + \theta h),$$

from which we obtain

$$A = \frac{h^u (1 - \theta)^u}{(n + 1 - u)n!} f^{(n+1)}(a + \theta h). \tag{5}$$

Substituting (5) into (1) yields (ii). $\square$

(4.38) **Remarks on (4.37)** (a) It is only by tradition that we use the name Taylor's Theorem for (4.37). In 1715 Brook Taylor first studied what we call the Taylor series of f. The notion of convergence was not well understood at that time. It was not until much later that Joseph Louis Lagrange gave the first precise estimate of the remainder. The form of the remainder given here is due to Schlömilch (1847).

(b) Taking $u = 0$ we obtain the form of the remainder due to Lagrange:

$$R_n(h) = \left[h^{n+1}/(n+1)! \right] f^{(n+1)}(a + \theta h).$$

If we put $a + h = x$, this becomes

$$R_n(h) = \left[(x - a)^{n+1}/(n+1)! \right] f^{(n+1)}(\xi)$$

where $\xi = (1 - \theta)a + \theta x$ is in the open interval between a and x.

(c) Taking $u = n$ gives Cauchy's form:

$$R_n = \left[h^{n+1}(1 - \theta)^n/n! \right] f^{(n+1)}(a + \theta h).$$

(d) Notice that if $n = u = 0$, then (4.37) is just a restatement of the Mean Value Theorem.

We conclude this section with some examples to show how (4.37) can be applied to prove that certain functions are the sums of their Taylor series.

(4.39)   Examples   (a) If f is a polynomial of degree n and $a \in \mathbb{R}$, then

$$f(x) = \sum_{k=0}^{n} \frac{f^{(k)}(a)}{k!} (x - a)^k$$

for all $x \in \mathbb{R}$. This is true because $f^{(n+1)} \equiv 0$ so $R_n = 0$.

(b) Let α be an arbitrary real number. Consider the function f defined on $]-1, \infty[$ by $f(x) = (1 + x)^\alpha$. Since

$$f^{(k)}(x) = \alpha(\alpha - 1)(\alpha - 2) \ldots (\alpha - k + 1) \cdot (1 + x)^{\alpha - k},$$

we have

$$f^{(k)}(0)/k! = \binom{\alpha}{k} \qquad (k \geq 0),$$

where, for $k > 0$, we define

$$\binom{\alpha}{0} = 1, \qquad \binom{\alpha}{k} = \frac{1}{k!} \prod_{j=0}^{k-1} (\alpha - j).$$

Notice that if α is a nonnegative integer and $k \leq \alpha$, then (1) agrees with the earlier definition of binomial coefficients. The nth Taylor polynomial for f at $a = 0$ is therefore

$$P_n(h) = \sum_{k=0}^{n} \binom{\alpha}{k} h^k. \tag{1}$$

Taking Cauchy's form of the remainder (4.38.c), we see that $h > -1$ and $n \in \mathbb{N}$ imply that

$$
\begin{aligned}
R_n(h) &= \frac{(1 - \theta)^n}{n!} h^{n+1} f^{(n+1)}(\theta h) \\
&= \frac{(1 - \theta)^n}{n!} h^{n+1} (1 + \theta h)^{\alpha - n - 1} \prod_{j=0}^{n} (\alpha - j) \\
&= \alpha h \cdot (1 + \theta h)^{\alpha - 1} \cdot \left(\frac{1 - \theta}{1 + \theta h} \right)^n \cdot \prod_{j=1}^{n} \left[\left(\frac{\alpha}{j} - 1 \right) h \right]
\end{aligned}
\tag{2}
$$

for some θ with $0 < \theta < 1$. Now fix any $h, \beta \in \mathbb{R}$ such that $|h| < \beta < 1$. Since $0 < 1 - \theta < 1 + \theta h$, we have

$$0 < (1 - \theta)/(1 + \theta h) < 1. \tag{3}$$

Choose $N \in \mathbb{N}$ such that

$$|(\alpha/j - 1)h| < \beta \quad \text{if } j > N. \tag{4}$$

Write $M = 2^{\alpha - 1}$ if $\alpha \geq 1$ and $M = (1 - \beta)^{\alpha - 1}$ if $\alpha < 1$. Then M does not depend on n and we have

$$0 < (1 + \theta h)^{\alpha - 1} \leq M \tag{5}$$

for all $n \in \mathbb{N}$ [θ depends on n]. Using (2)–(5), we see that

$$|R_n(h)| \leq |\alpha| M (|\alpha| + 1)^N \beta^{n-N}$$

for all $n > N$. It follows that $R_n(h) \to 0$ as $n \to \infty$. We conclude that

$$(1 + h)^\alpha = \sum_{k=0}^{\infty} \binom{\alpha}{k} h^k \tag{6}$$

if $h \in \mathbb{R}$, $\alpha \in \mathbb{R}$, and $|h| < 1$. If $\alpha \in \mathbb{N}$, this is the Binomial Theorem.

(c) Let $f(x) = \log(1 + x)$ for $x > -1$. Then $f^{(0)}(0) = 0$ and $f^{(k)}(0) = (-1)^{k-1}(k-1)!$ for $k > 0$. The corresponding Taylor polynomials and Cauchy remainders for f at $a = 0$ are given by

$$P_n(h) = \sum_{k=1}^{n} (-1)^{k-1} \frac{h^k}{k},$$

$$R_n(h) = (1 - \theta)^n h^{n+1} (-1)^n (1 + \theta h)^{-n-1}$$

where $0 < \theta < 1$. For $|h| < \beta < 1$, we can use (3) to see that

$$|R_n(h)| \leq \beta^{n+1}(1 - \beta)^{-1} \to 0 \quad \text{as } n \to \infty.$$

Thus

$$\log(1 + h) = \sum_{k=1}^{\infty} (-1)^{k-1} \frac{h^k}{k} \tag{7}$$

if $h \in \mathbb{R}$ and $|h| < 1$.

In Chapter 7 we shall find all complex numbers α and h for which formulas (6) and (7) obtain. The proofs there make no use of Taylor's Theorem.

Exercises

1. (a) Prove the following *mean value theorem of order k*. Let $a < b$ in $\mathbb{R}$ and let $f : [a, b] \to \mathbb{R}$ be continuous on $[a, b]$. If f is k times differentiable on $[a, b]$ for some $k \in \mathbb{N}$ and if $h = (b - a)/k$, then there exists some $\xi \in]a, b[$ such that

$$\sum_{j=0}^{k} (-1)^{k-j} \binom{k}{j} f(a + jh) = h^k f^{(k)}(\xi).$$

[Hints: Use induction on k. If $k > 1$ and we suppose the theorem true when k is replaced by $k - 1$, then apply that theorem to the function g defined on $[a, b - h]$ by $g(x) = f(x + h) - f(x)$ to obtain a $< \eta < b - h$ such that

$$\sum_{j=0}^{k-1} (-1)^{k-1-j} \binom{k-1}{j} g(a + jh) = h^{k-1} g^{(k-1)}(\eta).$$

Note that $(b - h - a)/(k - 1) = (b - a)/k = h$. Then apply the Mean Value Theorem to the function $f^{(k-1)}$ on the interval $[\eta, \eta + h]$. Write all g's in terms of f and use the formula

$$\binom{k-1}{j-1} + \binom{k-1}{j} = \binom{k}{j}.]$$

(b) Use (a) to prove the following fact. If $c \in \mathbb{R}$ and if n^c is an integer for every $n \in \mathbb{N}$, then c is a nonnegative integer. [Hints: Assume false. Let k be the least integer greater than c. Choose an integer a such that $c(c-1)\ldots(c-k+1)$ $< a^{k-c}$. Apply part (a) to the function $f(x) = x^c$ where $b = a + k$, $h = 1$, to obtain an integer in $]0, 1[$. Note that $\xi^{c-k} < a^{c-k}$ if $\xi > a$.]

2.   Let $f : \mathbb{R} \to \mathbb{R}$ be continuous and such that

$$\lim_{x \to \infty} f(x) = \lim_{x \to -\infty} f(x) = 0.$$

Then for each $\epsilon > 0$ there exists a function $g \in C^\infty(\mathbb{R})$ which is identically 0 outside some compact interval such that $|f(x) - g(x)| < \epsilon$ for all $x \in \mathbb{R}$. [Hint: Use the Stone–Weierstrass Theorem and (4.31). Consider the collection of all g's in $C^\infty(\mathbb{R})$ that vanish outside some compact interval.]

3.   Let f be a real-valued function on an open interval $I \subset \mathbb{R}$ and let $a \in I$.
(a) If f is twice differentiable at a, then

$$f''(a) = \lim_{h \to 0} \frac{f(a + h) + f(a - h) - 2f(a)}{h^2}.$$

[Hint: Use (4.22).]
(b) If $a = 0$, $f(x) = x^2$ $(x \geq 0)$, $f(x) = -x^2$ $(x \leq 0)$, then the limit in (a) exists, but $f''(0)$ does not exist.
(c) If f is thrice differentiable at a, then

$$\frac{1}{3} f'''(a) = \lim_{h \to 0} \frac{f(a + h) - f(a - h) - 2hf'(a)}{h^3}.$$

(d) Find an example of an f for which the limit in (c) exists, but $f'''(a)$ does not.

4.   (a) If P is a polynomial of degree $\leq n$ and $\lim_{h \to 0}[P(h)/h^n] = 0$, then all of P's coefficients are 0.
(b) Let a, f, and P_n be as in (4.34). If P is a polynomial of degree $\leq n$ such that

$$\lim_{h \to 0} \frac{f(a + h) - P(h)}{h^n} = 0,$$

then $P = P_n$.

5.   Suppose that I is an interval of $\mathbb{R}$, $a \in I$, and f is a twice differentiable real-valued function on I that satisfies $f'' + f = 0$ and $f(a) = f'(a) = 0$. Then $f = 0$. [Hint: First show that f is infinitely differentiable and then use Taylor's Theorem.]

6.   Evaluate
(a) $\lim_{t \downarrow 0} (1 + t)^{1/t}$,

(b) $\lim_{t \downarrow 0} \frac{1}{t} [e - (1 + t)^{1/t}]$,

(c) $\lim_{t \downarrow 0} \left\{ \frac{1}{t^2} [e - (1 + t)^{1/t}] - \frac{e}{2t} \right\}$,

(d) $\lim_{t \to 0} \frac{1}{t} [(1 + t)^{1/t} - \log(1 + t)^{e/t}]$.

[Hint: Use the log series (4.39.c) to obtain upper and lower estimates for $(1 + t)^{1/t}$; e.g., $\exp(1 - t/2) < (1 + t)^{1/t} < \exp(1 - t/2 + t^2/3)$ for $0 < t < 1$.]

Convex Functions

(4.40)   Definition Let $I \subset \mathbb{R}$ be an interval and $f : I \to \mathbb{R}$. Then f is said to be *convex on I* if

(i) $$f((1 - t)a + tb) \leqq (1 - t)f(a) + tf(b)$$

whenever $a, b \in I$ and $0 \leqq t \leqq 1$. We call f *concave* if $-f$ is convex.

(4.41)   Remark Writing $x = (1 - t)a + tb$, we see that inequality (4.40.i) is equivalent to

$$f(x) \leqq \frac{b - x}{b - a} f(a) + \frac{x - a}{b - a} f(b)$$

$$= \frac{f(b) - f(a)}{b - a}(x - a) + f(a) \leqq \max\{f(a), f(b)\}$$

whenever $a, b \in I$ and x is between a and b. That is, geometrically, f is convex on I if and only if the chord [line segment] having endpoints $(a, f(a))$ and $(b, f(b))$ is never below the graph of f.

Our next theorem shows that convex functions have very nice differentiability properties. First, a lemma.

(4.42)   Lemma *Let $I \subset \mathbb{R}$ be an interval and let $f : I \to \mathbb{R}$ be convex. If $a < b < c$ in I, then*

(i) $$\frac{f(b) - f(a)}{b - a} \leqq \frac{f(c) - f(a)}{c - a} \leqq \frac{f(c) - f(b)}{c - b}.$$

Proof Since

$$b = \frac{c - b}{c - a} a + \frac{b - a}{c - a} c,$$

we have

$$f(b) \leqq \frac{c - b}{c - a} f(a) + \frac{b - a}{c - a} f(c),$$

from which both inequalities in (i) follow easily. $\square$

(4.43)   Theorem *Let $I \subset \mathbb{R}$ be an* open *interval and let $f : I \to \mathbb{R}$ be convex. Then*
(i) *$f'_+(x)$ and $f'_-(x)$ exist and are finite at each $x \in I$,*
(ii) *f'_+ and f'_- are nondecreasing functions on I,*
(iii) *$f'_- \leqq f'_+$ on I,*
(iv) *f is differentiable except at countably many points of I,*
(v) *if $[\alpha, \beta] \subset I$ and $M = \max\{|f'_+(\alpha)|, |f'_-(\beta)|\}$, then*

$$|f(x) - f(y)| \leqq M|x - y|$$

for all $x, y \in [\alpha, \beta]$,
(vi) *f is continuous on I.*

Proof   For fixed $x \in I$, it follows from (4.42) that the function ϕ defined on a neighborhood of 0, except at $h = 0$, by

$$\phi(h) = (f(x+h) - f(x))/h$$

is nondecreasing. Therefore

$$\lim_{h\uparrow 0} \phi(h) = \sup\{\phi(h) : h < 0\}$$

$$\leq \inf\{\phi(h) : h > 0\} = \lim_{h\downarrow 0} \phi(h).$$

This proves (i) and (iii).

For $a < b$ in I, (4.42) shows that

$$f'_+(a) \leq \frac{f(b) - f(a)}{b - a} \leq \lim_{c\downarrow b} \frac{f(c) - f(b)}{c - b} = f'_+(b).$$

Similarly, f'_- is nondecreasing and so (ii) obtains.

The lemma shows that if $a < c$ in I, then $f'_+(a) \leq f'_-(c)$. Thus if a is a point of continuity of the monotone function f'_-, then

$$f'_-(a) \leq f'_+(a) \leq \lim_{c\downarrow a} f'_-(c) = f'_-(a).$$

This proves (iv).

Given $\alpha \leq x < y \leq \beta$ in I, define M as in (v) and obtain

$$-M \leq f'_+(\alpha) \leq f'_+(x) \leq \frac{f(x) - f(y)}{x - y} \leq f'_-(y) \leq f'_-(\beta) \leq M,$$

from which (v) is immediate. Finally, (vi) follows from (v).   □

It is evident geometrically that the graph of a convex function is never below any of its tangent lines. We make this remark analytically precise as follows.

(4.44)   Theorem   *Let I and f be as in (4.43) and let $c \in I$. Let m be any real number satisfying $f'_-(c) \leq m \leq f'_+(c)$. Then*

(i) $f(x) \geq m(x - c) + f(c)$

for all $x \in I$.

Proof   For $x > c$ we have

$$\frac{f(x) - f(c)}{x - c} \geq f'_+(c) \geq m$$

and for $x < c$ we have

$$\frac{f(x) - f(c)}{x - c} \leq f'_-(c) \leq m.$$

Therefore (i) obtains.   □

As a simple corollary we obtain the following inequality, which may also be proved by induction.

(4.45)   Jensen's Inequality   *Let I and f be as in (4.43). Suppose that $(x_j)_{j=1}^n \subset I$ and that $(p_j)_{j=1}^n \subset]0, \infty[$ satisfies $\sum\limits_{j=1}^n p_j = 1$. Then*

(i)
$$f\left(\sum_{j=1}^n p_j x_j \right) \leq \sum_{j=1}^n p_j f(x_j).$$

Moreover, equality obtains in (i) *if and only if f is linear on the smallest closed interval containing $(x_j)_{j=1}^n$.*

> **Proof**   We may suppose $x_1 \leqq x_2 \leqq \ldots \leqq x_n$. Let
> $$c = p_1 x_1 + p_2 x_2 + \ldots + p_n x_n.$$
> Plainly $x_1 \leqq c \leqq x_n$ and so $c \in I$. Let m be as in (4.44). Substituting x_j for x in (4.44.i), we have
> $$f(x_j) \geqq m(x_j - c) + f(c) \tag{1}$$
> for $1 \leqq j \leqq n$. Now multiply both sides of (1) by the positive number p_j and sum the resulting n inequalities to obtain (i). Clearly, equality holds in (i) if and only if equality holds in (1) for each j. But this is equivalent to the assertion that $f(x) = m(x - c) + f(c)$ for all $x \in [x_1, x_n]$.   □

We now give a useful criterion for convexity.

(4.46)   Theorem   *Let $I \subset \mathbb{R}$ be an open interval and let $f: I \to \mathbb{R}$ be differentiable on I. Then f is convex on I if and only if f' is nondecreasing on I.*

> **Proof**   Suppose that f fails to be convex on I. By (4.41) there exist $a < x < b$ in I such that
> $$f(x) > \frac{b-x}{b-a} f(a) + \frac{x-a}{b-a} f(b).$$
> This inequality is equivalent to
> $$\frac{f(x) - f(a)}{x - a} > \frac{f(b) - f(x)}{b - x} .$$
> Now apply the Mean Value Theorem to each of the intervals $[a, x]$ and $[x, b]$ to see that there exist $a < \xi < x < \eta < b$ such that $f'(\xi) > f'(\eta)$. Thus f' fails to be nondecreasing on I.
>
> Conversely, if f is convex on I, it follows from (4.43.ii) that f' is nondecreasing on I.   □

(4.47)   Corollary   *Let I and f be as in (4.46). Suppose that f is twice differentiable on I. Then f is convex on I if and only if $f''(x) \geqq 0$ for all $x \in I$.*

Our next theorem gives a useful property of convex functions, particularly for integration theory. First, a simple lemma.

(4.48)   Lemma   *Let I and f be as in (4.43). Then either f is monotone on I or there exists some $p \in I$ such that f is nonincreasing on $\{x \in I : x \leqq p\}$ and f is nondecreasing on $\{x \in I : p \leqq x\}$.*

Proof   Suppose that f is not monotone on I. Then there exist $x < y < z$ in I such that $f(x) > f(y)$ and $f(y) < f(z)$ [the other possibility is ruled out by convexity; see (4.42)]. By (4.43.vi) and (3.81), there is some $p \in [x, z]$ such that $f(p) \leqq f(t)$ for all $t \in [x, z]$. Take $b = x$ and $c = p$ in (4.42) to see that $f(a) \geqq f(p)$ if $a \in I$ and $a < x$. Take $a = p$ and $b = z$ in (4.42) to see that $f(c) \geqq f(p)$ if $c \in I$ and $c > z$. Then $f(p) \leqq f(t)$ for all $t \in I$. Now (4.42) shows that

$$a < b < p \text{ in } I \quad \text{implies} \quad f(a) \geqq f(b),$$
$$p < b < c \text{ in } I \quad \text{implies} \quad f(b) \leqq f(c). \quad \square$$

(4.49)   Theorem   *If $[a, b] \subset \mathbb{R}$ and $f : [a, b] \to \mathbb{R}$ is continuous and convex on $[a, b]$ then f is absolutely continuous on $[a, b]$.*

Proof   Recall the definition of absolute continuity (3.136). Let $\epsilon > 0$ be given. Choose an $\eta > 0$ such that

$$4\eta < b - a,$$
$$|f(x) - f(a)| < \epsilon/3 \qquad \text{if } a < x < a + 2\eta,$$
$$|f(x) - f(b)| < \epsilon/3 \qquad \text{if } b - 2\eta < x < b.$$

By (4.48), we may suppose that η is so small that f is monotone on both $[a, a + 2\eta]$ and $[b - 2\eta, b]$. Next, invoke (4.43.v) to obtain $M > 0$ such that

$$|f(x) - f(y)| \leqq M \cdot |x - y| \quad \text{if} \quad a + \eta \leqq x < y \leqq b - \eta.$$

Now let $\delta = \min\{\eta, \epsilon/(3M)\}$. Suppose that $\{]a_j, b_j[\}_{j=1}^n$ is a pairwise disjoint family of subintervals of $[a, b]$ such that $\sum_{j=1}^n (b_j - a_j) < \delta$. We may suppose that $b_j \leqq a_{j+1}$ for $1 \leqq j < n$ and, by adding two intervals if necessary, that $a_p < a + \eta \leqq a_{p+1} < b_{q-1} \leqq b - \eta < b_q$ for some $1 \leqq p < q \leqq n$. Then $a < b_p = (b_p - a_p) + a_p < \delta + a + \eta \leqq a + 2\eta$, so, using monotoneness, we obtain

$$\sum_{j=1}^p |f(b_j) - f(a_j)| = |\sum_{j=1}^p (f(b_j) - f(a_j))| \leqq |f(b_p) - f(a)| < \epsilon/3.$$

Similarly, at the other end we have

$$b > a_q = b_q - (b_q - a_q) > b - \eta - \delta \geqq b - 2\eta,$$
$$\sum_{j=q}^n |f(b_j) - f(a_j)| \leqq |f(b) - f(a_q)| < \epsilon/3.$$

Therefore,

$$\sum_{j=1}^{n} |f(b_j) - f(a_j)| < 2\epsilon/3 + \sum_{j=p+1}^{q-1} |f(b_j) - f(a_j)|$$

$$\leq 2\epsilon/3 + M \cdot \sum_{j=p+1}^{q-1} |b_j - a_j|$$

$$< 2\epsilon/3 + M\delta \leq \epsilon,$$

and so f satisfies (3.136). □

(4.50)   Corollary *If f is continuous and real-valued on $[a,b] \subset \mathbb{R}$ and if there exists a subdivision $a = x_0 < x_1 < \ldots < x_n = b$ such that f is differentiable and f' is monotone on each $]x_{k-1}, x_k[$, then f is absolutely continuous on $[a,b]$.*

Proof By (4.46), for each k, either f or $-f$ is convex on $[x_{k-1}, x_k]$. By (4.49), f is absolutely continuous on each $[x_{k-1}, x_k]$. □

Exercises

1.   There exists a function that is convex on $[0, 1]$ but not continuous at 0 or 1.

2.   If f and g are convex on an interval I, then so is $f + g$.

3.   Let F be a nonvoid set of convex functions on some interval I. Suppose

$$s(x) = \sup\{ f(x) : f \in F \} < \infty$$

for each $x \in I$. Then s is convex on I.

4.   For each of the functions f defined below, find all intervals on which f is convex.
(a) $f(x) = (1 + x^2)^{-1} \cdot (1 + x)$,
(b) $f(x) = e^x$,
(c) $f(x) = x^x$ $(x > 0)$,
(d) $(1 + x + x^2)^{-1}$,
(e) $f(x) = (\log x)^2$ $(x > 0)$.

5.   Let $(r_n)_{n=1}^{\infty}$ be an enumeration of the set of all positive rational numbers. Define f on $]0, \infty[$ by

$$f(x) = \sum_{r_n < x} \frac{x - r_n}{2^n}.$$

This notation means that the summation is to be extended over those $n \in \mathbb{N}$ for which $r_n < x$. Then
(a) f is convex and strictly increasing on $]0, \infty[$ [Hint: f is the sum of the functions $f_n(x) = \max\{x - r_n, 0\}/2^n$; see Exercises 2 and 3];
(b) $f'(x) = \sum_{r_n < x} (1/2^n)$ for every irrational $x > 0$ [Hint:

$$\lim_{h \downarrow 0} \sum_{x < r_n < x+h} (1/2^n) = 0];$$

(c) $f'_+ (r_p) = f'_- (r_p) + 1/2^p = \sum_{r_n \leq r_p} (1/2^n)$ for every $p \in \mathbb{N}$.

(d) There exists a convex function $g : \mathbb{R} \to \mathbb{R}$ which is differentiable at every irrational and at no rational.

6.   If f is convex and g is convex and nondecreasing, then the composite function $g \circ f$ is convex on any interval where it is defined. Show by an example that the monotoneness of g is needed.

7.   If f is convex on a bounded open interval I, then f is bounded below on I. [Compare (4.48).]

8.   If I is an open interval and $f: I \to \mathbb{R}$ satisfies

$$f((x + y)/2) \leq (1/2)[f(x) + f(y)]$$

for all $x, y \in I$, then f is said to be *midpoint convex* on I. Any continuous midpoint convex function is convex. [Hint: Inequality (4.40.i) obtains for any t of the form $t = m/2^n < 1$ where $m, n \in \mathbb{N}$.]

9.   If $x_j > 0$ and $p_j > 0$ for $1 \leq j \leq n$ and $\sum_{j=1}^{n} p_j = 1$, then

$$\prod_{j=1}^{n} x_j^{p_j} \leq \sum_{j=1}^{n} p_j x_j.$$

Moreover, equality obtains if and only if $x_1 = x_2 = \ldots = x_n$. What does this say when each $p_j = 1/n$? [Hint: $f(x) = -\log x$ is convex on $]0, \infty[.]$

10.   A *logarithmically convex* [*log convex*] function is a positive-valued function f defined on an open interval I such that the composite function $\log \circ f(x) = \log(f(x))$ is convex on I.

(a) Every log convex function is convex.

(b) If f is positive and twice differentiable on I, then f is log convex on I if and only if $(f')^2 \leq ff''$ on I.

(c) If f is linear and log convex on I, then f is a constant on I.

(d) A positive function f on I is log convex on I if and only if $f((1 - t)a + tb) \leq [f(a)]^{1-t} \cdot [f(b)]^t$ for every $a, b \in I$ and $0 \leq t \leq 1$.

(e) A positive function f on I is log convex on I if and only if f is continuous on I and $f((a + b)/2) \leq [f(a)f(b)]^{1/2}$ for all $a, b \in I$. [Hint: Use Exercise 8.]

(f) The sum and product of any two log convex functions on I are both log convex on I. [Hint: Use (e) and Cauchy's Inequality.]

(g) The product of two positive convex functions may fail to be convex.

(h) A positive function f on an open interval I is log convex on I if and only if for each $r \in \mathbb{R}$ the function $g_r(x) = e^{rx}f(x)$ is convex on I. [Hint: If $f((a + b)/2) > (f(a)f(b))^{1/2}$ for some $a, b \in I$, choose r such that $g_r(a) = g_r(b).]$

11.   A real-valued function f on an interval I is convex on I if and only if the determinant

$$\begin{vmatrix} a & f(a) & 1 \\ b & f(b) & 1 \\ c & f(c) & 1 \end{vmatrix}$$

is nonnegative whenever $a < b < c$ in I.

12. Let $P(x) = a_n x^n + \ldots + a_0$ $(n > 1, a_n \neq 0)$ be a polynomial with real coefficients that has only real roots. Then
    (a) the polynomials $P^{(k)}(x)$ $(1 \leqq k < n)$ have only real roots [use Rolle's Theorem],
    (b) the function $f(x) = -\log|P(x)|$ is convex on any open interval between two consecutive roots of $P(x)$,
    (c) $a_{k-1}a_{k+1} \leqq a_k^2$ for $1 \leqq k < n$. [Hints: Note that $a_k = P^{(k)}(0)/k!$ so we need only show that

$$P^{(k-1)}(0)P^{(k+1)}(0) \leqq \left[P^{(k)}(0) \right]^2.$$

    In view of (a), we need only consider the case $k = 1$.]

13. Suppose that f is a log convex function on $]0, \infty[$ [see Exercise 10] and that
    $$f(x + 1) = xf(x), \qquad f(1) = 1$$
    for all $x > 0$. Then
    (a) $f(n) = (n - 1)!$ for all $n \in \mathbb{N}$,
    (b) $f(x + n) = f(x) \displaystyle\prod_{k=0}^{n-1} (x + k)$ for $n \in \mathbb{N}$ and $x > 0$,
    (c) if $g = \log \circ f$, then

    $$g(n) - g(n - 1) \leqq \frac{g(x + n) - g(n)}{x} \leqq g(n + 1) - g(n)$$

    whenever $0 < x < 1$ and $n > 1$,
    (d) $(n - 1)^x \leqq \dfrac{f(x + n)}{(n - 1)!} \leqq n^x$

    whenever $0 < x < 1$ and $n \in \mathbb{N}$,
    (e) writing

    $$f_n(x) = \frac{n^x}{x} \prod_{k=1}^{n} \left(1 + \frac{x}{k}\right)^{-1},$$

    we have

    $$f_{n-1}(x) \leqq f(x) \leqq \left(1 + \frac{x}{n}\right) f_n(x)$$

    for $0 < x < 1$ and $n > 1$,
    (f) $\displaystyle\lim_{n \to \infty} f_n(x) = f(x)$ for $0 < x < 1$
    [show that $\overline{\lim} f_n \leqq f \leqq \underline{\lim} f_n$],
    (g) there is at most one function f that satisfies the hypotheses of this exercise.
    [Remark: The Gamma Function

    $$\Gamma(x) = \int_0^\infty t^{x-1} e^{-t} \, dt$$

    does satisfy the hypotheses (see Chapter 7).]

14. If x, y, a, and b are positive real numbers and $bx \neq ay$, then
    $$(x + y) \log \frac{x + y}{a + b} < x \log \frac{x}{a} + y \log \frac{y}{b} .$$

    [Hint: $u \to u \log u$ is convex on $[0, \infty[$.]

15. Let ϕ be any real-valued function on $]0, \infty[$. Write
    $$f(x) = x\phi(x) \qquad \text{and} \qquad g(x) = \phi(1/x)$$

for $x > 0$. If either f or g is convex on $]0, \infty[$, then so is the other one. [Hint:

$$\frac{2}{x+y} = \frac{x}{x+y} \cdot \frac{1}{x} + \frac{y}{x+y} \cdot \frac{1}{y}.$$

See Exercise 8.]

16. If $I \subset \mathbb{R}$ is a nonvoid open interval and if f is a continuous real-valued function on I such that

(i) $$\overline{\lim_{h \to 0}} \; \frac{f(x+h) + f(x-h) - 2f(x)}{h^2} \geqq 0$$

for all $x \in I$, then f is convex on I. [Hints: For $\epsilon > 0$, write $f_\epsilon(x) = f(x) + \epsilon x^2$. It suffices to show that each f_ϵ is convex. If f_ϵ is not convex, then there is some line $L(x) = ax + b$ such that the function $g = f_\epsilon - L$ has a local maximum at some $c \in I$. In this case,

$$0 \geqq \frac{g(c+h) + g(c-h) - 2g(c)}{h^2} = \frac{f(c+h) + f(c-h) - 2f(c)}{h^2} + 2\epsilon$$

for all sufficiently small $h > 0$.] Conversely, if f is convex on I, then (i) obtains for each $x \in I$. There exist discontinuous nonconvex functions $f: \mathbb{R} \to \mathbb{R}$ such that (i) holds at every $x \in \mathbb{R}$.

Differentiability Almost Everywhere

Let f be a real-valued function on $[a, b] \subset \mathbb{R}$. We have seen in (4.43) that if f is convex on $[a, b]$, then f is differentiable except at countably many points of $[a, b]$. In the converse direction, it is known that if f is differentiable except at countably many points of $[a, b]$ and λ is Lebesgue measure (2.75), then $E \subset [a, b]$ and $\lambda(E) = 0$ imply $\lambda(f(E)) = 0$ [see Exercise 6(a), p. 333]. Thus, if f is Lebesgue's singular function (3.92) and P is Cantor's ternary set, then

$$\lambda(P) = 0 \quad \text{and} \quad f(P) = [0, 1]$$

so it follows that f must be nondifferentiable at an uncountable set of points [all in P]. This example shows that a continuous monotone function can fail to be differentiable at a "large" set of points. In 1904, Henri Lebesgue showed that this set must be "small."

(4.51) Definition Let $P(x)$ be a proposition for $x \in \mathbb{R}$. We say that $P(x)$ is true *almost everywhere* on a set $E \subset \mathbb{R}$ and we write "$P(x)$ a.e. on E" if the set of $x \in E$ at which $P(x)$ is false is a λ-null set; i.e.,

$$\lambda(\{x \in E : P(x) \quad \text{is false}\}) = 0.$$

Lebesgue proved that every continuous monotone function is differentiable a.e. Seven years later, W. H. Young was able to show that Lebesgue's hypothesis of continuity was unnecessary. In view of Jordan's Decomposition Theorem (3.133), "monotone" can be replaced by "finite variation." The following proof does not use (3.133).

(4.52)   Lebesgue's Differentiation Theorem   *If $f:[a,b]\to\mathbb{R}$ satisfies $V_a^b f < \infty$, then f has a finite derivative almost everywhere on $[a,b]$.*

Proof   [Karl Stromberg]   The reader should review the definition of Dini derivates (4.4). Our proof proceeds as follows. Let

$$A = \{x : a < x < b, D_- f(x) < D^+ f(x)\},$$
$$B = \{x : a < x < b, D^- f(x) > D_+ f(x)\},$$
$$C = \{x : a < x < b, |D_+ f(x)| = \infty\}.$$

We then show, by contradiction, that

$$\lambda(A) = \lambda(B) = \lambda(C) = 0. \tag{1}$$

Supposing, for the moment, that this has been done, we complete the proof by writing $H = A \cup B \cup C \cup \{a,b\}$ and then noting that $\lambda(H) = 0$ [(2.76.v) and (2.79.ii)] and that if $x \in [a,b]\setminus H$, then

$$-\infty < D_+ f(x) \leq D^+ f(x) \leq D_- f(x) \leq D^- f(x) \leq D_+ f(x) < \infty$$

and so $f'(x)$ exists and is finite. Thus, we need only prove (1).

Assume that $\lambda(A) > 0$. For each $(c,n) \in \mathbb{Q} \times \mathbb{N}$ write

$$A_{c,n} = \{x \in A : D_- f(x) < c - 1/n < c + 1/n < D^+ f(x)\}.$$

Since $\{A_{c,n} : (c,n) \in \mathbb{Q} \times \mathbb{N}\}$ is a countable family of sets whose union is A, it follows from our assumption and (2.79.ii) that $\lambda(A_{c_0,n_0}) > 0$ for some $(c_0,n_0) \in \mathbb{Q} \times \mathbb{N}$. Write $E = A_{c_0,n_0}$, $\epsilon = 1/n_0$, and $g(x) = f(x) - c_0 x$ for $x \in [a,b]$. Then $V_a^b g \leq V_a^b f + |c_0|(b-a) < \infty$, $D_- g(x) = D_- f(x) - c_0$, and $D^+ g(x) = D^+ f(x) - c_0$ for $a < x < b$, and so

$$\lambda(E) > 0, \tag{2}$$

$$D_- g(x) < -\epsilon < \epsilon < D^+ g(x) \quad \text{for } x \in E. \tag{3}$$

Next choose a subdivision

$$P = \{a = x_0 < x_1 < \ldots < x_n = b\}$$

such that

$$\sum_{k=1}^{n} |g(x_k) - g(x_{k-1})| > V_a^b g - \frac{\epsilon}{6}\lambda(E). \tag{4}$$

Since P is a finite set, (2.76.v) and (2.79.iii) show that $\lambda(E\setminus P) = \lambda(E)$. Thus, we may replace E by $E\setminus P$ without affecting the validity of (2), (3), or (4), and so we suppose that

$$E \cap P = \varnothing. \tag{5}$$

For each $1 \leq k \leq n$ and each $x \in E \cap]x_{k-1}, x_k[$, use (3) to obtain $\alpha_x < \beta_x$

such that

$$x_{k-1} < \alpha_x \leqq x \leqq \beta_x < x_k, \tag{6}$$

$$\frac{g(\beta_x) - g(\alpha_x)}{\beta_x - \alpha_x} < -\epsilon \quad \text{if } g(x_{k-1}) \leqq g(x_k), \tag{7}$$

$$\frac{g(\beta_x) - g(\alpha_x)}{\beta_x - \alpha_x} > \epsilon \quad \text{if } g(x_{k-1}) > g(x_k). \tag{8}$$

[For (7), take $\beta_x = x$. For (8), take $\alpha_x = x$.] Now apply Lemma (2.80) to $\mathcal{G} = \{[\alpha_x, \beta_x] : x \in E\}$ to get a finite, pairwise disjoint, family $\mathcal{F} \subset \mathcal{G}$ such that

$$\sum_{I \in \mathcal{F}} |I| > \frac{1}{6} \lambda(E). \tag{9}$$

For each k such that $I \subset]x_{k-1}, x_k[$ for some $I \in \mathcal{F}$, use Lemma (3.135) and either (7) or (8) [whichever applies] to obtain

$$V_{x_{k-1}}^{x_k} g > |g(x_{k-1}) - g(x_k)| + 2\epsilon \cdot \sum |I|$$

where the sum is over all $I \in \mathcal{F}$ such that $I \subset]x_{k-1}, x_k[$. For all other k, we obviously have

$$V_{x_{k-1}}^{x_k} g \geqq |g(x_k) - g(x_{k-1})|.$$

Thus, using (3.131.ii), (4), and (9), we have

$$V_a^b g = \sum_{k=1}^{n} V_{x_{k-1}}^{x_k} g > \sum_{k=1}^{n} |g(x_k) - g(x_{k-1})| + 2\epsilon \sum_{I \in \mathcal{F}} |I|$$

$$> V_a^b g - (\epsilon/6)\lambda(E) + 2\epsilon \cdot (1/6)\lambda(E)$$

$$= V_a^b g + (\epsilon/6)\lambda(E) > V_a^b g.$$

This contradiction proves that our initial assumption that $\lambda(A) > 0$ was false. Therefore, $\lambda(A) = 0$.

   To prove that $\lambda(B) = 0$, we could give an argument similar to the preceding paragraph. However, it is easier to observe that $D^- f(x) = -D_-(-f)(x)$ and $D_+ f(x) = -D^+(-f)(x)$ and therefore the set B for f is just the set A for $-f$.

   Finally, assume that $\lambda(C) > 0$. Write $M = 6V_a^b f/(\lambda(C))$ and then, for each $x \in C$, choose $h_x > 0$ such that $x + h_x < b$ and

$$|f(x + h_x) - f(x)| > Mh_x.$$

Next, apply Lemma (2.80) to $\mathcal{G} = \{[x, x + h_x] : x \in C\}$ to get a finite pairwise disjoint family $\{[x_j, x_j + h_j]\}_{j=1}^{m} \subset \mathcal{G}$ for which

$$\sum_{j=1}^{m} h_j > \frac{1}{6}\lambda(C) = V_a^b f/M.$$

Then

$$V_a^b f < M \sum_{j=1}^{m} h_j < \sum_{j=1}^{m} |f(x_j + h_j) - f(x_j)| \leq V_a^b f.$$

This contradiction proves that $\lambda(C) = 0$ and so the proof is complete. $\qquad\square$

We next apply (4.52) to prove the following useful theorem on term-by-term differentiation.

(4.53)   Theorem [Fubini]   *Let* $[a,b] \subset \mathbb{R}$ *and let* $(f_n)_{n=1}^{\infty}$ *be a sequence of real-valued nondecreasing functions on* $[a,b]$. *Suppose that the series*

$$f(x) = \sum_{n=1}^{\infty} f_n(x)$$

converges for each $x \in [a,b]$. *Then*

$$f'(x) = \sum_{n=1}^{\infty} f_n'(x) \text{ is finite for almost every } x \in [a,b].$$

Proof   Without loss of generality, we suppose that $f_n(x) \geq 0$ for all n and x [otherwise, we could carry out the proof for the functions $f_n - f_n(a)$]. Then f is nonnegative and nondecreasing on $[a,b]$. By (4.52) and (2.79.ii), there is a set $E \subset]a,b[$ such that $\lambda([a,b] \setminus E) = 0$ and all of the functions f_n and f are differentiable at every point of E. Consider the partial sums

$$S_k = \sum_{n=1}^{k} f_n$$

and the remainders $r_k = f - S_k$. Clearly, every S_k and every r_k is differentiable at every $x \in E$ and

$$S_k'(x) = \sum_{n=1}^{k} f_n'(x), \qquad r_k'(x) = f'(x) - S_k'(x).$$

Now $r_k = \sum_{n=k+1}^{\infty} f_n$ is nondecreasing, so $r_k'(x) \geq 0$ and $f_n'(x) \geq 0$, whence

$$S_k'(x) \leq S_{k+1}'(x) \leq f'(x) \tag{1}$$

for all $x \in E$ and $k \in \mathbb{N}$. It follows that

$$\sum_{n=1}^{\infty} f_n'(x) = \lim_{k \to \infty} S_k'(x) \leq f'(x) < \infty$$

for all $x \in E$. We must show that $\lim_{k \to \infty} S_k'(x) = f'(x)$ a.e. on $[a,b]$. Because of (1), it is enough to find a subsequence of $(S_k')_{k=1}^{\infty}$ that converges to f' at almost every point of $[a,b]$.

Since $S_k(b) \to f(b) < \infty$ as $k \to \infty$, we can choose integers $1 \leq k_1$

$< k_2 < \cdots$ such that

$$\sum_{j=1}^{\infty} \left[f(b) - S_{k_j}(b) \right] < \infty$$

[say $S_{k_j}(b) > f(b) - 2^{-j}$]. Now

$$0 \leqq \left[f(x) - S_{k_j}(x) \right] = r_{k_j}(x) \leqq r_{k_j}(b) = \left[f(b) - S_{k_j}(b) \right]$$

for all $x \in [a,b]$, so the series

$$\sum_{j=1}^{\infty} \left[f - S_{k_j} \right]$$

of nondecreasing functions converges (uniformly) on $[a,b]$. Applying the reasoning of the preceding paragraph to this series, we see that

$$\sum_{j=1}^{\infty} \left[f'(x) - S_{k_j}'(x) \right] < \infty$$

and consequently

$$\lim_{j \to \infty} S_{k_j}'(x) = f'(x)$$

for almost every $x \in [a,b]$. $\square$

As a consequence of (4.53), we have the following remarkable fact.

(4.54) **Theorem** *There exists a strictly increasing continuous function $f : \mathbb{R} \to \mathbb{R}$ such that $f'(x) = 0$ for almost every $x \in \mathbb{R}$.*

Proof Let ψ be Lebesgue's singular function (3.92). The properties that ψ has which we use here are that ψ is continuous, ψ is nondecreasing, $0 \leqq \psi \leqq 1$ on $\mathbb{R}$, $\psi(0) < \psi(1)$, and $\psi' = 0$ a.e. [obviously, $\psi'(x) = 0$ if $x \in \mathbb{R} \backslash P$, where P is Cantor's ternary set]. Let $\{[a_n, b_n]\}_{n=1}^{\infty}$ be an enumeration of the set of all closed intervals of $\mathbb{R}$ having distinct rational endpoints. The idea is to place a copy of ψ on each $[a_n, b_n]$ and then add them. Precisely, for $n \in \mathbb{N}$, define f_n on $\mathbb{R}$ by

$$f_n(x) = 2^{-n} \psi \left(\frac{x - a_n}{b_n - a_n} \right)$$

and then define

$$f(x) = \sum_{n=1}^{\infty} f_n(x).$$

Since f_n is continuous and $0 \leqq f_n \leqq 2^{-n}$ on $\mathbb{R}$, Weierstrass's M-test shows that f is continuous on $\mathbb{R}$. Each f_n is nondecreasing on $\mathbb{R}$ so f is also. If $x < y$ in $\mathbb{R}$, we can choose n such that $x < a_n < b_n < y$, and so $f(y) - f(x) \geqq f_n(y) - f_n(x) \geqq f_n(b_n) - f_n(a_n) = [\psi(1) - \psi(0)]/2^n > 0$. Thus, f is strictly increasing on $\mathbb{R}$. Finally, (4.53) yields

$$f'(x) = \sum_{n=1}^{\infty} f_n'(x) = 0$$

a.e. on $\mathbb{R}$. $\square$

Exercises

1. If f, as in (4.8), is any continuous, nowhere differentiable function on $\mathbb{R}$, then $V_a^b f = \infty$ whenever $a < b$ in $\mathbb{R}$.

2. Let $E \subset \mathbb{R}$ be any set such that $\lambda(E) = 0$.
   (a) There exists a countable family $\mathcal{I}$ of open intervals such that each point of E is in infinitely many members of $\mathcal{I}$ and

   $$\sum_{I \in \mathcal{I}} |I| < 1.$$

   (b) There exists a bounded, continuous, nondecreasing function f on $\mathbb{R}$ such that $f'(x) = \infty$ for every $x \in E$. [Hints: Write $\mathcal{I} = \bigcup_{k=1}^{\infty} \mathcal{I}_k$ where

   $$E \subset \cup \mathcal{I}_k \quad \text{and} \quad \sum_{I \in \mathcal{I}_k} |I| < 2^{-k}.$$

   Define

   $$f(t) = \sum_{I \in \mathcal{I}} |] - \infty, t[\cap I|.$$

   Given $x \in E$ and $N \in \mathbb{N}$, choose $\delta > 0$ such that $]x - \delta, x + \delta[$ is contained in at least N members of $\mathcal{I}$. Then $0 < |h| < \delta$ implies $|f(x + h) - f(x)| \geq N \cdot |h|$.]

3. In contrast with (4.53), there exists a nondecreasing sequence $(f_n)_{n=1}^{\infty}$ of nondecreasing functions on $[0, 1]$ such that $f_n(x) \to x$ uniformly on $[0, 1]$ while $f_n'(x) \to 1$ for *no* $x \in [0, 1]$. [Hint: Write $f_n(x) = (k - 1) \cdot 2^{-n}$ if $(k - 1) \cdot 2^{-n} \leq x < k \cdot 2^{-n}$.]

4. (a) If $A \subset \mathbb{R}$ and $a < s < t$ in $\mathbb{R}$, then $\lambda(A \cap [a, t]) = \lambda(A \cap [a, s]) + \lambda(A \cap [s, t])$.
   (b) If $I \subset \mathbb{R}$ is an open interval, $x \in I$, and $\phi: I \to \mathbb{R}$ is a function which is differentiable at x, then

   $$\lim_{h + k \to 0} \frac{\phi(x + h) - \phi(x - k)}{h + k} = \phi'(x)$$

   where we require $h \geq 0$, $k \geq 0$, and $h + k > 0$. [Hint: The absolute difference between the quotient and $\phi'(x)$ is dominated by

   $$\left| \frac{\phi(x + h) - \phi(x)}{h} - \phi'(x) \right| + \left| \frac{\phi(x - k) - \phi(x)}{-k} - \phi'(x) \right|.$$

   (c) If $E \subset \mathbb{R}$, then almost every $x \in E$ has the property that

   $$\lim_{|I| \to 0} [\lambda(E \cap I)/|I|] = 1$$

   where the limit is taken over all intervals I with $x \in I$ and $|I| > 0$. Points $x \in \mathbb{R}$ having this property are called *points of density* for E. [Hints: Suppose that

$E \subset]a, b[$ for some $a < b$ in $\mathbb{R}$. Choose open sets V_n with $E \subset V_n \subset]a, b[$ and $\lambda(V_n) < \lambda(E) + 2^{-n}$ by letting V_n be the union of an appropriate cover of E by open intervals. Define $\phi(x) = \lambda(E \cap [a, x])$ and $\phi_n(x) = \lambda(V_n \cap [a, x])$ for $a \leq x \leq b$. Use (a) to show that the functions $f_n = \phi_n - \phi$ are each nondecreasing on $[a, b]$. Use (4.53) and (4.52) to show that $\phi_n'(x) \to \phi'(x)$ a.e. on $[a, b]$. Check that $\phi_n'(x) = 1$ for all $x \in E$ and all n. Conclude that $\phi'(x) = 1$ a.e. on E and use (b).]

5. *The Denjoy–Young–Saks Theorem.* Let f be an arbitrary real-valued function defined on an open interval $I \subset \mathbb{R}$. The theorem stated in (f) concerns the a.e. behavior of the four Dini derivates of f. The results are truly remarkable in view of the complete lack of hypotheses about f.

(a) At almost every x of the set $E = \{x \in I : f(t) \leq f(x)$ if $t \in I$ and $t < x\}$ we have $0 \leq D_- f(x) = D^+ f(x) < \infty$. [Hints: We suppose that $\lambda(E) > 0$ since there is nothing to prove otherwise. Write $a = \inf E$ and $b = \sup E$. For $a < u < b$, define $g(u) = \sup f(]a, u])$. Then $g = f$ on $E \cap]a, b[$ and g is nondecreasing. Fix any $c \in E$ at which g is differentiable and for which c is a point of density for E (see Exercise 4(c)). For $c < x < b$,

$$\frac{f(x) - f(c)}{x - c} \leq \frac{g(x) - g(c)}{x - c}$$

(because $f(x) \leq g(x)$ and $f(c) = g(c)$) so $D^+ f(c) \leq g'(c)$. Since

$$\lim_{n \to \infty} n\lambda(E \cap]c, c + 1/n]) = 1,$$

there is a sequence $(x_n)_{n=1}^{\infty} \subset E \cap]c, b[$ with $x_n \to c$ as $n \to \infty$. Now $g(x_n) = f(x_n)$ for all n, and so we have

$$D^+ f(c) \geq \lim_{n \to \infty} \frac{f(x_n) - f(c)}{x_n - c} = g'(c).$$

Therefore $D^+ f(c) = g'(c)$. Similar reasoning shows that $D_- f(c) = g'(c)$. Also, $0 \leq g'(c) < \infty]$.

(b) At almost every x of the set $A = \{x \in I : D_- f(x) > -\infty\}$ we have $D^+ f(x) = D_- f(x) < \infty$. [Hints: For $n \in \mathbb{N}$ and rational $r \in I$, write $f_n(x) = f(x) + nx$, $I_r = I \cap]r, \infty[$, and $E_{n,r} = \{x \in I_r : f_n(t) \leq f_n(x)$ if $r < t < x\}$. Apply (a) to see that $-n \leq D_- f(x) = D^+ f(x) < \infty$ for all $x \in E_{n,r} \setminus N_{n,r}$ where $\lambda(N_{n,r}) = 0$. Check that A is the union of the sets $E_{n,r}$, let N be the union of the sets $N_{n,r}$, and obtain $D_- f(x) = D^+ f(x) < \infty$ for all $x \in A \setminus N$.]

(c) At almost every x of the set $B = \{x \in I : D^+ f(x) < \infty\}$ we have $D^+ f(x) = D_- f(x) > -\infty$. [Hint: Apply (b) to the function g defined by $g(x) = f(-x)$ for $-x \in I$. Notice that $D^+ f(-x) = -D_- g(x)$.]

(d) At almost every $x \in I$ we have either $-\infty < D_- f(x) = D^+ f(x) < \infty$ or else $D_- f(x) = -\infty$ and $D^+ f(x) = \infty$. [Hint: Consider whether $x \in A \cup B$ or not.]

(e) At almost every $x \in I$ we have either $-\infty < D_+ f(x) = D^- f(x) < \infty$ or else $D_+ f(x) = -\infty$ and $D^- f(x) = \infty$. [Hint: Apply (d) to the function g defined by $g(x) = -f(x)$. Notice that $D_- g = -D^- f$ and $D^+ g = -D_+ f$.]

(f) There exists a set $Z \subset I$ with $\lambda(Z) = 0$ such that at each $x \in I \setminus Z$ we have just one of the four possibilities:

(1) $-\infty < D_- f(x) = D^+ f(x) = D_+ f(x) = D^- f(x) < \infty$, so f is differentiable at x;

(2) $D_- f(x) = D_+ f(x) = -\infty$ and $D^+ f(x) = D^- f(x) = \infty$;

(3) $D_- f(x) = -\infty$, $D^+ f(x) = \infty$, and $-\infty < D^- f(x) = D_+ f(x) < \infty$;

(4) $D_+ f(x) = -\infty$, $D^- f(x) = \infty$, and $-\infty < D_- f(x) = D^+ f(x) < \infty$.

(g) [A. N. Singh] For $x = \sum\limits_{n=1}^{\infty} (a_n/3^n)$ where each a_n is 0, 1, or 2 and $a_n \neq 2$ for

infinitely many n, define $f(x) = \sum\limits_{n=1}^{\infty} (b_n/2^n)$ where $b_n = 1$ if $a_n = 2$ and $b_n = 0$

otherwise. Then $D^+ f(x) = \infty$ for every $x \in [0, 1[$. Moreover, f is continuous except at those countably many x for which $a_n = 0$ for all sufficiently large n, and it is right continuous even at those x. [Hint: If $a_p \neq 2$, let x_p have the same expansion as x except that a_p is replaced by 2. Then $x_p > x$ and $[f(x_p) - f(x)]$ $/(x_p - x) \geq 3^p/2^{p+1}$.]

Termwise Differentiation of Sequences

Theorem (4.53) answers a rather special case of the general question: If $f_n \to f$, does it follow that $f_n' \to f'$? Before presenting a theorem that gives sufficient conditions for an affirmative answer, we give a few examples to illustrate the difficulties involved.

(4.55)   Examples (a) Let $f_n(x) = nx/(1 + n^2 x^2)$ and $f(x) = 0$ for all $n \in \mathbb{N}$ and $x \in \mathbb{R}$. Here $f_n(x) \to f(x)$ as $n \to \infty$ for every $x \in \mathbb{R}$, but the convergence is not uniform because $f_n(1/n) = 1/2$ for all n. We have $f_n'(x) = (n - n^3 x^2)/(1 + n^2 x^2)^2$, and so $f_n'(x) \to 0 = f'(x)$ for each $x \neq 0$, while $f_n'(0) = n \to \infty \neq 0 = f'(0)$.

(b) For this elegant example we anticipate some simple properties of the trigonometric functions which are studied in Chapter 5. Those readers who are concerned about possible logical circularity [there is none] may wish to omit this example until after they have studied the relevant facts of the next chapter.

Let $f_n(x) = n^{-1/2} \cdot \sin(nx)$ and $f(x) = 0$ for all $n \in \mathbb{N}$ and $x \in \mathbb{R}$. Since $|\sin t| \leq 1$ for all $t \in \mathbb{R}$, we have $|f(x) - f_n(x)| \leq n^{-1/2}$ for all n and x, and so $f_n \to f$ *uniformly* on $\mathbb{R}$. Notice that $f_n'(x) = n^{1/2} \cdot \cos(nx)$ and $f'(x) = 0$. We claim that for each $x \in \mathbb{R}$ the sequence $(f_n'(x))_{n=1}^{\infty}$ is unbounded and so cannot converge to $f'(x)$ or to any other finite limit. To see this, fix any $x \in \mathbb{R}$. If $p \in \mathbb{N}$ and $|\cos(px)| < 1/2$, then

$$|\cos(2px)| = |2\cos^2(px) - 1| > 1/2.$$

Thus, for each $p \in \mathbb{N}$, either

$$|\cos(px)| \geq 1/2 \quad \text{or} \quad |\cos(2px)| \geq 1/2.$$

Accordingly, $\{n \in \mathbb{N} : |f_n'(x)| \geq \sqrt{n}/2\}$ is infinite, and so the claim follows. To summarize, our sequence $(f_n) \subset C^\infty(\mathbb{R})$ converges *uniformly* on $\mathbb{R}$ but the sequence (f_n') of derivatives converges *nowhere* on $\mathbb{R}$.

To get a positive result, we need to know that the sequence of derivatives converges uniformly.

(4.56)　Theorem　*Let $a < b$ be in $\mathbb{R}$ and let $(f_n)_{n=1}^{\infty}$ be a sequence of real-valued functions that are differentiable on $[a, b]$. Suppose that $(f_n')_{n=1}^{\infty}$ converges uniformly on $[a, b]$ to some function g and that, for some $c \in [a, b]$, the sequence $(f_n(c))_{n=1}^{\infty}$ converges. Then there exists a function f that is differentiable on $[a, b]$ such that $f_n \to f$ uniformly on $[a, b]$ and $f'(x) = g(x) = \lim_{n \to \infty} f_n'(x)$ for all $x \in [a, b]$.*

Proof　For any function $h : [a, b] \to \mathbb{C}$, let us write

$$\|h\| = \sup\{|h(u)| : u \in [a, b]\}.$$

Now, given any x and t in $[a, b]$ and any $m, n \in \mathbb{N}$, we can apply the Mean Value Theorem to the function $f_m - f_n$ to obtain

$$|(f_m - f_n)(x) - (f_m - f_n)(t)| \leq \|f_m' - f_n'\| \cdot |x - t|. \tag{1}$$

If we take $t = c$ in (1), we have

$$|(f_m - f_n)(x)| \leq |(f_m - f_n)(x) - (f_m - f_n)(c)| + |(f_m - f_n)(c)|$$
$$\leq \|f_m' - f_n'\| \cdot (b - a) + |f_m(c) - f_n(c)| \tag{2}$$

for all $x \in [a, b]$. Our hypotheses show that the right side of (2), which does not depend on x, has limit 0 as $m, n \to \infty$. It follows that $(f_n)_{n=1}^{\infty}$ converges uniformly on $[a, b]$ to some function f.

To prove our differentiation assertions, let $x \in [a, b]$ and $\epsilon > 0$ be fixed and arbitrary. First choose $N \in \mathbb{N}$ such that

$$\|f_N' - g\| < \epsilon/3 \tag{3}$$

and then choose $\delta > 0$ so that

$$\left| \frac{f_N(t) - f_N(x)}{t - x} - f_N'(x) \right| < \frac{\epsilon}{3} \tag{4}$$

whenever $t \in [a, b]$ and $0 < |t - x| < \delta$. Now fix any such t. Using (1), we have

$$\left| \frac{f_m(t) - f_m(x)}{t - x} - \frac{f_N(t) - f_N(x)}{t - x} \right| \leq \|f_m' - f_N'\| \tag{5}$$

for any $m \in \mathbb{N}$. Letting $m \to \infty$ in (5) and using (3) yields

$$\left| \frac{f(t) - f(x)}{t - x} - \frac{f_N(t) - f_N(x)}{t - x} \right| \leq \|g - f_N'\| < \frac{\epsilon}{3}. \tag{6}$$

From (3) follows

$$|f_N'(x) - g(x)| < \epsilon/3. \tag{7}$$

Now add inequalities (6), (4), and (7) and use the triangle inequality to get

$$\left| \frac{f(t) - f(x)}{t - x} - g(x) \right| < \epsilon. \tag{8}$$

Since (8) holds whenever

$$t \in [a, b] \quad \text{and} \quad 0 < |t - x| < \delta,$$

we conclude that $f'(x) = g(x)$. $\square$

Exercises

1. Let (f_n) be a sequence of real-valued functions on $[a, b]$ which converges at each t in a certain dense subset D of $[a, b]$. Suppose that each f_n is differentiable on $[a, b]$ and that there exists some positive constant M such that $|f_n'(x)| \leq M < \infty$ for all n and x. Then (f_n) converges uniformly on $[a, b]$. [Hint: We have

$$|f_m(x) - f_n(x)| \leq |(f_m - f_n)(x) - (f_m - f_n)(t)| + |(f_m - f_n)(t)|$$
$$\leq 2M|x - t| + |(f_m - f_n)(t)|.$$

Given $\epsilon > 0$, first choose a finite set $T \subset D$ such that $\text{dist}(x, T) < \epsilon/(4M)$ for all $x \in [a, b]$.]

2. Write $f_n(x) = x(1 + nx^2)^{-1}$ for $n \in \mathbb{N}$ and $x \in \mathbb{R}$. Find the pointwise limits f and g of the sequences (f_n) and (f_n'). Also find all intervals on which these sequences converge uniformly. Find all x for which $f'(x) = g(x)$.

3. Let $I \subset \mathbb{R}$ be a nonvoid open interval and let $(u_n)_{n=0}^{\infty}$ be a sequence of real-valued functions that are differentiable on I. Suppose that the series $\sum_{n=0}^{\infty} u_n'$ converges uniformly on every compact subinterval of I and that there is some $c \in I$ such that $\sum_{n=0}^{\infty} u_n(c)$ converges. Then there exists a differentiable function $f : I \to \mathbb{R}$ such that $\sum_{n=0}^{\infty} u_n$ converges to f uniformly on every compact subinterval of I and $\sum_{n=0}^{\infty} u_n'(x) = f'(x)$ for all $x \in I$.

4. Use Exercise 3 to prove the special case of (4.63) (below) in which $a \in \mathbb{R}$, $(c_n)_{n=0}^{\infty} \subset \mathbb{R}$, $z = x \in \mathbb{R}$, and $D = I =]a - R, a + R[$. [Remark: We do not use this method to prove (4.63) itself because we have not proven any analog of (4.56) for complex derivatives. Recall that the Mean Value Theorem can fail for complex-valued functions.]

5. Write $u_n(x) = x^n/n - x^{n+1}/(n + 1)$ for $n \in \mathbb{N}$ and $x \in \mathbb{R}$. Define f and g by

$$f(x) = \sum_{n=1}^{\infty} u_n(x), \quad g(x) = \sum_{n=1}^{\infty} u_n'(x)$$

at all $x \in \mathbb{R}$ for which these series converge. For which x does the equality $f'(x) = g(x)$ obtain? Notice that $f_-'(1) \neq g(1)$ and that $f_+'(-1)$ is defined, while $g(-1)$ is not defined.

6. (a) Suppose that $I \subset \mathbb{R}$ is an open interval, $a \in I$, and $(\phi_n)_{n=1}^{\infty}$ is a sequence of real-valued functions on I that have derivatives $\phi_n'(a)$ (possibly infinite) at a and satisfy

$$0 \leq \frac{\phi_n(t) - \phi_n(a)}{t - a} \leq M\phi_n'(a)$$

for all $t \in I$ ($t \neq a$) and all $n \in \mathbb{N}$, where M is some positive real constant that depends neither on t nor on n. Suppose also that

$$g(t) = \sum_{n=1}^{\infty} \phi_n(t),$$

where this series converges for every $t \in I$. Then $g'(a)$ exists (possibly infinite) and

$$g'(a) = \sum_{n=1}^{\infty} \phi'_n(a). \tag{1}$$

[Hints: If series (1) diverges and its Nth partial sum exceeds b, then

$$\frac{g(t) - g(a)}{t - a} \geqq \sum_{n=1}^{N} \frac{\phi_n(t) - \phi_n(a)}{t - a} > b$$

for all t sufficiently near to a. If series (1) converges and $(M + 1) \sum_{N+1}^{\infty} \phi'_n(a) < \epsilon$, then

$$\left| \frac{g(t) - g(a)}{t - a} - \sum_{n=1}^{\infty} \phi'_n(a) \right| < \epsilon + \sum_{n=1}^{N} \left| \frac{\phi_n(t) - \phi_n(a)}{t - a} - \phi'_n(a) \right|.$$

(b) Suppose that $\phi : \mathbb{R} \to \mathbb{R}$ has a (possibly infinite) derivative $\phi'(a)$ at each $a \in \mathbb{R}$ and that there is some positive real constant M such that

$$0 \leqq \frac{\phi(t) - \phi(a)}{t - a} \leqq M\phi'(a)$$

for all $a, t \in \mathbb{R}$ with $a \neq t$. Then, for any three sequences $(r_n)_{n=1}^{\infty}$, $(c_n)_{n=1}^{\infty}$, $(\lambda_n)_{n=1}^{\infty}$ of real numbers with $c_n \geqq 0$ and $\lambda_n > 0$ for all n, the functions ϕ_n defined on $\mathbb{R}$ by $\phi_n(t) = (c_n/\lambda_n)\phi[\lambda_n(t - r_n)]$ satisfy all the conditions of the first sentence of (a) at every $a \in \mathbb{R} = I$.

(c) The conditions in (b) are fulfilled if $\phi(t) = t^{1/3}$ and $M = 4$.

(d) If $(r_n)_{n=1}^{\infty} \subset \mathbb{R}$, then the function h defined on $\mathbb{R}$ by

$$h(t) = t + \sum_{n=1}^{\infty} \frac{(t - r_n)^{1/3}}{n^2(1 + |r_n|)^{1/3}}$$

is a strictly increasing continuous function from $\mathbb{R}$ *onto* $\mathbb{R}$ which has a derivative at each $a \in \mathbb{R}$. Moreover, $h'(a) > 1$ for each $a \in \mathbb{R}$ and $h'(r_n) = \infty$ for each $n \in \mathbb{N}$. [Hints: Use (c), (b), and (a) with $c_n = 1$ and $\lambda_n = n^3(1 + |r_n|)^{1/2}$. Since $|\phi_n(t)| < n^{-2}(1 + |t|)^{1/3}$, the series converges uniformly on each bounded interval of $\mathbb{R}$.]

(e) There exists a strictly increasing differentiable function H from $\mathbb{R}$ *onto* $\mathbb{R}$ such that $H'(x) < 1$ for all $x \in \mathbb{R}$ and $\{x \in \mathbb{R} : H'(x) = 0\}$ is dense in $\mathbb{R}$. [Hint: Let $(r_n)_{n=1}^{\infty}$ be dense in $\mathbb{R}$, write $H = h^{-1}$, and note that $H'(h(r_n)) = 0$ for all n.]

(f) For any function H as in (e), the set $\{x \in \mathbb{R} : H'(x) > 0\}$ is dense in $\mathbb{R}$ and the function H' is discontinuous at every point of this set.

7. Let D denote the set of all *bounded* functions $f : \mathbb{R} \to \mathbb{R}$ for which there exists some differentiable function $F : \mathbb{R} \to \mathbb{R}$ such that $F'(x) = f(x)$ for all $x \in \mathbb{R}$. Supply the set D with the uniform metric:

$$d(f, g) = \|f - g\|_u = \sup\{|f(x) - g(x)| : x \in \mathbb{R}\}$$

for $f, g \in D$. We call D the *space of bounded derivatives*.

(a) The metric space D is complete. [Hint: If $(f_n)_{n=1}^{\infty}$ is a Cauchy sequence, we can choose F_n such that $F_n' = f_n$ and $F_n(0) = 0$. Apply (4.56).]

(b) For each $f \in D$, the set $Z(f) = \{x \in \mathbb{R} : f(x) = 0\}$ is a G_δ-set in $\mathbb{R}$. Recall that this means that $Z(f)$ is the intersection of some countable family of open subsets of $\mathbb{R}$. We call $Z(f)$ the *zero set* of f. [Hint: Let $f = F'$ and $f_n(x) = n[F(x + 1/n) - F(x)]$. Check that $f_n(x) \to f(x)$ for all x and that

$$Z(f) = \bigcap_{k=1}^{\infty} \bigcap_{m=1}^{\infty} \bigcup_{n=m}^{\infty} \{x \in \mathbb{R} : |f_n(x)| < 1/k\}.]$$

Now let $D_0 = \{f \in D : Z(f) \text{ is dense in } \mathbb{R}\}$.

(c) If f and g are in D_0 and $\alpha \in \mathbb{R}$, then $f + g$ and αf are in D_0. [Hint: $Z(f + g) \supset Z(f) \cap Z(g)$, and this intersection is dense in $\mathbb{R}$ by (b) and the Baire Category Theorem.]

(d) The set D_0 is a closed subset of D. [Hint: If $(f_n)_{n=1}^{\infty} \subset D_0, f \in D$ and $\|f - f_n\|_u \to 0$, then $Z(f) \supset \bigcap_{n=1}^{\infty} Z(f_n).]$

(e) If I is a nonvoid open interval of $\mathbb{R}$, then the sets $A = \{f \in D_0 : f(x) \geqq 0$ for all $R \in I\}$ and $B = \{f \in D_0 : f(x) \leqq 0$ for all $x \in I\}$ are closed and nowhere dense in the complete metric space D_0. [Hint: Let H be as in Exercise 6(e) and choose $a \in \mathbb{R}$ for which $H'(a) > 0$. Suppose $f \in A$ and $\epsilon > 0$. Choose $b \in I$ for which $f(b) = 0$ and define g by $g(x) = f(x) - \epsilon H'(x - b + a)$. Then $g \in D_0$, $g(b) < 0, g \notin A$, and $\|f - g\|_u \leqslant \epsilon$. Thus f is not an interior point of A.]

(f) The set of all $f \in D$ for which the three sets $Z(f)$, $P(f) = \{x \in \mathbb{R} : f(x) > 0\}$ and $N(f) = \{x \in \mathbb{R} : f(x) < 0\}$ are all dense in $\mathbb{R}$ is a dense subset of D_0. [Hint: Let $\{I_n\}_{n=1}^{\infty}$ be an enumeration of the family of all nonvoid open intervals of $\mathbb{R}$ having rational endpoints. For each n, let A_n and B_n correspond to I_n as in (e). By the Baire Category Theorem, the set $D_0 \backslash \bigcup_{n=1}^{\infty} (A_n \cup B_n)$ is dense in D_0.]

(g) There exist functions $F : \mathbb{R} \to \mathbb{R}$ such that $F'(x)$ exists with $-1 < F'(x) < 1$ at each $x \in \mathbb{R}$ but yet there is *no* nonvoid open interval of $\mathbb{R}$ on which F is monotone. We might say that F is reasonably smooth and wiggles everywhere. [Hint: Choose f as in (f) with $\|f\|_u < 1$ and then choose F so that $F' = f$.]

(h) Let F be as in (g) and let $a < b$ be real numbers. Then
   (1) $|F(b) - F(a)| < b - a$,
   (2) F is absolutely continuous on $[a, b]$,
   (3) $V_a^b F \leqq b - a$,
   (4) F has a local minimum at some $u \in \,]a, b[$,
   (5) F has a local maximum at some $v \in \,]a, b[$,
   (6) $Z(F')$ is a dense G_δ in $\mathbb{R}$,
   (7) if F' is continuous at x, then $F'(x) = 0$,
   (8) the sets $]a, b[\cap P(F')$ and $]a, b[\cap N(F')$ are both nonvoid.

   We remark that the sets in (8) both have positive Lebesgue measure, but it is most convenient to prove this fact by using (6.84). We are indebted to Professor Clifford E. Weil for suggesting to us this elegant category proof of the existence of such functions F. An explicit construction of such an F (due to Yitzhak Katznelson and the present author) is outlined in the following exercise.

8. Define ψ on $\mathbb{R}$ by $\psi(x) = 2(1 + x)^{1/2} - 2$ if $x \geqq 0$ and $\psi(x) = 2 - 2(1 - x)^{1/2}$ if $x \leqq 0$.

(a) We have $\psi'(x) = (1 + |x|)^{-1/2}$ for all $x \in \mathbb{R}$.

(b) If $t, a \in \mathbb{R}$ with $t \neq a$, then

$$[\psi(t) - \psi(a)]/(t - a) < 4\psi'(a).$$

[Hint: For the case that $ta < 0$, it may be useful to prove and use the fact that $(r + s - 2)/(r^2 + s^2 - 2) < 2/s$ whenever $r > 1$ and $s > 1$. This inequality is equivalent to $(r - s)^2 + (r - 1)(s - 1) + r^2 + r + 3s > 5$.]

(c) If ϕ is any function of the form

$$\phi(x) = c_0 + \sum_{j=1}^{m} \frac{c_j}{\lambda_j} \psi(\lambda_j(x - r_j))$$

where $c_1, \ldots, c_m, \lambda_1, \ldots, \lambda_m$ are positive real numbers and $c_0, r_1, \ldots, r_m$ are any real numbers, then

$$(\phi(t) - \phi(a))/(t - a) < 4\phi'(a)$$

for all $t, a \in \mathbb{R}$ with $t \neq a$.

(d) If $(\phi_n)_{n=1}^{\infty}$ is any sequence of functions as in (c) for which $g(x) = \sum_{n=1}^{\infty} \phi_n'(x) < \infty$ for all $x \in \mathbb{R}$ and $\sum_{n=1}^{\infty} |\phi_n(a)| < \infty$ for some $a \in \mathbb{R}$, then the series $G = \sum_{n=1}^{\infty} \phi_n$ converges uniformly on every bounded interval of $\mathbb{R}$, and $G'(x) = g(x)$ for every $x \in \mathbb{R}$. [Hints: If $b > |a|$ and $|x| \leq b$, then $|\phi_n(x)| < 8b\phi_n'(a) + |\phi_n(a)|$. Compare Exercise 6(a).]

(e) Let $I_1, \ldots, I_m$ be pairwise disjoint open intervals of $\mathbb{R}$ having midpoints $r_1, \ldots, r_m$, respectively, and let ϵ and $y_1, \ldots, y_m$ be positive real numbers. Then it is possible to select the c's and λ's so that the function ϕ in (c) satisfies: (1) $\phi'(r_j) > y_j$ for all j; (2) $\phi'(x) < y_j + \epsilon$ if $x \in I_j$; (3) $\phi'(x) < \epsilon$ if x is in no I_j; (4) $\phi(0) = 0$. [Hint: For $1 \leq j \leq m$, take $c_j = y_j + \epsilon/2$ and then fix λ_j so that $c_j \psi'(\lambda_j t) < \epsilon/(2m)$ if $|t| \geq |I_j|/2$. Finally choose c_0 so that (4) obtains.]

(f) If $\{r_j\}_{j=1}^{\infty}$ and $\{s_j\}_{j=1}^{\infty}$ are disjoint countable subsets of $\mathbb{R}$, then there exists an everywhere differentiable function $G : \mathbb{R} \to \mathbb{R}$ such that $G'(r_j) = 1$, $G'(s_j) < 1$ and $0 < G'(x) \leq 1$ for all $j \in \mathbb{N}$ and $x \in \mathbb{R}$. [Hints: Let G be as in (d) with partial sums $G_m = \sum_{n=1}^{m} \phi_n$ where (e) is used to inductively construct the sequence $(\phi_n)_{n=1}^{\infty}$ so that the three conditions $A_m : G_m'(r_j) > (m - 1)/m$ for $1 \leq j \leq m$, $B_m : G_m'(x) < m/(m + 1)$ for $x \in \mathbb{R}$, $C_m : \phi_m'(s_j) < 1/(2m \cdot 2^m)$ for $1 \leq j \leq m$ are satisfied for all $m \in \mathbb{N}$. For $m = 1$, choose I_1 (with midpoint r_1 to miss s_1 and obtain $\phi_1 = G_1$ to correspond to $y_1 = \epsilon = 1/4$. Supposing $\phi_1, \ldots, \phi_{m-1}$ have been constructed, choose the I_j to miss $s_1, \ldots, s_m$ and satisfy $G_{m-1}'(x) < G_{m-1}'(r_j) + [m(m + 1)]^{-1} - [2m \cdot 2^m]^{-1}$ for $x \in I_j$ $(1 \leq j \leq m)$ and then invoke (e) with $\epsilon = [2m \cdot 2^m]^{-1}$ and $y_j = 1 - m^{-1} - G_{m-1}'(r_j)$ to obtain ϕ_m.]

(g) There exists an everywhere differentiable function $F : \mathbb{R} \to \mathbb{R}$ which is monotone on *no* nonvoid open interval of $\mathbb{R}$ such that $|F'(x)| < 1$ for all $x \in \mathbb{R}$. [Hint: For disjoint dense subsets $\{r_j\}_{j=1}^{\infty}$ and $\{s_j\}_{j=1}^{\infty}$ of $\mathbb{R}$, let G be as in (f), let H be the same as G except that the roles of the r_j and the s_j are interchanged, and write $F = G - H$.]

Complex Derivatives

Until now, we have discussed derivatives only for real-valued functions defined on intervals of $\mathbb{R}$. The same formal definition of derivative, as a limit of difference quotients, can be used to extend our concept to complex-valued functions that are defined on intervals of $\mathbb{R}$ or on open subsets of $\mathbb{C}$. We therefore make the following two definitions.

(4.57) **Definition** Let $I \subset \mathbb{R}$ be an interval of positive length and let f be a complex-valued function defined on I. We say that f is *differentiable* at a point $c \in I$ if there exists a complex number $f'(c)$ such that

$$\lim_{h \to 0} \frac{f(c + h) - f(c)}{h} = f'(c) \qquad (c + h \in I).$$

If this holds, we call $f'(c)$ *the derivative of f at c* [it is obviously unique if it exists]. In case c is an endpoint of I, we refer to the right or left derivative at c and write $f'_+(c)$ or $f'_-(c)$ instead of $f'(c)$. If f is differentiable at every $c \in I$, we say that f is *differentiable on I*.

(4.58) **Definition** Let D be a nonvoid open subset of $\mathbb{C}$ and let f be a complex-valued function defined on D. For $c \in D$, we say that f is *complex differentiable at c* if there exists a complex number $f'(c)$ such that

$$\text{(i)} \qquad \lim_{h \to 0} \frac{f(c + h) - f(c)}{h} = f'(c) \qquad (c + h \in D).$$

In this case, we call $f'(c)$ *the complex derivative of f at c* [it is obviously unique if it exists]. In case f is complex differentiable at every $c \in D$, we say that f is *differentiable* [or *holomorphic*] *on D*. We use the adjective "complex" to emphasize that the limit in (i) is required to exist and equal $f'(c)$ as h tends to 0 through arbitrary *complex* values; i.e., for each $\epsilon > 0$ there exists some $\delta > 0$ such that

$$\left| \frac{(f(c + h) - f(c))}{h} - f'(c) \right| < \epsilon$$

whenever $h \in \mathbb{C}$, $c + h \in D$, and $0 < |h| < \delta$. The following example should clarify the distinction between this and the preceding definition.

(4.59) **Example** Let $D = \mathbb{C}$ and $f(z) = \text{Re}(z)$ for $z \in D$. For any $c \in D$ and any $h = u + iv \neq 0$ $(u, v \in \mathbb{R})$, we have

$$\frac{(f(c + h) - f(c))}{h} = u/(u + iv). \tag{1}$$

It is clear that if we consider only real h $(v = 0)$, then the expression in (1) is equal to 1 for all $h \in \mathbb{R}$, $h \neq 0$. In particular, the restriction of f to $\mathbb{R}$ has derivative 1 at every $c \in \mathbb{R}$. However, there is no $c \in \mathbb{C}$ at which f has a complex derivative. To see this, notice .that if $h \to 0$ along the real axis $(v = 0)$, the limit of (1) is 1, while if $h \to 0$ along the imaginary axis $(u = 0)$, the limit of (1) is 0. If $h \to 0$ along the line $v = mu (m \in \mathbb{R}$ fixed), the limit of (1) is $(1 + im)^{-1}$. If $h \to 0$ through the sequence $h_n = i^n/n$, the (1) has no limit.

The notion of derivative given in (4.57) reduces to that of real-valued functions because of the following easy theorem.

(4.60)   Theorem   *Let I, f, and c be as in (4.57). Write $f = f_1 + if_2$ where f_1 and f_2 are real-valued. Then f is differentiable at c if and only if both f_1 and f_2 are differentiable at c. In this case, $f'(c) = f_1'(c) + if_2'(c)$.*

Proof   Exercise. $\square$

(4.61)   Remarks   Many of the previous theorems of this chapter remain true for differentiation of complex-valued functions under either of the settings (4.57) and (4.58). Specifically, (4.6) and (4.9) remain valid with no change in the proofs. The reader can easily formulate three new versions of the chain rule for the differentiation of composite functions $[I \to J \to \mathbb{C}, I \to D \to \mathbb{C}, D_1 \to D_2 \to \mathbb{C}]$ and in each version the proof is the same as that given for (4.10). The definition of higher derivatives (4.24) and Theorems (4.26) and (4.27) remain the same with minor obvious changes. Taylor series are defined just as in (4.28). By using (4.60) and (4.52), one sees that any complex-valued function of finite variation on a closed interval is differentiable a.e. on that interval. Examples (4.5.ae) and (4.11.abdf) remain valid with the same proofs on the domain $\mathbb{C}$. In particular, $\exp' = \exp$ on $\mathbb{C}$.

The most notable exception to the foregoing remarks is that Rolle's Theorem, Taylor's Theorem, the Mean Value Theorems, and some results derived from them fail for complex-valued functions. For example, define f on $[0, 1]$ by $f(x) = (x - x^2) + i(x^2 - x^3)$. Then f satisfies the hypotheses of Rolle's Theorem (4.15) [except that it is *not* real-valued]. Indeed, $f'(x) = (1 - 2x) + i(2x - 3x^2)$ and $f(0) = f(1) = 0$. However, the conclusion fails because $f'(x) \neq 0$ for all x.

We do, however, have the following important fact, which will be used later.

(4.62)   Theorem   *Let D be an open subset of $\mathbb{C}$ and let $f : D \to \mathbb{C}$ satisfy $f'(z) = 0$ for every $z \in D$. If $a, b, \in D$ and if the line segment $[a, b] = \{a + (b - a)t : 0 \leq t \leq 1\}$ joining a to b lies entirely in D, then $f(a) = f(b)$. In particular, if there is some $a \in D$ such that $[a, b] \subset D$ for all $b \in D$ [such a D is said to be star-shaped from a], then f is a constant on D.*

Proof   Assume that there exists $a, b \in D$ such that $[a, b] \subset D$ and $|f(b) - f(a)| = \delta > 0$. Let $c = (a + b)/2$ be the midpoint of $[a, b]$. Since $\delta = |f(b) - f(a)| \leq |f(b) - f(c)| + |f(c) - f(a)|$, it follows that at least one of the segments $[a, c]$ or $[c, b]$— call it $[a_1, b_1]$— satisfies $|f(b_1) - f(a_1)| \geq \delta/2$ and $|b_1 - a_1| = |b - a|/2$. Similarly, let c_1 be the midpoint of $[a,_1, b_1]$ and choose one of the segments $[a_1, c_1]$ or $[c_1, b_1]$— call it $[a_2, b_2]$— such that $|f(b_2) - f(a_2)| \geq \delta/4$ and $|b_2 - a_2| = |b - a|/4$. Continuing in this way, we obtain a sequence $([a_n, b_n])_{n=1}^{\infty}$ of segments such that $[a_{n+1}, b_{n+1}] \subset [a_n, b_n]$, $|f(b_n) - f(a_n)| \geq \delta/2^n$, and $|b_n - a_n| = |b - a|/2^n$ for all n. Thus

$$|f(b_n) - f(a_n)| \geq \delta \cdot |b_n - a_n|/|b - a| \tag{1}$$

for all n. Using the Nested Interval Principle, we see that there is a (unique) $p \in \mathbb{C}$ such that $p \in [a_n, b_n]$ for all n. Then $p \in [a, b] \subset D$, so, because $f'(p) = 0$ and D is open, there is some $\eta > 0$ such that $z \in D$ and

$$|f(z) - f(p)| \leq (\delta/2) \cdot |z - p|/|b - a|$$

whenever $|z - p| < \eta$. Now fix an n such that $|z - p| < \eta$ for all $z \in [a_n, b_n]$. Then we have

$$\begin{aligned}
|f(b_n) - f(a_n)| &\leq |f(b_n) - f(p)| + |f(p) - f(a_n)| \\
&\leq (\delta/2) \cdot (|b_n - p| + |p - a_n|)/|b - a| \\
&= (\delta/2) \cdot |b_n - a_n|/|b - a|. \tag{2}
\end{aligned}$$

Combining (1) and (2) yields $\delta \leq \delta/2$. $\square$

The next theorem shows that complex power series may be differentiated term-by-term.

(4.63)   Theorem   *Let $a \in \mathbb{C}$, $(c_n)_{n=0}^{\infty} \subset \mathbb{C}$, and $R > 0$ be given. Suppose that*

(i) $$f(z) = \sum_{n=0}^{\infty} c_n(z - a)^n$$

converges on the disk $D = \{z \in \mathbb{C} : |z - a| < R\}$. Then f is infinitely differentiable on D and

(ii)

$$f^{(j)}(z) = \sum_{n=j}^{\infty} n(n-1) \ldots (n-j+1)c_n(z-a)^{n-j} = \sum_{n=j}^{\infty} j!\binom{n}{j}c_n(z-a)^{n-j}$$

for all $z \in D$ and $j \in \mathbb{N}$. Thus, in particular, $f^{(j)}(a) = j! \cdot c_j$ for $j \geq 0$, and so the series (i) is the Taylor series of f centered at a. That is, a function has at most one power series representation on a disk, namely, its Taylor series.

Proof   We lose nothing but notation complication by supposing that $a = 0$, so we do [otherwise substitute $w = z - a$]. If we can prove (ii) for the case $j = 1$, then the general case is obtained by repeated application of the theorem for that special case. Thus, we need only show that

$$f'(z) = \sum_{n=1}^{\infty} nc_n z^{n-1} \tag{1}$$

for all $z \in D[a = 0]$. Since $n^{1/n} \to 1$ as $n \to \infty$, we have

$$\varlimsup_{n \to \infty} |nc_n|^{1/n} = \left(\lim_{n \to \infty} n^{1/n} \right)\left(\varlimsup_{n \to \infty} |c_n|^{1/n} \right) = \varlimsup_{n \to \infty} |c_n|^{1/n},$$

and so the series (1) converges on D. Write

$$g(z) = \sum_{n=1}^{\infty} nc_n z^{n-1} \tag{2}$$

for $z \in D$. Now fix any $z \in D$ and choose $r \in \mathbb{R}$ such that $|z| < r < R$. If $w \ne z$ and $|w| \le r$, we have

$$\left| \frac{f(w) - f(z)}{w - z} - g(z) \right| = \left| \sum_{n=0}^{\infty} c_n \cdot \frac{w^n - z^n}{w - z} - \sum_{n=1}^{\infty} nc_n z^{n-1} \right|$$

$$\le \sum_{n=1}^{\infty} \left(|c_n| \cdot \left| \frac{w^n - z^n}{w - z} - nz^{n-1} \right| \right)$$

$$= \sum_{n=1}^{\infty} \left(|c_n| \cdot \left| \sum_{k=1}^{n} (w^{n-k}z^{k-1} - z^{n-1}) \right| \right) \tag{3}$$

and

$$|c_n| \left| \sum_{k=1}^{n} (w^{n-k}z^{k-1} - z^{n-1}) \right| \le |c_n| \cdot \sum_{k=1}^{n} (|w|^{n-k} \cdot |z|^{k-1} + |z|^{n-1})$$

$$< |c_n| \sum_{k=1}^{n} (r^{n-k}r^{k-1} + r^{n-1})$$

$$= |c_n| \cdot 2nr^{n-1} = M_n, \tag{4}$$

where the last equality defines M_n. Now $\sum_{n=1}^{\infty} rM_n < \infty$ by the Root Test, so (4) and the Weierstrass M-test show that the last series in (3) converges uniformly for $|w| \le r$, and so its sum is a function ϕ that is continuous on the disk $|w| \le r$. By the continuity of ϕ, we have $\phi(w) \to \phi(z) = 0$ as $w \to z$ and so (3) shows that

$$\lim_{w \to z} \frac{f(w) - f(z)}{w - z} = g(z). \tag{5}$$

Now (5) and (2) imply (1). $\square$

(4.64) Remark One of the many striking theorems of analytic function theory is the fact that if a complex-valued function f is defined and differentiable on some nonvoid open set $V \subset \mathbb{C}$, then f is infinitely differentiable on V and, for each $a \in V$, f has a representation of the form (4.63.i) on the largest open disk D centered at a such that $D \subset V$.

Exercises

1. If f is a *real*-valued function that is defined and differentiable on some nonvoid open set $D \subset \mathbb{C}$, then $f'(c) = 0$ for all $c \in D$. [Hint: Show that $f'(c)$ is both real and purely imaginary.]

2. If f is a complex-valued function on a nonvoid *connected* open set $D \subset \mathbb{C}$ and $f'(z) = 0$ for all $z \in D$, then f is a constant on D. [Hint: Use Exercise 19, p. 126]

3. If $c \in \mathbb{C}$ and f is a differentiable function on some open disk $D \subset \mathbb{C}$ such that $f'(z) = cf(z)$ for all $z \in D$, then there exists $a \in \mathbb{C}$ such that $f(z) = ae^{cz}$ for all $z \in D$. [Hint: Differentiate the function $g(z) = e^{-cz} \cdot f(z)$.]

4. *Pringsheim's Theorem.*
    (a) Let $(a_n)_{n=0}^{\infty} \subset \mathbb{C}$ satisfy $\mathrm{Re}(a_n) \geqq 0$ for all n and suppose that the two power series

    $$f(z) = \sum_{n=0}^{\infty} a_n z^n, \qquad g(z) = \sum_{n=0}^{\infty} \mathrm{Re}(a_n) z^n$$

    each have radius of convergence equal to 1. Then the number $z = 1$ is singular for the function f in the sense that for every $a \in [0, 1[$ the Taylor series

    $$\sum_{k=0}^{\infty} \frac{f^{(k)}(a)}{k!} (z - a)^k$$

    diverges if $|z - a| > 1 - a$; i.e., its open disk of convergence does not contain $z = 1$. [Hint: For $x > a$, use (2.50.ii) to obtain

    $$\sum_{k=0}^{\infty} \frac{\mathrm{Re}(f^{(k)}(a))}{k!} (x - a)^k = \sum_{k=0}^{\infty} \left(\sum_{n=k}^{\infty} \mathrm{Re}(a_n) \binom{n}{k} a^{n-k} \right)(x - a)^k$$

    $$= \sum_{n=0}^{\infty} \left(\mathrm{Re}(a_n) \sum_{k=0}^{n} \binom{n}{k} a^{n-k}(x - a)^k \right) = \sum_{n=0}^{\infty} \mathrm{Re}(a_n) x^n$$

    and note that the last series diverges if $x > 1$.]
    (b) The conclusion of (a) fails if $a_n = (-1)^n$ for all $n \geqq 0$.

5. Let $I \subset \mathbb{R}$ be a nonvoid open interval and let $f \colon I \to \mathbb{C}$ be infinitely differentiable on I. Then the following two statements are equivalent.
    (a) For each $a \in I$ there exists $R > 0$ depending on a such that if $x \in I$ and $|x - a| < R$, then

    $$f(x) = \sum_{k=0}^{\infty} \frac{f^{(k)}(a)}{k!} (x - a)^k.$$

    (b) For each $a \in I$ there exist two positive real numbers r and C depending on a such that

    $$|f^{(n)}(x)| \leqq C^{n+1} \cdot n!$$

    for all $x \in I$ with $|x - a| < r$ and all integers $n \geqq 0$. If either (and hence both) of (a) and (b) obtains, then f is said to be *real analytic on I*. [Hints: If (b) holds, choose $R \leqq r$ such that $0 < CR < 1$ and then apply Taylor's Theorem to the real and imaginary parts of f to obtain (a). Suppose (a) holds. Choose $0 < r < 1$

so that $2r < R$ and $[a - 2r, a + 2r] \subset I$. Series (a) converges when $x = a + 2r$, so there exists $N \in \mathbb{N}$ such that $|f^{(k)}(a)| < k!(2r)^{-k}$ for $k > N$. Now, for $n > N$ and $|x - a| < r$, use (4.63) and (4.39.b) to get

$$|f^{(n)}(x)| \leq n!(2r)^{-n} \cdot \sum_{k=n}^{\infty} \binom{k}{k-n}(1/2)^{k-n}$$

$$= n!(2r)^{-n} \cdot \sum_{j=0}^{\infty} \binom{-n-1}{j}(-1/2)^j$$

$$= n!(2r)^{-n}(1 - 1/2)^{-n-1} < n!(2/r)^{n+1}.$$

Now choose C so that $C > 2/r$ and $C > \max\{|f^{(n)}(x)| : |x - a| \leq r, \ 0 \leq n \leq N\}$.]

6.   Let $I \subset \mathbb{R}$ be a nontrivial compact interval and let $f : I \to \mathbb{C}$ be n times differentiable on I and satisfy

(i) $$f^{(n)}(x) = \sum_{k=0}^{n-1} \alpha_k f^{(k)}(x)$$

for all $x \in I$ (here $n \in \mathbb{N}$ and $\alpha_k \in \mathbb{C}$ for $0 \leq k < n$). Write

$$A = \max\{1, |\alpha_0|, \ldots, |\alpha_{n-1}|\}$$

and $B = 1 + \sum_{k=0}^{n-1} \|f^{(k)}\|_u$, where the uniform norm is taken over I. Then we have the following.

(a) The function f is infinitely differentiable on I and $\|f^{(n+j)}\|_u \leq 2^j A^{j+1} B$ for each integer $j \geq 0$. [Hint: Check by induction on j that

$$f^{(n+j)} = \sum_{k=0}^{n-1} \beta_{j,k} f^{(k)}$$

where $|\beta_{j,k}| \leq 2^j A^{j+1}$ for $0 \leq k < n$.]

(b) If $a \in I$ and $f^{(k)}(a) = 0$ for $0 \leq k < n$, then $f(x) = 0$ for all $x \in I$. [Hint: Check that $f^{(m)}(a) = 0$ for all $m \geq 0$, apply Taylor's Theorem to the real and imaginary parts of f, and then use (a) to estimate the remainders.]

(c) If f_1 and f_2 both satisfy (i) on I, $a \in I$, and $f_1^{(k)}(a) = f_2^{(k)}(a)$ for $0 \leq k < n$, then $f_1(x) = f_2(x)$ for all $x \in I$.

(d) If $(c_k)_{k=0}^{n-1} \subset \mathbb{C}$ and $a \in I$, then the differential equation (i) has at most one solution f on I such that $f^{(k)}(a) = c_k$ for $0 \leq k < n$.

7.   Let $I \subset \mathbb{R}$ be a nontrivial closed interval, let $\{\alpha, \beta, \gamma\} \subset \mathbb{C}$ with $\alpha \neq 0$, and let r_1 and r_2 be the roots of the equation $\alpha z^2 + \beta z + \gamma = 0$. Consider the differential equations

(i) $\alpha f' + \beta f = 0$,

(ii) $\alpha f'' + \beta f' + \gamma f = 0$.

Then we have the following.

(a) The only functions f on I that satisfy (i) are given by

$$f(x) = c_0 e^{-\beta x/\alpha}$$

where $c_0 \in \mathbb{C}$.

(b) If $r_1 \neq r_2$, then the only functions f on I that satisfy (ii) are given by

$$f(x) = b_1 e^{r_1 x} + b_2 e^{r_2 x}$$

where $b_1, b_2 \in \mathbb{C}$.

(c) If $r_1 = r_2 = r$, then the only solutions of (ii) on I are given by

$$f(x) = (b_1 + b_2 x)e^{rx}$$

where $b_1, b_2 \in \mathbb{C}$. [Hint: Check that all of these functions are solutions and then use (d) of Exercise 6 to prove uniqueness.]

8. In each of the following use Exercise 7 to find all functions f on $[0, 1]$ that satisfy the given conditions.
   (a) $f' = 0$, $f(0) = 0$;
   (b) $f' = f$, $f(0) = 1$;
   (c) $f'' = f$, $f(0) = 0$, $f'(0) = 1$;
   (d) $f'' = f$, $f(0) = 1$, $f'(0) = 0$;
   (e) $f'' + f = 0$, $f(0) = 0$, $f'(0) = 1$;
   (f) $f'' + f = 0$, $f(0) = 1$, $f'(0) = 0$;
   (g) $f'' + f = 2f'$, $f(0) = 0$, $f'(0) = 1$.

9. Show that L'Hospital's rule can fail for complex-valued functions by defining f and g on $\mathbb{R}$ by $f(x) = x$ and $g(x) = x \cdot \exp(-i/x)$ and considering $\lim_{x \to 0} (f(x)/g(x))$.

   [Remark: This exercise should probably be postponed until it is learned in Chapter 5 that $|\exp(it)| = 1$ for all $t \in \mathbb{R}$.]

5

THE ELEMENTARY
TRANSCENDENTAL FUNCTIONS

In previous chapters we have met the exponential function and studied some of its properties. We have also examined the real logarithm. Now that we have developed sufficient analytical machinery, it is possible to explore other functions of fundamental importance in analysis: trigonometric functions, the argument function, complex logarithms, and complex powers. As we shall see, these are all closely related to the exponential. Our first section is a repetition of earlier material, but we commit this redundancy for the sake of having the systematic presentation all in one place.

The Exponential Function

As in (2.74.a), we define the exponential function $\exp : \mathbb{C} \to \mathbb{C}$ by the formula

$$\exp(z) = \sum_{n=0}^{\infty} \frac{z^n}{n!}$$

where $0^0 = 1$. The Ratio Test shows that this series converges absolutely at every $z \in \mathbb{C}$.

(5.1)   **Theorem**   *If* $a, b \in \mathbb{C}$, *then*
(i) $\exp(a + b) = \exp(a)\exp(b)$,
(ii) $\exp(0) = 1$,
(iii) $\exp(a) \neq 0$,
(iv) $\exp(-a) = 1/\exp(a)$.

Proof   This is a repeat of (3.94). $\square$

(5.2)   **Theorem**   *The function* exp *is continuous and differentiable on* $\mathbb{C}$, *and* $\exp'(z) = \exp(z)$ *for all* $z \in \mathbb{C}$.

Proof   See (4.5.e). $\square$

(5.3)   **Theorem**   *We have*
(i) $\exp(x) > 0$ *for all* $x \in \mathbb{R}$;
(ii) $a < b$ *in* $\mathbb{R}$ *implies* $\exp(a) < \exp(b)$;
(iii) $\lim\limits_{x \to \infty} x^p \exp(x) = \infty$ *for every integer* p;
(iv) $\lim\limits_{x \to -\infty} \exp(x) = 0$;
(v) $\exp(\mathbb{R}) = \,]0, \infty[$.

Proof   See (3.95). $\square$

The Trigonometric Functions

As with exp, we define the trigonometric functions on $\mathbb{C}$. It is not obvious at all from our definitions that these functions have any relationship with triangles. Their application to geometry will be discussed later.

(5.4)   **Definitions**   We define the *sine function* sin and the *cosine function* cos from $\mathbb{C}$ to $\mathbb{C}$ by the formulas

(i) $$\sin z = \sum_{n=0}^{\infty} (-1)^n \frac{z^{2n+1}}{(2n+1)!},$$

(ii) $$\cos z = \sum_{n=0}^{\infty} (-1)^n \frac{z^{2n}}{(2n)!},$$

where $0^0 = 1$. It is clear from the Ratio Test that these two series converge absolutely at every $z \in \mathbb{C}$.

(5.5)   **Euler's Formulas**   *For all* $z \in \mathbb{C}$ *we have*
(i) $\exp(iz) = \cos z + i \sin z,$
(ii) $\exp(-iz) = \cos z - i \sin z,$
(iii) $\sin z = [1/(2i)][\exp(iz) - \exp(-iz)],$
(iv) $\cos z = (1/2)[\exp(iz) + \exp(-iz)].$

Proof   Since $(-1)^n z^{2n} = (iz)^{2n}$, $i(-1)^n z^{2n+1} = (iz)^{2n+1}$ for all integers $n \geq 0$, formula (i) is obtained by multiplying (5.4.i) by i and then adding term-by-term to (5.4.ii). Since (5.4.i) shows that $\sin(-z) = -\sin z$, (ii) follows from (i) by substituting $-z$ for z. To get (iii) and (iv), solve (i) and (ii) simultaneously. $\square$

In principle, all trigonometric identities can be proved by using only (5.5) and (5.1). The reader will have no trouble supplying such proofs for the following.

(5.6)   **Basic Identities**   *If $a, b \in \mathbb{C}$, then*
(i) $\sin(-b) = -\sin b$,
(ii) $\cos(-b) = \cos b$,
(iii) $\sin^2 a + \cos^2 a = 1$, *
(iv) $\sin(a + b) = \sin a \cos b + \cos a \sin b$,
(v) $\cos(a + b) = \cos a \cos b - \sin a \sin b$.

(5.7)   **Theorem**   *The functions* sin *and* cos *are continuous and differentiable on* $\mathbb{C}$. *Moreover,*

$$\sin' z = \cos z \quad and \quad \cos' z = -\sin z$$

for all $z \in \mathbb{C}$.

Proof   Apply (5.5), (5.2), and the Chain Rule. $\square$

The next theorem includes our *definition* of the number π.

(5.8)   **Theorem**   *There exists exactly one real number π having the following three properties:*

(i) $$\pi > 0,$$

(ii) $$\cos(\pi/2) = 0,$$

(iii) $$\cos x > 0 \quad whenever \quad 0 < x < \pi/2.$$

Moreover, $\pi < 4$.

Proof   Assuming that there are two such numbers π and π', say $\pi < \pi'$, we conclude from (i) and (iii) that $\cos(\pi/2) > 0$, contrary to (ii). Thus, at most one such number exists.
Since $2^{2n}/(2n)! > 2^{2n+2}/(2n+2)!$ for $n > 1$, it follows from (5.4.ii) and Leibnitz's Test (2.45) that

$$\cos 2 = 1 - 2^2/2! + 2^4/4! - 2^6/6! + \ldots$$
$$= -1 + 16/24 - + \ldots < -1 + 2/3 < 0. \tag{1}$$

Now cos is continuous and real-valued on $[0, 2]$ and $\cos 0 = 1 > 0$, so it follows from (1) and the Intermediate Value Theorem that the set

$$A = \{x \in [0, 2] : \cos x = 0\}$$

is nonvoid. Define $\pi/2 = \inf A$. Since cos is continuous, $\pi/2 \in A$, and

---

*We adopt the standard notation of writing $\sin^2 a$ for $(\sin a)^2$, etc.

since $\cos 0 = 1$, $\pi/2 > 0$. Thus π has properties (i) and (ii). Property (iii) also obtains because, otherwise, another application of the Intermediate Value Theorem would produce an element of A that is less than $\pi/2$. It follows from (1) that $\pi/2 < 2$. $\square$

(5.9) **Theorem** *Suppose $z \in \mathbb{C}$ and $x \in \mathbb{R}$. Then*
(i) sin *is strictly increasing and continuous on* $[-\pi/2, \pi/2]$ *with* $\sin(-\pi/2)$ $= -1$, $\sin 0 = 0$, *and* $\sin(\pi/2) = 1$;
(ii) $\sin(z + \pi/2) = \cos z$ *and* $\cos(z + \pi/2) = -\sin z$;
(iii) cos *is strictly decreasing and continuous on* $[0, \pi]$ *with* $\cos 0 = 1$, $\cos(\pi/2) = 0$, *and* $\cos \pi = -1$;
(iv) cos *is strictly increasing and continuous on* $[-\pi, 0]$ *with* $\cos(-\pi) = -1$, $\cos(-\pi/2) = 0$, *and* $\cos 0 = 1$;
(v) sin *is strictly decreasing and continuous on each of the intervals* $[-\pi,$ $-\pi/2]$ *and* $[\pi/2, \pi]$ *with* $\sin(-\pi) = 0$, $\sin(-\pi/2) = -1$, $\sin(\pi/2) = 1$, *and* $\sin \pi$ $= 0$;
(vi) $\sin(z + 2\pi) = \sin z$ *and* $\cos(z + 2\pi) = \cos z$;
(vii) $-1 \leqq \sin x \leqq 1$ *and* $-1 \leqq \cos x \leqq 1$.

Proof By (5.8.iii) and (5.7), $\sin' > 0$ on $]0, \pi/2[$, so the Mean Value Theorem shows that sin is strictly increasing on $[0, \pi/2]$. It is also strictly increasing on $[-\pi/2, 0]$ by (5.6.i). Since $0 = \sin 0 < \sin(\pi/2) = -\sin(-\pi/2)$ and $\sin^2(\pi/2) = 1 - \cos^2(\pi/2) = 1$, the proof of (i) is complete. To obtain (ii), use (i), (5.6.iv), and (5.6.v).

Now (iii) follows from (i) and the second formula in (ii). Next, (iv) follows from (iii) and (5.6.ii). We can again use $\sin' = \cos$ to obtain (v). To check (vi), use (ii) to write

$$\sin(z + 2\pi) = \cos(z + 3\pi/2)$$

$$= -\sin(z + \pi) = -\cos(z + \pi/2) = \sin z$$

and similarly for cos. Since (vii) holds for $-\pi \leqq x \leqq \pi$, (vi) shows that it holds for all $x \in \mathbb{R}$. $\square$

(5.10) **Remark** Nothing like (5.9.vii) is true for all complex numbers. For example, $\cos(iy) = [e^y + e^{-y}]/2$, and so cos maps the imaginary axis onto the interval $[1, \infty[$.

The next result gives a connection between the exponential and the unit circle.

(5.11) **Theorem** *For $\theta \in \mathbb{R}$, define $f(\theta) = \exp(i\theta)$. Then f is a continuous function from $\mathbb{R}$ onto the unit circle $\mathbb{T} = \{z \in \mathbb{C} : |z| = 1\}$. For each $z \in \mathbb{T}$ there exists exactly one $\theta \in \mathbb{R}$ such that $-\pi < \theta \leqq \pi$ and $f(\theta) = z$. For $\theta_1, \theta_2 \in \mathbb{R}$, we have $f(\theta_1) = f(\theta_2)$ if and only if $(\theta_2 - \theta_1) \in 2\pi\mathbb{Z} = \{2\pi n : n \in \mathbb{Z}\}$.*

Proof   It is clear from (5.2) that f is continuous. For $\theta \in \mathbb{R}$, we have

$$|f(\theta)| = |\cos\theta + i\sin\theta| = (\cos^2\theta + \sin^2\theta)^{1/2} = 1$$

because $\cos\theta$ and $\sin\theta$ are *real*, and so $f(\theta) \in \mathbb{T}$.

Now let $z \in \mathbb{T}$ be given. Then $z = a + bi$ where $a, b \in \mathbb{R}$ and $a^2 + b^2 = 1$. We wish to prove that there exists exactly one $\theta_1 \in \,]-\pi, \pi]$ such that $f(\theta_1) = z$; i.e., $\cos\theta_1 = a$ *and* $\sin\theta_1 = b$. Since $-1 \leq a \leq 1$, it follows from (5.9.iii) and the Intermediate Value Theorem that there is exactly one real number θ_0 such that $0 \leq \theta_0 \leq \pi$ and $\cos\theta_0 = a$. Using (5.9.iv), we see that the only solutions θ of the equation $\cos\theta = a$ that lie in the interval $[-\pi, \pi]$ are $\theta = \theta_0$ and $\theta = -\theta_0$. These two are the only candidates for θ_1. We will need (5.9.i) and (5.9.v). We have $|b| = \sqrt{b^2} = \sqrt{1 - a^2} = |\sin\theta_0|$. Since $\sin \geq 0$ on $[0, \pi]$ and $\sin < 0$ on $\,]-\pi, 0[$, we are forced to choose θ_1 as follows. If $b \geq 0$, take $\theta_1 = \theta_0$ [$-\theta_0$ must be ruled out unless $-\theta_0 = \theta_0 = 0$]. If $b < 0$, than $a < 1$, so we can and must take $\theta_1 = -\theta_0$. This proves all but our last assertion.

Let $\theta_1, \theta_2 \in \mathbb{R}$. If $\theta_2 - \theta_1 = 2\pi n$ for some $n \in \mathbb{Z}$, then it follows from (5.5.i), (5.9.vi) and induction that

$$f(\theta_2) = f(\theta_1 + 2\pi n) = f(\theta_1).$$

Conversely, suppose that $f(\theta_1) = f(\theta_2)$. Write $\theta_3 = \theta_2 - \theta_1$ and let n be the integer satisfying

$$n < (\theta_3 + \pi)/(2\pi) \leq n + 1.$$

Then

$$-\pi < \theta_3 - 2\pi n \leq \pi$$

and

$$f(\theta_3 - 2\pi n) = f(\theta_3) = f(\theta_2)/f(\theta_1) = 1,$$

and so the preceding paragraph shows that $\theta_3 - 2\pi n = 0$. That is, $\theta_2 = \theta_1 + 2\pi n$.   □

Recalling (3.98), we may write $\exp(z) = e^z$ for any $z \in \mathbb{C}$. The next theorem gives formulas that are very important in Fourier analysis. Its proof gives practice in manipulation.

(5.12)   **Theorem**   *Let n be a nonnegative integer. Define functions D_n and K_n on $\mathbb{R}$ by*

$$D_n(\theta) = \sum_{j=-n}^{n} e^{ij\theta}, \qquad K_n(\theta) = \frac{1}{n+1} \sum_{k=0}^{n} D_k(\theta).$$

Then $\theta \in \mathbb{R}\backslash 2\pi\mathbb{Z}$ and $n \in \mathbb{N}$ imply

(i) $$D_n(\theta) = 1 + 2\sum_{j=1}^{n} \cos(j\theta) = \frac{\sin\left[(n + 1/2)\theta\right]}{\sin\left[(1/2)\theta\right]}$$

and

(ii) $$K_n(\theta) = \sum_{j=-n}^{n} \left(1 - \frac{|j|}{n+1}\right) e^{ij\theta} = \frac{1}{n+1} \left(\frac{\sin[((n+1)/2)\theta]}{\sin[(1/2)\theta]}\right)^2.$$

Proof From (5.5.iv), the first equality in (i) is obvious. Summing a geometric progression, we have

$$D_n(\theta) = \frac{e^{i(n+1)\theta} - e^{-in\theta}}{e^{i\theta} - 1} = \frac{e^{i(n+1/2)\theta} - e^{-i(n+1/2)\theta}}{e^{i\theta/2} - e^{-i\theta/2}},$$

from which (i) follows upon application of (5.5.iii). We have

$$(n+1)K_n(\theta) = \sum_{k=0}^{n} \sum_{j=-k}^{k} e^{ij\theta} = \sum_{j=-n}^{n} \sum_{k=|j|}^{n} e^{ij\theta}$$

$$= \sum_{j=-n}^{n} (n+1-|j|)e^{ij\theta},$$

and we obtain the first equality in (ii) upon division by $n+1$. From (5.6.i) and (5.6.v), we have the identity $\cos(a-b) - \cos(a+b) = 2\sin a \sin b$, valid for all $a, b \in \mathbb{C}$. Therefore

$$(n+1)\sin^2(\theta/2)K_n(\theta) = \sum_{k=0}^{n} \sin[(k+1/2)\theta]\sin(\theta/2)$$

$$= \frac{1}{2} \sum_{k=0}^{n} [\cos k\theta - \cos[(k+1)\theta]]$$

$$= \frac{1}{2}[1 - \cos(n+1)\theta] = \sin^2[(n+1)\theta/2],$$

from which the second equality in (ii) follows. $\square$

The sequence $(D_n)_{n=0}^{\infty}$ is called the *Dirichlet kernel*. The sequence $(K_n)_{n=0}^{\infty}$ is called the *Fejér kernel*.

We close this section by giving the familiar definitions of the other four most popular trigonometric functions.

(5.13) **Definitions** If $z \in \mathbb{C}$ and $\cos z \neq 0$, define
$$\tan z = (\sin z)/(\cos z), \qquad \sec z = 1/(\cos z).$$

If $z \in \mathbb{C}$ and $\sin z \neq 0$, define
$$\cot z = (\cos z)/(\sin z), \qquad \csc z = 1/(\sin z).$$

The Argument

(5.14) **Definition** Let $z \in \mathbb{C}$. A real number θ is called *an argument* of z if $z = |z|e^{i\theta}$. If $z = 0$, then any real number is an argument of z. If $z \neq 0$, then, because $|z/|z|| = 1$, it follows from (5.11) that z has exactly one argument in the interval $]-\pi,\pi]$. We call this number the *principal argument* of z and denote it by $\operatorname{Arg} z$. Thus Arg is a real-valued function defined on $\mathbb{C}\backslash\{0\}$ which is characterized by the two properties

$$z = |z|\exp[i\operatorname{Arg} z], \qquad -\pi < \operatorname{Arg} z \leqq \pi$$

for all nonzero complex numbers z. It is clear from (5.11) that for $z \neq 0$, the arguments of z are just the numbers of the form $\operatorname{Arg} z + 2\pi n$, where n is any integer.

(5.15) **Theorem** *The function* Arg *is continuous on* $\mathbb{C}\backslash]-\infty,0]$.

Proof Assume that this is false. Then there exists a z_0 and a sequence $(z_n)_{n=1}^\infty$ all in $\mathbb{C}\backslash]-\infty,0]$ such that $z_n \to z_0$ but $\operatorname{Arg} z_n \not\to \operatorname{Arg} z_0$. Write $\theta_n = \operatorname{Arg} z_n$ for $n \geqq 0$. Since $(\theta_n)_{n=1}^\infty \subset]-\pi,\pi]$ and $\theta_n \not\to \theta_0$, there is some $\theta_0' \neq \theta_0$ and a subsequence (θ_{n_k}) such that $\theta_{n_k} \to \theta_0'$ [take $\theta_0' = \overline{\lim}\,\theta_n$ or $\theta_0' = \underline{\lim}\,\theta_n$, they cannot both be θ_0]. Then $-\pi \leqq \theta_0' \leqq \pi$ and

$$|z_0|e^{i\theta_0} = z_0 = \lim_{k\to\infty} z_{n_k} = \lim_{k\to\infty} |z_{n_k}|e^{i\theta_{n_k}} = |z_0|e^{i\theta_0'},$$

and so $e^{i\theta_0} = e^{i\theta_0'}$. It follows from (5.11) that $\theta_0 - \theta_0' = 2\pi n$ for some integer n. Since $\theta_0' \neq \theta_0$, $-\pi < \theta_0 \leqq \pi$, and $-\pi \leqq \theta_0' \leqq \pi$, it must be that $n = 1$, $\theta_0 = \pi$, and $\theta_0' = -\pi$. But then $z_0 = |z_0|e^{i\pi} = -|z_0|$, which contradicts the fact that $z_0 \notin]-\infty,0]$. $\square$

We next solve the equations $e^z = 1$, $\sin z = 0$, and $\cos z = 0$.

(5.16) **Theorem** *For* $z, w \in \mathbb{C}$, *we have*
(i) $e^z = e^w$ *if and only if* $w = z + 2\pi i n$ *for some integer* n;
(ii) $\sin z = 0$ *if and only if* $z = n\pi$ *for some integer* n;
(iii) $\cos z = 0$ *if and only if* $z = n\pi + \pi/2$ *for some integer* n.

Proof All of the "if" statements follow from (5.9), (5.5), and (5.1).
Suppose that $e^w = e^z$. Then $e^{w-z} = 1$ and, writing $w - z = x + iy$ with $x, y \in \mathbb{R}$, we have

$$e^x = e^x \cdot |e^{iy}| = |e^{x+iy}| = 1,$$

and so (5.3) shows that $x = 0$. Thus $e^{iy} = 1$. By (5.11), there is an integer n such that $y = 2\pi n$. Therefore $w = z + iy = z + 2\pi i n$, which proves (i).
Suppose $\sin z = 0$. Then (5.5.iii) yields $e^{iz} = e^{-iz}$. By (i) there is an integer n such that $iz = -iz + 2\pi i n$, and so $z = n\pi$. This proves (ii). Now (iii) follows from (ii) and (5.9.ii). $\square$

Exercises

1.   Prove the identities (5.6) and the following identities for $a, b \in \mathbb{C}$.
(a) $2\sin a \sin b = \cos(a-b) - \cos(a+b)$,
(b) $2\cos a \cos b = \cos(a-b) + \cos(a+b)$,

(c) $2 \sin a \cos b = \sin(a + b) + \sin(a - b)$,

(d) $\cos(2a) = 1 - 2 \sin^2 a = 2 \cos^2 a - 1$,

(e) $\sin(\pi/2 - a) = \cos a$.

2.   Prove that

(a) $\sin(\pi/4) = \cos(\pi/4) = \sqrt{2}/2$,

(b) $\sin(\pi/6) = \cos(\pi/3) = 1/2$,

(c) $\sin(\pi/3) = \cos(\pi/6) = \sqrt{3}/2$,

(d) $\cos(\pi/5) = (1 + \sqrt{5})/4$.

       [Hint: For (d), write $\theta = \pi/10$, $4\theta = \pi/2 - \theta$, $\cos \theta = \sin(4\theta) = 4 \sin \theta \cos \theta$ $\cos(2\theta)$, $1 = 4 \sin \theta \cos(2\theta) = 4 \cos(4\theta)\cos(2\theta) = 8 \cos^3(2\theta) - 4 \cos(2\theta)$.]

3.   Prove that $\cos(3/2) > 0$, $\cos x > 1 - x^2/2$ if $0 < x < 1$, and $\pi > 3$.

4.   (a) If $0 < x < \pi/2$, then $2x/\pi < \sin x < x < (\sin x)/(\cos x)$.

      (b) If $0 < a < b \leq \pi/2$, then $a/b < (\sin a)/(\sin b) < \pi a/(2b)$.

5.   (a) If $0 < x < \pi/2$, then $\cos x < ((\sin x)/x)^3$.

      (b) If $p > 3$ is a real number, then the equation $\cos x = ((\sin x)/x)^p$ has just one solution $x = x_0$ such that $0 < x_0 < \pi/2$. [Hint: Write $f(x) = x - (\sin x)$ $(\cos x)^{-1/p}$, compute f' and f'', then minimize f'.]

6.   If $a, b \in \mathbb{C}$, $a \neq 0$, satisfy $a + a^{-1} = 2 \cos b$, then $a^n + a^{-n} = 2 \cos(nb)$ for all integers n.

7.   (a) If $n \in \mathbb{N}$, $n > 1$, then

$$\sum_{k=1}^{n} e^{2k\pi i/n} = 0.$$

      [Hint: Call the sum S and then show that $e^{2\pi i/n}S = S$.]

      (b) If $n \in \mathbb{N}$, $n > 1$, $r > 0$, and $z \in \mathbb{C}$, then

$$\sum_{k=1}^{n} |z - re^{2k\pi i/n}|^2 = n(|z|^2 + r^2).$$

      (c) Give a geometric interpretation of (b) in terms of the distances from a fixed point to the vertices of a regular n-gon.

8.   (a) If $x \in \mathbb{R}$ and $n \in \mathbb{N}$, then

$$\cos(nx) = \sum_{j=0}^{[n/2]} (-1)^j \binom{n}{2j} \cos^{n-2j} x \sin^{2j} x$$

    and

$$\sin(nx) = \sum_{j=0}^{[(n-1)/2]} (-1)^j \binom{n}{2j+1} \cos^{n-2j-1} x \sin^{2j+1} x$$

    where $[t]$ is the integer satisfying $[t] \leq t < [t] + 1$. [Hint: Apply the Binomial Theorem to the right side of $\cos(nx) + i \sin(nx) = (\cos x + i \sin x)^n$ and then equate real and imaginary parts.]

    (b) If $x \in \mathbb{R}$ with $\sin x \neq 0$ and $m \in \mathbb{N}$, then

$$\sin((2m + 1)x) = \sin^{2m+1} x \cdot \sum_{j=0}^{m} (-1)^j \binom{2m + 1}{2j + 1} (\cot^2 x)^{m-j}.$$

(c) If $m \in \mathbb{N}$, then the m roots of the equation

$$\sum_{j=0}^{m} (-1)^j \binom{2m+1}{2j+1} t^{m-j} = 0$$

are $t = \cot^2 \dfrac{k\pi}{2m+1}$ $(k = 1, 2, \ldots, m)$.

(d) If $m \in \mathbb{N}$, then

$$\sum_{k=1}^{m} \cot^2 \frac{k\pi}{2m+1} = \frac{m(2m-1)}{3}.$$

(e) If $m \in \mathbb{N}$, then

$$\sum_{k=1}^{m} \csc^2 \frac{k\pi}{2m+1} = \frac{2m(m+1)}{3}.$$

[Hint: Use $\csc^2 z = 1 + \cot^2 z$.]

(f) If $m \in \mathbb{N}$, then

$$\frac{2m(2m-1)}{(2m+1)^2} \cdot \frac{\pi^2}{6} < \sum_{k=1}^{m} \frac{1}{k^2} < \frac{2m(2m+2)}{(2m+1)^2} \cdot \frac{\pi^2}{6}.$$

[Hint: Use (d) and (e) together with the inequalities $\sin x < x < \tan x$ and $\cot^2 x < 1/x^2 < \csc^2 x$ for $0 < x < \pi/2$.]

(g)

$$\sum_{k=1}^{\infty} \frac{1}{k^2} = \frac{\pi^2}{6}.$$

9.   Use (a) of the preceding exercise to prove that

$$\sin^2 \frac{\pi}{5} = \frac{5 - \sqrt{5}}{8}, \quad \cos^2 \frac{\pi}{5} = \frac{3 + \sqrt{5}}{8}, \quad \text{and} \quad \cos \frac{\pi}{5} = \frac{1 + \sqrt{5}}{4}.$$

10.   Which is larger e^π or π^e? [Hint: Minimize $x^{-e} e^x$ for $x > 0$.]

11.   For $0 < x \leq \pi/2$, we have $\cos x < e^{-x^2/2}$. [Hint: Show that $f(x) = -x^2/2 - \log \cos x$ is increasing on $[0, \pi/2[$.]

12.   (a) If $z = x + iy$ where $x, y \in \mathbb{R}$, then

$$\sin z = \sin x \cosh y + i \cos x \sinh y$$

$$\cos z = \cos x \cosh y - i \sin x \sinh y$$

where $\cosh y = (e^y + e^{-y})/2$ and $\sinh y = (e^y - e^{-y})/2$.

(b) For which $z \in \mathbb{C}$ is $\sin z$ a real number?

(c) For $x, y \in \mathbb{R}$,

$$|\sin(x + iy)|^2 = \sin^2 x + \sinh^2 y,$$

$$|\cos(x + iy)|^2 = \cos^2 x + \sinh^2 y.$$

(d) Sketch graphs of the equations

$$|\sin(x + iy)| = 1, \qquad |\cos(x + iy)| = 1.$$

13.   Let $a \in \mathbb{R}$. Then the equation $\tan z = az$ has infinitely many real roots. If $a \leq 0$ or $a \geq 1$, it has no other roots, while if $0 < a < 1$, it has exactly two nonreal roots and they are both purely imaginary. [Hint: Equate real and imaginary parts. See the preceding exercise.]

14.   Let $\phi : \mathbb{R} \to \mathbb{R}$ be continuous and not identically zero. Suppose that

(i) $\qquad\qquad\qquad \phi(s + t) = \phi(s)\phi(t)$ for all $s, t \in \mathbb{R}$.

Then there exists a unique $b \in \mathbb{R}$ such that

(ii) $\qquad\qquad\qquad \phi(x) = e^{bx}$ for all $x \in \mathbb{R}$.

Consider the following steps.
(a) $\phi(x) > 0$ for all $x \in \mathbb{R}$.
(b) $\phi(0) = 1$.
(c) $\phi(-x) = 1/(\phi(x))$ for all $x \in \mathbb{R}$.
(d) $\phi(mx) = (\phi(x))^m$ for all $m \in \mathbb{Z}, x \in \mathbb{R}$.
(e) $\phi(x/n) = (\phi(x))^{1/n}$ for all $n \in \mathbb{N}, x \in \mathbb{R}$.
(f) $\phi(rx) = (\phi(x))^r$ for all $r \in \mathbb{Q}, x \in \mathbb{R}$.
Write $b = \log \phi(1)$.

15.   Let χ be a continuous complex-valued function on $\mathbb{R}$ that satisfies

(i) $\qquad\qquad\qquad |\chi(x)| = 1$ for all $x \in \mathbb{R}$

and

(ii) $\qquad\qquad\qquad \chi(s + t) = \chi(s)\chi(t)$ for all $s, t \in \mathbb{R}$.

Then there exists a unique $a \in \mathbb{R}$ such that

(iii) $\qquad\qquad\qquad \chi(x) = e^{iax}$ for all $x \in \mathbb{R}$.

The following steps lead to a simple proof.
(a) If $x \in \mathbb{R}$ and $m \in \mathbb{Z}$, then $\chi(mx) = (\chi(x))^m$.
Writing $\theta_n = \mathrm{Arg}(\chi(2^{-n}))$, we have
(b) $\lim\limits_{n \to \infty} \theta_n = 0$,
(c) $(2\theta_{n+1} - \theta_n)/(2\pi) \in \mathbb{Z}$ for all $n \in \mathbb{N}$,
(d) there exists $p \in \mathbb{N}$ such that $2\theta_{n+1} = \theta_n$ for all $n \geq p$,
(e) $\theta_n = 2^{p-n}\theta_p$ for all $n \geq p$.
(f) The set $D = \{m \cdot 2^{-n} : m, n \in \mathbb{Z}, n \geq p\}$ is dense in $\mathbb{R}$.
Writing $a = 2^p\theta_p$, we have
(g) $\chi(x) = e^{iax}$ for all $x \in D$.

16.   If $\psi : \mathbb{R} \to \mathbb{C}$ is continuous and not identically zero and if ψ satisfies

(i) $\qquad\qquad\qquad \psi(s + t) = \psi(s)\psi(t)$ for all $s, t \in \mathbb{R}$,

then there exists a unique $c \in \mathbb{C}$ such that

(ii) $\qquad\qquad\qquad \psi(x) = e^{cx}$ for all $x \in \mathbb{R}$.

[Hint: Write $\phi = |\psi|$, $\chi = \psi/\phi$, and use the preceding two exercises.]

17.   If $\omega : \mathbb{C} \to \mathbb{C}$ is differentiable, not identically zero, and satisfies

(i) $\qquad\qquad\qquad \omega(z + w) = \omega(z)\omega(w)$ for all $z, w \in \mathbb{C}$,

then there exists a unique $c \in \mathbb{C}$ such that

$$\omega(z) = e^{cz} \quad \text{for all } z \in \mathbb{C}.$$

[Hint: Write $\psi_1(x) = \omega(x)$, $\psi_2(y) = \omega(iy)$, and use the preceding exercise.]

18.   If $0 < \alpha < \pi/2$ and $(c_n)_{n=1}^{\infty} \subset \mathbb{C}$ satisfy $-\alpha \leq \mathrm{Arg}\, c_n \leq \alpha$ for all n, then $\sum\limits_{n=1}^{\infty} c_n$ and

$\sum\limits_{n=1}^{\infty} |c_n|$ either both converge or both diverge.

[Hint: $|c_n|\cos \alpha \leq \mathrm{Re}\, c_n$.]

Complex Logarithms and Powers

The reader is urged to review our discussion of the real logarithm in (3.96), (3.97), and (4.11). We use its properties freely in discussing complex logarithms. The painful difficulty in extending the domain of the logarithm to the complex plane is that exp is not one-to-one on $\mathbb{C}$. The care that must be exercised here becomes clear by a simple example.

(5.17) Example   Since $e^{2\pi i} = 1$, if we "take logarithms," we obtain

$$2\pi i = 2\pi i \log e = \log e^{2\pi i} = \log 1 = 0,$$

a clear absurdity. The trouble here is the second equality; it equates two *different* values of "the complex logarithm of 1." After all, $e^0 = 1 = e^{2\pi i n}$ for all integers n. The formula $\log a^b = b \log a$, true for positive real a and *real* b by (3.98), is simply not true in general for complex b. The difference is that $\exp|_{\mathbb{R}}$ is one-to-one.

(5.18) Definition   Let $z \in \mathbb{C}$. A number $w \in \mathbb{C}$ is called *a complex logarithm of z* if $z = e^w$. It is clear from (5.1.iii) that 0 has *no* complex logarithm.

(5.19) Theorem   *Let $z \in \mathbb{C}$ with $z \neq 0$. Then the complex logarithms of z are precisely the numbers*

$$w_n = \log|z| + i(\operatorname{Arg} z + 2\pi n)$$

where n is any integer and $\log|z|$ is the real logarithm of the positive number $|z|$.

Proof   Each w_n is a complex logarithm of z because

$$e^{w_n} = \exp\big[\log|z|\big] \cdot \exp\big[i \operatorname{Arg} z\big] \cdot e^{2\pi i n}$$
$$= |z| \cdot \exp\big[i \operatorname{Arg} z\big] \cdot 1 = z$$

by (5.14). On the other hand, if w is any complex logarithm of z, then

$$e^w = z = e^{w_0}$$

and (5.16.i) shows that there exists an integer n such that

$$w = w_0 + 2\pi i n = w_n. \qquad \square$$

(5.20) Definition   For $z \in \mathbb{C}$, $z \neq 0$, define *the principal logarithm of z* to be

(i) $$\operatorname{Log} z = \log|z| + i \operatorname{Arg} z.$$

Thus, $\operatorname{Log} z$ is the complex logarithm w_0 of (5.19). It is the only complex logarithm having imaginary part in the interval $]-\pi, \pi]$, and any other complex logarithm of z differs from it by an integral multiple of $2\pi i$. In other words, $\operatorname{Log} z$ is the unique complex number such that

(ii) $$\exp[\operatorname{Log} z] = z, \qquad -\pi < \operatorname{Im} \operatorname{Log} z \leqq \pi.$$

If x is a positive real number, then $|x| = x$ and $\operatorname{Arg} x = 0$, so $\operatorname{Log} x = \log x$; i.e. the functions Log and log agree on the domain $]0, \infty[$ of the latter.

(5.21)   Theorem   *The function* Log *is continuous and differentiable on* $\mathbb{C}\backslash] - \infty, 0]$. *Moreover,*

$$\text{(i)} \qquad\qquad \operatorname{Log}' z = 1/z$$

for all $z \in \mathbb{C}\backslash] - \infty, 0]$.

> **Proof**   The continuity assertion follows from the fact that the functions $z \to \log |z|$ and $z \to \operatorname{Arg} z$ are both continuous on this domain [see (3.97.iii) and (5.15)]. Now fix any $z \in \mathbb{C}\backslash] - \infty, 0]$ and let $(z_n)_{n=1}^{\infty}$ be any sequence in this open set such that $z_n \to z$ and $z \neq z_n$ for all n. Writing $w = \operatorname{Log} z$ and $w_n = \operatorname{Log} z_n$, it follows from the continuity of Log that $w_n \to w$ and from (5.20) that $w_n \neq w$ for all n. Therefore, (5.2) and (2.8) yield
>
> $$\lim_{n \to \infty} \frac{\operatorname{Log} z_n - \operatorname{Log} z}{z_n - z} = \lim_{n \to \infty} \left(\frac{e^{w_n} - e^w}{w_n - w} \right)^{-1} = (e^w)^{-1} = z^{-1}.$$
>
> This proves (i).   □

(5.22)   Warning   It can easily happen that

$$\operatorname{Log} ab \neq \operatorname{Log} a + \operatorname{Log} b$$

for two nonzero complex numbers a and b. For example, take $a = b = -1 + i = \sqrt{2}\, e^{3\pi i/4}$. Then $ab = -2i = 2e^{-\pi i/2}$, $\operatorname{Log} a = \operatorname{Log} b = \log\sqrt{2} + 3\pi i/4$, $\operatorname{Log} ab = \log 2 - \pi i/2$, and $[\operatorname{Log} a + \operatorname{Log} b] - \operatorname{Log} ab = 2\pi i$. We do, however, have the following theorem.

(5.23)   Theorem   *Let* a, b *be nonzero complex numbers. If* $-\pi < \operatorname{Arg} a + \operatorname{Arg} b \leqq \pi$, *then*

$$\text{(i)} \qquad\qquad \operatorname{Log} ab = \operatorname{Log} a + \operatorname{Log} b.$$

If $-\pi < \operatorname{Arg} a - \operatorname{Arg} b \leqq \pi$, *then*

$$\text{(ii)} \qquad\qquad \operatorname{Log}(a/b) = \operatorname{Log} a - \operatorname{Log} b.$$

If $\operatorname{Re} a > 0$ *and* $\operatorname{Re} b > 0$, *then both* (i) *and* (ii) *obtain.*

> **Proof**   Suppose that $-\pi < \operatorname{Arg} a + \operatorname{Arg} b \leqq \pi$. Then the number $w = \operatorname{Log} a + \operatorname{Log} b = \log|a| + i\operatorname{Arg} a + \log|b| + i\operatorname{Arg} b = \log|ab| + i(\operatorname{Arg} a + \operatorname{Arg} b)$ satisfies $e^w = \exp(\operatorname{Log} a) \cdot \exp(\operatorname{Log} b) = ab$ and $-\pi < \operatorname{Im} w \leqq \pi$. These are the two requirements in (5.20.ii) that are both necessary and sufficient that $w = \operatorname{Log} ab$. Thus (i) obtains. The proof of (ii) is similar. Since $\operatorname{Re} z > 0$ implies $-\pi/2 < \operatorname{Arg} z < \pi/2$, the last assertion is a consequence of the first two.   □

We now imitate Definition (3.98) in defining complex powers.

(5.24)   Definition   Let a and b be complex numbers with $a \neq 0$. By *a value of the power* a^b, we mean any number of the form $\exp(bw)$ where w is a complex logarithm of a. By *the principal value of* a^b, we mean the number

(i) $$a^b = \exp[b \operatorname{Log} a].$$

We shall always use the symbol a^b to denote the right side of (i) unless the contrary is explicitly stated. In view of the last sentence of (5.20), (i) agrees with (3.98.i) in the case that a is a positive real number.

(5.25)   Examples   (a) Since $\operatorname{Log} i = \log|i| + i \operatorname{Arg} i = \pi i/2$, $i^i = e^{-\pi/2}$, a real number! The values of i^i are the real numbers $e^{-\pi/2 + 2\pi n}$, where n is any integer.
       (b) Since $\operatorname{Log}(-i) = -\pi i/2$ and $\operatorname{Log}(-1) = \pi i$, we have $(-1)^{1/2} = e^{\pi i/2} = i$ and $(-i)^{1/2} = e^{-\pi i/4}$. Thus

$$i = \left[(-i)(-i)\right]^{1/2} \neq (-i)^{1/2}(-i)^{1/2} = -i,$$

and so $(ab)^c = a^c b^c$ can fail.
       (c) Let $a = e^{2\pi}$, $b = i$, and $c = 1/2$. Then $\operatorname{Log} a = \log a = 2\pi$, so $a^b = e^{2\pi i} = 1$, $(a^b)^c = 1^c = \exp[(1/2)\operatorname{Log} 1] = 1$, and $a^{bc} = a^{i/2} = \exp[(i/2) \cdot 2\pi] = -1$. Thus

$$\left(a^b\right)^c \neq a^{bc}.$$

       (d) Let a be any nonzero complex number and let $n \in \mathbb{N}$. Using (5.24.i), we have

$$a^0 = \exp[0 \operatorname{Log} a] = 1,$$
$$a^{n-1} \cdot a = \exp[(n-1)\operatorname{Log} a] \cdot \exp[\operatorname{Log} a]$$
$$= \exp[n \operatorname{Log} a] = a^n,$$
$$a^{-n} = \exp[-n \operatorname{Log} a] = 1/(\exp[n \operatorname{Log} a])$$
$$= 1/a^n.$$

Thus, Definition (5.24) is consistent with our definition of powers with integral exponents given in (1.23).
       In the case of real exponents, we have the following important dichotomy.

(5.26)   Theorem   Let $a \in \mathbb{C} \setminus \{0\}$ and $b \in \mathbb{R}$ be given. Let a^b and $|a|^b$ denote the principal values of these powers, and for each integer k, let

$$z_k = a^b e^{2\pi i k b}.$$

Write $V = \{z_k : k \in \mathbb{Z}\}$ and $C = \{z \in \mathbb{C} : |z| = |a|^b\}$. Then V is the set of all values of a^b and $V \subset C$.

       (i) If $b \in \mathbb{Q}$ and n is the smallest natural number such that nb is an integer, then V has just n distinct elements $z_0, z_1, \ldots, z_{n-1}$. In fact, $z_k = z_j$ if and only if $(k - j)/n$ is an integer.
       (ii) If b is irrational, then $z_k \neq z_j$ whenever $k \neq j$. Also, V is a dense subset of the circle C.

Proof   The complex logarithms of a are just the numbers

$$w_k = \operatorname{Log} a + 2\pi i k \qquad (k \in \mathbb{Z}),$$

and so the values of a^b are the numbers

$$e^{bw_k} = e^{b \operatorname{Log} a} \cdot e^{2\pi i k b} = z_k$$

whether or not b is real. Since b is real and $|e^{i\theta}| = 1$ for all real θ, we have

$$|z_k| = |e^{b \operatorname{Log} a}| = |e^{b \operatorname{Log}|a|}| \cdot |e^{ib \operatorname{Arg} a}|$$

$$= e^{b \operatorname{Log}|a|} = |a|^b.$$

This proves our first assertion.

Now let b and n be as in (i). Given any two integers j and k, choose integers q and r such that $k - j = nq + r$ and $0 \leq r < n$. Then in view of (5.16.i) and since $a^b \neq 0$, the following statements are pairwise equivalent: $z_k = z_j$; $e^{2\pi i k b} = e^{2\pi i j b}$; $2\pi i (k - j) b \in 2\pi i \mathbb{Z}$; $(k - j) b \in \mathbb{Z}$; $(nbq + rb) \in \mathbb{Z}$; $rb \in \mathbb{Z}$; $r = 0$; $(k - j)/n = q$. This proves (i).

Now suppose that b is irrational and let $k \neq j$ be integers. Then $z_k / z_j = e^{2\pi i (k - j) b} \neq 1$ because $(k - j) b$ is not an integer, and therefore $z_k \neq z_j$. We need only show that V is dense in C. For $t \in \mathbb{R}$, let $\theta(t)$ denote the fractional part of t; i.e., $0 \leq \theta(t) < 1$ and $t - \theta(t)$ is an integer. We claim that the set $A = \{\theta(kb) : k \in \mathbb{Z}\}$ is dense in $]0, 1[$. In fact, let $0 < \alpha < \beta < 1$ be given. Choose $n \in \mathbb{N}$ such that $1/n < \beta - \alpha$. Dividing $[0, 1[$ into n disjoint subintervals of length $1/n$, we see that one of these subintervals contains at least two of the $n + 1$ numbers $\theta(b), \theta(2b), \ldots, \theta((n + 1)b)$.* Thus, there exist $j \neq k$ in $\mathbb{N}$ such that $|\theta(kb) - \theta(jb)| \leq 1/n < \beta - \alpha$. Taking $m = k - j$ or $j - k$ according as $\theta(kb) - \theta(jb)$ is positive or negative, we have $\theta(mb) = |\theta(kb) - \theta(jb)|$ and, since $m \neq 0$ and b is irrational, $\theta(mb) > 0$. Let r be the smallest natural number such that $r\theta(mb) > \alpha$. Then $r\theta(mb) = (r - 1)\theta(mb) + \theta(mb) < \alpha + (\beta - \alpha) = \beta$ and, since $0 < r\theta(mb) < 1, r\theta(mb) = \theta(rmb)$; hence, $\alpha < \theta(rmb) < \beta$. This establishes the above claim. Now let $z \in C$ be given. Write $z = a^b e^{2\pi i t}$ where $0 \leq t < 1$. By the claim, there exist integers $(k_j)_{j=1}^{\infty}$ such that $\theta(k_j b) \to t$ as $j \to \infty$. Then

$$z_{k_j} = a^b e^{2\pi i k_j b} = a^b e^{2\pi i \theta(k_j b)} \to a^b e^{2\pi i t} = z$$

as $j \to \infty$. Therefore V is dense in C. $\square$

Since we now know that every complex number has mth roots for every $m \in \mathbb{N}$, we can safely invoke Lemma (3.119) to give a simple proof of the following.

---

*This is an application of the Pigeon-hole Principle: If $n + 1$ letters are placed in n pigeon-holes, then some pigeon-hole contains at least two of the letters.

(5.27)   **Fundamental Theorem of Algebra**   *If $P(z)$ is any nonconstant complex polynomial as in (3.119), then $P(z_0) = 0$ for at least one $z_0 \in \mathbb{C}$.*

     Proof   Write $\alpha = \inf\{|P(z)|:z \in \mathbb{C}\}$. Since

$$|P(z)| \geq |z|^n \left[|c_n| - \sum_{k=0}^{n-1} |c_k z^{k-n}| \right],$$

$n > 0$, and $c_n \neq 0$, we have $|P(z)| \to \infty$ as $|z| \to \infty$. Thus there exists $0 < r < \infty$ such that $|P(z)| > \alpha + 1$ if $|z| \geq r$. Write $W = \{z \in \mathbb{C}:|z| < r\}$. Clearly $\alpha = \inf\{|P(z)| : z \in W\}$. Since $|P|$ is continuous and W^- is compact, there is a $z_0 \in W^-$ such that $|P(z_0)| = \alpha$. By our choice of r, $z_0 \in W$. It follows from (3.119.ii) that $P(z_0) = 0$. $\square$

Exercises

1.   Let $c \in \mathbb{C}$. Find all solutions $z \in \mathbb{C}$ of the equation $\cos z = c$ in terms of complex logarithms.

2.   (a) If $w \in \mathbb{C}\backslash\{0\}$ and $\operatorname{Arg} w \neq \pi$, then $\operatorname{Log}(1/w) = -\operatorname{Log} w$.
         (b) Let $a, b \in \mathbb{C}$. Describe geometrically the set of all $z \in \mathbb{C}$ such that

$$\operatorname{Log}((z - a)/(z - b)) = \operatorname{Log}(z - a) - \operatorname{Log}(z - b).$$

         (c) Do the same for the equation

$$\operatorname{Log}((z - a)/(z - b)) = \operatorname{Log}(1 - a/z) - \operatorname{Log}(1 - b/z).$$

3.   For $z \in \mathbb{C}\backslash\{0\}$, let $L(z) = \{w \in \mathbb{C}:w$ is a complex logarithm of $z\}$. Then $a, b \in \mathbb{C}\backslash\{0\}$ implies

$$L(a) + L(b) = L(ab)$$

where we define $L(a) + L(b) = \{w_1 + w_2 : w_1 \in L(a), w_2 \in L(b)\}$.

π Is Irrational

Since the square of any rational number is rational, the above statement follows from this theorem.

(5.28)   **Theorem** π^2 *is irrational.*

     Proof   [Ivan Niven]   Assume that $\pi^2 = a/b$ for some $a, b \in \mathbb{N}$. Since $\sum_{n=0}^{\infty} (a^n/n!) = e^a < \infty$, we can choose some $N \in \mathbb{N}$ such that

$$\pi a^N/N! < 1. \tag{1}$$

Define a polynomial of degree $2N$ by

$$f(x) = x^N(1 - x)^N/N!. \tag{2}$$

Expand the right side of (2) to obtain

$$f(x) = \frac{1}{N!} \sum_{n=N}^{2N} c_n x^n,$$

where the c_n are certain integers. Differentiating k times gives

$$f^{(k)}(x) = \frac{1}{N!} \sum_{n=k}^{2N} n(n-1) \cdots (n-k+1)c_n x^{n-k}$$

for $N \leq k \leq 2N$. Plainly (2) and induction yield $f^{(k)}(x) = (-1)^k f^{(k)}(1-x)$ for all $k \geq 0$. Therefore $f^{(k)}(1) = f^{(k)}(0) = 0$ for $0 \leq k < N$ and $f^{(k)}(0)$ and $f^{(k)}(1)$ are integers for $N \leq k \leq 2N$ $[f^{(k)}(0) = k! \, c_k/N!]$. Defining

$$F(x) = b^N \sum_{j=0}^{N} (-1)^j \pi^{2N-2j} f^{(2j)}(x) = \sum_{j=0}^{N} (-1)^j a^{N-j} b^j f^{(2j)}(x),$$

we see that

$$F(0) \text{ and } F(1) \text{ are integers.} \tag{3}$$

Now define a function g by

$$g(x) = F'(x)\sin(\pi x) - \pi F(x)\cos(\pi x).$$

Then, since $f^{(2N+2)}(x) = 0$ for all x,

$$g'(x) = \left[F''(x) + \pi^2 F(x) \right]\sin(\pi x)$$
$$= b^N \pi^{2N+2} f(x)\sin(\pi x)$$
$$= \pi^2 a^N f(x)\sin(\pi x).$$

By the Mean Value Theorem, there exists $0 < \xi < 1$ such that

$$\pi\left[F(1) + F(0) \right] = g(1) - g(0) = g'(\xi) = \pi^2 a^N f(\xi)\sin(\pi\xi). \tag{4}$$

It follows from (2) and (1) that

$$0 < a^N \pi f(\xi) < \pi a^N / N! < 1. \tag{5}$$

Obviously $0 < \sin(\pi\xi) \leq 1$, and so (4) and (5) show that

$$0 < \left[F(1) + F(0) \right] < 1. \tag{6}$$

Combining (3) and (6), we have found an integer between 0 and 1. This contradiction completes the proof. $\square$

(5.29) **Remark** Recall that an *algebraic number* is any complex number that is a root of some polynomial with integral (equivalently, rational) coefficients and that a *transcendental number* is any complex number that is not algebraic. It was proved by Hermite in 1873 that e is transcendental. In 1882, his technique was extended by Lindemann to show that π, too, is transcendental. It is known that the set A of all algebraic numbers is a subfield of $\mathbb{C}$. The Generalized Lindemann Theorem states that e^A is linearly independent over A; i.e., if $a_1, \ldots, a_n$ are distinct algebraic numbers and $c_1, \ldots, c_n$ are any nonzero algebraic numbers, then

$$c_1 e^{a_1} + c_2 e^{a_2} + \ldots + c_n e^{a_n} \neq 0.$$

It follows from this that both e and π are transcendental. Indeed, if $e \in A$, then take $a_1 = 0$, $a_2 = 1$, $c_1 = e$, $c_2 = -1$ to obtain $e \cdot e^0 + (-1)e^1 \neq 0$. Similarly, if $\pi \in A$, then $\pi i \in A$, and so $e^{\pi i} + e^0 \neq 0$. For proofs of these facts, see the delightful little book *Irrational Numbers* by Ivan Niven.

Lindemann's achievement was particularly outstanding because it finally proved the impossibility of squaring the circle, a problem that had been stated but unsettled for more than two thousand years! The problem is as follows: given a circle, construct, using only straight edge and compass, a square having the same area as the circle. This is equivalent to constructing a segment of length $\sqrt{\pi}$ from a given segment of length 1. It is fairly easy to show that all constructible lengths are algebraic and that if $\sqrt{\pi} \in A$, then so is π.

Incidentally, Niven's book also contains the Gelfond-Schneider Theorem: If $a, b \in A$, $0 \neq a \neq 1$, and $b \notin \mathbb{Q}$, then every value of a^b is transcendental.

Exercises

Deduce the following results from those cited in (5.29).

1.   If $a \in A$ and $a \neq 0$, then the numbers e^a, $\sin a$, $\cos a$, $\tan a$, $\sinh a$, $\cosh a$, and $\tanh a$ are all transcendental.

2.   For $a, c \in A$ with $a \neq 1$ and $z, w \in \mathbb{C}$ with $a = e^z$ and $c = e^w$, we have that z is transcendental and w/z is either rational or transcendental. [Hint: Put $b = w/z$ in the Gelfond-Schneider Theorem.]

3. e^π is transcendental.

Log Series and the Inverse Tangent

Here we give some very useful power series for the principal logarithm and the inverse tangent. Other power series for elementary functions are developed in Chapter 7.

(5.30)   Log Series   *For all $z \in \mathbb{C}$ such that $|z| < 1$, we have*

(i) $$\mathrm{Log}(1 + z) = \sum_{n=1}^{\infty} \frac{(-1)^{n-1} z^n}{n};$$

(ii) $$-\mathrm{Log}(1 - z) = \sum_{n=1}^{\infty} \frac{z^n}{n};$$

(iii) $$\mathrm{Log}\left(\frac{1+z}{1-z}\right) = 2 \cdot \sum_{n=0}^{\infty} \frac{z^{2n+1}}{2n+1}.$$

Proof It is obvious from the Ratio Test that each of these power series has radius of convergence $R = 1$. Let $f(z)$ denote the sum of the series on the right side of (i). According to (4.63), we may differentiate term-by-term to obtain

$$f'(z) = \sum_{n=1}^{\infty} (-1)^{n-1} z^{n-1} = (1+z)^{-1}$$

whenever $|z| < 1$. By (5.21),

$$\mathrm{Log}'(1+z) = (1+z)^{-1}$$

for all $z \in \mathbb{C}\backslash] - \infty, -1]$. Thus, the function g defined on the unit disk $D = \{z \in \mathbb{C} : |z| < 1\}$ by $g(z) = f(z) - \mathrm{Log}(1+z)$ satisfies $g'(z) = 0$ for all $z \in D$. It follows from (4.62) that $g(z) = g(0) = 0$ for all $z \in D$. This establishes (i). Now (ii) is obtained by replacing z with $-z$ in (i) and, since $z \in D$ implies $\mathrm{Re}(1+z) > 0$ and $\mathrm{Re}(1-z) > 0$, (iii) follows from (5.23) by adding (i) and (ii). $\square$

We next solve the equation $\tan w = z$ for w when z is given.

(5.31) **Theorem** *If $z \in \mathbb{C}$ and $z \neq \pm i$, then $\tan w = z$ if and only if $2iw$ is a complex logarithm of $(1 + iz)/(1 - iz)$. If $z = i$ or $-i$, then there is no $w \in \mathbb{C}$ such that $\tan w = z$.*

Proof For $z, w \in \mathbb{C}$, the following equalities are pairwise equivalent:

$$\sin w = z \cos w,$$
$$e^{iw} - e^{-iw} = iz(e^{iw} + e^{-iw}), \tag{1}$$

$$e^{2iw} - 1 = iz(e^{2iw} + 1),$$
$$(1 - iz)e^{2iw} = 1 + iz. \tag{2}$$

If $z = i$ or $-i$, it is clear that (2) fails for every w, and so $\tan w = z$ is impossible. Since $\sin w = \cos w = 0$ is impossible, the theorem follows from the equivalence of (1) and (2). $\square$

(5.32) **Definition** For $z \in \mathbb{C}$ with $z \neq \pm i$, we define the *principal inverse tangent* of z to be the number

(i) $$\mathrm{Arctan}\, z = \frac{1}{2i} \mathrm{Log} \frac{1+iz}{1-iz}.$$

This defines a function Arctan from $\mathbb{C}\backslash\{i, -i\}$ into $\mathbb{C}$. Some authors denote this function by Tan^{-1}.

We now list some important properties of Arctan.

(5.33) Theorem *Let* $D = \mathbb{C}\setminus\{iy : y \in \mathbb{R}, |y| \geq 1\}$ *and* $S = \{w \in \mathbb{C} : |\mathrm{Re}\,w| < \pi/2\}$.

(i) *If* $z \in \mathbb{C}\setminus\{i, -i\}$, *then* $\tan(\mathrm{Arctan}\,z) = z$.

(ii) *The restriction of* Arctan *to* D *is a one-to-one function from* D *onto* S, *and its inverse is the restriction of* tan *to* S.

(iii) *The function* Arctan *is differentiable, therefore continuous, on* D, *and*

$$\mathrm{Arctan}'\,z = (1 + z^2)^{-1} \quad \text{for } z \in D.$$

(iv) *If* $z \in \mathbb{C}$ *with* $|z| < 1$, *then*

$$\mathrm{Arctan}\,z = \sum_{n=0}^{\infty} (-1)^n \frac{z^{2n+1}}{2n+1}.$$

(v) *The restriction of* Arctan *to* $\mathbb{R}$ *is a strictly increasing continuous function from* $\mathbb{R}$ *onto* $]-\pi/2, \pi/2[$, $\lim_{x\to\infty} \mathrm{Arctan}\,x = \pi/2$, $\lim_{x\to-\infty} \mathrm{Arctan}\,x = -\pi/2$, *and* $\mathrm{Arctan}(-x) = -\mathrm{Arctan}\,x$ *for all* $x \in \mathbb{R}$.

Proof Assertion (i) follows from (5.31) and (5.32.i). Since

$$\mathrm{Re}[\mathrm{Arctan}\,z] = (1/2)\mathrm{Arg}((1 + iz)/(1 - iz)) \tag{1}$$

and since

$$(1 + iz)/(1 - iz) = (1 - |z|^2 + 2i\,\mathrm{Re}\,z)/|1 - iz|^2, \tag{2}$$

it follows that $(1 + iz)/(1 - iz)$ cannot be a nonpositive real number if $z \in D$; hence, $\mathrm{Arctan}(D) \subset S$. If $\mathrm{Arctan}\,z_1 = \mathrm{Arctan}\,z_2$, it follows from (i) that $z_1 = z_2$. Given $w_0 \in S$, define $z_0 = \tan w_0$ [note that $\cos w_0 \neq 0$] and then use (5.31) to observe that $e^{2iw_0} = (1 + iz_0)/(1 - iz_0)$ and $-\pi < \mathrm{Im}(2iw_0) < \pi$, which, with the aid of (5.20.ii), shows that

$$2iw_0 = \mathrm{Log}((1 + iz_0)/(1 - iz_0)) = 2i\,\mathrm{Arctan}\,z_0.$$

Moreover, $z_0 \in D$ because otherwise (1) and (2) imply that $\mathrm{Re}\,w_0 = \mathrm{Re}[\mathrm{Arctan}\,z_0] = \pi/2$. Therefore $\mathrm{Arctan}(D) = S$, and (ii) is established. According to (2), $(1 + iz)/(1 - iz)$ is in $\mathbb{C}\setminus]-\infty, 0]$ whenever $z \in D$, and so (iii) follows from (5.21) and the Chain Rule after a simple computation. To obtain (iv), simply replace z with iz in (5.30.iii) and use (5.32.i).

Finally, let $x \in \mathbb{R}$. Then

$$\mathrm{Im}[\mathrm{Arctan}\,x] = -(1/2)\log|(1 + ix)/(1 - ix)| = -(1/2)\log 1 = 0,$$

and so $\mathrm{Arctan}\,\mathbb{R} \subset \mathbb{R} \cap S =]-\pi/2, \pi/2[$. On the other hand, if $-\pi/2 < u < \pi/2$ and $x = \tan u$, then $x \in \mathbb{R} \subset D$ and $u \in S$, so it follows from (ii) that $\mathrm{Arctan}\,x = u$. Thus $\mathrm{Arctan}(\mathbb{R}) =]-\pi/2, \pi/2[$. From (iii) we see that Arctan is strictly increasing on $\mathbb{R}$. The preceding two sentences establish the asserted limit relations. The fact that $\tan(-u) = -\tan(u)$ for $-\pi/2 < u < \pi/2$ finishes the proof. $\square$

Rational Approximation to π

The power series (5.33.iv) is very useful for calculating rational approximants to π as follows.

(5.34)   **Theorem**   *We have*

$$\pi = 16 \cdot \sum_{n=0}^{\infty} \frac{(-1)^n}{(2n+1)5^{2n+1}} - 4 \sum_{n=0}^{\infty} \frac{(-1)^n}{(2n+1)(239)^{2n+1}} \,.$$

Proof   Write $x = \text{Arctan}(1/5)$. Then

$$\tan x = 1/5,$$
$$\tan 2x = (2 \tan x)/(1 - \tan^2 x) = 5/12,$$
$$\tan 4x = (2 \tan(2x))/(1 - \tan^2(2x)) = 120/119.$$

Next write $y = 4x - \pi/4$. Then, since $\tan(\pi/4) = 1$,

$$\tan y = (\tan(4x) - 1)/(1 + \tan(4x)) = 1/239.$$

It is clear that $0 < x < \pi/4$; hence $-\pi/4 < y < 3\pi/4$. But, since $\tan y > 0$, it follows that $0 < y < \pi/2$, and so $y = \text{Arctan}(1/239)$. Therefore

$$\pi = 4(4x - y) = 16 \,\text{Arctan}(1/5) - 4 \,\text{Arctan}(1/239),$$

and this yields the asserted formula upon substitution into (5.33.iv). $\square$

(5.35)   **Example**   Let

$$s_k = \sum_{n=0}^{k} \frac{(-1)^n \cdot 16}{(2n+1) \cdot 5^{2n+1}}, \qquad t_k = \sum_{n=0}^{k} \frac{(-1)^n \cdot 4}{(2n+1)(239)^{2n+1}}$$

$$s = \lim_{k \to \infty} s_k, \qquad\qquad t = \lim_{k \to \infty} t_k.$$

Then (5.34) says that $\pi = s - t$. By Leibnitz's Test, we have

$$s_{2k+1} < s < s_{2k}, \qquad t_{2k+1} < t < t_{2k}$$

for all $k \geq 0$. Using these facts, we can compute rational approximations to π as accurately as we desire. For example,

$$0 < s - s_3 < s_4 - s_3 = 16/(9 \cdot 5^9) < 1/10^6,$$
$$0 < t - t_1 < t_2 - t_1 = 4/\left(5 \cdot (239)^5\right)$$
$$< 4/\left(5 \cdot (200)^5\right) < 3/10^{12},$$
$$-3/10^{12} < \pi - (s_3 - t_1) = (s - s_3) - (t - t_1) < 1/10^6$$

A simple calculation yields $s_3 - t_1 = 3.1415926 \ldots$, and so the estimate

$$3.141592 < \pi < 3.141593$$

is obtained. An even simpler calculation yields the crude estimate

$$3.140 < s_1 - t_0 < s - t = \pi < s < s_0 = 3.200.$$

Exercises

1. If $a \in \mathbb{C}$, $(z_n)_{n=1}^{\infty} \subset \mathbb{C}$, and $|z_n| \to \infty$, then
    (a) $\lim_{n \to \infty} (1 + a/z_n)^{z_n} = e^a$,
    (b) $\lim_{n \to \infty} z_n(a^{1/z_n} - 1) = \text{Log}\, a$ if $a \neq 0$.
    Here, as usual, the powers have their principal values. [Hint: A power series is continuous on its open disk of convergence.]

2. If $|z - 1| \leq r < 1$, then
    (a) $|\text{Log}\, z| \leq \log[1/(1 - r)]$,
    (b) $2|\text{Arctan}(1 - z)| \leq \log[(1 + r)/(1 - r)]$.

3. If $z \in \mathbb{C}$ with $|z| < 1$, then

    $$|z|/(1 + |z|) \leq |\text{Log}(1 + z)| \leq |z|(1 + |z|)/|1 + z|.$$

    [Hints: For the second inequality, write a power series for $(1 + z)\text{Log}(1 + z)$ and estimate it. For the first, write $z = re^{i\theta}$ with $0 \leq r < 1$ and check that

    $$|\text{Log}(1 + z)| = |e^{-i\theta}\, \text{Log}(1 + z)| \geq \sum_{n=1}^{\infty} (-1)^{n-1} \frac{r^n}{n} \cos(n - 1)\theta \geq \frac{r}{1 + r}.$$

    The last inequality can be proved by differentiation.]

4. (a) If $t \in \mathbb{R}$ and $0 < |t| < 1$, then

    $$\frac{1}{2t} \log \frac{1 + t}{1 - t} < \frac{1}{1 - t^2}.$$

    (b) If $x > 0$ and $x \neq 1$, then

    $$(\log x)/(x - 1) < 1/\sqrt{x}.$$

    [Hint: Write $\sqrt{x} = (1 + t)/(1 - t)$.]
    (c) If $x > 1$, then

    $$(\log x)/(x - 1) > 2/(x + 1).$$

    (d) If $0 < a < b$, then

    $$2/(a + b) < (\log b - \log a)/(b - a) < 1/\sqrt{ab}.$$

The Sine Product
and Related Expansions

(5.36) **Definition** If $(u_n)_{n=p}^{\infty} \subset \mathbb{C}$, the symbol $\prod\limits_{n=p}^{\infty} u_n$ is called an *infinite product* and its *value* is $\lim\limits_{q \to \infty} \prod\limits_{n=p}^{q} u_n$ if this limit exists in $\mathbb{C}$. In this case, we write

$$\prod_{n=p}^{\infty} u_n = \lim_{q \to \infty} \prod_{n=p}^{q} u_n.$$

(5.37) **Example** Since

$$\prod_{n=2}^{q} \left(1 - \frac{2}{n(n+1)} \right) = \prod_{n=2}^{q} \frac{(n-1)(n+2)}{n(n+1)} = \frac{q+2}{3q},$$

$$\prod_{n=2}^{\infty} \left(1 - \frac{2}{n(n+1)} \right) = \frac{1}{3}.$$

We take up a more complete study of infinite products in Chapter 7. Our immediate goal is to obtain an infinite product expansion of the sine function. The following theorem, which has many applications, will be used to that end. It is a close relative of the Weierstrass M-test which allows us to interchange certain limiting operations.

(5.38) **Theorem** *For $n, k \in \mathbb{N}$ let $a_n(k) \in \mathbb{C}$; i.e., $(a_n)_{n=1}^{\infty}$ is a sequence of complex-valued functions on $\mathbb{N}$. Suppose that*

$$\lim_{k \to \infty} a_n(k) = A_n \in \mathbb{C} \quad \text{for each } n$$

and suppose that

$$|a_n(k)| \leq M_n < \infty \quad \text{for all } n \text{ and } k$$

where

$$\sum_{n=1}^{\infty} M_n < \infty.$$

Then

(i) $$\lim_{k \to \infty} \sum_{n=1}^{\infty} a_n(k) = \sum_{n=1}^{\infty} A_n$$

and

(ii) $$\lim_{k \to \infty} \prod_{n=1}^{\infty} (1 + a_n(k)) = \prod_{n=1}^{\infty} (1 + A_n),$$

where all sums and products appearing in (i) *and* (ii) *exist (as limits) in* $\mathbb{C}$.

Proof Plainly $|A_n| \leqq M_n$ for all n, and so all series in (i) converge absolutely in $\mathbb{C}$. Moreover, given $\epsilon > 0$, choose N such that $\sum\limits_{n=N+1}^{\infty} M_n < \epsilon/4$ and then choose K such that $|a_n(k) - A_n| < \epsilon/(2N)$ whenever $1 \leqq n \leqq N$ and $k > K$. Then $k > K$ implies that

$$\left| \sum_{n=1}^{\infty} a_n(k) - \sum_{n=1}^{\infty} A_n \right|$$

$$\leqq \sum_{n=1}^{N} |a_n(k) - A_n| + \sum_{n=N+1}^{\infty} |a_n(k)| + \sum_{n=N+1}^{\infty} |A_n|$$

$$\leqq N \cdot \frac{\epsilon}{2N} + 2 \sum_{n=N+1}^{\infty} M_n < \epsilon.$$

This proves (i).

To prove (ii), first choose m such that $M_n < 1/2$ for all $n > m$. Define $b_n(k) = \mathrm{Log}(1 + a_n(k))$ and $B_n = \mathrm{Log}(1 + A_n)$ for $n > m$ and all k. From the continuity of Log on the disk $|1 - z| < 1$ it follows that $\lim\limits_{k \to \infty} b_n(k) = B_n$ for $n > m$. Also, using (5.30.i), $n > m$ implies that

$$|b_n(k)| \leqq \sum_{j=1}^{\infty} \frac{|a_n(k)|^j}{j} \leqq M_n \sum_{j=0}^{\infty} \frac{M_n^j}{j+1} < M_n \sum_{j=0}^{\infty} 2^{-j} = 2M_n$$

for all k. Applying the first part of this theorem, we obtain

$$\lim_{k \to \infty} \sum_{n=m+1}^{\infty} b_n(k) = \sum_{n=m+1}^{\infty} B_n,$$

from which follows

$$\lim_{k \to \infty} \exp\left[\sum_{n=m+1}^{\infty} b_n(k) \right] = \exp\left[\sum_{n=m+1}^{\infty} B_n \right]. \tag{1}$$

But

$$\exp\left[\sum_{n=m+1}^{\infty} b_n(k) \right] = \lim_{q \to \infty} \exp\left[\sum_{n=m+1}^{q} b_n(k) \right]$$

$$= \lim_{q \to \infty} \prod_{n=m+1}^{q} (1 + a_n(k)) = \prod_{n=m+1}^{\infty} (1 + a_n(k)) \tag{2}$$

exists in $\mathbb{C}$ for each k and, similarly,

$$\exp\left(\sum_{n=m+1}^{\infty} B_n \right) = \prod_{n=m+1}^{\infty} (1 + A_n). \tag{3}$$

Therefore, the infinite products in (ii) all have values in $\mathbb{C}$ and equality (ii) follows easily from (2), (1), and (3) because

$$\lim_{k\to\infty} \prod_{n=1}^{\infty} (1 + a_n(k))$$

$$= \lim_{k\to\infty} \left[\left(\prod_{n=1}^{m} (1 + a_n(k)) \right) \cdot \left(\prod_{n=m+1}^{\infty} (1 + a_n(k)) \right) \right]$$

$$= \left(\lim_{k\to\infty} \prod_{n=1}^{m} (1 + a_n(k)) \right) \cdot \left(\lim_{k\to\infty} \prod_{n=m+1}^{\infty} (1 + a_n(k)) \right)$$

$$= \left(\prod_{n=1}^{m} (1 + A_n) \right) \cdot \left(\prod_{n=m+1}^{\infty} (1 + A_n) \right) = \prod_{n=1}^{\infty} (1 + A_n). \qquad \square$$

Here is a simple application.

(5.39) Theorem *For all $z \in \mathbb{C}$ we have*

$$\lim_{k\to\infty} (1 + z/k)^k = e^z.$$

Proof Fix $z \in \mathbb{C}$ and define $a_1(k) = 1 + z$,

$$a_n(k) = \begin{cases} \binom{k}{n} z^n / k^n, & 1 < n \leq k, \\ 0, & 1 \leq k < n. \end{cases}$$

Then $A_n = \lim_{k\to\infty} a_n(k) = z^n / n!$ for $n > 1$ and $A_1 = 1 + z$. Also, $|a_1(k)|$ $\leq 1 + |z| = M_1$, $|a_n(k)| \leq |z|^n / n! = M_n$ for $n > 1$, and $\sum_{n=1}^{\infty} M_n = e^{|z|} < \infty$.

Since

$$\sum_{n=1}^{\infty} a_n(k) = 1 + z + \sum_{n=2}^{k} \binom{k}{n} \left(\frac{z}{k} \right)^n = \left(1 + \frac{z}{k} \right)^k,$$

our conclusion follows from (5.38.i). $\square$

(5.40) Theorem [Euler, 1748] *For all $z \in \mathbb{C}$ we have*

(i) $$\sin(\pi z) = \pi z \cdot \prod_{n=1}^{\infty} (1 - z^2 / n^2).$$

Proof For $m \in \mathbb{N}$, let $P_m(z)$ be the polynomial given by

$$P_m(z) = [1/(2i)]((1 + \pi i z / m)^m - (1 - \pi i z / m)^m).$$

It follows from (5.39) that

$$\lim_{m\to\infty} P_m(z) = [1/(2i)][e^{\pi i z} - e^{-\pi i z}] = \sin(\pi z) \qquad (1)$$

for all $z \in \mathbb{C}$. The roots of $P_m(z)$ are the solutions of

$$\left(\frac{m + \pi i z}{m - \pi i z} \right)^m = 1 = e^{2j\pi i},$$

$$z = \frac{m(e^{2j\pi i/m} - 1)}{\pi i (e^{2j\pi i/m} + 1)} = \frac{m}{\pi} \tan \frac{j\pi}{m}$$

for integers j. If $m = 2k$ is even, then $P_m(z)$ is of degree $2k - 1$ and its (distinct) roots are

$$0, \quad \pm (2k/\pi)\tan(j\pi/(2k)) \qquad (1 \leq j < k).$$

Since the coefficient of z in the expansion of $P_{2k}(z)$ is π, we have

$$P_{2k}(z) = \pi z \cdot \prod_{j=1}^{k-1} \left(1 - \frac{\pi^2 z^2}{4k^2 \tan^2(j\pi/(2k))} \right) \tag{2}$$

for all $z \in \mathbb{C}$. Now fix $z \in \mathbb{C}$ and define

$$a_j(k) = \begin{cases} \dfrac{-\pi^2 z^2}{4k^2 \tan^2(j\pi/(2k))}, & 1 \leq j < k, \\ 0, & 1 \leq k \leq j. \end{cases}$$

It follows from (1) and (2) that

$$\sin(\pi z) = \pi z \cdot \lim_{k \to \infty} \prod_{j=1}^{\infty} (1 + a_j(k)). \tag{3}$$

Since $x < \tan x$ for $0 < x < \pi/2$ and $\lim_{u \to 0} [(\tan u)/u] = 1$ [Why?], we have

$$|a_j(k)| \leq |z|^2/j^2 = M_j, \qquad \lim_{k \to \infty} a_j(k) = -z^2/j^2 = A_j$$

for all j and k. Thus (i) follows from (3) and (5.38.ii). $\square$

(5.41) Wallis's Formulas   *We have*

(i) $$\frac{\pi}{2} = \prod_{n=1}^{\infty} \frac{2n}{2n - 1} \cdot \frac{2n}{2n + 1} = \frac{2}{1} \cdot \frac{2}{3} \cdot \frac{4}{3} \cdot \frac{4}{5} \cdot \frac{6}{5} \cdot \frac{6}{7} \cdots,$$

(ii) $$\lim_{k \to \infty} \frac{1}{\sqrt{k}} \prod_{n=1}^{k} \frac{2n}{2n - 1} = \sqrt{\pi}.$$

Proof   To obtain (i), simply put $z = 1/2$ in (5.40.i) and then take reciprocals. From (i) we have

$$\frac{\pi}{2} = \lim_{k \to \infty} \left(\frac{2}{1} \right)^2 \cdot \left(\frac{4}{3} \right)^2 \cdots \left(\frac{2k}{2k - 1} \right)^2 \cdot \frac{1}{2k + 1},$$

and so

$$\sqrt{\pi} = \lim_{k \to \infty} \frac{\sqrt{2}}{\sqrt{2k+1}} \prod_{n=1}^{k} \frac{2n}{2n-1}.$$

But

$$\lim_{k \to \infty} \left[\sqrt{2k+1} / \sqrt{k} \right] = \sqrt{2}$$

and so (ii) follows. ☐

(5.42) **Remark** The remarkable formula (5.41.i) was published by Wallis in 1656. It is perhaps the first known formula for π as the limit of an explicitly given sequence of rational numbers. The famous Leibnitz formula

$$\frac{\pi}{4} = 1 - \frac{1}{3} + \frac{1}{5} - \frac{1}{7} + - \cdots$$

was discovered in 1673. Both of these converge so slowly that they are of no practical use in computing decimal approximations to π.

Our next theorem gives the surprising "partial fraction decomposition" of the cotangent function.

(5.43) **Theorem** [Euler, 1748] *If $z \in \mathbb{C} \backslash \mathbb{Z}$, then*

(i) $$\pi z \cdot \cot(\pi z) = 1 + 2z^2 \cdot \sum_{n=1}^{\infty} \frac{1}{z^2 - n^2}.$$

Proof Let k be a nonnegative integer and let $m = 2k + 1$. We assert that

$$\sin z = \sum_{r=0}^{k} (-1)^r \binom{m}{2r+1} \cos^{m-2r-1} \left(\frac{z}{m} \right) \cdot \sin^{2r+1} \left(\frac{z}{m} \right) \qquad (1)$$

for all $z \in \mathbb{C}$. In fact, for $z \in \mathbb{R}$, this follows by applying the Binomial Theorem to the right side of

$$e^{iz} = (e^{iz/m})^m = (\cos(z/m) + i \sin(z/m))^m$$

and then equating imaginary parts. Since each side of (1) is the sum of an everywhere convergent power series [Merten's Theorem], the extension to complex z follows from the fact that if a power series [the difference of two] converges to 0 at each $z \in \mathbb{R}$, then all of its coefficients are 0 [Why?]. Now define

$$P_m(w) = \sum_{r=0}^{k} (-1)^r \binom{m}{2r+1} w^{2r+1}.$$

From (1) we have

$$\sin z = \cos^m(z/m) P_m(\tan(z/m)) \qquad (2)$$

for $|z| < m\pi/2 = k\pi + \pi/2$. Therefore the roots of the polynomial $P_m(w)$ are the m numbers

$$w_n = \tan(n\pi/m), \qquad -k \le n \le k, \tag{3}$$

and so

$$P_m(w) = (-1)^k \cdot \prod_{n=-k}^{k} (w - w_n)$$

for all $w \in \mathbb{C}$. Using a formula for differentiating a finite product [proved by induction on the number of factors], we conclude that

$$\frac{P_m'(w)}{P_m(w)} = \sum_{n=-k}^{k} \frac{1}{w - w_n} \tag{4}$$

for $w \ne w_n$. Next differentiate both sides of (2) with respect to z to obtain

$$\cos z = (1/m)\left[\cos^{m-2}(z/m) P_m'(\tan(z/m)) \right.$$
$$\left. - m \cos^{m-1}(z/m)\sin(z/m) P_m(\tan(z/m)) \right].$$

Dividing this equation by (2), we obtain

$$\cot z = \frac{1}{m \cos^2(z/m)} \frac{P_m'(\tan(z/m))}{P_m(\tan(z/m))} - \tan(z/m).$$

Now we substitute (3) and (4) and juggle to get

$$(\cot z + \tan(z/m))\cos^2(z/m)$$

$$= \frac{1}{m \tan(z/m)} + \sum_{n=1}^{k} \frac{2\tan(z/m)}{m(\tan^2(z/m) - \tan^2(n\pi/m))} \tag{5}$$

whenever $|z| < m\pi/2$ and z is not an integral multiple of π.

Now fix $z \in \mathbb{C}$, not in $\pi\mathbb{Z}$, and let $m = 2k + 1 > |z|$. Define

$$a_n(m) = \frac{2\tan(z/m)}{m(\tan^2(z/m) - \tan^2(n\pi/m))}$$

for $1 \le n \le k$ and $a_n(m) = 0$ for $n > k$. Now $\lim_{m\to\infty} m\tan(z/m) = z$, and so there exists an m_0 such that $m|\tan(z/m)| < 2|z|$ for $m > m_0$. Also, $\tan^2(n\pi/m) > n^2\pi^2/m^2$ for $1 \le n \le k$. It follows that

$$|a_n(m)| \le 4|z|/(n^2\pi^2 - 4|z|^2) = M_n$$

for $m > m_0$ and $n > 2|z|$. Since $\lim_{n\to\infty} n^2 M_n = 4|z|/\pi^2 = c$, there exists N such that $M_n < 2c/n^2$ for $n > N$ and so $\sum M_n < \infty$. Also, $\lim_{m\to\infty} a_n(m) = 2z/(z^2 - n^2\pi^2)$. According to (5) and (5.38.i), we now have

$$\cot z = \frac{1}{z} + \lim_{m\to\infty} \sum_{n=1}^{\infty} a_n(m) = \frac{1}{z} + \sum_{n=1}^{\infty} \frac{2z}{z^2 - n^2\pi^2}.$$

But this is equivalent to (i) if we replace z by πz and then require only that $z \notin \mathbb{Z}$. $\square$

Stirling's Formula

We next make use of Wallis's Formula and the Log Series to obtain the following important estimates of the size of $n!$.

(5.44)   **Theorem**   *If $n > 1$ is an integer, then*

(i) $$\exp\left[\frac{1}{12n + 1/4}\right] < \frac{n!}{\sqrt{2\pi n}\, e^{-n} n^n} < \exp\left[\frac{1}{12n}\right].$$

In particular,

(ii) $$\lim_{n \to \infty}\left[n!/(\sqrt{2\pi n}\, e^{-n} n^n) \right] = 1.$$

Proof   For any $n \in \mathbb{N}$, we can take $z = 1/(2n + 1)$ in (5.30.iii) to get

$$\left(n + \tfrac{1}{2}\right)\log\left(1 + \frac{1}{n}\right) - 1 = \frac{2n + 1}{2} \log \frac{1 + 1/(2n + 1)}{1 - 1/(2n + 1)} - 1$$

$$= \sum_{k=1}^{\infty} \frac{1}{(2k + 1)(2n + 1)^{2k}}. \qquad (1)$$

This implies that

$$\left(n + \tfrac{1}{2}\right)\log\left(1 + \frac{1}{n}\right) - 1 < \frac{1}{3} \sum_{k=1}^{\infty} \frac{1}{(2n + 1)^{2k}}$$

$$= \frac{1}{3} \cdot \frac{1/(2n + 1)^2}{1 - 1/(2n + 1)^2} = \frac{1}{12n} - \frac{1}{12(n + 1)}. \qquad (2)$$

Since induction shows that $(3/5)^{k-1} \le 3/(2k + 1)$ for $k \ge 1$ and since, for $n \ge 2, 6n + 49/16 > 15 > 72/5$, we also deduce from (1) that

$$\left(n + \tfrac{1}{2}\right)\log\left(1 + \frac{1}{n}\right) - 1 = \frac{1}{3(2n + 1)^2} \sum_{k=1}^{\infty} \frac{3}{(2k + 1)(2n + 1)^{2k-2}}$$

$$\ge \frac{1}{3(2n + 1)^2} \sum_{k=1}^{\infty} \left(\frac{3}{5(2n + 1)^2}\right)^{k-1}$$

$$= \frac{1}{3(2n + 1)^2} \cdot \frac{1}{1 - 3/\left[5(2n + 1)^2\right]}$$

$$= \frac{1}{12n^2 + 12n + 6/5}$$

$$= \frac{12}{144n^2 + 144n + 72/5}$$

$$> \frac{12}{144n^2 + 150n + 49/16}$$

$$= \frac{1}{12n + 1/4} - \frac{1}{12(n + 1) + 1/4}. \qquad (3)$$

For the sake of brevity, write

$$x_n = \exp\left[\frac{-1}{12n + 1/4}\right], \quad y_n = \exp\left[\frac{-1}{12n}\right], \quad a_n = n! e^n n^{-(n+1/2)}. \quad (4)$$

If we apply exp to inequalities (2) and (3), we find that

$$\frac{x_{n+1}}{x_n} < \frac{1}{e}\left(1 + \frac{1}{n}\right)^{n+1/2} = \frac{a_n}{a_{n+1}} < \frac{y_{n+1}}{y_n} \quad (5)$$

for all $n \geq 2$. Now $x_{n+1}/x_n > e^0 = 1$, so the sequence (a_n) is decreasing to some limit $a \geq 0$. We also see, using (5) and (4), that $a_n x_n {\downarrow} a$ and $a_n y_n {\uparrow} a$ as $n \to \infty$. Therefore

$$0 < a_n y_n < a < a_n x_n \quad (6)$$

for all $n \geq 2$. Using (4) to write $n! = a_n e^{-n} n^{n+1/2}$ and using Wallis's Formula (5.41.ii), we can evaluate a as follows:

$$\sqrt{\pi} = \lim_{n\to\infty} \frac{2^{2n}(n!)^2}{(2n)!\sqrt{n}} = \lim_{n\to\infty} \frac{2^{2n} a_n^2 n^{2n+1} e^{-2n}}{a_{2n}(2n)^{2n+1/2} e^{-2n}\sqrt{n}}$$

$$= \lim_{n\to\infty} \frac{a_n^2}{a_{2n} 2^{1/2}} = \frac{a}{\sqrt{2}}.$$

Thus, $a = \sqrt{2\pi}$. Substitution of this value into (6) yields

$$x_n^{-1} < a_n/\sqrt{2\pi} < y_n^{-1},$$

which is the same as (i). □

(5.45) Remark The limit statement (5.44.ii) was proved by James Stirling in 1764. Many bounds of the type (5.44.i) have since been obtained. The ones given here are due to E. Cesàro (1922). They can be improved by making sharper estimates of the series (5.44.1).

Exercises

1.   (a) Prove that $z \cdot \prod_{n=1}^{\infty} \cos(z/2^n) = \sin z$ for all $z \in \mathbb{C}$. [Hint: Let P_k denote the kth partial product and use the identity $2 \sin w \cos w = \sin(2w)$ to prove that $P_k \cdot 2^k \sin(z/2^k) = \sin z$.]

(b) If $x_1 = \sqrt{1/2}$ and $x_{n+1} = \sqrt{(1 + x_n)/2}$ for $n \in \mathbb{N}$, then

$$\frac{2}{\pi} = \prod_{n=1}^{\infty} x_n = \sqrt{\frac{1}{2}} \cdot \sqrt{\frac{1}{2} + \frac{1}{2}\sqrt{\frac{1}{2}}} \cdot \sqrt{\frac{1}{2} + \frac{1}{2}\sqrt{\frac{1}{2} + \frac{1}{2}\sqrt{\frac{1}{2}}}} \cdots$$

[Remark: This infinite product representation of $2/\pi$ was given by F. Vieta in 1646.]

2. $(e^{\pi a} - e^{-\pi a})/(2\pi a) = \displaystyle\prod_{n=1}^{\infty} (1 + a^2/n^2)$ for any $a \in \mathbb{R}$.

3. $\cos(\pi z) = \displaystyle\prod_{k=1}^{\infty} (1 - 4z^2/(2k-1)^2)$ for all $z \in \mathbb{C}$. [Hint: Use $\sin(2w) = 2\sin w \cos w$.]

4. $\pi/4 = \displaystyle\prod_{n=1}^{\infty} (1 - 1/(2n+1)^2)$. [Hint: Use Exercise 3 and (5.38.ii). Let $z_k \to 1/2$.]

5. $\displaystyle\lim_{n\to\infty} \left\{ \sum_{k=0}^{2n} \left[\binom{2n}{k} \middle/ \left[\binom{2n}{n} \sqrt{n\pi} \right] \right] \right\} = 1$. [Hint: Use Wallis's second formula.]

6. Evaluate $\displaystyle\prod_{n=1}^{\infty} (1 + z^4/n^4)$. [Hint: $(1 + w^4) = (1 - (\alpha w)^2)(1 - (\beta w)^2)$ where $\alpha^2 = i$ and $\beta^2 = -i$.]

7. Evaluate $\displaystyle\lim_{z\to 0} \prod_{n=1}^{\infty} (1 + (\cos z)/n^2)$.

8. $\pi \tan(\pi z/2) = \displaystyle\sum_{n=0}^{\infty} [4z/((2n+1)^2 - z^2)]$ unless z is an odd integer. [Hint: Use $2\cot w = \cot(w/2) - \tan(w/2)$ and check that the sum is 0 if z is an even integer.]

9. $\pi z \csc(\pi z) = 1 + 2z^2 \cdot \displaystyle\sum_{n=1}^{\infty} [(-1)^{n-1}/(n^2 - z^2)]$ unless z is an integer. [Hint: $\csc z = \cot z + \tan(z/2)$.]

10. $\pi \sec(\pi z) = \displaystyle\sum_{n=0}^{\infty} \frac{(-1)^n(2n+1)}{[(2n+1)/2]^2 - z^2}$ unless $2z$ is an odd integer. [Hint: Replace z by $1/2 - z$ in Exercise 9, then factor each term. Finally regroup the resulting terms in pairs. Be sure to justify all steps.]

11. (a)
$$\pi = a\tan\frac{\pi}{a}\left[1 - \frac{1}{a-1} + \frac{1}{a+1} - \frac{1}{2a-1} + \frac{1}{2a+1} - + \cdots \right.$$
$$\left. - \frac{1}{na-1} + \frac{1}{na+1} - + \cdots \right]$$

whenever $a > 0$ and $\cos(2\pi/a) \neq 0$,

(b)
$$\pi = a\sin\frac{\pi}{a} \cdot \left[1 + \frac{1}{a-1} - \frac{1}{a+1} - \frac{1}{2a-1} + \frac{1}{2a+1} + \frac{1}{3a-1} - - + + \cdots \right]$$

where $a > 0$ and $1/a$ is not an integer. In particular it must be shown that these series converge. Try some particular values of a.

12. Let $P_k = \displaystyle\prod_{n=1}^{k} [4n^2/(4n^2 - 1)]$ be the kth partial product in Wallis's Formula for $\pi/2$ and let $Q_k = \pi/(2P_k)$. Prove that
$$\frac{1}{3k+3} < \frac{\pi}{2} - P_k < \frac{1}{k}.$$

[Hint:

$$Q_k > \prod_{k+1}^{\infty} \left(1 + \frac{1}{4n^2}\right) > 1 + \frac{1}{4} \sum_{k+1}^{\infty} \frac{1}{n(n+1)} = 1 + \frac{1}{4(k+1)}$$

and

$$Q_k < \exp\left(\frac{1}{4} \sum_{k+1}^{\infty} \frac{1}{n(n-1)}\right) = \exp\left(\frac{1}{4k}\right) < 1 + \frac{1}{2k} \, .$$

Also

$$\frac{4}{3}(Q_k - 1) \leq \frac{\pi}{2} - P_k < \frac{\pi}{2}(Q_k - 1).]$$

13.   For $n \in \mathbb{N}$ with $n > 1$, we have
$$\exp\left(-\frac{1}{6n}\right) < \binom{2n}{n} \cdot 2^{-2n} \cdot \sqrt{\pi n} < \exp\left(-\frac{1}{12n+1}\right).$$

14.   (a) If $0 \leq x_j \leq 1$ for $1 \leq j \leq n$, then, by induction,
$$\prod_{j=1}^{n} (1 - x_j) \geq 1 - \sum_{j=1}^{n} x_j.$$

(b) If $n \in \mathbb{N}$ and $z \in \mathbb{C}$, then
$$|e^z - (1 + z/n)^n| \leq |e^{|z|} - (1 + |z|/n)^n| \leq |z|^2 e^{|z|}/(2n).$$

[Hint: For $n > 1$, we have

$$e^z - \left(1 + \frac{z}{n}\right)^n = \sum_{k=n+1}^{\infty} \frac{z^k}{k!} + \sum_{k=2}^{n} \left[1 - \left(1 - \frac{1}{n}\right)\left(1 - \frac{2}{n}\right) \cdots \left(1 - \frac{k-1}{n}\right)\right] \frac{z^k}{k!} \, .\right]$$

6

INTEGRATION

In this chapter we give an elementary development of the theory of the Lebesgue integral on $\mathbb{R}$ and $\mathbb{R}^n$. Our approach is essentially due to F. Riesz. It assumes no prior knowledge of any kind of integration theory or of any measure theory [aside from those contained earlier in this book]. We begin with the simple matter of integrating a step function.

Step Functions

Recall that if E is a subset of a set X, then the characteristic function of E is the function ξ_E defined on X by

$$\xi_E(x) = \begin{cases} 1 & \text{if } x \in E, \\ 0 & \text{if } x \in X \setminus E. \end{cases}$$

(6.1) **Definition** A real-valued function ϕ on $\mathbb{R}$ is called a *step function* if ϕ has the form

(i) $$\phi = \sum_{j=1}^{n} \alpha_j \xi_{I_j}$$

where $\{I_j\}_{j=1}^{n}$ is a pairwise disjoint family of bounded intervals [open, closed, or neither; singletons allowed] and $(\alpha_j)_{j=1}^{n} \subset \mathbb{R}$. That is, there exist real numbers $a = x_0 < x_1 < \ldots < x_m = b$ such that $\phi(x) = 0$ if $x < a$ or $x > b$ and ϕ is constant on each $]x_{k-1}, x_k[$. We denote the family of all step functions on $\mathbb{R}$ by $M_0(\mathbb{R})$ or simply by M_0.

The following proposition allows us to define an integral unambiguously on M_0.

(6.2) **Proposition** *Suppose that $\phi \in M_0$ and that ϕ is expressed in two different ways as in (6.1.i):*

(i) $$\phi = \sum_{j=1}^{n} \alpha_j \xi_{I_j}, \qquad \phi = \sum_{k=1}^{m} \beta_k \xi_{J_k}.$$

Then

(ii) $$\sum_{j=1}^{n} \alpha_j |I_j| = \sum_{k=1}^{m} \beta_k |J_k|.$$

Proof We may suppose that each family $\{I_j\}_{j=1}^{n}$ and $\{J_k\}_{k=1}^{m}$ covers $\mathbb{R}$ and is pairwise disjoint; otherwise we can adjoin a finite number of intervals to each family and, in (i), affix a zero coefficient to the characteristic function of each new interval. This adjunction does not affect (ii) [recall: $0 \cdot \infty = 0$]. This supposed, we have

$$I_j = \bigcup_{k=1}^{m} I_j \cap J_k, \qquad J_k = \bigcup_{j=1}^{n} I_j \cap J_k,$$

$$|I_j| = \sum_{k=1}^{m} |I_j \cap J_k|, \qquad |J_k| = \sum_{j=1}^{n} |I_j \cap J_k|.$$

Therefore

$$\sum_{j=1}^{n} \alpha_j |I_j| = \sum_{j=1}^{n} \sum_{k=1}^{m} \alpha_j |I_j \cap J_k|$$

$$= \sum_{j=1}^{n} \sum_{k=1}^{m} \beta_k |I_j \cap J_k|$$

$$= \sum_{k=1}^{m} \sum_{j=1}^{n} \beta_k |I_j \cap J_k| = \sum_{k=1}^{m} \beta_k |J_k|,$$

where the second equality obtains because if $|I_j \cap J_k| > 0$ and $x \in I_j \cap J_k$, then

$$\alpha_j = \alpha_j \xi_{I_j}(x) = \phi(x) = \beta_k \xi_{J_k}(x) = \beta_k. \qquad \square$$

(6.3) **Definition** If $\phi \in M_0$ is as in (6.1.i), we define the *Lebesgue integral* of ϕ to be the number

$$\int \phi = \sum_{j=1}^{n} \alpha_j |I_j|.$$

Thus $\int$ is a real-valued function on M_0.

We now prove three important lemmas concerning M_0 and the integral. Much of what follows will depend only on these lemmas and not on the specific form of the elements of M_0 or on the specific way that we have defined the integral. Of course we must refer to (6.1) and (6.3) to prove the lemmas.

(6.4)   **Lemma I**   *The set M_0, equipped with pointwise operations, is a real vector lattice of functions and $\int$ is a monotone linear functional on M_0. That is, if $\phi, \psi \in M_0$ and $\alpha \in \mathbb{R}$, then*

(i) $$\phi + \psi \in M_0,$$

(ii) $$\alpha\phi \in M_0,$$

(iii) $$\phi \vee \psi \in M_0,$$

(iv) $$\int (\phi + \psi) = \int \phi + \int \psi,$$

(v) $$\int \alpha\phi = \alpha \int \phi, \text{ and}$$

(vi) $$\phi \leq \psi \quad \text{implies} \quad \int \phi \leq \int \psi.$$

Proof   Write $\phi = \sum\limits_{j=1}^{n} \alpha_j \xi_{I_j}$ and $\psi = \sum\limits_{k=1}^{m} \beta_k \xi_{J_k}$, where the families $\{I_j\}_{j=1}^{n}$ and $\{J_k\}_{k=1}^{m}$ each cover $\mathbb{R}$ and are each pairwise disjoint [see the proof of (6.2)]. Then

$$\phi + \psi = \sum_{j=1}^{n} \sum_{k=1}^{m} (\alpha_j + \beta_k) \xi_{I_j \cap J_k}$$

and

$$\int \phi + \int \psi = \sum_{j=1}^{n} \left(\alpha_j \sum_{k=1}^{m} |I_j \cap J_k| \right) + \sum_{k=1}^{m} \left(\beta_k \sum_{j=1}^{n} |I_j \cap J_k| \right)$$

$$= \sum_{j=1}^{n} \sum_{k=1}^{m} (\alpha_j + \beta_k) |I_j \cap J_k|$$

$$= \int (\phi + \psi),$$

and so (i) and (iv) obtain. Statements (ii) and (v) are obviously true. To prove (iii), notice that

$$|\phi| = \sum_{j=1}^{n} |\alpha_j| \xi_{I_j} \in M_0$$

and

$$\phi \vee \psi = (1/2)(\phi + \psi + |\phi - \psi|)$$

and use (i) and (ii). Finally, (vi) obtains because if $\phi \leq \psi$, then $\psi - \phi \geq 0$, and so $\int \psi - \int \phi = \int (\psi - \phi) \geq 0$. $\square$

(6.5)   **Corollary**   *If $\phi, \psi \in M_0$, then the functions $\phi \wedge \psi$, $|\phi|$, ϕ^+, and ϕ^- are all in M_0.*

Proof   Just use (6.4) and the relations $\phi \wedge \psi = -[(-\phi) \vee (-\psi)]$, $\phi^+ = \phi \wedge 0$, $\phi^- = -(\phi \vee 0)$, and $|\phi| = \phi^+ + \phi^-$. $\square$

(6.6)   **Lemma II**   *Let* $\phi_1 \leqq \phi_2 \leqq \ldots$ *be a nondecreasing sequence in* M_0 *and suppose that* $\lim\limits_{n \to \infty} \int \phi_n < \infty$. *Then* $\lim\limits_{n \to \infty} \phi_n(x) < \infty$ *a.e.*

Proof   Let $E = \{x \in \mathbb{R} : \lim\limits_{n \to \infty} \phi_n(x) = \infty\}$. We are to show that $\lambda(E) = 0$. That is, given $\epsilon > 0$, we must find a countable family $\mathcal{G}$ of open intervals such that $E \subset \cup \mathcal{G}$ and $\sum\limits_{I \in \mathcal{G}} |I| < \epsilon$.

Let $\epsilon > 0$ be given. Since the (monotone) sequence $(\int \phi_n)_{n=1}^{\infty}$ converges, it is a Cauchy sequence. Choose $n_1 \in \mathbb{N}$ such that $n > n_1$ implies $\int \phi_n - \int \phi_{n_1} < 1/4$. If $k \geqq 2$ and n_{k-1} has been chosen, choose $n_k > n_{k-1}$ such that

$$n > n_k \quad \text{implies} \quad \int \phi_n - \int \phi_{n_k} < 1/4^k.$$

Inductively, we have selected a subsequence such that the sequence $\psi_k = \phi_{n_{k+1}} - \phi_{n_k}$ satisfies

$$0 \leqq \int \psi_k < 1/4^k \qquad (k = 1, 2, \ldots). \tag{1}$$

Writing $A_k = \{x \in \mathbb{R} : \psi_k(x) > 1/2^k\}$, we see that A_k is the union of a finite pairwise disjoint family $\mathcal{G}_k$ of intervals. Since $\xi_{A_k} \leqq 2^k \psi_k$, it follows from (1) that

$$\sum_{J \in \mathcal{G}_k} |J| = \int \xi_{A_k} \leqq 2^k \int \psi_k < 1/2^k \tag{2}$$

for all k [void sums are 0]. Choose $m \in \mathbb{N}$ such that $2^{-m+2} < \epsilon$ and write $\mathcal{G} = \bigcup\limits_{k=m}^{\infty} \mathcal{G}_k$. Then $\cup \mathcal{G} = \bigcup\limits_{k=m}^{\infty} A_k$ and, using (2), we have

$$\sum_{J \in \mathcal{G}} |J| < \sum_{k=m}^{\infty} 1/2^k < \epsilon/2.$$

Enumerate $\mathcal{G}$ as $\{J_j\}_{j=1}^{\infty}$ and, for each j, choose an *open* interval $I_j \supset J_j$ with $|I_j| < \epsilon/2^{j+1} + |J_j|$. Write $\mathcal{G} = \{I_j\}_{j=1}^{\infty}$. Then

$$\sum_{I \in \mathcal{G}} |I| < \epsilon.$$

We complete the proof by showing that $E \subset \cup \mathcal{G} \subset \cup \mathcal{G}$.

Let $x \in \mathbb{R} \setminus \cup \mathcal{G}$. Then we have $x \in \mathbb{R} \setminus A_k$ for all $k \geqq m$, and so $\psi_k(x) \leqq 2^{-k}$ for all such k. Thus, given $p > m$, we obtain

$$\phi_{n_p}(x) - \phi_{n_m}(x) = \sum_{k=m}^{p-1} (\phi_{n_{k+1}}(x) - \phi_{n_k}(x)) = \sum_{k=m}^{p-1} \psi_k(x) < 1,$$

and so the monotone sequence $(\phi_{n_p}(x))_{p=m+1}^{\infty}$ is bounded above by the real number $\phi_{n_m}(x) + 1$. Therefore $x \notin E$. $\square$

Lemma II suggests an alternative definition of "measure zero." The following theorem makes this explicit.

(6.7) **Theorem** *Let $E \subset \mathbb{R}$. Then a necessary and sufficient condition that $\lambda(E)$ = 0 is that there exists a nondecreasing sequence $(\phi_n)_{n=1}^{\infty} \subset M_0$ such that*

(i)
$$\lim_{n \to \infty} \int \phi_n < \infty$$

and

(ii)
$$\lim_{n \to \infty} \phi_n(x) = \infty \quad \textit{for all } x \in E.$$

Proof The sufficiency follows from (6.6). To prove necessity, suppose that $\lambda(E) = 0$. For $k \in \mathbb{N}$, choose a family $\{I_{k,j}\}_{j=1}^{\infty}$ of open intervals that covers E such that

$$\sum_{j=1}^{\infty} |I_{k,j}| < 1/2^k.$$

For $n \in \mathbb{N}$, define

$$\phi_n = \sum_{k=1}^{n} \sum_{j=1}^{n} \xi_{I_{k,j}}$$

Then $(\phi_n)_{n=1}^{\infty}$ is nondecreasing and

$$\int \phi_n = \sum_{k=1}^{n} \sum_{j=1}^{n} |I_{k,j}| \le \sum_{k=1}^{n} 1/2^k < 1$$

for each n. Thus (i) obtains. For $x \in E$, there are infinitely many pairs (k, j) with $x \in I_{k,j}$ [at least one for each k], and so

$$\lim_{n \to \infty} \phi_n(x) = \sum_{k=1}^{\infty} \sum_{j=1}^{\infty} \xi_{I_{k,j}}(x) = \infty.$$

Therefore (ii) obtains. $\square$

(6.8) **Lemma III** *If $(\phi_n)_{n=1}^{\infty} \subset M_0$ is a nonincreasing sequence of nonnegative step functions such that $\lim_{n \to \infty} \phi_n(x) = 0$ a.e., then $\lim_{n \to \infty} \int \phi_n = 0$.*

Proof Choose $a < b$ in $\mathbb{R}$ such that $\phi_1(x) = 0$ if $x < a$ or $x > b$ and choose a positive real number β such that $\phi_1(x) < \beta$ for all $x \in \mathbb{R}$. Let $D = \{x \in \mathbb{R} : \phi_n$ is discontinuous at x for some $n\}$ and $E = \{x \in \mathbb{R} : \lim_{n \to \infty} \phi_n(x) = 0$ is false$\}$. Then $\lambda(D \cup E) = 0$.

Let $\epsilon > 0$ be given. Choose a countable family $\mathcal{G}$ of open intervals that covers $D \cup E$ and satisfies

$$\sum_{I \in \mathcal{G}} |I| < \epsilon/(2\beta). \tag{1}$$

Write $C = [a, b] \setminus (D \cup E)$. Then $\lim_{n \to \infty} \phi_n(x) = 0$ for all $x \in C$. Thus, given $x \in C$, there is an $n(x) \in \mathbb{N}$ such that $\phi_{n(x)}(x) < \epsilon/(2(b - a))$. Since $\phi_{n(x)}$ is continuous at x, there is an open interval $J(x)$ containing x such that

$$\phi_{n(x)}(t) < \epsilon/(2(b - a)) \quad \text{for } t \in J(x). \tag{2}$$

The family $\mathcal{I} \cup \{J(x) : x \in C\}$ covers $[a, b]$, and so the Heine–Borel Theorem yields $I_1, \ldots, I_p \in \mathcal{I}$ and $x_1, \ldots, x_q \in C$ such that

$$[a, b] \subset \bigcup_{j=1}^{p} I_j \cup \bigcup_{k=1}^{q} J(x_k)$$

Write $N = \max\{n(x_k) : 1 \leq k \leq q\}$. Then it follows from (2) that

$$\phi_N(t) < \epsilon/(2(b - a)) \quad \text{for } t \in \bigcup_{k=1}^{q} J(x_k). \tag{3}$$

The set $[a, b] \backslash \bigcup_{j=1}^{p} I_j$ can be expressed as a finite disjoint union $\bigcup_{k=1}^{r} L_k$ of intervals. Define

$$\psi = \sum_{j=1}^{p} \beta \xi_{I_j} + \sum_{k=1}^{r} \frac{\epsilon}{2(b - a)} \xi_{L_k}.$$

It follows from (3) and our choice of β that $\phi_N \leq \psi$, and so, with the use of (1), $n \geq N$ implies

$$\int \phi_n \leq \int \phi_N \leq \int \psi = \beta \sum_{j=1}^{p} |I_j| + \sum_{k=1}^{r} \frac{\epsilon}{2(b - a)} |L_k|$$

$$< \epsilon/2 + \epsilon/2 = \epsilon. \qquad \square$$

For use in extending the domain of our integral to a larger set of functions, we prove the following.

(6.9)   Theorem *Suppose that the sequences $(\phi_n)_{n=1}^{\infty} \subset M_0$ and $(\psi_m)_{m=1}^{\infty} \subset M_0$ are nondecreasing and that*

$$\lim_{n \to \infty} \phi_n(x) \leq \lim_{m \to \infty} \psi_m(x)$$

a.e. on $\mathbb{R}$. Then

$$\lim_{n \to \infty} \int \phi_n \leq \lim_{m \to \infty} \int \psi_m.$$

Proof   Let $n \in \mathbb{N}$ be fixed. Then the sequence $(\phi_n - \psi_m)_{m=1}^{\infty}$ is nonincreasing and

$$\lim_{m \to \infty} [\phi_n(x) - \psi_m(x)] = \phi_n(x) - \lim_{m \to \infty} \psi_m(x) \leq 0$$

a.e. Therefore the sequence $([\phi_n - \psi_m]^+)_{m=1}^{\infty}$ is nonincreasing and has limit 0 a.e. It follows from (6.8) that

$$\int \phi_n - \lim_{m \to \infty} \int \psi_m = \lim_{m \to \infty} \int [\phi_n - \psi_m] \leq \lim_{m \to \infty} \int [\phi_n - \psi_m]^+ = 0.$$

Thus the inequality

$$\int \phi_n \leq \lim_{m \to \infty} \int \psi_m$$

obtains for each $n \in \mathbb{N}$. Now let $n \to \infty$. $\square$

The First Extension

(6.10)   Definition   If f is a real-valued function that is defined a.e. on $\mathbb{R}$, then we write $f \in M_1$ provided that there exists a nondecreasing sequence $(\phi_n)_{n=1}^{\infty} \subset M_0$ such that $\lim_{n\to\infty} \phi_n(x) = f(x)$ a.e. and $\lim_{n\to\infty} \int \phi_n < \infty$. This defines the set M_1. In this case, we define the *Lebesgue integral* of f to be the real number $\int f = \lim_{n\to\infty} \int \phi_n$.

(6.11)   Remarks   (a) The above definition of $\int f$ is unambiguous: it does not depend on the particular choice of the sequence $(\phi_n)_{n=1}^{\infty}$. Indeed, suppose that $(\psi_m)_{m=1}^{\infty} \subset M_0$ is another nondecreasing sequence such that $\lim_{m\to\infty} \psi_m(x) = f(x) = \lim_{n\to\infty} \phi_n(x)$ a.e. It follows from (6.9) that

$$\lim_{n\to\infty} \int \phi_n = \lim_{m\to\infty} \int \psi_m.$$

   (b) If $f \in M_0$, then $f \in M_1$, and $\int f$ as defined in (6.10) agrees with $\int f$ as defined in (6.3), for we can take $\phi_n = f$ for all n.

   (c) If $f \in M_1$ and $g = f$ a.e., then $g \in M_1$ and $\int g = \int f$. This is clear since we can use the same sequence $(\phi_n)_{n=1}^{\infty}$ for g that we did for f.

(6.12)   Examples   (a) Let $f : \mathbb{R} \to [0, \infty]$ be such that the set $C = \{x \in \mathbb{R} : f(x) < \infty, f \text{ is continuous at } x\}$ satisfies $\lambda(\mathbb{R} \setminus C) = 0$. We claim that there is a nondecreasing sequence $(\phi_n)_{n=1}^{\infty} \subset M_0$ for which $\lim_{n\to\infty} \phi_n(x) = f(x)$ a.e. on $\mathbb{R}$. In fact, write $N_n = n \cdot 2^n$, $I_{n,k} =](k-1)/2^n, k/2^n]$, and define

$$\phi_n = \sum_{k=1-N_n}^{N_n} \left[\inf f(I_{n,k}) \right] \cdot \xi_{I_{n,k}}.$$

Then $\phi_n \in M_0$ because C is dense in $\mathbb{R}$. Since $I_{n,k} = I_{n+1,2k-1} \cup I_{n+1,2k}$ and $\phi_n = 0$ off $]-n, n]$, we have $0 \leq \phi_n \leq \phi_{n+1} \leq f$. The set $E = (\mathbb{R} \setminus C)$ has measure 0. Given $x \in C$ and any $\alpha < f(x)$, choose $\delta > 0$ such that $f(t) > \alpha$ whenever $|t - x| < \delta$, and then observe that $\phi_n(x) \geq \alpha$ whenever $n > |x|$ and $2^{-n} < \delta$. Therefore, $\lim_{n\to\infty} \phi_n(x) = f(x)$ for $x \in \mathbb{R} \setminus E$. Note that f need not be in M_1 since we could have $\lim_{n\to\infty} \int \phi_n = \infty$.

   (b) Let f be as in (a). Suppose also that $f(\mathbb{R})$ is bounded above by some real number β and that f vanishes identically outside $[a, b]$. Then $f \in M_1$ because

$$\phi_n \leq \beta \xi_{[a,b]}, \qquad \int \phi_n \leq \beta(b-a) < \infty$$

for all n.

   (c) Define $f(x) = 1/\sqrt{x}$ for $0 < x \leq 1$ and $f(x) = 0$ otherwise. Then $f \in M_1$ because, with ϕ_n as in (a), we have

$$\int \phi_n = \sum_{k=1}^{2^n} \left(\frac{2^n}{k} \right)^{1/2} \cdot \frac{1}{2^n} = \frac{1}{2^{n/2}} \sum_{k=1}^{2^n} \frac{1}{\sqrt{k}}$$

and so [with the use of Exercise 12, Chapter 1]

$$\frac{2\sqrt{2^n + 1} - 2}{2^{n/2}} < \int \phi_n < \frac{2\sqrt{2^n} - 1}{2^{n/2}} < 2.$$

It follows that

$$\int f = \lim_{n \to \infty} \int \phi_n = 2.$$

Our next theorem lists a few simple properties of M_1 and the integral.

(6.13)   Theorem   *If $f, g \in M_1$ and $0 \leqq \alpha < \infty$, then $\alpha f, f + g, f \wedge g$, and $f \vee g$ are all in M_1; $\int \alpha f = \alpha \int f, \int (f + g) = \int f + \int g$. If $f \leqq g$, then $\int f \leqq \int g$.*

Proof   Choose nondecreasing sequences $(\phi_n)_{n=1}^\infty, (\psi_n)_{n=1}^\infty \subset M_0$ that converge a.e. to f and g, respectively, such that

$$\lim_{n \to \infty} \int \phi_n < \infty, \qquad \lim_{n \to \infty} \int \psi_n < \infty.$$

If $f \leqq g$, then (6.9) yields $\int f \leqq \int g$. One checks immediately that the nondecreasing sequences $(\alpha \phi_n)_{n=1}^\infty, (\phi_n + \psi_n)_{n=1}^\infty, (\phi_n \wedge \psi_n)_{n=1}^\infty, (\phi_n \vee \psi_n)_{n=1}^\infty$ converge a.e. to the respective functions $\alpha f, f + g, f \wedge g, f \vee g$. To see that their corresponding sequences of integrals have finite limits, observe that

$$\lim_{n \to \infty} \int \alpha \phi_n = \alpha \lim_{n \to \infty} \int \phi_n = \alpha \int f,$$

$$\lim_{n \to \infty} \int (\phi_n + \psi_n) = \lim_{n \to \infty} \int \phi_n + \lim_{n \to \infty} \int \psi_n = \int f + \int g,$$

and

$$\int (\phi_n \wedge \psi_n) \leqq \int \phi_n \leqq \int f,$$

$$\int (\phi_n \vee \psi_n) \leqq \int [\phi_n + |\phi_1| + \psi_n + |\psi_1|] \leqq \int f + \int |\phi_1| + \int g + \int |\psi_1|$$

for all n.   □

Integrable Functions

(6.14)   Definition   A real-valued function defined a.e. on $\mathbb{R}$ is said to be *Lebesgue integrable* [or *Lebesgue summable*] and we write $f \in L_1^r(\mathbb{R})$ or $f \in L_1^r$ in case there exist $g, h \in M_1$ such that $f = g - h$ a.e. In this case we define the *Lebesgue integral* of f to be the real number

$$\int f = \int g - \int h.$$

(6.15)   Remarks   (a) The above definition of $\int f$ does not depend on the particular choice of g and h. In fact, if $f = g_1 - h_1 = g_2 - h_2$ a.e. where $g_j, h_j \in M_1$ $(j = 1, 2)$, then $g_1 + h_2 = g_2 + h_1$ a.e., and so

$$\int g_1 + \int h_2 = \int (g_1 + h_2) = \int (g_2 + h_1) = \int g_2 + \int h_1,$$

$$\int g_1 - \int h_1 = \int g_2 - \int h_2.$$

(b) If $f \in M_1$, then $f \in L_1^r$ and the above definition of $\int f$ agrees with that given in (6.10) because we can take $g = f$ and $h = 0$.

(c) If $f_1 \in L_1^r$ and $f_2 = f_1$ a.e., then $f_2 \in L_1^r$ and $\int f_2 = \int f_1$ because we can use the same g and h to represent both f_1 and f_2.

Our first theorem for L_1^r is similar to (6.4) and (6.5) for M_0.

(6.16) **Theorem** *The set L_1^r, equipped with pointwise (a.e.) operations, is a real vector lattice of functions, and $\int$ is a monotone linear functional on L_1^r. That is, if $f_1, f_2 \in L_1^r$ and $\alpha \in \mathbb{R}$, then*

(i) $\qquad\qquad\qquad\qquad f_1 + f_2 \in L_1^r,$

(ii) $\qquad\qquad\qquad\qquad \alpha f_1 \in L_1^r,$

(iii) $\qquad\qquad\qquad\qquad f_1 \vee f_2 \in L_1^r,$

(iv) $\qquad\qquad\qquad\qquad f_1 \wedge f_2 \in L_1^r,$

(v) $\qquad\qquad\qquad\qquad |f_1| \in L_1^r,$

(vi) $\qquad\qquad\qquad\qquad f_1^+, f_1^- \in L_1^r$

(vii) $\qquad\qquad\qquad\qquad \int (f_1 + f_2) = \int f_1 + \int f_2$

(viii) $\qquad\qquad\qquad\qquad \int \alpha f_1 = \alpha \int f_1,$

(ix) $\qquad\qquad f_1 \leqq f_2 \ \ implies \ \ \int f_1 \leqq \int f_2.$

Proof Choose $g_j, h_j \in M_1$ such that $f_j = g_j - h_j$ a.e. $(j = 1, 2)$. Then $f_1 + f_2 = (g_1 + g_2) - (h_1 + h_2)$ a.e. and $\int (f_1 + f_2) = \int (g_1 + g_2) - \int (h_1 + h_2) = \int g_1 - \int h_1 + \int g_2 - \int h_2 = \int f_1 + \int f_2$, so (i) and (vii) obtain. If $\alpha < 0$, we have $\alpha f_1 = (-\alpha) h_1 - (-\alpha) g_1$ a.e. and $\int \alpha f_1 = \int (-\alpha) h_1 - \int (-\alpha) g_1 = -\alpha \int h_1 + \alpha \int g_1 = \alpha \int f_1$. If $\alpha \geqq 0$, the verifications of (ii) and (viii) are even simpler.

If $f_1 \leqq f_2$, then $g_1 + h_2 \leqq g_2 + h_1$, $\int g_1 + \int h_2 \leqq \int g_2 + \int h_1$, $\int f_1 = \int g_1 - \int h_1 \leqq \int g_2 - \int h_2 = \int f_2$, and so (ix) obtains. To check (v), notice that

$$|f_1| = (g_1 \vee h_1) - (g_1 \wedge h_1) \ \text{a.e.}$$

The rest follow from

$$f_1 \vee f_2 = (1/2)[f_1 + f_2 + |f_1 - f_2|],$$
$$f_1 \wedge f_2 = (1/2)[f_1 + f_2 - |f_1 - f_2|],$$
$$f_1^+ = f_1 \vee 0,$$
$$f_1^- = -(f_1 \wedge 0). \quad \square$$

(6.17)   Corollary   *If* $f \in L_1^r$, *then* $|\int f| \leq \int |f|$.

Proof   Since $\pm f \leq |f|$, we have $\pm \int f \leq \int |f|$. $\square$

The step functions are dense in L_1^r in the following sense.

(6.18)   Theorem   *If* $f \in L_1^r$, *then there exists a sequence* $(\phi_n)_{n=1}^\infty \subset M_0$ *such that*

(i) $$\lim_{n \to \infty} \int |f - \phi_n| = 0$$

and

(ii) $$\lim_{n \to \infty} \phi_n(x) = f(x) \quad a.e.$$

Proof   Choose $g, h \in M_1$ such that $f = g - h$ a.e. and then choose nondecreasing sequences $(\psi_n)_{n=1}^\infty$, $(\chi_n)_{n=1}^\infty$ in M_0 that converge a.e. to g and h respectively. If we set $\phi_n = \psi_n - \chi_n$, (ii) is obvious and (i) follows from

$$\int |f - \phi_n| \leq \int (g - \psi_n) + \int (h - \chi_n).   \square$$

Two Limit Theorems

In this section we prove two important theorems which each have the features that they allow us to construct members of L_1^r as a.e. limits and they allow us to interchange the operations of integration and taking limits. Improved versions of these theorems appear later in this chapter, but we need the present versions first. We begin with a lemma.

(6.19)   Lemma   *Let* $f \in L_1^r$ *be nonnegative a.e. and let* $\epsilon > 0$. *Then there exist nonnegative functions* $g, h \in M_1$ *such that* $f = g - h$ *a.e. and* $\int h < \epsilon$.

Proof   Choose $g_0, h_0 \in M_1$ such that $f = g_0 - h_0$ a.e. Next choose a nondecreasing sequence $(\phi_n)_{n=1}^\infty \subset M_0$ such that $\phi_n \to h_0$ a.e. Then $\int \phi_n \to \int h_0$, and so we can choose $p \in \mathbb{N}$ such that $\int \phi_p > \int h_0 - \epsilon$. Writing $h = h_0 - \phi_p$, $g = g_0 - \phi_p$, we have $h \geq 0$ a.e., $g = f + h \geq 0$ a.e., and $\int h < \epsilon$. By redefining g and h on sets of measure 0, we can make them nonnegative everywhere. $\square$

(6.20)   Monotone Convergence Theorem for L_1^r *Let* $(f_n)_{n=1}^\infty$ *be a nondecreasing sequence in* L_1^r *such that* $\lim_{n \to \infty} \int f_n < \infty$. *Define* $f(x) = \lim_{n \to \infty} f_n(x)$ *a.e. on* $\mathbb{R}$. *Then* $f \in L_1^r$ *and*

$$\int f = \lim_{n \to \infty} \int f_n.$$

Proof *Case* 1. Suppose that $f_n \in M_1$ for all n. For each n, choose a nondecreasing sequence $(\phi_{n,k})_{k=1}^\infty \subset M_0$ such that $\lim_{k \to \infty} \phi_{n,k}(x) = f_n(x)$ a.e. Next define

$$\phi_n(x) = \max\{\phi_{j,k}(x) : 1 \le j \le n, 1 \le k \le n\}.$$

Then $(\phi_n)_{n=1}^\infty \subset M_0$ is nondecreasing and, since $\phi_{j,k} \le f_j \le f_n$ a.e. for $j \le n$, $\phi_n \le f_n$ a.e. Thus

$$\lim_{n \to \infty} \int \phi_n \le \lim_{n \to \infty} \int f_n < \infty,$$

and so the function g defined by $g(x) = \lim_{n \to \infty} \phi_n(x)$ is in M_1 $[g(x) < \infty$ a.e. by (6.6)] and

$$\lim_{n \to \infty} \int \phi_n = \int g.$$

Since $\phi_n \le f_n \le f$ a.e., we have $g \le f$ a.e. and

$$\int g \le \lim_{n \to \infty} \int f_n. \tag{1}$$

On the other hand, for fixed n we have $k \ge n$ implies

$$\phi_{n,k} \le \phi_k \le g,$$

and so, letting $k \to \infty$, we have

$$f_n \le g \quad \text{a.e.}$$

Thus $f \le g$ a.e., $f = g$ a.e., $f \in M_1$. Now (1) yields

$$\int f \le \lim_{n \to \infty} \int f_n \le \int f.$$

Case 2. Suppose $(f_n)_{n=1}^\infty \subset L_1'$ and $\lim_{n \to \infty} \int f_n = \beta < \infty$. For $k \in \mathbb{N}$, write $F_k = f_{k+1} - f_k \ge 0$. Apply (6.19) to F_k to obtain nonnegative functions $G_k, H_k \in M_1$ such that $F_k = G_k - H_k$ a.e. and $\int H_k < 2^{-k}$. Write $g_n = \sum_{k=1}^n G_k$ and $h_n = \sum_{k=1}^n H_k$. Then $(g_n)_{n=1}^\infty$ and $(h_n)_{n=1}^\infty$ are nondecreasing sequences in M_1 that satisfy

$$\int h_n = \sum_{k=1}^n \int H_k < 1 < \infty$$

and

$$\int g_n = \sum_{k=1}^n \int (F_k + H_k) = \sum_{k=1}^n \int F_k + \sum_{k=1}^n \int H_k$$
$$< \int f_{n+1} - \int f_1 + 1 \le \beta + 1 - \int f_1 < \infty$$

for all n. If we write $g(x) = \lim_{n \to \infty} g_n(x)$ and $h(x) = \lim_{n \to \infty} h_n(x)$ for $x \in \mathbb{R}$, it follows from Case 1 that $g, h \in M_1$ and

$$\int g = \lim_{n \to \infty} \int g_n, \qquad \int h = \lim_{n \to \infty} \int h_n.$$

But

$$f_{n+1} = f_1 + \sum_{k=1}^{n} F_k = f_1 + g_n - h_n$$

for all n. Therefore

$$f = f_1 + g - h \in L_1^r$$

and

$$\int f = \int f_1 + \int g - \int h = \int f_1 + \lim_{n\to\infty} \int g_n - \lim_{n\to\infty} \int h_n$$

$$= \lim_{n\to\infty} \int (f_1 + g_n - h_n)$$

$$= \lim_{n\to\infty} \int f_{n+1}. \qquad \square$$

(6.21)   Remark   Theorem (6.20) also holds for nonincreasing sequences $(f_n)_{n=1}^{\infty} \subset L_1^r$ such that $\lim_{n\to\infty} \int f_n > -\infty$. We need only apply (6.20) to the sequence $(-f_n)_{n=1}^{\infty}$.

(6.22)   Lebesgue's Dominated Convergence Theorem   For $(f_n)_{n=1}^{\infty} \subset L_1^r$, suppose that there is a $g \in L_1^r$ such that $|f_n(x)| \leq g(x)$ a.e. for all n. Suppose also that there is a function f such that $\lim_{n\to\infty} f_n(x) = f(x)$ a.e. Then $f \in L_1^r$ and $\int f = \lim_{n\to\infty} \int f_n$.

Proof   For $1 \leq k \leq m$, define

$$g_{k,m}(x) = \min\{f_n(x) : k \leq n \leq m\} \quad a.e.$$

and

$$g_k(x) = \inf\{f_n(x) : n \geq k\} \quad a.e.$$

Since L_1^r is a lattice $g_{k,m} \in L_1^r$. Also, $g_{k,m} \geq -g$ a.e., and so

$$\int g_{k,m} \geq \int (-g) > -\infty.$$

Notice that the nonincreasing sequence $(g_{k,m})_{m=k}^{\infty}$ converges to g_k a.e. It follows from (6.21) that $g_k \in L_1^r$ and

$$\int g_k = \lim_{m\to\infty} \int g_{k,m} \leq \int g < \infty$$

for all $k \geq 1$. By definition of limit inferior, we have

$$\lim_{k\to\infty} g_k(x) = \varliminf_{n\to\infty} f_n(x) = f(x) \quad a.e.$$

Applying (6.20) to the nondecreasing sequence $(g_k)_{k=1}^{\infty} \subset L_1^r$ and noting that $g_k \leq f_k$ a.e. for all k, we obtain $f \in L_1^r$ and

$$\int f = \lim_{k\to\infty} \int g_k \leq \varliminf_{k\to\infty} \int f_k.$$

We can apply this reasoning to $(-f_n)_{n=1}^{\infty}$ and get

$$-\int f = \int (-f) \leq \varliminf_{k\to\infty} -\int f_k = -\varlimsup_{k\to\infty} \int f_k.$$

Therefore

$$\int f \leq \varliminf_{k\to\infty} \int f_k \leq \varlimsup_{k\to\infty} \int f_k \leq \int f,$$

and so $\lim_{k\to\infty} \int f_k$ exists and is equal to $\int f$. $\square$

The presence of the dominating function g in (6.22) is all important as the following examples show.

(6.23)   **Examples**   (a) Let $f_n = \xi_{[n,n+1]}$ and $f = 0$. Then $\lim_{n\to\infty} f_n(x) = f(x)$ for all $x \in \mathbb{R}$, but $\lim_{n\to\infty} \int f_n = 1 \neq 0 = \int f$.

(b) Let $f_n = n\xi_{]0,1/n]}$ and $f = 0$. Again $\lim_{n\to\infty} f_n(x) = f(x)$ for all $x \in \mathbb{R}$, but $\lim_{n\to\infty} \int f_n \neq \int f$.

We may now prove the following important fact.

(6.24)   **Theorem**   *Let $f \in L_1^r$. Then $\int |f| = 0$ if and only if $f = 0$ a.e.*

Proof   If $f = 0$ a.e., then $|f| = 0$ a.e., and so (6.15.c) shows that $\int |f| = \int 0 = 0$.

Conversely, suppose that $\int |f| = 0$. Define $A = \{x \in \mathbb{R}: f(x) \neq 0\}$. Write $f_n = n|f|$ for all $n \in \mathbb{N}$. Then $\lim_{n\to\infty} f_n(x) = \infty \cdot \xi_A(x)$ for all $x \in \mathbb{R}$. Since $f_1 \leq f_2 \leq \ldots$ and $\int f_n = 0$ for all n, it follows from (6.20) that $\infty \cdot \xi_A \in L_1^r$, and so $\infty \cdot \xi_A$ is finite a.e. Therefore A has measure zero. $\square$

The Riemann Integral

The Riemann integral is just the "definite integral" studied in elementary calculus. It was set forth by the great German geometer and analyst Bernard Riemann (1826–66) many years before Lebesgue's time. Aside from historical interest, perhaps its greatest utility for us lies in its definition. We shall show in this section that the Riemann integral is the restriction of the Lebesgue integral to a certain subset of L_1^r and we shall characterize this subset. First, a definition.

(6.25)   **Definition**   Let $-\infty < a < b < \infty$ and let f be a bounded real-valued function on $[a,b]$. For each subdivision $P = \{a = x_0 < \ldots < x_n = b\}$ and choice of $t_k \in [x_{k-1}, x_k]$, define the corresponding *Riemann sum* by

$$S(f, P, t_k) = \sum_{k=1}^{n} f(t_k)\Delta x_k$$

where $\Delta x_k = x_k - x_{k-1}$. Define the *mesh* of P to be the number

$$|P| = \max\{\Delta x_k : 1 \leq k \leq n\}.$$

We say that f is *Riemann integrable* over $[a, b]$ if there exists $A \in \mathbb{R}$ such that to each $\epsilon > 0$ corresponds a $\delta > 0$ such that whenever $|P| < \delta$ we have

$$|A - S(f, P, t_k)| < \epsilon$$

for all choices of the t_k's. [It is easy to see that at most one such A can exist: $|A_1 - A_2| < 2\epsilon$ for all $\epsilon > 0$.] If A exists, we call it the *Riemann integral* of f over $[a, b]$ and write

$$A = R \int_a^b f.$$

(6.26)   Lemma *If* $\phi \in M_0$ *and* $\phi = 0$ *outside* $[a, b]$, *then* ϕ *is Riemann integrable over* $[a, b]$ *and*

$$R \int_a^b \phi = \int \phi.$$

Proof   Choose $\beta \in \mathbb{R}$ such that $|\phi(x)| < \beta$ for all $x \in \mathbb{R}$ and then choose a subdivision $Q = \{a = y_0 < \ldots < y_m = b\}$ and real numbers $\alpha_1, \ldots, \alpha_m$ such that $\phi(x) = \alpha_j$ whenever $y_{j-1} < x < y_j$. Then

$$\int \phi = \sum_{j=1}^m \alpha_j (y_j - y_{j-1}). \qquad (1)$$

Let $\epsilon > 0$ be given. Choose $0 < \delta < \epsilon/(4\beta m)$ and let P and $\{t_k\}_{k=1}^n$ be as in (6.25) with $|P| < \delta$. If we sum over only those k's for which $[x_{k-1}, x_k]$ contains some y_j, we get

$$S_0 = |\sum \phi(t_k)\Delta x_k| < 2m\beta\delta.$$

For fixed j, if we sum only over those k's for which $y_{j-1} < x_{k-1} < x_k < y_j$, we get

$$S_j = |\alpha_j(y_j - y_{j-1}) - \sum \phi(t_k)\Delta x_k| = |\alpha_j|\,|(y_j - y_{j-1}) - \sum \Delta x_k| < 2\beta\delta.$$

Therefore

$$|\int \phi - S(\phi, P, t_k)| \leq S_0 + \sum_{j=1}^m S_j < 4m\beta\delta < \epsilon. \qquad \square$$

(6.27)   Definition   Let $-\infty < a < b < \infty$, let f be a bounded real-valued function on $\mathbb{R}$ such that $f(x) = 0$ if $x < a$ or $x > b$, and let $P = \{a = x_0 < \ldots < x_n = b\}$ be any subdivision of $[a, b]$. For $1 \leq k \leq n$, define

$$\alpha_k = \inf f([x_{k-1}, x_k]), \qquad \beta_k = \sup f([x_{k-1}, x_k]).$$

Let ϕ [resp. ψ] be the step function with value α_k [resp. β_k] on $]x_{k-1}, x_k[$ $(1 \leq k$

$\leq n$) that agrees with f elsewhere. Plainly $\phi \leq f \leq \psi$. The functions ϕ and ψ are called the *lower* and *upper functions* for f relative to P. Clearly

$$\int \phi \leq S(f, P, t_k) \leq \int \psi$$

for any choice of $(t_k)_{k=1}^n$.

Next, an important criterion for Riemann integrability.

(6.28) **Theorem** *Let $[a, b]$ and f be as in (6.27). Suppose that $(P_j)_{j=1}^\infty$ is any sequence of subdivisions of $[a, b]$ such that $\lim_{j \to \infty} |P_j| = 0$. For each j, let ϕ_j and ψ_j be the lower and upper functions for f relative to P_j (6.27).*

(i) *If f is Riemann integrable over $[a, b]$, then*

$$\lim_{j \to \infty} \int \phi_j = \lim_{j \to \infty} \int \psi_j = R \int_a^b f.$$

(ii) *If $\lim_{j \to \infty} \int \phi_j = \lim_{j \to \infty} \int \psi_j = A$, then f is Riemann integrable over $[a, b]$ and*

$R \int_a^b f = A$.

Proof (i) Suppose that f is Riemann integrable over $[a, b]$ and let $\epsilon > 0$ be given. Let $\delta > 0$ correspond to ϵ as in (6.25). Choose j_0 such that $|P_j| < \delta$ for all $j \geq j_0$. Then, for fixed $j \geq j_0$, we have

$$\left| R \int_a^b f - S(f, P_j, t_{j,k}) \right| < \epsilon$$

for all choices of $t_{j,k}$ in the kth interval $[x_{j,k-1}, x_{j,k}]$ of P_j. By appropriate choice of $t_{j,k}$ we can make the value $f(t_{j,k})$ as near as we wish to the infimum [or to the supremum] of f over this interval. We conclude that $j \geq j_0$ implies

$$\left| R \int_a^b f - \int \phi_j \right| \leq \epsilon, \qquad \left| R \int_a^b f - \int \psi_j \right| \leq \epsilon.$$

This proves (i).

(ii) Suppose that $\lim_{j \to \infty} \int \phi_j = \lim_{j \to \infty} \int \psi_j = A$. Let $\epsilon > 0$ be given. Fix j such that

$$\int \phi_j > A - \epsilon/2, \qquad \int \psi_j < A + \epsilon/2. \tag{1}$$

Since ϕ_j and ψ_j are Riemann integrable (6.26), there exists a $\delta > 0$ such that $|P| < \delta$ implies

$$\int \phi_j - \epsilon/2 = R \int_a^b \phi_j - \epsilon/2 < S(\phi_j, P, t_k),$$

$$\int \psi_j + \epsilon/2 = R \int_a^b \psi_j + \epsilon/2 > S(\psi_j, P, t_k). \tag{2}$$

Combining (1) and (2), we see that $|P| < \delta$ implies

$$A - \epsilon < S(\phi_j, P, t_k) \leq S(f, P, t_k) \leq S(\psi_j, P, t_k) < A + \epsilon,$$
$$|A - S(f, P, t_k)| < \epsilon.$$

This proves (ii). □

(6.29) Theorem *Let $[a, b]$ and f be as in (6.27). Then f is Riemann integrable over $[a, b]$ if and only if f is continuous a.e. on $[a, b]$. In this case, $f \in L_1^r$ and*

$$R \int_a^b f = \int f.$$

Proof Let $(P_j)_{j=1}^\infty$, $(\phi_j)_{j=1}^\infty$, and $(\psi_j)_{j=1}^\infty$ be as in (6.28). For each $\delta > 0$ define functions m_δ and M_δ on $\mathbb{R}$ by

$$m_\delta(x) = \inf f([x - \delta, x + \delta]), \qquad M_\delta(x) = \sup f([x - \delta, x + \delta])$$

and then define m and M on $\mathbb{R}$ by

$$m(x) = \lim_{\delta \downarrow 0} m_\delta(x), \qquad M(x) = \lim_{\delta \downarrow 0} M_\delta(x).$$

It is clear that $m(x) \leq M(x)$ for all x. Moreover, it is an easy exercise to show that f is continuous at x if and only if $m(x) = M(x)$.
 We next show that

$$\lim_{j \to \infty} \phi_j(x) = m(x) \quad \text{a.e.,} \tag{1}$$

$$\lim_{j \to \infty} \psi_j(x) = M(x) \quad \text{a.e.} \tag{2}$$

It is clear that equalities (1) and (2) obtain if $x \notin [a, b]$ since everything is zero. Let $x \in [a, b] \backslash \bigcup_{j=1}^\infty P_j$ and $\epsilon > 0$ be given. Choose $\delta > 0$ such that $m_\delta(x) > m(x) - \epsilon$ and $M_\delta(x) < M(x) + \epsilon$. Next select j_0 such that $|P_j| < \delta$ for $j > j_0$. Given $j > j_0$, we can select the closed interval I_j determined by P_j that contains x in its interior. Then $I_j \subset [x - \delta, x + \delta]$, and so $m(x) - \epsilon < m_\delta(x) \leq \inf f(I_j) = \phi_j(x) \leq m(x)$ [the last inequality because there is some $\delta' > 0$ such that $[x - \delta', x + \delta'] \subset I_j$]. Similarly, $j > j_0$ implies

$$M(x) \leq \psi_j(x) < M(x) + \epsilon.$$

This establishes (1) and (2) since the countable set $\bigcup_{j=1}^\infty P_j$ has measure zero.
 The sequences $(\phi_j)_{j=1}^\infty$ and $(\psi_j)_{j=1}^\infty$ are both dominated by $g = \beta \xi_{[a,b]} \in L_1^r$ if β is a bound for $|f|$. Invoking (6.22), (1), and (2), we see that m and M are in L_1^r and that

$$\lim_{j \to \infty} \int \phi_j = \int m, \tag{3}$$

$$\lim_{j \to \infty} \int \psi_j = \int M. \tag{4}$$

Back to the theorem at hand! Suppose that f is Riemann integrable. Then (6.28.i), (3), and (4) imply that

$$\int m = \int M = R \int_a^b f. \tag{5}$$

Thus $\int (M - m) = 0$. Since $m \leq M$, we see from (6.24) that $m = M$ a.e., i.e., f is continuous a.e. Since $m \leq f \leq M$ and $m = M$ a.e., we have $f = M$ a.e. This implies that $f \in L_1^r$ and, by (5),

$$\int f = \int M = R \int_a^b f.$$

Conversely, if f is continuous a.e., then $m = M$ a.e. Thus, by (3) and (4),

$$\lim_{j \to \infty} \int \phi_j = \lim_{j \to \infty} \int \psi_j.$$

Therefore, (6.28.ii), f is Riemann integrable. $\square$

There exist sets $P \subset [0, 1]$ such that $\xi_P \in L_1^r$, but such that there is no function f that is Riemann integrable over $[0, 1]$ for which $f = \xi_P$ a.e.

(6.30)   Example   Recall the discussion of Cantor sets given in (2.81) and (2.82). We take P to be the Cantor set corresponding to the sequence $a_n = 2^{-n-1} + 2^{-1} \cdot 3^{-n}$ $(n \geq 0)$. Write

$$\psi_n = \xi_{P_n} = \sum_{k=1}^{2^n} \xi_{J_{n,k}}.$$

Then

$$\int \psi_n = \sum_{k=1}^{2^n} |J_{n,k}| = (1/2)(1 + 2^n/3^n)$$

and

$$\psi_n(x) \to \xi_P(x) \quad \text{as } n \to \infty$$

for all $x \in \mathbb{R}$. Since $\psi_n \leq \xi_{[0,1]}$ for all n,   (6.22) shows that $\xi_P \in L_1^r$ and $\int \xi_P = 1/2 = \lambda(P)$. Assume that f is a function that is Riemann integrable over $[0, 1]$ and $f = \xi_P$ a.e. Write $N = \{x \in [0, 1]: f(x) \neq \xi_P(x)$ or f is discontinuous at $x\}$. Then $\lambda(N) = 0$ (6.29). By (2.79.iii), $\lambda(P \setminus N) = \lambda(P) = 1/2$, and so there exists $p \in P \setminus N$, $0 < p < 1$. Then $f(p) = \xi_P(p) = 1$ and f is continuous at p. If I is any open interval with $p \in I \subset]0, 1[$, then the open set $V = I \cap (]0, 1[\setminus P)$ is nonvoid (2.85.b); hence, $\lambda(V \setminus N) = \lambda(V) > 0$ (2.76.vi), and so there exists $x \in I$ such that $f(x) = \xi_P(x) = 0$. This shows that we can choose a sequence $(x_k)_{k=1}^{\infty}$ such that $\lim_{k \to \infty} x_k = p$ and $\lim_{k \to \infty} f(x_k) = 0 \neq 1 = f(p)$. But f is continuous at p—a contradiction.

We hasten to remind the reader that we have defined the Riemann integral only for *bounded* functions on *bounded* intervals. It can happen that a function has an "improper integral," in the sense of elementary calculus, without being Lebesgue integrable. The next example should suffice to illustrate this point.

(6.31) Example Define $f = \sum_{n=1}^{\infty} \frac{(-1)^{n-1}}{n} \xi_{[n-1,n[}$. Then f is Riemann integrable on $[0,b]$ for all $0 < b < \infty$ and

$$\lim_{b \to \infty} R \int_0^b f = \lim_{b \to \infty} \int f \cdot \xi_{[0,b]} = \sum_{n=1}^{\infty} \frac{(-1)^{n-1}}{n},$$

which converges. But $f \notin L_1^r$, for if it were, so would $|f|$ be, and then $|f|$ would serve as a dominating function for the sequence $f_k = \sum_{n=1}^{k} \frac{1}{n} \xi_{[n-1,n[}$ $(k = 1, 2, \dots)$.

But $\lim_{k \to \infty} \int f_k = \sum_{n=1}^{\infty} \frac{1}{n} = \infty$.

The trouble is that $f \in L_1^r$ implies $|f| \in L_1^r$. Of course, we are at liberty to define "improper Lebesgue integrals," but we must be cautious not to attribute to them the properties of the integral on L_1^r. The situation is analogous to the contrast between absolutely convergent series and conditionally convergent series.

(6.32) Notation Let $I \subset \mathbb{R}$ be an interval having endpoints $a, b \in \mathbb{R}^{\#}$ $(a \leq b)$ and let f be a real-valued function that is defined a.e. on I. We write $f\xi_I$ for the function that agrees with f on I and is equal to 0 everywhere on $\mathbb{R} \backslash I$. If $f\xi_I \in L_1^r$, we write $f \in L_1^r(I)$, and

$$\int_a^b f = \int f\xi_I.$$

In this case, we also denote the number $\int_a^b f$ by

$$\int_a^b f(x)\, dx \quad \text{or} \quad \int_a^b f(t)\, dt \quad \text{or} \quad \int_a^b f(\beta)\, d\beta$$

or any similar symbol obtained from the first one by replacing x by any letter at all that does not already appear in this first symbol. In view of (6.29) and this notational convention, we are free to drop the prefix R in the symbol for the Riemann integral and we shall usually do so. Thus, if $a, b \in \mathbb{R}$ and f is defined on and Riemann integrable over $[a,b]$, then we have

$$R \int_a^b f = \int_a^b f = \int_a^b f(y)\, dy.$$

We also *define*, if $a < b$,

$$\int_b^a f = - \int_a^b f.$$

Notice that if $a = b$, we have $\int_a^a f = 0$ because $f\xi_I = 0$ a.e. on $\mathbb{R}$.

Exercises

1. *Fundamental Theorem of Calculus for Riemann Integrals.* If F is a real-valued function that is defined and differentiable on a closed interval $[a,b] \subset \mathbb{R}$ and if the derivative F' is Riemann integrable over $[a,b]$ (in particular, F' is bounded on

[a, b]), then

$$\int_a^b F' = F(b) - F(a).$$

[Hint: Letting P be as in (6.25), invoke the Mean Value Theorem to obtain t_k with $x_{k-1} < t_k < x_k$, for which $F(x_k) - F(x_{k-1}) = F'(t_k)\Delta x_k$ and conclude that $S(F', P, t_k) = F(b) - F(a)$.]

2. (a) If f is Riemann integrable over $[a, b]$, then so is $|f|$ and

$$\left| \int_a^b f \right| \leq \int_a^b |f|.$$

(b) There exist functions f on $[0, 1]$ that take only the values -1 and 1 (so $|f| = 1$) and are not Riemann integrable over $[0, 1]$. [It is more difficult to find such an f that is not even in $L_1'([0, 1])$ (see (6.68)).]

3. The function F defined on $\mathbb{R}$ by $F(0) = 0$ and $F(x) = x^2 \sin(1/x)$ for $x \neq 0$ is differentiable on $\mathbb{R}$, and F' is Riemann integrable over $[-1, 1]$, but F' is not continuous at 0.

4. The function F defined on $\mathbb{R}$ by $F(0) = 0$ and $F(x) = x^2 \sin(1/x^2)$ for $x \neq 0$ is differentiable on $\mathbb{R}$, F' is continuous except at 0, and $F'(I) = \mathbb{R}$ for every interval I of positive length containing 0. Thus F' is not Riemann integrable over $[0, 1]$ but we do have

$$\lim_{a \downarrow 0} \int_a^b F' = F(b) \quad \text{for all } b > 0.$$

However, the function f that is equal to F' on $[0, 1]$ and equal to 0 on $\mathbb{R}\setminus[0, 1]$ is not Lebesgue integrable. [Hints: Use Exercise 1 and the fact that $|f| \in L_1'$ if $f \in L_1'$.]

5. If f is Riemann integrable over $[a, b]$, then the function F defined on $[a, b]$ by

$$F(x) = \int_a^x f \quad (a \leq x \leq b)$$

is continuous on $[a, b]$. Moreover, if f is continuous at $c \in [a, b]$, then F is differentiable at c and $F'(c) = f(c)$. [Hints: Use Exercise 2 to estimate the size of integrals. Use the definition of derivative and obtain

$$\frac{F(c + h) - F(c)}{h} - f(c) = \frac{1}{h} \int_c^{c+h} [f - f(c)]$$

where $\int_c^{c+h}$ means $-\int_{c+h}^c$ if $h < 0$.]

6. *Change of Variable Theorem for Riemann Integrals.* Let $f \in C'([a, b])$ and let $\phi : [\alpha, \beta] \to [a, b]$ be such that ϕ has a continuous derivative on $[\alpha, \beta]$. Then

$$\int_{\phi(\alpha)}^{\phi(\beta)} f = \int_\alpha^\beta [(f \circ \phi)\phi'].$$

[Hints: Use Exercise 5 and the Chain Rule to obtain $(F \circ \phi)' = (f \circ \phi)\phi'$, then use Exercise 1.]

7. *Integration by Parts for Riemann Integrals.* Suppose that u and v are real-valued functions that have derivatives u' and v' that are Riemann integrable over $[a, b]$. Then

$$\int_a^b uv' = u(b)v(b) - u(a)v(a) - \int_a^b vu'.$$

[Hint: Apply Exercise 1 to $F = uv$.]

8.   Let $(f_n)_{n=1}^{\infty}$ be a uniformly bounded sequence of real-valued functions that are each defined and Riemann integrable over $[a, b]$. Suppose that $f(x) = \lim_{n\to\infty} f_n(x)$ exists for each $x \in [a, b]$.
      (a) If f is Riemann integrable over $[a, b]$, then

$$\lim_{n\to\infty} \int_a^b f_n = \int_a^b f.$$

[Attempts to prove this without using (6.22) should be enough to convince one of the power of the Lebesgue theory. However, such proofs are possible and the result itself predates Lebesgue's work.]
      (b) It can happen that f is not Riemann integrable even if every f_n is continuous. [Hint: Refer to (6.30). Let $f = \xi_P$, let $f_n = 1$ on P_n, and let $f_n(x) = 0$ if $\mathrm{dist}(x, P_n) \geq 1/n$.]
      (c) If $f_n \to f$ uniformly on $[a, b]$, then f is Riemann integrable over $[a, b]$.

9.   Let $c < d$ be in $\mathbb{R}^{\#}$ and let f be a real-valued function defined on $]c, d[$.
      (a) If $f \in L_1^r(]c, d[)$, then

$$\int_c^d f = \lim_{a\downarrow c}\lim_{b\uparrow d} \int_a^b f = \lim_{b\uparrow d}\lim_{a\downarrow c} \int_a^b f.$$

(b) If f is Riemann integrable (or only Lebesgue integrable) over every $[a, b] \subset \,]c, d[$ and if

$$\lim_{n\to\infty} \int_{a_n}^{b_n} |f| < \infty$$

for some two strictly monotone sequences $(a_n)_{n=1}^{\infty}$ and $(b_n)_{n=1}^{\infty}$ with $a_n \downarrow c$ and $b_n \uparrow d$, then $f \in L_1^r(]c, d[)$ and equalities (a) obtain. [Hints: For (a), use (6.22) by considering arbitrary sequences as in (b). For (b), use (6.20).] This exercise, in concert with Exercise 1 and Theorem (6.29), is very useful indeed in proving that certain functions are Lebesgue integrable and in the evaluation of their integrals. (See the next exercise.)

10.   *Limit Comparison Tests.* Let c, d, and f be as in Exercise 9 and let g be a positive function in $L_1^r(]c, d[)$.
      (a) If it is known that $f \in L_1^r(]c, b[)$ whenever $c < b < d$ and

$$\varlimsup_{x\uparrow d} [|f(x)|/g(x)] < \infty,$$

then $f \in L_1(]c, d[)$ and $\int_c^d f = \lim_{b\uparrow d} \int_c^b f$.
      (b) If it is known that $f \in L_1^r(]a, d[)$ whenever $c < a < d$ and

$$\varlimsup_{x\downarrow c} [|f(x)|/g(x)] < \infty,$$

then $f \in L_1^r(]c, d[)$ and $\int_c^d f = \lim_{a\downarrow c} \int_a^d f$.
      (c) The functions g defined by $g(x) = x^{-p}$ are in $L_1^r(]c, \infty[)$ whenever $p > 1$ and $c > 0$ and they are in $L_1^r(]0, d[)$ whenever $p < 1$ and $0 < d < \infty$.
      (d) The functions g defined by $g(x) = x^{-1}|\log x|^{-p}$ are in $L_1^r(]c, \infty[)$ whenever $p > 1$ and $c > 1$ and they are in $L_1^r(]0, d[)$ whenever $p > 1$ and $0 < d < 1$. [Hint: Use Exercise 6 with $\phi(t) = e^{-t}$.]
      (e) The functions g defined by $g(x) = |\log x|^p$ are in $L_1^r(]0, 1[)$ if $0 \leq p < \infty$.

11. Define

$$F(x) = \int_0^x \frac{\sin t}{t}\, dt \quad \text{for } x \in \mathbb{R}.$$

(a) The function F is strictly increasing on $[2n\pi, (2n+1)\pi]$ and strictly decreasing on $[(2n+1)\pi, (2n+2)\pi]$ for each integer $n \geq 0$.

(b) We have

$$F(0) < F(2\pi) < F(4\pi) < F(6\pi) < \cdots,$$
$$F(\pi) > F(3\pi) > F(5\pi) > F(7\pi) > \cdots,$$

and each term of the first sequence is less than each term of the second sequence.

(c) $\lim_{n \to \infty} [F((2n+1)\pi) - F(2n\pi)] = 0$.

(d) $\lim_{x \to \infty} \int_0^x [(\sin t)/t]\, dt$ exist in $\mathbb{R}$.

(e) $\lim_{x \to \infty} \int_0^x [|\sin t|/t]\, dt = \infty$.

(f) $F(-x) = -F(x)$ for all $x \in \mathbb{R}$.

(g) $|\int_a^b [(\sin t)/t]\, dt| \leq 2F(\pi) < 2\pi$ for all $a, b \in \mathbb{R}$.

12. We have for $a \in \mathbb{R}$ that $\lim_{n \to \infty} \int_0^n (1 + x/n)^n e^{-ax}\, dx = 1/(a-1)$ if $a > 1$,

$$\lim_{n \to \infty} \int_0^n \left(1 - \frac{x}{n}\right)^n e^{ax}\, dx = \frac{1}{1-a} \quad \text{if } a < 1,$$

and these limits are both ∞ in all other cases. [Hint: Check that $(1 + x/n)^n \leq (1 + x/(n+1))^{n+1}$ if $n > -x$.]

13. (a) If $f \in L_1^r$ and $\phi \in M_0$, then $f\phi \in L_1^r$.

(b) There exists $f \in M_1$ such that $f^2 \notin M_1$. [Hint: Write $f(x) = x^{-1/2}$ if $0 < x \leq 1$ and $f(x) = 0$ if $x \leq 0$ or $x > 1$.]

14. If $f \in L_1^r$ and $\int_0^x f = 0$ for all $x \in \mathbb{R}$, then $f = 0$ a.e. [Hints: First show that $\int f\phi = 0$ for all $\phi \in M_0$. Then choose $(\phi_n)_{n=1}^\infty \subset M_0$ such that $\phi_n \to f$ a.e., define $f_n = f\phi_n(1 + |\phi_n|)^{-1}$, and then use Exercise 13(a) and (6.22) to show that $f^2(1 + |f|)^{-1}$ is in L_1^r and has integral 0.]

15. Suppose that f is a real-valued function defined on the plane $\mathbb{R}^2$ such that, for each fixed x, the function $y \to f(x, y)$ is in L_1^r and for each fixed y, the function $x \to f(x, y)$ is continuous on $\mathbb{R}$. Define ϕ on $\mathbb{R}$ by

$$\phi(x) = \int_{-\infty}^\infty f(x, y)\, dy.$$

(a) If there is some function $g \in L_1^r$ such that $|f(x, y)| \leq |g(y)|$ for all $y \in \mathbb{R}$ and all x in some open interval $I \subset \mathbb{R}$, then ϕ is continuous on I.

(b) The function f defined by

$$f(x, y) = \begin{cases} |x| & \text{if } |xy| \leq 1, \\ |x|(2 - |xy|) & \text{if } 1 \leq |xy| \leq 2, \\ 0 & \text{if } |xy| \geq 2, \end{cases}$$

is continuous on $\mathbb{R}^2$ and satisfies our hypotheses. However, $\phi(x) = 3$ for all $x \neq 0$ and $\phi(0) = 0$, so ϕ is *not* continuous at 0.

16. The sequence of functions f_n defined by $f_n(x) = \sin(nx)$ has *no* subsequence which converges a.e. on $\mathbb{R}$. [Hint: If it had such a subsequence $(f_{n_k})_{k=1}^\infty$, then the sequence

$g_k = (f_{n_k} - f_{n_{k+1}})^2$ would converge to 0 a.e. on $[0, 2\pi]$, while each g_k has integral 2π over $[0, 2\pi]$. Incidentally, for topology students, this is a simple proof of the fact that the product of a continuum of closed intervals is *not* sequentially compact.]

17.   Using Exercises 7, 10, and 11(e), we have

$$\lim_{b \to \infty} \cdot \int_{\pi/2}^{b} \frac{\sin x}{x} \, dx = -\int_{\pi/2}^{\infty} \frac{\cos x}{x^2} \, dx,$$

where the integrand on the right is in $L_1^i([\pi/2, \infty[)$ while the one on the left is not.

18.   For $x > 0$ and $y > 0$ define

$$f(x, y) = \int_{x}^{y} \sin(t^2) \, dt.$$

Then we have
(a)

$$f(x, y) = \frac{\cos(x^2)}{2x} - \frac{\cos(y^2)}{2y} - \frac{1}{4} \int_{x^2}^{y^2} \frac{\cos s}{s^{3/2}} \, ds,$$

(b) $|f(x, y)| < 1/x$ if $y > x > 0$,
(c) $\lim_{y \to \infty} f(x, y)$ exists in $\mathbb{R}$ for each x.

19.   (a) If $a < b$ in $\mathbb{R}^{\#}$ and if $(u_n)_{n=1}^{\infty}$ is a sequence of functions in $L_1^i(]a, b[)$ satisfying

$$\sum_{n=1}^{\infty} \int_{a}^{b} |u_n(x)| \, dx < \infty,$$

then the series

$$f(x) = \sum_{n=1}^{\infty} u_n(x)$$

converges absolutely for almost every $x \in]a, b[$, and the function f thereby defined is in $L_1^i(]a, b[)$ and satisfies

$$\int_{a}^{b} f(x) \, dx = \sum_{n=1}^{\infty} \int_{a}^{b} u_n(x) \, dx.$$

[Hints: Use (6.20) to show that the function $g = \sum_{n=1}^{\infty} |u_n|$ is in $L_1^i(]a, b[)$ and then apply (6.22) to the partial sums $f_N = \sum_{n=1}^{N} u_n$.]

(b) Let $f_n = n\xi_{]0,1/n]}$ for $n \in \mathbb{N}$ as in (6.23.b) and let $u_n = f_n - f_{n+1}$. Then

$$\sum_{n=1}^{\infty} u_n(x) = f_1(x),$$

and this series converges absolutely for each $x \in \mathbb{R}$, but

$$\sum_{n=1}^{\infty} \int_{0}^{1} u_n = 0 \neq 1 = \int_{0}^{1} f_1.$$

Notice that the hypothesis in (a) is not met because $\int_{0}^{1} |u_n| = 2/(n+1)$ for each $n \in \mathbb{N}$.

20. If $p \in \mathbb{R}$, $p > -1$, and n is an integer with $0 \le n \le p + 1$, then

$$\int_0^1 x^p (\log x)^n \, dx = \frac{(-1)^n \cdot n!}{(p+1)^{n+1}} \, .$$

[Hint: Use Exercises 7, 9, and 1.]

21. From Exercises 19(a) (if necessary) and 20, we have

$$\int_0^1 x^{-x} \, dx = \sum_{n=1}^{\infty} \frac{1}{n^n}$$

and

$$\int_0^1 x^x \, dx = \sum_{n=1}^{\infty} \frac{(-1)^{n-1}}{n^n} \, .$$

22. (a) There exist unbounded continuous functions $f : \mathbb{R} \to \mathbb{R}$ such that $f \in L_1^r$.
     (b) If $f \in L_1^r$ is defined everywhere on $\mathbb{R}$ and is uniformly continuous on $\mathbb{R}$, then
     $$\lim_{x \to \infty} f(x) = \lim_{x \to -\infty} f(x) = 0, \text{ and so } f \text{ is bounded.}$$

23. [Casper Goffman, 1977] In this exercise we give a very simple example of a function $F : [0, 1] \to \mathbb{R}$ such that $F'(x)$ exists with $0 \le F'(x) \le 1$ for each $x \in [0, 1]$, yet F' is *not* Riemann integrable over $[0, 1]$.
     Let $\{I_n\}_{n=1}^{\infty}$ be a pairwise disjoint family of open subintervals of $[0, 1]$ for which $\sum_{n=1}^{\infty} |I_n| = 1/2$ and the closed set $E = [0, 1] \setminus \bigcup_{n=1}^{\infty} I_n$ has neither isolated points nor interior points. (For example, take E to be the Cantor set corresponding to the sequence $a_n = 2^{-n-1} + 3^{-n} \cdot 2^{-1}$ as in (2.82) and let the I_n's be an enumeration of the component intervals of $[0, 1] \setminus E$.) Let J_n be the open interval concentric with I_n for which $|J_n| = |I_n|^2$. Let $f_n : [0, 1] \to [0, 1]$ be a function that is continuous on $[0, 1]$, is equal to 1 at the midpoint c_n of I_n, and is identically 0 on $[0, 1] \setminus J_n$. Define F_n, f, and F on $[0, 1]$ by $F_n(x) = \int_0^x f_n$, $f(x) = \sum_{n=1}^{\infty} f_n(x)$, and $F(x) = \sum_{n=1}^{\infty} F_n(x)$.
     (a) If $P = \{0 = x_0 < x_1 < \dots < x_p = 1\}$ and ϕ and ψ correspond to f and P as in (6.27), then $\int \psi - \int \phi \ge 1/2$. [Hint: If $]x_{k-1}, x_k[$ meets E, then it contains some I_n, so $\beta_k = 1$ and $\alpha_k = 0$. The sum of the Δx_k for which this fails cannot exceed $1/2$.]
     (b) The function f is continuous on each I_n, discontinuous at each point of E, and is *not* Riemann integrable over $[0, 1]$. [Hint: Each point of E is a limit point of the set $\{c_n\}_{n=1}^{\infty}$.]
     (c) If $I \subset [0, 1]$ is an interval with $I \cap E \ne \varnothing$, then
     $$\sum_{n=1}^{\infty} |I \cap J_n| \le 16 |I|^2.$$

     [Hints: This is obvious if the set $A = \{n \in \mathbb{N} : I \cap J_n \ne \varnothing\}$ is void. Suppose $A \ne \varnothing$. For $n \in A$, we have $|I \cap I_n| \ge (1/2)(|I_n| - |J_n|) \ge (1/4)|I_n|$ (because $|I_n| \le (1/2)$), so $|I \cap J_n| \le |J_n| = |I_n|^2 \le 16 |I \cap I_n|^2$.
     (d) If $x \in E$, then $F'(x) = 0 = f(x)$. [Hint: If $0 \le s \le x \le t \le 1$, then $F(t) - F(s) \le 16(t-s)^2$ because $\int_s^t f_n \le |[s, t] \cap J_n|$.]

(e) If $x \in I_n$ for some n, then $F'(x) = f_n(x) = f(x)$. [Hint: For $s \leq x \leq t$ in I_n, we have $F(t) - F(s) = F_n(t) - F_n(s) = \int_s^t f_n$.

(f) We have $f \in L_1^r([0, 1])$ and $F(x) = \int_0^x f$ for all $x \in [0, 1]$.

24.   [L. M. Levine] Let $f : [a, b] \to \mathbb{R}$ be any function.

(a) If A is the set of $x \in]a, b]$ at which $f(x-) = \lim_{t \uparrow x} f(t)$ exists in $\mathbb{R}$ and yet f is discontinuous at x, then A is countable. [Hints: Define m and M as in the proof of (6.29). For $n \in \mathbb{N}$ let A_n be the set of $x \in A$ for which $M(x) - m(x) \geq 1/n$. For fixed n, show that each point of A_n is the right endpoint of some open interval that is disjoint from A_n and so deduce that A_n is countable.]

(b) If f is bounded, then f is Riemann integrable over $[a, b]$ if and only if the left hand limit $f(x-)$ exists for almost every $x \in]a, b]$.

25.   Let $\{r_n\}_{n=1}^{\infty}$ be an enumeration of $\mathbb{Q}$. Define $f_n(x) = (x - r_n)^{-1/2}$ if $r_n < x < r_n + 1$ and $f_n(x) = 0$ for all other $x \in \mathbb{R}$. Also define $f(x) = \sum_{n=1}^{\infty} 2^{-n} f_n(x)$ for $x \in \mathbb{R}$. Then

(a) $f \in L_1^r$ and $\int f = 2$,

(b) for each $k \in \mathbb{N}$, the set $V_k = \{x \in \mathbb{R} : f(x) > k\}$ is open and dense in $\mathbb{R}$,

(c) the set $F = \{x \in \mathbb{R} : f(x) < \infty\}$ is of first Baire category in $\mathbb{R}$ and $\lambda(\mathbb{R} \setminus F) = 0$. [Hints: For (a), use Exercise 19(a). For (b), notice that $f(x) = \infty$ for every x in the set

$$\bigcap_{j=1}^{\infty} \bigcup_{n=j}^{\infty} \{x \in \mathbb{R} : 0 < x - r_n < 4^{-n}\}$$

and that this set is dense in $\mathbb{R}$ by the Baire Category Theorem.]

26.   (a) If $[a, b] \subset \mathbb{R}$, $(f_n)_{n=1}^{\infty} \subset L_1^r([a, b])$ and $f_n \to f$ uniformly on $[a, b]$, then

$$f \in L_1^r([a, b]) \quad \text{and} \quad \int_a^b f_n \to \int_a^b f.$$

(b) Defining $f_n(x) = n^{-1} e^{-x/n}$, we have $(f_n)_{n=1}^{\infty} \subset L_1^r([0, \infty[)$, $f_n \to 0$ uniformly on $[0, \infty[$, and $\int_0^{\infty} f_n = 1$ for all $n \in \mathbb{N}$.

27.   *The Wallis Formula.* If $I_n = \int_0^{\pi/2} \sin^n x \, dx$ for $n \geq 0$, then we have

(a) $I_{n+1} = [n/(n + 1)] I_{n-1}$ for $n \geq 1$,

(b) $0 < I_{n+1} < I_n < I_{n-1}$ for $n \geq 1$,

(c) $1 < I_n \cdot I_{n+1}^{-1} < 1 + (1/n)$ for $n \geq 1$,

(d) $I_{2k+1} = \dfrac{2k}{2k + 1} \cdot \dfrac{2k - 2}{2k - 1} \cdot \dfrac{2k - 4}{2k - 3} \cdots \cdot \dfrac{2}{3}$,

(e) $I_{2k} = \dfrac{2k - 1}{2k} \cdot \dfrac{2k - 3}{2k - 2} \cdot \dfrac{2k - 5}{2k - 4} \cdots \cdot \dfrac{1}{2} \cdot \dfrac{\pi}{2}$ if $k \in \mathbb{N}$,

(f) $\dfrac{\pi}{2} \cdot \dfrac{I_{2k+1}}{I_{2k}} = \dfrac{2 \cdot 2}{1 \cdot 3} \cdot \dfrac{4 \cdot 4}{3 \cdot 5} \cdot \dfrac{6 \cdot 6}{5 \cdot 7} \cdots \cdot \dfrac{2k \cdot 2k}{(2k - 1)(2k + 1)}$,

(g) $\dfrac{\pi}{2} = \lim_{k \to \infty} \prod_{n=1}^{k} \dfrac{4n^2}{4n^2 - 1}$,

(h) $\int_0^1 (1 - t^2)^k \, dt = I_{2k-1}$ if $k \in \mathbb{N}$,

(i) $\int_0^{\infty} \dfrac{dt}{(1 + t^2)^k} = I_{2k-2}$ if $k \in \mathbb{N}$,

(j) $1 - u < e^{-u} < 1/(1 + u)$ if $u > 0$,

(k) $I_{2k+1} < \int_0^1 e^{-kt^2} dt = \dfrac{1}{\sqrt{k}} \int_0^{\sqrt{k}} e^{-x^2} dx,$

(l) $\dfrac{1}{\sqrt{k}} \int_0^\infty e^{-x^2} dx = \int_0^\infty e^{-kt^2} dt < I_{2k-2},$

(m) $\dfrac{\pi}{4(k+1)} = I_{2k+1} \cdot I_{2k+2} < I_{2k+1}^2,$

(n) $I_{2k-2}^2 < I_{2k-2} \cdot I_{2k-3} = \pi/(4(k-1)),$

(o) $\int_0^\infty e^{-x^2} dx = \dfrac{\sqrt{\pi}}{2}.$

28. *The First Mean Value Theorem for Integrals.* Let $f : [a, b] \to \mathbb{R}$ be continuous.

(a) If $w \in L_1'([a, b])$ is nonnegative on $[a, b]$, then there is some ξ with $a < \xi < b$ such that

$$\int_a^b f(x) w(x) \, dx = f(\xi) \int_a^b w(x) \, dx.$$

(b) There is some ξ with $a < \xi < b$ such that

$$\int_a^b f(x) \, dx = f(\xi)(b - a).$$

[Hints: If $\alpha = \min f([a, b])$ and $\beta = \max f([a, b])$, then $\alpha \int_a^b w \le \int_a^b fw \le \beta \int_a^b w$, so use the Intermediate Value Theorem to obtain $\xi \in [a, b]$. It is harder to show that ξ can always be chosen in $]a, b[$.]

29. *Taylor's Theorem Again.* Let $n \in \mathbb{N}$, let $I \subset \mathbb{R}$ be a closed interval, and let $f : I \to \mathbb{R}$ be a function that is $n + 1$ times differentiable on I such that $f^{(n+1)}$ is Riemann integrable over I. Given $a, a + h \in I$, write

$$\phi_k(t) = (a + h - t)^k / k! \qquad (k = 0, 1, \ldots, n).$$

(a) For $1 \le k \le n$, Exercise 7 yields

$$\int_a^{a+h} \phi_{k-1}(t) f^{(k)}(t) \, dt - \int_a^{a+h} \phi_k(t) f^{(k+1)}(t) \, dt = \dfrac{f^{(k)}(a)}{k!} h^k.$$

(b) Using Exercise 1, we obtain

$$f(a + h) = \sum_{k=0}^n \dfrac{f^{(k)}(a)}{k!} h^k + R_n$$

where

$$R_n = \int_a^{a+h} \phi_n(t) f^{(n+1)}(t) \, dt.$$

(c) If $f^{(n+1)}$ is continuous on I, we can apply Exercise 28 to obtain

$$R_n = \dfrac{f^{(n+1)}(a + \theta h)}{(n+1)!} h^{n+1} = \dfrac{f^{(n+1)}(a + \theta' h)}{n!} (1 - \theta')^n h^{n+1}$$

for some $\theta, \theta' \in]0, 1[$.

(d) The Darboux Theorem can be used to remove the continuity hypothesis in (c).

30. *The Riemann-Stieltjes Integral.* Let a, b, f, P, and $(t_k)_{k=1}^n$ be just as in (6.25) where the Riemann integral is defined, except that we allow the *bounded* function f to be complex-valued. Let g be any other *bounded* complex-valued function on $[a, b]$ and define the corresponding *Riemann-Stieltjes sum* to be the complex number

$$S(f, g, P, t_k) = \sum_{k=1}^{n} f(t_k)[g(x_k) - g(x_{k-1})]$$

We say that f is *Riemann-Stieltjes integrable with respect to g over* $[a, b]$ and we write $f \in RS(g, [a, b])$ if there exists $A \in \mathbb{C}$ such that to each $\epsilon > 0$ corresponds a $\delta > 0$ such that whenever $|P| < \delta$ we have

$$|A - S(f, g, P, t_k)| < \epsilon$$

for all choices of the t_k's.

(a) There can exist at most one such A. If A exists, we call it the *Riemann-Stieltjes integral of f with respect to g over* $[a, b]$ and we denote this number A by

$$A = \int_a^b f\, dg = \int_a^b f(x)\, dg(x).$$

(b) If $g(x) = 0$ for $0 \le x \le 1$ and $g(x) = 1$ for $1 < x \le 2$, then $f \in RS(g, [0, 2])$ if and only if f is continuous at $x = 1$. In this case, $\int_0^2 f\, dg = f(1)$.

(c) If ϕ is a complex-valued step function such that g is continuous at every y where ϕ is discontinuous, then $\phi \in RS(g, [a, b])$. More precisely, suppose that

$$Q = \{a = y_0 < y_1 < \ldots < y_m = b\}.$$

$$(\alpha_j)_{j=1}^{m} \subset \mathbb{C}, \qquad \phi(x) = \alpha_j$$

if $y_{j-1} < x < y_j$, $\phi(a) = \alpha_1$, $\phi(b) = \alpha_m$, and g is continuous at y_j for $0 < j < m$. Then

$$\int_a^b \phi\, dg = \sum_{j=1}^{m} \alpha_j [g(y_j) - g(y_{j-1})].$$

[Hints: Imitate the proof of (6.26) with $g(x_k) - g(x_{k-1})$ in place of Δx_k. Given $\epsilon > 0$, write $\eta = \epsilon/(6\beta m)$ and choose $\delta < |Q|$ so that $|g(x) - g(y_j)| < \eta$ if $|x - y_j| < \delta$ and $0 < j < m$. In S_0 omit $k = 1$ and $k = n$. Include $k = 1$ in S_1 and $k = n$ in S_m.]

(d) If $V_a^b g < \infty$ and $f \in RS(g, [a, b])$, then

$$\left| \int_a^b f\, dg \right| \le \|f\|_u \cdot V_a^b g$$

where $\|f\|_u = \sup\{|f(x)| : a \le x \le b\}$.

(e) If $f_1, f_2 \in RS(g, [a, b])$ and $\alpha, \beta \in \mathbb{C}$, then $\alpha f_1 + \beta f_2 \in RS(g, [a, b])$ and

$$\int_a^b (\alpha f_1 + \beta f_2)\, dg = \alpha \int_a^b f_1\, dg + \beta \int_a^b f_2\, dg.$$

(f) If $V_a^b g < \infty$, $(f_m)_{m=1}^{\infty} \subset RS(g, [a, b])$, and $f_m \to f$ uniformly on $[a, b]$, then $f \in RS(g, [a, b])$ and

$$\int_a^b f\, dg = \lim_{m \to \infty} \int_a^b f_m\, dg.$$

[Hints: Use (d) and (e) to show that the integrals of the f_m form a Cauchy sequence so converge to some $A \in \mathbb{C}$. Given $\epsilon > 0$, fix m such that $\|f - f_m\|_u < \epsilon$ and $|\int_a^b f_m\, dg - A| < \epsilon$. Let δ correspond to ϵ as in the above definition of $\int_a^b f_m\, dg$ and show that $|A - S(f, g, P, t_k)| < 2\epsilon + \epsilon V_a^b g$ whenever $|P| < \delta$.]

(g) If f is continuous on $[a, b]$ and $V_a^b g < \infty$, then $f \in RS(g, [a, b])$. [Hint: Find a sequence of functions of the type ϕ described in (c) that converges uniformly to f on $[a, b]$ and use (f).]

(h) If $f \in RS(g, [a, b])$, then $g \in RS(f, [a, b])$ and

$$\int_a^b g \, df = f(b) g(b) - f(a) g(a) - \int_a^b f \, dg.$$

[Hint: Let P and the choice of the t_k's be arbitrary. Check that $S(g, f, P, t_k)$ $= f(b)g(b) - f(a)g(a) - S(f, g, Q, x_{k-1})$, where $Q = \{a = t_0 \leq t_1 \leq \ldots \leq t_n \leq t_{n+1} = b\}$ and $t_{k-1} \leq x_{k-1} \leq t_k$ for $1 \leq k \leq n + 1$. Note also that $|Q| \leq 2|P|$.]

(i) If g is continuous on $[a, b]$ and $V_a^b f < \infty$, then $f \in RS(g, [a, b])$.

(j) If $f \in RS(g_j, [a, b])$ for $j = 1$ and 2 and if $\alpha, \beta \in \mathbb{C}$, then $f \in RS(\alpha g_1 + \beta g_2, [a, b])$ and $\int_a^b f \, d(\alpha g_1 + \beta g_2) = \alpha \int_a^b f \, dg_1 + \beta \int_a^b f \, dg_2$.

(k) If f is continuous a.e. on $[a, b]$ and g is absolutely continuous on $[a, b]$, then $fg' \in L_1([a, b])$, $f \in RS(g, [a, b])$, and $\int_a^b f \, dg = \int_a^b fg'$. (This part is out of place since its proof requires (6.85). The next part is a special case which can easily be proved here.) [Hint: By (e), (j), and Exercise 6 at the end of Chapter 3, we may suppose that f and g are real-valued and that g is nondecreasing on $[a, b]$. Using the notation of (6.28), the fact that $g(d) - g(c) = \int_c^d g'$ for all $c, d \in [a, b]$ leads to

$$\int_a^b \phi_j g' \leq S(f, g, P_j, t_{j,k}) \leq \int_a^b \psi_j g'$$

for all j. As $j \to \infty$ these last two integrals converge to $\int_a^b fg'$.]

(l) If g is differentiable on $[a, b]$ and f and g' are both continuous on $[a, b]$, then $V_a^b g < \infty$ and $\int_a^b f \, dg = \int_a^b fg'$. [Hint: In case g is real-valued, choose t_k so that

$$g(x_k) - g(x_{k-1}) = g'(t_k)\Delta x_k.]$$

(m) Given f and g on $[a, b]$, for $\delta > 0$ let A_δ denote the set of all complex numbers obtainable as Riemann-Stieltjes sums of f with respect to g in which $|P| < \delta$. Then $f \in RS(g, [a, b])$ if and only if $\lim_{\delta \downarrow 0} \text{diam} A_\delta = 0$. In this case, $\int_a^b f \, dg$ is the unique complex number that is in the closure of every A_δ. (Here $\text{diam} A_\delta = \sup\{|\alpha - \beta| : \alpha, \beta \in A_\delta\}$.)

(n) If $f \in RS(g, [a, b])$ and $a < c < b$, then

$$f \in RS(g, [a, c]) \cap RS(g, [c, b])$$

and

$$\int_a^c f \, dg + \int_c^b f \, dg = \int_a^b f \, dg.$$

[Hint: Use (m). Show that the difference of any two R–S sums over $[a, c]$ for subdivisions having meshes $< \delta$ is equal to the difference of two such sums over $[a, b]$ by extending the two subdivisions of $[a, c]$ to subdivisions of $[a, b]$ which agree on $[c, b]$.]

(o) Given $a < c < b$, define f and g by $f(x) = g(x) = 0$ if $a \leq x < c$, $f(x) = g(x) = 1$ if $c < x \leq b$, $f(c) = 0$, $g(c) = 1$. Then f is R–S integrable with respect to g over both $[a, c]$ and $[c, b]$ but *not* over $[a, b]$.

31.   *The Lebesgue–Stieltjes Integral.* Let g be a real-valued (monotone) nondecreasing

function on $\mathbb{R}$. For $E \subset \mathbb{R}$, define

$$\lambda_g(E) = \inf \sum_{n=1}^{\infty} [g(b_n) - g(a_n)]$$

where the infimum is taken over all countable families $\{]a_n, b_n[\}_{n=1}^{\infty}$ of open intervals that cover E for which g is continuous at each a_n and each b_n. A proposition about $x \in \mathbb{R}$ is said to hold "g-a.e." if the set E of those x for which it fails satisfies $\lambda_g(E) = 0$. Let $M_0(g)$ be the set of all step functions whose points of discontinuity are each points at which g is continuous. If I is any bounded interval having left endpoint a and right endpoint b, write $|I|_g = g(b) - g(a)$. Making these replacements (λ_g for λ, "g-a.e." for "a.e.", $M_0(g)$ for M_0, and $|I|_g$ for $|I|$), reformulate all the definitions, lemmas and theorems in (6.2)–(6.24) and make any required alterations to make the proofs go through. (Actually, this project can be carried out successfully for nearly all of (6.33)–(6.80) as well.) In particular, obtain the class $L_1'(g)$ of functions f that are integrable with respect to g and the Lebesgue–Stieltjes integral $f \to \int f \, dg$ on it.

32. *Abstract Integration Theory.* Let X be a set, let M_0 be a set of real-valued functions defined on X, and let I be a real-valued function defined on M_0. Suppose that (6.4) is true when $I(f)$ is written in place of $\int f$. With an eye on (6.7), we *define* a set $E \subset X$ to be an *I-null set* if there exists a nondecreasing sequence $(\phi_n)_{n=1}^{\infty} \subset M_0$ such that $\lim_{n \to \infty} I(\phi_n) < \infty$ and $\lim_{n \to \infty} \phi_n(x) = \infty$ for every $x \in E$. A proposition $P(x)$ about $x \in X$ is said to hold *I-a.e.* if the set E of x for which it fails is an I-null set. We also suppose that the analog of (6.8) holds for our I and M_0: If $(\phi_n)_{n=1}^{\infty} \subset M_0$, $\phi_n(x) \geq \phi_{n+1}(x) \geq 0$ for all $x \in X$ and all $n \in \mathbb{N}$, and $\lim_{n \to \infty} \phi_n(x) = 0$ *I-a.e.*, then $\lim_{n \to \infty} I(\phi_n) = 0$. Now write out the analogs of all the definitions and theorems of (6.9)–(6.24) and prove the theorems. In this way, extend the domain of I to the space $L_1'(I)$ on which the two main convergence theorems hold good. This abstract theory can be carried further through analogy with much of the ongoing concrete theory of this chapter. One can develop the theory of absolutely convergent numerical series this way. In fact, take $X = \mathbb{N}$, take M_0 to be the set of all real-valued functions ϕ on $\mathbb{N}$ (sequences) such that $\phi(n) \neq 0$ for only finitely many $n \in \mathbb{N}$, and define

$$I(\phi) = \sum_{n=1}^{\infty} \phi(n)$$

(this is a finite sum in disguise). One easily checks that, in this case, the above hypotheses are met, the resulting space $L_1'(I)$ is the set of all $f : \mathbb{N} \to \mathbb{R}$ such that

$$\sum_{n=1}^{\infty} |f(n)| < \infty,$$

and that the "integral" of f is $\sum_{n=1}^{\infty} f(n)$.

33. *The Number e Is Transcendental.* Assume, to the contrary, that there exist integers $(a_j)_{j=0}^n$ with $a_0 \neq 0$, $a_n \neq 0$, and $\sum_{j=0}^{n} a_j e^j = 0$. Fix any $\beta \in \mathbb{R}$ such that

$$\prod_{j=1}^{n} |x - j| \leq \beta$$

whenever $0 \le x \le n$ and then fix any prime number p such that $p > n$, $p > |a_0|$, and

$$\frac{n^{p-1}\beta^p e^n}{(p-1)!} \cdot \sum_{j=1}^{n} |a_j| < 1.$$

(a) Such numbers β and p exist.

(b) If $m \ge 0$ is an integer, then $\int_0^\infty u^m e^{-u}\, du = m!$.

Define the polynomial f by

$$f(x) = x^{p-1}\left[\prod_{j=1}^{n} (x-j) \right]^p.$$

(c) We have $f(x) = \sum_{k=0}^{np} b_k x^{k+p-1}$, where the b_k are integers and b_0 is *not* divisible by p. For $0 \le j \le n$, write

$$c_j = \frac{e^j}{(p-1)!} \int_j^\infty f(x)e^{-x}\, dx$$

and

$$\alpha_j = \frac{e^j}{(p-1)!} \int_0^j f(x)e^{-x}\, dx.$$

(d) The number c_0 is an integer that is *not* divisible by p.

(e) For $1 \le j \le n$, we have

$$(p-1)!\, c_j = \int_0^\infty u^p g_j(u)e^{-u}\, du,$$

where g_j is a polynomial with integer coefficients, and so c_j is an integer divisible by p.

(f) The integer $a_0 c_0$ is not divisible by p.

(g) The integer $c = \sum_{j=0}^{n} a_j c_j$ is *not* 0.

(h) For $1 \le j \le n$, we have $|\alpha_j| < n^{p-1}\beta^p e^n/(p-1)!$.

(i) The number $\alpha = \sum_{j=1}^{n} a_j \alpha_j$ satisfies $|\alpha| < 1$.

(j) $\alpha + c = \sum_{j=0}^{n} a_j e^j c_0 = 0$.

(k) The integer c satisfies $0 < |c| < 1$.

Measurable Functions

Lebesgue's Dominated Convergence Theorem tells us that any a.e. limit of a sequence of integrable functions is itself integrable *provided* the sequence is dominated by an integrable function. In this section we study a larger class of functions that is closed under the formation of a.e. limits of sequences, with no provisos.

(6.33) **Definition** An extended real-valued function f that is defined a.e. on $\mathbb{R}$ is said to be *Lebesgue measurable*, or simply *measurable*, if there exists a sequence

$(\phi_n)_{n=1}^{\infty} \subset M_0$ such that $\lim_{n \to \infty} \phi_n(x) = f(x)$ a.e. We denote the class of all such f's by M^r.

(6.34)   Remarks   (a) It follows from (6.18) that $L_1^r \subset M^r$.

(b) Example (6.12.a) shows that if f is a real-valued function that is continuous a.e. on $\mathbb{R}$, then f^+ and f^- are in M^r. It follows that $f \in M^r$. Thus M^r contains many functions that are not integrable.

(c) Although it is not at all obvious, we will show later in this chapter that there are sets $E \subset \mathbb{R}$ such that $\xi_E \notin M^r$.

The next three theorems are important but easy. We leave their proofs to the reader [cf. (6.13)].

(6.35)   Theorem   *Let* $f, g \in M^r$ *and* $\alpha \in \mathbb{R}$. *Then* $\alpha f, f \vee g, f \wedge g, f^+, f^-,$ *and* $|f|$ *are all in* M^r. *Moreover, if* $f + g$ *is defined a.e., then* $f + g \in M^r$. *[Recall that* $\infty + (-\infty)$ *and* $(-\infty) + \infty$ *are not defined.]*

(6.36)   Theorem   *Let* $n \in \mathbb{N}$, *let* $f_1, \ldots, f_n \in M^r$ *be finite a.e., and let* F *be a continuous function from an open subset of* $\mathbb{R}^n$ *into* $\mathbb{R}$. *Suppose that* $(f_1(x), \ldots, f_n(x))$ *is in the domain of* F *a.e. Then the function* h *defined a.e. by*

$$h(x) = F(f_1(x), \ldots, f_n(x))$$

is in M^r.

Using (6.36) we obtain

(6.37)   Theorem   *Let* $f, g \in M^r$ *be finite a.e. Then the functions* $fg, |f|^p$ $(p > 0)$, *and* f/g $(g \neq 0 \ a.e.)$ *are all in* M^r.

The next theorem shows that all "small" members of M^r are integrable.

(6.38)   Theorem   *Let* $f \in M^r, g \in L_1^r$ *and suppose that* $|f| \leq g$ *a.e. Then* $f \in L_1^r$.

Proof   Choose any sequence $(\phi_n)_{n=1}^{\infty} \subset M_0$ such that $\phi_n(x) \to f(x)$ a.e. Define $f_n = (\phi_n \wedge g) \vee (-g)$ a.e. Then $(f_n)_{n=1}^{\infty} \subset L_1^r$ and $|f_n| \leq g$ a.e. Since $f_n(x) = \phi_n(x)$ if $|\phi_n(x)| \leq g$, we have $f_n(x) \to f(x)$ a.e. Thus (6.22) shows that $f \in L_1^r$. $\square$

We can now prove that M^r is closed under a.e. limits.

(6.39)   Theorem   *Suppose that* $(f_n)_{n=1}^{\infty} \subset M^r$ *and let* f *be an* $\mathbb{R}^{\#}$-*valued function defined a.e. on* $\mathbb{R}$ *such that* $f_n(x) \to f(x)$ *a.e. Then* $f \in M^r$.

Proof   *Case 1:* $f_n \in L_1^r$ for all n. Using (6.18), we can choose $(\phi_k)_{k=1}^{\infty} \subset M_0$ such that

$$\int |f_k - \phi_k| < \frac{1}{2^k}$$

for all k. Define

$$g_n = \sum_{k=1}^{n} |f_k - \phi_k|.$$

Then $(g_n)_{n=1}^{\infty} \subset L_1^r$ is a nondecreasing sequence and

$$\lim_{n \to \infty} \int g_n = \lim_{n \to \infty} \sum_{k=1}^{n} \int |f_k - \phi_k| < \sum_{k=1}^{\infty} \frac{1}{2^k} = 1.$$

From (6.20), we conclude that

$$\sum_{k=1}^{\infty} |f_k - \phi_k| = \lim_{n \to \infty} g_n \in L_1^r;$$

hence

$$\sum_{k=1}^{\infty} |f_k - \phi_k| < \infty \quad \text{a.e.}$$

It follows that $\lim_{k \to \infty} |f_k - \phi_k| = 0$ a.e., and so, since

$$\phi_k = f_k - (f_k - \phi_k),$$

we have $\phi_k \to f$ a.e. Therefore $f \in M^r$.

Case 2: $(f_n)_{n=1}^{\infty} \subset M^r$. Define

$$h_n = [(f_n \wedge n) \vee (-n)] \cdot \xi_{[-n,n]} \in M^r.$$

That is,

$$h_n(x) = 0 \quad \text{if } |x| > n$$

and, for $|x| \le n$,

$$h_n(x) = f_n(x) \quad \text{if } |f_n(x)| \le n,$$
$$h_n(x) = n \quad \quad \text{if } f_n(x) > n,$$
$$h_n(x) = -n \quad \text{if } f_n(x) < -n.$$

Since $|h_n| \le n\xi_{[-n,n]}$, it follows from (6.38) that $h_n \in L_1^r$. Plainly $h_n \to f$ a.e. By Case 1, $f \in M^r$. □

(6.40) Corollary *Let* $(f_n)_{n=1}^{\infty} \subset M^r$. *Then* $\inf_{n \ge 1} f_n$, $\sup_{n \ge 1} f_n$, $\lim_{n \to \infty} f_n$, *and* $\overline{\lim}_{n \to \infty} f_n$ *are all in* M^r. [*As usual, these four functions are defined pointwise a.e.*]

Proof Define $g_m = f_1 \wedge f_2 \wedge \ldots \wedge f_m \in M^r$. Then $\inf_{n \ge 1} f_n = \lim_{m \to \infty} g_m \in M^r$.
Similarly, $\sup_{n \ge 1} f_n \in M^r$. Define $h_k = \inf_{n \ge k} f_n \in M^r$. Then $\underline{\lim}_{n \to \infty} f_n = \sup_{k \ge 1} h_k = \lim_{k \to \infty} h_k \in M^r$. Similarly, $\overline{\lim}_{n \to \infty} f_n \in M^r$. □

We next extend the domain of our integral to a subset of M^r larger than L_1^r.

(6.41) **Definition** Let $f \in M'$. If $f \in L_1'$ we define $\int f$ as before. If $f \notin L_1'$ and $f \geq 0$ a.e., define $\int f = \infty$. If these two cases fail, define $\int f = \int f^+ - \int f^-$ *provided* that one of the latter two integrals is finite. Otherwise, we *do not* define $\int f$. Note that in every case where $\int f$ is defined, we always have

$$\int f = \int f^+ - \int f^-.$$

(6.42) **Theorem** *Suppose that $f, g \in M'$ and $\int f$ and $\int g$ are defined. If $f \leq g$, then $\int f \leq \int g$.*

Proof We may obviously suppose that $\int f > -\infty$ and $\int g < \infty$. Then $\int f^- < \infty$ and $\int g^+ < \infty$. Thus $f^-, g^+ \in L_1'$. From (6.38) and the evident inequalities $f^+ \leq g^+$, $g^- \leq f^-$, we conclude that $f^+, g^- \in L_1'$, and so

$$0 \leq \int f^+ \leq \int g^+ < \infty, \qquad 0 \leq \int g^- \leq \int f^- < \infty.$$

Therefore

$$\int f = \int f^+ - \int f^- \leq \int g^+ - \int g^- = \int g. \quad \square$$

We now give an improved version of the Monotone Convergence Theorem (MCT).

(6.43) **Monotone Convergence Theorem** *Let $(f_n)_{n=1}^\infty \subset M'$ be a nondecreasing sequence with $f_1^- \in L_1'$ [for example, $f_1 \geq 0$]. Then*

$$\lim_{n \to \infty} \int f_n = \int \left(\lim_{n \to \infty} f_n \right).$$

Proof Write $f(x) = \lim_{n \to \infty} f_n(x)$ for $x \in \mathbb{R}$. Since $\int f_1^- < \infty$ and $f^- \leq f_n^- \leq f_1^-$, we have $\int f^- \leq \int f_n^- < \infty$, and so $\int f$ and $\int f_n$ are defined for all n. Since $f_n \leq f$, we have $\int f_n \leq \int f$ for all n. Thus $\lim_{n \to \infty} \int f_n \leq \int f$. If $\lim_{n \to \infty} \int f_n = \infty$, there is nothing left to prove. If $\lim_{n \to \infty} \int f_n < \infty$, then $(f_n)_{n=1}^\infty \subset L_1'$ and the theorem follows from (6.20). $\square$

(6.44) **Examples** (a) Let $f(x) = 1/x$ for $0 < x \leq 1$ and $f(x) = 0$ otherwise. Define

$$f_n = 2^n \cdot \sum_{k=1}^{2^n} \frac{1}{k} \xi_{](k-1)/2^n, k/2^n]}.$$

One checks that $f_1 \leq f_2 \leq \ldots$ and $f_n \to f$. Thus

$$\int f = \lim_{n \to \infty} \int f_n = \lim_{n \to \infty} \sum_{k=1}^{2^n} \frac{1}{k} = \infty.$$

(b) The hypothesis in (6.43) that f_1^- be integrable cannot be dispensed with. In fact, let f be as in (a) and define $g_n = -f/n$. Then $g_n(x) \to 0$ for all x, and

$g_1 \leqq g_2 \leqq \ldots$. But $\int g_n = -\infty$ for all n. Thus

$$\lim_{n \to \infty} \int g_n = -\infty \neq 0 = \int 0.$$

The integral is not linear on this extended domain. However, it is still homogeneous and it is *countably* additive for nonnegative summands.

(6.45)   Theorem   *If $\alpha \in \mathbb{R}$ and $f \in M'$ with $\int f$ defined, then $\int \alpha f$ is defined and equal to $\alpha \int f$.*

We omit the trivial proof.

(6.46)   Theorem [Lebesgue]   *Let $(f_n)_{n=1}^{\infty} \subset M'$ with $f_n \geqq 0$ for all n; then*

(i) $$\int \left(\sum_{n=1}^{\infty} f_n \right) = \sum_{n=1}^{\infty} \int f_n.$$

Proof   If all f_n are in L_1^r, then

$$\int \left(\sum_{n=1}^{k} f_n \right) = \sum_{n=1}^{k} \int f_n$$

for all k, and we can apply (6.43) to the sequence $\displaystyle\sum_{n=1}^{k} f_n$ $(k = 1, 2, \ldots)$. Otherwise, both sides of (i) equal ∞. $\square$

The following inequality is often useful.

(6.47)   Fatou's Lemma   *Let $(f_n)_{n=1}^{\infty} \subset M'$ with $f_n \geqq 0$ for all n. Then*

$$\int \left(\varliminf_{n \to \infty} f_n \right) \leqq \varliminf_{n \to \infty} \int f_n.$$

Proof   Apply (6.43) to the sequence $g_k = \inf_{n \geqq k} f_n$ to obtain

$$\int \varliminf_{n \to \infty} f_n = \int \lim_{k \to \infty} g_k = \lim_{k \to \infty} \int g_k \leqq \varliminf_{k \to \infty} \int f_k,$$

where the last inequality obtains because $g_k \leqq f_k$ for all k. $\square$

(6.48)   Remark   Strict inequality can obtain in (6.47). For example, let

$$f_n = (1/n)\xi_{[0,n]}$$

Complex-Valued Functions

In this short section we extend the notions of measurability, integrability, and the integral to complex-valued functions. Most of this is rather routine.

(6.49)   Definition   Let f be a complex-valued [never infinite] function that is defined a.e. on $\mathbb{R}$. We call f *Lebesgue measurable* if Re f and Im f are measurable, and we write $f \in M$. We call f *Lebesgue integrable* if Re f and Im f are integrable, and we write $f \in L_1$. If $f \in L_1$, we define

$$\int f = \int \operatorname{Re} f + i \int \operatorname{Im} f.$$

(6.50)   Remark   It is obvious that $L_1^r \subset L_1$ and that the above definition of $\int$ extends that given on L_1^r (6.14). Also, if $f \in M^r$ is never infinite, then $f \in M$.

(6.51)   Theorem   *If $f, g \in M$ and $\alpha \in \mathbb{C}$, then $\alpha f, f + g, fg, f/g$ $(g \neq 0$ a.e.$), \bar{f}$, and $|f|$ are all in M.*

> **Proof**   Apply (6.35) and (6.37).   □

(6.52)   Theorem   *If $f, g \in L_1$ and $\alpha \in \mathbb{C}$, then $\alpha f, f + g, \bar{f}$, and $|f|$ are all in L_1. Moreover,*

(i) $$\int \alpha f = \alpha \int f,$$

(ii) $$\int (f + g) = \int f + \int g,$$

(iii) $$\int \bar{f} = \overline{\int f},$$

(iv) $$\left| \int f \right| \leq \int |f|.$$

> **Proof**   All but the assertions about $|f|$ are immediate if we use (6.16). Write $f = f_1 + if_2$, where $f_1, f_2 \in L_1^r$. Since $|f| \in M^r$ and $|f| \leq |f_1| + |f_2|$, it follows from (6.38) that $|f| \in L_1^r \subset L_1$. To prove (iv), write $|\int f| = u \int f$, where $u \in \mathbb{C}$ with $|u| = 1$. [If $\int f \neq 0$, solve for u. If $\int f = 0$, take $u = 1$.] Write $uf = h = h_1 + ih_2$ where $h_1, h_2 \in L_1^r$. Then
>
> $$\left| \int f \right| = u \int f = \int uf = \int h = \int h_1 + i \int h_2 = \int h_1,$$
>
> where the last equality obtains because its left side, like $|\int f|$, is a real number. Thus, using (6.17), we conclude
>
> $$\left| \int f \right| = \int h_1 \leq \int |h_1| \leq \int |h| = \int |f|.   \square$$

Our next result gives a very useful and simple criterion that a measurable function be integrable.

(6.53)   Theorem   *Let $f \in M$. Then $f \in L_1$ if and only if $|f| \in L_1^r$, i.e., $\int |f| < \infty$.*

Proof Write $f = f_1 + if_2$, where $f_1, f_2 \in M'$. If $|f| \in L_1'$, then the inequalities $|f_1| \leq |f|$ and $|f_2| \leq |f|$ along with (6.38) show that $f_1, f_2 \in L_1'$; hence, $f \in L_1$. The converse is part of (6.52). $\square$

The following, though important, is a simple consequence of (6.39).

(6.54) Theorem *Let $(f_n)_{n=1}^{\infty} \subset M$ and let f be a complex-valued function defined a.e. on $\mathbb{R}$. Suppose that $f_n \to f$ a.e. Then $f \in M$.*

We can now give our most general form of

(6.55) Lebesgue's Dominated Convergence Theorem *Let $(f_n)_{n=1}^{\infty} \subset M$ and let f be a complex-valued function defined a.e. on $\mathbb{R}$ such that $f_n \to f$ a.e. Suppose that there is a $g \in L_1'$ such that*

(i) $$|f_n| \leq g \quad a.e. \quad \text{for all } n.$$

Then $f \in L_1$ and

(ii) $$\lim_{n \to \infty} \int f_n = \int f.$$

Proof Separate each function into real and imaginary parts and apply (6.38), (6.39), and (6.22) to these two sequences of real-valued functions. We leave the details as an exercise for the reader. $\square$

One crucial advantage of the Lebesgue integral over the Riemann integral is that L_1 is *complete* in the following sense.

(6.56) Theorem *Let $(f_n)_{n=1}^{\infty} \subset L_1$ satisfy*

$$\lim_{m,n \to \infty} \int |f_m - f_n| = 0.$$

Then there exists $f \in L_1$ such that

$$\lim_{n \to \infty} \int |f - f_n| = 0.$$

This function f is unique in the sense that if $h \in L_1$ and $\int |h - f_n| \to 0$, then $f = h$ a.e. Moreover, there is a subsequence $(f_{n_k})_{k=1}^{\infty}$ such that

$$\lim_{k \to \infty} f_{n_k}(x) = f(x) \quad a.e.$$

Proof Choose $n_1 < n_2 < \ldots$ such that

$$\int |f_m - f_{n_k}| < \frac{1}{2^k} \qquad (m \geq n_k).$$

This can be done inductively [cf. proof of (6.6)]. Define

$$g_k = |f_{n_1}| + \sum_{j=2}^{k} |f_{n_j} - f_{n_{j-1}}|.$$

Then $(g_k)_{k=2}^{\infty} \subset L_1^r$ is a nondecreasing sequence with

$$\int g_k = \int |f_{n_1}| + \sum_{j=2}^{k} \int |f_{n_j} - f_{n_{j-1}}| < \int |f_{n_1}| + \sum_{j=2}^{k} \frac{1}{2^{j-1}}$$

$$< \int |f_{n_1}| + 1 < \infty$$

for all $k > 1$. It follows from (6.20) that the function g defined by $g(x) = \lim_{k \to \infty} g_k(x)$ a.e. is in L_1^r, and so

$$|f_1(x)| + \sum_{j=2}^{\infty} |f_{n_j}(x) - f_{n_{j-1}}(x)| = g(x) < \infty \quad \text{a.e.}$$

Thus the series

$$f_{n_1} + \sum_{j=2}^{\infty} (f_{n_j} - f_{n_{j-1}}),$$

being absolutely convergent, is convergent a.e. The kth partial sum of this series is f_{n_k}, and so there is a complex-valued function f defined a.e. on $\mathbb{R}$ such that $f(x) = \lim_{k \to \infty} f_{n_k}(x)$ a.e. By (6.54), $f \in M$.

Let $\epsilon > 0$ be given. Choose $r \in \mathbb{N}$ such that

$$\int |f_m - f_n| < \epsilon \quad \text{if } m, n > n_r.$$

Then

$$\int |f_{n_k} - f_n| < \epsilon \quad \text{if } k > r, n > n_r.$$

Applying Fatou's Lemma (6.47), we have

$$\int |f - f_n| = \int \left(\lim_{k \to \infty} |f_{n_k} - f_n| \right) \leq \varliminf_{k \to \infty} \int |f_{n_k} - f_n| \leq \epsilon$$

for all $n > n_r$. Thus, for any $n > n_r$, $f - f_n \in L_1$, and so $f = (f - f_n) + f_n \in L_1$. Since ϵ was arbitrary,

$$\lim_{n \to \infty} \int |f - f_n| = 0.$$

Supposing that $\int |h - f_n| \to 0$, we have

$$\int |f - h| \leq \int |f - f_n| + \int |f_n - h| \to 0 \quad \text{as } n \to \infty$$

and so $\int |f - h| = 0$. By (6.24), $f = h$ a.e. $\quad\square$

(6.57) **Remark** It is possible to find sequences $(f_n)_{n=1}^{\infty}$ that are Riemann integrable over $[0, 1]$ such that

(i)
$$\lim_{m,n\to\infty} R\int_0^1 |f_n - f_m| = 0,$$

but there is no Riemann integrable function f on $[0, 1]$ satisfying

(ii)
$$\lim_{n\to\infty} R\int_0^1 |f - f_n| = 0.$$

In fact, let $f_n = \psi_n$ as in (6.30). Then f_n and $|f_n - f_m|$ are Riemann integrable over $[0, 1]$ for all m, n since each is continuous except at finitely many points (6.29). Also $f_n \to \xi_P$ everywhere and $0 \le f_n - \xi_P \le \xi_{[0,1]} \in L'_1$ everywhere. It follows from (6.22) and (6.29) that

$$R\int_0^1 |f_n - f_m| = \int |f_n - f_m| \le \int |f_n - \xi_P| + \int |\xi_P - f_m| \to 0$$

as $m, n \to \infty$. Thus (i) obtains. Assume that (ii) obtains for some Riemann integrable f. Then, defining $f = 0$ off $[0, 1]$, we have

$$\lim_{n\to\infty} \int |f - f_n| = 0.$$

Since $\int |\xi_P - f_n| \to 0$, it follows that $f = \xi_P$ a.e., contrary to (6.30).

Measurable Sets

Lebesgue's original definition of his integral was based upon his notion of measurable sets and measurable functions which he studied first. We have done just the opposite in starting with the integral, moving on to measurable functions, and now arriving at measurable sets.

(6.58) **Definition** A set $E \subset \mathbb{R}$ is said to be *Lebesgue measurable* [or simply *measurable*], and we write $E \in \mathfrak{M}$, if $\xi_E \in M$.

(6.59) **Theorem** *The family $\mathfrak{M}$ has the following properties.*
  (i) $\mathbb{R} \in \mathfrak{M}$ *and* $\varnothing \in \mathfrak{M}$.
  (ii) $E \in \mathfrak{M}$ *implies* $E' = \mathbb{R}\setminus E \in \mathfrak{M}$.
  (iii) $(E_n)_{n=1}^{\infty} \subset \mathfrak{M}$ *implies* $\bigcup_{n=1}^{\infty} E_n \in \mathfrak{M}$.
  (iv) $(E_n)_{n=1}^{\infty} \subset \mathfrak{M}$ *implies* $\bigcap_{n=1}^{\infty} E_n \in \mathfrak{M}$.
  (v) $A, B \in \mathfrak{M}$ *implies* $A \setminus B \in \mathfrak{M}$.
  (vi) $E \in \mathfrak{M}$ *if* E *is* λ-*null.*
  (vii) $U \in \mathfrak{M}$ *for every open set* $U \subset \mathbb{R}$.
  (viii) $C \in \mathfrak{M}$ *for every closed set* $C \subset \mathbb{R}$.

Proof If E is λ-null, then $\xi_E = 0$ a.e. So $\xi_E \in M_1 \subset M$; hence $E \in \mathfrak{M}$. Thus (vi) obtains. Since $\xi_{E'} = 1 - \xi_E$, we have (ii). Assertion (iii) follows from

$$\xi_E = \sup_{n \geq 1} \xi_{E_n}, \quad \text{where } E = \bigcup_{n=1}^{\infty} E_n$$

and (6.40). From (ii) and (iii) and

$$\bigcap_{n=1}^{\infty} E_n = \left(\bigcup_{n=1}^{\infty} E_n' \right)',$$

we get (iv). Using (iv), (ii), and $A \setminus B = A \cap B'$, we have (v).

Since $\xi_I \in M_0 \subset M$ for each bounded interval I and since each open set is a countable union of bounded intervals, we can use (iii) to obtain (vii). Now (i) is trivial. Finally, (viii) follows from (ii) and (vii). $\square$

In the following theorem, λ denotes, as usual, Lebesgue [outer] measure as defined in (2.75).

(6.60) **Theorem** *If $E \in \mathfrak{M}$, then*

$$\lambda(E) = \int \xi_E.$$

Proof If $\mathcal{I}$ is a countable family of open intervals with $E \subset \mathcal{I}$, then

$$\xi_E \leq \sum_{I \in \mathcal{I}} \xi_I,$$

and so (6.46) yields

$$\int \xi_E \leq \sum_{I \in \mathcal{I}} \int \xi_I = \sum_{I \in \mathcal{I}} |I|.$$

[If I is unbounded, then I contains bounded intervals J of arbitrarily great length, $\xi_J \leq \xi_I$, $|J| = \int \xi_J \leq \int \xi_I$; hence $\int \xi_I = \infty = |I|$.] Taking the infimum over all such $\mathcal{I}$, we obtain

$$\int \xi_E \leq \lambda(E). \tag{1}$$

It is our job to prove the reversed inequality. We may suppose that $\int \xi_E < \infty$; i.e., $\xi_E \in L_1$.

Case 1: $E \subset [-a, a]$ where $0 < a < \infty$. By (6.18), there is some sequence $(\phi_n)_{n=1}^{\infty} \subset M_0$ such that $\phi_n \to \xi_E$ a.e. We suppose, as we obviously may, that each ϕ_n vanishes off $[-a, a]$. Define

$$U_n = \{ x \in \mathbb{R} : \phi_n(x) > 1/2, \phi_n \text{ is continuous at } x \}.$$

Then U_n is the union of finitely many disjoint open subintervals of $[-a, a]$. Write

$$A = \{ x \in \mathbb{R} : \phi_n(x) \nrightarrow \xi_E(x) \text{ or } \phi_n \text{ is discontinuous at } x \text{ for some } n \}.$$

Then A, being a union of countably many λ-null sets, is λ-null. Thus

$$\lambda(E \setminus A) = \lambda(E). \tag{2}$$

Plainly

$$\lim_{n \to \infty} \xi_{U_n}(x) = \xi_E(x) \tag{3}$$

for all $x \in \mathbb{R} \setminus A$ [consider the cases $x \in E$ and $x \notin E$]. Let

$$V_k = \bigcup_{n=k}^{\infty} U_n.$$

It follows from (3) that $\xi_{V_k} \to \xi_E$ a.e. [except on A]. Since each V_k is open, $\xi_{V_k} \in M$ for all k. Clearly $\xi_{V_k} \leq \xi_{[-a,a]} \in L_1^r$ for all k. By (6.55),

$$\lim_{k \to \infty} \int \xi_{V_k} = \int \xi_E. \tag{4}$$

Let $\epsilon > 0$ be given. By (4), we can fix a k such that

$$\int \xi_{V_k} < \int \xi_E + \epsilon.$$

Let $\mathcal{I}$ be a countable pairwise disjoint family of intervals such that $\cup \mathcal{I} = V_k$. If $x \in E \setminus A$, then $x \in U_n$ for all large n, by (3), and so $x \in V_k$. Thus $E \setminus A \subset \cup \mathcal{I} = V_k$. Since $\mathcal{I}$ is pairwise disjoint, $\xi_{V_k} = \sum_{I \in \mathcal{I}} \xi_I$. Therefore

$$\lambda(E) = \lambda(E \setminus A) \leq \sum_{I \in \mathcal{I}} |I| = \sum_{I \in \mathcal{I}} \int \xi_I$$

$$= \int \sum_{I \in \mathcal{I}} \xi_I = \int \xi_{V_k} < \int \xi_E + \epsilon.$$

Since ϵ was arbitrary,

$$\lambda(E) \leq \int \xi_E$$

and the proof is complete for Case 1.

Case 2: $E \in \mathfrak{M}$. Write $E = \bigcup_{p=1}^{\infty} E_p$ where $\{E_p\}_{p=1}^{\infty} \subset \mathfrak{M}$ is pairwise disjoint and each E_p is bounded [e.g., $E_p = E \cap (]-p, p[\,\backslash\,]-p+1, p-1[)$]. Then, using Case 1 and (6.46),

$$\lambda(E) \leq \sum_{p=1}^{\infty} \lambda(E_p) = \sum_{p=1}^{\infty} \int \xi_{E_p} = \int \sum_{p=1}^{\infty} \xi_{E_p} = \int \xi_E. \quad \square$$

(6.61) **Remark** If we had not defined λ previously [except for the notion of a λ-null set (cf. (6.7))], we could use (6.60) as a definition of λ on $\mathfrak{M}$. We could then extend λ to arbitrary sets $A \subset \mathbb{R}$ by defining

$$\lambda(A) = \inf\{\lambda(E) : A \subset E, E \in \mathfrak{M}\}.$$

(6.62) **Theorem** *The measure λ has the following properties.*

(i) $\qquad\qquad\qquad\qquad \lambda(\varnothing) = 0.$

(ii) *If* $\{E_n\}_{n=1}^\infty \subset \mathfrak{M}$ *is pairwise disjoint, then*

$$\lambda\left(\bigcup_{n=1}^\infty E_n\right) = \sum_{n=1}^\infty \lambda(E_n).$$

(iii) *If* $A, B \in \mathfrak{M}$, $B \subset A$, *and* $\lambda(B) < \infty$, *then*

$$\lambda(A \setminus B) = \lambda(A) - \lambda(B).$$

(iv) *If* $(E_n)_{n=1}^\infty \subset \mathfrak{M}$ *and* $E_n \subset E_{n+1}$ *for all* n, *then*

$$\lambda\left(\bigcup_{n=1}^\infty E_n\right) = \lim_{n\to\infty} \lambda(E_n).$$

(v) *If* $(E_n)_{n=1}^\infty \subset \mathfrak{M}$, $E_n \supset E_{n+1}$ *for all* n, *and* $\lambda(E_p) < \infty$ *for some* p, *then*

$$\lambda\left(\bigcap_{n=1}^\infty E_n\right) = \lim_{n\to\infty} \lambda(E_n).$$

Proof   Assertion (ii) follows from (6.60), (6.46), and

$$\xi_E = \sum_{n=1}^\infty \xi_{E_n}, \quad \text{where } E = \bigcup_{n=1}^\infty E_n.$$

To prove (iii), use (i), (ii), and $A = B \cup (A \setminus B) \cup \emptyset \cup \emptyset \cup \ldots$. For (iv), use (6.60), (6.43), and

$$\xi_E = \lim_{n\to\infty} \xi_{E_n}, \quad \text{where } E = \bigcup_{n=1}^\infty E_n.$$

To prove (v), use (6.60), (6.55), $\xi_{E_n} \leq \xi_{E_p} \in L_1^r$ $(n \geq p)$, and

$$\xi_E = \lim_{n\to\infty} \xi_{E_n}, \quad \text{where } E = \bigcap_{n=1}^\infty E_n. \quad \square$$

(6.63)   Remarks on (6.62)   (a) The requirement in (v) that $\lambda(E_p) < \infty$ for some p cannot be dispensed with. For example, let $E_n = [n, \infty[$.

(b) Assertions (iv) and (v) can be deduced directly from (i) and (ii) as (iii) was. For (iv), write

$$\bigcup_{n=1}^\infty E_n = E_1 \cup \bigcup_{n=2}^\infty (E_n \setminus E_{n-1}).$$

For (v), write

$$\bigcap_{n=1}^\infty E_n = \bigcap_{n=p}^\infty E_n = E_p \setminus \bigcup_{n=p}^\infty (E_n \setminus E_{n+1}).$$

We leave the details to the reader.

The next two theorems connect the topology of $\mathbb{R}$ with the measure λ.

(6.64)   Theorem [Outer regularity]   *If* $E \subset \mathbb{R}$, *then*

(i) $\lambda(E) = \inf\{\lambda(U) : E \subset U \subset \mathbb{R}, U \text{ is open}\}.$

Proof Let α denote the right side of (i). Since $\lambda(E) \leq \lambda(U)$ whenever $E \subset U$, we have $\lambda(E) \leq \alpha$. Assume that $\lambda(E) < \alpha$. By Definition (2.75), there is a countable family $\mathcal{I}$ of open intervals such that $E \subset \cup \mathcal{I}$ and

$$\sum_{I \in \mathcal{I}} |I| < \alpha.$$

Let $U = \cup \mathcal{I}$. By (2.76.iv),

$$\lambda(U) < \alpha.$$

But $E \subset U$. This contradicts the definition of α. $\square$

(6.65) **Theorem** [Inner regularity] *If $E \in \mathfrak{M}$, then*

$$\lambda(E) = \sup\{\lambda(C) : C \subset E, C \text{ is compact}\}.$$

Proof For $n \geq 1$, write $E_n = [-n, n] \cap E$ and $D_n = [-n, n] \backslash E$. Choose, using (6.64), open $U_n \supset D_n$ such that

$$\lambda(U_n) < \lambda(D_n) + 1/n.$$

Write $C_n = [-n, n] \backslash U_n$. Then C_n is compact and $C_n \subset E_n$. By (6.62.iii) we have

$$\lambda(C_n) = 2n - \lambda([-n, n] \cap U_n)$$
$$\geq 2n - \lambda(U_n) > 2n - \lambda(D_n) - 1/n = \lambda(E_n) - 1/n.$$

From this and (6.62.iv), we deduce

$$\lambda(E) = \lim_{n \to \infty} \lambda(E_n)$$

$$\leq \lim_{n \to \infty} \left[\lambda(C_n) + 1/n\right] \leq \varlimsup_{n \to \infty} \lambda(C_n) \leq \lambda(E).$$

Thus $\lim_{n \to \infty} \lambda(C_n) = \lambda(E)$. $\square$

We can use the preceding two theorems to give a simple proof of a theorem due to Steinhaus. First, a definition.

(6.66) **Definition** For $A, B \subset \mathbb{R}$ and $x \in \mathbb{R}$ define

$$x + A = A + x = \{x + a : a \in A\},$$

$$A + B = \{a + b : a \in A, b \in B\},$$

$$A - B = \{a - b : a \in A, b \in B\}.$$

(6.67) **Steinhaus's Theorem** *If $E \in \mathfrak{M}$ and $\lambda(E) > 0$, then there is a $\delta > 0$ such that $]-\delta, \delta[\subset E - E$.*

Proof By (6.65), we can choose a compact set $C \subset E$ such that $\lambda(C) > 0$. Next use (6.64) to obtain an open set $U \supset C$ such that

$$\lambda(U) < 2\lambda(C).$$

Write $\delta = \text{dist}(C, \mathbb{R} \setminus U) = \inf\{|x - y| : x \in C, y \in \mathbb{R} \setminus U\}$. Then $\delta > 0$ and $t + C \subset U$ whenever $|t| < \delta$. Then

$$(t + C) \cap C \neq \emptyset \quad \text{if } |t| < \delta, \tag{1}$$

otherwise $(t + C) \cup C \subset U$ and $2\lambda(C) = \lambda(t + C) + \lambda(C) = \lambda((t + C) \cup C) \leq \lambda(U) < 2\lambda(C)$ by (2.76.vii). It follows from (1) that if $|t| < \delta$, then there exist $x, y \in C$ such that $t + x = y$, and so $t = y - x \in C - C \subset E - E$. $\square$

The reader may very well ask whether or not there exist nonmeasurable subsets of $\mathbb{R}$. One can make the following somewhat vague statement. Any subset of $\mathbb{R}$ whose elements can be designated precisely by "elementary set theory" is Lebesgue measurable. It has been demonstrated in recent years that any proof of the existence of non-Lebesgue measurable subsets of $\mathbb{R}$ requires, at some stage, *uncountably* many arbitrary choices of elements of $\mathbb{R}$. Roughly speaking: if you have a nonmeasurable set, then you can't tell me which numbers are in it [although you might be able to tell me some of them]. The reader should be attentive to the arbitrary choices in the next proof.

(6.68) **Theorem** *If* $A \subset \mathbb{R}$ *and* $\lambda(A) > 0$, *then there exists* $E \subset A$ *such that* $E \notin \mathfrak{M}$.

Proof We may suppose that $A \in \mathfrak{M}$, otherwise take $E = A$. Since $0 < \lambda(A) = \lim_{n \to \infty} \lambda(A \cap [-n, n])$, we may suppose that $A \subset [-p, p]$ for some $p \in \mathbb{N}$ [otherwise we locate $E \subset A \cap [-p, p]$, where $\lambda(A \cap [-p, p]) > 0$].

Clearly A is infinite [$\lambda(A) > 0$], so A has a countably infinite subset D. Let H be the additive subgroup of $\mathbb{R}$ generated by D. That is, H consists of all numbers of the form

$$\sum_{k=1}^{m} n_k d_k$$

where $m \in \mathbb{N}$, $n_k \in \mathbb{Z}$, and $d_k \in D$. For $x, y \in A$, write $x \sim y$ if $x - y \in H$. Then $\sim$ is an equivalence relation on A; i.e., A is the union of a pairwise disjoint family $\{A_\gamma\}_{\gamma \in \Gamma}$ of nonvoid sets satisfying, for $x \in A_\gamma$, $A_\gamma = \{y \in A : x \sim y\}$. For each $\gamma \in \Gamma$, fix one $x_\gamma \in A_\gamma$ and then let $E = \{x_\gamma : \gamma \in \Gamma\}$. Then

$$x \in A \text{ implies } x - y \in H \quad \text{for some } y \in E, \tag{1}$$

$$y_1, y_2 \in E, y_1 \neq y_2 \quad \text{implies} \quad y_1 - y_2 \notin H. \tag{2}$$

Define $B = (A - A) \cap H$. Then $D - D \subset B \subset H$, and so, since $D - D$ is infinite and H is countable, it follows that B is countably infinite.

Suppose that $b_1, b_2 \in B$ and $(b_1 + E) \cap (b_2 + E) \neq \emptyset$. Then there exist $y_1, y_2 \in E$ such that $b_1 + y_1 = b_2 + y_2$; whence, $y_1 - y_2 = b_2 - b_1 \in H$. It follows from (2) that $y_1 = y_2$ and therefore $b_1 = b_2$. This argument shows that the countably infinite family $\{b + E : b \in B\}$ is pairwise disjoint. Accordingly, if we assume that $E \in \mathfrak{M}$,

$$\lambda(B + E) = \lambda\left(\bigcup_{b \in B} (b + E)\right) = \sum_{b \in B} \lambda(b + E) \tag{3}$$

By (2.76.vii), all summands on the right side of (3) are equal to $\lambda(E)$. Thus, if we assume that $E \in \mathfrak{M}$, either $\lambda(B + E) = \infty$ or $\lambda(B + E) = 0$. We shall complete the proof that $E \notin \mathfrak{M}$ by showing that neither of these alternatives is possible.

Since $B + E \subset [A - A] + E \subset [A - A] + A \subset [-3p, 3p]$, it is impossible that $\lambda(B + E) = \infty$.

To show that $\lambda(B + E) = 0$ cannot obtain, we show that $A \subset B + E$. In fact, if $x \in A$, then (1) shows that $x - y \in H$ for some $y \in E \subset A$. Then $x - y \in (A - A) \cap H = B$, and so $x \in B + E$. $\quad\square$

(6.69) Example To illustrate (6.68) and its proof more concretely, let $A = [0, 1]$, $p = 1$, and $D = \mathbb{Q} \cap [0, 1]$. Then $H = \mathbb{Q}$, $D - D = B = \mathbb{Q} \cap [-1, 1]$, and the A_γ's are the intersections with $[0, 1]$ of the distinct cosets of the subgroup $\mathbb{Q}$ of the additive group $\mathbb{R}$. The set E consists of precisely one element from each such A_γ.

Though λ is countably additive on $\mathfrak{M}$ (6.62.ii), the next theorem shows that it is not even finitely additive for arbitrary subsets of $\mathbb{R}$.

(6.70) Theorem *Let $A \in \mathfrak{M}$, $\lambda(A) < \infty$, and $E \subset A$. If*

$$\lambda(A) = \lambda(E) + \lambda(A \backslash E),$$

then $E \in \mathfrak{M}$.

Proof For $n \in \mathbb{N}$, use (6.64) to obtain an open $U_n \supset E$ such that $\lambda(U_n) \leq \lambda(E) + 1/n$. Write $G = A \cap \bigcap_{n=1}^{\infty} U_n$. Then $G \in \mathfrak{M}$, and since $E \subset G \subset U_n$ for all n, $\lambda(E) \leq \lambda(G) \leq \lambda(U_n) < \lambda(E) + 1/n$ for all n. Therefore, $\lambda(G) = \lambda(E)$. Similarly we can find $H \in \mathfrak{M}$ such that $A \backslash E \subset H \subset A$ and $\lambda(H) = \lambda(A \backslash E)$. Since $G, H \in \mathfrak{M}$ and $G \cup H = A$, we have

$$\lambda(A) = \lambda(E) + \lambda(A \backslash E) = \lambda(G) + \lambda(H)$$
$$= \lambda(G \backslash H) + \lambda(G \cap H) + \lambda(H \backslash G) + \lambda(G \cap H)$$
$$= \lambda(G \cup H) + \lambda(G \cap H)$$
$$= \lambda(A) + \lambda(G \cap H),$$

and so, since $\lambda(A) < \infty$, $\lambda(G \cap H) = 0$. But $G \backslash E \subset G \cap H$, so $\lambda(G \backslash E) = 0$ and hence $G \backslash E \in \mathfrak{M}$. We conclude that $E \in \mathfrak{M}$ because $E = G \backslash (G \backslash E)$. $\quad\square$

‚0.71)   Note   It follows from (6.70) that if A and E are as in (6.68) with $A \in \mathfrak{M}$ and $\lambda(A) < \infty$, then $\lambda(A) = \lambda(E \cup (A \setminus E)) < \lambda(E) + \lambda(A \setminus E)$.

Structure of Measurable Functions

(6.72)   Theorem   *Let f be an $\mathbb{R}^{\#}$-valued function defined on $\mathbb{R}$. Then the following five statements are equivalent.*

(i) *f is measurable.*
(ii) *$\{x \in \mathbb{R} : f(x) \leqq \alpha\} \in \mathfrak{M}$ for all $\alpha \in \mathbb{R}$.*
(iii) *$\{x \in \mathbb{R} : f(x) > \alpha\} \in \mathfrak{M}$ for all $\alpha \in \mathbb{R}$.*
(iv) *$\{x \in \mathbb{R} : f(x) \geqq \alpha\} \in \mathfrak{M}$ for all $\alpha \in \mathbb{R}$.*
(v) *$\{x \in \mathbb{R} : f(x) < \alpha\} \in \mathfrak{M}$ for all $\alpha \in \mathbb{R}$.*

Moreover, if (i) *obtains and $f \geqq 0$ on $\mathbb{R}$, then there exists a nondecreasing sequence $0 \leqq s_1 \leqq s_2 \leqq \ldots$ of functions of the form*

(vi)
$$s = \sum_{j=1}^{m} \alpha_j \xi_{E_j},$$

[where $m \in \mathbb{N}$, $\alpha_1, \ldots, \alpha_m$ are (nonnegative) real numbers, and $\{E_j\}_{j=1}^{m}$ is a pairwise disjoint family of measurable sets] such that

$$\lim_{n \to \infty} s_n(x) = f(x) \quad \text{for all } x \in \mathbb{R}.$$

Such functions s are called (nonnegative) *measurable simple functions*.

Proof   That (ii) implies (iii) and that (iv) implies (v) follow from the fact that $\mathfrak{M}$ is closed under complementation (6.59.ii).

If (i) holds, and $\alpha \in \mathbb{R}$, the functions

$$f_n = n\left[\left\{(f \wedge n) \vee \left(\alpha + \frac{1}{n}\right)\right\} - \{(f \wedge n) \vee \alpha\}\right]$$

($n = 1, 2, \ldots$) are all measurable (6.35), and, since

$$\lim_{n \to \infty} f_n(x) = \xi_A(x) \quad \text{for all } x \in \mathbb{R}$$

where A is the set in (ii), it follows from (6.39) that ξ_A is measurable, and so $A \in \mathfrak{M}$. Thus (i) implies (ii).

To see that (iii) implies (iv), write

$$\{x \in \mathbb{R} : f(x) \geqq \alpha\} = \bigcap_{n=1}^{\infty} \{x \in \mathbb{R} : f(x) > \alpha - 1/n\}$$

and use (6.59.iv). Similarly (v) implies (ii).

Thus far, we have seen that (i) implies (ii) and that the latter four are equivalent. Next, suppose that any, and hence all, of the latter four obtain. We must prove (i). For $n \in \mathbb{N}$ and $-n \cdot 2^n < k \leq n \cdot 2^n$, write

$$E_{n,k} = \{x \in \mathbb{R} : (k-1)/2^n \leq f(x) < k/2^n\},$$

$$A_n = \{x \in \mathbb{R} : f(x) < -n\},$$

$$B_n = \{x \in \mathbb{R} : f(x) \geq n\}.$$

Then our hypothesis shows that all of these sets are measurable [$E_{n,k}$ is the intersection of two measurable sets], and so the functions

$$s_n = -n\xi_{A_n} + n\xi_{B_n} + \sum_{k=-n \cdot 2^n + 1}^{n \cdot 2^n} \frac{k-1}{2^n} \xi_{E_{n,k}}$$

are all measurable. We have

$$0 \leq f(x) - s_n(x) < 1/2^n \quad \text{if } |f(x)| < n,$$

$$s_n(x) = -n \qquad\qquad \text{if } f(x) \leq -n,$$

$$s_n(x) = n \qquad\qquad\; \text{if } f(x) \geq n.$$

Therefore

$$f(x) = \lim_{n \to \infty} s_n(x) \quad \text{for all } x \in \mathbb{R},$$

and so (i) follows from (6.39).

        Finally, suppose that $f \geq 0$ and that (i) obtains. For $n \geq 1$ and $k \leq 0$, the sets A_n and $E_{n,k}$ are void, so the functions s_n are nonnegative and have the form (vi). To see that $s_n \leq s_{n+1}$, consider any fixed $x \in \mathbb{R}$. Note that $E_{n,k} = E_{n+1,2k-1} \cup E_{n+1,2k}$. If $x \in E_{n,k}$ where $1 \leq k \leq n \cdot 2^n$, then $s_n(x) = (k-1)2^{-n} \leq s_{n+1}(x)$ because $s_{n+1}(x)$ is either $(k-1)2^{-n}$ or $(2k-1)2^{-n-1}$. If x is in no such $E_{n,k}$, then $x \in B_n$ (so $s_n(x) = n$) and x is in either B_{n+1} or some $E_{n+1,j}$ with $j > 2n \cdot 2^n$ (so $s_{n+1}(x) \geq n$). In all cases $s_n(x) \leq s_{n+1}(x)$. $\square$

(6.73) Theorem *Let f be a complex-valued function defined on $\mathbb{R}$. Then f is measurable if and only if $f^{-1}(V) \in \mathfrak{M}$ for each open set $V \subset \mathbb{C}$.*

Proof Write $f = f_1 + if_2$, where f_1 and f_2 are real-valued. Suppose that $f^{-1}(V) \in \mathfrak{M}$ for all open $V \subset \mathbb{C}$. Given $\alpha \in \mathbb{R}$, write $U = \{z \in \mathbb{C} : \text{Re}(z) < \alpha\}$ and $V = \{z \in \mathbb{C} : \text{Im}(z) < \alpha\}$. Then U and V are open, and so $\{x \in \mathbb{R} : f_1(x) < \alpha\} = f^{-1}(U) \in \mathfrak{M}$, $\{x \in \mathbb{R} : f_2(x) < \alpha\} = f^{-1}(V) \in \mathfrak{M}$. Therefore f_1 and f_2 are measurable by (6.72), and so f is measurable.

        Next, suppose that f is measurable. Let $I =]a, b[$ be any open interval [rectangle] in $\mathbb{C}$: $a = a_1 + ia_2$, $b = b_1 + ib_2$, $a_1 \leq b_1$, $a_2 \leq b_2$. Since

f_1 and f_2 are measurable, it follows from (6.72) that

$$f^{-1}(I) = \left[\{ x \in \mathbb{R} : f_1(x) > a_1 \} \right.$$
$$\cap \{ x \in \mathbb{R} : f_1(x) < b_1 \}$$
$$\cap \{ x \in \mathbb{R} : f_2(x) > a_2 \}$$
$$\left. \cap \{ x \in \mathbb{R} : f_2(x) < b_2 \} \right] \in \mathfrak{M}.$$

If $V \subset \mathbb{C}$ is open, we can find a countable family $\{ I_n \}_{n=1}^{\infty}$ of open intervals of the form I $[a_j, b_j \in \mathbb{Q}]$ such that $V = \bigcup_{n=1}^{\infty} I_n$. Then (6.59.iii) yields

$$f^{-1}(V) = \bigcup_{n=1}^{\infty} f^{-1}(I_n) \in \mathfrak{M}. \quad \square$$

The following remarkable theorem shows that pointwise convergence a.e. [on a set of *finite* measure] is "almost uniform."

(6.74) Egorov's Theorem [1911] *Let $E \in \mathfrak{M}$ with $\lambda(E) < \infty$. Suppose that $(f_n)_{n=1}^{\infty} \subset M$ and that $f_n \to f$ a.e. on E, where f is a complex-valued function defined a.e. on E. Then for each $\delta > 0$ there exists a compact set C such that $C \subset E$, $\lambda(E \setminus C) < \delta$, and $f_n \to f$ uniformly on C.*

Proof Let A be the set of all $x \in E$ for which $f(x)$ and $f_n(x)$ $(n \in \mathbb{N})$ are all defined and $f_n(x) \to f(x)$. Then $\lambda(E \setminus A) = 0$, and so $A \in \mathfrak{M}$. Define $g_n = f_n \cdot \xi_A$ and $g = f$ on A, $g = 0$ off A. Then $(g_n)_{n=1}^{\infty} \subset M$ and $g_n(x) \to g(x)$ for all $x \in \mathbb{R}$; hence, $g \in M$. For $n, k \in \mathbb{N}$, write

$$B_{k,n} = \bigcup_{j=n}^{\infty} \{ x \in A : |g(x) - g_j(x)| \geq 1/k \}.$$

It follows from (6.72) and (6.59.iii) that $B_{k,n} \in \mathfrak{M}$. Since $B_{k,1} \supset B_{k,2} \supset \cdots$, $\bigcap_{n=1}^{\infty} B_{k,n} = \varnothing$, and $\lambda(A) < \infty$, (6.62.v) yields

$$\lim_{n \to \infty} \lambda(B_{k,n}) = 0$$

for each $k \in \mathbb{N}$. Given k, choose $n(k)$ such that

$$\lambda(B_{k,n(k)}) < \delta/3^k$$

and write

$$B = \bigcup_{k=1}^{\infty} B_{k,n(k)}.$$

Then

$$\lambda(B) \leq \sum_{k=1}^{\infty} \lambda(B_{k,n(k)}) < \sum_{k=1}^{\infty} \delta/3^k = \delta/2.$$

and

$$A \setminus B = \bigcap_{k=1}^{\infty} \bigcap_{j=n(k)}^{\infty} \{x \in A : |g(x) - g_j(x)| < 1/k\}.$$

Thus, given $k \in \mathbb{N}$, we have that $j \geq n(k)$ implies

$$|f(x) - f_j(x)| = |g(x) - g_j(x)| < 1/k$$

for all $x \in A \setminus B$; i.e., $f_j \to f$ uniformly on $A \setminus B$. Now use (6.65) to choose a compact set $C \subset A \setminus B$ such that

$$\lambda(C) > \lambda(A \setminus B) - \delta/2$$
$$= \lambda(A) - \lambda(B) - \delta/2$$
$$= \lambda(E) - \lambda(B) - \delta/2 > \lambda(E) - \delta.$$

Plainly $f_n \to f$ uniformly on C. $\square$

(6.75) Remarks (a) The requirement that $\lambda(E) < \infty$ in (6.74) is important. Indeed, if $E = \mathbb{R}$, $f_n = \xi_{[n,n+1]}$, and $f = 0$, then $f_n(x) \to f(x)$ for all $x \in \mathbb{R}$; but if $C \in \mathfrak{M}$ [not necessarily compact] and $\lambda(\mathbb{R} \setminus C) < 1$, then $C \cap [n, n+1] \neq \emptyset$ for all n, and so

$$\sup_{x \in C} |f(x) - f_n(x)| = 1$$

for all n; hence, $f_n \not\to f$ uniformly on C.

(b) One cannot replace δ by 0 in (6.74). In fact, take $E = [0, 1]$, $f_n(x) = x^n$ for $0 \leq x < 1$, $f_n(x) = 0$ otherwise, and $f(x) = 0$ for all x. Then $f_n(x) \to f(x)$ for all $x \in \mathbb{R}$; but if $C \subset E$ and $\lambda(E \setminus C) = 0$, then C is dense in E and so

$$\sup_{x \in C} |f(x) - f_n(x)| = 1$$

for all n.

We next show that measurable functions are "almost continuous" even though they need not be continuous anywhere [consider $\xi_{\mathbb{Q}}$].

(6.76) Luzin's Theorem *Let $f \in M$. Then for each $\delta > 0$ there exists a continuous complex-valued function g on $\mathbb{R}$ such that*

$$\lambda(\{x \in \mathbb{R} : f(x) \neq g(x)\}) < \delta.$$

Moreover, if $|f(x)| \leq \beta$ a.e., we can choose g such that $|g(x)| \leq \beta$ for all x.

Proof *Case* 1: f is real-valued. Choose a sequence $(\phi_k)_{k=1}^{\infty} \subset M_0$ such that $\phi_k \to f$ a.e. Let D denote the set of all $x \in \mathbb{R}$ such that $f(x)$ is not defined or $\phi_k(x) \not\to f(x)$ or some ϕ_k is discontinuous at x. Then $\lambda(D) = 0$. For each $n \in \mathbb{N}$, let

$$E_n = \{x \in \mathbb{R} : n - 1 < |x| < n\} \setminus D$$

and apply (6.74) to obtain a compact set $C_n \subset E_n$ such that $\lambda(E_n \setminus C_n) < \delta/2^n$ and $\phi_k \to f$ uniformly on C_n. Since each ϕ_k is continuous on C_n, it follows from (3.107) that f, when restricted to C_n, is continuous in the relative topology of C_n. Let $C = \bigcup_{n=1}^{\infty} C_n$. Then C is closed, $\lambda(\mathbb{R}\backslash C) = \sum_{n=1}^{\infty} \lambda(E_n \backslash C) < \delta$, and f, when restricted to C, is continuous. Define $g(x) = f(x)$ for $x \in C$ and then extend g to be continuous on $\mathbb{R}$ [for example, define g linearly on each component interval of $\mathbb{R}\backslash C$].

Case 2: f is complex-valued. Write $f = f_1 + if_2$, where f_1 and f_2 are real-valued. Apply Case 1 to f_j to obtain a continuous g_j such that

$$\lambda(\{x \in \mathbb{R} : f_j(x) \neq g_j(x)\}) < \delta/2$$

$(j = 1, 2)$ and then define $g = g_1 + ig_2$.

Finally, suppose that $|f(x)| \leq \beta$ a.e. Define $\phi : \mathbb{C} \to \{z \in \mathbb{C} : |z| \leq \beta\}$ by $\phi(z) = z$ if $|z| \leq \beta$, $\phi(z) = \beta z / |z|$ if $|z| > \beta$. Then ϕ is continuous and so the composite function $h = \phi \circ g$ is continuous. Also, $|h(x)| \leq \beta$ for all x and

$$\{x \in \mathbb{R} : h(x) \neq f(x)\} \subset \{x \in \mathbb{R} : g(x) \neq f(x)\} \cup \{x \in \mathbb{R} : |f(x)| > \beta\}.$$

$\square$

(6.77) Corollary *Let f be a complex-valued function defined a.e. on $\mathbb{R}$. Then $f \in M$ if and only if there exists a sequence $(g_n)_{n=1}^{\infty}$ of continuous functions on $\mathbb{R}$ such that $g_n \to f$ a.e.*

Proof If the latter condition obtains, then $f \in M$ by (6.73) and (6.54).
Conversely, suppose that $f \in M$. For each $n \in \mathbb{N}$, apply (6.76) to obtain a continuous g_n such that the set

$$E_n = \{x \in \mathbb{R} : f(x) \neq g_n(x)\}$$

satisfies $\lambda(E_n) < 2^{-n}$. Writing $F_k = \bigcup_{n=k}^{\infty} E_n$, we have $\lambda(F_k) < \sum_{n=k}^{\infty} 2^{-n} = 2^{-k+1}$. Then $F_1 \supset F_2 \supset \dots$, and so the set $F = \bigcap_{k=1}^{\infty} F_k$ satisfies

$$\lambda(F) = \lim_{k \to \infty} \lambda(F_k) = 0.$$

Now

$$\mathbb{R}\backslash F = \bigcup_{k=1}^{\infty} \bigcap_{n=k}^{\infty} \{x \in \mathbb{R} : f(x) = g_n(x)\},$$

and so

$$\lim_{n \to \infty} g_n(x) = f(x)$$

for all $x \subset \mathbb{R}\backslash F$. $\square$

Integration Over Measurable Sets

(6.78)   Definition   Let $E \subset \mathbb{R}$ and let f be a function defined a.e. on E. We interpret $f \cdot \xi_E$ to be that function that agrees with f on E and vanishes everywhere on $\mathbb{R} \backslash E$, whether or not f is defined on $\mathbb{R} \backslash E$. If $E \in \mathfrak{M}$, then we say that f is *integrable over* E and we write $f \in L_1(E)$ provided that $f \cdot \xi_E \in L_1(\mathbb{R})$. In this case, we define

(i)
$$\int_E f = \int f \cdot \xi_E$$

Similarly, we say that f is *measurable on* E $(E \in \mathfrak{M})$ if $f \cdot \xi_E$ is measurable. In this case, we use (i) to define its left side if its right side is defined. In particular,

$$\int_{\mathbb{R}} f = \int f.$$

For $a \leq b$ in $\mathbb{R}^\#$, we write

$$\int_a^b f = \int_{]a,b[} f$$

and we also define

$$\int_b^a f = -\int_a^b f.$$

As is customary, and sometimes helpful, we also write

$$\int_E f(x)\, dx, \quad \int_E f(t)\, dt, \quad \int_E f(\theta)\, d\theta, \quad \text{etc}.$$

for $\int_E f$.

(6.79)   Theorem   *Let $f \in L_1$. Then for each $\epsilon > 0$ there is a $\delta > 0$ such that*

$$\int_E |f| < \epsilon$$

whenever $E \in \mathfrak{M}$ and $\lambda(E) < \delta$.

Proof   For $n \in \mathbb{N}$, write

$$A_n = \{ x \in \mathbb{R} : |f(x)| \leq n \}.$$

Then $A_n \in \mathfrak{M}$, and so

$$g_n = |f| \xi_{A_n} \in M.$$

Plainly $g_1 \leq g_2 \leq \ldots$, $g_n \to |f|$ a.e., and $g_n \leq |f| \in L_1^r$ for all n. Therefore

$$\lim_{n \to \infty} \int g_n = \int |f|.$$

Given $\epsilon > 0$, choose n such that

$$\int |f| - \int g_n < \epsilon/2$$

and let $\delta = \epsilon/(2n)$. Then $E \in \mathfrak{M}$, $\lambda(E) < \delta$ implies

$$\int_E |f| = \int_E (|f| - g_n) + \int_E g_n$$

$$\leq \int (|f| - g_n) + \int n\xi_E < \epsilon/2 + n\delta = \epsilon. \quad \square$$

(6.80)   **Theorem**   *Let $f \in L_1$ or $f \in M'$ with $f \geq 0$. If $\{E_n\}_{n=1}^{\infty} \subset \mathfrak{M}$ is pairwise disjoint and $E = \bigcup_{n=1}^{\infty} E_n$, then*

$$\int_E f = \sum_{n=1}^{\infty} \int_{E_n} f.$$

Proof   Since

$$\sum_{n=1}^{k} f\xi_{E_n} \to f\xi_E \quad \text{a.e.}$$

as $k \to \infty$, this theorem follows from (6.55) and (6.46). $\square$

Exercises

1.   Recall that the function sgn is defined on $\mathbb{C}$ by sgn $0 = 0$ and sgn $z = z/|z|$ for $z \neq 0$. If $f \in M$, then sgn $\bar{f} \in M$ and $f \cdot$ sgn $\bar{f} = |f|$ where sgn $\bar{f}$ is the composite function defined where f is defined by (sgn $\bar{f})(x) = \text{sgn}(\overline{f(x)})$, the bar denoting complex conjugation.

2.   (a) If $f \in L_1([0, 1])$, $|f| \leq 1$ a.e., and $\int_0^1 f = 1$, then $f = 1$ a.e. on $[0, 1]$. [Hint: Check that $\int_0^1 (1 - \text{Re } f) = 0$ and use (6.24) to get Re $f = 1$ a.e.]
     (b) If $g \in L_1([0, 1])$, $\int_0^1 g = \alpha$, and $|g| \leq |\alpha|$ a.e., then $g = \alpha$ a.e. on $[0, 1]$.

3.   If $f \in L_1$ and $\int_E f = 0$ for every measurable set E, then $f = 0$ a.e.

4.   For $f \in M$, define the *essential supremum* of $|f|$ to be the number

$$\|f\|_{\infty} = \inf\{ \beta \in \mathbb{R}^{\#} : |f(x)| \leq \beta \text{ a.e.} \}$$

and write $f \in L_{\infty}$ if $\|f\|_{\infty} < \infty$. For $\phi \in L_1$, write $\|\phi\|_1 = \int |\phi|$. If $f, g \in L_{\infty}$, $h \in L_1$, and $\alpha \in \mathbb{C}$ are given, then
     (a) $\|f\|_{\infty} = 0$ if and only if $f = 0$ a.e.,
     (b) $\|\alpha f\|_{\infty} = |\alpha| \cdot \|f\|_{\infty}$,
     (c) $\|f + g\|_{\infty} \leq \|f\|_{\infty} + \|g\|_{\infty}$,
     (d) $fh \in L_1$ and $|\int fh| \leq \|fh\|_1 \leq \|f\|_{\infty} \cdot \|h\|_1$.
     (e) If $(f_n)_{n=1}^{\infty} \subset L_{\infty}$ and $\lim_{m,n \to \infty} \|f_m - f_n\|_{\infty} = 0$, then there exists $f \in L_{\infty}$ such that $\lim_{n \to \infty} \|f - f_n\|_{\infty} = 0$. [Hint: Let E be the set of $x \in \mathbb{R}$ such that $|f_m(x) - f_n(x)| > \|f_m - f_n\|_{\infty}$ for some m and n. Show that $\lambda(E) = 0$ and that $(f_n)_{n=1}^{\infty}$ is uniformly convergent on $\mathbb{R}\backslash E$.]
     (f) If $f \in L_{\infty}$, then $\|f\|_{\infty} = \sup\{|\int fh| : h \in L_1, \|h\|_1 \leq 1\}$. [Hint: Given $\beta < \|f\|_{\infty}$, choose a measurable set A such that $0 < \lambda(A) < \infty$ and $|f(x)| > \beta$ for all $x \in A$. Then put $h = (\xi_A \cdot \text{sgn } \bar{f})/\lambda(A)$ and obtain $\int fh \geq \beta$.]

(g) If $h \in L_1$, then $\|h\|_1 = \sup\{|\int fh| : f \in L_\infty, \|f\|_\infty \leq 1\}$. [Hint: Put $f = \operatorname{sgn} \bar{h}$.]

(h) If $f \in M$ and $fh \in L_1$ for every $h \in L_1$, then $f \in L_\infty$. [Hints: Assuming $f \notin L_\infty$, inductively define positive numbers $\beta_1 < \beta_2 < \dots$ such that $\sum_{n=1}^{\infty} 1/\beta_n < \infty$ and $\lambda(\{x \in \mathbb{R} : \beta_n < |f(x)| \leq \beta_{n+1}\}) > 0$. Then choose measurable sets B_n such that $0 < \lambda(B_n) < \infty$ and $\beta_n < |f(x)| \leq \beta_{n+1}$ for $x \in B_n$. Consider

$$h = \sum_{n=1}^{\infty} \frac{1}{\beta_n \lambda(B_n)} \xi_{B_n}.]$$

(i) If $h \in M$ and $fh \in L_1$ for every $f \in L_\infty$, then $h \in L_1$.

5.   (a) If $E \subset \mathbb{R}$ is a measurable set and $0 < \alpha < \beta < \lambda(E)$, then there exists a compact set A which is nowhere dense in $\mathbb{R}$ such that $A \subset E$ and $\alpha < \lambda(A) < \beta$. [Hints: Choose a compact $C \subset E \backslash \mathbb{Q}$ such that $\lambda(C) > \beta$. Cover C by open sets $V_1, \dots, V_n$, each having measure $< \beta - \alpha$. Then let $A = C\backslash(V_1 \cup \dots \cup V_m)$ where m is the smallest integer for which this set has measure $< \beta$.]

   (b) If $E \subset \mathbb{R}$ is a measurable set and $0 < \alpha < \lambda(E)$, then there exists a compact set A that is nowhere dense in $\mathbb{R}$ such that $A \subset E$ and $\lambda(A) = \alpha$. [Hint: Use (a) to choose $A_1 \supset A_2 \supset \dots$ such that $\alpha < \lambda(A_n) < \alpha + 1/n$ and let A be their intersection.]

6.   There exists a measurable set $A \subset \mathbb{R}$ such that $0 < \lambda(A \cap I) < \lambda(I)$ for each nonvoid open interval $I \subset \mathbb{R}$. [Hints: Let $\{I_n\}_{n=1}^{\infty}$ be an enumeration of the family of nonvoid open intervals having rational endpoints. Write $A_0 = \emptyset$ and $r_0 = 1$. After compact nowhere dense sets $A_0, \dots, A_{n-1}$ and positive numbers $r_0, \dots, r_{n-1}$ have been chosen, let $r_n = \min\{r_0, \dots, r_{n-1}, \lambda(I_n\backslash(A_0 \cup \dots \cup A_{n-1}))\}$ and use Exercise 5 to choose a compact nowhere dense set $A_n \subset I_n\backslash(A_0 \cup \dots \cup A_{n-1})$ with $0 < \lambda(A_n) < r_n/2^n$. Let $A = \bigcup_{j=1}^{\infty} A_j$ and note that

$$0 < \lambda(A_n) \leq \lambda(I_n \cap A) < \lambda(I_n) - r_n + \sum_{j=n}^{\infty} \lambda(A_j) < \lambda(I_n).]$$

7.   If $E \subset \mathbb{R}$, $0 < \alpha < 1$, and $\lambda(E \cap I) \leq \alpha\lambda(I)$ for every interval I, then $\lambda(E) = 0$. [Hint: Show that $\lambda(E) \leq \alpha\lambda(V)$ for every open set V with $E \subset V$.]

8.   Suppose that $E \subset \mathbb{R}$, $a \in \mathbb{R}$, $\delta > 0$, and, for every $t \in \mathbb{R}$ with $|t| < \delta$, either $a + t$ or $a - t$ is in E. Then $\lambda(E) \geq \delta$. [Hint: $]-\delta, \delta[\subset (E - a) \cup (a - E)$.]

9.   Suppose that $f : \mathbb{R} \to \mathbb{R}$ satisfies $f(x + y) = f(x) + f(y)$ for all $x, y \in \mathbb{R}$.

   (a) If $x \in \mathbb{R}$ and $r \in \mathbb{Q}$, then $f(rx) = rf(x)$. [Hint: Consider the cases $r = 0, r = -1$, $r = n$, and $r = 1/n$ where $n \in \mathbb{N}$.]

   (b) If f is continuous on $\mathbb{R}$ and $a = f(1)$, then $f(x) = ax$ for all $x \in \mathbb{R}$.

   (c) If f is continuous at 0, then f is continuous on $\mathbb{R}$.

   (d) If f is bounded on some neighborhood of 0, then f is continuous at 0.

   (e) If f is measurable, then $f(x) = ax$ for all $x \in \mathbb{R}$. [Hint: Show that the set $E_n = f^{-1}([-n, n])$ has positive measure for some $n \in \mathbb{N}$ and then use (6.67) to obtain $\delta > 0$ such that $|f(t)| \leq 2n$ if $-\delta < t < \delta$.]

10.   *Midpoint Convex Functions*. Let $I \subset \mathbb{R}$ be an open interval and let $f : I \to \mathbb{R}$ satisfy

$$f((x + y)/2) \leq (1/2)(f(x) + f(y))$$

for all $x, y \in I$.

(a) If $n \in \mathbb{N}$, $\{x_1, x_2, \ldots, x_n\} \subset I$, and $r_1, r_2, \ldots, r_n$ are positive rational numbers with $\sum_{j=1}^{n} r_j = 1$, then

$$f\left(\sum_{j=1}^{n} r_j x_j\right) \leq \sum_{j=1}^{n} r_j f(x_j).$$

[Hints: First, prove this by induction on k in the case that $n = 2^k$ and $r_j = 2^{-k}$ for all j. Next, if n is arbitrary and $r_j = 1/n$ for all j, put $a = \dfrac{1}{n} \sum_{j=1}^{n} x_j$, fix any k with $2^k > n$, and apply the preceding case to the 2^k numbers $y_1, \ldots, y_{2^k}$ where $y_j = x_j$ for $1 \leq j \leq n$ and $y_j = a$ for $n < j \leq 2^k$. In the general case, choose $q \in \mathbb{N}$ such that $q r_j = p_j \in \mathbb{N}$ for all j and then apply the second case to q numbers, p_j of which equal x_j for each j.]

(b) If f is continuous on I, then $f(\alpha x + \beta y) \leq \alpha f(x) + \beta f(y)$ whenever $x, y \in I$, $0 \leq \alpha \leq 1$, and $\beta = 1 - \alpha$. That is, f is convex on I.

(c) If $x \in I$ and $\overline{\lim}_{y \to x} f(y) < \infty$, then f is continuous at x. [Hints: Write $\bar{f}(x) = \overline{\lim}_{y \to x} f(y)$ and $\underline{f}(x) = \underline{\lim}_{y \to x} f(y)$. From $2f(x + t) \leq f(x + 2t) + f(x)$ obtain $\bar{f}(x) \leq f(x)$. From $2f(x) - f(x + t) \leq f(x - t)$ obtain $2f(x) - \bar{f}(x) \leq \underline{f}(x)$.]

(d) If f is measurable, then f is continuous on I and so f is convex. [Hints: Assume that f is discontinuous at some fixed $x \in I$. Choose $\delta > 0$ such that $]x - 2\delta$, $x + 2\delta[\subset I$ and apply (c) to obtain $x_n \to x$ with $|x - x_n| < \delta$ and $f(x_n) > n$ for all n. Write $A_n = \{y : |y - x| < 2\delta, f(y) > n\}$ and $E_n = \{y : |y - x_n| < \delta, f(y) > n\}$. Note that $E_n \subset A_n$ and $A_{n+1} \subset A_n$ for all n. Use Exercise 8 to show that $\lambda(E_n) \geq \delta$ for all n. Conclude that the set $A = \bigcap_{n=1}^{\infty} A_n$ is nonvoid to get a contradiction.]

11. *Sigma Algebras.* Let X be a set. A family $\mathcal{F}$ of subsets of X is called a *σ-algebra* (or *σ-field*) of subsets of X if

    (1) $X \in \mathcal{F}$,

    (2) $X \setminus A \in \mathcal{F}$ whenever $A \in \mathcal{F}$, and

    (3) $\bigcup_{n=1}^{\infty} A_n \in \mathcal{F}$ whenever $(A_n)_{n=1}^{\infty} \subset \mathcal{F}$.

(a) If $\mathcal{F}$ is as above, then (4) $\varnothing \in \mathcal{F}$, (5) $\bigcap_{n=1}^{\infty} B_n \in \mathcal{F}$ whenever $(B_n)_{n=1}^{\infty} \subset \mathcal{F}$, and

    (6) $A \cup B, A \cap B$, and $A \setminus B$ are all in $\mathcal{F}$ whenever $A, B \in \mathcal{F}$.

(b) If $\mathcal{E}$ is any family of subsets of X, then there exists a unique σ-algebra $\mathcal{S}(\mathcal{E})$ of subsets of X such that $\mathcal{E} \subset \mathcal{S}(\mathcal{E}) \subset \mathcal{F}$ for every $\mathcal{F}$ as above for which $\mathcal{E} \subset \mathcal{F}$. [Hint: The intersection of any collection of σ-algebras is a σ-algebra.]

(c) If a σ-algebra $\mathcal{F}$ has infinitely many members, then $\mathcal{F}$ is uncountable. [Hints: Produce a countably infinite, pairwise disjoint family $\mathcal{D} \subset \mathcal{F}$ and note that $\bigcup \mathcal{C} \in \mathcal{F}$ for all $\mathcal{C} \subset \mathcal{D}$. If there exists a sequence $(E_n)_{n=1}^{\infty} \subset \mathcal{F}$ with $E_{n+1} \subsetneqq E_n$ for all n, take $D_n = E_n \setminus E_{n+1}$. Otherwise, $\mathcal{F}$ contains infinitely many nonvoid sets D such that if $F \in \mathcal{F}$ and $F \subsetneqq D$, then $F = \varnothing$.]

12. *Borel Sets.* If X is a topological space and $\mathcal{E}$ is the family of all open subsets of X, then the σ-algebra $\mathcal{S}(\mathcal{E})$, as in Exercise 11(b), is denoted by $\mathcal{B}(X)$ and its members are called the *Borel sets* of X.

(a) Every Borel set of $\mathbb{R}$ is Lebesgue measurable: $\mathcal{B}(\mathbb{R}) \subset \mathfrak{M}$.

(b) If $f: \mathbb{R} \to \mathbb{C}$ is measurable, then $f^{-1}(B) \in \mathfrak{M}$ for every $B \in \mathcal{B}(\mathbb{C})$. [Hint: The family $\mathcal{F}$ of all $B \subset \mathbb{C}$ for which $f^{-1}(B) \in \mathfrak{M}$ is a σ-algebra containing every open set.]

(c) There exists a continuous one-to-one function $f: \mathbb{R} \to \mathbb{R}$ and a set $M \in \mathfrak{M}$ such that $f^{-1}(M)$ is not in $\mathfrak{M}$. [Hints: Let P be any Cantor set of measure zero as in (2.81)–(2.84) and let ψ be the associated continuous nondecreasing function from $\mathbb{R}$ onto $[0, 1]$ for which $\psi(P) = [0, 1]$ as in (3.91). Write $g(x) = (\psi(x) + x)/2$ for $x \in \mathbb{R}$ and check that g is strictly increasing and continuous from $\mathbb{R}$ onto $\mathbb{R}$. Note that

$$g\left(\sum_{j=1}^{\infty} \epsilon_j r_j \right) = \sum_{j=1}^{\infty} \epsilon_j r_j'$$

where $r_j' = (2^{-j} + r_j)/2$ so that $g(P)$ is the Cantor set P' constructed from the numbers

$$a_n' = \sum_{j=n+1}^{\infty} r_j' = \frac{1}{2^{n+1}} + \frac{a_n}{2}.$$

Then $\lambda(P') = \lim_{n \to \infty} 2^n a_n' = 1/2$, so we can choose a nonmeasurable $S \subset P'$ and let $M = g^{-1}(S)$. Take $f = g^{-1}$.]

(d) There exist measurable sets that are not Borel sets: $\mathfrak{M} \setminus \mathcal{B}(\mathbb{R}) \neq \varnothing$.

(e) If $E \in \mathfrak{M}$, then there exists $B \in \mathcal{B}(\mathbb{R})$ such that $B \subset E$ and $\lambda(E \setminus B) = 0$.

13. *Baire Class* 1. Let X be a topological space. A function $f: X \to \mathbb{C}$ is said to be of *Baire class* 1 on X if there exists some sequence of complex-valued functions which are continuous on X that converges to f at every point of X.

(a) If f is as above, then the set D of all points of X at which f is discontinuous is of first Baire category in X and $f^{-1}(V)$ is an F_σ-set in X for every open set $V \subset \mathbb{C}$. [Hint: See Osgood's Theorem given in Exercise 7 immediately following (3.77).]

(b) If f is as above and there is a $\beta \in \mathbb{R}$ such that $|f(x)| \leq \beta$ for all $x \in X$, then the sequence $(f_n)_{n=1}^{\infty}$ of continuous functions that converges to f at every $x \in X$ can be chosen so that $|f_n(x)| \leq \beta$ for all n and x. [Hint: See the last part of the proof of (6.76).]

(c) There exists a bounded measurable function g on $\mathbb{R}$ for which *no* function f of Baire class 1 on $\mathbb{R}$ satisfies $f = g$ a.e. on $\mathbb{R}$. [Hints: Let $g = \xi_A$, where A is as in Exercise 6, and assume that some f as above exists. Obtain a contradiction to the Baire Category Theorem by using (a) to show that the disjoint sets $f^{-1}(0)$ and $f^{-1}(1)$ are both dense G_δ-sets.]

(d) If g is as in (c), then there exists a uniformly bounded sequence of continuous functions that converges to g a.e. on $\mathbb{R}$, but no such sequence can converge everywhere on $\mathbb{R}$.

(e) If g is any bounded measurable function on $\mathbb{R}$, then there is a function f on $\mathbb{R}$ which is the pointwise (everywhere) limit of a sequence of functions of Baire class 1 on $\mathbb{R}$ such that $f = g$ a.e. Such functions f are said to be of *Baire class* 2. [Hint: Suppose that g is real-valued. Choose a uniformly bounded sequence (g_n) of real-valued continuous functions that converges to g a.e. Define $f_n = \lim_{k \to \infty} (g_n \wedge g_{n+1} \wedge \cdots \wedge g_{n+k}).$]

(f) If $I \subset \mathbb{R}$ is an open interval and $f: I \to \mathbb{R}$ is differentiable on I, then f' is of Baire class 1 on I.

14.   If $f \in L_1$, and $\epsilon > 0$, then there exists a continuous function $g : \mathbb{R} \to \mathbb{C}$ which vanishes identically outside some bounded interval and satisfies $\int |f - g| < \epsilon$. [Hint: For $f \in L_1^r$, use (6.18) to get $\phi \in M_0$ with $\int |f - \phi| < \epsilon/2$ and then let $g = \phi$ except on small intervals containing discontinuity points of ϕ.]

15.   Let g and h be in M^r. Then the sets $\{x \in \mathbb{R} : g(x) < h(x)\}$, $\{x \in \mathbb{R} : g(x) \geq h(x)\}$, $\{x \in \mathbb{R} : g(x) > h(x)\}$, $\{x \in \mathbb{R} : g(x) \leq h(x)\}$, and $\{x \in \mathbb{R} : g(x) = h(x)\}$ are all measurable. [Hint: The first set is the union of the sets $\{x \in \mathbb{R} : g(x) < a\} \cap \{x \in \mathbb{R} : b < h(x)\}$ over all $(a, b) \in \mathbb{Q} \times \mathbb{Q}$ with $a < b$.]

16.   (a) If $(f_n)_{n=1}^{\infty} \subset M^r$ and A is the set of $x \in \mathbb{R}$ at which $f(x) = \lim_{n \to \infty} f_n(x)$ exists in $\mathbb{R}^{\#}$, then A is measurable. Also, if we define $f(x) = 0$ for $x \in \mathbb{R} \backslash A$, then f is in M^r. [Hint: Use Exercise 15 noting that A is the set on which $\underline{\lim} f_n = \overline{\lim} f_n$.]
        (b) Part (a) remains true if we replace M^r by M and $\mathbb{R}^{\#}$ by $\mathbb{C}$.

17.   *Simple Functions.* A *simple function* is a function that takes on only a finite number of values.
        (a) A measurable function $s : \mathbb{R} \to \mathbb{C}$ is a simple function if and only if

$$ s = \sum_{j=1}^{m} \alpha_j \xi_{E_j} $$

        for some $m \in \mathbb{N}$, $(\alpha_j)_{j=1}^{m} \subset \mathbb{C}$, and $(E_j)_{j=1}^{m} \subset \mathfrak{M}$.
        (b) If $f : \mathbb{R} \to [0, \infty]$ is measurable, then there exist measurable simple functions $s_n : \mathbb{R} \to [0, \infty[$ such that $s_1 \leq s_2 \leq \ldots$ and $\lim_{n \to \infty} s_n(x) = f(x)$ for all $x \in \mathbb{R}$. Moreover, if f is bounded on $\mathbb{R}$, we can make this convergence uniform on $\mathbb{R}$. [See (6.72).]
        (c) If $f : \mathbb{R} \to \mathbb{C}$ is measurable, then there exist measurable simple functions $s_n : \mathbb{R} \to \mathbb{C}$ such that $|s_1| \leq |s_2| \leq \ldots$ and $\lim_{n \to \infty} s_n(x) = f(x)$ for all $x \in \mathbb{R}$. In case f is bounded on $\mathbb{R}$, this convergence can be made uniform on $\mathbb{R}$.
        (d) If f is as in (b), then there exist positive real numbers $(c_k)_{k=1}^{\infty}$ and measurable sets $(A_k)_{k=1}^{\infty}$ such that

$$ f(x) = \sum_{k=1}^{\infty} c_k \xi_{A_k}(x) $$

        for all $x \in \mathbb{R}$. [Hint: $f = \sum_{n=1}^{\infty} (s_n - s_{n-1})$ where $s_0 = 0$. Each $s_n - s_{n-1}$ is a nonnegative simple function. Rearrange a double series.]

18.   *Vitali–Carathéodory Theorem.* If $f \in L_1^r$ and $\epsilon > 0$, then there exist functions ϕ and ψ defined everywhere on $\mathbb{R}$ which satisfy
                (i) $\phi, \psi \in L_1^r$
                (ii) $\phi(x) \leq f(x) \leq \psi(x)$ at every x where f is defined,
                (iii) $\int (\psi - \phi) < \epsilon$,
                (iv) ϕ is upper semicontinuous and bounded above,
                (v) ψ is lower semicontinuous and bounded below.
        (Recall our discussion of semicontinuity in Exercise 4 just after (3.93).) [Hints: First suppose $f \geq 0$ is defined everywhere on $\mathbb{R}$. Write f as in Exercise 17(d). Choose compact F_k and open V_k with $F_k \subset A_k \subset V_k$ and $\lambda(V_k \backslash F_k) < 4^{-k} c_k^{-1} \epsilon$. Choose k_0

so that $\sum_{k=k_0}^{\infty} c_k \lambda(F_k) < \epsilon/2$. Write

$$\phi = \sum_{k=1}^{k_0} c_k \xi_{F_k}, \qquad \psi = \sum_{k=1}^{\infty} c_k \xi_{V_k}.$$

In the general case, write $f = f^+ - f^-$.]

19.   *H. Bohr's Proof of Tietze's Theorem.* Let X be a metric space with metric ρ, let A be a closed subset of X, and let $f : A \to [0, 1]$ be continuous on A. Define d on X by $d(x) = \text{dist}(x, A) = \inf\{\rho(x, a) : a \in A\}$. For $x \in X$ and $r > 0$, write $B_r(x) = \{y \in X : \rho(x, y) < r\}$ and $\psi(x, r) = \sup f(A \cap B_r(x))$ where $\sup \emptyset = 0$. Define g on $X \backslash A$ by

$$g(x) = \frac{1}{d(x)} \int_{d(x)}^{2d(x)} \psi(x, r)\, dr.$$

    (a) For fixed $x \in X \backslash A$, we have $d(x) > 0$ and the function $r \to \psi(x, r)$ is nondecreasing, so $g(x)$ is well defined and $0 \leq g(x) \leq 1$.

    (b) The function g is continuous at each $x_0 \in X \backslash A$. [Hints: For $0 < \delta < d(x_0) \leq r$ and $\rho(x, x_0) < \delta$ we have $d(x_0) - \delta < d(x) < d(x_0) + \delta$ and $\psi(x_0, r - \delta) \leq \psi(x, r) \leq \psi(x_0, r + \delta)$, so

$$\frac{d(x_0)}{d(x_0) + \delta}\left(g(x_0) - \frac{3\delta}{d(x_0)} \right) \leq g(x) \leq \frac{d(x_0)}{d(x_0) - \delta}\left(g(x_0) + \frac{3\delta}{d(x_0)} \right).]$$

    (c) If $x_0 \in A$, $x \in X \backslash A$, and $\rho(x, x_0) < \delta$, then $\inf f(A \cap B_{3\delta}(x_0)) \leq g(x) \leq \psi(x_0, 3\delta)$. [Hints: Since $B_{2d(x)}(x) \subset B_{3\delta}(x_0)$, we have $\psi(x, 2d(x)) \leq \psi(x_0, 3\delta)$. Also, $d(x) < r < 2d(x)$ implies $B_r(x) \cap A \cap B_{3\delta}(x_0) \neq \emptyset$ so $\psi(x, r) \geq \inf f(A \cap B_{3\delta}(x_0))$.]

    (d) The function $F : X \to [0, 1]$ defined by $F(x) = f(x)$ if $x \in A$ and $F(x) = g(x)$ if $x \in X \backslash A$ is continuous on X.

20.   (a) Let $g : [a, b] \to \mathbb{R}$ be continuous on $[a, b]$ and let D be a countable subset of $[a, b]$. Suppose that for each $x \in [a, b[\backslash D$ there exists some $\delta = \delta(x)$ with $0 < \delta \leq b - x$ such that $g(x) < g(t)$ whenever $x < t < x + \delta$. Then g is strictly increasing on $[a, b]$. [Hints: Assume $g(c) > g(d)$ for some $c < d$ in $[a, b]$. For each y with $g(c) > y > g(d)$, define $x_y = \sup\{x : c < x < d, g(x) = y\}$. Check that $g(x_y) = y$ so the mapping $y \to x_y$ is one-to-one, and then obtain a contradiction by showing that $x_y \in D$. This proves that g is nondecreasing.]

    (b) Assertion (a) is false if the continuity hypothesis is omitted.

    (c) Assertion (a) is false if the countability hypothesis is replaced by $\lambda(D) = 0$. [Hint: Consider Lebesgue's singular function ψ as in (3.92) and let $g(x) = x - \psi(x)$.]

21.   (a) *Fundamental Theorem of Calculus.* Let $F : [a, b] \to \mathbb{R}$ be continuous on $[a, b]$ and have a finite right derivative $F'_+(x)$ at every $x \in [a, b[\backslash D$, where D is some countable set. Suppose also that $F'_+ \in L_1([a, b])$. Then $F(b) - F(a) = \int_a^b F'_+(x)\, dx$. [Hints: Write $f = F'_+$ on $[a, b[\backslash D$ and $f = 0$ elsewhere on $\mathbb{R}$. Given $\epsilon > 0$, use Exercise 18 to obtain a lower semicontinuous $\psi > f$ on $\mathbb{R}$ (add ce^{-x^2} to ψ if necessary to obtain *strict* inequality) such that $\int \psi < \int f + \epsilon$ and then define g by

$$g(x) = \int_a^x \psi(u)\, du - [F(x) - F(a) - (x - a)\epsilon].$$

For $x \in [a, b[\backslash D$, select $\delta(x) > 0$ so that

$$\psi(t) > f(x) \quad \text{and} \quad F(t) - F(x) < (t - x)(f(x) + \epsilon)$$

whenever $x < t < x + \delta(x)$. Check the hypotheses of Exercise 20 for g and obtain $g(b) > g(a) = 0$. Deduce that $\int_a^b f + \epsilon + (b - a)\epsilon > F(b) - F(a)$. Conclude that $\int_a^b f \geq F(b) - F(a)$. Finally, replace F by $- F$.]
(b) Theorem (a) is false if the continuity hypothesis is omitted.
(c) Theorem (a) is false if the countability of D is replaced by $\lambda(D) = 0$.

22.   [Vito Volterra, 1881] Let P be any Cantor set constructed as in (2.81) for which $\lambda(P) > 0$. Let $\phi(t) = t^2 \sin(1/t)$ for $t \neq 0$. We define a function F on $[0, 1]$ as follows. For $x \in P$, put $F(x) = 0$. If $]a, b[$ is a component interval of $[0, 1]\backslash P$, let $c = \sup\{t : 0 < t \leq (b - a)/2, \ \phi'(t) = 0\}$ and define $F(a + t) = F(b - t) = \phi(t)$ if $0 < t \leq c$ and $F(x) = \phi(c)$ if $a + c \leq x \leq b - c$. Then F has the following properties.
(a) F is differentiable on $[0, 1]$.
(b) $F'(x) = 0$ for every $x \in P$.
(c) $|F'(x)| < 3$ for every $x \in [0, 1]$.
(d) F' is discontinuous at every point of P.
(e) F' is *not* Riemann integrable over any interval $[0, x]$ for $0 < x \leq 1$. However (see Exercise 21),
(f) $F(x) = \int_0^x F'$ for all $x \in [0, 1]$.
     We remark that Lebesgue said in his thesis (1902) that it was this example of Volterra's that motivated him to devise a method of integration by which functions with bounded derivatives can be reconstructed from their derivatives.

23.   Let P be the Cantor set constructed from a given sequence $(a_n)_{n=0}^\infty$ as in (2.81).
(a) If $3a_{n+1} \geq a_n$ for all n, then $P - P = [-1, 1]$, where $P - P$ is as in (6.66). [Hint: Prove by induction that $P_n - P_n = [-1, 1]$ for all n. Notice that $P_n - P_n$ is the union of the 3^n intervals $[x - a_n, x + a_n]$ for x of the form $\sum_{j=1}^n \delta_j r_j$, where $r_j = a_{j-1} - a_j$ as in (2.83) and each δ_j is -1 or 0 or 1.]
(b) If $\lim_{n\to\infty} 3^n a_n = 0$, then $\lambda(P - P) = 0$, so $P - P$ contains no nonvoid open interval.
(c) If $0 < \xi < 1/2$ and $a_n = \xi^n$ for all n, then $\lambda(P - P) = 0$ for $\xi < 1/3$ and $P - P = [-1, 1]$ for $\xi \geq 1/3$.
(d) Let G be the additive subgroup of $\mathbb{R}$ generated by P; that is, G is the set of all sums of finitely many elements of $P - P$. If $\lim_{n\to\infty} k^n a_n = 0$ for every $k > 0$, then $\lambda(G) = 0$. [Hint: The set G_N of all sums of N elements of $P - P$ is, for each n, a subset of the union of $(2N + 1)^n$ intervals of length $2Na_n$.]
(e) If $\lim_{n\to\infty} a_n^{1/n} = 0$ or $\lim_{n\to\infty} (a_{n+1}/a_n) = 0$, then $\lambda(G) = 0$.

24.   (a) *Fejér's Lemma.* Let g be a bounded measurable function defined on $\mathbb{R}$ that is p-periodic for some positive real number p: $g(x + p) = g(x)$ for all $x \in \mathbb{R}$. Then, for every $f \in L_1$, we have

(i) $$\lim_{t\to\infty} \int_{\mathbb{R}} f(x)g(tx)\, dx = \left(\int_{\mathbb{R}} f\right)\left(\frac{1}{p}\int_0^p g\right)$$

[Hints: First, consider $f = \xi_{[a,b]}$, write the integral on the left as $I(t)$, and obtain (for fixed t with $t(b - a) > p$)

$$I(t) = \frac{1}{t} \int_{at}^{bt} g = \sum_{k=1}^{n} \frac{1}{t} \int_{x_{k-1}}^{x_k} g + \frac{1}{t} \int_{x_n}^{bt} g$$

where $x_k = at + kp$ and $n = n(t)$ satifies $0 \leq (b - a)t - np < p$. The last integral is a bounded function of t, each integral in the sum equals $\int_0^p g$, and $n/t \to (b - a)/p$ as $t \to \infty$. For $f \in L_1^r$ and $\epsilon > 0$, choose $\phi \in M_0$ so that $\int |f - \phi| < \epsilon$, check that (i) holds with ϕ in place of f, and note that

$$\left| \int f(x)g(tx)\,dx - \int \phi(x)g(tx)\,dx \right| < \beta\epsilon$$

and

$$\left| \left(\int \phi \right)\left(\frac{1}{p} \int_0^p g \right) - \left(\int f \right)\left(\frac{1}{p} \int_0^p g \right) \right| < \beta\epsilon$$

where β is a bound for $|g|$.]
(b) *Riemann–Lebesgue Lemma.* If $f \in L_1$, then

$$\lim_{t \to \infty} \int_{-\infty}^{\infty} f(x)\cos(tx)\,dx = 0$$

and

$$\lim_{t \to \infty} \int_{-\infty}^{\infty} f(x)\sin(tx)\,dx = 0.$$

(c) If $f \in L_1$ and $\theta_t \in \mathbb{R}$ for $t \in \mathbb{R}$, then

$$\lim_{t \to \infty} \int_{-\infty}^{\infty} f(x)\cos^2(tx + \theta_t)\,dx = \frac{1}{2} \int_{-\infty}^{\infty} f(x)\,dx.$$

[Hint: $2\cos^2 u = 1 + \cos(2u)$.]
(d) *Cantor–Lebesgue Theorem.* If $a_n, b_n \in \mathbb{R}$ for $n \in \mathbb{N}$ and if $\lim_{n \to \infty} [a_n \cos(nx) + b_n \sin(nx)] = 0$ for all $x \in E$, where E is measurable and $\lambda(E) > 0$, then $\lim_{n \to \infty} a_n = \lim_{n \to \infty} b_n = 0$. [Hint: Write

$$a_n \cos(nx) + b_n \sin(nx) = r_n \cos(nx + \theta_n), \quad \text{where } r_n = \left(a_n^2 + b_n^2 \right)^{1/2}.$$

Assuming $r_{n_k} \geq \epsilon > 0$ for all k, obtain a contradiction to (c) by taking $f = \epsilon\xi_C$ for a compact $C \subset E$.]

25. (a) Let g be as in Exercise 24 and write $a = (1/p)\int_0^p g$. Suppose there is some sequence $(t_n)_{n=1}^{\infty}$ of real numbers with $t_n \to \infty$ such that the corresponding set $A = \{x \in [0, p] : \lim_{n \to \infty} g(t_n x) \text{ exists}\}$ satisfies $\lambda(A) > 0$. Then $g = a$ a.e. on $\mathbb{R}$.
[Hints: First check that the set A is measurable because each function $x \to g(t_n x)$ is measurable. Write $h(x) = \lim_{n \to \infty} g(t_n x)$ for $x \in A$. Use Exercise 24 to show that $\int_E (h - a) = 0$ for every measurable set $E \subset A$ and conclude that $h = a$ a.e. on A. Use Exercise 24 again to obtain

$$0 = \lim_{n \to \infty} \int_A |g(t_n x) - a|\,dx = \frac{\lambda(A)}{p} \int_0^p |g - a|.]$$

(b) If the sequence $(t_n)_{n=1}^{\infty} \subset \mathbb{R}$ has limit ∞, then the series $\sum\limits_{n=1}^{\infty} \cos(t_n x)$ and $\sum\limits_{n=1}^{\infty} \sin(t_n x)$ both diverge a.e.

(c) Let b and $p_n < q_n \leq p_{n+1}$ $(n = 1, 2, 3, \dots)$ be natural numbers such that $\sum\limits_{n=1}^{\infty} b^{p_n - q_n} < \infty$. Let E be the set of all $x \in [0, 1]$ having a base b expansion

$$x = \sum_{j=1}^{\infty} x_j b^{-j} \qquad (x_j \in \mathbb{Z}, 0 \leq x_j < b)$$

that satisfies $x_j = 0$ if $p_n < j < q_n$ for some n. Then E is an uncountable closed set. If $t_n = \pi b^{p_n}$ for all n, then $\sum\limits_{n=1}^{\infty} \sin(t_n x)$ converges absolutely and uniformly on E. Of course $\lambda(E) = 0$. For example let $p_n = n^2 = q_n - n$.

26.   There exists a sequence $(f_n)_{n=1}^{\infty}$ of real-valued continuous functions on $[0, 1]$ that converges to 0 at every $x \in [0, 1]$ but converges uniformly on *no* subinterval of $[0, 1]$ having positive length. Compare this fact with Egorov's Theorem. Find a set of measure $1/2$ on which your sequence does converge uniformly. [Hints: For $p \in \mathbb{N}$, let

$$D_p = \{(2k - 1)2^{-p} : k = 1, 2, \dots, 2^{p-1}\}.$$

Given n, define f_n so that $f_n(a) = f_n(0) = 0$ and $f_n([a, a + 2^{-n}]) = [0, 2^{-p}]$ whenever $a \in D_p$ and $1 \leq p \leq n$. If $|I| > 2^{1-p}$, then $\sup f_n(I) \geq 2^{-p}$ for all $n \geq p$. If $x > 0$ is in no D_p and $\epsilon > 0$, then there is a p such that $2^{-p} < \epsilon$ and $x \in [a, a + 2^{-p}]$ for some $a \in D_p$; hence $f_n(x) < \epsilon$ for all $n \geq p$.]

27.   (a) If $(E_n)_{n=1}^{\infty}$ is a sequence of measurable subsets of $[0, 1]$ for which $\varlimsup\limits_{n\to\infty} \lambda(E_n) = 1$ and if $\alpha < 1$, then there exists a subsequence for which $\lambda(\bigcap\limits_{k=1}^{\infty} E_{n_k}) > \alpha$. [Hint: Let $F_n = [0, 1] \setminus E_n$ and make $\sum\limits_{k=1}^{\infty} \lambda(F_{n_k}) < 1 - \alpha$.]

(b) For any $\alpha < 1$, there is a sequence $(E_n)_{n=1}^{\infty}$ of closed subsets of $[0, 1]$ such that $\lambda(E_n) > \alpha$ for all n and yet each subsequence satisfies $\lambda(\bigcap\limits_{k=1}^{\infty} E_{n_k}) = 0$. [Hint: Let $b > (1 - \alpha)^{-1}$ be an integer and let E_n consist of those x in $[0, 1]$ having a base b expansion whose nth digit is not 0.]

28.   Let $(E_n)_{n=1}^{\infty}$ be any sequence of measurable sets. Define

$$\varliminf_{n\to\infty} E_n = \bigcup_{k=1}^{\infty} \bigcap_{n=k}^{\infty} E_n \quad \text{and} \quad \varlimsup_{n\to\infty} E_n = \bigcap_{k=1}^{\infty} \bigcup_{n=k}^{\infty} E_n.$$

(a) $\varliminf\limits_{n\to\infty} E_n \subset \varlimsup\limits_{n\to\infty} E_n$.

(b) $\varliminf\limits_{n\to\infty} E_n$ consists of those x such that $x \in E_n$ for all but finitely many n.

(c) $\varlimsup\limits_{n\to\infty} E_n$ consists of those x such that $x \in E_n$ for infinitely many n.

(d) $\lambda(\varliminf\limits_{n\to\infty} E_n) \leq \varliminf\limits_{n\to\infty} \lambda(E_n)$.

(e) If $\lambda(\bigcup\limits_{n=p}^{\infty} E_n) < \infty$ for some p, then $\lambda(\varlimsup\limits_{n\to\infty} E_n) \geq \varlimsup\limits_{n\to\infty} \lambda(E_n)$.

(f) If $E_n = [n, n+1]$, then inequality (d) is strict and inequality (e) is false.

(g) If $\sum_{n=1}^{\infty} \lambda(E_n) < \infty$, then $\lambda(\overline{\lim_{n\to\infty}} E_n) = 0$.

Compare (d) with Fatou's Lemma.

29. *Convergence in Measure.* Let $(f_n)_{n=1}^{\infty} \subset M$ be given.

(a) [F. Riesz] Suppose that

$$\lim_{m,n\to\infty} \lambda(\{x \in \mathbb{R} : |f_m(x) - f_n(x)| \geq \delta\}) = 0$$

for every $\delta > 0$. In this case we say that $(f_n)_{n=1}^{\infty}$ is *Cauchy in measure.* Then there exists a subsequence $(f_{n_k})_{k=1}^{\infty}$ and an $f \in M$ such that $f(x) = \lim_{k\to\infty} f_{n_k}(x)$ a.e. Moreover, the original sequence *converges in measure* to f which means that

$$\lim_{n\to\infty} \lambda(\{x \in \mathbb{R} : |f(x) - f_n(x)| \geq \delta\}) = 0$$

for every $\delta > 0$. [Hints: The proof is analogous to that of (6.56). Obtain the subsequence so that the sets $E_k = \{x \in \mathbb{R} : |f_{n_{k+1}}(x) - f_{n_k}(x)| \geq 2^{-k}\}$ satisfy $\lambda(E_k) < 2^{-k}$. Use Exercise 28 to see that the set $E = \overline{\lim_{k\to\infty}} E_k$ satisfies $\lambda(E) = 0$. Show that this subsequence converges at every $x \in \mathbb{R}\backslash E$. Notice that

$$\{x : |f(x) - f_n(x)| \geq \delta\} \subset \underline{\lim_{k\to\infty}} \{x : |f_{n_k}(x) - f_n(x)| \geq \delta/2\}$$

and use Exercise 28(d) to see that the latter set has measure $< \epsilon$ for all sufficiently large n.]

(b) If $f_n = \xi_{I_n}$ with $I_n = [2^{-k}j, 2^{-k}(j+1)]$ where $n = 2^k + j, k \geq 0, 0 \leq j < 2^k$, then $(f_n)_{n=1}^{\infty}$ converges nowhere on $[0, 1]$ but does converge in measure to $f = 0$ and $f_{2^k}(x) \to 0$ for all $x \neq 0$.

(c) [Lebesgue] Suppose that $X \subset \mathbb{R}$, $\lambda(X) < \infty$, and every f_n is identically 0 on $\mathbb{R}\backslash X$. If $f \in M$ and $f_n \to f$ a.e., then $f_n \to f$ in measure. [Hint: Use Egorov's Theorem.]

(d) If $f_n = \xi_{[n, n+1]}$, then $f_n \to 0$ everywhere on $\mathbb{R}$ but (f_n) is *not* Cauchy in measure.

(e) If $\lim_{m,n\to\infty} \int |f_m - f_n| = 0$, then (f_n) is Cauchy in measure.

(f) If $f_n = n \cdot \xi_{[0, 1/n]}$, then (f_n) is Cauchy in measure, but it is false that $\lim_{m,n\to\infty} \int |f_m - f_n| = 0$.

30. *Lebesgue's Definitions.*

(a) A set $E \subset \mathbb{R}$ is Lebesgue measurable if and only if $\lambda([a, b] \cap E) + \lambda([a, b]\backslash E) = b - a$ whenever $a < b$ in $\mathbb{R}$. [Hint: The condition implies that $\lambda(V \cap E) + \lambda(V\backslash E) = \lambda(V)$ for all open $V \subset \mathbb{R}$. In case $\lambda(E) < \infty$, obtain $V_1 \supset V_2 \supset \cdots$ $\supset E$ with $\lambda(V_n) \to \lambda(E)$ and show that E differs from $\bigcap_{n=1}^{\infty} V_n$ by a null set.]

(b) Let $f : [a, b] \to \mathbb{R}$ be measurable and bounded with $-\infty < c \leq f(x) < d < \infty$ for all $x \in [a, b]$. For each subdivision $P = \{c = y_0 < y_1 < \cdots < y_n = d\}$ write

$$E_k = \{x \in [a, b] : y_{k-1} \leq f(x) < y_k\}, L(P) = \sum_{k=1}^{n} y_{k-1}\lambda(E_k), \text{ and } U(P)$$

$$= \sum_{k=1}^{n} y_k \lambda(E_k). \text{ Then}$$

$$\lim_{|P|\to 0} L(P) = \int_a^b f = \lim_{|P|\to 0} U(P).$$

We remark that Lebesgue's rich theory of integration flowed from his simple, but brilliant, idea of subdivising the y-axis instead of the x-axis.

31.   *Tonelli's Definition.* A function $f:[a,b]\to\mathbb{R}$ is called *quasi-continuous on $[a,b]$* if there exists a sequence $(f_n)_{n=1}^\infty$ of continuous functions $f_n:[a,b]\to\mathbb{R}$ and a sequence $(V_n)_{n=1}^\infty$ of open subsets of $[a,b]$ such that $f_n = f$ on $[a,b]\backslash V_n$ and $\lim_{n\to\infty}\lambda(V_n)=0$.

(a) A function f is quasi-continuous on $[a,b]$ if and only if it is measurable.

(b) Let f be quasi-continuous on $[a,b]$ with $|f(x)|\leqq\beta<\infty$ for all $x\in[a,b]$. Then the functions f_n can be chosen as above such that $|f_n(x)|\leqq\beta$ for all $x\in[a,b]$ and $n\in\mathbb{N}$. Once this is done we have

$$\int_a^b f= \lim_{n\to\infty}\int_a^b f_n.$$

(c) It is possible to find f and $(f_n)_{n=1}^\infty$ as in (b) such that $(f_n(x))_{n=1}^\infty$ converges for *no* $x\in[a,b]$. [Hint: See Exercise 29(b).]

32.   *Sak's Definition.* Let $f:\mathbb{R}\to[0,\infty]$ be measurable. For each finite, pairwise disjoint family $\mathcal{D} = \{E_1,\ldots,E_n\}$ of nonvoid measurable sets that covers $\mathbb{R}$, write

$$s(\mathcal{D}) = \sum_{k=1}^n (\inf f(E_k))\lambda(E_k)$$

(recall that $0\cdot\infty = \infty\cdot 0 = 0$). Then

$$\int f= \sup\{s(\mathcal{D}) : \mathcal{D} \text{ is as above}\}.$$

33.   Let $f:[a,b]\to[0,\infty[$ be measurable. For $y\geqq 0$, define $\phi(y)=\lambda(\{x\in[a,b]:f(x)\geqq y\})$. Then

$$\int_a^b f= \int_0^\infty \phi= \lim_{d\to\infty}\int_0^d\phi.$$

[Hint: For a subdivision P of $[0,d]$ as in Exercise 30(b) (with $c=0$), rewrite the Riemann sum $S(\phi,P,y_k)$ as a Lebesgue lower sum $L(P)$ for the function f_d that agrees with f when $f<d$ and is 0 otherwise.]

34.   Suppose that $f:\mathbb{R}\to\mathbb{C}$ is measurable, that $(s_n)_{n=1}^\infty$ is a sequence of integrable simple functions that converges to f a.e. on $\mathbb{R}$ (or in measure) and that $\lim_{m,n\to\infty}\int|s_m - s_n| = 0$. Then $f\in L_1$ and $\lim_{n\to\infty}\int s_n = \int f$.

35.   [J. Marcinkiewicz, 1935] Let $I\subset\mathbb{R}$ be a closed interval and let $(h_n)_{n=1}^\infty$ be any fixed sequence of nonzero real numbers having limit 0. Then there exists a continuous function $F:I\to\mathbb{R}$ having the following property. If $\phi:I\to\mathbb{R}$ is any measurable function, then there is a subsequence $(h_{n_j})_{j=1}^\infty$ of $(h_n)_{n=1}^\infty$ such that

$$\lim_{j\to\infty}\frac{F(x + h_{n_j}) - F(x)}{h_{n_j}} = \phi(x)$$

for almost every $x\in I$. One and the same F works for all ϕ. Of course the subsequence depends on ϕ.*

This exercise consists of verifying the following bite-size steps leading to a proof of this remarkable theorem. For $f\in C'(I)$, we write $\|f\| = \sup\{|f(x)| : x\in I\}$.

---

*We are grateful to Professor R. B. Burckel for having brought Marcinkiewicz's paper to our attention.

(a) If $G \in C'(I)$ and $\epsilon > 0$, then there exists $H \in C'(I)$ such that $H' = 0$ a.e. and $\|G - H\| < \epsilon$. [Hint: Divide I into subintervals I_k such that $|G(s) - G(t)| < \epsilon$ if $s, t \in I_k$ and define H on each I_k to be monotone and agree with G at the endpoints. Remember (3.92).]

(b) If $Q, G_1 \in C'(I)$, Q is differentiable a.e., and $\epsilon > 0$, then there exists $G_2 \in C'(I)$ such that $G_2' = Q'$ a.e. and $\|G_1 - G_2\| < \epsilon$. [Hint: Take $G = G_1 - Q$ in (a) and let $G_2 = Q + H$.]

(c) There is an enumeration $\{P_k\}_{k=1}^{\infty}$ of the set of all polynomials (in one variable) having rational coefficients for which $P_1 = 0$.

　　　Now let $F_1 = 0$, $E_1 = \emptyset$, $n_1 = 1$, and write $t_1 = h_1$. Suppose that $k > 1$, that $F_{k-1} \in C'(I)$, $E_{k-1} \subset I$, and $n_{k-1} \in \mathbb{N}$ have been chosen and write $t_{k-1} = h_{n_{k-1}}$.

(d) There exists $F_k \in C'(I)$ such that $F_k' = P_k$ a.e. and $\|F_k - F_{k-1}\| < |t_{k-1}|/(k - 1)$. [Hint: Use (b).]

(e) There exist a measurable set $E_k \subset I$ with $\lambda(E_k) < 2^{-k}$ and a natural number $n_k > n_{k-1}$ such that $t_k = h_{n_k}$ satisfies $|t_k| < |t_{k-1}|/2$ and

$$\left| \frac{F_k(x + t_k) - F_k(x)}{t_k} - P_k(x) \right| < \frac{1}{k}$$

for all $x \in I \setminus E_k$. [Hint: Apply (6.74) to the sequence

$$f_n(x) = h_n^{-1}[F_k(x + h_n) - F_k(x)].]$$

(f) This defines F_k, E_k, n_k, and t_k for all $k \in \mathbb{N}$.

(g) For $n, k \in \mathbb{N}$, we have $\|F_{k+n} - F_k\| < 2|t_k|/k$.

(h) There is a function $F \in C'(I)$ such that $\|F - F_k\| \leqq 2|t_k|/k$ for all $k \in \mathbb{N}$.

(i) If $k \in \mathbb{N}$ and $x \in I \setminus E_k$, then

$$\left| \frac{F(x + t_k) - F(x)}{t_k} - P_k(x) \right| < \frac{5}{k} \, .$$

Now let $\phi : I \to \mathbb{R}$ be a measurable function.

(j) There exist natural numbers $k_1 < k_2 < \ldots$ and a sequence $(B_j)_{j=1}^{\infty}$ of measurable subsets of I such that for each $j \in \mathbb{N}$ we have $\lambda(B_j) < 2^{-j}$ and $|\phi(x) - P_{k_j}(x)| < j^{-1}$ for all $x \in I \setminus B_j$. [Hint: Use (6.76) and (3.118).]

(k) Writing $s_j = t_{k_j}$, we have

$$\left| \frac{F(x + s_j) - F(x)}{s_j} - \phi(x) \right| < \frac{6}{j}$$

for all $j \in \mathbb{N}$ and $x \in I \setminus (E_{k_j} \cup B_j)$.

(l) The set $A = \bigcap_{r=1}^{\infty} \bigcup_{j=r}^{\infty} (E_{k_j} \cup B_j)$ satisfies $\lambda(A) = 0$. [Hint: (6.62.v).]

(m) The sequence (s_j) is a subsequence of (h_n) and

$$\lim_{j \to \infty} \frac{F(x + s_j) - F(x)}{s_j} = \phi(x)$$

for all $x \in I \setminus A$.

　　　We remark that this proof can be amended somewhat in order to show that the set of functions F having the desired property is residual in $C'(I)$ in the sense of Baire category.

The Fundamental Theorem of Calculus

It is our purpose in this section to investigate the famous formula

(i) $$F(b) - F(a) = \int_a^b F'(x)\,dx.$$

Certainly this formula is not always true even if $F' \in L_1([a,b])$. In fact, if F is Lebesgue's singular function (3.92), then $F' = 0$ a.e. on $[0,1]$, and so

$$F(1) - F(0) = 1 > 0 = \int_0^1 F'.$$

We need to know that the integral is translation invariant.

(6.81) Theorem *Let $f \in L_1$ and $t \in \mathbb{R}$. Then*

(i) $$\int f(x + t)\,dx = \int f(x)\,dx.$$

In particular, if $a \leqq b$ in $\mathbb{R}$, then

(ii) $$\int_a^b f(x + t)\,dx = \int_{a+t}^{b+t} f(x)\,dx.$$

Proof To prove (i), consider successively the cases $f \in M_0$, $f \in M_1$, $f \in L_1'$, and $f \in L_1$, We leave the details to the reader.
To prove (ii), simply note that

$$f(x + t)\xi_{]a,b[}(x) = (f \cdot \xi_{]a+t,b+t[})(x + t)$$

for almost every x. □

(6.82) Theorem *Let F be a nondecreasing function on $[a,b] \subset \mathbb{R}$. Then*

$$F' \in L_1([a,b])$$

and

(i) $$\int_a^b F' \leqq F(b) - F(a).$$

Proof For $x > b$ define $F(x) = F(b)$. Define F_n on $[a,b]$ by

$$F_n(x) = n[F(x + 1/n) - F(x)].$$

It follows from Lebesgue's Differentiation Theorem (4.52) that $F_n(x) \to F'(x)$ a.e. on $[a,b]$. Since each F_n is measurable [being continuous except at a countable set of points], it follows that $F' \in M$. We need only prove (i). By Fatou's Lemma (6.47), we have

$$\int_a^b F' \leqq \varliminf_{n \to \infty} \int_a^b F_n.$$ (1)

But

$$\int_a^b F_n = n\int_a^b F\left(x + \frac{1}{n}\right)dx - n\int_a^b F(x)\,dx$$

$$= n\left[\int_{a+1/n}^{b+1/n} F(x)\,dx - \int_a^b F(x)\,dx\right]$$

$$= n\int_b^{b+1/n} F - n\int_a^{a+1/n} F$$

$$\leqq n\int_b^{b+1/n} F(b) - n\int_a^{a+1/n} F(a) = F(b) - F(a), \qquad (2)$$

where the second equality uses (6.81). Combine (1) and (2) to get (i). $\square$

(6.83) Corollary *If F is a complex-valued function on $[a,b]$ and $V_a^b F < \infty$, then $F' \in L_1([a,b])$.*

Proof Use Jordan's Decomposition Theorem (3.133) to write

$$F = (F_1 - F_2) + i(F_3 - F_4)$$

where each F_j is nondecreasing on $[a,b]$. $\square$

(6.84) Theorem *Let $f \in L_1([a,b])$ and $c \in \mathbb{C}$. Define F on $[a,b]$ by*

$$F(x) = c + \int_a^x f(t)\,dt.$$

Then F is absolutely continuous on $[a,b]$ and $F' = f$ a.e. on $[a,b]$. In particular, $F'(x) = f(x)$ if f is continuous at x.

Proof Suppose that f is continuous at x. Given $\epsilon > 0$, we can choose $\delta > 0$ such that

$$|f(t) - f(x)| < \epsilon \quad \text{if } |t - x| < \delta.$$

Let $h \neq 0$ and $|h| < \delta$. Then

$$\left|\frac{F(x+h) - F(x)}{h} - f(x)\right| = \left|\frac{1}{h}\int_a^{x+h} f(t)\,dt - \frac{1}{h}\int_a^x f(t)\,dt - f(x)\right|$$

$$= \left|\frac{1}{h}\int_x^{x+h} [f(t) - f(x)]\,dt\right|$$

$$\leqq \frac{1}{|h|}\int_I |f(t) - f(x)|\,dt < \frac{1}{|h|}\int_I \epsilon\,dt = \epsilon,$$

where I is the open interval having endpoints x and $x + h$. Therefore

$$F'(x) = \lim_{h\to 0} \frac{F(x+h) - F(x)}{h} = f(x).$$

To see that F is absolutely continuous, let $\epsilon > 0$ be given and choose δ as in (6.79). If $\{]a_j, b_j[\}_{j=1}^n$ is a pairwise disjoint family of subintervals of $[a,b]$ such that

$$\sum_{j=1}^{n} (b_j - a_j) < \delta,$$

write E for the union of this family, note that $\lambda(E) < \delta$, and obtain

$$\sum_{j=1}^{n} |F(b_j) - F(a_j)| = \sum_{j=1}^{n} \left| \int_{a_j}^{b_j} f(t)\, dt \right|$$

$$\leq \sum_{j=1}^{n} \int_{a_j}^{b_j} |f| = \int_{E} |f| < \epsilon.$$

We prove that $F' = f$ a.e. by considering five cases. We suppose that $f = 0$ off $[a,b]$.

Case 1: $f = \xi_I$, where $I =]\alpha, \beta[\subset [a,b]$. Then

$$F(x) = |I \cap [a,x]| + c = \begin{cases} c, & a \leq x \leq \alpha, \\ x - \alpha + c, & \alpha \leq x \leq \beta, \\ \beta - \alpha + c, & \beta \leq x \leq b, \end{cases}$$

and so the conclusion is obvious.

Case 2: $f \in M_0$. This case follows at once from Case 1.

Case 3: $f \in M_1$. Choose a nondecreasing sequence $(\phi_n)_{n=1}^{\infty} \subset M_0$ such that $\phi_n \to f$ a.e. Define

$$\Phi_n(x) = c + \int_{a}^{x} \phi_n(t)\, dt.$$

Then

$$F(x) = \lim_{n \to \infty} \Phi_n(x) = \Phi_1(x) + \sum_{k=2}^{\infty} \left[\Phi_k(x) - \Phi_{k-1}(x) \right]$$

for all $x \in [a,b]$. Since each summand in this series is a nondecreasing function of x, we can apply Case 2 and Fubini's Theorem (4.53) to write

$$F'(x) = \Phi_1'(x) + \sum_{k=2}^{\infty} \left[\Phi_k'(x) - \Phi_{k-1}'(x) \right]$$

$$= \phi_1(x) + \sum_{k=2}^{\infty} \left[\phi_k(x) - \phi_{k-1}(x) \right] = \lim_{n \to \infty} \phi_n(x) = f(x) \text{ a.e.}$$

Case 4: $f \in L_1^r$. Write $f = g - h$, where $g, h \in M_1$, and apply Case 3.

Case 5: $f \in L_1$. Write $f = f_1 + if_2$, where $f_1, f_2 \in L_1^r$, and apply Case 4. $\square$

(6.85)   **Fundamental Theorem of Calculus**   *Let F be a complex-valued function that is absolutely continuous on $[a,b]$. Then $F' \in L_1([a,b])$ and*

(i) $$F(x) = F(a) + \int_{a}^{x} F'(t)\, dt \quad \text{for all } x \in [a,b].$$

Our proof of this theorem requires four lemmas.

(6.86) **Lemma** *Let $E_1 \subset E_2 \subset \ldots$ be subsets of $\mathbb{R}$. Then*

$$\lambda\left(\bigcup_{n=1}^{\infty} E_n\right) = \lim_{n \to \infty} \lambda(E_n).$$

[*Note that these sets need not be measurable.*]

Proof Write $E = \bigcup_{n=1}^{\infty} E_n$. Plainly, $\lambda(E_n) \le \lambda(E)$ for all n, and so

$$\lim_{n \to \infty} \lambda(E_n) \le \lambda(E). \tag{1}$$

To prove the reversed inequality, we may suppose that $\lim_{n \to \infty} \lambda(E_n) < \infty$. Let $\epsilon > 0$ be given. For each n, choose an open set U_n such that $E_n \subset U_n$ and

$$\lambda(U_n) < \lambda(E_n) + \epsilon/2^n.$$

Writing $V_n = \bigcup_{k=1}^{n} U_k$, we have

$$V_n \subset V_{n+1}, \qquad V_{n+1} \backslash V_n = U_{n+1} \backslash (U_{n+1} \cap V_n),$$

$$\lambda(V_{n+1}) - \lambda(V_n) = \lambda(U_{n+1}) - \lambda(U_{n+1} \cap V_n)$$

$$\le \lambda(U_{n+1}) - \lambda(E_n) < \lambda(E_{n+1}) - \lambda(E_n) + \epsilon/2^{n+1}.$$

Sum these inequalities for $n = 0, 1, \ldots, p-1$ [$V_0 = E_0 = \varnothing$] and obtain

$$\lambda(V_p) < \lambda(E_p) + \epsilon$$

for all $p \ge 1$. Therefore

$$\lambda(E) \le \lambda\left(\bigcup_{p=1}^{\infty} V_p\right) = \lim_{p \to \infty} \lambda(V_p) \le \lim_{p \to \infty} \lambda(E_p) + \epsilon.$$

Since ϵ was arbitrary, we have proved the reverse of (1). $\square$

(6.87) **Lemma** *If F is a real-valued function that is absolutely continuous on $[a,b]$, $A \subset [a,b]$, and $\lambda(A) = 0$, then $\lambda(F(A)) = 0$.*

Proof Let $\epsilon > 0$ be given and choose δ as in the definition of absolute continuity (3.136). Select a pairwise disjoint family $\{]a_j, b_j[\}_{j=1}^{\infty}$ of subintervals of $[a,b]$ that covers $A \backslash \{a,b\}$ and satisfies

$$\sum_{j=1}^{\infty} (b_j - a_j) < \delta.$$

Choose $c_j, d_j \in [a_j, b_j]$ such that

$$F([a_j, b_j]) = [F(c_j), F(d_j)].$$

Then $F(A \backslash \{a,b\})$ is covered by these intervals, and so

$$\lambda(F(A)) = \lambda(F(A \backslash \{a,b\})) \le \sum_{j=1}^{\infty} |F(d_j) - F(c_j)| \le \epsilon. \quad \square$$

(6.88)   **Lemma**   *Let F be a real-valued function on $[a,b]$, Suppose that $B \subset]a,b[$, $\beta \geq 0$, and $D^+F(x) \leq \beta$ and $D_-F(x) \geq -\beta$ for all $x \in B$. Then $\lambda(F(B)) \leq \beta\lambda(B)$.*

Proof   Let $\epsilon > 0$ be given. For $n \in \mathbb{N}$, define

$$B_n = \{ x \in B : F(x+h) - F(x) < (\beta + \epsilon)|h|$$
$$\text{when } h \neq 0, a < x+h < b, |h| < 1/n \}.$$

Then $B_1 \subset B_2 \subset \ldots$ and $\bigcup_{n=1}^{\infty} B_n = B$. By (6.86) we have

$$\lambda(F(B)) = \lim_{n \to \infty} \lambda(F(B_n)).$$

Therefore, since ϵ is arbitrary, it suffices to show that

$$\lambda(F(B_n)) < (\beta + \epsilon)(\lambda(B) + \epsilon)$$

for each n.

Fix n and choose a sequence $(I_k)_{k=1}^{\infty}$ of subintervals of $]a,b[$ that covers B_n such that $|I_k| < 1/n$ for all k and $\sum_{k=1}^{\infty} |I_k| < \lambda(B_n) + \epsilon$. Then, from the definition of B_n we have

$$|F(s) - F(t)| < (\beta + \epsilon)|I_k|$$

whenever $s, t \in B_n \cap I_k$. Therefore

$$\lambda(F(B_n)) \leq \sum_{k=1}^{\infty} \lambda(F(B_n \cap I_k))$$

$$\leq \sum_{k=1}^{\infty} \operatorname{diam} F(B_n \cap I_k)$$

$$\leq (\beta + \epsilon) \sum_{k=1}^{\infty} |I_k| < (\beta + \epsilon)(\lambda(B_n) + \epsilon). \quad \square$$

(6.89)   **Lemma**   *Let F be a complex-valued function that is absolutely continuous on $[a,b]$. Suppose that $F'(x) = 0$ a.e. on $[a,b]$. Then $F(x) = F(a)$ for all $x \in [a,b]$.*

Proof   It suffices to consider the case that F is real-valued. Let $B = \{ x : a < x < b, F'(x) = 0 \}$ and $A = [a,b] \setminus B$. Then $\lambda(A) = 0$, and so $\lambda(F(A)) = 0$ (6.87). Using (6.88) with $\beta = 0$, we see that $\lambda(F(B)) = 0$. Since the interval $F([a,b])$ satisfies

$$\lambda(F([a,b])) \leq \lambda(F(A)) + \lambda(F(B)) = 0,$$

it follows that this interval, the range of F, is a single point. $\quad \square$

Proof of (6.85)   By (6.83), $F' \in L_1([a,b])$. Define

$$G(x) = \int_a^x F'.$$

According to (6.84), G is absolutely continuous on $[a,b]$ and $G' = F'$ a.e. Thus the absolutely continuous function $H = F - G$ satisfies $H' = 0$ a.e. By (6.89), $F(x) - G(x) = H(x) = H(a) = F(a) - G(a) = F(a)$ for all $x \in [a,b]$. $\square$

Integration by Parts

(6.90) Theorem *Let F and G be absolutely continuous on $[a,b]$. Then*

$$\int_a^b F \cdot G' + \int_a^b F' \cdot G = F(b)G(b) - F(a)G(a).$$

Proof Since

$$\sum_{j=1}^n |F(b_j)G(b_j) - F(a_j)G(a_j)|$$

$$\leq \sum_{j=1}^n |G(b_j)| \cdot |F(b_j) - F(a_j)| + \sum_{j=1}^n |F(a_j)| \cdot |G(b_j) - G(a_j)|$$

$$\leq \alpha \sum_{j=1}^n |F(b_j) - F(a_j)| + \beta \sum_{j=1}^n |G(b_j) - G(a_j)|,$$

where $|G(x)| \leq \alpha, |F(x)| \leq \beta$ for all $x \in [a,b]$, it is easy to see that $F \cdot G$ is absolutely continuous on $[a,b]$. Now apply (6.85) to $F \cdot G$ and use the formula

$$(F \cdot G)' = F \cdot G' + F' \cdot G. \quad \square$$

Integration by Substitution

In this section we wish to investigate the validity of the formula

$$\int_{\phi(\alpha)}^{\phi(\beta)} f(x)\, dx = \int_\alpha^\beta f(\phi(t))\phi'(t)\, dt.$$

Of course, this does not always obtain. Indeed, if $f = 1$, it reduces to (6.85.i).

This question has a long history [see (6.99)]. We believe that Theorem (6.95), due to J. Serrin and D. E. Varberg (1969), gives the best account of the matter to date.

(6.91) Notation Let $a < b$ and $c < d$ in $\mathbb{R}$. Suppose that ϕ is a function from $[a,b]$ into $[c,d]$ and that F is a real-valued function that is differentiable a.e. on $[c,d]$. We write $F' = f$ a.e. on $[c,d]$.

(6.92) Lemma *Let ϕ be as in (6.91). Suppose that $E \subset [a,b]$, that $\phi'(x)$ exists [possibly infinite] for each $x \in E$, and that $\lambda(\phi(E)) = 0$. Then $\phi' = 0$ a.e. on E.*

Proof Write

$$B = \{t \in E : \phi'(t) \neq 0\},$$
$$B_n = \{t \in B : |\phi(s) - \phi(t)| > |s - t|/n$$

whenever $a \leqq s \leqq b$ and $0 < |s - t| < 1/n\}$. Then $B = \bigcup_{n=1}^{\infty} B_n$, and so we need only show that $\lambda(B_n) = 0$ for all n. Fix $n \in \mathbb{N}$. Since B_n can be covered by finitely many intervals of length less than $1/n$, it suffices to show that $\lambda(I \cap B_n) = 0$ for every such interval I. Let I be given and write $A = I \cap B_n$. We want to prove $\lambda(A) = 0$. Let $\epsilon > 0$ be given. Since $\lambda(\phi(A)) = 0$, there exists a family $\{J_k\}_{k=1}^{\infty}$ of open intervals that covers $\phi(A)$ and satisfies

$$\sum_{k=1}^{\infty} |J_k| < \frac{\epsilon}{n}.$$

Let $A_k = A \cap \phi^{-1}(J_k)$. Then $A = \bigcup_{k=1}^{\infty} A_k$ and, by definition of B_n, $\operatorname{diam} A_k \leqq n \cdot \operatorname{diam} \phi(A_k) \leqq n \cdot |J_k|$ for each k. Therefore

$$\lambda(A) \leqq \sum_{k=1}^{\infty} \lambda(A_k) \leqq \sum_{k=1}^{\infty} \operatorname{diam} A_k \leqq n \cdot \sum_{k=1}^{\infty} |J_k| < \epsilon. \qquad \square$$

(6.93) **Theorem** [Chain Rule] *With the notation of* (6.91), *suppose also that* ϕ *and* $F \circ \phi$ *are both differentiable a.e. on* $[a,b]$ *and that* $\lambda(F(A)) = 0$ *whenever* $A \subset [c,d]$ *and* $\lambda(A) = 0$. *Then*

(i) $$(F \circ \phi)' = (f \circ \phi) \cdot \phi'$$

a.e. on $[a,b]$.

Proof Write

$$D = \{x : c \leqq x \leqq d, F'(x) \text{ exists in } \mathbb{R}, F'(x) = f(x)\},$$
$$B = [c,d] \backslash D, \qquad S = \phi^{-1}(B),$$
$$T = \phi^{-1}(D) = [a,b] \backslash S.$$

Let $t \in T$ be such that ϕ is differentiable at t. Then ϕ is continuous at t and $F'(\phi(t)) = f(\phi(t))$; hence,

$$F(\phi(t + h)) - F(\phi(t)) = [f(\phi(t)) + \delta(h)] \cdot [\phi(t + h) - \phi(t)], \quad (1)$$

where $\delta(h) \to 0$ as $h \to 0$ [put $\delta(h) = 0$ if $\phi(t + h) = \phi(t)$, otherwise solve for $\delta(h)$]. Now divide (1) by h and let $h \to 0$ to see that (i) holds a.e. on T.

Since $\lambda(\phi(S)) \leqq \lambda(B) = 0$, our hypothesis on F yields $\lambda(F(\phi(S))) = 0$. Let

$$E = \{s \in S : (F \circ \phi)'(s) \text{ and } \phi'(s) \text{ both exist}\}.$$

Then $\lambda(S \setminus E) = 0 = \lambda(\phi(E)) = \lambda(F \circ \phi(E))$. Applying (6.92) to ϕ and to $F \circ \phi$, we obtain $(F \circ \phi)' = 0 = \phi'$ a.e. on E. Therefore (i) holds a.e. on S.

$\square$

(6.94)   Example   The hypothesis in (6.93) that F maps λ-null sets to λ-null sets is important. In fact, let ϕ be a strictly increasing continuous function on $[0,1]$ such that $\phi' = 0$ a.e. (4.54) and let $F = \phi^{-1}$. Then F, being monotone, is differentiable a.e. and $(F \circ \phi)'(x) = 1$ for all $x \in [0,1]$. Therefore (6.93.i) fails a.e.

(6.95)   Change of Variable Theorem   *With the notation of (6.91), we suppose also that ϕ is differentiable a.e. on $[a,b]$ and that F is absolutely continuous on $[c,d]$. Then the following two statements are equivalent.*
   (i) *$F \circ \phi$ is absolutely continuous on $[a,b]$.*
   (ii) *$(f \circ \phi) \cdot \phi' \in L_1([a,b])$ and*

$$\int_{\phi(\alpha)}^{\phi(\beta)} f(x)\,dx = \int_{\alpha}^{\beta} f(\phi(t))\phi'(t)\,dt$$

for all α, $\beta \in [a,b]$.

Proof   Suppose that (ii) obtains. Then (6.85) shows that

$$F(\phi(s)) - F(\phi(a)) = \int_{\phi(a)}^{\phi(s)} f(x)\,dx = \int_{a}^{s} f(\phi(t))\phi'(t)\,dt$$

for all $s \in [a,b]$, and so (i) follows from (6.84).
   Conversely, suppose that (i) obtains. Our hypotheses and (6.87) show that the hypotheses of (6.93) are satisfied. Since $(F \circ \phi)' \in L_1([a,b])$ (6.85),   (6.93.i) yields

$$(f \circ \phi) \cdot \phi' \in L_1([a,b]).$$

Also, (6.85) and (6.93.i) give

$$\int_{\phi(\alpha)}^{\phi(\beta)} f(x)\,dx = F(\phi(\beta)) - F(\phi(\alpha))$$

$$= \int_{\alpha}^{\beta} (F \circ \phi)'(t)\,dt = \int_{\alpha}^{\beta} f(\phi(t))\phi'(t)\,dt$$

for all α, $\beta \in [a,b]$. $\square$

(6.96)   Examples   (a) It is not necessary that ϕ be absolutely continuous in order that (6.95.ii) should obtain. For example, let $\phi(t) = t\sin(1/t)$ for $t \neq 0$, $\phi(0) = 0$, and $F(x) = x^2$. Then ϕ is differentiable except at $t = 0$ and ϕ is not absolutely continuous on $[0,1]$ because $V_0^1 \phi = \infty$. However, F and $F \circ \phi$ are both absolutely continuous on $[0,1]$ because their derivatives are both bounded there.
   (b) It can happen that $F \circ \phi$ fails to be absolutely continuous even though both F and ϕ are. For instance, let $\phi(t) = t^2 \sin^2(1/t)$ for $t \neq 0$, $\phi(0) = 0$, and $F(x) = \sqrt{x}$ for $x \geq 0$. Then $V_0^1 F \circ \phi = \infty$ so $F \circ \phi$ is not absolutely continuous on

[0, 1]. But ϕ is absolutely continuous on [0, 1] since ϕ' is bounded there. Also, F is absolutely continuous on [0, 1] as can be seen, using (6.84), from the fact that

$$F(x) = \int_0^x f(t)\, dt \qquad (x \geqq 0),$$

where $f(t) = 1/(2\sqrt{t}\,)$ for $t > 0$. In fact, $f \in L_1([0, 1])$ because

$$\int_0^1 |f(t)|\, dt = \lim_{n \to \infty} \int_{1/n}^1 \frac{dt}{2\sqrt{t}}$$

$$= \lim_{n \to \infty} \left(1 - 1/\sqrt{n}\,\right) = 1,$$

where we have used the fact that $F' = f$ is bounded on $[1/n, 1]$, (6.85), and the monotone convergence theorem.

(6.97)   Corollary *Let ϕ be absolutely continuous on $[a, b]$ into $[c, d]$ and let $f \in L_1'([c, d])$. Then any one of the following three conditions is sufficient to imply (6.95.ii).*
   (i)   *ϕ is monotone on $[a, b]$.*
   (ii)   *f is bounded on $[c, d]$.*
   (iii) *$(f \circ \phi) \cdot \phi' \in L_1'([a, b])$.*

Proof Define F on $[c, d]$ by $F(x) = \int_c^x f(u)\, du$. Then F is absolutely continuous on $[c, d]$ and $F' = f$ a.e. on $[c, d]$. Certainly ϕ is differentiable a.e. on $[a, b]$. Thus, it suffices to show that (6.95.i) obtains.
    Suppose (i) obtains. Given $\epsilon > 0$, choose $\eta > 0$ such that

$$\sum_{j=1}^n |F(d_j) - F(c_j)| < \epsilon \tag{1}$$

whenever the $]c_j, d_j[$ are pairwise disjoint in $[c, d]$ and the sum of their lengths is less than η. Next choose $\delta > 0$ such that

$$\sum_{j=1}^n |\phi(b_j) - \phi(a_j)| < \eta \tag{2}$$

whenever the $]a_j, b_j[$ are pairwise disjoint in $[a, b]$ and the sum of their lengths is less than δ. If such $]a_j, b_j[$ are given, then, owing to (i), the open intervals with endpoints $\phi(a_j)$ and $\phi(b_j)$ are pairwise disjoint, and so (1) yields

$$\sum_{j=1}^n |F \circ \phi(b_j) - F \circ \phi(a_j)| < \epsilon.$$

Thus $F \circ \phi$ is absolutely continuous on $[a, b]$.
    Suppose (ii) obtains. Choose $C \in \mathbb{R}$ such that $|f(x)| \leq C$ for all $x \in [c, d]$. Given $\epsilon > 0$, choose $\eta = \epsilon/C$ and then choose $\delta > 0$ so that (2)

holds. Then

$$\sum_{j=1}^{n} |F \circ \phi(b_j) - F \circ \phi(a_j)| = \sum_{j=1}^{n} \left| \int_{\phi(a_j)}^{\phi(b_j)} f(u)\, du \right|$$

$$\leq \sum_{j=1}^{n} C|\phi(b_j) - \phi(a_j)| < C\eta = \epsilon$$

whenever the $]a_j, b_j[$ are as described following (2). Thus again $F \circ \phi$ satisfies (6.95.i).

Suppose (iii) obtains. For $n \in \mathbb{N}$, write $f_n(x) = f(x)$ if $|f(x)| \leq n$ and $f_n(x) = 0$ otherwise. Then $f_n \to f$ a.e. on $[c, d]$. Also, for any $t \in [a, b]$ for which $f \circ \phi(t)\phi'(t)$ is defined and finite [i.e. by (6.92), almost every t] we have $f_n \circ \phi(t)\phi'(t) \to f \circ \phi(t)\phi'(t)$. Since $|f_n| \leq |f| \in L_1([c, d])$ and $|(f_n \circ \phi) \cdot \phi'| \leq |(f \circ \phi) \cdot \phi'| \in L_1([a, b])$, we can use Lebesgue's Dominated Convergence Theorem to write

$$\int_{\phi(\alpha)}^{\phi(\beta)} f(x)\, dx = \lim_{n \to \infty} \int_{\phi(\alpha)}^{\phi(\beta)} f_n(x)\, dx \tag{3}$$

and

$$\int_{\alpha}^{\beta} f(\phi(t))\phi'(t)\, dt = \lim_{n \to \infty} \int_{\alpha}^{\beta} f_n(\phi(t))\phi'(t)\, dt \tag{4}$$

for all $\alpha, \beta \in [a, b]$. It follows from (ii) that the right sides of (3) and (4) are equal. $\square$

(6.98)   Remark   It is a simple matter to check, by separation into real and imaginary parts, that (6.95) and (6.97) remain true for the case that f and F are complex-valued.

(6.99)   History   Corollary (6.97.i) is due to Lebesgue (1909). Both (6.97.iii) and (6.95) [with ϕ absolutely continuous] were stated by de la Vallée Poussin (1914). The latter work was criticized and (6.95) [with ϕ continuous and monotone] was enunciated by Fichtenholz (1922). The author is unaware of any correct proof of these latter two assertions prior to 1969.

Two Mean Value Theorems

If one interprets the number

$$\frac{1}{b - a} \int_{a}^{b} f(x)\, dx$$

to be the "average value" of f on $[a, b]$, then it is a consequence of the next theorem $[g = 1]$ that, for continuous integrands, this average value is attained.

(6.100) First Mean Value Theorem for Integrals *Let f be a continuous real-valued function on* $[a,b]$ *and let* $g \in L_1'([a,b])$ *be nonnegative. Then there exists* $\xi \in [a,b]$ *such that*

(i)
$$\int_a^b f(x)g(x)\,dx = f(\xi)\int_a^b g(x)\,dx.$$

Proof   Write $f([a,b]) = [c,d]$. Then $cg(x) \le f(x)g(x) \le dg(x)$ a.e. on $[a,b]$. Integrating, we obtain

$$c\int_a^b g(x)\,dx \le \int_a^b f(x)g(x)\,dx \le d\int_a^b g(x)\,dx.$$

If $\int_a^b g > 0$, we may divide by it and realize that the quotient of the two integrals in (i) is in $[c,d]$ and so has the form $f(\xi)$ for some $\xi \in [a,b]$. Otherwise, $g = 0$ a.e. and both sides of (i) are zero for any choice of ξ.   □

(6.101) Second Mean Value Theorem for Integrals *Let* $f \in L_1'([a,b])$ *and let* α *be a monotone function on* $[a,b]$. *Then there exists* $\xi \in [a,b]$ *such that*

(i)
$$\int_a^b f(x)\alpha(x)\,dx = \alpha(a)\int_a^\xi f(x)\,dx + \alpha(b)\int_\xi^b f(x)\,dx.$$

Proof   We may suppose that α is nondecreasing since if the theorem were true in this case, then the result for α nonincreasing would follow by replacing α by $-\alpha$.

     Case 1: α is absolutely continuous on $[a,b]$. Defining

$$F(x) = \int_a^x f(t)\,dt,$$

we can apply (6.84) and (6.90) to obtain

$$\int_a^b f(x)\alpha(x)\,dx = F(b)\alpha(b) - F(a)\alpha(a) - \int_a^b F(x)\alpha'(x)\,dx. \tag{1}$$

Since F is continuous on $[a,b]$ and $\alpha' \ge 0$, it follows from (6.100) and (6.85) that

$$\int_a^b F(x)\alpha'(x)\,dx = F(\xi)\int_a^b \alpha'(x)\,dx = F(\xi)\big[\alpha(b) - \alpha(a)\big] \tag{2}$$

for some $\xi \in [a,b]$.

     Combining (1) and (2) yields

$$\int_a^b f(x)\alpha(x)\,dx = \alpha(a)\big[F(\xi) - F(a)\big] + \alpha(b)\big[F(b) - F(\xi)\big]$$

$$= \alpha(a)\int_a^\xi f(x)\,dx + \alpha(b)\int_\xi^b f(x)\,dx,$$

as desired.

Case 2: α is nondecreasing. We want to choose a sequence $(\alpha_n)_{n=1}^{\infty}$ of functions that are nondecreasing and absolutely continuous on $[a,b]$ such that $\alpha_n(x) \to \alpha(x)$ a.e. on $[a,b]$, $\alpha_n(a) = \alpha(a)$, $\alpha_n(b) = \alpha(b)$. For instance, let $h_n = (b-a)/n$ and $x_{n,k} = a + kh_n$ for $n \in \mathbb{N}$ and $0 \leq k \leq n$. Let α_n be the function that agrees with α at $x_{n,k}$ $(0 \leq k \leq n)$ and is linear on each interval $[x_{n,k-1}, x_{n,k}]$ $(1 \leq k \leq n)$. We leave it to the reader to check that this sequence $(\alpha_n)_{n=1}^{\infty}$ has the stated properties $[\alpha_n(x) \to \alpha(x)$ if α is continuous at x since $|\alpha(x) - \alpha_n(x)| \leq |\alpha(x_{n,k-1}) - \alpha(x_{n,k})|$ if $x_{n,k-1} \leq x \leq x_{n,k}]$.

Apply Case 1 to α_n to obtain $\xi_n \in [a,b]$ such that

$$\int_a^b f(x)\alpha_n(x)\,dx = \alpha(a)\int_a^{\xi_n} f(x)\,dx + \alpha(b)\int_{\xi_n}^b f(x)\,dx. \tag{3}$$

Passing to a subsequence, if need be, we may suppose that $\xi_n \to \xi \in [a,b]$. We check that Lebesgue's Dominated Convergence Theorem applies to each of the three integrals in (3), and so, letting $n \to \infty$ in (3), we obtain (i).

$\square$

Arc Length

Recall that a curve in the plane is, by definition, a complex-valued function γ that is defined and continuous on some interval of $\mathbb{R}$. If f is a real-valued function that is defined and continuous on an interval $I \subset \mathbb{R}$, then the graph of f can be regarded as a curve γ by defining γ on I by $\gamma(t) = t + if(t)$.

In this section we define the length of a curve and obtain an important formula for computing such lengths.

(6.102) Definition If γ is a curve in the plane whose domain is a closed interval $[a,b]$, we call γ an *arc* and define the *length* of γ to be the number $V_a^b\gamma$.

(6.103) Theorem *Let γ be as in (6.102). If γ is absolutely continuous on $[a,b]$, then*

$$V_a^b\gamma = \int_a^b |\gamma'(t)|\,dt.$$

Proof For any subdivision $P = \{a = x_0 < \ldots < x_n = b\}$ we have

$$\sum_{k=1}^n |\gamma(x_k) - \gamma(x_{k-1})| = \sum_{k=1}^n \left| \int_{x_{k-1}}^{x_k} \gamma'(t)\,dt \right|$$

$$\leq \sum_{k=1}^n \int_{x_{k-1}}^{x_k} |\gamma'(t)|\,dt = \int_a^b |\gamma'(t)|\,dt.$$

Therefore,

$$V_a^b\gamma \leq \int_a^b |\gamma'(t)|\,dt. \tag{1}$$

To prove that (1) is actually an equality, let us define f a.e. on $[a,b]$ by $f(t) = 0$ if $\gamma'(t) = 0$, $f(t) = \gamma'(t)/|\gamma'(t)|$ if $\gamma'(t) \neq 0$. One checks that $f \in M([a,b])$ [γ' is measurable and the sets $\{t \in [a,b] : \gamma'(t) = 0\}$ and $\{t \in [a,b] : \gamma'(t) \text{ exists}, \gamma'(t) \neq 0\}$ are both measurable]. Thus there exists a sequence $(\phi_n)_{n=1}^\infty$ of complex-valued step functions such that

$$\lim_{n \to \infty} \phi_n(t) = f(t) \quad \text{a.e.} \tag{2}$$

Each ϕ_n has the form

$$\phi = \sum_{k=1}^m \alpha_k \xi_{[x_{k-1}, x_k[}, \tag{3}$$

where $a = x_0 < \ldots < x_m = b$ and $(\alpha_k)_{k=1}^m \subset \mathbb{C}$. Since $|f| \leq 1$ a.e. we may suppose that $|\phi_n| \leq 1$ for all n [if $|\alpha_k| > 1$, replace α_k in (3) by $\alpha_k/|\alpha_k|$]. We have

$$\lim_{n \to \infty} |\gamma'(t)[f(t) - \phi_n(t)]| = 0 \quad \text{a.e.},$$

$$|\gamma' \cdot [f - \phi_n]| \leq 2|\gamma'| \in L_1([a,b])$$

for all n. Thus, given any $\epsilon > 0$, we appeal to Lebesgue's Dominated Convergence Theorem to see that there exists a ϕ as in (3) such that

$$\int_a^b |\gamma'[f - \phi]| < \epsilon.$$

Since, as planned, $|\gamma'| = \gamma' f$ a.e., we have

$$\int_a^b |\gamma'| = \int_a^b \gamma' f$$

$$\leq \int_a^b |\gamma'[f - \phi]| + \left| \int_a^b \gamma' \phi \right| < \epsilon + \left| \sum_{k=1}^m \alpha_k \int_{x_{k-1}}^{x_k} \gamma' \right|$$

$$\leq \epsilon + \sum_{k=1}^m \left| \int_{x_{k-1}}^{x_k} \gamma' \right|$$

$$= \epsilon + \sum_{k=1}^m |\gamma(x_k) - \gamma(x_{k-1})|$$

$$\leq \epsilon + V_a^b \gamma.$$

Therefore,

$$\int_a^b |\gamma'| < \epsilon + V_a^b \gamma$$

for all $\epsilon > 0$. This proves the reverse of inequality (1). $\square$

We can now compute lengths of circular arc very easily.

(6.104) Example   Let $\alpha < \beta$ be real numbers and let γ be the arc defined on $[\alpha, \beta]$ by $\gamma(t) = e^{it}$. Then $\gamma'(t) = ie^{it}$, and so the length of γ is

$$\int_\alpha^\beta |ie^{it}| \, dt = \int_\alpha^\beta 1 \, dt = \beta - \alpha.$$

As t "runs" from α to β, this arc γ "traces along" the unit circle in a "counterclockwise direction." If $\beta - \alpha > 2\pi$ this arc "traces" part of the circle more than once. If $\beta - \alpha = 2\pi$, the length of γ is just the circumference of the circle.

It should be noted that the length of an arc cannot always be determined from an examination of its range [a "geometrical curve"]. For instance, the arcs

$$\gamma_1(t) = e^{it},$$
$$\gamma_2(t) = e^{2it},$$
$$\gamma_3(t) = \exp(4\pi it \cdot \sin(t^{-1})), \qquad t \neq 0,$$
$$\gamma_3(0) = 1$$

defined on $[0, 2\pi]$ all have the unit circle as range, but their lengths are 2π, 4π, and ∞, respectively.

Exercises

1. Suppose that $F : [a, b] \to \mathbb{C}$ is continuous on $[a, b]$, is absolutely continuous on $[\alpha, \beta]$ whenever $a < \alpha < \beta < b$, and that $V_a^b F < \infty$. Then F is absolutely continuous on $[a, b]$. [Hint: Use (6.85), (6.83), (6.55), and (6.84).]

2. *Lebesgue's Decomposition Theorem.* If $f : [a, b] \to \mathbb{C}$ satisfies $V_a^b f < \infty$, then there exists a unique pair of functions g and h on $[a, b]$ such that
   1. $f(x) = g(x) + h(x)$ for $a \leq x \leq b$,
   2. g is absolutely continuous on $[a, b]$,
   3. $h' = 0$ a.e. on $[a, b]$,
   4. $g(a) = 0$.
   Moreover, if f is real-valued and nondecreasing on $[a, b]$, then so are g and h. [Hint: Define $g(x) = \int_a^x f'$.]

3. Let $f : [a, b] \to \mathbb{R}$ be measurable and let B be a measurable subset of $[a, b]$ at each point of which f is differentiable.
   (a) The function $f' \xi_B$ is measurable and
   $$\lambda(f(B)) \leq \int_B |f'|. \tag{*}$$
   [Hints: Put $f(x) = 0$ if $x > b$ or $x < a$. For $x \in B$ and $x < b$, we have $f'(x) = \overline{\lim}_{n \to \infty} n[f(x + 1/n) - f(x)]$ and the function on the right is measurable (on $\mathbb{R}$). Given $\epsilon > 0$, let $B_n = \{x \in B : (n - 1)\epsilon \leq |f'(x)| < n\epsilon\}$ and then apply (6.88) to obtain $\lambda(f(B_n)) \leq n\epsilon\lambda(B_n) \leq \int_{B_n} |f'| + \epsilon\lambda(B_n)$.]
   (b) If f is nondecreasing on $[a, b]$, then equality obtains in (*). [Hints: Let I be any interval for which $J = f^{-1}(I) \neq \emptyset$ and choose sequences (α_k) and (β_k) in J such that $\alpha_k \leq \beta_k$, $\alpha_k \downarrow \inf J$, and $\beta_k \uparrow \sup J$. Employ (6.82) to obtain $\int_J f' = \lim_{k \to \infty} \int_{\alpha_k}^{\beta_k} f' \leq |I|$. For any sequence of intervals I_n covering $f(B)$, the intervals $J_n = f^{-1}(I_n)$ cover B, and so $\int_B f' \leq \sum |I_n|$.]

4. Let $f : [a, b] \to \mathbb{C}$ satisfy $V_a^b f < \infty$ and define v on $[a, b]$ by $v(x) = V_a^x f$.
   (a) Then $v' = |f'|$ a.e. on $[a, b]$. [Hints: Write $f = g + h$ as in Exercise 2. Define $u(x) = V_a^x g$ and $s(x) = V_a^x h$. Use (6.103) and (6.84) to get $u' = |f'|$ a.e. Show that u'

$\le v' \le u' + s'$ a.e. To show that $s' = 0$ a.e., first suppose h is real-valued. If $P_n = \{a = x_0 < \ldots < x_m = b\}$, let s_n be the unique function on $[a, b]$ for which $s_n(a) = 0$, $s_n(x_k) - s_n(x_{k-1}) = |h(x_k) - h(x_{k-1})|$, and either $s_n - h$ or $s_n + h$ is constant on $[x_{k-1}, x_k]$. Then $s - s_n$ is nondecreasing on $[a, b]$, and we may suppose $s(b) - s_n(b) < 2^{-n}$, so (4.53) yields $s' = \lim_{n\to\infty}(s - s_n)' = 0$ a.e.]

(b) If E is a measurable subset of $[a, b]$, then

$$\int_E |f'(x)|\,dx \le \lambda(v(E)).$$

Moreover, equality is obtained if f is absolutely continuous on $[a, b]$. [Hints: Use (a) and Exercise 3(b) applied to v and $B = \{x \in E: v$ is differentiable at $x\}$. If f is a.c., so is v and (6.87) yields $\lambda(v(B)) = \lambda(v(E))$.]

(c) If E is as in (b) and $\lambda(v(E)) = 0$, then $f' = 0$ a.e. on E.
(d) If f is real-valued and monotone on $[a, b]$ and E is as in (b), then $\lambda(f(E)) = 0$ if and only if $f' = 0$ a.e. on E.

5.   Let f be real-valued and *continuous* on $[a, b]$. For $y \in \mathbb{R}$, let $N(y)$ be the number of values of x in $[a, b]$ for which $f(x) = y$ if this number is finite and let $N(y) = \infty$ otherwise. The function N is called the *Banach indicatrix of f on* $[a, b]$. This exercise yields Banach's useful formula: $V_a^b f = \int N$. For any subdivision $P = \{a = x_0 < x_1 < \ldots < x_n = b\}$, write $I_1 = [x_0, x_1]$, $I_k = \,]x_{k-1}, x_k]$ for $2 \le k \le n$, and then define

$$N_P = \sum_{k=1}^n \xi_{f(I_k)} \quad \text{and} \quad \omega_P = \sum_{k=1}^n |f(I_k)|.$$

We then have
(a) $N_P(y) \le N(y)$ for all y and P;
(b) $\int N_P = \omega_P$ for all P;
(c) if $P_1 \subset P_2 \subset \ldots$ is an increasing sequence of subdivisions of $[a, b]$ with meshes tending to 0, then the functions N_{P_n} form a nondecreasing sequence which converges to N at each y. [Hint: If $1 \le m \le N(y)$, m an integer, and $t_1, \ldots, t_m$ are distinct roots of $f(t) = y$, then $N_{P_n}(y) \ge m$ whenever the mesh of P_n is less than $\min\{|t_i - t_j| : i \ne j\}$.];
(d) N is measurable and $\int N = \sup\{\int N_P : P$ is a subdivision of $[a, b]\}$;
(e) $V_a^b f = \sup\{\omega_P : P$ is a subdivision of $[a, b]\}$. [Hint: $|f(x_k) - f(x_{k-1})| \le |f(I_k)|$ for all k, so the sum V_P of these left sides does not exceed ω_P. By adjoining to P points at which f takes its max and min in each $[x_{k-1}, x_k]$ we obtain P^* with $V_{P^*} \ge \omega_P$.];
(f) $V_a^b f = \int_{-\infty}^{\infty} N(y)\,dy$;
(g) if $V_a^b f < \infty$, then the set of $y \in \mathbb{R}$ such that $f(x) = y$ for infinitely many $x \in [a, b]$ is a set of measure 0;
(h) [Varberg] if $V_a^b f < \infty$ and $v(x) = V_a^x f$ for $x \in [a, b]$, then we have $\lambda(v(E)) = 0$ whenever $E \subset [a, b]$ and $\lambda(f(E)) = 0$. [Hints: Let W be any open set containing $f(E)$ and let $]a_n, b_n[$ $(n = 1, 2, \ldots)$ be the component intervals of $f^{-1}(W)$. Check that $\lambda(v(E)) \le \sum V_{a_n}^{b_n} f = \sum \int N_n = \int_W N$, where N_n is the indicatrix of f for $[a_n, b_n]$. Use the integrability of N to make the last integral small.];
(i) if $V_a^b f < \infty$, $E \subset [a, b]$, and $\lambda(f(E)) = 0$, then $f' = 0$ a.e. on E. [Use (h) and Exercise 4(c).];
(j) give an example to show that (f) may fail if f is discontinuous at some point. The status of (i) for discontinuous f seems to be unknown.

6. *N-functions.* Let $f:[a,b] \to \mathbb{R}$ be given. We call f an *N-function* if $\lambda(f(A)) = 0$ whenever $A \subset [a,b]$ and $\lambda(A) = 0$.
   (a) If f is differentiable at all but countably many points of $[a,b]$ then f is an *N-function.* [Hint: Let $A_n = \{x \in A : -n \leqq f'(x) \leqq n\}$ and use (6.88).]
   (b) *Banach–Zarecki Theorem.* In order that f be absolutely continuous on $[a,b]$, it is necessary and sufficient that (1) f is continuous on $[a,b]$, (2) $V_a^b f < \infty$, and (3) f is an *N-function.* [Hints: If (1)–(3) obtain and $a_j < b_j$ in $[a,b]$, write $B = \{x : a_j < x < b_j, f$ is differentiable at $x\}$ and $A = [a_j, b_j] \backslash B$ and then use Exercise 3 to get $|f(b_j) - f(a_j)| \leqq \lambda(f([a_j, b_j])) \leqq \lambda(f(A)) + \lambda(f(B)) = \lambda(f(B)) \leqq \int_{a_j}^{b_j} |f'|$. Since $f' \in L_1([a,b])$, we can use (6.79).]
   (c) If f is continuous and convex (or concave) on $[a,b]$, then f is absolutely continuous on $[a,b]$.
   (d) If f is continuous on $[a,b]$, then f is an *N-function* if and only if $f(E)$ is measurable whenever $E \subset [a,b]$ is measurable. [Hints: If E is measurable, then $E = A \cup C$ where $\lambda(A) = 0$ and C is a countable union of compact sets. If $\lambda(A) = 0$ and $\lambda(f(A)) > 0$, then $f(E)$ is nonmeasurable for some $E \subset A$.]
   (e) If f is absolutely continuous on $[a,b]$, $f([a,b]) \subset [c,d]$, and $g:[c,d] \to \mathbb{R}$ is absolutely continuous on $[c,d]$, then the composite function $g \circ f$ is absolutely continuous on $[a,b]$ if and only if $V_{ag}^b \circ f < \infty$.
   (f) If f is continuous on $[a,b]$, f is differentiable except at countably many points of $[a,b]$, and $f' \in L_1([a,b])$, then f is absolutely continuous on $[a,b]$.

7. Let P be any Cantor set constructed in $[0,1]$ as in (2.81). Define f on $[0,1]$ by $f(x) = 0$ if $x \in P$ and $f(x) = (x - a)(b - x)$ if $x \in [a,b]$ where $]a,b[$ is a component interval of $[0,1] \backslash P$. Next define F on $[0,1]$ by $F(x) = \int_0^x f$. Then
   (a) f is continuous on $[0,1]$.
   (b) F is absolutely continuous and strictly increasing on $[0,1]$,
   (c) $F'(x) = f(x)$ for all $x \in [0,1]$,
   (d) $\{x : F'(x) = 0\} = P$,
   (e) $\lambda(F(P)) = 0$ [Hint: Show that $\sum (F(b) - F(a)) = \sum \int_a^b f = F(1)$, where the sums are over all $[a,b]$ as above.],
   (f) the inverse function F^{-1} is absolutely continuous on $[0, F(1)]$ if and only if $\lambda(P) = 0$ [Use Exercise 6(b) if necessary.],
   (g) f is absolutely continuous on $[0,1]$.

8. Let P be as in Exercise 7, let $\{]s_n, t_n[\}_{n=1}^\infty$ be any enumeration of the component intervals of $[0,1] \backslash P$, let $m_n = (s_n + t_n)/2$ be the midpoint of $]s_n, t_n[$, and let $(h_n)_{n=1}^\infty$ be any sequence of positive numbers converging to 0. Define f on $[0,1]$ by $f(x) = 0$ if $x \in P$, $f(m_n) = h_n$, and f is linear on each of $[s_n, m_n]$ and $[m_n, t_n]$; that is, $f(x) = h_n - h_n|x - m_n|/(t_n - m_n)$ for $s_n \leqq x \leqq t_n$. Then we have
   (a) f is continuous on $[0,1]$;
   (b) f is an *N-function* (see Exercise 6);
   (c) $V_0^1 f = 2 \cdot \sum_{n=1}^{\infty} h_n$;
   (d) f is absolutely continuous on $[0,1]$ if and only if $\sum_{n=1}^{\infty} h_n < \infty$ (see Exercise 6(b));
   (e) if r_n is the length of the longest of the $n + 1$ component intervals of $]0,1[$ $\backslash \bigcup_{j=1}^{n} [s_j, t_j]$ and $h_n = r_n + t_n - m_n$, then $h_n \to 0$ and $f'(x)$ exists for no $x \in P$

[Hints: Fix $x \in P$ with $x \neq 1$ and $x \neq s_n$ for all n. Then $D_+ f(x) = 0$. Let $n_1 = \min\{n : m_n > x\}$ and, inductively, $n_k = \min\{n : x < m_n < m_{n_{k-1}}\}$. Show that $n_1 < n_2 < \ldots,]x, s_{n_1}[$ is contained in one of the $n_k + 1$ intervals that define $r_{n_k}, m_{n_k} - x \leq h_{n_k}, [f(m_{n_k}) - f(x)]/(m_{n_k} - x) \geq 1$, and so $D^+ f(x) \geq 1$. Similarly, if $x \in P$, $x \neq 0$, and $x \neq t_n$ for all n, then $D^- f(x) = 0$ and $D_- f(x) \leq -1$.];

(f) there exists a continuous N-function on $[0, 1]$ which is not differentiable a.e.

9. *Convex Functions as Integrals.* Let $I \subset \mathbb{R}$ be an open interval and let $F : I \to \mathbb{R}$ be given. Then F is convex on I if and only if there exist an $a \in I$ and a nondecreasing $f : I \to \mathbb{R}$ such that

$$F(x) = F(a) + \int_a^x f$$

for all $x \in I$. [Hints: For the "if" part see the proof of (4.46). For the "only if" use (4.49), (4.43), and (6.85).]

10. Suppose that $f : \mathbb{R} \to \mathbb{C}$ is bounded, measurable, not identically 0, and $f(x + y) = f(x)f(y)$ for all $x, y \in \mathbb{R}$. Then there exists a unique $\alpha \in \mathbb{R}$ such that $f(x) = e^{i\alpha x}$ for all $x \in \mathbb{R}$. [Hints: Show that f is never 0 and $f(0) = 1$. Write $F(x) = \int_0^x f$, fix a with $F(a) \neq 0$, and show that $F(x + a) - F(x) = f(x)F(a)$. Conclude successively that F is continuous, f is continuous, F has a continuous derivative, and so does f. Obtain $f'(x + y) = f(x)f'(y)$ for all $x, y \in \mathbb{R}$ so that $f'(x) = i\alpha f(x)$, where $\alpha i = f'(0)$ ($\alpha \in \mathbb{C}$). Check that the function $x \to f(x)e^{-i\alpha x}$ has derivative 0 at every x, so is identically equal to 1 (its value at 0). Since f is bounded, α is real.]

11. Theorems (6.100) and (6.101) both become false if we omit the requirement that f be *real*-valued. Show this by considering $[a, b] = [0, 1]$, $f(x) = 3x^2 + 2ix$, and $g(x) = \alpha(x) = x$.

12. *Bonnet's Mean Value Theorems.* Let f and α be as in (6.101) and suppose that $\alpha \geq 0$ on $[a, b]$.
(a) If α is nondecreasing, then there exists $\xi \in [a, b]$ such that $\int_a^b f(x)\alpha(x)\,dx = \alpha(b-)\int_\xi^b f(x)\,dx$.
(b) If α is nonincreasing, then there exists $\xi \in [a, b]$ such that $\int_a^b f(x)\alpha(x)\,dx = \alpha(a+)\int_a^\xi f(x)\,dx$. [Hint: The left sides of these equalities are unaffected by altering the values of α at a and b.]
(c) If $[a, b] = [-1, 1]$ and $f(x) = \alpha(x) = x$, then both of these formulas fail for every $\xi \in \mathbb{R}$.

13. (a) If $0 \leq a < b < \infty$, then

$$\left| \int_a^b \frac{\sin x}{x}\,dx \right| < 3.$$

[Hint: If $b \leq \pi/2$, the left side is less than $\pi/2$. For $a \geq \pi/2$, use Exercise 12(b) to dominate the left side by $2/a < 4/\pi$.]
(b) If $0 \leq \xi < \delta < \pi/2$ and $m \in \mathbb{R}$, then

$$\left| \int_\xi^\delta \frac{\sin(mu)}{\sin u}\,du \right| \leq \frac{3\delta}{\sin \delta} \leq \frac{3\pi}{2}.$$

[Use Exercise 12(a) with $\alpha(u) = u/(\sin u)$ and use (a) after putting $mu = x$.]
(c) If $0 < \delta \leq a \leq b \leq \pi/2$ and $m > 0$, then

$$\left| \int_a^b \frac{\sin mu}{\sin u}\,du \right| \leq \frac{2}{m \sin \delta}.$$

(d) If $\alpha : [0, \pi/2] \to \mathbb{R}$ is nondecreasing with $\alpha(0+) \geq 0$, $0 < \delta < \pi/2$, and $m > 0$, then

$$\left| \frac{2}{\pi} \int_0^{\pi/2} \alpha(u) \frac{\sin(mu)}{\sin u} \, du \right| \leq 3\alpha(\delta) + \frac{4\alpha(\pi/2)}{m\pi \sin \delta} .$$

(e) If $x \in \mathbb{R}$, $\phi : [x - \pi, x + \pi] \to \mathbb{R}$ is nondecreasing, $n \in \mathbb{N}$, D_n is Dirichlet's Kernel as in (5.12), and $0 < \delta < \pi/2$, then

$$\left| \frac{1}{2\pi} \int_{-\pi}^{\pi} \phi(x - t) D_n(t) \, dt - \frac{\phi(x-) + \phi(x+)}{2} \right|$$

$$\leq \frac{3}{2} \left\{ [\phi(x-) - \phi(x - 2\delta)] + [\phi(x + 2\delta) - \phi(x+)] \right\}$$

$$+ \frac{\phi(x + \pi) - \phi(x - \pi)}{n\pi \sin \delta} .$$

[Hints: Use (d) twice with $m = 2n + 1$. (1) $\alpha(u) = \phi(x + 2u) - \phi(x+)$ and $u = -t/2$; (2) $\alpha(u) = \phi(x-) - \phi(x - 2u)$ and $u = t/2$. Notice that $\int_0^{\pi} D_n = \pi$.]

Now let $f : \mathbb{R} \to \mathbb{R}$ be 2π-periodic with $V_0^{2\pi} f < \infty$. Write

$$s_n(x) = \frac{1}{2\pi} \int_{-\pi}^{\pi} f(x - t) D_n(t) \, dt.*$$

(f) For all $x \in \mathbb{R}$ and $n \geq 0$, we have

$$|s_n(x)| < 2V_0^{2\pi} f + \|f\|_u$$

where $\|f\|_u = \sup\{|f(x)| : x \in \mathbb{R}\}$.

(g) For all $x \in \mathbb{R}$, we have

$$\lim_{n \to \infty} s_n(x) = \frac{f(x-) + f(x+)}{2} .$$

(h) If $[a, b] \subset \mathbb{R}$ and f is continuous at every $x \in [a, b]$ (two-sided continuity at a and b), then $s_n \to f$ uniformly on $[a, b]$. [Hints: In (f)–(h), we may suppose $x \geq 0$, $a \geq 0$. Write $v(t) = V_{-\pi}^t f$, $\phi_1 = \frac{1}{2}(v + f)$, and $\phi_2 = \frac{1}{2}(v - f)$. Check that v is continuous at all points where f is, ϕ_1 and ϕ_2 are nondecreasing on $[-\pi, \infty[$, $f = \phi_1 - \phi_2$, and $v = \phi_1 + \phi_2$. For (h), choose δ so that $v(x + 2\delta) - v(x - 2\delta) < \epsilon/3$ for all $x \in [a, b]$.]

14. *Vitali's Covering Theorem.* Let E be an arbitrary nonvoid subset of $\mathbb{R}$ (not necessarily measurable) and let $\mathcal{V}$ be a family of closed intervals of positive length such that for each $x \in E$ and each $\epsilon > 0$ there exists an $I \in \mathcal{V}$ for which $x \in I$ and $|I| < \epsilon$. In this case, we say that $\mathcal{V}$ is a *Vitali cover* of E. Then there exists a countable, pairwise disjoint, family $\mathcal{C} \subset \mathcal{V}$ whose union $A = \cup\mathcal{C}$ satisfies $\lambda(E \backslash A) = 0$. [Hints: We sketch Banach's brilliant proof, leaving many details to be filled in by the reader. It suffices to consider the case that $E \subset]a, b[$ and $I \subset]a, b[$ for all $I \in \mathcal{V}$, where $a < b$ in $\mathbb{R}$. Let $r_1 = \sup\{|I| : I \in \mathcal{V}\}$ and choose I_1 with $2|I_1| > r_1$. Suppose $I_1, \ldots, I_{n-1}$ have been selected, are pairwise disjoint, and A_{n-1} is their union. If $E \subset A_{n-1}$, then we are done. Otherwise, let $r_n = \sup\{|I| : I \in \mathcal{V}, I \cap A_{n-1} = \emptyset\}$ and choose I_n with

---

*As shown in (8.27), the function s_n is the nth partial sum of the Fourier series of f. Exercise 13 is essentially a recapitulation of Jordan's proof of the important convergence theorem (8.47).

$2|I_n| > r_n$. Supposing $E \setminus A_n \neq \emptyset$ for all n, we obtain $\mathcal{C} = \{I_n\}_{n=1}^{\infty}$. To see that this does the job, let J_n be the closed interval with the same midpoint as I_n and $|J_n| = 5|I_n|$. Now $\lambda(\bigcup_{n=p}^{\infty} J_n) \leqq 5 \sum_{n=p}^{\infty} |I_n| \to 0$ as $p \to \infty$ since $\sum_{n=1}^{\infty} |I_n| \leqq b - a$. We show that $E \setminus A \subset \bigcup_{n=p}^{\infty} J_n$ for every p. Fix p and $x \in E \setminus A$. Choose $I \in \mathcal{V}$ such that $x \in I$ and $I \cap A_p = \emptyset$. Since $r_n \downarrow 0$ and $|I| \leqq r_{n+1}$ whenever $I \cap A_n = \emptyset$, there is q such that $I \cap A_q \neq \emptyset$ and $I \cap A_{q-1} = \emptyset$. Then $q > p$, $I \cap I_q \neq \emptyset$, and $|I| \leqq r_q < 2|I_q|$, so $I \subset J_q$ and $x \in \bigcup_{n=p}^{\infty} J_n$.]

15. Let $f: [a, b] \to \mathbb{R}$ be nondecreasing.

(a) If $B \subset [a, b]$, $\beta \in \mathbb{R}$, and, for each $x \in B$, either $D_+ f(x) \leqq \beta$ or $D_- f(x) \leqq \beta$, then $\lambda(f(B)) \leqq \beta\lambda(B)$. [Hints: Let $B_0 = \{x \in B : a < x < b, f(x) \neq f(t)$ if $t \in [a, b]$ and $t \neq x\}$ write $E = f(B_0)$ and check that $f(B) \setminus E$ is countable. Let V be any open set with $B_0 \subset V \subset]a, b[$ and let $\beta' > \beta$. Apply Exercise 14 to the family $\mathcal{V}$ of intervals $[f(s), f(t)]$ for which $[s, t] \subset V$, s or t is in B_0, and $f(t) - f(s) < \beta'(t - s)$. Obtain $\lambda(E) \leqq \lambda(A) \leqq \beta'\lambda(V)$.]

(b) If $C \subset [a, b]$, $\gamma \in \mathbb{R}$, and, for each $x \in C$, either $D^+ f(x) \geqq \gamma$ or $D^- f(x) \geqq \gamma$, then $\lambda(f(C)) \geqq \gamma\lambda(C)$. [Hints: Suppose $\gamma > \gamma' > 0$ and that W is an open set containing $f(C)$. Let $E = \{x \in C : f$ is continuous at $x\}$ and note that $C \setminus E$ is countable. Apply Exercise 14 to the family $\mathcal{V}$ of intervals $[s, t]$ for which s or t is in E, $[f(s), f(t)] \subset W$, and $f(t) - f(s) > \gamma'(t - s)$. Obtain $\gamma'\lambda(E) \leqq \gamma'\lambda(A) \leqq \lambda(W)$.]

(c) If $E = \{x \in [a, b] : D^+ f(x) = \infty$ or $D^- f(x) = \infty\}$, then $\lambda(E) = 0$. [Hint: By (b), $f(b) - f(a) \geqq \lambda(f(E)) \geqq \gamma\lambda(E)$ for all $\gamma \in \mathbb{R}$.]

(d) It follows from (a) and (b) that $f'(x)$ exists a.e. on $[a, b]$. [Hint: For *rational* numbers $\beta < \gamma$, the set $E_{\beta, \gamma}$ where $D_+ f \leqq \beta$ and $D^- f \geqq \gamma$ has measure 0, so $D_+ f \geqq D^- f$ a.e.]

Now let $I = \{x \in]a, b[: f'(x) = \infty\}$ and $Z = \{x \in]a, b[: f'(x) = 0\}$.

(e) The function f is an N-function (see Exercise 6) if and only if $\lambda(F(I)) = 0$.

(f) We always have $\lambda(f(Z)) = 0$.

In the remaining parts, suppose that f is continuous as well as nondecreasing on $[a, b]$.

(g) The function f is absolutely continuous on $[a, b]$ if and only if $\lambda(f(I)) = 0$. [Hint: Use Exercise 6(b).]

(h) We have $f' = 0$ a.e. on $[a, b]$ if and only if $\lambda(f(I)) = f(b) - f(a)$. [Hint: The set $[a, b] \setminus (I \cup Z)$ is the union of sets $A_n = B \cap C$, where B and C are as in (a) and (b) with $\beta = n$ and $\gamma = 1/n$. That is, $(1/n)\lambda(A_n) \leqq \lambda(f(A_n)) \leqq n\lambda(A_n)$.]

(i) If f is strictly increasing from $[a, b]$ onto $[c, d]$ and $g = f^{-1}$, then g is absolutely continuous on $[c, d]$ if and only if $\lambda(Z) = 0$.

(j) If f and g are as in (i), then $g' = 0$ a.e. on $[c, d]$ if and only if $f' = 0$ a.e. on $[a, b]$.

(k) Assertion (b) fails for certain continuous (but not monotone) functions $f: [0, 1] \to \mathbb{R}$. [Hint: See Exercise 8(e).]

16. Let $F: [a, b] \to \mathbb{R}$ be continuous, let $\beta \geqq 0$, and let C be a countable set. Suppose that $D^+ F(x) \leqq \beta$ and $D_- F(x) \geqq -\beta$ for all $x \in]a, b[\setminus C$. Then $|F(x) - F(y)| \leqq \beta|x - y|$ for all $x, y \in [a, b]$. [Hint: Apply (6.88) to $B =]x, y[\setminus C$.]

17. Let $F: [c, d] \to \mathbb{C}$ be given. If F has any of the following six properties, then all six obtain.

(a) For some $\beta \in \mathbb{R}$, we have $|F(x) - F(y)| \leq \beta|x - y|$ for all x, y in $[c, d]$.

(b) F is absolutely continuous on $[c, d]$ and $|F'| \leq \beta$ a.e. on $[c, d]$ for some $\beta \in \mathbb{R}$.

(c) For each $\epsilon > 0$, there is a $\delta > 0$ such that

$$\sum_{k=1}^{n} |F(d_k) - F(c_k)| < \epsilon \quad \text{if} \quad \sum_{k=1}^{n} (d_k - c_k) < \delta$$

and $c \leq c_k < d_k \leq d$ for $1 \leq k \leq n$.

(d) If $[a, b] \subset \mathbb{R}$ and $\phi : [a, b] \to [c, d]$ is absolutely continuous on $[a, b]$, then $F \circ \phi$ is absolutely continuous on $[a, b]$.

(e) If $[a, b] \subset \mathbb{R}$ and $\phi : [a, b] \to [c, d]$ satisfies $|\phi(s) - \phi(t)| \leq |s - t|$ for all $s, t \in [a, b]$, then $V_a^b F \circ \phi < \infty$.

(f) There exists some $\delta > 0$ such that $|F(x) - F(y)| < 1/n$ whenever $n \in \mathbb{N}, c \leq x < y \leq d$, and $y - x < \delta/n$. [Hints: If F is unbounded, then all six fail, so suppose $|F(x)| \leq C < \infty$ for all $x \in [c, d]$. Assuming (a) fails, show that (f) fails as follows. Given $0 < \delta < (2C + 1)^{-1}$, choose x, y with $|F(x) - F(y)| > \delta^{-2}|x - y|$ and consider the smallest n such that $1/n \leq |F(x) - F(y)|$. Notice that $n = (n - 1) + 1 < |F(x) - F(y)|^{-1} + (\delta^{-1} - 1)/(2C) < \delta^2|x - y|^{-1} + (\delta^{-1} - 1)\delta^2|x - y|^{-1} = \delta|x - y|^{-1}$. Thus (f) implies (a). To show that the failure of (f) implies the failure of (e), suppose that $2^k n_k(y_k - x_k) < 1 \leq n_k|F(x_k) - F(y_k)|$ and (passing to subsequences if need be) that $|x_{k+1} - x_k| < 2^{-k}$. Consider closed intervals $I_1 < I_2 < \ldots$ in $[0, 3]$ with $|I_k| = n_k(y_k - x_k)$ and $\text{dist}(I_k, I_{k+1}) = 2^{1-k}$. Divide I_k into n_k subintervals of length $y_k - x_k$ and define ϕ at the division points to be alternately x_k and y_k. Make ϕ piecewise linear.]

18. Give direct proofs that the function F defined on $[0, 1] = [c, d]$ by $F(x) = \sqrt{x}$ has none of the properties (a)–(f) of Exercise 17.

19. Let α and β be real numbers and define F on $[0, \infty[$ by $F(0) = 0$ and $F(x) = x^\alpha \sin(x^{-\beta})$ if $x > 0$. Then we have

(a) F is continuous on $[0, \infty[$ if and only if $\alpha > 0$ or $\beta < \alpha \leq 0$;

(b) F is bounded on $[0, \infty[$ if and only if $\beta \geq \alpha > 0$ or $\alpha = 0$ or $\beta \leq \alpha < 0$;

(c) $V_0^1 F < \infty$ if and only if $\beta < \alpha$ or $\beta = \alpha \leq 0$;

(d) F is absolutely continuous on $[0, 1]$ if and only if $\beta < \alpha$;

(e) F' is bounded on $[0, 1]$ if and only if $\alpha \geq \beta + 1$;

(f) $F'_+(0)$ exists and is finite if and only if $\beta > 0$ and $\alpha > 1$ or $\beta \leq 0$ and $\alpha \geq \beta + 1$.

20. Let $F : \mathbb{R} \to \mathbb{R}$ be any everywhere differentiable function such that F' is bounded on $\mathbb{R}$ but F is monotone on *no* nonvoid open interval of $\mathbb{R}$ (see Exercises 7 and 8, p. 216–218, for the existence of such functions).

(a) If $a < b$ in $\mathbb{R}$, then F is absolutely continuous on $[a, b]$.

(b) The function F' is of Baire class 1 on $\mathbb{R}$ (see Exercise 13, p. 309).

(c) The sets $A = \{x \in \mathbb{R} : F'(x) > 0\}$ and $B = \{x \in \mathbb{R} : F'(x) < 0\}$ satisfy $\lambda(A \cap I) > 0$ and $\lambda(B \cap I) > 0$ for every nonvoid open interval $I \subset \mathbb{R}$. [Hint: If not, then F is monotone on I by (6.85).]

(d) If F' is continuous at x, then $F'(x) = 0$.

(e) If $a < b$ in $\mathbb{R}$, then F' is not Riemann integrable over $[a, b]$.

(f) The set of points at which F attains a local maximum (resp., minimum) value is dense in $\mathbb{R}$.

21.   Suppose that $F: \mathbb{R} \to \mathbb{C}$ is absolutely continuous on $[-b, b]$ for all $b > 0$, that
        $F \in L_1(\mathbb{R})$, and that $F' \in L_1(\mathbb{R})$. Then
        (a) $F(x) = \int_{-\infty}^{x} F'(t)\, dt$ for all $x \in \mathbb{R}$,
        (b) $\lim_{x \to -\infty} F(x) = 0 = \lim_{x \to \infty} F(x)$,
        (c) $\int_{-\infty}^{\infty} F'(t)\, dt = 0$.
        [Hint: First find $a_n \to -\infty$ with $F(a_n) \to 0$ and then use (6.85).]

22.   In order that a function $F: \mathbb{R} \to \mathbb{C}$ be expressible in the form

$$F(x) = \int_{-\infty}^{x} f(t)\, dt \qquad (x \in \mathbb{R})$$

        for some $f \in L_1(\mathbb{R})$, it is necessary and sufficient that (1) F is absolutely continu-
        ous on $[-b, b]$ for all $b > 0$, (2) $\lim_{x \to -\infty} F(x) = 0$, and (3) $\lim_{b \to \infty} V_{-b}^{b} F < \infty$. In
        this event, we have $f = F'$ a.e.

23.   Let $u: \mathbb{R} \to \mathbb{R}$ satisfy (1) $u(-x) = u(x)$ for all x, (2) $u(x) = 0$ if $x \geq 1$, (3) u is
        absolutely continuous on $[-1, 1]$, (4) u is nonincreasing on $[0, 1]$, and (5) $\int_{-1}^{1} u$
        $= 1$. For each $n > 0$ define u_n by $u_n(x) = nu(nx)$.
        (a) If $f \in L_1([a, b])$, $a < c < b$, and

$$\lim_{h \downarrow 0} \frac{1}{2h} \int_{c-h}^{c+h} f(t)\, dt = s,$$

        where $s \in \mathbb{C}$, then

$$\lim_{n \to \infty} \int_{a}^{b} f(x) u_n(c - x)\, dx = s.$$

        [Hints: Write $F(x) = \int_{a}^{x} f$ and show that

$$\int_{a}^{b} f(x) u_n(c - x)\, dx = -\int_{0}^{1/n} \frac{F(c + h) - F(c - h)}{2h} \cdot 2hu_n'(h)\, dh$$

        and

$$s = -\int_{0}^{1/n} s \cdot 2hu_n'(h)\, dh.$$

        Then subtract, take absolute values, and note that $-2hu_n'(h) \geq 0$ a.e.]
        (b) Let $I \subset \mathbb{R}$ be an open interval and suppose that $f: I \to \mathbb{C}$ is measurable and
        satisfies $\int_{a}^{b} |f| < \infty$ whenever $a < b$ in I. Then

$$\lim_{n \to \infty} \int_{I} f(t) u_n(x - t)\, dt = f(x)$$

        for almost every $x \in I$. [Hint: Use (a) and (6.84).]
        (c) There exists a function u in $C^{\infty}(\mathbb{R})$ which has properties (1)–(5) above. [Hint:
        Let g be as in the proof of (4.31) and take $u(x) = cg(x)g(-x)$ for an appropri-
        ate constant c.]
        (d) Let I and f be as in (b) and let g be another function having the same properties
        as f. Suppose that

$$\int_{I} f(t) \phi'(t)\, dt = -\int_{I} g(t) \phi(t)\, dt \qquad (*)$$

        for all $\phi \in C^{\infty}(\mathbb{R})$ that are identically 0 outside some closed subinterval of I.
        Then there exists $a \in I$ such that

$$f(x) - f(a) = \int_{a}^{x} g(t)\, dt$$

for almost every $x \in I$. In particular, there is a function f_0 on I which is absolutely continuous on each closed subinterval of I such that $f_0 = f$ and $f_0' = g$ a.e. on I. If (∗) obtains for all ϕ as above, we say that g is the *distributional derivative* of f on I. [Hints: Fix any $a, x \in I$ at each of which the conclusion of (b) obtains (where u is as in (c)). Let $\phi_n(y) = \int_{-\infty}^{y} [u_n(x - t) - u_n(a - t)]\, dt$. Considering the cases $x < a$ and $x > a$, find $\lim_{n \to \infty} \phi_n(y)$. Put ϕ_n into (∗) and let $n \to \infty$.]

24.   Let $f \in L_1([a, b])$ be given and write

$$A = \lim_{h \downarrow 0} \int_a^{b-h} |f(x + h) - f(x)|\, dx, \qquad B = \varlimsup_{h \downarrow 0} \int_a^{b-h} |f(x + h) - f(x)|\, dx.$$

(a) If $A = 0$, then there exists $c \in \mathbb{C}$ such that $f(x) = c$ a.e. on $[a, b]$. [Hint: Write $F(x) = \int_a^x f$ and suppose that $a < s < t < b$ satisfy $F'(s) = f(s)$ and $F'(t) = f(t)$. Check that

$$\int_s^t [f(x + h) - f(x)]\, dx = [F(t + h) - F(t)] - [F(s + h) - F(s)]$$

for small h and conclude that

$$|f(t) - f(s)| \leq \lim_{h \downarrow 0} \int_s^t |f(x + h) - f(x)|\, dx.]$$

(b) If $A < \infty$, then there exists a function $f_0 : [a, b] \to \mathbb{C}$ such that $f = f_0$ a.e. on $[a, b]$ and $V_a^b f_0 \leq A$. [Hints: Let $E = \{t : a < t < b, F'(t) = f(t)\}$, show that

$$\sum_{k=1}^n |f(t_k) - f(t_{k-1})| \leq A \text{ if } \{t_0 < t_1 < \cdots < t_n\} \subset E. \text{ Let } f_0 = f \text{ on } E.]$$

(c) If $V_a^b f < \infty$, then $B < \infty$. [Hint: First, suppose f is nondecreasing on $[a, b]$.]

25.   Suppose that $f : \mathbb{R} \to \mathbb{C}$ is measurable and has arbitrarily small periods; that is, for each $\epsilon > 0$ there exists $0 < h < \epsilon$ such that $f(x + h) = f(x)$ for all $x \in \mathbb{R}$. Then there is some $c \in \mathbb{C}$ such that $f(x) = c$ a.e. on $\mathbb{R}$. [Hint: Consider truncates of f. See Exercise 24(a).]

Hölder's and Minkowski's Inequalities

These two inequalities are indispensable in the study of many of the function spaces of vital importance in modern analysis. Before presenting them, we prove a simple lemma.

(6.105) Lemma   *Let $1 < p < \infty$ be a real number and write $p' = p/(p - 1)$. If $a \geq 0$ and $b \geq 0$ are real numbers, then*

$$ab \leq a^p/p + b^{p'}/p'.$$

Moreover, equality obtains if and only if $b = a^{p-1}$.

Proof   For $t \geq 0$, define

$$f(t) = 1/p + t^{p'}/p' - t.$$

Then $f(1) = 0$, $f'(t) > 0$ for $t > 1$, and $f'(t) < 0$ for $0 \leq t < 1$, so $f(t) > 0$

unless $t = 1$. Given $a > 0$ and $b \geqq 0$, we have

$$0 \leqq f(a^{1-p}b) = \frac{1}{p} + a^{-p}b^{p'}/p' - a^{1-p}b,$$

where the inequality is strict if and only if $b \neq a^{p-1}$. $\square$

(6.106) Hölder's Inequality   *Let* $1 < p < \infty$ *be a real number and write* p'
$= p/(p-1)$. *If* $f, g \in M$, *then*

(i)
$$\int |fg| \leqq \left(\int |f|^p \right)^{1/p} \cdot \left(\int |g|^{p'} \right)^{1/p'},$$

where ∞^a *means* ∞ *for* $a > 0$. *In case* $p = 2$, (i) *is called* Schwarz's Inequality.

Proof   Given f and g, write

$$A = \left(\int |f|^p \right)^{1/p} \quad \text{and} \quad B = \left(\int |g|^{p'} \right)^{1/p'}. \tag{1}$$

If $A = 0$ or $B = 0$, then $fg = 0$ a.e., and hence both sides of (i) are 0. Thus
we suppose $AB > 0$. If A or B is ∞, then (i) is obvious, so we also suppose
$A < \infty$ and $B < \infty$.

Given $x \in \mathbb{R}$ at which both f and g are defined, put $a = |f(x)|/A$
and $b = |g(x)|/B$ and then apply (6.105) to obtain

$$\frac{|f(x)g(x)|}{AB} \leqq \frac{|f(x)|^p}{pA^p} + \frac{|g(x)|^{p'}}{p'B^{p'}}. \tag{2}$$

Since (2) obtains a.e., we have

$$\frac{1}{AB} \int |fg| \leqq \frac{1}{pA^p} \cdot \int |f|^p + \frac{1}{p'B^{p'}} \cdot \int |g|^{p'} = \frac{1}{p} + \frac{1}{p'} = 1$$

by (1). Therefore $\int |fg| \leqq AB$. $\square$

(6.107) Minkowski's Inequality   *Let* $1 \leqq p < \infty$ *be a real number. If* $f, g \in M$, *then*

$$\left(\int |f + g|^p \right)^{1/p} \leqq \left(\int |f|^p \right)^{1/p} + \left(\int |g|^p \right)^{1/p},$$

where ∞^a *means* ∞ *for* $a > 0$.

Proof   This is obvious if either integral on the right is ∞ or the integral on
the left is 0. Thus we suppose that $\int |f|^p < \infty, \int |g|^p < \infty$, and $\int |f + g|^p > 0$.
Since

$$|f + g|^p \leqq (|f| + |g|)^p \leqq (2[|f| \vee |g|])^p$$

$$= 2^p [|f|^p \vee |g|^p] \leqq 2^p [|f|^p + |g|^p]$$

a.e., we also have $\int |f + g|^p < \infty$. We already know this theorem if $p = 1$, so we suppose $p > 1$ and use (6.106). Since $(p - 1)p' = p$, we have

$$\int |f + g|^p \le \int (|f| \cdot |f + g|^{p-1}) + \int (|g| \cdot |f + g|^{p-1})$$

$$\le \left(\int |f|^p\right)^{1/p} \cdot \left(\int |f + g|^p\right)^{1/p'} + \left(\int |g|^p\right)^{1/p} \cdot \left(\int |f + g|^p\right)^{1/p'}$$

Since $1 - 1/p' = 1/p$, the proof is completed by dividing by $(\int |f + g|^p)^{1/p'}$.
$\square$

The L_p Spaces

(6.108) Definition For a positive real number p, we let $L_p(\mathbb{R})$, or simply L_p, denote the set of all $f \in M$ for which $|f|^p \in L_1$, and we write

$$\|f\|_p = \left(\int |f|^p\right)^{1/p}.$$

The number $\|f\|_p$ is called the L_p-*norm* of f. For $E \in \mathfrak{M}$, $L_p(E)$ denotes the set of all f that are measurable on E for which $f\xi_E \in L_p$; that is,

$$\int_E |f|^p < \infty$$

[see (6.78)].

(6.109) Theorem *Let $1 \le p < \infty$, let $f, g \in L_p$, and let $\alpha \in \mathbb{C}$. Then*
    (i)   *αf and $f + g$ are in L_p,*
    (ii) *$\|\alpha f\|_p = |\alpha| \cdot \|f\|_p$,*
    (iii) *$\|f\|_p > 0$ unless $f = 0$ a.e.,*
    (iv) *$\|f + g\|_p \le \|f\|_p + \|g\|_p$.*

Proof Apply (6.108) and (6.107). $\square$

The L_p spaces, like L_1, are complete in the following sense.

(6.110) Theorem *Let $1 \le p < \infty$ be given and suppose that $(f_n)_{n=1}^{\infty} \subset L_p$ satisfies*

    (i)
$$\lim_{m,n \to \infty} \|f_m - f_n\|_p = 0.$$

Then there exists $f \in L_p$ such that

    (ii)
$$\lim_{n \to \infty} \|f - f_n\|_p = 0.$$

This function f is unique in the sense that if $h \in L_p$ satisfies $\|h - f_n\|_p \to 0$, then $h = f$ a.e. Moreover, there is a subsequence

$$\left(f_{n_k}\right)_{k=1}^{\infty} \text{ such that}$$

(iii) $$\lim_{k\to\infty} f_{n_k}(x) = f(x) \quad a.e.$$

Proof This proof is almost the same as that of (6.56), so we give it with less detail than we did there.

Choose $n_1 < n_2 < \cdots$ in $\mathbb{N}$ such that $\| f_m - f_{n_k} \|_p < 2^{-k}$ whenever $m > n_k$. Writing $f_{n_0} = 0$, (6.109.iv) yields

$$\int \left(\sum_{j=1}^{k} |f_{n_j} - f_{n_{j-1}}| \right)^p = \left\| \sum_{j=1}^{k} |f_{n_j} - f_{n_{j-1}}| \right\|_p^p$$

$$\leqq \left(\sum_{j=1}^{k} \| f_{n_j} - f_{n_{j-1}} \|_p \right)^p$$

$$< \left(\| f_{n_1} \|_p + \sum_{j=2}^{\infty} 2^{-j+1} \right)^p = (\| f_{n_1} \|_p + 1)^p = C < \infty$$

for all $k \geqq 1$. By (6.20) [or (6.43)], the function $\left(\sum_{j=1}^{\infty} |f_{n_j} - f_{n_{j-1}}| \right)^p$ is in L_1, so it is finite a.e. Thus, at almost every $x \in \mathbb{R}$, the series $\sum_{j=1}^{\infty} (f_{n_j}(x) - f_{n_{j-1}}(x))$, whose kth partial sum is $f_{n_k}(x)$, converges [absolutely], and so there exists $f \in M$ such that (iii) obtains [use (6.54)].

Given $\epsilon > 0$, use (i) to get $r \in \mathbb{N}$ such that

$$\| f_{n_k} - f_n \|_p < \epsilon \quad \text{if} \quad k > r \quad \text{and} \quad n > n_r.$$

By (6.47), $n > n_r$ implies

$$\| f - f_n \|_p \leqq \varliminf_{k\to\infty} \| f_{n_k} - f_n \|_p \leqq \epsilon.$$

Therefore $f = f_n + (f - f_n) \in L_p$ and (ii) obtains. $\square$

Our next theorem establishes that the complex-valued step functions and the continuous functions vanishing outside bounded intervals both form dense subsets of each L_p space.

(6.111) Theorem Let $p \geqq 1$ and $\epsilon > 0$ be real numbers, and let $f \in L_p$. Then there exist a function $\phi : \mathbb{R} \to \mathbb{C}$ having real and imaginary parts in M_0 and a function $g : \mathbb{R} \to \mathbb{C}$ which is continuous on $\mathbb{R}$ and is identically 0 outside some bounded interval such that

$$\| f - \phi \|_p < \epsilon \quad \text{and} \quad \| f - g \|_p < \epsilon.$$

Proof First, suppose that f is real-valued. For $n \in \mathbb{N}$, define f_n on $\mathbb{R}$ by $f_n(x) = f(x)$ if $f(x)$ is defined and neither $|x|$ nor $|f(x)|$ exceeds n, and

$f_n(x) = 0$ for all other x. Since $\lim_{n\to\infty} |f(x) - f_n(x)|^p = 0$ a.e. and $|f(x) - f_n(x)|^p \le |f(x)|^p$ a.e. for all n, (6.22) implies that $\lim_{n\to\infty} \|f - f_n\|_p = 0$. Thus, we can find $N \in \mathbb{N}$ such that $\|f - f_N\|_p < \epsilon/6$. Now $f_N \in L_1^r$ and $|f_N| \le N\xi_{[-N,N]}$, so (6.18) provides a sequence $(\phi_k)_{k=1}^\infty \subset M_0$ such that $\lim_{k\to\infty} |f_N(x) - \phi_k(x)|^p = 0$ a.e. We may suppose that $|\phi_k| \le N\xi_{[-N,N]}$ and so $|f_N - \phi_k|^p \le 2^p N^p \xi_{[-N,N]}$ for all k. Another application of (6.22) yields some k such that $\phi = \phi_k$ satisfies $\|f_N - \phi\|_p < \epsilon/6$. By (6.109.iv) we have $\|f - \phi\|_p \le \|f - f_N\|_p + \|f_N - \phi\|_p < \epsilon/3$. We next alter ϕ slightly to make it continuous. Choose an open set $V \subset [-N-1, N+1]$ that contains all of the (finitely many) points at which ϕ is discontinuous and satisfies $2^p N^p \lambda(V) < (\epsilon/6)^p$. Now let g be any continuous function on $\mathbb{R}$ such that $g = \phi$ on $\mathbb{R}\setminus V$ and $g(V) \subset [-N, N]$ (e.g., we can take g to be linear on each component interval of V). Then g, like ϕ, is identically 0 outside $[-N-1, N+1]$ and

$$\|\phi - g\|_p^p = \int_V |\phi - g|^p \le \int_V (2N)^p < \left(\frac{\epsilon}{6}\right)^p.$$

Thus $\|f - g\|_p < \epsilon/3 + \epsilon/6 = \epsilon/2$.

In case f is complex-valued, write $f = f_1 + if_2$, where f_1 and f_2 are real-valued, and then apply the preceding paragraph to each f_j to obtain ϕ_j and g_j such that $\|f_j - \phi_j\|_p < \epsilon/2$ and $\|f_j - g_j\|_p < \epsilon/2$ $(j = 1, 2)$. Finally, write $\phi = \phi_1 + i\phi_2$ and $g = g_1 + ig_2$. $\square$

Exercises

1. Let $0 < r < s < \infty$ be given.
    (a) If $f \in M$ and $E \in \mathfrak{M}$, then

    $$\left(\int_E |f|^r\right)^{1/r} \le \left(\int_E |f|^s\right)^{1/s} (\lambda(E))^{1/r - 1/s}.$$

    [Hint: Use Hölder's Inequality with $p = s/r$.]
    (b) If $E \in \mathfrak{M}$ and $\lambda(E) < \infty$, then $L_s(E) \subset L_r(E)$.
    (c) $L_r([0,1])\setminus L_s([0,1]) \ne \varnothing$.

2. Let $0 < p < \infty$ be given and define $f(x) = 0$ for $x \le 0$, $f(x) = x^{-1/p}(1 + |\log x|)^{-2/p}$ for $x > 0$. Then $f \in L_p\setminus L_q$ if $0 < q \ne p$.

3. (a) If $0 < r < s < t < \infty$ and $f \in L_r \cap L_t$, then $f \in L_s$ and $\|f\|_s^s \le \|f\|_r^{r\alpha} \cdot \|f\|_t^{t(1-\alpha)}$, where $s = \alpha r + (1 - \alpha)t$.
    (b) Suppose $f \in L_p \cap L_q$ and $\|f\|_p > 0$ for some $q > p > 0$. Let $a = \inf\{p > 0: f \in L_p\}$ and $b = \sup\{p > 0: f \in L_p\}$. Then the formula $\phi(p) = p\log(\|f\|_p)$ defines a function ϕ that is convex on $]a, b[$.

4. Let $E \in \mathfrak{M}$ have finite measure and let f be measurable on E. Then

    $$\lim_{p\to\infty} \left(\int_E |f|^p\right)^{1/p} = \|f\|_\infty,$$

where $\|f\|_\infty$ is defined by

$$\|f\|_\infty = \inf\{t \in \mathbb{R} : t > 0, \lambda(\{x \in E : |f(x)| > t\}) = 0\}$$

and $\|f\|_\infty = \infty$ if no such t exists.
[Hint: If $0 < s < \|f\|_\infty$ and $A = \{x \in E : |f(x)| > s\}$, then $(\int_E |f|^p)^{1/p} \geq s(\lambda(A))^{1/p} \to s$.]

5.   Let $E \in \mathfrak{M}$ with $0 < \lambda(E) < \infty$ and let $f \in M$. For $0 < p < \infty$, write

$$m(p) = \left(\frac{1}{\lambda(E)} \int_E |f|^p\right)^{1/p}$$

[$m(p) = \infty$ if $f \notin L_p(E)$].
(a) The function m is nondecreasing on $]0, \infty[$ and $\lim\limits_{p \to \infty} m(p) = \|f\|_\infty$. [See Exercises 1 and 4.]
(b) If $y \geq 0$ is a real number, then $p^{-1}(y^p - 1) \downarrow \log y$ as $p \downarrow 0$, where we define $\log 0 = -\infty$. [Hint: Check monotonicity of functions by differentiation.]
(c) If $m(p_0) < \infty$ for some $p_0 > 0$, then $\int_E \log|f|$ is defined and

$$\lim_{p \downarrow 0} \frac{1}{p} \int_E (|f|^p - 1) = \int_E \log|f|.$$

[Hint: Consider negatives and use the Monotone Convergence Theorem.]
(d) Defining $\log \infty = \infty$ and $\log 0 = -\infty$, we have

$$\frac{1}{\lambda(E)} \int_E \log|f| \leq \log(m(p))$$

for all $p > 0$ if this integral is defined. [Hints: Check the inequality $\log t \leq t - 1$, substitute $t = (|f(x)|/(m(p)))^p$, and then integrate over E.]
(e) If $m(p_0) < \infty$ for some $p_0 > 0$, then

$$\lim_{p \downarrow 0} m(p) = \exp\left[\frac{1}{\lambda(E)} \int_E \log|f|\right].$$

The right side of this equality is called the *geometric mean* of $|f|$ over E while $m(1)$ is called the *arithmetic mean* of $|f|$ over E. [Hints: Use $\log t \leq t - 1$ to see that

$$\log(m(p)) \leq \frac{1}{p\lambda(E)} \int_E (|f|^p - 1).]$$

(f) If $E =]0, 1]$ and $f(x) = \exp(x^{-1/2})$, then the equalities in (c) and (e) both fail.

6.   *Generalized Hölder Inequality.* If $f_1, f_2, \ldots, f_n$ are in M and $\alpha_1, \alpha_2, \ldots, \alpha_n$ are positive real numbers whose sum is 1, then

$$\int\left(\prod_{j=1}^n |f_j|^{\alpha_j}\right) \leq \prod_{j=1}^n \left(\int |f_j|\right)^{\alpha_j}.$$

7.   Let X be a nonvoid set. For any function $\phi : X \to [0, \infty]$ define

$$\sum_{x \in X} \phi(x) = \sup\left\{\sum_{j=1}^n \phi(x_j) : n \in \mathbb{N}, \ \{x_1, \ldots, x_n\} \subset X\right\}.$$

For any real number $p > 0$, let $l_p(X)$ denote the set of all functions $f : X \to \mathbb{C}$ for which $\sum\limits_{x \in X} |f(x)|^p < \infty$ and write

$$\|f\|_p = \left(\sum_{x \in X} |f(x)|^p \right)^{1/p}.$$

(a) If $1 < p < \infty$, $p' = p/(p-1)$, $f \in l_p(X)$, and $g \in l_{p'}(X)$, then $fg \in l_1(X)$ and $\|fg\|_1 \leqq \|f\|_p \cdot \|g\|_{p'}$.

(b) If $1 \leqq p < \infty$, $f \in l_p(X)$, and $g \in l_p(X)$, then $f + g \in L_p(X)$ and $\|f + g\|_p \leqq \|f\|_p + \|g\|_p$.

(c) For $1 \leqq p < \infty$, the formula $d(f, g) = \|f - g\|_p$ defines a metric d on $l_p(X)$ and makes $l_p(X)$ a complete metric space. This space is separable if and only if X is countable. The set $B = \{f \in l_p(X) : \|f\|_p \leqq 1\}$ is compact if and only if X is finite.

(d) If $0 < p < q < \infty$, then $l_p(X) \subset l_q(X)$ and $\|f\|_q \leqq \|f\|_p$ for all $f \in l_p(X)$. [Hint: Notice that $t^p \geqq t^q$ if $0 \leqq t \leqq 1$ and put $t = |f(x)|/\|f\|_q$.]

8. (a) If $f : \mathbb{R} \to \mathbb{C}$ is uniformly continuous on $\mathbb{R}$ and $f \in L_p$ for some $0 < p < \infty$, then $\lim\limits_{|x| \to \infty} f(x) = 0$.

(b) For any $0 < p < \infty$, there exist continuous unbounded functions $f \in L_p$.

9. (a) Find necessary and sufficient conditions on f and g that equality obtains in Hölder's Inequality.

(b) Do the same for Minkowski's Inequality. [Hint: Keeping (6.105) in mind, trace through the proofs to determine how strict inequality can creep in.]

10. Let $0 < p < 1$ be given.

(a) There exist $f, g \in L_p([0, 1])$ such that $\|f + g\|_p > \|f\|_p + \|g\|_p$.

(b) For all $f, g \in L_p$ we have $\|f + g\|_p^p \leqq \|f\|_p^p + \|g\|_p^p$. [Hint: Check that $(a + b)^p \leqq a^p + b^p$ for $a, b \geqq 0$ by showing that $1 + t^p - (1 + t)^p$ increases with t and putting $t = b/a$.]

(c) If we define $d(f, g) = \|f - g\|_p^p$, then d is a metric for L_p (except that $d(f, g) = 0$ only implies $f = g$ a.e.) and L_p is a complete metric space.

(d) If $f, g \in L_p$, then $\|f + g\|_p \leqq 2^{(1-p)/p}[\|f\|_p + \|g\|_p]$. [Hint: By maximizing $\phi(t) = (1 + t)^{1/p}(1 + t^{1/p})^{-1}, t \geqq 0$, show that $\phi(t) \leqq 2^{(1-p)/p}$ and then put $t = \|g\|_p^p/\|f\|_p^p$.]

(e) Theorem (6.111) is true for $0 < p < 1$.

Integration on $\mathbb{R}^n$

Much of what appears earlier in this chapter can be repeated *verbatim* to develop the theory of Lebesgue integration on $\mathbb{R}^n$. In this section we give the necessary definitions for this extension of the theory and we point out precisely which ones of our earlier theorems carry over to this setting. In addition we list those few nonobvious amendments that must be made to the proofs for $\mathbb{R}$ in order to obtain proofs for $\mathbb{R}^n$. We implore the reader to carry out the valuable exercise of turning back and carefully checking that what we claim for $\mathbb{R}^n$ is actually correct.

(6.112) Definitions   Let $n \in \mathbb{N}$ be fixed. An *interval* of $\mathbb{R}^n$ is a set of the form

$$I = \mathop{\times}\limits_{j=1}^{n} I_j = \{ x = (x_1, \ldots, x_n) \in \mathbb{R}^n : x_j \in I_j \quad \text{for} \quad 1 \leq j \leq n \}$$

where each I_j is an interval of $\mathbb{R}$. We call I an *open* [resp., *closed*] *interval* if each I_j is open [resp., closed]. If $a_j \leq b_j$ in $\mathbb{R}^{\#}$ $(1 \leq j \leq n)$, we use the notation

$$\begin{aligned}
]a, b[&= \{ x \in \mathbb{R}^n : a_j < x_j < b_j \quad \text{for} \quad 1 \leq j \leq n \}, \\
[a, b] &= \{ x \in \mathbb{R}^n : a_j \leq x_j \leq b_j \quad \text{for} \quad 1 \leq j \leq n \}, \\
[a, b[&= \{ x \in \mathbb{R}^n : a_j \leq x_j < b_j \quad \text{for} \quad 1 \leq j \leq n \}, \\
]a, b] &= \{ x \in \mathbb{R}^n : a_j < x \leq b_j \quad \text{for} \quad 1 \leq j \leq n \}.
\end{aligned}$$

For I as above, we define the *n-dimensional volume* of I to be the number

$$|I| = \prod_{j=1}^{n} |I_j|.$$

Notice that $|I| = 0$ if $|I_j| = 0$ for some j and that $|I| = \infty$ if $|I| \neq 0$ and $|I_j| = \infty$ for some j.

For $E \subset \mathbb{R}^n$, we define the *n-dimensional Lebesgue* [*outer*] *measure* of E to be the number

$$\lambda^n(E)$$

$$= \inf \left\{ \sum_{I \in \mathscr{I}} |I| : \mathscr{I} \text{ is a countable family of open intervals of } \mathbb{R}^n \text{ that covers } E \right\}.$$

A set $E \subset \mathbb{R}^n$ is called a *null set* of $\mathbb{R}^n$ if $\lambda^n(E) = 0$. [Compare (2.75) and (2.78).]

If $P(x)$ is a proposition about elements $x \in \mathbb{R}^n$, then "$P(x)$ almost everywhere" and "$P(x)$ a.e." mean that

$$\lambda^n(\{ x \in \mathbb{R}^n : P(x) \text{ is false} \}) = 0.$$

One can now prove the following theorem just as we did for the corresponding assertions in (2.76) and (2.79).

(6.113) Theorem
(i) $E \subset \mathbb{R}^n$ *implies* $0 \leq \lambda^n(E) \leq \infty$.
(ii) $\lambda^n(\emptyset) = 0$.
(iii) $E \subset F \subset \mathbb{R}^n$ *implies* $\lambda^n(E) \leq \lambda^n(F)$.
(iv) $(E_k)_{k=1}^{\infty} \subset \mathscr{P}(\mathbb{R}^n)$ *implies* $\lambda^n(\bigcup\limits_{k=1}^{\infty} E_k) \leq \sum\limits_{k=1}^{\infty} \lambda^n(E_k)$.
(v) *If* $C \subset \mathbb{R}^n$ *is countable, then* $\lambda^n(C) = 0$.
(vi) *If* $E \subset \mathbb{R}^n$ *and* $x \in \mathbb{R}^n$, *then* $\lambda^n(E + x) = \lambda^n(E)$, *where*

$$E + x = \{ y + x : y \in E \}.$$

(vii) *If* $E, N \subset \mathbb{R}^n$ *and* $\lambda^n(N) = 0$, *then* $\lambda^n(E \cup N) = \lambda^n(E)$.

(6.114) Remark Observe that we did *not* include in the above theorem the assertion that $\lambda^n(I) = |I|$ for intervals $I \subset \mathbb{R}^n$. This statement is true, but a direct proof from (6.112) is quite complicated. It follows though as a corollary of our extension to $\mathbb{R}^n$ of (6.60).

In order to carry over our proof of (6.60) to the present setting we need to know that each open subset of $\mathbb{R}^n$ is expressible as the union of a countable pairwise disjoint family of (not necessarily open) intervals of positive volume. This fact is a corollary of our next theorem.

(6.115) Definition Let m be any integer. A *dyadic n-cube of order m* is an interval of the form

$$I = \{x \in \mathbb{R}^n : k_j \leq 2^m x_j < k_j + 1 \quad \text{for } 1 \leq j \leq n\}$$

where $k = (k_1, k_2, \ldots, k_n) \in \mathbb{Z}^n$ is an *n*-tuple of integers. Thus $I = [a, b[$, where $a_j = k_j/2^m$ and $b_j = a_j + 1/2^m$ for $1 \leq j \leq n$. Let $\mathcal{C}_m$ or, more precisely, $\mathcal{C}_m^n$ denote the family of all such I [for fixed m and n]. Plainly $I \in \mathcal{C}_m$ means $I = [a, b[$, where $2^m b_j - 1 = 2^m a_j \in \mathbb{Z}$ for $1 \leq j \leq n$. For example, $[2, 4[$ is a dyadic 1-cube of order -1, but $[1, 3[$ is not.

Given $x \in I \in \mathcal{C}_m$, we have $k_j = [2^m x_j]$ for all j [here $[t]$ is the integer for which $[t] \leq t < [t] + 1$], so each $\mathcal{C}_m$ is a pairwise disjoint family covering $\mathbb{R}^n$.

(6.116) Theorem *If V is an open subset of $\mathbb{R}^n$, then V is the union of a pairwise disjoint family $\mathcal{V}$ of dyadic n-cubes.*

Proof Recalling that the union of the void family is void, we suppose that $V \neq \emptyset$. Let $\mathcal{V}_1 = \{I \in \mathcal{C}_1 : I \subset V\}$. After $\mathcal{V}_m$ has been defined for $m = 1, 2, \ldots, p$, write $V_m = \cup \mathcal{V}_m$ for $1 \leq m \leq p$ and define $\mathcal{V}_{p+1} = \{I \in \mathcal{C}_{p+1} : I \subset V \setminus \bigcup_{m=1}^{p} V_m\}$. It is clear that the family $\mathcal{V} = \bigcup_{m=1}^{\infty} \mathcal{V}_m$ is pairwise disjoint and $\cup \mathcal{V} \subset V$. To see that $\cup \mathcal{V} = V$, let $x \in V$ be fixed. For each $m \in \mathbb{N}$, let $I_m \in \mathcal{C}_m$ be the cube containing x. Then $I_1 \supset I_2 \supset \ldots$ and, since V is open, $I_m \subset V$ for all sufficiently large m. If $I_1 \subset V$, then $I_1 \in \mathcal{V}_1$ and $x \in V_1 \subset \cup \mathcal{V}$. Otherwise, let p be the positive integer for which $I_p \not\subset V$ and $I_{p+1} \subset V$. Then $I_{p+1} \in \mathcal{V}_{p+1}$, so $x \in \cup \mathcal{V}_{p+1} \subset \cup \mathcal{V}$. □

In order to carry over items (6.2)–(6.8) to our present setting, we need a few facts about intervals of $\mathbb{R}^n$ which we collect here in a single theorem.

(6.117) Theorem (i) *If $\{I_k\}_{k=1}^{m}$ is a finite pairwise disjoint family of intervals of $\mathbb{R}^n$ and if $I = \bigcup_{k=1}^{m} I_k$ is an interval, then*

$$|I| = \sum_{k=1}^{m} |I_k|.$$

(ii) *If I is an interval of $\mathbb{R}^n$, then ξ_I is continuous a.e. on $\mathbb{R}^n$.*

(iii) *If $\mathcal{Q}$ is the family of all subsets of $\mathbb{R}^n$ that can be expressed as a finite disjoint union of intervals, then $\mathcal{Q}$ is an algebra of sets: $\mathbb{R}^n \in \mathcal{Q}$ and, whenever $E, F \in \mathcal{Q}$, we also have that $E \cap F$, $E \cup F$, and $E \setminus F$ are all in $\mathcal{Q}$.*

Proof (i) This is obvious if $m = 1$. Suppose, for an inductive proof, that $m > 1$ and that we know the assertion to be true whenever the number of I_k's is less than m. We may suppose that $I_1 \neq \varnothing \neq I_m$. Write $I_k = \underset{j=1}{\overset{n}{\times}} I_{k,j}$ as a product of intervals of $\mathbb{R}$. Since $\varnothing = I_1 \cap I_m = \underset{j=1}{\overset{n}{\times}} (I_{1,j} \cap I_{m,j})$, there is some j_0 for which $I_{1,j_0} \cap I_{m,j_0} = \varnothing$. We may suppose (renumbering if necessary) that I_{1,j_0} is to the left of I_{m,j_0}. Let a be the left endpoint of I_{m,j_0} and consider the intervals

$$J = \{ x \in I : x_{j_0} < a \}, \qquad K = \{ x \in I : x_{j_0} > a \}.$$

Then

$$J = \bigcup_{k=1}^{m-1} J \cap I_k \quad \text{and} \quad K = \bigcup_{k=2}^{m} K \cap I_k$$

are expressions of J and K as disjoint unions of fewer than m intervals, and so our induction hypothesis yields

$$|I| = |J| + |K| = \sum_{k=1}^{m-1} |J \cap I_k| + \sum_{k=2}^{m} |K \cap I_k|$$

$$= \sum_{k=1}^{m} [|J \cap I_k| + |K \cap I_k|] = \sum_{k=1}^{m} |I_k|,$$

where we have used the fact that $|\varnothing| = 0$.

(ii) It is clear that ξ_I is continuous at each point of either of the open sets I^0 or $\mathbb{R}^n \setminus I^-$. Thus it suffices to show that $\lambda^n(\partial I) = 0$, where ∂I, the boundary of I, is the set of remaining points of $\mathbb{R}^n$. But ∂I is a subset of the union of $2n$ sets of the form

$$E = \{ x \in \mathbb{R}^n : x_{j_0} = c \},$$

where $1 \leq j_0 \leq n$ and $c \in \mathbb{R}$. By (6.113), it suffices to show that $\lambda^n(E) = 0$. To this end, let $\epsilon > 0$ be given. For each $k \in \mathbb{N}$, let

$$I_k = \{ x \in \mathbb{R}^n : |x_{j_0} - c| < \delta_k, |x_j| < k \quad \text{for } j \neq j_0 \},$$

where $(2k)^{n-1} \cdot 2^{k+2} \cdot \delta_k = \epsilon$. Then $E \subset \bigcup_{k=1}^{\infty} I_k$ and $\sum_{k=1}^{\infty} |I_k| = \epsilon/2 < \epsilon$.

(iii) Let $E = \bigcup_{k=1}^{p} I_k$ and $F = \bigcup_{m=1}^{q} J_m$ each be disjoint unions of intervals. Then

$$E \cap F = \bigcup_{k=1}^{p} \bigcup_{m=1}^{q} I_k \cap J_m$$

is a finite disjoint union and $I_k \cap J_m$ is an interval for each k and m, so $E \cap F \in \mathcal{Q}$. For any interval J, we have $\mathbb{R}^n \backslash J \in \mathcal{Q}$ [e.g., $\mathbb{R}^n \backslash [a,b] = \bigcup_{j=1}^{n} \{x \in \mathbb{R}^n : a_i \leq x_i \text{ for } 1 \leq i < j, \ x_j < a_j\} \cup \bigcup_{j=1}^{n} \{x \in \mathbb{R}^n : a_i \leq x_i \text{ for } 1 \leq i \leq n, \ x_i \leq b_i \text{ for } 1 \leq i < j, \ b_j < x_j\}$]. Therefore

$$E \backslash F = E \cap \bigcap_{m=1}^{q} (\mathbb{R}^n \backslash J_m)$$

is a finite intersection of members of $\mathcal{Q}$, so is in $\mathcal{Q}$. Finally,

$$E \cup F = \mathbb{R}^n \backslash \left[(\mathbb{R}^n \backslash E) \cap (\mathbb{R}^n \backslash F) \right]$$

is in $\mathcal{Q}$. $\square$

We now define step functions just as we did in (6.1).

(6.118) Definition A real-valued function ϕ on $\mathbb{R}^n$ is called a *step function* if ϕ has the form

$$\phi = \sum_{k=1}^{m} \alpha_k \xi_{I_k}$$

where $\{I_k\}_{k=1}^{m}$ is a pairwise disjoint family of *bounded* intervals of $\mathbb{R}^n$ and $(\alpha_k)_{k=1}^{m} \subset \mathbb{R}$. We denote the set of all step functions on $\mathbb{R}^n$ by $M_0(\mathbb{R}^n)$ or simply by M_0.

We now develop the theory of Lebesgue integration and Lebesgue measurability of functions on $\mathbb{R}^n$ by repeating all of (6.2)–(6.110) *mutatis mutandis* with the exceptions of (6.25)–(6.31) and (6.82)–(6.104) and of examples (6.12.c), (6.23), (6.44), (6.48), (6.57), (6.63.a), (6.69), and (6.75). Since we have reserved n for the dimension of our underlying space $\mathbb{R}^n$, n must be replaced by some other letter [ν, say]. In the proof of Luzin's Theorem (6.76) for $\mathbb{R}^n$ ($n > 1$), we must invoke Tietze's Extension Theorem [consult the index]. In Steinhaus's Theorem (6.67), $]-\delta, \delta[$ can be replaced by the open ball of radius δ centered at 0. Actually, our treatment of the Riemann integral (6.25)–(6.31) can be carried over to closed intervals $[a,b] \subset \mathbb{R}^n$ without too much effort, but we shall not need this extension. We hereby suppose that all of these extensions have been made and we regard them as known. Thus we cite the definitions and theorems of (6.2)–(6.110) as if they were phrased in terms of $\mathbb{R}^n$. For instance, $L_1(\mathbb{R}^n)$ is the space of Lebesgue integrable functions on $\mathbb{R}^n$, $M(\mathbb{R}^n)$ is the set of Lebesgue measurable functions on $\mathbb{R}^n$, and $\mathfrak{M}(\mathbb{R}^n)$ is the family of Lebesgue measurable subsets of $\mathbb{R}^n$. For $f \in L_1(\mathbb{R}^n)$, we write $\int f, \int_{\mathbb{R}^n} f, \int f(x)\,dx$, $\int_{\mathbb{R}^n} f(t)\,dt$, etc., for the Lebesgue integral of f.

Iteration of Integrals

Throughout this section m and n are fixed positive integers. For $x = (x_1, \ldots, x_m) \in \mathbb{R}^m$ and $y = (y_1, \ldots, y_n) \in \mathbb{R}^n$, we write (x, y) for the ordered $m + n$-tuple whose jth coordinate is x_j if $1 \leq j \leq m$ and is y_{j-m} if $m + 1 \leq j \leq m + n$. Thus we make the natural identification of $\mathbb{R}^m \times \mathbb{R}^n$ with $\mathbb{R}^{m+n}$ by writing $(x, y) = (x_1, x_2, \ldots, x_m, y_1, y_2, \ldots, y_n)$. If f is a measurable function on $\mathbb{R}^{m+n}$ whose Lebesgue integral is defined, we write

$$\int_{\mathbb{R}^{m+n}} f(x, y) \, d(x, y) \quad \text{or} \quad \int_{\mathbb{R}^{m+n}} f(z) \, dz \qquad (1)$$

for that integral [here, z and (x, y) are regarded as generic elements of $\mathbb{R}^{m+n}$ with $z = (x_1, \ldots, x_m, y_1, \ldots, y_n)$].

Let f be a function [either complex or extended real-valued] defined a.e. on $\mathbb{R}^{m+n}$. If, for a fixed $y \in \mathbb{R}^n$, the function $x \to f(x, y)$ [whose value at $x \in \mathbb{R}^m$ is $f(x, y)$] is defined a.e. on $\mathbb{R}^m$ and has a Lebesgue integral, we write

$$g(y) = \int_{\mathbb{R}^m} f(x, y) \, dx$$

for the value of that integral. Thus we define a function g on a (possibly void) subset of $\mathbb{R}^n$. If, in addition, it should happen that g is defined a.e. on $\mathbb{R}^n$ and has a Lebesgue integral, we write

$$\int_{\mathbb{R}^n} \int_{\mathbb{R}^m} f(x, y) \, dx \, dy = \int_{\mathbb{R}^n} g \qquad (2)$$

for the value of that integral, and we call this number an *iterated integral* of f. In an analogous way, we define another iterated integral

$$\int_{\mathbb{R}^m} \int_{\mathbb{R}^n} f(x, y) \, dy \, dx \qquad (3)$$

provided, of course, that the inside integral defines a function on $\mathbb{R}^m$ that has an integral. These iterated integrals, if they exist, are extraordinarily useful. For instance, they can be used to reduce the evaluation of integrals over higher-dimensional spaces to integrals over $\mathbb{R}$ which in turn can often be evaluated by use of the Fundamental Theorem of Calculus.

It is our purpose in this section to prove the Fubini and Tonelli Theorems which give two circumstances under which the integrals (2) and (3) exist, are equal, and are both equal to the integral (1). We begin with some preliminaries.

(6.119) Theorem *Let $\phi \in M_0(\mathbb{R}^{m+n})$. Then we have the following conclusions.*
   (i) *For each $a \in \mathbb{R}^m$, the function $y \to \phi(a, y)$ is in $M_0(\mathbb{R}^n)$.*
   (ii) *For each $b \in \mathbb{R}^n$, the function $x \to \phi(x, b)$ is in $M_0(\mathbb{R}^m)$.*
   (iii) *The function*

$$x \to \int_{\mathbb{R}^n} \phi(x, y) \, dy$$

is in $M_0(\mathbb{R}^m)$.
   (iv) *The function*

$$y \to \int_{\mathbb{R}^m} \phi(x, y) \, dx$$

is in $M_0(\mathbb{R}^n)$.
   (v)

$$\int_{\mathbb{R}^m} \int_{\mathbb{R}^n} \phi(x, y) \, dy \, dx = \int_{\mathbb{R}^{m+n}} \phi(x, y) \, d(x, y) = \int_{\mathbb{R}^n} \int_{\mathbb{R}^m} \phi(x, y) \, dx \, dy.$$

Proof Since each bounded interval of $\mathbb{R}^{m+n}$ is of the form $I \times J$, where I and J are bounded intervals of $\mathbb{R}^m$ and $\mathbb{R}^n$, respectively, and since ϕ is a finite linear combination of characteristic functions of such intervals, it suffices (by invoking (6.4)) to consider the case that $\phi = \xi_{I \times J}$. Then $\phi(x, y) = \xi_I(x)\xi_J(y)$ for all $(x, y) \in \mathbb{R}^{m+n}$, so the functions in (i)–(iv) are

$$\xi_I(a)\xi_J, \quad \xi_J(b)\xi_I, \quad |J| \cdot \xi_I, \quad \text{and} \quad |I|\xi_J,$$

respectively. Also, (v) is just the obvious assertion

$$|J| \cdot |I| = |I \times J| = |I| \cdot |J|.$$

☐

We next prove an important lemma which will be improved in (6.124).

(6.120) Lemma *Let $E \subset \mathbb{R}^{m+n}$ be a null set. This means that $\lambda^{m+n}(E) = 0$. For $a \in \mathbb{R}^m$ and $b \in \mathbb{R}^n$, write*

$$E_a = \{ y \in \mathbb{R}^n : (a, y) \in E \},$$
$$E^b = \{ x \in \mathbb{R}^m : (x, b) \in E \}.$$

Then

(i) *$\lambda^n(E_a) = 0$ for almost every $a \in \mathbb{R}^m$,*
(ii) *$\lambda^m(E^b) = 0$ for almost every $b \in \mathbb{R}^n$.*

Proof According to (6.7), there exists a nondecreasing sequence $(\phi_k)_{k=1}^{\infty}$ $\subset M_0(\mathbb{R}^{m+n})$ such that

$$\lim_{k \to \infty} \int_{\mathbb{R}^{m+n}} \phi_k(x, y) d(x, y) < \infty \tag{1}$$

and

$$\lim_{k \to \infty} \phi_k(x, y) = \infty \quad \text{if } (x, y) \in E. \tag{2}$$

By (6.119) and (1), the functions Φ_k on $\mathbb{R}^m$ defined by

$$\Phi_k(x) = \int_{\mathbb{R}^n} \phi_k(x, y) \, dy \tag{3}$$

form a nondecreasing sequence in $M_0(\mathbb{R}^m)$ whose sequence of integrals converges. Thus (6.6) provides a set $A \subset \mathbb{R}^m$ for which $\lambda^m(\mathbb{R}^m \setminus A) = 0$ and

$$\lim_{k \to \infty} \Phi_k(a) < \infty \quad \text{if } a \in A. \tag{4}$$

Fix any $a \in A$. The functions $y \to \phi_k(a, y)$ form a nondecreasing sequence in $M_0(\mathbb{R}^n)$, so (3), (4), and (6.6) furnish $F \subset \mathbb{R}^n$ (depending on a) such that $\lambda^n(\mathbb{R}^n \setminus F) = 0$ and

$$\lim_{k \to \infty} \phi_k(a, y) < \infty \quad \text{if } y \in F.$$

Using (2), we conclude that $E_a \cap F = \emptyset$, so $\lambda^n(E_a) = 0$. This completes the proof of (i). The proof of (ii) is similar. ☐

(6.121) Fubini's Theorem *Let $f \in L_1(\mathbb{R}^{m+n})$. Then we have the following conclusions.*

   (i) *For almost every $a \in \mathbb{R}^m$, the function $y \to f(a, y)$ is defined a.e. on $\mathbb{R}^n$ and is in $L_1(\mathbb{R}^n)$.*
   (ii) *For almost every $b \in \mathbb{R}^n$, the function $x \to f(x, b)$ is defined a.e. on $\mathbb{R}^m$ and is in $L_1(\mathbb{R}^m)$.*
   (iii) *The function*

$$x \to \int_{\mathbb{R}^n} f(x, y)\, dy$$

is defined a.e. on $\mathbb{R}^m$ and is in $L_1(\mathbb{R}^m)$.
   (iv) *The function*

$$y \to \int_{\mathbb{R}^m} f(x, y)\, dx$$

is defined a.e. on $\mathbb{R}^n$ and is in $L_1(\mathbb{R}^n)$.
   (v)

$$\int_{\mathbb{R}^m}\int_{\mathbb{R}^n} f(x, y)\, dy\, dx = \int_{\mathbb{R}^{m+n}} f(x, y)\, d(x, y) = \int_{\mathbb{R}^n}\int_{\mathbb{R}^m} f(x, y)\, dx\, dy.$$

Moreover, if $f \in M_1(\mathbb{R}^{n+n})$, then the functions in (i)–(iv) are in M_1.

Proof *Case* 1: Suppose that $f \in M_1(\mathbb{R}^{m+n})$. Then (6.10) provides a nondecreasing sequence $(\phi_k)_{k=1}^\infty \subset M_0(\mathbb{R}^{m+n})$ and a set $E \subset \mathbb{R}^{m+n}$ such that

$$\int_{\mathbb{R}^{n+n}} f(x, y)\, d(x, y) = \lim_{k\to\infty} \int_{\mathbb{R}^{m+n}} \phi_k(x, y)\, d(x, y) < \infty, \qquad (1)$$

$\lambda^{m+n}(E) = 0$, and, for all $(x, y) \in \mathbb{R}^{m+n} \setminus E$, $f(x, y)$ is defined and

$$\lim_{k\to\infty} \phi_k(x, y) = f(x, y) < \infty. \qquad (2)$$

Defining Φ_k as in (3) of the preceding proof, we see as there that (6.6) applies to yield

$$\lim_{k\to\infty} \int_{\mathbb{R}^n} \phi_k(x, y)\, dy < \infty \qquad (3)$$

for almost every $x \in \mathbb{R}^m$. Define $A = \{x \in \mathbb{R}^m : (3) \text{ obtains}, \lambda^n(E_x) = 0\}$. Then, using (6.120), we infer that $\lambda^m(\mathbb{R}^m \setminus A) = 0$. For $x \in A$, let $F(x)$ denote the limit in (3). Thus

$$F(x) = \lim_{k\to\infty} \Phi_k(x) \quad \text{a.e. on } \mathbb{R}^m$$

and, by (6.119) and (1),

$$\lim_{k\to\infty} \int_{\mathbb{R}^m} \Phi_k(x)\, dx = \int_{\mathbb{R}^{m+n}} f(x, y)\, d(x, y) < \infty.$$

Invoking (6.10), we have $F \in M_1(\mathbb{R}^m)$ and

$$\int_{\mathbb{R}^m} F(x)\, dx = \int_{\mathbb{R}^{m+n}} f(x, y)\, d(x, y). \qquad (4)$$

Now fix any $a \in A$. Noting (2), the functions $y \to \phi_k(a, y)$ form a nondecreasing sequence in $M_0(\mathbb{R}^n)$ which converges except for $y \in E_a$ [therefore a.e.] to the function $y \to f(a, y)$ and, by (3), their integrals converge to $F(a) < \infty$. It follows that $y \to f(a, y)$ is in $M_1(\mathbb{R}^n)$ and that

$$\int_{\mathbb{R}^n} f(a, y)\, dy = F(a).$$

Combining this with (4) and the fact that A is almost all of $\mathbb{R}^m$, we deduce (i), (iii), and the first equality in (v) in Case 1. The remaining three assertions are proved similarly.

Case 2: Suppose that $f \in L_1'(\mathbb{R}^{m+n})$. According to (6.14), there exist g and h in $M_1(\mathbb{R}^{m+n})$ such that $f = g - h$ a.e. and $\int f = \int g - \int h$. The theorem is now proved for f by applying Case 1 and (6.16) to g and to h.

Case 3. Suppose that $f \in L_1(\mathbb{R}^{m+n})$. All we need do in this case is apply Definition (6.39) and Case 2 to Re f and Im f. $\square$

In order to safely apply the obviously useful conclusions of Fubini's Theorem, we need to know in advance that $f \in L_1(\mathbb{R}^{m+n})$. Fortunately, our first corollary of the following theorem allows us to reduce (all but the measurability part of) the checking of this hypothesis to the consideration of integrals over spaces of lower dimension.

(6.122) Tonelli's Theorem *Let f be a nonnegative $[0, \infty]$-valued Lebesgue measurable function defined a.e. on $\mathbb{R}^{m+n}$. Then we have the following conclusions.*

(i) For almost every $a \in \mathbb{R}^m$, the function $y \to f(a, y)$ is defined a.e. on $\mathbb{R}^n$ and is in $M'(\mathbb{R}^n)$.

(ii) For almost every $b \in \mathbb{R}^n$, the function $x \to f(x, b)$ is defined a.e. on $\mathbb{R}^m$ and is in $M'(\mathbb{R}^m)$.

(iii) The function

$$x \to \int_{\mathbb{R}^n} f(x, y)\, dy$$

is defined a.e. on $\mathbb{R}^m$ and is in $M'(\mathbb{R}^m)$.

(iv) The function

$$y \to \int_{\mathbb{R}^m} f(x, y)\, dx$$

is defined a.e. on $\mathbb{R}^n$ and is in $M'(\mathbb{R}^n)$.

(v) We have

$$\int_{\mathbb{R}^m}\int_{\mathbb{R}^n} f(x, y)\, dy\, dx = \int_{\mathbb{R}^{m+n}} f(x, y)\, d(x, y) = \int_{\mathbb{R}^n}\int_{\mathbb{R}^m} f(x, y)\, dx\, dy$$

independently of whether or not these integrals are finite.

Proof For each $k \in \mathbb{N}$ consider the cube $I_k = \{(x, y) \in \mathbb{R}^{m+n} : |x_i| \le k$ for $1 \le i \le m, |y_j| \le k$ for $1 \le j \le n\}$ and define the truncate f_k of f by $f_k = (k \wedge f)\xi_{I_k}$. That is, for $(x, y) \in I_k$, $f_k(x, y) = \min\{k, f(x, y)\}$ and $f_k(x, y) = 0$ for other (x, y). Since $0 \le f_k \le k\xi_{I_k}$, it follows from (6.35), (6.37), and (6.38) that $f_k \in L_1(\mathbb{R}^{m+n})$ for each k. Thus all of the conclusions of (6.121) are true for each f_k. Plainly $0 \leq f_1 \leq f_2 \leq \ldots$ and $f_k \to f$ a.e. on $\mathbb{R}^{m+n}$ Thus (i)–(v) follow from several applications of (6.120), (6.121), (6.39), and (6.43). We leave the (straightforward) details to the reader. $\square$

(6.123) Corollary *Let f be a complex-valued Lebesgue measurable function defined a.e. on $\mathbb{R}^{m+n}$. That is, $f \in M(\mathbb{R}^{m+n})$. Suppose that one of the two iterated integrals*

$$\int_{\mathbb{R}^m} \int_{\mathbb{R}^n} |f(x, y)| \, dy \, dx, \qquad \int_{\mathbb{R}^n} \int_{\mathbb{R}^m} |f(x, y)| \, dx \, dy$$

is finite. Then $f \in L_1(\mathbb{R}^{m+n})$ and all of the conclusions of (6.121) are true.

Proof It follows from (6.51) and (6.122) that the two iterated integrals are meaningful and equal and each is equal to $\int |f(x, y)| d(x, y)$. We conclude from (6.53) that $f \in L_1$. $\square$

(6.124) Corollary *Let $E \subset \mathbb{R}^{m+n}$ be a Lebesgue measurable set. That is, $E \in \mathfrak{M}(\mathbb{R}^{m+n})$. Define E_a and E^b as in (6.120). Then*
  (i) $E_a \in \mathfrak{M}(\mathbb{R}^n)$ *for almost every $a \in \mathbb{R}^m$*,
  (ii) $E^b \in \mathfrak{M}(\mathbb{R}^m)$ *for almost every $b \in \mathbb{R}^n$*.
Moreover, the following three statements are equivalent.
  (iii) $\lambda^{m+n}(E) = 0$.
  (iv) $\lambda^n(E_a) = 0$ *for almost every $a \in \mathbb{R}^m$*.
  (v) $\lambda^m(E^b) = 0$ *for almost every $b \in \mathbb{R}^n$*.

Proof Write $f = \xi_E$. Then $f \in M(\mathbb{R}^{m+n})$ and $f \geq 0$, so the conclusions of (6.122) are true of f. We have

$$f(a, y) = \xi_{E_a}(y) \quad \text{and} \quad f(x, b) = \xi_{E^b}(x),$$

so (i) and (ii) follow from (6.122.i) and (6.122.ii) respectively. For this f, (6.122.v) reads

$$\int_{\mathbb{R}^m} \lambda^n(E_x) \, dx = \lambda^{m+n}(E) = \int_{\mathbb{R}^n} \lambda^m(E^y) \, dy,$$

and so the equivalence of (iii), (iv), and (v) follows from this and (6.24). [Of course we proved that (iii) implies (iv) and (v) in (6.120).] $\square$

(6.125) Remark By using a well-ordering and transfinite induction argument, it is possible to prove the existence of a subset S of the plane $\mathbb{R}^2$ which meets each horizontal line and each vertical line in at most one point and also meets each compact subset of $\mathbb{R}^2$ having positive λ^2 measure in at least one point [see *Real and Abstract Analysis* by Edwin Hewitt and Karl Stromberg, (21.27), p. 393]. No such S can be in $\mathfrak{M}(\mathbb{R}^2)$ [Exercise!]. For $E = S$ and $m = n = 1$, conclusions (i), (ii), (iv), and (v) of (6.124) are true while conclusion (iii) is false. For $f = \xi_S$ and $m = n = 1$, the two iterated integrals in (6.123) are 0, but $f \notin L_1(\mathbb{R}^2)$, and the middle integral in (6.122.v) is not defined, but the other two are 0. This remark is meant to emphasize that the measurability hypotheses in (6.121)–(6.124) are important and must be checked before the conclusions may be inferred.

To illustrate how one can check measurability of certain functions defined on product spaces, we prove the following theorem which answers a (commonly overlooked) question that arises fairly often in analysis.

(6.126) Theorem *Let $f \in M(\mathbb{R}^n)$ and define F on $\mathbb{R}^n \times \mathbb{R}^n = \mathbb{R}^{2n}$ by*

$$F(x, y) = f(x - y)$$

for those (x, y) for which f is defined at $x - y$. Then $F \in M(\mathbb{R}^{2n})$.

Proof We may suppose that f is real-valued: $f \in M'(\mathbb{R}^n)$. By (6.33), there exists a set $E \subset \mathbb{R}^n$ and a sequence $(\phi_k)_{k=1}^\infty \subset M_0(\mathbb{R}^n)$ such that $\lambda^n(E) = 0$ and, for $t \in \mathbb{R}^n \backslash E$, $f(t)$ is defined and $\phi_k(t) \to f(t)$ as $k \to \infty$. Since the boundary of any interval is a null set [see (6.117)], we may suppose that each ϕ_k is a finite linear combination of *open* intervals [adjoin to E all of the (countably many) boundaries of the intervals that appear in the representations (6.118) of the ϕ_k and replace each such interval by its interior]. Thus the functions Φ_k defined on $\mathbb{R}^{2n}$ by $\Phi_k(x, y) = \phi_k(x - y)$ are finite linear combinations of characteristic functions of sets of the form $\{(x, y) : x - y \in I\}$, where I is an open interval of $\mathbb{R}^n$. Since $(x, y) \to x - y$ is a continuous mapping, all such sets are open, and so (6.58), (6.59.vii), and (6.35) show that each Φ_k is in $M(\mathbb{R}^{2n})$. Plainly, $\lim_{k \to \infty} \Phi_k(x, y) = F(x, y)$ unless $(x, y) \in A$, where

$$A = \{(x, y) \in \mathbb{R}^{2n} : x - y \in E\}.$$

If we can show that $\lambda^{2n}(A) = 0$, it will follow from (6.39) that F is measurable. We complete the proof by showing that $\lambda^{2n}(A) = 0$.

Using (6.64), we obtain a G_δ set $G \supset E$ such that $\lambda^n(G) = 0$ $[G = \bigcap_{k=1}^\infty V_k$, where each V_k is open in $\mathbb{R}^n$, $E \subset V_k$, and $\lambda^n(V_k) < 1/k]$. The set $B = \{(x, y) : x - y \in G\}$ is a G_δ set in $\mathbb{R}^{2n}$, so, by (6.59), $B \in M(\mathbb{R}^{2n})$ and (6.124) applies to B. For each $b \in \mathbb{R}^n$, we have

$$B^b = \{x \in \mathbb{R}^n : x - b \in G\} = b + G,$$

so

$$\lambda^n(B^b) = \lambda^n(b + G) = \lambda^n(G) = 0,$$

where the second equality is (6.113.vi). Thus $\lambda^{2n}(B) = 0$. Since $A \subset B$, the proof is complete. $\square$

We close this section by giving a simple example to show that interchange of the order of integration can give well defined and unequal answers.

(6.127) Example   Define f on $\mathbb{R}^2$ by $f(x, y) = (y^2 - x^2)/[(x^2 + y^2)^2]$ if $0 < x < 1$ and $0 < y < 1$ and $f(x, y) = 0$ for all other $(x, y) \in \mathbb{R}^2$. Then

$$\int_{\mathbb{R}} \int_{\mathbb{R}} f(x, y) \, dx \, dy = \int_0^1 \int_0^1 \frac{y^2 - x^2}{(x^2 + y^2)^2} \, dx \, dy.$$

For fixed $y \neq 0$, the last integrand is continuous on $[0, 1]$ and is the derivative (with respect to x) of $x(x^2 + y^2)^{-1}$, so the Fundamental Theorem of Calculus (for Riemann integrals) shows that

$$\int_0^1 \frac{y^2 - x^2}{(x^2 + y^2)^2} \, dx = (1 + y^2)^{-1}.$$

Since the derivative of Arctan y is $(1 + y^2)^{-1}$ [see (5.33.iii)], we have

$$\int_{\mathbb{R}} \int_{\mathbb{R}} f(x, y) \, dx \, dy = \int_0^1 (1 + y^2)^{-1} \, dy = \frac{\pi}{4}.$$

Plainly, $f(x, y) = -f(y, x)$ for all $(x, y) \in \mathbb{R}^2$, so

$$\int_{\mathbb{R}} \int_{\mathbb{R}} f(x, y) \, dy \, dx = -\int_{\mathbb{R}} \int_{\mathbb{R}} f(y, x) \, dy \, dx = -\frac{\pi}{4}.$$

It is trivial, using continuity and (6.72), to show that f is measurable on $\mathbb{R}^2$. Since the two iterated integrals are not equal, it follows from (6.123) that

$$\int_{\mathbb{R}} \int_{\mathbb{R}} |f(x, y)| \, dx \, dy = \infty.$$

Of course, this equality can easily be checked by direct computation.

Exercises

1.   Define f on $\mathbb{R}^2$ by $f(x, y) = x^{-3}$ if $0 < y < |x| < 1$ and $f(x, y) = 0$ otherwise. Then
      (a) f is measurable on $\mathbb{R}^2$,
      (b) for all $a, b \in \mathbb{R}$, the functions $y \to f(a, y)$ and $x \to f(x, b)$ are in $L_1(\mathbb{R})$ and are Riemann integrable over $[0, 1]$ and $[-1, 1]$ respectively,
      (c) the function $x \to \int_{\mathbb{R}} f(x, y) \, dy$ is in $M'(\mathbb{R})$ but its integral does not exist,
      (d) the function $y \to \int_{\mathbb{R}} f(x, y) \, dx$ is identically 0,
      (e) $\int_{\mathbb{R}} \int_{\mathbb{R}} f(x, y) \, dx \, dy = 0$ and $\int_{\mathbb{R}} \int_{\mathbb{R}} f(x, y) \, dy \, dx$ is a meaningless symbol,
      (f) $f \notin L_1(\mathbb{R}^2)$.

2.   The function f defined on $\mathbb{R}^2$ by $f(0, 0) = 0$ and $f(x, y) = xy(x^2 + y^2)^{-2}$ if $(x, y) \neq (0, 0)$ is *not* in $L_1(\mathbb{R}^2)$ even though its two iterated integrals over $\mathbb{R}^2$ both exist and equal 0. Moreover, for each fixed $y \in \mathbb{R}$, the function $x \to f(x, y)$ is continuous and in $L_1(\mathbb{R})$.

3.   If $f \in L_1(\mathbb{R}^n)$ and $a \in \mathbb{R}^n$, then

$$\int_{\mathbb{R}^n} f(x - a) \, dx = \int_{\mathbb{R}^n} f(x) \, dx = \int_{\mathbb{R}^n} f(-x) \, dx.$$

[Hint: Consider successively the cases $f \in M_0, f \in M_1, f \in L_1^+$, and $f \in L_1$.]

4.   Let p be a positive real number.

(a) If $0 < p < 1$ and $\{c_1, \ldots, c_N\} \subset \mathbb{C}$, then $\left| \sum_{k=1}^{N} c_k \right|^p \leq \sum_{k=1}^{N} |c_k|^p$. [Hint: Show that $(1 + t)^p \leq 1 + t^p$ for all $t \geq 0$ and use induction on N.]

(b) If I is a bounded interval of $\mathbb{R}^n$ and $\delta > 0$, then there exist infinitely differentiable functions $g_j : \mathbb{R} \to [0, 1]$ $(1 \leq j \leq n)$, where each g_j vanishes identically outside some bounded interval such that the function g defined by

$$g(x) = \prod_{j=1}^{n} g_j(x_j) \qquad (x \in \mathbb{R}^n)$$

satisfies $g(x) = 0$ for $x \in \mathbb{R}^n \setminus I$ and

$$\int_{\mathbb{R}^n} |\xi_I - g|^p < \delta.$$

[Hint: Let ϕ be as in (4.31). Given $a < b$ in $\mathbb{R}$ and $0 < \eta < (b - a)/2$ choose an odd positive integer m such that $(1 - 2^{-1/m})(b - a) < 2\eta$ and write $\psi(t) = \phi((2t - a - b)^m(b - a)^{-m})$ to obtain $\psi \in C^\infty(\mathbb{R})$ satisfying $0 \leq \psi \leq 1$, $\psi = 1$ on $[a + \eta, b - \eta]$, and $\psi = 0$ off $[a, b]$.]

(c) If $f \in L_p(\mathbb{R}^n)$ and $\epsilon > 0$, then there exists a complex-valued step function ϕ on $\mathbb{R}^n$ (this means that $\mathrm{Re}\,\phi$ and $\mathrm{Im}\,\phi$ are in $M_0(\mathbb{R}^n)$) such that $\|f - \phi\|_p < \infty$. [Hint: For $k \in \mathbb{N}$, write $f_k(x) = f(x)$ if $|f(x)| < k$ and $|x_j| < k$ for all j and write $f_k(x) = 0$ for all other x. Use dominated convergence to obtain k for which $\|f - f_k\|_p < \epsilon$ and then find an appropriate sequence of ϕ's that converges to f_k a.e.]

(d) If $f \in L_p(\mathbb{R}^n)$ and $\epsilon > 0$, then there exists a function h which is a finite linear combination of functions of the type g described in (b) such that $\|f - h\|_p < \epsilon$.

5.   If $1 \leq p < \infty$ and $f \in L_p(\mathbb{R}^n)$, then

$$\lim_{a \to 0} \|f_a - f\|_p = 0$$

where $f_a(x) = f(x - a)$ for $a, x \in \mathbb{R}^n$. [Hints: Given $\epsilon > 0$, choose a continuous h on $\mathbb{R}^n$ that vanishes off some compact interval J for which $\|f - h\|_p < \epsilon/3$ (see Exercise 4(d)) and then use the uniform continuity of h to obtain $\delta > 0$ such that $\|h_a - h\|_p < \epsilon/3$ if $|a_j| < \delta$ for $1 \leq j \leq n$.]

6.   Let $f \in L_1(\mathbb{R}^n)$, $g \in M(\mathbb{R}^n)$, and suppose that g is bounded. Then the formula

$$f * g(x) = \int_{\mathbb{R}^n} f(x - y)g(y) \, dy$$

defines a function $f * g$ (called the *convolution* of f with g) that is bounded and uniformly continuous on $\mathbb{R}^n$. Moreover, if $A, B \subset \mathbb{R}^n$, f vanishes on $\mathbb{R}^n \setminus A$, and g vanishes on $\mathbb{R}^n \setminus B$, then $f * g$ vanishes on $\mathbb{R}^n \setminus (A + B)$, where $A + B = \{a + b : a \in A, b \in B\}$. Also,

$$f * g(x) = \int_{\mathbb{R}^n} f(t)g(x - t) \, dt$$

for all $x \in \mathbb{R}^n$. [Hint: Use Exercises 5 and 3.]

7.   If $A, B \in \mathfrak{M}(\mathbb{R}^n)$ are both of positive measure, then the set $A + B$ contains a nonvoid open subset of $\mathbb{R}^n$. [Hint: Use Exercise 6. Suppose $\lambda^n(A) < \infty$, write $f = \xi_A$, $g = \xi_B$, and show that $f * g$ is not identically 0 by integrating it.]

8.   There exists a closed set $E \subset \mathbb{R}^{2n}$ with $\lambda^{2n}(E) > 0$ such that if $A, B \in \mathfrak{M}(\mathbb{R}^n)$ and

$A \times B \subset E$, then $\lambda^n(A) = 0$ or $\lambda^n(B) = 0$. [Hint: Use Exercise 7. Take $E = \{(x, y) \in \mathbb{R}^{2n} : x + y \in P\}$, where P is a closed nowhere dense subset of $\mathbb{R}^n$ with $\lambda^n(P) > 0$. Draw a picture for $n = 1$.]

9. A function $u : \mathbb{R}^n \to \mathbb{C}$ is said to be *infinitely differentiable on* $\mathbb{R}^n$ and we write $u \in C^\infty(\mathbb{R}^n)$ if, for each ordered n-tuple $m = (m_1, \ldots, m_n)$ of nonnegative integers, the function $D^m u$ obtained by successively differentiating u m_1 times with respect to x_1, m_2 times with respect to $x_2, \ldots,$ and m_n times with respect to x_n exists and is continuous on $\mathbb{R}^n$. Now let ϕ be as in (4.31), write $c = \int_{-1}^1 \phi(t) \, dt$, and, for each real number $\alpha > 0$, define u_α on $\mathbb{R}^n$ by

$$u_\alpha(x) = \left(\frac{\alpha}{c} \right)^n \cdot \prod_{j=1}^n \phi(\alpha x_j).$$

Then we have

(a) $u_\alpha(-x) = u_\alpha(x) \geq 0$ for all α and x,

(b) $u_\alpha(x) = 0$ of $|x_j| \geq 1/\alpha$ for some j,

(c) $u_\alpha \in C^\infty(\mathbb{R}^n)$ and $D^m u_\alpha(x) = c^{-n} \alpha^{n+|m|} \cdot \prod_{j=1}^n \phi^{(m_j)}(\alpha x_j)$ where $|m| = m_1 + m_2 + \ldots + m_n$,

(d) $\int_{\mathbb{R}^n} u_\alpha(y) \, dy = 1$,

(e) if $f \in L_1(\mathbb{R}^n)$, then $f * u_\alpha \in C^\infty(\mathbb{R}^n)$ and $D^m(f * u_\alpha) = f * (D^m u_\alpha)$ [Hints: See Exercise 6. Use induction on $|m|$, the definition of derivative, the mean value theorem, and dominated convergence to prove the equality.],

(f) if $f \in L_1(\mathbb{R}^n)$, then $f * u_\alpha \in L_1(\mathbb{R}^n)$ and $\lim_{\alpha \to \infty} \| f * u_\alpha - f \|_1 = 0$ [Hint: Write $f * u_\alpha(x) - f(x) = \int [f(x - y) - f(x)] u_\alpha(y) \, dy$. Use Exercise 5 and (b).],

(g) if $A, B \subset \mathbb{R}^n$, A is compact, B is closed, and $A \cap B = \emptyset$, then there exists a function $\phi \in C^\infty(\mathbb{R}^n)$ such that $\phi(x) = 1$ for $x \in A$, $\phi(x) = 0$ for $x \in B$, and $0 \leq \phi(x) \leq 1$ for $x \in \mathbb{R}^n$. [Hint: Put $V_\alpha = \{ y \in \mathbb{R}^n : |y_j| < \alpha^{-1} \text{ for all } j \}$, choose α so large that $(A + V_\alpha) \cap (B + V_\alpha) = \emptyset$, and let $\phi = f * u_\alpha$, where $f = 1$ on $A + V_\alpha$ and $f = 0$ elsewhere.]

10. If $f, g \in L_1(\mathbb{R}^n)$, then the formula

$$f * g(x) = \int_{\mathbb{R}^n} f(x - y) g(y) \, dy$$

defines a function $f * g \in L_1(\mathbb{R}^n)$. This function $f * g$ is called the *convolution* of f with g. Moreover, $f * g = g * f$, $\| f * g \|_1 \leq \| f \|_1 \cdot \| g \|_1$, and, for any $h \in L_1(\mathbb{R}^n)$, $f * (g * h) = (f * g) * h$ and $f * (g + h) = f * g + f * h$. [Hint: Check that the integrand is in $L_1(\mathbb{R}^{2n})$ and apply Fubini's Theorem.]

11. Let $\phi \in L_1(\mathbb{R}^n)$ satisfy $\int_{\mathbb{R}^n} \phi = 1$ and, for $\alpha > 0$, define ϕ_α on $\mathbb{R}^n$ by $\phi_\alpha(t) = \alpha^n \phi(\alpha t)$.

(a) If $f \in L_1(\mathbb{R}^n)$, then $\lim_{\alpha \to \infty} \| f * \phi_\alpha - f \|_1 = 0$. [Hints: See Exercise 10. Check that $\int \phi_\alpha = 1$ and

$$\| f * \phi_\alpha - f \|_1 \leq \int_{\mathbb{R}^n} \int_{\mathbb{R}^n} |f(x - t) - f(x)| \, dx \, |\phi_\alpha(t)| \, dt$$

$$= \int_{\mathbb{R}^n} \left\{ \int_{\mathbb{R}^n} \left| f\left(x - \frac{1}{\alpha} y\right) - f(x) \right| dx \, |\phi(y)| \right\} dy.$$

Notice that the function in braces is dominated by $2\| f \|_1 \cdot |\phi|$ and converges to 0 a.e. as $\alpha \to \infty$ by Exercise 5.]

(b) If $f : \mathbb{R}^n \to \mathbb{C}$ is bounded and uniformly continuous, then $\lim_{\alpha \to \infty} \| f * \phi_\alpha - f \|_u = 0$.

12. Let $1 < p < \infty$ be a real number and let $p' = p/(p - 1)$. If $f \in L_p(\mathbb{R}^n)$ and $g \in L_{p'}(\mathbb{R}^n)$, then the *convolution* $f*g$ of f with g defined by the formula

$$f*g(x) = \int_{\mathbb{R}^n} f(x - y)g(y)\,dy$$

is in $C_0(\mathbb{R}^n)$ and $\|f*g\|_u \le \|f\|_p \cdot \|g\|_{p'}$ (see (3.124) for the definition of C_0). [Hints: Consider the truncates f_k and g_k of f and g as in Exercise 4(c). Use Exercise 6 to see that each f_k*g_k is continuous and vanishes outside of a compact set. Use Hölder's Inequality to show that $f*g$ is defined at each x and that $(f_k*g_k)_{k=1}^\infty$ converges to $f*g$ uniformly on $\mathbb{R}^n$.]

13. For $x, y \in \mathbb{R}^n$, write

$$xy = \sum_{j=1}^{n} x_j y_j.$$

(a) If $f \in L_1(\mathbb{R}^n)$, then the function $\hat{f}$ defined on $\mathbb{R}^n$ by

$$\hat{f}(y) = \int_{\mathbb{R}^n} f(x)e^{-2\pi i xy}\,dx$$

is in $C_0(\mathbb{R}^n)$ and $\|\hat{f}\|_u \le \|f\|_1$. [Hints: See (3.124) for the definition of C_0. Suppose first that $f = \xi_I$ where I is a bounded interval of $\mathbb{R}^n$.] The function $\hat{f}$ is called the *Fourier transform* of f.

(b) If $f, g \in L_1(\mathbb{R}^n)$ and $f*g$ is as in Exercise 10, then $\widehat{f*g} = \hat{f}\hat{g}$.

(c) For $x \in \mathbb{R}^n$, write $|x|^2 = \sum_{j=1}^{n} x_j^2$ and define the *Gauss Kernel* G on $\mathbb{R}^n$ by

$G(x) = \exp(-\pi |x|^2)$. Then $G \in L_1(\mathbb{R}^n)$ and $\hat{G} = G$. [Hints: Consider first the case $n = 1$. We have $\|G\|_1 = \hat{G}(0)$ and we show that $\hat{G}(0) = 1$ by using the hoary trick of changing to polar coordinates (see the next section) as follows:

$$\hat{G}(0)^2 = \int_{-\infty}^{\infty}\int_{-\infty}^{\infty} e^{-\pi(x^2+y^2)}\,dx\,dy$$

$$= \int_0^{\infty}\int_0^{2\pi} e^{-\pi r^2} r\,d\theta dr = \int_0^{\infty} e^{-\pi r^2} 2\pi r\,dr = 1.$$

Write $g = \hat{G}$ and justify the interchange of differentiation and integration (using dominated convergence) to obtain

$$g'(y) = i\int_{-\infty}^{\infty} (-2\pi x e^{-\pi x^2})e^{-2\pi i xy}\,dx.$$

Next use integration by parts (combined with dominated convergence) to see that $g'(y) = -2\pi y\, g(y)$ for all $y \in \mathbb{R}$. The function $\phi(y) = g(y)e^{\pi y^2}$ satisfies $\phi'(y) = 0$ for all y and $\phi(0) = 1$, so $g(y) = e^{-\pi y^2}$ for all y.]

(d) There is no function $f \in L_1(\mathbb{R}^n)$ such that $f*G = G$. [Hint: Assuming f exists, apply (b), (c), and (a).]

(e) If $f, g \in L_1(\mathbb{R}^n)$, $\alpha \in \mathbb{R}$, $\alpha > 0$, and $t \in \mathbb{R}^n$, then

$$\int_{\mathbb{R}^n} \hat{f}(y)g\left(\frac{1}{\alpha}y\right)e^{2\pi i ty}\,dy = \alpha^n \int_{\mathbb{R}^n} f(x)\hat{g}(\alpha(x - t))\,dx.$$

(f) For $\alpha > 0$, write $G_\alpha(u) = \alpha^n G(\alpha u)$ $(u \in \mathbb{R}^n)$. If $f \in L_1(\mathbb{R}^n)$, $\alpha > 0$, and $t \in \mathbb{R}^n$, then

$$\int_{\mathbb{R}^n} \hat{f}(y)G\left(\frac{1}{\alpha}y\right)e^{2\pi i ty}\,dy = \int_{\mathbb{R}^n} f(x)G_\alpha(x - t)\,dx = f*G_\alpha(t).$$

(g) If $f \in L_1(\mathbb{R}^n)$, then

$$\lim_{\alpha \to \infty} \| f * G_\alpha - f \|_1 = 0.$$

[Hint: Use Exercise 11.]

(h) *Uniqueness Theorem.* If $f_1, f_2 \in L_1(\mathbb{R}^n)$ and $\hat{f}_1 = \hat{f}_2$, then $f_1 = f_2$ a.e. [Hint: Write $f = f_1 - f_2$ and apply (f) and (g).]

(i) If $f \in L_1(\mathbb{R}^n)$ and $f * f = f$ a.e., then $f = 0$ a.e.

(j) *Inversion Theorem.* If $f \in L_1(\mathbb{R}^n)$ and also $\hat{f} \in L_1(\mathbb{R}^n)$, then

$$f(t) = \int_{\mathbb{R}^n} \hat{f}(y) e^{2\pi i t y}\, dy$$

for almost every $t \in \mathbb{R}^n$. That is, $f(t) = \hat{\hat{f}}(-t)$ a.e. If, in addition, f is continuous on $\mathbb{R}^n$, then these equalities hold for every $t \in \mathbb{R}^n$. [Hints: In view of (g) and (6.56), there is a sequence $\alpha_k \to \infty$ such that $f * G_{\alpha_k} \to f$ a.e. Now put $\alpha = \alpha_k$ in (f) and use Lebesgue's dominated convergence theorem.]

(k) If $f \in L_1(\mathbb{R}^n)$, $\hat{f} \geqq 0$ on $\mathbb{R}^n$, and f is continuous at 0, then $\hat{f} \in L_1(\mathbb{R}^n)$ and

$$\int_{\mathbb{R}^n} \hat{f}(y)\, dy = f(0),$$

so the conclusion of (j) obtains. [Hint: By (f), with $t = 0$, we have

$$\int_{\mathbb{R}^n} \hat{f}(y) G\left(\frac{1}{\alpha} y\right) dy = \int_{\mathbb{R}^n} f(x) G_\alpha(x)\, dx.$$

The integral on the left has limit $\| \hat{f} \|_1$ as $\alpha \to \infty$ by (6.43) and the one on the right has limit $f(0)$ since, given $\epsilon > 0$, there exists $\delta > 0$ such that $|f(x) - f(0)| < \epsilon$ if $|x| \leqq \delta$, and so

$$\int_{|x| \leqq \delta} |f(x) - f(0)| G_\alpha(x)\, dx < \epsilon$$

for all α, while

$$\int_{|x| > \delta} |f(x) - f(0)| G_\alpha(x)\, dx$$

$$\leqq \sup_{|x| > \delta} G_\alpha(x) \| f \|_1 + |f(0)| \cdot \int_{|x| > \delta} G_\alpha(x)\, dx$$

$$\leqq \alpha^n \exp(-\pi \alpha^2 \delta^2) \| f \|_1 + |f(0)| \int_{|u| > \alpha \delta} G(u)\, du \to 0$$

as $\alpha \to \infty$ by dominated convergence.]

(l) For all $x \in \mathbb{R}$ we have

$$e^{-|x|} = \frac{1}{\pi} \int_{-\infty}^{\infty} \frac{e^{ixt}}{1 + t^2}\, dt = \frac{2}{\pi} \int_0^\infty \frac{\cos(xt)}{1 + t^2}\, dt.$$

[Hint: Apply (j) when $n = 1$ and $f(x) = e^{-|x|}$.]

(m) For all $x \in \mathbb{R}$ we have

$$\frac{2}{\pi} \int_0^\infty \left(\frac{\sin t}{t} \right)^2 \cos(2xt)\, dt = \max\{0, 1 - |x|\}.$$

(n) For each real number $\alpha \geqq 0$ we have

$$e^{-2\pi\alpha} = \frac{1}{\sqrt{\pi}} \int_0^\infty u^{-1/2} e^{-u - \alpha^2\pi^2/u}\, du.$$

[Hints: Put $x = 2\pi\alpha$ in (l), write

$$\frac{1}{1 + t^2} = \int_0^\infty e^{-(1+t^2)u}\, du,$$

apply Fubini's Theorem, write $ut^2 = \pi x^2$, and then use (c) for $n = 1$.]

(o) The *Poisson kernel* is defined on $\mathbb{R}^n$ by $P(x) = c_n(1 + |x|^2)^{-(n+1)/2}$, where

$$c_n = \pi^{-(n+1)/2} \int_0^\infty e^{-v} v^{(n-1)/2}\, dv.$$

We have $P = \hat{f}$, where $f(y) = e^{-2\pi|y|}$ for $y \in \mathbb{R}^n$. [Hints: Use (n) with $\alpha = |y|$ to obtain

$$\hat{f}(x) = \frac{1}{\sqrt{\pi}} \int_0^\infty u^{-1/2} e^{-u} \int_{\mathbb{R}^n} G\left(\sqrt{\pi/u}\ y\right) e^{-2\pi ixy}\, dy\, du,$$

put $t = \sqrt{\pi/u}\ y$, and then use (c).]

(p) For $\alpha > 0$, write $P_\alpha(u) = \alpha^n P(\alpha u)$ $(u \in \mathbb{R}^n)$. If $f \in L_1(\mathbb{R}^n)$, then $\lim\limits_{\alpha \to \infty} \| f * P_\alpha - f \|_1 = 0$. [Hint: Use (n), (j), and Exercise 11.]

(q) If $f \in L_1(\mathbb{R}^n)$, $\alpha > 0$, and $t \in \mathbb{R}^n$, then

$$\int_{\mathbb{R}^n} \hat{f}(y) e^{-2\pi|y|/\alpha} e^{2\pi ity}\, dy = f * P_\alpha(t).$$

(r) If f is in both $L_1(\mathbb{R}^n)$ and $L_2(\mathbb{R}^n)$ [we write $f \in L_1 \cap L_2(\mathbb{R}^n)$], then $\hat{f} \in L_2(\mathbb{R}^n)$ and $\| \hat{f} \|_2 = \| f \|_2$. [Hints: Write $g(x) = \overline{f(-x)}$ and $h = f * g$. Check that $\hat{g} = \overline{\hat{f}}$ and $\hat{h} = |\hat{f}|^2$. Use Exercise 12 and (k) to see that $\hat{h} \in L_1$ and that $\| \hat{f} \|_2^2 = h(0) = \int f(x) g(-x)\, dx = \| f \|_2^2.$]

(s) If $f \in L_2(\mathbb{R}^n)$, then there exists a function $g \in L_2(\mathbb{R}^n)$ such that, whenever a sequence $(f_k)_{k=1}^\infty \subset L_1 \cap L_2(\mathbb{R}^n)$ satisfies $\lim\limits_{k \to \infty} \| f - f_k \|_2 = 0$, we also have $\lim\limits_{k \to \infty} \| \hat{f}_k - g \|_2 = 0$. Moreover, if h is any other function having the same two properties as g, then $h = g$ a.e. on $\mathbb{R}^n$. In case, $f \in L_1 \cap L_2(\mathbb{R}^n)$, then $g = \hat{f}$ a.e. on $\mathbb{R}^n$. [Hints: To produce one such g, fix a particular sequence of f_k's so that $\| f - f_k \|_2 \to 0$, use (r) to see that $\| \hat{f}_j - \hat{f}_k \|_2 \to 0$, and invoke (6.110). If another sequence f_k' produces some h, then the interlaced sequence $f_1, f_1', f_2, f_2', \dots$ produces a ϕ for which $g = \phi$ a.e. and $h = \phi$ a.e.]

Let us agree in the remaining parts of this exercise to not distinguish between two functions in $L_p(\mathbb{R}^n)$ that are equal a.e. on $\mathbb{R}^n$. For $f \in L_2(\mathbb{R}^n)$, we denote by $\hat{f}$ the function g described in (s) and we call $\hat{f}$ the *Fourier* [or *Plancherel*] *transform* of f. For ϕ and $(\phi_k)_{k=1}^\infty$ in $L_2(\mathbb{R}^n)$, write $\phi(x) = \text{l.i.m.}\limits_{k \to \infty}\ \phi_k(x)$ to mean that $\| \phi - \phi_k \|_2 \to 0$ as $k \to \infty$ and read l.i.m. as *limit in mean*.

(t) For $k > 0$, let $B_k = \{ x \in \mathbb{R}^n : |x| \le k \}$. If $f \in L_2(\mathbb{R}^n)$, then

$$f \xi_{B_k} \in L_1 \cap L_2(\mathbb{R}^n) \quad \text{for all } k \in \mathbb{N}$$

and

$$\hat{f}(y) = \text{l.i.m.}\limits_{k \to \infty} \int_{B_k} f(x) e^{-2\pi ixy}\, dx.$$

(u) If $f \in L_2(\mathbb{R}^n)$, then $\hat{f} \in L_2(\mathbb{R}^n)$ and $\| \hat{f} \|_2 = \| f \|_2$. [Hint: Apply (r) to any sequence $(f_k)_{k=1}^\infty$ as in (s).]

(v) Let ϕ and ψ be in $L_1 \cap L_2(\mathbb{R}^n)$ and write $f(x) = \hat{\phi}(-x)$, $g(x) = \hat{\psi}(-x)$ for

$x \in \mathbb{R}^n$. Then $fg \in L_1 \cap L_2(\mathbb{R}^n)$ and $\widehat{fg} = \phi * \psi$. [Hint: Apply (j) to compute $\widehat{\phi * \psi}$ after noting that $\widehat{\phi * \psi} = \hat{\phi}\hat{\psi} \in L_1(\mathbb{R}^n)$.]

(w) If $\phi \in L_1 \cap L_2(\mathbb{R}^n)$ and $f(x) = \hat{\phi}(-x)$ for $x \in \mathbb{R}^n$, then $f \in L_2(\mathbb{R}^n)$ and $\hat{f} = \phi$. [Hints: For $\alpha > 0$, write $g_\alpha(x) = G((1/\alpha)x)$. Use dominated convergence to see that $\|f - fg_\alpha\|_2 \to 0$ as $\alpha \to \infty$ and conclude that $\|\hat{f} - \widehat{fg_\alpha}\|_2 \to 0$. Use (v) to see that $\widehat{fg_\alpha} = \phi * G_\alpha$ and then invoke (g) and (6.110) to obtain $\hat{f} = \phi$ a.e.]

(x) If $\phi \in L_2(\mathbb{R}^n)$, then there exists $f \in L_2(\mathbb{R}^n)$ such that $\hat{f} = \phi$. [Hints: Let $\phi_k \in L_1 \cap L_2$ satisfy $\|\phi - \phi_k\|_2 \to 0$. Use (w) to get $f_k \in L_2$ such that $\hat{f_k} = \phi_k$ and then use (u) and (6.110) to produce f with $\|f - f_k\|_2 \to 0$.]

(y) *Plancherel's Theorem.* The restriction of the Fourier transform (defined in (a)) to $L_1 \cap L_2(\mathbb{R}^n)$ has a unique extension to a linear isometry of $L_2(\mathbb{R}^n)$ onto $L_2(\mathbb{R}^n)$.

(z) *M. Riesz's Theorem.* We have $\{\phi * \psi : \phi, \psi \in L_2(\mathbb{R}^n)\} = \{\hat{h} : h \in L_1(\mathbb{R}^n)\}$. [Hints: Given $\phi, \psi \in L_2$, let $h = fg$, where $f(x) = \hat{\phi}(-x)$, $g(x) = \hat{\psi}(-x)$. Approximate ϕ and ψ by sequences from $L_1 \cap L_2$ and use (v) to show that $\hat{h} = \phi * \psi$. Given $h \in L_1(\mathbb{R}^n)$, write $h = fg$ where $f, g \in L_2(\mathbb{R}^n)$.]

14. *Weyl's Theorem on Uniform Distribution.* Let $\xi = (\xi_1, \dots, \xi_n)$ be a fixed vector in $\mathbb{R}^n$ having the property that if $m = (m_1, \dots, m_n) \in \mathbb{Z}^n$ is an n-tuple of integers for which $m\xi = \sum\limits_{j=1}^{n} m_j \xi_j$ is an integer, then $m_j = 0$ for all j $(1 \leq j \leq n)$.

(a) The set of all $x \in \mathbb{R}^n$ which fail to have the above property of ξ has Lebesgue measure 0. [Hint: The described set is the union of the hyperplanes $H_{m,k} = \{x \in \mathbb{R}^n : mx = k\}$, where $m \in \mathbb{Z}^n \setminus \{0\}$ and $k \in \mathbb{Z}$.] Let $I = \{x \in \mathbb{R}^n : 0 \leq x_j < 1$ for $1 \leq j \leq n\}$ and let F be the set of all complex-valued functions f that are defined everywhere on $\mathbb{R}^n$ and satisfy $f(x + m) = f(x)$ whenever $x \in \mathbb{R}^n$ and $m \in \mathbb{Z}^n$ (these functions are periodic of period 1 in each variable and so are completely determined by their values at points of I). For $f \in F$ and $p \in \mathbb{N}$, write

$$A_p(f) = \frac{1}{p} \sum_{k=1}^{p} f(k\xi)$$

where $k\xi = (k\xi_1, \dots, k\xi_n)$. Let F^* be the set of all $f \in F$ for which f is Lebesgue integrable over I and

$$\lim_{p \to \infty} A_p(f) = \int_I f. \qquad (*)$$

(b) For $m \in \mathbb{Z}^n$, the function $\mathcal{X}_m$ defined on $\mathbb{R}^n$ by

$$\mathcal{X}_m(x) = e^{2\pi i m x} = \exp\left(2\pi i \sum_{j=1}^{n} m_j x_j \right)$$

is in F^*. [Hint: If $m \neq 0$, one can use the independence property of ξ to compute $A_p(\mathcal{X}_m)$ explicitly by summing a geometric progression.]

(c) With pointwise operations, the set F^* is a complex linear space; that is, if $f, g \in F^*$ and $\alpha \in \mathbb{C}$, then $\alpha f, f + g \in F^*$.

(d) If $(f_k)_{k=1}^{\infty} \subset F^*$ and $f_k \to f$ uniformly on $\mathbb{R}^n$, then $f \in F^*$.

(e) If $f \in F$ is continuous on $\mathbb{R}^n$, then $f \in F^*$. [Hint: Use the Stone–Weierstrass Theorem (3.122) to show that f is a uniform limit of finite linear combinations of functions $\mathcal{X}_m$ as in (b); compare (3.129). Specifically, let X be the compact space $\{z \in \mathbb{C}^n : z = (z_1, \dots, z_n), |z_j| = 1$ for $1 \leq j \leq n\}$, show that the set of all linear combinations of the functions given by $\widetilde{\mathcal{X}}_m(z) = z_1^{m_1} z_2^{m_2} \dots z_n^{m_n}$ is dense in $C(X)$, and approximate the function $\tilde{f}$ given by

$$\tilde{f}(e^{2\pi i t_1}, \ldots, e^{2\pi i t_n}) = f(t_1, \ldots, t_n).]$$

(f) If a real-valued $f \in F$ is such that the sets $G = \{g \in F^* : g \text{ is real-valued}, g \leq f$ on $\mathbb{R}^n\}$ and $H = \{h \in F^* : h \text{ is real-valued}, f \leq h \text{ on } \mathbb{R}^n\}$ are both nonvoid and satisfy

$$\sup\left\{\int_I g : g \in G\right\} = \inf\left\{\int_I h : h \in H\right\},$$

then $f \in F^*$. [Hints: Choose $(g_k)_{k=1}^{\infty} \subset G$ and $(h_k)_{k=1}^{\infty} \subset H$ whose integrals over I have the same limit α. Write $g = \overline{\lim} g_k$, $h = \underline{\lim} h_k$, and apply (6.47) to $(h_1 - g_k)$ and to $(h_k - g_1)$ to get $\alpha \leq \int_I g$ and $\int_I h \leq \alpha$. Infer that $g = f = h$ a.e. For fixed k, obtain

$$\int_I g_k \leq \lim_{p \to \infty} A_p(f) \leq \overline{\lim}_{p \to \infty} A_p(f) \leq \int_I h_k$$

and then let $k \to \infty$.]
(g) If $J \subset I$ is an interval of $\mathbb{R}^n$ and f is the function in F such that $f(x) = 1$ for $x \in J$ and $f(x) = 0$ for $x \in I \setminus J$, then $f \in F^*$. [Hints: Show that f satisfies the hypothesis of (f) where the g's and h's can be chosen to be continuous.]
(h) Let J be as in (g). For each $p \in \mathbb{N}$, let $N_p(J)$ be the number of integers k such that $1 \leq k \leq p$ and $k\xi - m \in J$ for some $m \in \mathbb{N}^n$. (Necessarily $m_j = [k\xi_j]$ for all j, where $[t]$ is the largest integer not exceeding t). Then

$$\lim_{p \to \infty} \frac{1}{p} N_p(J) = |J|.$$

It is because of this result that we say that the sequence of vectors $(k\xi)_{k=1}^{\infty}$ is *uniformly distributed in I modulo $\mathbb{Z}^n$*.
(i) If $f \in F$ is bounded and continuous a.e. on $\mathbb{R}^n$, then $f \in F^*$. [Hints: For each $k \in \mathbb{N}$, define functions g_k and h_k on $\mathbb{R}^n$ by $g_k(x) = \inf\{f(y) : y \in C_k(x)\}$ and $h_k(x) = \sup\{f(y) : y \in C_k(x)\}$, where $C_k(x)$ is the dyadic n-cube of order k that contains x (see (6.115)). Then g_k and h_k are each constant on every dyadic n-cube of order k, so they are in F^*. Also, (g_k) and (h_k) are monotone sequences that both converge to f at every x where f is continuous. Compare the proof of (6.29).]
(j) There exist open sets (not intervals) $J \subset I$ such that if $N_p(J)$ is defined as in (h), then $N_p(J) = p$ for all $p \in \mathbb{N}$ but $\lambda^n(J) < 1$.
(k) There exist bounded lower semicontinuous functions $f \in F \setminus F^*$.

15. *Kronecker's Approximation Theorems.*
(a) Let ξ be as in Exercise 14. Then for each $\epsilon > 0$, each $x \in \mathbb{R}^n$, and each $k_0 \in \mathbb{N}$ there exists $m \in \mathbb{Z}^n$ and $k \in \mathbb{N}$ with $k > k_0$ such that

$$|x_j - m_j - k\xi_j| < \epsilon$$

for all j $(1 \leq j \leq n)$. [Hint: Use Exercise 14(h). Suppose $x \in I$ and choose a dyadic n-cube J of order $r > 0$ with $x \in J$ and $2^{-r} < \epsilon$. If no such m and k exist, then $N_p(J) \leq k_0$ for all p.]
(b) Let $\theta = (\theta_1, \ldots, \theta_n) \in \mathbb{R}^n$ be such that if $m \in \mathbb{Z}^n$ and $m\theta = 0$, then $m_j = 0$ for all j $(1 \leq j \leq n)$. Then for each $\epsilon > 0$, each $x \in \mathbb{R}^n$, and each real number $t_0 > 0$, there exist $m \in \mathbb{Z}^n$ and a real number $t > t_0$ such that

$$|x_j - m_j - t\theta_j| < \epsilon$$

for all j $(1 \leq j \leq n)$. [Hint: Choose any real number $\alpha > 0$ that is not in the countable set of numbers $\{q\theta : q \in Q^n\}$, apply (a) with $\xi_j = \theta_j / \alpha$ and $k_0 > t_0 \alpha$ to obtain k and m, then take $t = k / \alpha$.]

(c) The set of all $\theta \in \mathbb{R}^n$ that fail to satisfy the hypothesis of (b) is of measure 0.

(d) If $p_1, p_2, \ldots, p_n$ are n distinct prime numbers, then the vector θ with $\theta_j = \log p_j$ $(1 \leq j \leq n)$ satisfies the hypothesis of (b). [We remark that this θ also satisfies the more stringent hypothesis made on ξ in (a). This follows from the fact that e is a transcendental number.]

Some Differential Calculus
in Higher Dimensions

In order to present the second major theorem about integration over $\mathbb{R}^n$ [Fubini's Theorem is the first], namely, the Change of Variable Theorem (6.148), we find it necessary to digress for one section from the theme of this chapter. In particular, we need to become familiar with the meaning and significance of the Jacobian of a "change function" ϕ and we need to know the Inverse Function Theorem (6.140). The subject of calculus in $\mathbb{R}^n$ is much too large to be done properly in the present volume. A reasonably thorough treatment, including differential forms, Grassmann algebras, and integration on manifolds, really needs a separate book. We recommend the book *Functions of Several Variables* by Wendell H. Fleming as an excellent introduction.

Our presentation in this and the following section does presuppose the most elementary facts about matrices and determinants such as can be found in nearly any book on matrix theory or elementary linear algebra. Specifically, we need to know about elementary row and column operations, the fact that the determinant of the product of square matrices is the product of their determinants, and Cramer's Rule on the solution of systems of equations by determinants. We have decided to leave it to the reader to consult his favorite book on these topics, if necessary, but we suggest reading our two sections first because the prerequisites needed from linear algebra are small indeed.

We begin with several definitions.

(6.128) Definition   Let n and p be natural numbers. A mapping $A : \mathbb{R}^n \rightarrow \mathbb{R}^p$ is called a *linear mapping* [or linear transformation] if

$$A(s + t) = A(s) + A(t) \quad \text{and} \quad A(\alpha t) = \alpha A(t)$$

for all $s, t \in \mathbb{R}^n$ and $\alpha \in \mathbb{R}$. For $1 \leq j \leq n$, let $e^j \in \mathbb{R}^n$ be the vector having 1 in the jth coordinate and 0 in all other coordinates. The set $\{e^1, \ldots, e^n\}$ is the *standard basis* for $\mathbb{R}^n$. Thus, for $h = (h_1, \ldots, h_n) \in \mathbb{R}^n$, we have

$$h = \sum_{j=1}^{n} h_j e^j$$

and, for A as above, we have

$$A(h) = \sum_{j=1}^{n} h_j A(e^j).$$

Writing $a_{ij} = A_i(e^j)$ for the ith coordinate of the vector $A(e^j)$ $(1 \leq i \leq p)$, we see that the ith coordinate of $A(h)$ is given by

$$A_i(h) = \sum_{j=1}^{n} a_{ij} h_j \tag{1}$$

Thus the vector $A(h)$ is obtained by the familiar matrix multiplication of the p by n rectangular matrix (a_{ij}) by the column vector (n by 1 matrix) h. This matrix (a_{ij}), which has the real number a_{ij} in its ith row and jth column, is called the *matrix of A with respect to the standard basis* [or simply *the matrix of A*]. [Note that any p by n matrix having real entries defines a linear mapping of $\mathbb{R}^n$ into $\mathbb{R}^p$ via (1)]. In case $p = n$, we denote by $\det A$ the determinant of the matrix (a_{ij}) of A and we call this number the determinant of A.

It is clear from (1) and Cauchy's Inequality (1.44) that if A is as above, then

$$|A_i(h)|^2 \leq \left(\sum_{j=1}^{n} |a_{ij}|^2 \right) |h|^2$$

for $1 \leq i \leq p$ [as usual, $|h| = \left(\sum_{j=1}^{n} h_j^2 \right)^{1/2}$ for $h = (h_1, \ldots, h_n) \in \mathbb{R}^n$], and so

$$|A(h)| = \left(\sum_{i=1}^{p} |A_i(h)|^2 \right)^{1/2} \leq \left(\sum_{i=1}^{p} \sum_{j=1}^{n} |a_{ij}|^2 \right)^{1/2} \cdot |h| = \|A\| \cdot |h| \tag{2}$$

for all $h \in \mathbb{R}^n$, where the last equality defines the number $\|A\|$.

If A is as above and $B : \mathbb{R}^p \to \mathbb{R}^q$ is a linear mapping, then, letting $f^1, \ldots, f^p$ be the standard basis in $\mathbb{R}^p$, we see that the entry in the hth row and jth column $(1 \leq h \leq q, 1 \leq j \leq n)$ of the matrix of the composite mapping $B \circ A$ is given by

$$c_{hj} = (B \circ A)_h(e^j) = B_h(A(e^j)) = B_h\left(\sum_{i=1}^{p} A_i(e^j) f^i \right) = \sum_{i=1}^{p} A_i(e^j) B_h(f^i)$$

$$= \sum_{i=1}^{n} b_{hi} a_{ij}.$$

Thus the matrix (c_{hj}) of $B \circ A$ is the ordinary matrix product $(b_{hi})(a_{ij})$ of the matrices of B and A respectively.

(6.129) Definitions Let n and p be natural numbers, let Ω be an open subset of $\mathbb{R}^n$, let $\phi : \Omega \to \mathbb{R}^p$ be a mapping, and let $a \in \Omega$. We say that ϕ is *differentiable* at a if there exists a linear mapping $A : \mathbb{R}^n \to \mathbb{R}^p$ such that

$$\lim_{h \to 0} \frac{1}{|h|} |\phi(a + h) - \phi(a) - A(h)| = 0 \tag{1}$$

in $\mathbb{R}^p$ (here $h \to 0$ in $\mathbb{R}^n$ and $|h| = \left(\sum\limits_{j=1}^{n} h_j^2 \right)^{1/2}$). That is, to each $\epsilon > 0$ there

corresponds a neighborhood U of a, $U \subset \Omega$, such that

$$\left| \phi(t) - [\phi(a) + A(t - a)] \right| \leq \epsilon |t - a| \tag{2}$$

for all $t \in U$ [write $h = t - a$] (here $|x| = \left(\sum\limits_{i=1}^{p} x_i^2 \right)^{1/2}$ for $x = (x_1, \ldots, x_p) \in \mathbb{R}^p$).

It is clear that at most one such A can exist [if A_1 and A_2 both satisfy the condition and if for some $b \in \mathbb{R}^n$ we have $|A_1(b) - A_2(b)| = c > 0$, choose ϵ so that $0 < 2|b|\epsilon < c$, take $t = a + \alpha b$, and use (2) to obtain the contradiction

$$\alpha c = |A_1(t - a) - A_2(t - a)| \leq 2|t - a|\epsilon = 2\alpha |b|\epsilon < \alpha c$$

for all sufficiently small positive α]. If A exists, we call it the *differential of ϕ at a* and we denote it by $d\phi(a)$. In this way we see that $d\phi$ is a linear transformation-valued function defined at those $a \in \Omega$ at which ϕ is differentiable. If ϕ is differentiable at each point of Ω, we say that ϕ is *differentiable on Ω*.

If $p = 1$ [so that ϕ is real-valued] and j is fixed with $1 \leq j \leq n$, then we (unsurprisingly) define the *jth partial derivative of ϕ at a* to be the derivative at a_j of the function

$$t_j \to \phi(a_1, \ldots, a_{j-1}, t_j, a_{j+1}, \ldots, a_n)$$

if this derivative exists and, if so, we denote this number by $D_j\phi(a)$ [or $(\partial\phi/\partial t_j)(a)$]. Thus, using the notation of (6.128),

$$D_j\phi(a) = \lim_{\alpha \to 0} \frac{1}{\alpha} \left[\phi(a + \alpha e^j) - \phi(a) \right]$$

if this limit exists. We regard $D_j\phi$ as an extended real-valued function defined at those $a \in \Omega$ at which this limit does exist in $\mathbb{R}^{\#}$.

Supposing that $p = n$ and that ϕ is differentiable at a, we define the *Jacobian of ϕ at a* to be the determinant of the linear mapping $d\phi(a)$ and we denote this real number by $J_\phi(a)$:

$$J_\phi(a) = \det(d\phi(a)).$$

In this way we obtain J_ϕ as a real-valued function defined at those $a \in \Omega$ at which ϕ is differentiable.

The following theorem relates the notions just defined.

(6.130) Theorem *Let n, p, Ω, and a be as in (6.129). For $1 \leq i \leq p$, define the coordinate functions $\phi_i : \Omega \to \mathbb{R}$ of ϕ by letting $\phi_i(t)$ be the ith coordinate of $\phi(t)$:*

$$\phi(t) = (\phi_1(t), \ldots, \phi_p(t)), \quad t \in \Omega.$$

(a) Suppose that ϕ is differentiable at a. Then all of the pn partial derivatives $D_j\phi_i(a)$ exist in $\mathbb{R}$ and the matrix of the linear mapping $d\phi(a) = A$ is the p by n matrix having the number $D_j\phi_i(a)$ in its ith row and jth column. The ith coordinate of $A(h)$ is

given by

(i) $$A_i(h) = \sum_{j=1}^{n} D_j\phi_i(a)h_j \qquad (1 \leq i \leq p).$$

Each ϕ_i is differentiable at a and $d\phi_i(a) = A_i$, where A_i is defined from $\mathbb{R}^n$ to $\mathbb{R}$ by (i).
   (b) *If each ϕ_i is differentiable at a and $d\phi_i(a) = A_i$, then ϕ is differentiable at a and $d\phi(a) = A$, where A is defined from $\mathbb{R}^n$ to $\mathbb{R}^p$ by*

$$A(h) = (A_1(h), \ldots, A_p(h)).$$

   (c) *If all of the pn partial derivatives $D_j\phi_i(t)$ $(1 \leq i \leq p, 1 \leq j \leq n)$ exist in $\mathbb{R}$ at every $t \in \Omega$ and if the pn functions thereby defined are continuous on Ω [in this case we say that ϕ is of class $C^{(1)}$ on Ω], then ϕ is differentiable on Ω.*

Proof   (a) Suppose that ϕ is differentiable at a and write $A = d\phi(a)$. By definition (6.128), the (i, j) entry in the matrix of A is the ith coordinate $A_i(e^j)$ of the vector $A(e^j) \in \mathbb{R}^p$. We have

$$\varlimsup_{\alpha \to 0} |(1/\alpha)[\phi_i(a + \alpha e^j) - \phi_i(a)] - A_i(e^j)|$$

$$= \varlimsup_{\alpha \to 0} (1/|\alpha|)|\phi_i(a + \alpha e^j) - \phi_i(a) - A_i(\alpha e^j)|$$

$$\leq \lim_{\alpha \to 0} \frac{|\phi(a + \alpha e^j) - \phi(a) - A(\alpha e^j)|}{|\alpha e^j|} = 0$$

where we have used the evident inequality $|x_i| \leq |x|$ for $x \in \mathbb{R}^p$ and Definition (6.129). Therefore $D_j\phi_i(a)$ exists and equals $A_i(e^j)$. Because $A_i(h) = A_i\left(\sum_{j=1}^{n} h_j e^j\right) = \sum_{j=1}^{n} h_j A_i(e^j)$, this also proves (i). To see that $d\phi_i(a) = A_i$, write

$$|\phi_i(a + h) - \phi_i(a) - A_i(h)| \leq |\phi(a + h) - \phi(a) - A(h)|,$$

divide by $|h|$, and let $h \to 0$.
   (b) For any $x \in \mathbb{R}^p$ we have

$$|x| = \left(\sum_{i=1}^{p} x_i^2\right)^{1/2} \leq \sum_{i=1}^{p} |x_i|.$$

Apply this inequality with $x = \phi(a + h) - \phi(a) - A(h)$, where A is defined as in (b), divide by $|h|$ and let $h \to 0$.
   (c) In view of (b) we need only show that each ϕ_i is differentiable on Ω. Thus we suppose that $p = 1$. We prove (c) by induction on n. If $n = 1$ and $t \in \Omega \subset \mathbb{R}$, we have

$$\lim_{h \to 0} \frac{|\phi(t + h) - \phi(t) - \phi'(t)h|}{|h|} = \lim_{h \to 0} \left|\frac{\phi(t + h) - \phi(t)}{h} - \phi'(t)\right| = 0,$$

so $d\phi(t) = A$, where $A(h) = \phi'(t)h$ for $h \in \mathbb{R}$. Now suppose that $n > 1$ is fixed and that the theorem is true when n is replaced by $n - 1$ and $p = 1$. For $t \in \mathbb{R}^n$, write $t' = (t_1, \ldots, t_{n-1}) \in \mathbb{R}^{n-1}$. Let $a \in \Omega$ be given and define the function ψ on the set $\Omega' = \{t \in \mathbb{R}^{n-1} : (t, a_n) = (t_1, \ldots, t_{n-1}, a_n) \in \Omega\}$ by $\psi(t) = \phi(t, a_n)$. One checks that Ω' is open in $\mathbb{R}^{n-1}$, $D_j\psi(t) = D_j\phi(t, a_n)$ for $t \in \Omega'$ and $1 \le j \le n - 1$, and that these partial derivatives of ψ are continuous on Ω'. Therefore our induction hypothesis shows that ψ is differentiable at $a' = (a_1, \ldots, a_{n-1})$ and (a) shows that

$$d\psi(a')(h) = \sum_{j=1}^{n-1} D_j\phi(a)h_j$$

for all $h \in \mathbb{R}^{n-1}$. Thus, given $\epsilon > 0$, Definition (6.127) provides $\delta_1 > 0$ such that $h' \in \mathbb{R}^{n-1}$ and $|h'| < \delta_1$ imply that $a' + h' \in \Omega'$ and

$$\left| \psi(a' + h') - \psi(a') - \sum_{j=1}^{n-1} D_j\phi(a)h_j \right| \le \frac{\epsilon|h'|}{2} . \tag{1}$$

Since $D_n\phi$ is continuous at a, we can choose δ so small that $0 < \delta \le \delta_1$ and that if $t \in \mathbb{R}^n$ and $|t - a| < \delta$, then $t \in \Omega$ and $|D_n\phi(t) - D_n\phi(a)| < \epsilon/2$. Now fix any $h \in \mathbb{R}^n$ with $|h| < \delta$. Then $|h'| < \delta_1$, (1) obtains, and the Mean Value Theorem supplies θ such that $0 < \theta < 1$ and

$$\phi(a + h) - \psi(a' + h') = \phi(a' + h', a_n + h_n) - \phi(a' + h', a_n)$$
$$= D_n\phi(a' + h', a_n + \theta h_n)h_n.$$

Writing $t = (a' + h', a_n + \theta h_n)$, we have $|t - a| = |(h', \theta h_n)| < \delta$, and so our choice of δ yields

$$|\phi(a + h) - \psi(a' + h') - D_n\phi(a)h_n| = |D_n\phi(t)h_n - D_n\phi(a)h_n|$$
$$\le \epsilon|h_n|/2 \le \epsilon|h|/2. \tag{2}$$

Now combine (1) and (2), define A as in (i), and recall that $\phi(a) = \psi(a')$ to obtain

$$|\phi(a + h) - \phi(a) - A(h)| \le |\phi(a + h) - \psi(a' + h') - D_n\phi(a)h_n|$$
$$+ \left| \psi(a' + h') - \psi(a') - \sum_{j=1}^{n-1} D_j\phi(a)h_j \right|$$
$$\le \epsilon|h|/2 + \epsilon|h'|/2 \le \epsilon|h|. \tag{3}$$

Since (3) holds whenever $|h| < \delta$, this proves that ϕ is differentiable at a with A as its differential. Since $a \in \Omega$ was arbitrary, the proof is complete.

□

(6.131) Examples (a) Let $n = 2$, $p = 1$, $\Omega = \mathbb{R}^2$, and define $\phi : \mathbb{R}^2 \to \mathbb{R}$ by $\phi(s, t) = t^3/(s^2 + t^2)$ if $(s, t) \ne (0, 0)$ and $\phi(0, 0) = 0$. Since $|\phi(s, t)| \le |t| \le \sqrt{s^2 + t^2} = |(s, t)|$ for all $(s, t) \in \mathbb{R}^2$, ϕ is continuous on $\mathbb{R}^2$. A direct calculation shows that the partial derivatives $D_1\phi$ and $D_2\phi$ exist everywhere on $\mathbb{R}^2$ and are continuous except at $(0, 0)$.

Thus ϕ is differentiable except at $(0,0)$ by (6.130.c). Moreover, one checks that

$$|D_1\phi(s,t)| \leq 3\sqrt{3}/8 \quad \text{and} \quad |D_2\phi(s,t)| \leq 9/8$$

everywhere [by making the substitutions $s = r\cos\theta$, $t = r\sin\theta$]. However, ϕ is *not* differentiable at $(0,0)$ because, if it were, (6.130.a) would yield $d\phi(0,0) = A$, where $A(s,t) = t$ $[D_1\phi(0,0) = 0, \quad D_2\phi(0,0) = 1]$ and hence

$$0 = \lim_{(s,t)\to(0,0)} \frac{|\phi(s,t) - \phi(0,0) - A(s,t)|}{|(s,t)|}$$

$$= \lim_{r\to 0} \frac{|\phi(r\cos\theta, r\sin\theta) - r\sin\theta|}{|r|}$$

$$= |\sin^3\theta - \sin\theta|$$

for all real θ—a palpable contradiction. It is also interesting to notice that if $\gamma: \mathbb{R} \to \mathbb{R}^2$ is differentiable at 0 and satisfies $\gamma(0) = (0,0), \gamma'(0) \neq (0,0)$ [this means $\gamma(x) = (\gamma_1(x), \gamma_2(x))$ and $\gamma'(0) = (\gamma_1'(0), \gamma_2'(0))$], then the composite function $f = \phi \circ \gamma$ [the "restriction of ϕ to the curve $\gamma(\mathbb{R})$"] is differentiable at 0 and

$$f'(0) = \lim_{x\to 0} \frac{f(x)}{x} = \frac{(\gamma_2'(0))^3}{(\gamma_1'(0))^2 + (\gamma_2'(0))^2} .$$

(b) *Polar Coordinates in* $\mathbb{R}^2$. Let $n = p = 2$ and define $\phi: \mathbb{R}^2 \to \mathbb{R}^2$ by $\phi(r,\theta) = (r\cos\theta, r\sin\theta) = (x, y)$ so that $x = \phi_1(r,\theta) = r\cos\theta$ and $y = \phi_2(r,\theta) = r\sin\theta$. Plainly ϕ is of class $C^{(1)}$ on $\mathbb{R}^2$. Thus ϕ is differentiable on $\mathbb{R}^2$ and the matrix of the linear mapping $d\phi(r,\theta)$ is

$$\begin{pmatrix} D_1\phi_1(r,\theta) & D_2\phi_1(r,\theta) \\ D_1\phi_2(r,\theta) & D_2\phi_2(r,\theta) \end{pmatrix} = \begin{pmatrix} \cos\theta & -r\sin\theta \\ \sin\theta & r\cos\theta \end{pmatrix}$$

It follows that the Jacobian of ϕ is given by $J_\phi(r,\theta) = r$. If $\theta_0 \in \mathbb{R}$, then one checks that the restriction of ϕ to the open set $\Omega = \{(r,\theta) \in \mathbb{R}^2 : r > 0, \theta_0 < \theta < \theta_0 + 2\pi\}$ is a one-to-one $C^{(1)}$ mapping of Ω onto the open set $\Delta = \mathbb{R}^2 \setminus \{(r\cos\theta_0, r\sin\theta_0) : r \geq 0\}$.

(c) *Spherical Coordinates in* $\mathbb{R}^n$. For integers $n \geq 2$, we define $C^{(1)}$ mappings $\phi^n: \mathbb{R}^n \to \mathbb{R}^n$ inductively as follows:

$$\phi^2(\rho,\theta_1) = (\rho\cos\theta_1, \rho\sin\theta_1) \tag{1}$$

and, for $n > 2$,

$$\phi^n(\rho,\theta_1, \ldots, \theta_{n-1}) = (x_1, \ldots, x_n) \tag{2}$$

where

$$x_1 = \rho\cos\theta_1,$$
$$(x_2, \ldots, x_n) = \phi^{n-1}(\rho\sin\theta_1, \theta_2, \ldots, \theta_{n-1}). \tag{3}$$

Notice that ϕ^2 is the mapping ϕ of Example (b). It is evident, by induction on n, that if (2) obtains, then $|x|^2 = \rho^2$ and

$$
\begin{aligned}
x_1 &= \rho \cos \theta_1 \\
x_2 &= \rho \sin \theta_1 \cos \theta_2 \\
&\vdots \\
x_{n-1} &= \rho \sin \theta_1 \ldots \sin \theta_{n-2} \cos \theta_{n-1} \\
x_n &= \rho \sin \theta_1 \ldots \sin \theta_{n-2} \sin \theta_{n-1}.
\end{aligned}
$$

That is,

$$
x_i = \rho \cos \theta_i \prod_{j=1}^{i-1} \sin \theta_j \qquad (1 \le i < n)
$$

and

$$
x_n = \rho \prod_{j=1}^{n-1} \sin \theta_j
$$

where, as usual, $\prod_{j=1}^{0} \sin \theta_j = 1$. Defining the open sets

$$
\Omega_n = \{(\rho, \theta_1, \ldots, \theta_{n-1}) \in \mathbb{R}^n : \rho > 0, 0 < \theta_j < \pi
$$

$$
\text{if} \quad 1 \le j < n - 1, -\pi < \theta_{n-1} < \pi\}
$$

and

$$
\Delta_n = \mathbb{R}^n \setminus \{x \in \mathbb{R}^n : x_n = 0 \quad \text{and} \quad x_{n-1} \le 0\}
$$
$$
= \{x \in \mathbb{R}^n : x_n \neq 0 \quad \text{or} \quad x_{n-1} > 0\},
$$

we next see by induction that ϕ^n is a one-to-one mapping of Ω_n onto Δ_n. For $n = 2$, this is obvious, so suppose that $n > 2$ is given and that the assertion is true for $n - 1$ in place of n. Letting $x \in \Delta_n$ be given, we write $\rho = |x| > 0$ and, noting that $-1 < x_1/\rho < 1$, we choose the unique θ_1 for which $0 < \theta_1 < \pi$ and $\rho \cos \theta_1 = x_1$. Since $(x_2, \ldots, x_n) \in \Delta_{n-1}$, our inductive hypothesis provides a unique point $(r, \theta_2, \ldots, \theta_{n-1}) \in \Omega_{n-1}$ for which $\phi^{n-1}(r, \theta_2, \ldots, \theta_{n-1}) = (x_2, \ldots, x_n)$. As noted above, this implies that $r = (x_2^2 + \cdots + x_n^2)^{1/2} = (\rho^2 - \rho^2 \cos^2 \theta_1)^{1/2} = \rho \sin \theta_1$. Thus $(\rho, \theta_1, \ldots, \theta_{n-1}) \in \Omega_n$ and ϕ^n maps it to x. The uniqueness statements along with (1)–(3) show that ϕ^n is one-to-one on Ω_n.

If one writes out the matrix of the linear mapping $d\phi^n(\rho, \theta_1, \ldots, \theta_{n-1})$ for an $n > 2$ and then expands the determinant of this matrix according to the cofactors of the first row, being careful to factor out common factors of columns that have such, one obtains the formula

$$
J_{\phi^n}(\rho, \theta_1, \ldots, \theta_{n-1}) = \rho(\sin \theta_1)^{n-2} J_{\phi^{n-1}}(\rho, \theta_2, \ldots, \theta_{n-1}).
$$

Using this and the fact that $J_{\phi^2}(\rho, \theta_1) = \rho$, one finds that the Jacobian of ϕ^n is given by

$$
J_{\phi^n}(\rho, \theta_1, \ldots, \theta_{n-1}) = \rho^{n-1} \cdot \prod_{j=1}^{n-2} (\sin \theta_j)^{n-j-1}
$$

and that it is always positive on Ω_n.

The next theorem is simple but important.

(6.132) Theorem   *Let* $n, p, \Omega, a,$ *and* ϕ *be as in (6.129). Suppose that* ϕ *is differentiable at* a. *Then* ϕ *is continuous at* a. *In fact, to each* $\epsilon > 0$ *corresponds some* $\delta > 0$ *such that* $t \in \Omega$ *and*

$$|\phi(t) - \phi(a)| \leq (\|d\phi(a)\| + \epsilon)|t - a|$$

for all $t \in \mathbb{R}^n$ *for which* $|t - a| < \delta$ *[see (6.128) for the definition of* $\|A\|$*]. If* ϕ *is of class* $C^{(1)}$ *on* Ω, *then* ϕ *is continuous on* Ω.

   Proof   Let $\epsilon > 0$ be given and write $d\phi(a) = A$. By (6.129.2) there is some $\delta > 0$ such that $|t - a| < \delta$ implies that $t \in \Omega$ and

$$|\phi(t) - \phi(a) - A(t - a)| \leq \epsilon |t - a|$$

   and hence, by (6.128.2),

$$|\phi(t) - \phi(a)| \leq |\phi(t) - \phi(a) - A(t - a)| + |A(t - a)|$$
$$\leq \epsilon |t - a| + \|A\| \cdot |t - a|.$$

   If ϕ is of class $C^{(1)}$ on Ω, then (6.130.c) shows that ϕ is differentiable at each point of Ω. $\square$

(6.133) Example   Of course, the mere existence of finite partial derivatives does *not* imply continuity. The function $\phi : \mathbb{R}^2 \to \mathbb{R}$ defined by $\phi(0,0) = 0$ and $\phi(s,t) = st/(s^2 + t^2)$ if $s^2 + t^2 \neq 0$ has finite partial derivatives everywhere on $\mathbb{R}^2$ but is discontinuous at $(0,0)$. Evidently ϕ is of class $C^{(1)}$ on $\mathbb{R}^2 \backslash \{(0,0)\}$.

   We next show that composites of differentiable mappings are differentiable and we obtain a useful formula for the partial derivatives of composite maps.

(6.134) Chain Rule   *Let* $n, p,$ *and* q *be natural numbers, let* $\Omega \subset \mathbb{R}^n$ *and* $\Delta \subset \mathbb{R}^p$ *be open sets, and let* $\phi : \Omega \to \Delta$ *and* $\psi : \Delta \to \mathbb{R}^q$ *be given mappings. Suppose that* ϕ *is differentiable at a point* $a \in \Omega$ *and that* ψ *is differentiable at the point* $b = \phi(a) \in \Delta$. *Then the composite mapping* $\mathcal{X} = \psi \circ \phi$ *defined from* Ω *into* $\mathbb{R}^q$ *by* $\mathcal{X}(t) = \psi(\phi(t))$ *is differentiable at* a *and its differential there is the composite of the linear mappings* $d\phi(a)$ *and* $d\psi(b)$:

   (i) $d(\psi \circ \phi)(a) = d\psi(\phi(a)) \circ d\phi(a).$

Furthermore, the coordinate functions $\mathcal{X}_h$ $(1 \leq h \leq q)$ *defined by* $\mathcal{X}(t) = (\mathcal{X}_1(t), \ldots, \mathcal{X}_q(t))$ *satisfy*

   (ii) $D_j \mathcal{X}_h(a) = \displaystyle\sum_{i=1}^{p} D_i \psi_h(b) D_j \phi_i(a)$ *for* $1 \leq j \leq n.$

   Proof   Write $A = d\phi(a)$, $B = d\psi(b)$ and $\alpha = \max\{\|A\|, \|B\|\} + 1$. We are to prove that $d\mathcal{X}(a)$ exists and is equal to $B \circ A$. We have, using the linearity of B, that

$$\mathcal{X}(a + h) - \mathcal{X}(a) - B \circ A(h) = \left[\psi(\phi(a + h)) - \psi(b) - B(\phi(a + h) - b) \right]$$
$$+ \left[B(\phi(a + h) - \phi(a) - A(h)) \right] \qquad (1)$$

whenever $a + h \in \Omega$. Let $\epsilon > 0$ be given. Use (6.129) and (6.132) to obtain $\eta > 0$ such that if $h \in \mathbb{R}^n$, $x \in \mathbb{R}^p$, $|h| < \eta$, $|x - b| < \eta$, then $a + h \in \Omega$, $x \in \Delta$,

$$|\phi(a + h) - \phi(a) - A(h)| \leq (\epsilon/(2\alpha))|h|, \tag{2}$$

$$|\phi(a + h) - \phi(a)| \leq \alpha|h|, \tag{3}$$

and

$$|\psi(x) - \psi(b) - B(x - b)| \leq (\epsilon/(2\alpha))|x - b|. \tag{4}$$

Write $\delta = \eta/\alpha \leq \eta$. For any $h \in \mathbb{R}^n$ with $|h| < \delta$, (3) yields $|\phi(a + h) - b| \leq \alpha|h| < \eta$, so we may take $x = \phi(a + h)$ in (4) to get

$$|\psi(\phi(a + h)) - \psi(b) - B(\phi(a + h) - b)| \leq (\epsilon/(2\alpha))|\phi(a + h) - b|$$
$$\leq (\epsilon/(2\alpha)) \cdot \alpha|h|$$
$$= (\epsilon/2)|h|. \tag{5}$$

and (2) gives

$$|B(\phi(a + h) - \phi(a) - A(h))| \leq \|B\| \cdot (\epsilon/(2\alpha))|h| \leq (\epsilon/2)|h|. \tag{6}$$

Combining (1), (5), and (6), we have

$$|\mathcal{X}(a + h) - \mathcal{X}(a) - B \circ A(h)| \leq \epsilon|h|$$

whenever $h \in \mathbb{R}^n$ and $|h| < \delta$. Therefore (i) obtains. According to (6.130.a), formulas (ii) simply express the fact noted in (6.128) that the matrix of $B \circ A$ is the product of the matrices of B and A in that order. $\square$

(6.135) Example Let $n = q = 1$, $p = 2$, $\Omega = \mathbb{R}$, $\Delta = \mathbb{R}^2$. Define $\psi(0, 0) = 0$, $\psi(x, y) = y^3/(x^2 + y^2)$ if $x^2 + y^2 > 0$, and $\phi(t) = (t \cos \theta, t \sin \theta)$, where $\theta \in \mathbb{R}$ is fixed. Then $\mathcal{X}(t) = \psi \circ \phi(t) = t \sin^3\theta$ for all $t \in \mathbb{R}$, so $D_1\mathcal{X}_1(0) = \mathcal{X}'(0) = \sin^3\theta$ while

$$\sum_{i=1}^{2} D_i\psi_1(0, 0)D_1\phi_i(0) = 0 \cos \theta + 1 \sin \theta = \sin \theta.$$

Thus (6.134.ii) fails when $h = j = 1$, $a = 0$ and $b = (0, 0)$ unless $\sin \theta = \sin^3\theta$. The trouble is that ψ is *not* differentiable at $(0, 0)$. [Compare (6.131.a).]

(6.136) Remark If in (6.134) we write $x = \phi(t)$ and $y = \psi(x)$, then (6.134.ii) can be written in the somewhat imprecise but familiar form

$$\frac{\partial y_h}{\partial t_j} = \sum_{i=1}^{p} \frac{\partial y_h}{\partial x_i} \cdot \frac{\partial x_i}{\partial t_j}.$$

(6.137) Corollary *Let everything be as in (6.134) and suppose, in addition, that* $q = p = n$. *Then*

$$J_{\psi \circ \phi}(a) = J_{\psi}(\phi(a))J_{\phi}(a).$$

Proof   This follows from (6.134) and the well-known fact that the determinant of the product of two n by n matrices is equal to the product of their determinants.   □

As another useful corollary of (6.134), we have the following.

(6.138) **Mean Value Theorem for $\mathbb{R}^n$**   *Let $V \subset \mathbb{R}^n$ be an open set and let $\psi : V \to \mathbb{R}$ be a function that is differentiable on V. Let $a, b \in V$, $a \neq b$, be such that the segment $S = \{a + \alpha(b - a) : \alpha \in \mathbb{R}, 0 < \alpha < 1\}$ lies entirely in V. Then there exists a vector $v \in S$ such that*

$$\psi(b) - \psi(a) = \sum_{j=1}^{n} D_j \psi(v)(b_j - a_j).$$

Proof   Define $\phi : [0, 1] \to V$ by $\phi(\alpha) = a + \alpha(b - a)$. Then ϕ is continuous on $[0, 1]$, $\phi(]0, 1[) = S \subset V$, and ϕ is differentiable on $]0, 1[$ with $D_1 \phi_j(\alpha) = \phi_j'(\alpha) = b_j - a_j$ for $1 \leq j \leq n$ and $0 < \alpha < 1$. By (6.134), we see that the composite function $\mathcal{X} = \psi \circ \phi$ is differentiable on $]0, 1[$ with

$$\mathcal{X}'(\alpha) = D_1 \mathcal{X}_1(\alpha) = \sum_{j=1}^{n} D_j \psi_1(\phi(\alpha)) D_1 \phi_j(\alpha) = \sum_{j=1}^{n} D_j \psi(\phi(\alpha))(b_j - a_j)$$

whenever $0 < \alpha < 1$. By (6.132), ψ is continuous on V, and so $\mathcal{X}$ is continuous on $[0, 1]$. Therefore the Mean Value Theorem provides some ξ with $0 < \xi < 1$ such that

$$\psi(b) - \psi(a) = \mathcal{X}(1) - \mathcal{X}(0) = \mathcal{X}'(\xi).$$

Let $v = \phi(\xi)$.   □

We now prove two inequalities that will be useful in proving the Inverse Function Theorem.

(6.139) **Lemma**   *Let n, p, Ω, ϕ, and a be as in (6.129). Suppose that ϕ is of class $C^{(1)}$ on Ω and that $\epsilon > 0$. Write $A = d\phi(a)$. Then there exists a neighborhood V of a such that*

(i) $|\phi(s) - \phi(t)| \leq (\|A\| + \epsilon)|s - t|$ *for all $s, t \in V$.*

If, in addition, $p = n$ and A is invertible [i.e., $J_\phi(a) \neq 0$], then for the same V we have

(ii) $|\phi(s) - \phi(t)| \geq (\|A^{-1}\|^{-1} - \epsilon)|s - t|$ *for all $s, t \in V$.*

Proof   Define $\psi = \phi - A$ on Ω. According to (6.130.a), $D_j \psi_i(t) = D_j \phi_i(t) - D_j \phi_i(a)$ for all $t \in \Omega$, $1 \leq j \leq n$, $1 \leq i \leq p$. Therefore each $D_j \psi_i$ is continuous on Ω and is 0 at a. So we can find $\delta > 0$ such that $V = \{t \in \mathbb{R}^n : |t - a| < \delta\} \subset \Omega$ and $|D_j \psi_i(t)| < \epsilon/(pn)$ for all $t \in V$ and all i and j. Now, given any $s, t \in V$ and $1 \leq i \leq p$, (6.138) supplies $v \in V$ for which

$$\psi_i(s) - \psi_i(t) = \sum_{j=1}^{n} D_j \psi_i(v)(s_j - t_j),$$

and so

$$|\psi_i(s) - \psi_i(t)| \leq \sum_{j=1}^{n} \frac{\epsilon}{pn} |s_j - t_j| \leq \frac{\epsilon}{p} |s - t|.$$

Summing these inequalities gives

$$|\psi(s) - \psi(t)| \leq \sum_{i=1}^{p} |\psi_i(s) - \psi_i(t)| \leq \epsilon |s - t|.$$

Since

$$|A(s) - A(t)| = |A(s - t)| \leq \|A\| \cdot |s - t|,$$

the triangle inequality yields (i).

Now suppose that $p = n$ and A is invertible. Since

$$\|A^{-1}\| \cdot |A(u)| \geq |A^{-1}A(u)| = |u|$$

for all $u \in \mathbb{R}^n$, we have

$$|\phi(s) - \phi(t)| \geq |A(s - t)| - |\psi(s) - \psi(t)| \geq \|A^{-1}\|^{-1} \cdot |s - t| - \epsilon |s - t|$$

for all $s, t \in V$, so (ii) obtains. $\square$

The next theorem is exceedingly important. It asserts the existence of local $C^{(1)}$ inverses for certain $C^{(1)}$ mappings in $\mathbb{R}^n$.

(6.140) Inverse Function Theorem *Let n be a natural number and let ϕ be a mapping of class $C^{(1)}$ from an open set $\Omega \subset \mathbb{R}^n$ into $\mathbb{R}^n$ such that $J_\phi(t) \neq 0$ for all $t \in \Omega$. Then each $a \in \Omega$ has an open neighborhood $V \subset \Omega$ such that*
   (i) *the set $W = \phi(V)$ is open in $\mathbb{R}^n$,*
   (ii) *the restriction $\phi|_V$ of ϕ to V is one-to-one on V,*
   (iii) *the inverse ψ of $\phi|_V$ is of class $C^{(1)}$ on W and*

$$d\psi(w) = [d\phi(\psi(w))]^{-1} \quad \text{for all } w \in W.$$

Moreover, the set $\Delta = \phi(\Omega)$ is open in $\mathbb{R}^n$.

Proof Let $a \in \Omega$, write $A = d\phi(a)$, define α by $2\alpha = \|A^{-1}\|^{-1}$, and apply (6.139) with $\epsilon = \alpha$ to obtain an open neighborhood V of a such that $V \subset \Omega$ and

$$|\phi(s) - \phi(t)| \geq \alpha |s - t| \tag{1}$$

for all $s, t \in V$. Clearly (ii) obtains for this V. To prove (i), let $w \in W$ be given. We must find a neighborhood N of w such that $N \subset W$. Let v be the *unique* point of V such that $\phi(v) = w$. Choose $\delta > 0$ such that $B = \{t \in \mathbb{R}^n :$

$|t - v| \leq \delta\} \subset V$, and let $S = \{t \in \mathbb{R}^n : |t - v| = \delta\}$. Then $\phi(S)$ is compact [S is compact and ϕ is continuous (6.132)] and $w \notin \phi(S)$, so the number ϵ defined by

$$2\epsilon = \inf\{|\phi(s) - w| : s \in S\} \tag{2}$$

is positive. We now show that the ball $N = \{x \in \mathbb{R}^n : |x - w| < \epsilon\}$ is a subset of W. Fix any $x \in N$ and define a function f on Ω by

$$f(t) = |\phi(t) - x|^2 = \sum_{i=1}^{n} (\phi_i(t) - x_i)^2.$$

To show that $x \in W$, we will find $u \in V$ such that $f(u) = 0$. Since f is continuous on Ω, $B \subset \Omega$ and B is compact, there is some $u \in B$ such that $f(u) \leq f(t)$ for all $t \in B$. For any $s \in S$, we use (2) to see that

$$|\phi(s) - x| > |\phi(s) - x| + |x - w| - \epsilon \geq |\phi(s) - w| - \epsilon \geq \epsilon$$

and conclude that $f(s) > \epsilon^2$. But $f(u) \leq f(v) = |w - x|^2 < \epsilon^2$, so $u \notin S$, $u \in B \setminus S$. It follows that for each j, $1 \leq j \leq n$, the function $t_j \to f(u_1, \ldots, u_{j-1}, t_j, u_{j+1}, \ldots, u_n)$ has a local minimum at u_j, and so

$$0 = D_j f(u) = \sum_{i=1}^{n} 2(\phi_i(u) - x_i) D_j \phi_i(u).$$

Therefore, $y_i = \phi_i(u) - x_i$ $(1 \leq i \leq n)$ is a solution for the system of equations

$$\sum_{i=1}^{n} D_j \phi_i(u) y_i = 0 \quad (1 \leq j \leq n).$$

By (6.130.a), the determinant of the matrix of coefficients of this system is $J_\phi(u) \neq 0$, and so it has the unique solution $y_1 = y_2 = \cdots = y_n = 0$ [the linear mapping $d\phi(u)$ is one-to-one]. This proves that $\phi_i(u) = x_i$ for all i, $\phi(u) = x$, and concludes the proof of (i).

To prove (iii), we first show that ψ is differentiable on W. Let $w \in W$ and $\epsilon > 0$ be given, write $v = \psi(w)$ and $M = d\phi(v)$, let α be as in (1), and define $\beta = \epsilon\alpha\|M^{-1}\|^{-1}$. By definition of M, there is a neighborhood V_1 of v such that $V_1 \subset V$ and

$$|\phi(t) - \phi(v) - M(t - v)| \leq \beta|t - v| \tag{3}$$

for all $t \in V_1$. Applying (i) to $v \in \Omega$, we may also suppose that $W_1 = \phi(V_1)$ is open. Thus W_1 is a neighborhood of w. Given $x \in W_1$, write $t = \psi(x)$ and invoke (3) to obtain

$$|\psi(x) - \psi(w) - M^{-1}(x - w)| = |t - v - M^{-1}(\phi(t) - \phi(v))|$$

$$= |M^{-1}[M(t - v) - (\phi(t) - \phi(v))]|$$

$$\leq \|M^{-1}\| \cdot |\phi(t) - \phi(v) - M(t - v)|$$

$$\leq \epsilon\alpha|t - v| \leq \epsilon|\phi(t) - \phi(v)| = \epsilon|x - w|,$$

where the last inequality follows from (1). This proves that ψ is differentiable at w and that $d\psi(w) = M^{-1} = [d\phi(\psi(w))]^{-1}$. Thus, for all $w \in W$ the product of the matrix of $d\phi(\psi(w))$ and the matrix of $d\psi(w)$ is the identity matrix. In the language of partial derivatives [see (6.130.a)], this says that

$$\sum_{j=1}^{n} D_j\phi_i(\psi(w))D_k\psi_j(w) = \delta_{ik} \qquad (4)$$

for $w \in W$, $1 \leq i \leq n$, $1 \leq k \leq n$, where $\delta_{ik} = 1$ or 0 according as $i = k$ or $i \neq k$. Note that ψ is continuous on W (6.132) and each $D_j\phi_i$ is continuous on V. In view of (4) and Cramer's Rule, each function $D_k\psi_j$ is the quotient of two linear combinations of products [actually determinants] of the continuous functions $(D_j\phi_i) \circ \psi$ of which the denominator is $J_{\phi \circ \psi}$, which is never 0 on W. Therefore each $D_k\psi_j$ is continuous on W and the proof of (iii) is complete.

Finally, for each $a \in \Omega$ choose a neighborhood $V_a \subset \Omega$ of a such that $W_a = \phi(V_a)$ is open in $\mathbb{R}^n$. Then $\Delta = \phi(\Omega) = \phi(\cup\{V_a : a \in \Omega\}) = \cup\{W_a : a \in \Omega\}$ is open in $\mathbb{R}^n$. $\square$

(6.141) Examples (a) Let $n = 2$, $\Omega = \mathbb{R}^2$, and define ϕ by $\phi(s, t) = (e^s \cos t, e^s \sin t)$. [That is, if we write $z = (s, t) = s + it$, then $\phi(z) = e^z = \exp(z) = (\mathrm{Re}(e^z), \mathrm{Im}(e^z)) \in \mathbb{R}^2 = \mathbb{C}$.] Then ϕ satisfies the hypotheses of (6.140) with $J_\phi(a, b) = e^{2a} > 0$ for all $(a, b) \in \Omega$ and $\Delta = \phi(\Omega) = \mathbb{R}^2 \setminus \{(0, 0)\}$. Plainly ϕ is *not* one-to-one on Ω since $\phi(s, t) = \phi(s, t + 2\pi)$ for all $(s, t) \in \Omega$. However, if $(a, b) \in \Omega$ is given, we can choose $V = \{(s, t) \in \Omega : |t - b| < \pi\}$ and have $\phi|_V$ be one-to-one with $W = \phi(V) = \mathbb{R}^2 \setminus \{(u, v) : u = r\cos(b + \pi), v = r\sin(b + \pi) \text{ for some } r \geq 0\}$. In this case, the mapping $\psi : W \to V$ inverse to $\phi|_V$ is given by $\psi(u, v) = (s, t)$, where $s = \log\sqrt{u^2 + v^2}$ and t is the unique value of $\arg(u + iv)$ lying in $]b - \pi, b + \pi[$. [Thus, ψ is a branch of the complex logarithm.]

(b) The mapping $\phi : \mathbb{R}^2 \to \mathbb{R}^2$ defined by $\phi(s, t) = (s^2 - t^2, 2st)$ satisfies the hypotheses of (6.140) except that $J_\phi(0, 0) = 0$, because $J_\phi(s, t) = 4(s^2 + t^2)$ for all $(s, t) \in \mathbb{R}^2$. Since $\phi(s, t) = \phi(-s, -t)$, there is *no* neighborhood of $(0, 0)$ on which ϕ is one-to-one. One checks that ϕ is one-to-one on any half plane $V = \{(s, t) \in \mathbb{R}^2 : as + bt > 0\}$, where $a, b \in \mathbb{R}$ are not both 0.

Exercises

1.   Define $\phi : \mathbb{R}^2 \to \mathbb{R}$ by $\phi(s, t) = f(s) + f(t)$, where $f(0) = 0$ and $f(x) = x^2 \sin(1/x)$ for $x \neq 0$. Then $D_1\phi$ and $D_2\phi$ exist and are finite everywhere on $\mathbb{R}^2$, they are both discontinuous at $(0, 0)$, but yet ϕ is differentiable at $(0, 0)$.

2.   (a) Let $\Omega = \mathbb{R}^2 \setminus \{(0, t) : t \geq 0\}$ and define ϕ on Ω by $\phi(s, t) = t^2$ if $s > 0$ and $t \geq 0$ and $\phi(s, t) = 0$ if $s < 0$ or $t < 0$. Then $D_1\phi = 0$ everywhere on Ω but ϕ is *not* independent of s.

   (b) Suppose that $\Omega \subset \mathbb{R}^2$ is an open set having the property that for each $b \in \mathbb{R}$ the set $\{s \in \mathbb{R} : (s, b) \in \Omega\}$ is an interval. If $\phi : \Omega \to \mathbb{R}$ is a function such that $D_1\phi = 0$ everywhere on Ω, then ϕ is independent of s.

3.   (a) Define $\phi : \mathbb{R}^2 \to \mathbb{R}$ by $\phi(0, 0) = 0$ and $\phi(s, t) = st(s^2 - t^2)/(s^2 + t^2)$ if $s^2 + t^2 > 0$. Then $\phi \in C^{(1)}(\mathbb{R}^2)$, the partial derivatives of $D_1\phi$ and of $D_2\phi$ all exist and are finite everywhere on $\mathbb{R}^2$, but

$$D_2 D_1\phi(0, 0) = -1 \neq 1 = D_1 D_2\phi(0, 0).$$

(b) Let $\Omega \subset \mathbb{R}^2$ be an open set and let $\phi : \Omega \to \mathbb{R}$ be a function for which $D_1\phi$, $D_2\phi$, and $D_2 D_1\phi$ all exist and are finite at every point of Ω. Suppose also that $D_2 D_1\phi$ is continuous at a point $(a, b) \in \Omega$. Then $D_1 D_2\phi$ exists at (a, b) and

$$D_1 D_2\phi(a, b) = D_2 D_1\phi(a, b).$$

[Hints: Given $\epsilon > 0$, choose $\delta > 0$ such that $S = \{(s, t) : |s - a| < \delta, \; |t - b| < \delta\} \subset \Omega$, and $|D_2 D_1\phi(s, t) - D_2 D_1\phi(a, b)| < \epsilon$ for all $(s, t) \in S$. Fixing $(s, t) \in S$ with $s \neq a$, $t \neq b$, apply the Mean Value Theorem (for $\mathbb{R}$) twice to obtain $(\sigma, \tau) \in S$ such that

$$\Delta(s, t) = [\phi(s, t) - \phi(s, b)] - [\phi(a, t) - \phi(a, b)]$$
$$= D_2 D_1\phi(\sigma, \tau)(t - b)(s - a)$$

and conclude that

$$\left| \frac{\Delta(s, t)}{(s - a)(t - b)} - D_2 D_1\phi(a, b) \right| < \epsilon.$$

Now let $t \to b$.]

(c) Check directly that if ϕ is the function in (a), then $D_2 D_1\phi$ is *not* continuous at $(0, 0)$.

4.   *Partial Derivatives of Higher Order.* Let $\Omega \subset \mathbb{R}^n$ be an open set, let $\phi : \Omega \to \mathbb{R}$ be a function, let $r > 1$ be an integer, and let $j = (j_1, \ldots, j_r)$ be an ordered r-tuple of integers with $1 \leq j_k \leq n$ for $1 \leq k \leq r$. Write j' for the ordered $(r - 1)$-tuple $(j_1, \ldots, j_{r-1})$. Supposing that $D_{j'}\phi$ is defined (and real-valued) on Ω, we define $D_j\phi$ on Ω by

$$D_j\phi(t) = D_{j_r} D_{j'}\phi(t)$$

at those $t \in \Omega$ where $D_{j'}\phi$ has a partial derivative with respect to its j_rth variable. This inductively defines the n^r partial derivatives of ϕ or *order r*. Thus

$$D_j\phi = D_{j_r} D_{j_{r-1}} \cdots D_{j_2} D_{j_1}\phi.$$

We also write

$$D_j\phi = \frac{\partial^r \phi}{\partial t_{j_r} \partial t_{j_{r-1}} \cdots \partial t_{j_2} \partial t_{j_1}}.$$

For example, if $n = 4$ and $r = 5$, $D_{(2,1,4,4,1)}\phi = D_1 D_4 D_4 D_1 D_2\phi$ is one of the 1024 partial derivatives of ϕ of order 5. We say that ϕ is *of class $C^{(r)}$ on Ω* if all n^r of the functions $D_j\phi$ are defined, finite, and continuous on Ω.

(a) If ϕ is of class $C^{(r)}$ on Ω, then ϕ is of class $C^{(r-1)}$ on Ω.

(b) If ϕ is of class $C^{(r)}$ on Ω, j is as above, and i is another r-tuple obtained from j by permuting its coordinates, then $D_i\phi = D_j\phi$ on Ω. [See Exercise 3(b).] For example, $D_{(2, 1, 4, 4, 1)}\phi = D_{(1, 1, 2, 4, 4)}\phi$.

    A mapping $\phi : \Omega \to \mathbb{R}^p$ is said to be *of class $C^{(r)}$* on Ω if its p coordinate functions $\phi_1, \ldots, \phi_p$ are of class $C^{(r)}$ on Ω.

(c) Let ϕ, ψ, and $\mathcal{X}$ be as in (6.134). If ϕ and ψ are of class $C^{(r)}$ on Ω and Δ respectively, then $\mathcal{X}$ is of class $C^{(r)}$ on Ω.

(d) Let everything be as in (6.140). If ϕ is of class $C^{(r)}$ on Ω, then ψ is of class $C^{(r)}$ on W.

5. For a given $n \in \mathbb{N}$, a *multiindex* (of *rank n*) is an ordered n-tuple $\alpha = (\alpha_1, \alpha_2, \ldots, \alpha_n)$ of nonnegative integers α_j. The *order* of α is the sum $|\alpha| = \alpha_1 + \alpha_2 + \ldots + \alpha_n$ of the coordinates of α. For functions ϕ of class $C^{(|\alpha|)}$ defined on open subsets of $\mathbb{R}^n$ we define the *differential operator* D^α by

$$D^\alpha \phi = D_1^{\alpha_1} D_2^{\alpha_2} \ldots D_n^{\alpha_n} \phi$$

where $D_j^{\alpha_j}$ differentiates α_j times with respect to the jth variable. In view of Exercise 4(b), the order in which these differentiations are carried out is not important for such ϕ. The multiindex superscript tells us how many times to differentiate with respect to each variable. Thus, in the notation of Exercise 4,

$$D_{(2,1,4,4,1)} \phi = D^{(2,1,0,2)} \phi$$

for functions ϕ of class $C^{(5)}$ on open subsets Ω of $\mathbb{R}^4$. We define the *monomial* x^α, for α as above and $x \in \mathbb{R}^n$, by

$$x^\alpha = \prod_{j=1}^{n} x_j^{\alpha_j} = x_1^{\alpha_1} x_2^{\alpha_2} \ldots x_n^{\alpha_n}.$$

If α and β are two multiindices, we write $\beta \leq \alpha$ if $\beta_j \leq \alpha_j$ for all j. We define

$$\alpha! = \prod_{j=1}^{n} \alpha_j! = (\alpha_1!)(\alpha_2!) \ldots (\alpha_n!)$$

and

$$\binom{\alpha}{\beta} = \frac{\alpha!}{\beta!(\alpha - \beta)!}$$

when $\beta \leq \alpha$ and $(\alpha - \beta)_j = \alpha_j - \beta_j$ for all j.

(a) *Leibnitz's Formula.* If ϕ and ψ are two real-valued functions of class $C^{(r)}$ on an open set $\Omega \subset \mathbb{R}^n$ and if α is a multiindex with $|\alpha| \leq r$, then

$$D^\alpha(\phi\psi) = \sum_{\beta=0}^{\alpha} \binom{\alpha}{\beta}(D^{\alpha-\beta}\phi)(D^\beta\psi)$$

everywhere on Ω where the summation is extended over all multiindices β for which $\beta \leq \alpha$. [Hint: Use induction on $|\alpha|$.]

(b) *Taylor's Theorem for $\mathbb{R}^n$.* Let n, r, and u be integers with $n > 0$, $r \geq 0$, and $u < r + 1$. Let $\Omega \subset \mathbb{R}^n$ be an open set and let ψ be a real-valued function of class $C^{(r+1)}$ on Ω. Then, for each $a, h \in \mathbb{R}^n$ $(h \neq 0)$ for which the line segment $\{a + th : 0 \leq t \leq 1\}$ is entirely contained in Ω, there exists $0 < \theta < 1$ such that

$$\psi(a + h) = \sum_{k=0}^{r} \sum_{|\alpha|=k} \frac{1}{\alpha!} D^\alpha \psi(a) h^\alpha + R_r(h),$$

where

$$R_r(h) = \frac{(1 - \theta)^u (r + 1)}{r + 1 - u} \cdot \sum_{|\alpha|=r+1} \frac{1}{\alpha!} D^\alpha \psi(a + \theta h) h^\alpha$$

(here $\sum_{|\alpha|=k}$ denotes the summation over *all* multiindices α for which $|\alpha| = k$).

[Hints: Apply (4.37) (with r in place of n, $a = 0$, $h = 1$) to the function f defined on [0, 1] by $f(t) = \psi(a + th)$. Apply (6.134) k times to find $f^{(k)}(t)$.] Notice that when $r = u = 0$, this is just (6.138).

6. *Extrema on $\mathbb{R}^n$.* Let $\Omega \subset \mathbb{R}^n$ be an open set, let $a \in \Omega$, and let $\phi : \Omega \to \mathbb{R}$ be a function. We say that ϕ has a *local minimum* [resp., *absolute local minimum*] at a if there exists a neighborhood V of a such that $\phi(x) \geqq \phi(a)$ [resp., $\phi(x) > \phi(a)$] for all $x \in V \cap \Omega$ with $x \neq a$. We define *local maximum* and *absolute local maximum* the same way except that the inequalities are reversed. Many authors say "relative" in place of "local." There is no need that Ω be open in these definitions.

(a) If ϕ has a local minimum (or local maximum) at a and if $D_j\phi(a)$ exists for some j, then $D_j\phi(a) = 0$.

Now suppose that ϕ is of class $C^{(2)}$ on Ω. For $x \in \Omega$ and $h \in \mathbb{R}^n$, define

$$Q(x, h) = \sum_{i=1}^{n} \sum_{j=1}^{n} D_i D_j\phi(x)h_i h_j$$

and

$$\beta(x) = \inf\{\, Q(x, u) : u \in \mathbb{R}^n, |u| = 1 \,\}.$$

(b) We have $Q(x, h) \geqq \beta(x)|h|^2$ for all $x \in \Omega$ and $h \in \mathbb{R}^n$.

(c) The function β is continuous at a and so on Ω. [Hints: Given $\epsilon > 0$ choose open V with $a \in V \subset \Omega$ such that $|D_i D_j\phi(x) - D_i D_j\phi(a)| < \epsilon/(2n)$ for all $x \in V$ and all i and j. Then $x \in V$ and $|u| = 1$ imply $|Q(x, u) - Q(a, u)| < \epsilon/2$, so $Q(x, u) > \beta(a) - \epsilon/2$, which implies $\beta(x) > \beta(a) - \epsilon$. Also $Q(a, v) < \beta(a) + \epsilon/2$ for some v with $|v| = 1$, so, for $x \in V$, $\beta(x) \leqq Q(x, v) < \beta(a) + \epsilon$.]

(d) If $h \in \mathbb{R}^n$ is such that $\{a + th : 0 \leqq t \leqq 1\} \subset \Omega$, then for some $0 < \theta < 1$ we have

$$\phi(a + h) = \phi(a) + \sum_{j=1}^{n} D_j\phi(a)h_j + \frac{1}{2} Q(a + \theta h, h).$$

[Hint: Use Taylor's Theorem, Exercise 5(b).]

(e) $\beta(a) > 0$ if and only if $Q(a, h) > 0$ for all $h \in \mathbb{R}^n$ with $|h| > 0$.

(f) $\beta(a) \geqq 0$ if and only if $Q(a, h) \geqq 0$ for all $h \in \mathbb{R}^n$.

(g) If ϕ has a local minimum at a, then $\beta(a) \geqq 0$. [Hint: Assume false and use (f), (a), (d), and continuity to obtain a contradiction.]

(h) If $D_j\phi(a) = 0$ for all j and $\beta(a) > 0$, then ϕ has an absolute local minimum at a. [Hint: Use (d), (b), and (c) to find $\delta > 0$ so that $\phi(a + h) - \phi(a) \geqq (1/4)\beta(a)|h|^2$ whenever $|h| < \delta$.]

(i) In case $n = 2$, the condition $\beta(a) \geqq 0$ is equivalent to

$$\phi_{11}(a) \geqq 0 \quad \text{and} \quad \phi_{11}(a)\phi_{22}(a) \geqq [\phi_{12}(a)]^2,$$

where $\phi_{ij} = D_i D_j\phi$.

(j) Assertion (i) obtains if $\geqq$ is replaced by $>$ throughout.

(k) In case $n = 1$, we have $\beta(a) = \phi''(a)$.

(l) The function $\phi : \mathbb{R}^2 \to \mathbb{R}$ given by $\phi(x, y) = (x - y^2)(x - 3y^2)$ has the property that its restriction to any line $\{(\alpha t, \beta t) : t \in \mathbb{R}\}$ ($\alpha, \beta \in \mathbb{R}$ not both 0) through $(0, 0)$ has an absolute local minimum at $(0, 0)$ but yet ϕ fails to have a local minimum at $(0, 0)$ or anywhere else.

7. If $\Omega \subset \mathbb{R}^n$ is an open set, $a \in \Omega$, and $\phi : \Omega \to \mathbb{R}$ is continuous on Ω, of class $C^{(1)}$ on

$\Omega \backslash \{a\}$, and satisfies

$$\lim_{t \to a} D_j \phi(t) = d_j \in \mathbb{R}$$

for $1 \leqq j \leqq n$, then ϕ is of class $C^{(1)}$ on Ω and $D_j \phi(a) = d_j$ for each j.

8.   (a) If $\Omega \subset \mathbb{R}^n$ is a nonvoid open set and $\phi : \Omega \to \mathbb{R}$ is such that $D_j \phi$ is a bounded function on Ω for $1 \leqq j \leqq n$, then ϕ is continuous on Ω. [Hint: For $a \in \Omega$ and $h \in \mathbb{R}^n$ with $|h|$ sufficiently small, write

$$\phi(a+h) - \phi(a) = \sum_{j=1}^{n} \left[\phi(a+h^j) - \phi(a+h^{j-1}) \right],$$

where $h^0 = (0, 0, \ldots, 0)$ and $h^j = (h_1, \ldots, h_j, 0, \ldots, 0)$ for $1 \leqq j \leqq n$.]
    (b) Theorem (6.138) need not apply in (a), as Example (6.131.a) shows.

9.   *Derivatives of Integrals.* Let $E \subset \mathbb{R}^n$ be a Lebesgue measurable set and let $I \subset \mathbb{R}$ be an open interval. Suppose that $f : I \times E \to \mathbb{C}$ is a function such that (1) for each $t \in I$ the function $y \to f(t, y)$ is in $L_1(E)$, (2) for each $y \in E$ the function $t \to f(t, y)$ has a derivative $D_1 f(n, y)$ that is continuous on I, and (3) there exists a function $g \in L_1(E)$ satisfying $|D_1 f(t, y)| \leqq |g(y)|$ for all $t \in I$ and $y \in E$. Then we have the following.
    (a) The function F defined on I by

$$F(t) = \int_E f(t, y)\, dy$$

has a continuous derivative on I that is given by

$$F'(t) = \int_E D_1 f(t, y)\, dy.$$

[Hint: Supposing that f is real-valued, use the Mean Value Theorem and Lebesgue's Dominated Convergence Theorem.]
    (b) If $n = 1$, E is an interval of $\mathbb{R}$, f is continuous on $I \times E$, and α and β are differentiable functions from I into E, then the function g defined on I by

$$g(t) = \int_{\alpha(t)}^{\beta(t)} f(t, y)\, dy$$

is differentiable on I and we have *Leibnitz's Rule:*

$$g'(t) = \int_{\alpha(t)}^{\beta(t)} D_1 f(t, y)\, dy + f(t, \beta(t)) \beta'(t) - f(t, \alpha(t)) \alpha'(t)$$

for all $t \in I$. If α' and β' are continuous on I, then so is g'. [Hint: Apply the Chain Rule (6.134) to the function F defined on I^3 by

$$F(t, u, v) = \int_u^v f(t, y)\, dy.]$$

    (c) Consider the example where $n = 1$, $E =]0, \infty[$, $I = \mathbb{R}$, and $f(t, y) = t^3 e^{-t^2 y}$. Here hypotheses (1) and (2) obtain while (3) and assertion (a) both fail. In fact, $D_1 f(t, y) = (3t^2 - 2t^4 y) e^{-t^2 y}$, $D_1 f(y^{-1/2}, y) = (ey)^{-1}$, $F(t) = t$, and $F'(0) = 1 \neq 0 = \int_0^\infty D_1 f(0, y)\, dy$.

10.   *The Laplacian in Polar Coordinates.* Let $\Delta \subset \mathbb{R}^2$ be an open set not containing $(0, 0)$ and let ψ be a function of class $C^{(2)}$ on Δ (see Exercise 4). Write $\Omega = \{(r, \theta) \in \mathbb{R}^2 : (r \cos \theta, r \sin \theta) \in \Delta\}$ and define $\mathcal{X}$ on Ω by $\mathcal{X}(r, \theta) = \psi(r \cos \theta, r \sin \theta)$. Then $\mathcal{X}$ is of class $C^{(2)}$ on Ω and $\psi_{11}(r \cos \theta, r \sin \theta) + \psi_{22}(r \cos \theta, r \sin \theta) = \mathcal{X}_{11}(r, \theta) +$

$(1/r^2)\mathfrak{X}_{22}(r,\theta) + (1/r)\mathfrak{X}_1(r,\theta)$ for all $(r,\theta) \in \Omega$. Here $f_{ij} = D_j D_i f$ and $f_i = D_i f$. The function $\psi_{11} + \psi_{22}$ is called the *Laplacian* of ψ. [Hint: Use (6.134) and Exercise 3(b).]

11. The function ϕ defined on $\mathbb{R}$ by $\phi(0) = 0$ and $\phi(t) = t + 2t^2 \cos(1/t)$ for $t \neq 0$ is differentiable on $\mathbb{R}$ and $\phi'(0) = 1 \neq 0$, but there is *no* neighborhood V of 0 on which ϕ is one-to-one. Why does this not contradict (6.140)?

12. Let $K \subset \mathbb{R}^n$ be a compact set having nonvoid interior Ω. Suppose that $\phi : K \to \mathbb{R}$ is continuous on K, differentiable on Ω, and $\phi(t) = 0$ for all $t \in K \backslash \Omega$. Then there exists $a \in \Omega$ such that $d\phi(a) = 0$. [Hint: Compare the proof of Rolle's Theorem.]

13. In this exercise we identify $\mathbb{R}^2$ with the complex plane $\mathbb{C}$ in the usual way: $(x, y) = x + iy = z$. Let $\Omega \subset \mathbb{C}$ be an open set and let $f : \Omega \to \mathbb{C}$ be a function with $u = \mathrm{Re}(f)$ and $v = \mathrm{Im}(f)$. Regard f also as a mapping of Ω into $\mathbb{R}^2$ by writing $f(x, y) = f(z) = u(z) + iv(z) = (u(x, y), v(x, y))$ for $x + iy = z \in \Omega$. We write $u_x = D_1 u$, $u_y = D_2 u$, $v_x = D_1 v$, $v_y = D_2 v$, $f_x = u_x + iv_x$, and $f_y = u_y + iv_y$ at all points of Ω where these partial derivatives exist (and are finite).

(a) The mapping f is differentiable at a point $z_0 = (x_0, y_0) \in \Omega$ if and only if there exist complex numbers α and β such that

$$\lim_{z \to z_0} \frac{f(z) - f(z_0) - \alpha(x - x_0) - \beta(y - y_0)}{|z - z_0|} = 0.$$

In this case, $\alpha = f_x(z_0)$, $\beta = f_y(z_0)$, and the matrix of the linear mapping $df(z_0)$ is

$$\begin{pmatrix} u_x(z_0) & u_y(z_0) \\ v_x(z_0) & v_y(z_0) \end{pmatrix}.$$

(b) If the partial derivatives u_x, u_y, v_x, and v_y exist (and are finite) everywhere on Ω and if all four of them are continuous at some $z_0 \in \Omega$, then f is differentiable at z_0. [Hint: Compare the proof of (6.130.c).]

(c) If $f(z) = |z|^2 \sin(1/|z|)$ for $z \neq 0$ and $f(0) = 0$, then f is differentiable at every point of $\mathbb{C}$, but the partial derivatives f_x and f_y are *not* continuous at $z_0 = 0$.

(d) The function f has a complex derivative

$$f'(z_0) = \lim_{\Delta z \to 0} \frac{f(z_0 + \Delta z) - f(z_0)}{\Delta z}$$

at a point $z_0 \in \Omega$ if and only if the mapping f is differentiable at z_0 ($df(z_0)$ exists) *and* the *Cauchy–Riemann equations* $u_x(z_0) = v_y(z_0)$ and $v_x(z_0) = -u_y(z_0)$ are satisfied. In this case, the value of the linear mapping $df(z_0)$ at a complex number Δz is $f'(z_0)\Delta z$ and we have $f'(z_0) = f_x(z_0) = -if_y(z_0)$ and $J_f(z_0) = |f'(z_0)|^2$.

(e) If $f(z) = |xy|^{1/2}$, then the Cauchy–Riemann equations are satisfied at $z_0 = 0$, but neither $df(0)$ nor $f'(0)$ exists.

14. *Implicit Function Theorem.* For $x = (x_1, \ldots, x_n) \in \mathbb{R}^n$ and $t = (t_1, \ldots, t_p) \in \mathbb{R}^p$, we write $(x, t) = (x_1, \ldots, x_n, t_1, \ldots, t_p) \in \mathbb{R}^{n+p}$. Let f be a mapping of class $C^{(1)}$ from an open set $\Omega \subset \mathbb{R}^{n+p}$ into $\mathbb{R}^n$. Suppose that $(a, b) \in \Omega$ is such that $f(a, b) = 0$, and the n by n matrix $(D_j f_i(a, b))$ $(1 \leq i \leq n, 1 \leq j \leq n)$ has a nonzero determinant. [This is the matrix of the differential $d\tilde{f}(a)$, where $\tilde{f}(x) = f(x, b)$ so our hypothesis is that $\tilde{f}(a) = 0$ and $J_{\tilde{f}}(a) \neq 0$.] Then there exists an open set $T \subset \mathbb{R}^p$ with $b \in T$ for which there is exactly one mapping $g : T \to \mathbb{R}^n$ satisfying (1) g is

continuous on T,   (2) $g(b) = a$, and (3) $(g(t), t) \in \Omega$ and $f(g(t), t) = 0$ for every $t \in T$. Moreover, g is of class $C^{(1)}$ on T. [Hints: Apply (6.140) to the mapping $\phi : \Omega \to \mathbb{R}^{n+p}$ defined by $\phi(x, t) = (f(x, t), t)$ to obtain open neighborhoods V and W of (a, b) and $(0, b)$, respectively, and a one-to-one $C^{(1)}$ mapping ψ of W onto V which is the inverse of the restriction of ϕ to V. Let T be any *connected* open neighborhood of $b \in \mathbb{R}^p$ such that $(0, t) \in W$ for all $t \in T$ (for example, T can be an open interval of $\mathbb{R}^p$). Define g on T by $\psi(0, t) = (g(t), t)$. If h is any mapping of T into $\mathbb{R}^n$ satisfying (1)–(3) (with h in place of g), then the set $\{t \in T : (h(t), t) \in V\}$ $= \{t \in T : h(t) = g(t)\}$ contains b and is both open and closed relative to T, so it is all of T.] This theorem furnishes a local solution to the problem of solving the system of equations

$$f_j(x_1, \ldots, x_n, t_1, \ldots, t_p) = 0 \qquad (1 \leq j \leq n)$$

for $x_1, \ldots, x_n$ in terms of $t_1, \ldots, t_p : x_j = g_j(t_1, \ldots, t_p)$.

15. *The Lagrange Multiplier Method.*   Let m and n be natural numbers with $n < m$ and let Ω be an open subset of $\mathbb{R}^m$. Suppose that $\phi : \Omega \to \mathbb{R}$ and $f : \Omega \to \mathbb{R}^n$ are of class $C^{(1)}$ on Ω and that $c \in M = \{u \in \Omega : f(u) = 0 \in \mathbb{R}^n\}$ is such that the restriction $\phi|_M$ of ϕ to M has a local extremum at c (this means that there is some neighborhood V of c such that either $\phi(u) \leq \phi(c)$ for all $u \in V \cap M$ or $\phi(u) \geq \phi(c)$ for all $u \in V \cap M$). Suppose also that the n by n matrix $(D_k f_i(c))$ $(1 \leq i \leq n, 1 \leq k \leq n)$ has a nonzero determinant. Then there exists a (unique) vector $\lambda = (\lambda_1, \ldots, \lambda_n) \in \mathbb{R}^n$ such that

$$D_j \phi(c) + \sum_{i=1}^{n} \lambda_i D_j f_i(c) = 0 \qquad (1 \leq j \leq m). \qquad (*)$$

[Hints: Apply Cramer's Rule to obtain the unique solution λ of the system

$$\sum_{i=1}^{n} \lambda_i D_k f_i(c) = - D_k \phi(c) \qquad (1 \leq k \leq n). \qquad (1)$$

The task is to show that $(*)$ also obtains for $n < j \leq m$. Write $p = m - n$ and denote elements of $\mathbb{R}^m$ as $(x, t) \in \mathbb{R}^n \times \mathbb{R}^p$ just as in Exercise 14. In particular, write $c = (a, b)$. Use Exercise 14 to obtain T and g. Define $F(t) = f(g(t), t)$ and $\Phi(t) = \phi(g(t), t)$ for $t \in T$. Since $F = 0$ on T, $dF(b) = 0$, so the Chain Rule yields

$$0 = D_j F_i(b) = \sum_{k=1}^{n} D_k f_i(c) D_j g_k(b) + D_{n+j} f_i(c)$$

for $1 \leq i \leq n$ and $1 \leq j \leq p$. Multiply by λ_i, sum over all i, and use (1) to obtain

$$0 = - \sum_{k=1}^{n} D_k \phi(c) D_j g_k(b) + \sum_{i=1}^{n} \lambda_i D_{n+j} f_i(c) \qquad (2)$$

for $1 \leq j \leq p$. Now $(g(t), t) \in M$ for all $t \in T$, so Φ has a local extremum at b, $d\Phi(b) = 0$, and

$$0 = D_j \Phi(b) = \sum_{k=1}^{n} D_k \phi(c) D_j g_k(b) + D_{n+j} \phi(c) \qquad (3)$$

for $1 \leq j \leq p$. Now combine (2) and (3).] The virtue of this result is that if one seeks points c at which $\phi|_M$ has local extrema, then he need only search among those $c \in M$ for which the system $(*)$ has a solution λ. That is, he can confine his search by considering solutions $c_1, \ldots, c_m, \lambda_1, \ldots, \lambda_n$ of the system of $m + n$ equations given by $(*)$ and $f_i(c) = 0$ $(1 \leq i \leq n)$. The variables $\lambda_1, \ldots, \lambda_n$ are known as *Lagrange multipliers*.

16.   As applications of Exercise 15, solve the following problems.

(a) Let $a_1, a_2, \ldots, a_m$, and b be given real numbers with $a_j \neq 0$ for some j. Find the point of the hyperplane $M = \left\{ x \in \mathbb{R}^m : \sum_{j=1}^{m} a_j x_j = b \right\}$ that is nearest the origin of $\mathbb{R}^m$. [Hint: Consider $\phi(x) = |x|^2$.]

(b) Prove that if $x_1, x_2, \ldots, x_m$ are real numbers for which $\sum_{j=1}^{m} x_j^2 = 1$, then

$$\prod_{j=1}^{m} x_j^2 \leq m^{-m},$$ and equality obtains if and only if $x_j^2 = 1/m$ for all j.

(c) Use (b) to prove that if $a_1, \ldots, a_m$ are nonnegative real numbers, then

$$\left(\prod_{j=1}^{m} a_j \right)^{1/m} \leq \frac{1}{m} \sum_{j=1}^{m} a_j,$$

and equality obtains if and only if $a_j = a_1$ for all j. [Hint: Put $x_j^2 = a_j/(mA)$, where A is the right side.]

(d) Find the maximum and the minimum values of $Ax + By + Cz$ on the ellipsoid

$$M = \{ (x, y, z) \in \mathbb{R}^3 : x^2/a^2 + y^2/b^2 + z^2/c^2 = 1 \}.$$

(e) Prove that if $a \in \mathbb{R}^m$, then

$$\max \left\{ \sum_{j=1}^{m} a_j x_j : x \in \mathbb{R}^m, |x| = 1 \right\} = |a|.$$

(f) Use (e) to prove Cauchy's Inequality.

(g) Let $a, c \in \mathbb{R}^m$ with $a \neq 0$ and $b \in \mathbb{R}$ be given. Prove that

$$\min \left\{ |x - c| : x \in \mathbb{R}^m, \sum_{j=1}^{m} a_j x_j = b \right\} = \frac{1}{|a|} \left| \sum_{j=1}^{m} a_j c_j - b \right|.$$

(h) Prove *Hadamard's Inequality*, which states that if $x = (x_{ij})$ is an n by n real matrix and if $r_i(x) = \left(\sum_{j=1}^{n} x_{ij}^2 \right)^{1/2}$ is the length of its ith row, then the determinant $\det(x)$ of x satisfies

$$|\det(x)| \leq \prod_{i=1}^{n} r_i(x).$$

[Hints: Given positive real numbers $R_1, \ldots, R_n$, let $M = \{ x \in \mathbb{R}^{n^2} : r_i(x) = R_i$ for $1 \leq i \leq n \}$. Let $\phi = \det$, $f_i(x) = r_i(x)^2 - R_i^2$, and $f(x) = (f_1(x), \ldots, f_n(x))$. Choose $a \in M$ such that $\phi(a) \geq \phi(x)$ for all $x \in M$. Check that $(\partial \phi / \partial x_{ij})(a) = A_{ij}$, where A_{ij} is the cofactor of a_{ij}, and recall that $\phi(a) = \sum_{j=1}^{n} a_{ij} A_{ij}$ for all i. Use Exercise 15 to obtain $\lambda_1, \ldots, \lambda_n$ such that $A_{ij} + 2\lambda_i a_{ij} = 0$ for all i and j. Deduce that $\phi(a) + 2\lambda_i R_i^2 = 0$ for all i, so that $a_{ij} \phi(a) = A_{ij} R_i^2$. Let $b = (b_{ij})$, where $b_{ij} = A_{ji}$, and check that $ab = \phi(a)I$, where I is the identity matrix. Note that $\phi(a)^n = \phi(ab) = \phi(a)\phi(b) = \phi(a) \left(\prod_{i=1}^{n} \frac{\phi(a)}{R_i^2} \right) \phi(a)$, so that $\phi(a) = \prod_{i=1}^{n} R_i$.]

17.   *Diagonalization of Symmetric Matrices.*   In this exercise the *transpose* of an $r \times c$ matrix $D = (d_{ij})$ having r rows and c columns with the entry d_{ij} in the ith row and jth column is the $c \times r$ matrix $D' = (d'_{ji})$ having c rows and r columns with the entry $d'_{ji} = d_{ij}$ in the jth row and ith column. Thus the ith row of D is the ith column of D'

and the jth column of D is the jth row of D^t. We shall always regard a vector $x \in \mathbb{R}^m$ as an $m \times 1$ matrix (column vector), so that x^t is a $1 \times m$ matrix (row vector).

Let $A = (a_{ij})$ be a given $m \times m$ real matrix that is symmetric ($A = A^t$, $a_{ji} = a_{ij} \in \mathbb{R}$). Define the linear mapping $L : \mathbb{R}^m \to \mathbb{R}^m$ and the quadratic form $Q : \mathbb{R}^m \to \mathbb{R}$ by

$$L(x) = Ax \quad \text{and} \quad Q(x) = x^t A x,$$

where the right sides of these equalities denote matrix products. Thus the ith coordinate of $L(x)$ is the number

$$L_i(x) = \sum_{j=1}^{m} a_{ij} x_j$$

and the number $Q(x)$ (1×1 matrix) is

$$Q(x) = \sum_{i=1}^{m} \sum_{j=1}^{m} a_{ij} x_i x_j.$$

(a) For $1 \leq k \leq m$, we have $D_k \, Q(x) = 2L_k(x)$.

For $n = 1, 2, \ldots, m$, inductively select sets M_n, real numbers β_n, and vectors v_n (having ith coordinate v_{in}) according to the prescription

$$M_n = \left\{ x \in \mathbb{R}^m : |x| = 1, \sum_{i=1}^{m} v_{ij} x_i = 0 \quad \text{if} \quad 1 \leq j < n \right\},$$

$$\beta_n = \sup\{ Q(x) : x \in M_n \}, \quad v_n \in M_n \quad \text{and} \quad Q(v_n) = \beta_n.$$

(b) The sets M_n are compact and nonvoid, so such selections are possible. [Hint: Use induction starting with $M_1 = \{x \in \mathbb{R}^m : |x| = 1\}$.]

Let $V = (v_{ij})$ be the $m \times m$ matrix having the $m \times 1$ matrix v_j as its jth column.

(c) We have $\beta_1 \geq \beta_2 \geq \ldots \geq \beta_m$ and $V^t V = I = VV^t$, where I is the $m \times m$ identity matrix. [Use the well-known theorem of linear algebra that $CB = I$ if $BC = I$.]

(d) For each n, the vector $L(v_n)$ is a linear combination of the vectors $v_1, \ldots, v_n$:

$$L(v_n) = \sum_{r=1}^{n} \lambda_r v_r \quad \text{for some real numbers } \lambda_1, \ldots, \lambda_n. \text{ [Hint: Suppose } n < m \text{ is}$$

given. Use Exercise 15 to obtain $\lambda_1, \ldots, \lambda_n$ such that the function g defined on $\mathbb{R}^m$ by

$$g(x) = Q(x) + \lambda_n(1 - |x|^2) - 2 \sum_{j=1}^{n-1} \sum_{i=1}^{m} \lambda_j v_{ij} x_i$$

has all partial derivatives equal to 0 at v_n (if $n = 1$, the double sum means 0).]

(e) In fact, $L(v_n) = \beta_n v_n$ for all n. [Hint: Given n, write $L(v_n)$ as in (d) and check that

$$\lambda_k = v_k^t A v_n = v_n^t A v_k$$

for $1 \leq k \leq n$. Conclude that $\lambda_n = \beta_n$ and that if $L(v_k) = \beta_k v_k$ for $k < n$, then $\lambda_k = 0$ for such k. Use induction.]

(f) We have $V^t A V = B$, where $B = (b_{ij})$ is the diagonal matrix with $b_{ij} = 0$ if $i \neq j$ and $b_{jj} = \beta_j$.

(g) The numbers $\beta_1, \ldots, \beta_m$ are solutions of the mth degree polynomial equation

$\det(A - \beta I) = 0$, and there are no other solutions. [Hint: Write $B - \beta I$ $= V^t(A - \beta I)V$ and take determinants.]

(h) If $x \in \mathbb{R}^m$ and $\xi = V^t x$, then the coordinates of ξ are given by $\xi_i = x^t v_i = v_i^t x$, and we have

$$x = \sum_{i=1}^m \xi_i v_i, \quad L(x) = \sum_{i=1}^m \beta_i \xi_i v_i, \quad Q(x) = \sum_{i=1}^m \beta_i \xi_i^2.$$

The numbers $\xi_1, \ldots, \xi_m$ are called the *coordinates of x with respect to the basis* $\{v_1, v_2, \ldots, v_m\}$.

(i) We have $\beta_m = \inf\{Q(x) : x \in M_1\}$.

(j) We have $\det A = \prod_{i=1}^m \beta_i$.

(k) We call A and Q *positive definite* if $Q(x) > 0$ for $x \neq 0$. This is so if and only if $\beta_i > 0$ for all i.

Transformations of Integrals on $\mathbb{R}^n$

In this section we investigate the "change of variable formula" on $\mathbb{R}^n$:

$$\int_\Delta f(x)\,dx = \int_\Omega f(\phi(t)|J_\phi(t)|)\,dt,$$

which is valid when Ω and Δ are suitable subsets of $\mathbb{R}^n$, $\phi : \Omega \to \Delta$ is a suitable mapping, and f is a suitable function on Δ. In the case $n = 1$, Theorem (6.95) and Corollary (6.97) provide very general "suitable" conditions that such a formula obtain. For general n, we shall content ourselves with much more restrictive conditions on ϕ in order that we may apply the Chain Rule and the Inverse Function Theorem as presented in the preceding section. Throughout this section, we adhere to the following notation and hypotheses.

(6.142) Notation   Let n be a natural number, let Ω be a nonvoid open subset of $\mathbb{R}^n$, and let $\phi : \Omega \to \mathbb{R}^n$ be a mapping that is one-to-one, of class $C^{(1)}$ on Ω, and satisfies $J_\phi(t) \neq 0$ for all $t \in \Omega$. It follows from (6.140) that the set $\Delta = \phi(\Omega)$ is open in $\mathbb{R}^n$, that the mapping $\psi = \phi^{-1}$ is of class $C^{(1)}$ on Δ, and that

$$J_\phi(\psi(x))J_\psi(x) = 1 \tag{1}$$

for all $x \in \Delta$.

We begin our investigation with some lemmas.

(6.143) Lemma   Let $A : \mathbb{R}^n \to \mathbb{R}^n$ be any linear mapping and let E be any subset of $\mathbb{R}^n$. Then

(i) $$\lambda^n(A(E)) = |\det A|\lambda^n(E),$$

where, as usual, λ^n is Lebesgue measure on $\mathbb{R}^n$.

Proof   First, suppose that A is invertible [$\det A \neq 0$] and that the lemma obtains when E is the unit cube $Q = \{x \in \mathbb{R}^n : 0 \le x_j < 1 \text{ for } 1 \le j \le n\}$.

That is,

$$\lambda^n(A(Q)) = |\det A|. \tag{1}$$

If, for some natural number m, $I = \{x \in \mathbb{R}^n : 0 \leq x_j < 2^{-m}$ for $1 \leq j \leq n\}$, then Q is the disjoint union of 2^{mn} translates of I and each such translate $v + I$ satisfies $\lambda^n(A(v + I)) = \lambda^n(A(v) + A(I)) = \lambda^n(A(I))$. The set $A(Q)$ is the disjoint union of the 2^{mn} sets $A(v + I)$ [A is one-to-one], and each of these sets is measurable [being the intersection of an open set with a closed set], so, by (1), $|\det A| = 2^{mn}\lambda^n(A(I))$. Since $\lambda^n(I) = 2^{-mn}$, this proves (i) if E is I or any translate of I. Thus (i) obtains if E is any countable disjoint union of dyadic n-cubes (6.115). But then (6.116) ensures that (i) is valid for any open set $E \subset \mathbb{R}^n$. Since the open supersets U of $A(E)$ are just the sets $(A(V)$, where V is an open superset of E, the outer regularity of λ^n (6.64) allows us to deduce (i) for all E from its validity for $E = V$. In summary, we now know that, for invertible A, the lemma follows if we can prove (1).

We now prove the lemma for three special types of A's.

Type 1. Suppose that A simply interchanges pth and qth coordinates. That is, for some fixed natural numbers p and q not exceeding n, $A_p(x) = x_q$, $A_q(x) = x_p$, and $A_j(x) = x_j$ for all other j. Then A is invertible, $A^{-1} = A$, the matrix of A is obtained from the identity matrix by interchanging its pth and qth rows, $\det A = -1$, and $A(Q) = Q$. Thus (1) and hence the lemma are true for this A.

Type 2. Suppose that for some $\beta \in \mathbb{R}$ and some p $(1 \leq p \leq n)$ we have $A_p(x) = \beta x_p$ and $A_j(x) = x_j$ for all other j. Then the matrix of A is obtained from the identity matrix by multiplying its pth row by β and det $A = \beta$. If $\beta = 0$, then $A(E) \subset A(\mathbb{R}^n) = \{x \in \mathbb{R}^n : x_p = 0\}$, so $\lambda^n(A(E)) = 0$ and (i) obtains for all E. If $\beta \neq 0$, then A is invertible (with A^{-1} of the same type as A), $A(Q) = \{x \in \mathbb{R}^n : 0 \leq x_j < 1$ if $j \neq p$, $0 \leq x_p < \beta$ if $\beta > 0$, $\beta < x_p \leq 0$ if $\beta < 0\}$, and $\lambda^n(A(Q)) = |\beta|$, so (1) and the lemma obtain.

Type 3. Suppose that for some $p \neq q$ $(1 \leq p \leq n, 1 \leq q \leq n)$ we have $A_p(x) = x_p + x_q$ and $A_j(x) = x_j$ for all $j \neq p$. Then A is invertible (with $A^{-1} = B \circ A \circ B$, where B is the mapping of Type 2 which multiplies the qth coordinate by -1), the matrix of A is obtained from the identity matrix by replacing its pth row by the sum of its pth and qth rows, det $A = 1$, and $A(Q) = \{x \in \mathbb{R}^n : 0 \leq x_j < 1$ if $j \neq p$, $0 \leq x_p - x_q < 1\}$. Writing $S = \{x \in A(Q) : x_p < 1\}$, $T = A(Q) \setminus S$, and e^p for the vector with $e_p^p = 1$ and $e_j^p = 0$ for other j, we have

$$S = \{x \in Q : x_q \leq x_p\} \quad \text{and} \quad T - e^p = \{x \in Q : x_p < x_q\},$$

so

$$\lambda^n(A(Q)) = \lambda^n(S) + \lambda^n(T) = \lambda^n(S) + \lambda^n(T - e^p)$$
$$= \lambda^n(Q) = 1 = \det A.$$

Thus (1) and the lemma are true for this A.

Next notice that if the lemma is true for two linear mappings A_1 and A_2, then it is true for the composite mapping $A_1 \circ A_2$ because

$$\lambda^n(A_1 \circ A_2(E)) = |\det A_1|\lambda^n(A_2(E)) = |\det A_1| \cdot |\det A_2|\lambda^n(E)$$
$$= |\det A_1 \circ A_2|\lambda^n(E).$$

Finally, let A be arbitrary. As is well known, A can be expressed as the composite $A = A_1 \circ A_2 \circ \ldots \circ A_m$ of finitely many mappings of the three types considered above.* Since the lemma is true for each of these A_k's, it follows from the preceding paragraph and induction that it is true for A. □

(6.144) Lemma   *Suppose (6.142) obtains. Let $f: \Delta \to \mathbb{R}$ be a nonnegative continuous function. Then*

(i)
$$\int_{\phi(I)} f(x)\, dx \leqq \int_I f(\phi(t))|J_\phi(t)|\, dt$$

for every dyadic n-cube I (see (6.115)) whose closure I^- is a subset of Ω.

Proof   Write $\mu(I)$ for the left side of (i) and $\nu(I)$ for the right side. Assume that (i) fails for some I_0:

$$\mu(I_0) = \nu(I_0) + c \tag{1}$$

where $c > 0$. Let m_0 be the order of I_0 (see (6.115)) and choose one of the 2^n dyadic n-cubes of order $m_0 + 1$ whose union is I_0 (call it I_1) such that $\mu(I_1) \geqq \nu(I_1) + 2^{-n}c$. [If this choice were impossible, then the 2^n subcubes $J_1, J_2, \ldots, J_{2^n}$ of I_0 would each satisfy $\mu(J_i) < \nu(J_i) + 2^{-n}c$, and summing these 2^n inequalities would yield $\mu(I_0) < \nu(I_0) + c$, contrary to (1).] Next, subdivide I_1 in a similar way to obtain $I_2 \subset I_1$ such that $\mu(I_2) \geqq \nu(I_2) + 2^{-2n}c$. Proceeding inductively we obtain a sequence $I_0 \supset I_1 \supset \ldots \supset I_p \supset \ldots$ of dyadic n-cubes such that

$$\mu(I_p) \geqq \nu(I_p) + 2^{-pn}c \tag{2}$$

(and the order of I_p is $m_0 + p$) for all $p \geqq 0$. Let a be the unique point that is in every I_p^-. Then $a \in I_0^- \subset \Omega$. Write $b = \phi(a) \in \Delta$, $\alpha = f(b)$, $\beta = |J_\phi(a)|$, and choose positive numbers ϵ and δ such that $2^{-m_0 n}\epsilon < c$, $\delta < \beta$,

$$(\alpha + \delta)(1 + \delta)^n \beta < \epsilon + (\alpha - \delta)(\beta - \delta). \tag{3}$$

Choose a neighborhood W of b such that $W \subset \Delta$ and

$$|f(x) - f(b)| < \delta \quad \text{if } x \in W. \tag{4}$$

---

*Perhaps it is better known that the matrix of A is a finite product of matrices of mappings of the three types—usually called elementary matrices or matrices of elementary row (or column) operations. See any book on matrix theory. The usual theorem states that if M is any square matrix, then there exist matrices P and Q—each a product of invertible elementary matrices—such that $PMQ = I_r$ for some r $(0 \leqq r \leqq n)$, where I_r is obtained from the identity matrix by multiplying its last $n - r$ rows by 0. Plainly, then, $M = P^{-1}I_r Q^{-1}$ is a product of elementary matrices.

Since $J_\phi(a) \neq 0$, the linear mapping $A = d\phi(a)$ is invertible, and we have $|A^{-1}(x)| \leq M|x|$ for all $x \in \mathbb{R}^n$, where $M = \|A^{-1}\|$. By the definition of A (6.129) and the continuity of ϕ and J_ϕ (6.142), we choose a neighborhood V of a with $V \subset \Omega$ such that if $t \in V$, then $\phi(t) \in W$,

$$|\phi(t) - [b + A(t - a)]| \leq \delta|t - a|/(2Mn), \qquad (5)$$

and

$$|J_\phi(t)| > |J_\phi(a)| - \delta = |\det A| - \delta = \beta - \delta. \qquad (6)$$

Next, fix p so large that $I_p \subset V$ and write $I = I_p$. By (4), we have $\alpha - \delta < f(x) < \alpha + \delta$ whenever $x \in \phi(I)$ and, by (6), $|J_\phi(t)| > \beta - \delta$ whenever $t \in I$. Thus $\mu(I) < (\alpha + \delta)\lambda^n(\phi(I))$ and $\nu(I) > (\alpha - \delta)(\beta - \delta)\lambda^n(I) > [(\alpha + \delta)(1 + \delta)^n \beta - \epsilon]\lambda^n(I)$, where the last inequality is (3). Therefore

$$\epsilon\lambda^n(I) = 2^{-(m_0+p)n}\epsilon < 2^{-pn}c \leq \mu(I) - \nu(I)$$

$$< (\alpha + \delta)\lambda^n(\phi(I)) - [(\alpha + \delta)(1 + \delta)^n \beta - \epsilon]\lambda^n(I)$$

$$= \lambda^n(I)\epsilon + (\alpha + \delta)[\lambda^n(\phi(I)) - (1 + \delta)^n \beta\lambda^n(I)],$$

and so, since $\alpha + \delta > f(b) \geq 0$, we have

$$\lambda^n(\phi(I)) > (1 + \delta)^n \beta\lambda^n(I). \qquad (7)$$

Let I' be the closed n-cube that is concentric with I and has side length $(1 + \delta)\sigma$, where $\sigma = 2^{-(m_0+p)}$ is the side length of I. Then

$$\lambda^n(I') = (1 + \delta)^n \sigma^n = (1 + \delta)^n \lambda^n(I),$$

and so (7) yields

$$\lambda^n(\phi(I)) > \beta\lambda^n(I') = |\det A|\lambda^n(I'). \qquad (8)$$

We will show that

$$\phi(I) \subset b + A(I') - A(a). \qquad (9)$$

If this could be done, then we could conclude from (6.142) that

$$\lambda^n(\phi(I)) \leq \lambda^n(A(I')) = |\det A|\lambda^n(I'),$$

contrary to (8), and this contradiction would complete our proof that no violator I_0 of (i) can exist. Thus, we need only establish (9).

Let $x \in \phi(I)$ be given and define $s = A^{-1}(x - b) + a$. Then

$$b + A(s - a) = x = \phi(t) \qquad (10)$$

for some $t \in I \subset V$, and so

$$s - t = A^{-1}(A(s - a) - A(t - a)) = A^{-1}(\phi(t) - b - A(t - a)).$$

Our definitions of M and σ, (5), and the fact that $t, a \in I^-$ now yield

$$\max_{1 \leq j \leq n} |s_j - t_j| \leq |s - t| \leq M|\phi(t) - b - A(t - a)|$$

$$\leq \frac{\delta|t - a|}{2n} \leq \frac{\delta}{2n}\sqrt{n} \max_{1 \leq j \leq n} |t_j - a_j| \leq \frac{\delta\sigma}{2}.$$

Our choice of I' then shows that $s \in I'$, and so by (10), x is in the right side of (9). This establishes (9). $\square$

(6.145) Lemma *Suppose that (6.142) obtains and that $f : \Delta \to \mathbb{R}$ is a nonnegative continuous function. Then*

(i) $$\int_{\phi(K)} f(x)\, dx = \int_K f(\phi(t))|J_\phi(t)|\, dt$$

for every compact set $K \subset \Omega$.

Proof Let K be given and define g on Ω by $g(t) = f(\phi(t))|J_\phi(t)|$. For each integer m, let K_m be the union of all dyadic n-cubes of order m that meet K (see (6.115)). Then it is evident that

$$K_1 \supset K_2 \supset K_3 \supset \cdots \quad \text{and} \quad \bigcap_{m=1}^{\infty} K_m = K.$$

Since K is compact and Ω is open, we can select $m_0 \geq 1$ such that $K_{m_0}^- \subset \Omega$ [the distance from K to the complement of Ω is positive]. By (6.144), $m \geq m_0$ implies that

$$\int_{\phi(K)} f \leq \int_{\phi(K_m)} f = \sum \int_{\phi(I)} f \leq \sum \int_I g = \int_{K_m} g, \tag{1}$$

where the summations are over those cubes I of order m whose union is K_m. By Lebesgue's Dominated Convergence Theorem we have

$$\int_K g = \int g\xi_K = \lim_{m \to \infty} \int g\xi_{K_m} = \lim_{m \to \infty} \int_{K_m} g \tag{2}$$

[the dominating function is $g\xi_{K_{m_0}}$, which is integrable because $K_{m_0}^-$ is compact and g is continuous]. Combining (1) and (2), we see that

$$\int_{\phi(K)} f \leq \int_K g \tag{3}$$

which is (i) with "$=$" replaced by "$\leq$".

We next apply the partial result (3) with Ω and Δ interchanged, g in place of f, $\psi = \phi^{-1}$ in place of ϕ, and $L = \phi(K)$ in place of K. Note from (6.142) that Δ is open, ψ is one-to-one of class $C^{(1)}$ on Δ, ψ maps Δ onto Ω, J_ψ is never 0 on Δ, g is nonnegative and continuous on Ω, and L is compact. Our partial result for these choices reads

$$\int_{\psi(L)} g(t)\, dt \leq \int_L g(\psi(x))|J_\psi(x)|\, dx. \tag{4}$$

But $\psi(L) = K$ and, using (6.142.1), we have $g(\psi(x)) = f(x)|J_\phi(\psi(x))| = f(x)|J_\psi(x)|^{-1}$ for all $x \in \Delta$. Therefore (4) becomes

$$\int_K g(t)\, dt \leq \int_{\phi(K)} f(x)\, dx,$$

which, combined with (3), gives (i). $\square$

(6.146) Lemma *Suppose that (6.142) obtains and that* $N \subset \Omega$ *satisfies* $\lambda^n(N) = 0$. *Then* $\lambda^n(\phi(N)) = 0$.

Proof First suppose also that the closure N^- of N is a compact subset of Ω. Choose $\delta > 0$ so that $M^- \subset \Omega$, where M is the open set $\{t \in \Omega : |t - s| < \delta$ for some $s \in N\}$. Since M^- is compact and J_ϕ is continuous, $\alpha = \sup\{|J_\phi(t)| : t \in M^-\} < \infty$. Given $\epsilon > 0$, use the definition of λ^n and (6.116) to obtain a family $\{I_k\}_{k=1}^\infty$ of dyadic n-cubes that covers N and satisfies $I_k \subset M$ for all k and $\sum_{k=1}^\infty |I_k| < \epsilon/\alpha$. Notice that $\{\phi(I_k)\}_{k=1}^\infty$ covers $\phi(N)$ and apply (6.144) with $f = 1$ to deduce that

$$\lambda^n(\phi(N)) \le \sum_{k=1}^\infty \lambda^n(\phi(I_k)) \le \sum_{k=1}^\infty \int_{I_k} |J_\phi(t)| \, dt \le \sum_{k=1}^\infty \alpha |I_k| < \epsilon.$$

Therefore $\lambda^n(\phi(N)) = 0$.

In the general case, let $\{K_i\}_{i=1}^\infty$ be any countable family of compact sets whose union is Ω, write $N_i = N \cap K_i$, and apply the preceding paragraph to each N_i to obtain

$$\lambda^n(\phi(N)) = \lambda^n\left(\bigcup_{i=1}^\infty \phi(N_i)\right) \le \sum_{i=1}^\infty \lambda^n(\phi(N_i)) = 0. \quad \square$$

(6.147) Theorem *Suppose that (6.142) obtains and that E is a Lebesgue measurable subset of Ω. Then $\phi(E)$ is a Lebesgue measurable set and*

(i) $$\lambda^n(\phi(E)) = \int_E |J_\phi(t)| \, dt.$$

Proof First, suppose that $\lambda^n(E) < \infty$. By the inner regularity of λ^n (6.65), for each natural number p there is a compact set $C_p \subset E$ such that $\lambda^n(C_p) > \lambda^n(E) - 1/p$. Writing $K_p = C_1 \cup C_2 \cup \cdots \cup C_p$, we see that each K_p is compact, $K_1 \subset K_2 \subset \cdots$, $K = \bigcup_{p=1}^\infty K_p \subset E$, and the set $E \setminus K = N$ satisfies $\lambda^n(N) = 0$ because $N \subset E \setminus K_p \subset E \setminus C_p$ and $\lambda^n(E \setminus C_p) < 1/p$ for all p. Accordingly, each $\phi(K_p)$ is compact, $\phi(K_1) \subset \phi(K_2) \cdots$, $\phi(E) = \phi(N) \cup \bigcup_{p=1}^\infty \phi(K_p)$, and, invoking (6.146), we get $\lambda^n(\phi(N)) = 0$. By (6.59), $\phi(E)$ is measurable. Also, $\xi_{K_1} \le \xi_{K_2} \le \cdots$, and this sequence converges to ξ_E a.e. [except on N], so (6.62.iv), (6.145) with $f = 1$ and the Monotone Convergence Theorem imply that

$$\lambda^n(\phi(E)) = \lambda^n(\phi(K)) = \lim_{p \to \infty} \lambda^n(\phi(K_p))$$

$$= \lim_{p \to \infty} \int_{K_p} |J_\phi(t)| \, dt = \int_E |J_\phi(t)| \, dt.$$

This proves the theorem if E has finite measure.

In the general case, write $E_m = \{t \in E : |t| < m\}$ and note that E_m is measurable and $\lambda^n(E_m) < \infty$ for all $m \geq 1$. By the preceding paragraph, $\phi(E) = \bigcup_{m=1}^{\infty} \phi(E_m)$ is measurable and the equality (i) obtains when E is replaced by E_m. When we apply the convergence theorems as above to the resulting equality, we obtain (i). $\quad\square$

(6.148) Change of Variable Theorem for $\mathbb{R}^n$ *Suppose that (6.142) obtains and let $f : \Delta \to \mathbb{C}$ be a Lebesgue measurable function. Then we have the following.*
   (i) *The composite function $f \circ \phi : \Omega \to \mathbb{C}$ is Lebesgue measurable.*
   (ii) $f \in L_1(\Delta)$ *if and only if* $(f \circ \phi)|J_\phi| \in L_1(\Omega)$.
   (iii) *If $f \in L_1(\Delta)$ or $f \geq 0$, then*

$$\int_\Delta f(x)\,dx = \int_\Omega f(\phi(t))|J_\phi(t)|\,dt.$$

Proof   If V is any open subset of $\mathbb{C}$, then $f^{-1}(V)$ is a measurable set by (6.73), and so $(f \circ \phi)^{-1}(V) = \phi^{-1}(f^{-1}(V)) = \psi(f^{-1}(V))$ is a measurable set by (6.147) with ψ in place of ϕ. This proves (i). We prove (ii) and (iii) by a standard method of considering special cases and then using convergence theorems.

    Case 1.   Suppose that $f = \xi_S$, where S is a measurable subset of Δ. Write $E = \psi(S)$. Then (6.147) (with ψ in place of ϕ) shows that E is a measurable subset of Ω and

$$\int_\Delta f(x)\,dx = \lambda^n(S) = \lambda^n(\phi(E)) = \int_E |J_\phi(t)|\,dt.$$

Since $\xi_E(t) = \xi_{\phi(E)}(\phi(t)) = f(\phi(t))$ for all $t \in \Omega$, the last integral is the right side of equality (iii), so (iii) obtains in this case. For nonnegative measurable functions, membership in L_1 is equivalent to finiteness of the integral, so (ii) follows from (iii) here (notice that J_ϕ, being continuous, is measurable).

    Case 2.   If f is a finite linear combination of characteristic functions of measurable sets:

$$f = \sum_{k=1}^{m} \alpha_k \xi_{S_k},$$

then (ii) and (iii) follow from Case 1 by the linearity of the integral and of L_1.

    Case 3.   Suppose that $f \geq 0$. By (6.72), there exists a nondecreasing sequence $(f_m)_{m=1}^{\infty}$ of nonnegative measurable simple functions that converges to f at each point of Δ. Plainly, $f_m \circ \phi|J_\phi| \to f \circ \phi|J_\phi|$ as $m \to \infty$ at each point of Ω. By Case 2, formula (iii) obtains for f_m in place of f, and so two applications of the Monotone Convergence Theorem show that it also obtains for f.

    Case 4.   Finally, suppose that f is complex-valued. Since Case 3 shows that formula (iii) holds good when f is replaced by $|f|$, we see that (ii) is true for f. Suppose that $f \in L_1(\Delta)$. Choose nonnegative real-valued functions f_1, f_2, f_3, f_4 in $L_1(\Delta)$ such that $f = \sum_{k=1}^{4} i^k f_k$ [for example, f_4 $= (\operatorname{Re} f)^+$, $f_2 = (\operatorname{Re} f)^-$, $f_1 = (\operatorname{Im} f)^+$, $f_3 = (\operatorname{Im} f)^-$]. By Case 3, formula (iii) obtains for each f_k, and so linearity again shows that it also obtains for f. $\square$

(6.149) Remark   It is instructive to look back over this section to discover where the real difficulties in the proofs of the preceding two theorems lie. There are, aside from our convergence theorems, just three troublesome steps. First, we needed a little matrix theory in (6.143) in order to see how linear mappings transform measure. Second, in order to prove inequality (6.144.i) (or something like it), we needed the differential theory of the preceding section in order to approximate ϕ locally by linear maps and so take advantage of the first step. Third, in order to reverse the inequality (6.145.3), we needed the Inverse Function Theorem and the Chain Rule so as to reverse the roles of ϕ and ψ. The rest was relatively easy.

   If we had known only that ϕ was a continuous one-to-one mapping of the open set Ω onto the open set Δ which had continuous partial derivatives and, *in addition*, satisfied

$$\text{(i)} \qquad \lambda^n(\phi(I)) = \int_I |J_\phi(t)| \, dt$$

for all dyadic n-cubes I having $I^- \subset \Omega$, then we could have deduced (6.147) and a slightly restricted version of (6.148) while completely avoiding the three difficulties just cited. Indeed, we would first obtain (i) for arbitrary compact $K \subset \Omega$ in place of I by using the sets K_m as in the proof of (6.145), and then the proofs of (6.146) and (6.147) could be repeated verbatim. In order to avoid, in the proof of (6.148), the phrases "(6.147) with ψ in place of ϕ" [which used the Inverse Function Theorem as cited in (6.142)], we could simply add the hypothesis that $f \circ \phi$ is measurable.

   For some specific cases (i) is easily checked. The following is an important example.

(6.150) Polar Coordinates in the Plane   Let $n = 2$, $\Omega = \{(r, \theta) \in \mathbb{R}^2 : r > 0, -\pi < \theta < \pi\}$, and $\phi(r, \theta) = (r \cos \theta, r \sin \theta)$. Then $\Delta = \phi(\Omega) = \mathbb{R}^2 \backslash \{(x, 0) : x \leq 0\}$ and $J_\phi(r, \theta) = r > 0$ on Ω [see (6.131.b)]. It is easy to check that (6.149.i) holds in this case. Since $\lambda^2(\mathbb{R}^2 \backslash \Delta) = 0$, (6.148), (6.121), and (6.122) show that

$$\int_{-\infty}^{\infty}\int_{-\infty}^{\infty} f(x, y) \, dx \, dy = \int_\Delta f(x, y) d(x, y) = \int_\Omega f(r \cos \theta, r \sin \theta) r d(r, \theta)$$

$$= \int_0^\infty \int_{-\pi}^{\pi} f(r \cos \theta, r \sin \theta) r \, d\theta \, dr$$

$$= \int_{-\pi}^{\pi}\int_0^\infty f(r \cos \theta, r \sin \theta) r \, dr \, d\theta$$

whenever $f \in L_1(\mathbb{R}^2)$ or $f \geq 0$ is measurable on $\mathbb{R}^2$.

Exercises

1. Given $n \in \mathbb{N}$, define $\phi : \mathbb{R}^n \to \mathbb{R}^n$ by $\phi(t) = x$, where $x_i = \phi_i(t) = (1 - t_i) \prod_{j=1}^{i-1} t_j$ [here,

as usual, $\prod_{j=1}^{0} t_j = 1$, so $x_1 = 1 - t_1$]. We have the following.

(a) $J_\phi(t) = (-1)^n \prod_{j=1}^{n} t_j^{n-j}$, $[0^0 = 1]$.

(b) If $\phi(t) = x$, then

$$\sum_{j=1}^{i} x_j = 1 - \prod_{j=1}^{i} t_j \quad \text{for } 1 \leq i \leq n.$$

(c) The set $\Omega = \{t \in \mathbb{R}^n : t_j \neq 0 \text{ if } 1 \leq j < n\}$ is open in $\mathbb{R}^n$ and so is the set

$$\Delta = \phi(\Omega) = \{x \in \mathbb{R}^n : \sum_{j=1}^{i} x_j \neq 1 \text{ if } 1 \leq i < n\}.$$

(d) If $t \in \Omega$ and $x = \phi(t)$, then

$$t_i = \left(1 - \sum_{j=1}^{i} x_j\right)\left(1 - \sum_{j=1}^{i-1} x_j\right)^{-1}$$

[here, as usual, $\sum_{j=1}^{0} x_j = 0$].

(e) The mapping ϕ is one-to-one on Ω.

(f) If $I = \{t \in \mathbb{R}^n : 0 < t_j < 1 \text{ for all } j\}$ is *the open unit n-cube* and $S = \{x \in \mathbb{R}^n : \sum_{i=1}^{n} x_i < 1 \text{ and } x_i > 0 \text{ for all } i\}$ is *the open unit n-simplex*, then $\phi(I) = S$ and the "volume" of S is given by $\lambda^n(S) = 1/n!$.

2. Let $v^j = (v_1^j, \ldots, v_n^j)$, $1 \leq j \leq n + 1$, be $n + 1$ given vectors (points) in $\mathbb{R}^n$ and let the *open n-simplex* determined by these vectors be the set

$$E = \left\{ \sum_{j=1}^{n+1} \alpha_j v^j : \sum_{j=1}^{n+1} \alpha_j = 1 \text{ and } \alpha_j > 0 \text{ for all } j \right\}$$

where $\alpha v = (\alpha v_1, \ldots, \alpha v_n)$ for $\alpha \in \mathbb{R}$ and $v \in \mathbb{R}^n$. Then $n! \, \lambda^n(E)$ is equal to the determinant of the $n \times n$ matrix whose jth column is the vector $v^j - v^{n+1}$: $\lambda^n(E) = (1/n!) \cdot \det(v_i^j - v_i^{n+1})$. [Hint: Use Exercise 1. Check that $\phi(S) = E$, where $\phi(t) = v^{n+1} + \sum_{j=1}^{n} t_j(v^j - v^{n+1})$.] The set E is open in $\mathbb{R}^n$ if and only if $\lambda^n(E) > 0$.

3. Using the definition of λ^n (6.112), prove the following fact. If $a \in \mathbb{R}^n$, $E \subset \mathbb{R}^n$, and $E_a = \{(a_1 x_1, a_2 x_2, \ldots, a_n x_n) : x \in E\}$, then $\lambda^n(E_a) = |a_1 a_2 \ldots a_n| \lambda^n(E)$.

4. *Integrals of Radial Functions.* A function f on $\mathbb{R}^n$ is called a *radial function* if $f(x) = f(y)$ whenever $|x| = |y|$, so that the value of f at a point depends only on its distance from the origin. That is, f is radial if there is some function ϕ defined on $[0, \infty[$ such that $f(x) = \phi(|x|)$ for all $x \in \mathbb{R}^n$. Let $V_n = \lambda^n(\{x \in \mathbb{R}^n : |x| \leq 1\}$ be the

"volume" of the closed unit ball in $\mathbb{R}^n$. Prove the following facts without recourse to any of the results of (6.119)–(6.150).

(a) Let ϕ be a nonnegative extended real-valued Lebesgue measurable function on $[0, \infty[$ and let $n \in \mathbb{N}$. Then

(i) $$\int_{\mathbb{R}^n} \phi(|x|)\, dx = nV_n \int_0^\infty \phi(r) r^{n-1}\, dr.$$

[Hints: Use Exercise 3 to show that $\lambda^n(\{x \in \mathbb{R}^n : |x| \le R\}) = \lambda^n(\{x \in \mathbb{R}^n : |x| < R\}) = V_n R^n$, noting that the open ball is the countable union of closed balls. Consider successively the cases $\phi = \xi_{]a,\,b[}$, $\phi = \xi_W$ where W is open, $\phi = \xi_N$ where $\lambda(N) = 0$, $\phi = \xi_E$ where E is a measurable set, ϕ is a simple function. Use (6.72).]

(b) Let ϕ be any complex-valued Lebesgue measurable function on $[0, \infty[$. If either of the two integrands in (i) is Lebesgue integrable (in L_1) over its domain, then so is the other one and equality (i) obtains.

5.   *The Volume of the n-ball.*   For each $n \in \mathbb{N}$ let V_n be as in Exercise 4 and define

$$I_n = \int_{\mathbb{R}^n} e^{-|x|^2/2}\, dx, \qquad J_n = \int_0^\infty r^{n-1} e^{-r^2/2}\, dr.$$

(a) By Tonelli's Theorem, $I_n = I_1^n$.

(b) By Exercise 4, $I_n = nV_n J_n$.

(c) $J_2 = 1$, $I_2 = 2\pi$, $I_1 = \sqrt{2\pi}$, $J_1 = \sqrt{\pi/2}$.

(d) For all n, $I_n = (2\pi)^{n/2}$.

(e) If $n > 2$, then $J_n = (n-2)J_{n-2}$.

(f) If $n = 2k$ is even, then

$$V_n = \frac{(2\pi)^{n/2}}{n(n-2)\cdot \ldots \cdot 4 \cdot 2} = \frac{\pi^k}{k!}.$$

(g) If $n = 2k - 1$ is odd, then

$$V_n = \frac{2(2\pi)^{(n-1)/2}}{n(n-2)\cdot \ldots \cdot 3 \cdot 1} = \frac{k! \cdot 2^{2k} \cdot \pi^{k-1}}{(2k)!}.$$

(h) We have $V_1 < V_2 < V_3 < V_4 < V_5$, $V_n > V_{n+1}$ if $n \ge 5$, and

$$\sum_{n=1}^\infty V_n t^n < \infty$$

for all $t > 0$.

6.   *The Gamma and Beta Functions.*   For $a > 0$, define

$$\Gamma(a) = \int_0^\infty t^{a-1} e^{-t}\, dt.$$

(a) We have $0 < \Gamma(a) < \infty$ for all $a > 0$.

(b) Integration by parts yields $\Gamma(a+1) = a\Gamma(a)$ for $a > 0$.

(c) If n is a nonnegative integer, then

$$\Gamma(n+1) = n! \,.$$

(d) For $a > 0$, we have

$$\Gamma(a) = 2^{1-a} \int_0^\infty x^{2a-1} e^{-x^2/2}\, dx.$$

For $a, b > 0$, define

$$B(a, b) = 2 \int_0^{\pi/2} (\cos \theta)^{2a-1} (\sin \theta)^{2b-1} d\theta.$$

(e) For $a, b > 0$, we have

$$\Gamma(a)\Gamma(b) = \Gamma(a + b)B(a, b).$$

[Hints: Use (d) and Tonelli's Theorem to write

$$\Gamma(a)\Gamma(b) = 2^{2-a-b} \int_Q x^{2a-1} y^{2b-1} e^{-(x^2+y^2)/2} d(x, y),$$

where $Q = \{(x, y) \in \mathbb{R}^2 : x > 0, y > 0\}$ is the first quadrant. Then transform to polar coordinates.]

(f) $\Gamma(1/2) = \sqrt{\pi}$.

(g) If k is a nonnegative integer, then

$$\Gamma(k + 1/2) = (2k)! \sqrt{\pi} / (2^{2k} \cdot k!).$$

(h) If $a, b > 0$, then

$$B(a, b) = \int_0^1 t^{a-1} (1 - t)^{b-1} dt.$$

(i) The function Γ is convex and therefore continuous on $]0, \infty[$.

(j) There exists a number a_0 with $1 < a_0 < 2$ such that Γ is strictly decreasing on $]0, a_0]$ and strictly increasing on $[a_0, \infty[$.

(k) $\lim_{a \to \infty} p^{-a} \Gamma(a) = \infty$ for all $p > 0$.

(l) $\lim_{a \downarrow 0} a\Gamma(a) = 1$, so $\lim_{a \downarrow 0} \Gamma(a) = \infty$.

(m) The function Γ is infinitely differentiable on $]0, \infty[$ and

$$\Gamma^{(n)}(a) = \int_0^\infty t^{a-1} (\log t)^n e^{-t} dt$$

for each $a > 0$ and each nonnegative integer n.

(n) If $V_n = \lambda^n(\{x \in \mathbb{R}^n : |x| \leq 1\})$ (as in Exercise 4), then

$$\frac{V_n}{V_{n-1}} = \int_{-1}^1 (1 - x_n^2)^{(n-1)/2} dx_n = \int_0^1 t^{-1/2} (1 - t)^{(n-1)/2} dt$$

$$= \sqrt{\pi} \, \Gamma\left(\frac{n+1}{2}\right) / \Gamma\left(\frac{n+2}{2}\right),$$

where $V_0 = 1$, and so

$$V_n = \frac{\pi^{n/2}}{(n/2)\Gamma(n/2)}.$$

We study the functions Γ and B more extensively in the next chapter.

7.   Using Exercises 4 and 6, we see that

$$\int_{\mathbb{R}^n} \frac{dx}{(1 + |x|^2)^p} = \frac{\pi^{n/2} \Gamma(p - n/2)}{\Gamma(p)}$$

whenever $p \in \mathbb{R}$, $n \in \mathbb{N}$, and $p > n/2$. If $p \leq n/2$, this integral equals ∞.

8.   *Centroids.*   (a) Let E be a measurable subset of $\mathbb{R}^n$ with $\lambda^n(E) > 0$. Then there

exists a unique $c \in \mathbb{R}^n$ such that

$$\int_{E-c} x_i dx = 0 \qquad (1 \leq i \leq n)$$

where $E - c = \{x - c : x \in E\}$. Moreover, the coordinates of c satisfy

$$\lambda^n(E)c_i = \int_E x_i dx \qquad (1 \leq i \leq n).$$

The vector c is called the *centroid* of E.

(b) If E is the half-ball $\{x \in \mathbb{R}^n : |X| \leq 1, x_n \geq 0\}$, then the centroid c of E has $c_i = 0$ for $1 \leq i < n$ and

$$c_n = 2V_{n-1}/[(n+1)V_n]$$

where V_n is as in Exercises 4 and 6.

(c) Using Exercise 1, we see that the centroid c of the open unit n-simplex S has $c_i = (n+1)^{-1}$ for $1 \leq i \leq n$.

(d) If E and c are as in (a) and $\phi : \mathbb{R}^n \to \mathbb{R}^n$ is a nonsingular affine mapping (that is, $\phi(t) = a + A(t)$ for some $a \in \mathbb{R}^n$ and some linear mapping A with $\det A \neq 0$), then the centroid of $\phi(E)$ is $\phi(c)$.

(e) If E is as in Exercise 2 and $\lambda^n(E) > 0$, then the centroid of E is

$$c = \frac{1}{n+1} \sum_{j=1}^{n+1} v^j.$$

9. *Solids of Revolution.* Let $E \subset \mathbb{R}^2$ be a measurable set such that $r \geq 0$ whenever $(r, z) \in E$. Write $S = \{(x, y, z) \in \mathbb{R}^3 : (\sqrt{x^2 + y^2}, z) \in E\}$. Then S is a measurable subset of $\mathbb{R}^3$ and

$$\lambda^3(S) = 2\pi \int_0^\infty r\lambda(E_r)\, dr,$$

where $E_r = \{z \in \mathbb{R} : (r, z) \in E\}$. If $\lambda^2(E) > 0$, then $\lambda^3(S) = 2\pi \bar{r}\lambda^2(E)$ where $(\bar{r}, \bar{z})$ is the centroid of E (see Exercise 8).

10. Let A be an $m \times m$ real, symmetric, positive definite matrix and let Q be its quadratic form just as in Exercise 17 of the preceding section. Then

$$\int_{\mathbb{R}^m} \phi(Q(x))\, dx = (\det A)^{-1/2} \int_{\mathbb{R}^m} \phi(|t|^2)\, dt$$

for any nonnegative measurable function ϕ defined on $]0, \infty[$. [Hint: Use that Exercise 17. Make two changes of variable, the first of which is $\xi = V^t x$.] In particular, using Exercise 5,

$$\int_{\mathbb{R}^m} e^{-(1/2)Q(x)}\, dx = (2\pi)^{m/2}(\det A)^{-1/2}.$$

11. Let $0 < a < b$ and $0 \leq c < d$ be real numbers. Show that the area of the region bounded by the curves having equations $y = ax^2$, $y = bx^2$, $xy = c$, and $xy = d$ is $(1/3)(d-c)\log(b/a)$ by considering the mapping ϕ that is inverse to the mapping ψ given by $\psi(x, y) = (s, t)$, where $s = x^{-2}y$ and $t = xy$.

12. Let $a < b$ be positive real numbers. Find the area of $\Delta = \{(x, y) \in \mathbb{R}^2 : x^2 - y^2 > a, x^2 + y^2 < b, x > 0, y > 0\}$ by considering the mapping ϕ that is inverse to the mapping ψ such that $\psi(x, y) = (s, t)$, where $s = x^2 + y^2$ and $t = x^2 - y^2$.

13. Let $a < b$ and $c < d$ be positive numbers. Find the area of $\{(x, y) \in \mathbb{R}^2 : ax < y$

$< bx, \ c < x + y < d\}$ by considering the mapping defined by $(x, y) = \phi(s, t) = (t/(1 + s), st/(1 + s))$.

14. Let $v^1, v^2, \ldots, v^n$ be n vectors in $\mathbb{R}^n$ and let

$$P = \left\{ \sum_{j=1}^{n} t_j v^j : 0 \leqq t_j \leqq 1 \quad \text{for} \quad 1 \leqq j \leqq n \right\}$$

be the *n-parallelogram* determined by these vectors. Then $\lambda^n(P)$ equals the absolute value of the determinant of the matrix (v_i^j) having the vector v^j as its jth column.

15. For $x \in \mathbb{R}^{n-1}$ and $t \in \mathbb{R}$, write $(x, t) = (x_1, \ldots, x_{n-1}, t) \in \mathbb{R}^n$. Let B be a measurable subset of $\mathbb{R}^{n-1}$ $(n > 1)$ and let h be a positive real number. Then the *cone*

$$C = \{(\alpha b, \alpha h) : b \in B, 0 \leqq \alpha \leqq 1\}$$

having *base B, height h,* and *vertex* at the origin 0 satisfies $\lambda^n(C) = (h/n)\lambda^{n-1}(B)$. [Hint: Check the set $C_t = \{x \in \mathbb{R}^{n-1} : (x, t) \in C\}$ is the set $(t/h)B = \{(t/h)b : b \in B\}$ if $0 \leqq t \leqq h$ and is void otherwise. Use Tonelli's Theorem and Exercise 3.]

16. Theorems (6.147) and (6.148) remain valid even if in hypothesis (6.142) we replace "$J_\phi(t) \neq 0$ for all $t \in \Omega$" by "$\lambda^n(N) = 0$, where $N = \{t \in \Omega : J_\phi(t) = 0\}$."

7

INFINITE SERIES AND
INFINITE PRODUCTS

Series Having Monotone Terms

We begin this chapter with two extremely useful tests which apply, however, only to series whose terms [from some point onward] form a monotone sequence.

(7.1) **Cauchy's Condensation Test** (1821) *Let $(a_n)_{n=0}^{\infty}$ be a monotone nonincreasing sequence of positive numbers. Then the two series*

$$\sum_{n=0}^{\infty} a_n \quad and \quad \sum_{k=0}^{\infty} 2^k a_{2^k}$$

either both converge or both diverge.

Proof First we have

$$\frac{1}{2} \sum_{k=0}^{m} 2^k a_{2^k} = \frac{1}{2} a_1 + a_2 + 2a_4 + 4a_8 + \ldots + 2^{m-1} a_{2^m}$$

$$< a_1 + a_2 + (a_3 + a_4) + (a_5 + a_6 + a_7 + a_8) + \ldots$$

$$+ (a_{2^{m-1}+1} + a_{2^{m-1}+2} + \ldots + a_{2^m}) < \sum_{n=0}^{\infty} a_n$$

for all $m > 3$, and so

$$\sum_{k=0}^{\infty} 2^k a_{2^k} \leq 2 \sum_{n=0}^{\infty} a_n. \tag{1}$$

Next, given any $m \geq 1$, choose k such that $2^k > m$, and we have

$$\sum_{n=0}^{m} a_n < a_0 + a_1 + (a_2 + a_3) + (a_4 + a_5 + a_6 + a_7) + \ldots$$

$$+ (a_{2^k} + a_{2^k+1} + \ldots + a_{2^{k+1}-1})$$

$$\leq a_0 + a_1 + 2a_2 + 4a_4 + \ldots + 2^k a_{2^k} < a_0 + \sum_{k=0}^{\infty} 2^k a_{2^k},$$

and so

$$\sum_{n=0}^{\infty} a_n \leq a_0 + \sum_{k=0}^{\infty} 2^k a_{2^k}. \tag{2}$$

Inequalities (1) and (2) prove the theorem. $\square$

(7.2)   **Remark**   In the preceding theorem the sums could begin with any integer $n_0 > 0$. The doubtful reader could define

$$a_0 = a_1 = \ldots = a_{n_0}$$

if need be. Actually (2.40) shows that the lower index of summation can never affect the convergence behavior of an infinite series.

(7.3)   **Examples**   (a) If p is a real number, then

$$\sum_{n=1}^{\infty} \frac{1}{n^p} < \infty$$

if and only if $p > 1$ because

$$\sum_{k=1}^{\infty} 2^k \frac{1}{(2^k)^p} = \sum_{k=1}^{\infty} (2^{1-p})^k$$

is a geometric series which converges if and only if $|2^{1-p}| < 1$; i.e., $p > 1$.
     (b) If p is a real number, then

$$\sum_{n=2}^{\infty} \frac{1}{n(\log n)^p} < \infty \quad \text{if and only if} \quad p > 1$$

because

$$\sum 2^k \frac{1}{2^k (\log 2^k)^p} = \frac{1}{(\log 2)^p} \sum \frac{1}{k^p}.$$

(7.4)   **Integral Test**   *Let a be an integer and let f be a nonnegative real-valued function that is defined and nonincreasing on $[a, \infty[$. Write*

$$S_n = \sum_{k=a}^{n} f(k) \quad \text{and} \quad I_n = \int_a^n f(x)\, dx \quad \text{for } n \geq a.$$

Then the sequence $(S_n - I_n)_{n=a}^{\infty}$ is nonincreasing and converges to a limit $\alpha \in [0, f(a)]$. In fact,

$$(i) \qquad \alpha = \lim_{x \to \infty} f(x) + \sum_{n=a}^{\infty} \left[f(n) - \int_n^{n+1} f(x)\, dx \right].$$

Moreover, unless f is constant on each interval $]n, n+1[$ $(n = a, a+1, \ldots)$, we have $0 < \alpha < f(a)$. Writing

$$S = \lim S_n = \sum_{k=1}^{\infty} f(k), \qquad I = \lim I_n = \int_a^{\infty} f(x)\, dx,$$

we have $S = I + \alpha$. In particular, $S < \infty$ if and only if $I < \infty$. If $S < \infty$, then

(ii) $$\int_{n+1}^{\infty} f(x)\,dx \leq S - S_n \leq \int_{n}^{\infty} f(x)\,dx \quad \text{for all } n \geq a.$$

Proof By the monotoneness of f, we have

$$f(n + 1) \leq f(x) \leq f(n) \quad \text{for } a \leq n \leq x \leq n + 1, \text{and so}$$

(1)

$$f(n + 1) \leq \int_{n}^{n+1} f(x)\,dx \leq f(n).$$

Thus

$$(S_n - I_n) - (S_{n+1} - I_{n+1}) = (I_{n+1} - I_n) - (S_{n+1} - S_n)$$

$$= \int_{n}^{n+1} f(x)\,dx - f(n + 1) \geq 0, \quad (2)$$

and so $(S_n - I_n)_{n=a}^{\infty}$ is nonincreasing. From (1) we see that

$$S_k - I_k = \sum_{n=a}^{k} f(n) - \sum_{n=a}^{k-1} \int_{n}^{n+1} f(x)\,dx$$

$$= f(k) + \sum_{n=a}^{k-1} \left[f(n) - \int_{n}^{n+1} f(x)\,dx \right] \quad (3)$$

is a sum of nonnegative terms, so is nonnegative. This proves that our (monotone) sequence has a limit $\alpha \geq 0$. Also, (i) follows from (3) by letting $k \to \infty$. Again with the use of (1), (3) also shows that

$$S_k - I_k \leq f(k) + \sum_{n=a}^{k-1} \left[f(n) - f(n + 1) \right] = f(a), \quad (4)$$

and so $\alpha \leq f(a)$.

Suppose that for some $n_0 \geq a$, f is not constant on $]n_0, n_0 + 1[$. Choose $x_0 \in]n_0, n_0 + 1[$ such that $f(n_0) > f(x_0)$. Then

$$f(n_0) > f(n_0)(x_0 - n_0) + f(x_0)(n_0 + 1 - x_0)$$

$$\geq \int_{n_0}^{x_0} f(x)\,dx + \int_{x_0}^{n_0+1} f(x)\,dx = \int_{n_0}^{n_0+1} f(x)\,dx;$$

whence (i) shows that $\alpha > 0$. Similar reasoning shows that $\int_{n_0}^{n_0+1} f(x)\,dx$ $> f(n_0 + 1)$, and therefore a glance at (3) shows that if $k > n_0$, then strict inequality holds in (4), and so $\alpha \leq S_k - I_k < f(a)$.

Writing $S_n = (S_n - I_n) + I_n$ and letting $n \to \infty$, we see that $S = \alpha + I$ [these limits exist in $\mathbb{R}^{\#}$ because (S_n) and (I_n) are monotone]. Finally, if $S < \infty$ and $k \geq a$, then, using (i), we have

$$S_k - I_{k+1} = \sum_{n=a}^{k} \left[f(n) - \int_{n}^{n+1} f(x)\,dx \right] \leq \alpha = S - I \leq S_k - I_k,$$

from which follows

$$\int_{k+1}^{\infty} f(x)\, dx = I - I_{k+1} \le S - S_k \le I - I_k = \int_{k}^{\infty} f(x)\, dx,$$

and this is (ii). $\square$

(7.5) **Examples** (a) Take $f(x) = 1/x$ on $[1, \infty[$. Then

$$S_n = \sum_{k=1}^{n} \frac{1}{k}, \qquad I_n = \log n,$$

$$S = I = \infty,$$

$$0 < \alpha = \lim_{n \to \infty} (S_n - I_n) < f(1) = 1.$$

In this particular case the number α is usually denoted by γ and is called *Euler's constant*. Its decimal expansion to six places is $\gamma = .577215 \ldots$. It is not known whether or not γ is rational.

(b) Write $f(x) = x^{-p}$ for $x \ge 1$. Then $I_n = \int_{1}^{n} x^{-p}\, dx = n^{1-p}/(1-p) - 1/(1-p)$ if $p \ne 1$ and $I_n = \log n$ if $p = 1$. Thus $\lim I_n < \infty$ if and only if $p > 1$. Inequalities (7.4.ii) also show that if $p > 1$, then

$$\frac{1}{(p-1)(n+1)^{p-1}} \le \sum_{k=n+1}^{\infty} \frac{1}{k^p} \le \frac{1}{(p-1)n^{p-1}}$$

for all $n \ge 1$. For instance, if $p = 2$, this tells us that the error in estimating $\sum_{n=1}^{\infty} \frac{1}{n^2}$ by the sum of its first twenty terms is between $1/21$ and $1/20$.

(c) We shall show that

$$\sum_{n=3}^{\infty} \frac{1}{n \log n (\log \log n)^p} < \infty$$

if and only if $p > 1$. Take

$$f(x) = \frac{1}{x \log x (\log \log x)^p}, \qquad x \ge 3.$$

Since log is strictly increasing it is easy to see that f is strictly decreasing and positive for $x \ge 3 > e$. Also, making the substitution $t = \log \log x$, we have

$$I_n = \int_{3}^{n} f(x)\, dx = \int_{\log \log 3}^{\log \log n} t^{-p}\, dt,$$

and so, as in (b), $\lim I_n < \infty$ if and only if $p > 1$.

Limit Comparison Tests

The Comparison Test (2.48) is useful for testing series of positive terms against other such series whose convergence behavior is known. For instance, if it is known that $\sum c_n < \infty$, then to prove that $\sum a_n < \infty$ by this test, we must show that for some n_0 we have $0 \leq a_n \leq c_n$ for all $n \geq n_0$. It is frequently difficult or nearly impossible to discover such an n_0. In such cases, the following theorem may prove useful.

(7.6) **Theorem** *Let (a_n) and (b_n) be sequences of positive real numbers.*
(i) *If $\overline{\lim}(a_n/b_n) < \infty$ and $\sum b_n < \infty$, then $\sum a_n < \infty$.*
(ii) *If $\underline{\lim}(a_n/b_n) > 0$, and $\sum b_n = \infty$, then $\sum a_n = \infty$.*

Proof (i) Choose β such that $\overline{\lim}(a_n/b_n) < \beta < \infty$. Then there exists n_0 such that $a_n/b_n < \beta$ for all $n \geq n_0$; hence,

$$\sum_{n=n_0}^{\infty} a_n < \beta \sum_{n=n_0}^{\infty} b_n < \infty.$$

(ii) Choose α satisfying $0 < \alpha < \underline{\lim}(a_n/b_n)$. Then, for some n_0, we have $a_n/b_n > \alpha$ for all $n \geq n_0$; hence,

$$\sum_{n=n_0}^{\infty} a_n \geq \alpha \sum_{n=n_0}^{\infty} b_n = \infty. \quad \square$$

(7.7) **Examples** (a) Let $a_n = 1/n - \log(1 + 1/n)$ for $n \geq 1$. We shall show that $\sum a_n < \infty$ by comparing it with $\sum(1/n^2)$. Let $b_n = 1/n^2$. Using l'Hospital's Rule, we have

$$\lim_{n \to \infty} \frac{a_n}{b_n} = \lim_{n \to \infty} n^2\left(\frac{1}{n} - \log\left(1 + \frac{1}{n}\right)\right) = \lim_{x \to 0} \frac{x - \log(1 + x)}{x^2}.$$

$$= \lim_{x \to 0} \frac{1}{2(1 + x)} = \frac{1}{2} < \infty$$

and our result follows.

(b) For $x > 0$, write $a_n = \sqrt[n]{x} - 1$ and $b_n = 1/n$. Then

$$\lim_{n \to \infty} (a_n/b_n) = \lim_{n \to \infty} n(\sqrt[n]{x} - 1) = \lim_{t \to 0} \frac{x^t - 1}{t} = \lim_{t \to 0} x^t \log x = \log x.$$

If $x > 1$, then $a_n > 0$ and $\log x > 0$, so $\sum a_n = \infty$. If $0 < x < 1$, then $-a_n > 0$ and $\lim_{n \to \infty} -(a_n/b_n) = -\log x > 0$, so $\sum - a_n = \infty$; hence, $\sum a_n = -\infty$.

(c) Let α and β be real numbers, $a_n = n^\alpha \exp\left[-\beta \sum_{k=1}^{n} \frac{1}{k}\right]$, $b_n = n^{\alpha - \beta}$.
Then $a_n > 0$, $b_n > 0$, and

$$\frac{a_n}{b_n} = n^\beta \exp\left[-\beta \sum_{k=1}^{n} \frac{1}{k}\right] = \exp\left[\beta\left(\log n - \sum_{k=1}^{n} \frac{1}{k}\right)\right] \to e^{-\beta\gamma},$$

where γ is Euler's constant (7.5.a). Since this limit is positive and finite, $\sum_{n=1}^{\infty} a_n$ and $\sum_{n=1}^{\infty} b_n$ either both converge or both diverge. Thus $\sum a_n < \infty$ if and only if $\beta - \alpha > 1$.

The next two theorems are included here to dramatically demonstrate the limitations of the Limit Comparison Test: given any series $\sum b_n < \infty$ $(b_n > 0)$, one can find a series $\sum a_n < \infty$ $(a_n > 0)$ such that $\lim(a_n/b_n) = \infty$; given any series $\sum b_n = \infty$ $(b_n > 0)$, one can find a series $\sum a_n = \infty$ $(a_n > 0)$ such that $\lim(a_n/b_n) = 0$.

(7.8)   **Theorem** [Dini] *Let* $\displaystyle\sum_{n=1}^{\infty} b_n$ *be any convergent series of positive numbers and let* $\alpha \in \mathbb{R}$. *Write* $r_n = \displaystyle\sum_{k=n}^{\infty} b_k$ $(n = 1, 2, \ldots)$ *and* $a_n = b_n/r_n^{\alpha}$. *Then*

(i) $\sum a_n < \infty$ *if* $\alpha < 1$,

(ii) $\sum a_n = \infty$ *if* $\alpha \geq 1$.

Proof   (i) Choose an integer p such that $0 < 1/p < 1 - \alpha$. For $0 \leq x \leq 1$, we have the obvious inequality

$$1 - x^p = (1 + x + x^2 + \ldots + x^{p-1})(1 - x) \leq p(1 - x).$$

Since $r_{n+1} \leq r_n$, we may replace x by $(r_{n+1}/r_n)^{1/p}$ to obtain

$$1 - r_{n+1}/r_n \leq p(1 - r_{n+1}^{1/p}/r_n^{1/p}).$$

Multiplying by $r_n^{1/p}$, we have

$$\frac{b_n}{r_n^{1-1/p}} = \frac{r_n - r_{n+1}}{r_n} \cdot r_n^{1/p} \leq p(r_n^{1/p} - r_{n+1}^{1/p}) \tag{1}$$

for all n. Since $r_n \to 0$, there is an n_0 such that $r_n < 1$ for all $n \geq n_0$. Thus, since $1 - 1/p > \alpha$, we have

$$a_n = b_n/r_n^{\alpha} < b_n/r_n^{1-1/p}, \tag{2}$$

for all $n \geq n_0$. Combining (1) and (2) and summing yields

$$\sum_{n=n_0}^{\infty} a_n \leq \sum_{n=n_0}^{\infty} p(r_n^{1/p} - r_{n+1}^{1/p}) = \lim_{k \to \infty} p \sum_{n=n_0}^{k} (r_n^{1/p} - r_{n+1}^{1/p})$$

$$= \lim_{k \to \infty} p(r_{n_0}^{1/p} - r_{k+1}^{1/p}) = p r_{n_0}^{1/p} < \infty.$$

Thus $\sum a_n < \infty$.

(ii) Since $\alpha \geq 1$ and $1 > r_n > r_{n+1} > 0$ for all $n \geq n_0$ [hence $r_n^{\alpha} \leq r_n$] it follows that if $q > p > n_0$, then

$$\sum_{n=p}^{q} a_n \geq \sum_{n=p}^{q} \frac{b_n}{r_n} > \frac{1}{r_p} \sum_{n=p}^{q} b_n = \frac{1}{r_p}(r_p - r_{q+1}) = 1 - \frac{r_{q+1}}{r_p}. \tag{3}$$

But $r_q \to 0$, so, given any p, there exists $q_p > p$ such that $r_{q_p+1}/r_p < 1/2$. From (3) it follows that $p > n_0$ implies

$$\sum_{n=p}^{q_p} a_n > \frac{1}{2},$$

and so Cauchy's Criterion (2.39) shows that $\sum a_n = \infty$. $\square$

(7.9)   **Theorem**   [Abel–Dini]   *Let* $\sum\limits_{n=1}^{\infty} b_n = \infty$ $(b_n > 0$ *for all* $n)$ *and let* $\alpha \in \mathbb{R}$.

Write $S_n = \sum\limits_{k=1}^{n} b_k$ *and* $a_n = b_n/S_n^{\alpha}$. *Then*

(i) $\sum a_n = \infty$ *if* $\alpha \leq 1$,

(ii) $\sum a_n < \infty$ *if* $\alpha > 1$.

Proof   (i) The case $\alpha = 1$ was proved in (2.41). It follows that if $\alpha < 1$ and n_0 is chosen so that $S_n > 1$ for all $n \geq n_0$, then

$$\sum_{n=n_0}^{\infty} \frac{b_n}{S_n^{\alpha}} \geq \sum_{n=n_0}^{\infty} \frac{b_n}{S_n} = \infty,$$

and so $\sum a_n = \infty$.

(ii) For $\alpha > 1$ choose an integer p such that $1 + 1/p < \alpha$. In the inequality $1 - x^p \leq p(1 - x)$ of the preceding proof, take $x = (S_{n-1}/S_n)^{1/p}$ and then divide by $S_{n-1}^{1/p}$ to obtain

$$\frac{b_n}{S_n S_{n-1}^{1/p}} = \left(1 - \frac{S_{n-1}}{S_n}\right) \frac{1}{S_{n-1}^{1/p}} \leq p\left(\frac{1}{S_{n-1}^{1/p}} - \frac{1}{S_n^{1/p}}\right). \tag{1}$$

For $n \geq n_0 > 1$, we have $S_n > 1$, and so $S_n^{\alpha} > S_n^{1+1/p} > S_n S_{n-1}^{1/p}$. Combining this with (1) yields

$$a_n = \frac{b_n}{S_n^{\alpha}} < p\left(\frac{1}{S_{n-1}^{1/p}} - \frac{1}{S_n^{1/p}}\right) \tag{2}$$

for all $n \geq n_0$. Now sum inequalities (2) for $n = n_0,\ n_0 + 1, \ldots$ and use $\lim\limits_{k \to \infty} S_k^{1/p} = \infty$ to get

$$\sum_{n=n_0}^{\infty} a_n < \sum_{n=n_0}^{\infty} p\left(\frac{1}{S_{n-1}^{1/p}} - \frac{1}{S_n^{1/p}}\right) = \frac{p}{S_{n_0-1}^{1/p}} < \infty.$$

Therefore $\sum a_n < \infty$. $\square$

Notice what this theorem says if we take $b_n = 1$ for all $n \geq 1$.

Two Log Tests

The Root Test and the Ratio Test are very weak tests. Both have the properties that they can prove divergence only if the terms do not tend to 0 and that they can prove convergence only if the series in question is dominated by a (rapidly convergent) geometric series. In this section we present two more powerful tests. They are more powerful because they rely ultimately on comparison with series of the form $\sum(1/n^p)$ and $\sum[1/n(\log n)^p]$, respectively.

(7.10) **First Log Test** *Let (a_n) be a sequence of positive real numbers. Write*
$$L_n = \left[\log(1/a_n)\right]/(\log n).$$

Then

(i) $$\underline{\lim} L_n > 1 \quad implies \quad \sum a_n < \infty$$

and

(ii) $$L_n \leq 1 \quad for \ all \ n \geq N \quad implies \quad \sum a_n = \infty.$$

Proof (i) Select p such that $1 < p < \underline{\lim} L_n$. Then there exists n_0 such that $L_n > p$ for all $n \geq n_0$. Thus $n \geq n_0$ implies
$$\log(1/a_n) > \log(n^p)$$
$$1/a_n > n^p,$$
$$0 < a_n < 1/n^p.$$

Since $\sum 1/n^p < \infty$, we have $\displaystyle\sum_{n=n_0}^{\infty} a_n < \infty$.

(ii) The hypothesis $L_n \leq 1$ is equivalent to $a_n \geq 1/n$. Therefore
$$\sum_{n=N}^{\infty} a_n \geq \sum_{n=N}^{\infty} 1/n = \infty. \quad \square$$

(7.11) **Examples** (a) For $x > 0$, write $a_n = x^{\log n}$. Then $L_n = -\log x$ for all $n > 1$, and so $\sum a_n < \infty$ if and only if $-\log x > 1$; i.e., $0 < x < 1/e$.

(b) For $x > 0$, write $a_n = x^{\log(\log n)}$. Then $L_n = -(\log x)(\log(\log n))(\log n)^{-1}$ for $n > 2$ and $\displaystyle\lim_{n\to\infty} L_n = 0$. Thus, $\sum a_n = \infty$ for all $x > 0$.

(c) If $a_n = (\log n)^{-\log n}$, then $L_n = \log(\log n) \to \infty > 1$, and so $\sum a_n < \infty$.

(7.12) **Second Log Test** *Let (a_n) be a sequence of positive real numbers and write*
$$M_n = \left[\log(1/(na_n))\right]/\left[\log(\log n)\right]$$

Then

(i) $\underline{\lim} M_n > 1 \quad implies \quad \sum a_n < \infty,$

(ii) $M_n \leq 1 \quad for \ all \quad n \geq N \quad implies \quad \sum a_n = \infty.$

Proof The proof is so similar to that of the First Log Test that we leave the details to the reader. It depends only on the facts that $\sum 1/n\log n = \infty$, $\sum 1/n(\log n)^p < \infty$ if $p > 1$, and the Comparison Test. $\quad \square$

(7.13) **Example** If $a_n = n^{-x_n}$, where $x_n = 1 + 1/(\log n)$ $(n > 1)$, then $M_n = 1/[\log(\log n)] < 1$ for all $n > e^e$, and so $\sum a_n = \infty$.

Other Ratio Tests

When one attempts the Ratio Test on a series $\sum a_n$ of positive terms and finds that $\underline{\lim}(a_{n+1}/a_n) \leqq 1 \leqq \overline{\lim}(a_{n+1}/a_n)$, he gains no information regarding the convergence of the series. There are, however, more sophisticated "ratio tests" which sometimes succeed where the old one failed. We obtain some of these as corollaries of the following very general test.

(7.14) Dini–Kummer Test*   *Let (a_n) and (b_n) be sequences of positive real numbers and write*

$$D_n = b_n - b_{n+1} \frac{a_{n+1}}{a_n} \ .$$

Then

(i) $\underline{\lim} \, D_n > 0$ *implies* $\sum a_n < \infty$,

(ii) $D_n \leqq 0$ *for all* $n \geqq N$ *and* $\sum \dfrac{1}{b_n} = \infty$ *imply* $\sum a_n = \infty$.

Proof   (i) Choose β such that $0 < \beta < \underline{\lim} \, D_n$. Then there exists n_0 such that $D_n > \beta$ for all $n \geqq n_0$. This implies that

$$0 < a_n < (1/\beta)(\,a_n b_n - a_{n+1} b_{n+1}) \tag{1}$$

for all $n \geqq n_0$. Thus the sequence $(a_n b_n)_{n=n_0}^{\infty}$ is a strictly decreasing sequence of positive numbers, and so it converges to a limit $\gamma \geqq 0$. Accordingly,

$$\sum_{n=n_0}^{\infty} \frac{1}{\beta}(a_n b_n - a_{n+1} b_{n+1}) = \lim_{p \to \infty} \sum_{n=n_0}^{P} \frac{1}{\beta}(a_n b_n - a_{n+1} b_{n+1})$$

$$= \lim_{p \to \infty} \frac{1}{\beta}(a_{n_0} b_{n_0} - a_{p+1} b_{p+1})$$

$$= \frac{1}{\beta}(a_{n_0} b_{n_0} - \gamma) < \infty.$$

Now apply (1) and the Comparison Test to see that $\sum a_n < \infty$.

(ii) For $n \geqq N$ we have

$$a_n b_n - a_{n+1} b_{n+1} \leqq 0,$$

and so $(a_n b_n)_{n=N}^{\infty}$ is a nondecreasing sequence. Thus

$$a_n \geqq a_N b_N / b_n \qquad (n \geqq N),$$

and so the Comparison Test shows that $\sum a_n$ diverges. $\square$

---

*Part (i) of this theorem is often called Kummer's Test. It was published by Kummer in 1835 under the additional hypothesis $\lim a_n b_n = 0$. Dini, in 1867, showed this restriction to be superfluous and also proved (ii).

(7.15) **Remarks** (a) If we choose $b_n = 1$ for all n, then $D_n = 1 - a_{n+1}/a_n$ and $\underline{\lim} D_n = 1 - \overline{\lim}(a_{n+1}/a_n)$. Thus, in this case, (7.14) reduces to the Ratio Test (2.61).

(b) If we take $b_n = 2^n$ and $a_n = 1/n^2$, then $\lim_{n \to \infty} D_n = -\infty$ and $\sum a_n < \infty$. This shows that the hypothesis $\sum 1/b_n = \infty$ in (7.14.ii) cannot be entirely dispensed with.

Plainly each choice of a fixed sequence (b_n) gives rise to a test for convergence that involves the ratio a_{n+1}/a_n. We call each of these *ratio tests*. We content ourselves here with three well-known tests of this kind that are sometimes useful.

(7.16) **Raabe's Test** (1832) *If $a_n > 0$ for all n and*

$$R_n = n(1 - a_{n+1}/a_n),$$

then

(i) $$\underline{\lim} R_n > 1 \quad \text{implies} \quad \sum a_n < \infty$$

and

(ii) $$R_n \leqq 1 \quad \text{for all } n \geqq N \quad \text{implies} \quad \sum a_n = \infty.$$

Proof In (7.14), take $b_n = n - 1$ for $n > 1$ and $b_1 = 1$. Then $\sum \dfrac{1}{b_n} = \infty$

and

$$R_n - 1 = (n - 1) - n \cdot \frac{a_{n+1}}{a_n} = D_n \text{ for } n > 1. \quad \square$$

(7.17) **Note** In Raabe's Test, the hypothesis $\underline{\lim} R_n > 1$ is equivalent to the existence of some $\beta > 1$ such that $R_n > \beta$ for all large n; i.e. $a_{n+1}/a_n < 1 - \beta/n$. Similarly, $R_n \leqq 1$ is equivalent to $a_{n+1}/a_n \geqq 1 - 1/n$. This shows that if $a_{n+1}/a_n < \alpha < 1$ for all large n [whence $\sum a_n < \infty$ by the Ratio Test], then $\sum a_n < \infty$ by Raabe's Test; and if $a_{n+1}/a_n \geqq 1$ for all large n [whence $\sum a_n = \infty$ by the Ratio Test], then $\sum a_n = \infty$ by Raabe's Test. Thus Raabe's Test successfully tests any series that the Ratio Test does. That Raabe's Test is stronger [tests more series] than the Ratio Test is shown by the following example.

(7.18) **Example** We shall test the binomial series $\sum_{n=0}^{\infty} \binom{\alpha}{n} x^n$ at $x = -1$, where α

is a fixed real number but not a nonnegative integer. Thus set

$$a_n = \binom{\alpha}{n}(-1)^n = \frac{\alpha(\alpha - 1) \ldots (\alpha - n + 1)}{n!}(-1)^n$$

for $n > 0$. Then

$$a_{n+1}/a_n = (n - \alpha)/(n + 1) \to 1, \tag{1}$$

and so the Ratio Test fails to apply to $\sum a_n$. It does, however, follow from (1) that all terms a_n have the same sign if $n > \max\{\alpha, 0\}$. Thus, multiplying every term by -1 if necessary [which affects neither convergence nor the ratios], we may suppose $a_n > 0$ for all large n. We have

$$R_n = n(1 - a_{n+1}/a_n) = n(1 + \alpha)/(n + 1) \to 1 + \alpha.$$

Therefore Raabe's Test shows that $\sum a_n$ converges if $\alpha > 0$ and diverges if $\alpha < 0$. Obviously it converges if $\alpha = 0$.

(7.19)   Bertrand's Test (1842)   *If $a_n > 0$ for all n and*

$$B_n = (R_n - 1)\log n = (n - 1)\log n - (a_{n+1}/a_n)n \log n,$$

then

(i) $\underline{\lim} B_n > 1$ *implies* $\sum a_n < \infty$

and

(ii) $\overline{\lim} B_n < 1$ *implies* $\sum a_n = \infty$.

Proof   In (7.14) take $b_{n+1} = n \log n$. Then $\sum 1/b_n = \infty$ and
$$D_n = (n - 1)\log(n - 1) - (n \log n)a_{n+1}/a_n = (n - 1)\log((n - 1)/n) + B_n.$$

Since

$$\lim_{n \to \infty} (n - 1)\log((n - 1)/n) = \lim_{n \to \infty} n \log(n/(n + 1))$$
$$= \lim_{n \to \infty} \log(1 + 1/n)^{-n} = \log(1/e) = -1,$$

we have

$$\overline{\lim} B_n = 1 + \overline{\lim} D_n,$$

and so the theorem follows from (7.14).   □

Bertrand's Test, though of infrequent practical use, does offer a simple proof of the next theorem.

(7.20)   Gauss' Test (1812)   *If $a_n > 0$ for all n and if there exist a real number α, a positive number ϵ, and a* bounded *sequence $(x_n) \subset \mathbb{R}$ such that*

(i) $a_{n+1}/a_n = 1 - \alpha/n - x_n/n^{1+\epsilon}$

for all n, then $\sum a_n < \infty$ if and only if $\alpha > 1$.

Proof   We have
$$R_n = n(1 - a_{n+1}/a_n) = \alpha + x_n/n^\epsilon \to \alpha$$

because (x_n) is bounded. Thus, for $\alpha \neq 1$, the theorem follows from Raabe's Test. For $\alpha = 1$ we use Bertrand's Test. We have
$$B_n = (R_n - 1)\log n = x_n(\log n)/n^\epsilon \to 0,$$

and so $\sum a_n = \infty$.   □

We remark that it is always possible to find *some* sequence (x_n) so that (i) holds. Just choose any α and $\epsilon > 0$ and then solve (i) for x_n. The issue is that we insist that (x_n) be *bounded*.

(7.21) Example Let α, β, γ be real numbers, none of which is a negative integer or 0. Define $a_0 = 1$ and, for $n > 0$,

$$a_n = \frac{\alpha(\alpha + 1)(\alpha + 2) \ldots (\alpha + n - 1)}{1 \cdot 2 \cdot 3 \cdot \ldots \cdot n} \frac{\beta(\beta + 1)(\beta + 2) \ldots (\beta + n - 1)}{\gamma(\gamma + 1)(\gamma + 2) \ldots (\gamma + n - 1)}$$

The series $\sum\limits_{n=0}^{\infty} a_n$ is called the *hypergeometric series*. Plainly

$$\frac{a_{n+1}}{a_n} = \frac{(\alpha + n)(\beta + n)}{(1 + n)(\gamma + n)} = \frac{n^2 + (\alpha + \beta)n + \alpha\beta}{n^2 + (\gamma + 1)n + \gamma}.$$

Since $\lim\limits_{n \to \infty} (a_{n+1}/a_n) = 1$, we see that, for all large n, the terms a_n have the same sign, and so Gauss' Test may be applied. Solving the equation

$$\frac{n^2 + (\alpha + \beta)n + \alpha\beta}{n^2 + (\gamma + 1)n + \gamma} = 1 - \frac{(\gamma + 1) - (\alpha + \beta)}{n} - \frac{x_n}{n^2}$$

for x_n, one finds that (x_n) converges and hence is bounded. Therefore Gauss' Test shows that $\sum a_n$ converges if and only if $\alpha + \beta < \gamma$.

Exercises

1. (a) If $p > 1$ and $n \in \mathbb{N}$, then

$$\frac{1}{(p - 1)(n + 1)^{p-1}} < \sum_{k=n+1}^{\infty} \frac{1}{k^p} < \frac{1}{(p - 1)n^{p-1}}.$$

(b) If $p > 0$, $p \neq 1$, and $n > 1$, then

$$\frac{1}{1 - p}\left(\frac{1}{(n + 1)^{p-1}} - \frac{1}{2^{p-1}}\right) < \sum_{k=2}^{n} \frac{1}{k^p} < \frac{1}{1 - p}\left(\frac{1}{n^{p-1}} - 1\right).$$

2. Let $x \in \mathbb{R}$, $x > 0$. Test each series $\sum a_n$ for convergence, where a_n is given by

(a) $\pi/2 - \text{Arctan } n$,

(b) $\left(1 - \dfrac{1}{\sqrt{n}}\right)^n$,

(c) $\sqrt[n]{x} - 1 - \dfrac{1}{n}$,

(d) $\dfrac{\sqrt{n + 1} - \sqrt{n}}{n^x}$,

(e) $[\log(\log n)]^{-\log n}$ $(n > 2)$,

(f) $x^{\sqrt{n}}$,

(g) $\left(\dfrac{\log n}{n}\right)^x$,

(h) $(\sqrt[n]{n} - 1)^x$,

(i) $\left(e - \left(1 + \dfrac{1}{n}\right)^n\right)^x$,

(j) $2^{-n^2} \cdot x^n$,

(k) $\left(\dfrac{1}{2} \cdot \dfrac{3}{4} \cdot \dfrac{5}{6} \cdot \ldots \cdot \dfrac{2n - 1}{2n}\right)^x$,

(l) $\left(\dfrac{1}{2} \cdot \dfrac{3}{4} \cdot \dfrac{5}{6} \cdot \ldots \cdot \dfrac{2n - 1}{2n}\right) \cdot \dfrac{1}{2n + 1}$,

(m) $\left(1 - \dfrac{\log n}{n}\right)^n$,

(n) $n^n e^{-n^2}$.

3.   Let $(p_n)_{n=1}^{\infty}$ be an increasing sequence of positive real numbers. What further conditions on this sequence will assure the convergence of $\sum a_n$, where a_n is given by

(a) n^{-p_n},   (b) p_n^{-n},   (c) $p_n^{-\log n}$,

(d) $p_n^{-\log(\log n)}$,   (e) $(\log n)^{-p_n}$

for $n > 1$?

4.   Let $f(x) = 1 - e^{-x}$. Define $a_1 = 1$ and $a_{n+1} = f(a_n)$ for $n \geq 1$. Then $\sum a_n = \infty$. [Hint: $f(x) > x/(1 + x)$ for $x > 0$.]

5.   Let F be a positive function defined and differentiable on $[a, \infty[$ for some $a \in \mathbb{N}$. Suppose also that F' is nonnegative and nonincreasing on $[a, \infty[$. Then

(a) $\displaystyle\sum_{n=a}^{\infty} F'(n) < \infty$ if and only if $\displaystyle\sum_{n=a}^{\infty} \frac{F'(n)}{F(n)} < \infty$,

(b) if $\displaystyle\sum_{n=a}^{\infty} F'(n) = \infty$, then $\displaystyle\sum_{n=a}^{\infty} \frac{F'(n)}{[F(n)]^p}$ converges for $p > 1$ and diverges for

$p \leq 1$. [Hint: Use the Integral Test.]

6.   (a) If (ϵ_n) is a strictly decreasing sequence with limit 0, then there is some divergent series $\sum d_n$, $d_n > 0$, such that $\sum \epsilon_n d_n < \infty$.

     (b) If (M_n) is a strictly increasing sequence with limit ∞, then there is a convergent series $\sum c_n$, $c_n > 0$, such that $\sum M_n c_n = \infty$. [Hint: Use (7.9) and (7.8):

$$s_n = 1/\epsilon_n, \qquad r_n = 1/M_n.]$$

7.   (a) If $t_n = \displaystyle\sum_{k=1}^{n} \frac{1}{2k-1}$, then

$$\lim_{n \to \infty} (t_n - (1/2)\log n) = (1/2)\gamma + \log 2$$

where γ is Euler's constant.

(b) Use (a) to prove that $\displaystyle\sum_{k=1}^{\infty} \frac{1}{4k^3 - k} = 2\log 2 - 1$.

(c) If $s_n = \displaystyle\sum_{k=1}^{n} \frac{1}{k}$, then

$$\sum_{k=1}^{n} \frac{1}{36k^3 - k} = 3t_{3n+1} - t_n - s_n - 3$$

$$\to \frac{3}{2}\log 3 + 2\log 2 - 3 \quad \text{as } n \to \infty.$$

8.   Let $\sum d_n$ be a divergent series of positive terms. What can be said about the series

(a) $\displaystyle\sum \frac{d_n}{1 + d_n}$,   (b) $\displaystyle\sum \frac{d_n}{1 + nd_n}$,

(c) $\displaystyle\sum \frac{d_n}{1 + n^2 d_n}$,   (d) $\displaystyle\sum \frac{d_n}{1 + d_n^2}$?

9.   Let $\sum a_n$ be a series of positive terms and write

$$T_n = n \log(a_{n+1}/a_n).$$

If $\overline{\lim} \, T_n < -1$, then the series converges. If $T_n \geq -1$ for all $n \geq N$, then the series diverges. [Hint: If $0 < \alpha < \beta$, then $1 - \beta/n < e^{-\beta/n} < 1 - \alpha/n$ for all large n.]

10. Let $(a_n)_{n=1}^{\infty}$ be a nonincreasing sequence of positive terms.

(a) If $1 \leq n_1 < n_2 < \ldots$ and, for some $\beta \in \mathbb{R}$,

$$n_{k+1} - n_k \leq \beta(n_k - n_{k-1})$$

for all $k > 1$, then $\sum a_n < \infty$ if and only if $\sum(n_{k+1} - n_k)a_{n_k} < \infty$.

(b) If $p > 1$ is an integer, then the sequences $n_k = p^k$ and $n_k = k^p$ satisfy the hypothesis of (a).

(c) The conclusion of (a) can fail if $n_k = k!$.

11. Let $\sum c_n$ be any convergent series of positive terms. Then there exists a monotone sequence (d_n) of positive terms such that $\sum d_n = \infty$ and $\underline{\lim}(d_n/c_n) = 0$. [Hint: Choose $1 = n_1 < n_2 < \ldots$ such that $c_{n_{k+1}} < c_{n_k}$ and $(n_{k+1} - n_k)c_{n_k} > k$ for all k. Then put $d_n = c_{n_k}/k$ for $n_k \leq n < n_{k+1}$.]

12. (a) If $(c_n)_{n=1}^{\infty}$ is a monotone sequence of positive terms such that $\sum c_n < \infty$, then $\lim_{n \to \infty} nc_n = 0$. [Hint: $nc_{2n} \leq c_{n+1} + \ldots + c_{2n}$.]

(b) Assertion (a) is best possible in the sense that if $p_n \to \infty$, then there is a series $\sum c_n$ as in (a) such that $\overline{\lim}_{n \to \infty} np_nc_n = \infty$. Thus, the conclusion of (a) *cannot* be changed to read $\lim_{n \to \infty} (n \log n)c_n = 0$. [Hint: Write $p_0 = 0$ and choose $0 = n_0 < n_1 < n_2 < \ldots, p_{n_{k-1}} < p_{n_k}$ and $p_{n_k} > 4^k$ for all $k \geq 1$. Let $c_n = \left(n_k\sqrt{p_{n_k}}\right)^{-1}$ for $n_{k-1} < n \leq n_k$.]

Infinite Products

In (5.36) we defined the *value* of an infinite product $\prod_{n=p}^{\infty} u_n$ with factors $u_n \in \mathbb{C}$ to be the limit $\lim_{q \to \infty} \prod_{n=p}^{q} u_n$ if this limit exists in $\mathbb{C}$, but we made no mention of the notion of convergence for infinite products. Because of the peculiar multiplicative role of the number 0 and in an effort to make the theory of infinite products parallel to that of infinite series, we make the following definition. Though it may seem strange at first, the theorems that we obtain are ample justification.

(7.22) **Definition** An infinite product $\prod_{n=1}^{\infty} u_n$, $u_n \in \mathbb{C}$, is said to *converge* if (i) there exists a $p \in \mathbb{N}$ such that $u_n \neq 0$ for $n \geq p$ *and* (ii) $\prod_{n=p}^{\infty} u_n$ has a *nonzero* value in $\mathbb{C}$ [$\lim_{q \to \infty} \prod_{n=p}^{q} u_n$ exists in $\mathbb{C}\backslash\{0\}$]. In this case, the value of $\prod_{n=1}^{\infty} u_n$ is the number

$$\left(\prod_{n=1}^{p-1} u_n\right)\left(\lim_{q \to \infty} \prod_{n=p}^{q} u_n\right) = \left(\prod_{n=1}^{p-1} u_n\right)\left(\prod_{n=p}^{\infty} u_n\right) = \prod_{n=1}^{\infty} u_n.$$

Note that neither the convergence nor the value is affected by the particular choice of p satisfying (i). All other infinite products are said to *diverge*.

If a divergent product has the value 0, we say that it *diverges* to 0. This occurs if $u_n = 0$ for infinitely many n or else $u_n \neq 0$ for $n \geq p$ but $\lim\limits_{q \to \infty} \prod\limits_{n=p}^{q} u_n = 0$. A convergent product has the value zero if and only if at least one [but only finitely many] of its factors is 0.

(7.23)   Examples   (a) The product $\prod\limits_{n=1}^{\infty} \dfrac{n}{n+1}$ has no zero factor and $\prod\limits_{n=1}^{q} \dfrac{n}{n+1}$

$= \dfrac{1}{2} \cdot \dfrac{2}{3} \cdot \ldots \cdot \dfrac{q}{q+1} = \dfrac{1}{q+1} \to 0$, so the value is 0 and the product diverges to 0.

(b) The product $\prod\limits_{n=1}^{\infty} \dfrac{n^2 - 1}{n^2}$ has just one zero factor and

$$\prod_{n=2}^{q} \frac{n^2 - 1}{n^2} = \frac{1 \cdot 3}{2 \cdot 2} \cdot \frac{2 \cdot 4}{3 \cdot 3} \cdot \frac{3 \cdot 5}{4 \cdot 4} \cdot \ldots \cdot \frac{(q-1)(q+1)}{q \cdot q} = \frac{q+1}{2q} \to \frac{1}{2} \neq 0.$$

Thus this product converges to the value 0.

(c) The product $\prod\limits_{n=1}^{\infty} \dfrac{n-1}{n}$ has only one zero factor, but, for any $p > 1$,

$\prod\limits_{n=p}^{q} \dfrac{n-1}{n} = \dfrac{p-1}{q} \to 0$ as $q \to \infty$. Thus, this product diverges to 0.

(d) The product $\prod\limits_{n=1}^{\infty} [1 + (-1)^n]$ diverges to 0 because infinitely many factors are 0.

(e) The products $\prod\limits_{n=1}^{\infty} (-1)^n$ and $\prod\limits_{n=1}^{\infty} \dfrac{n+1}{n}$ both diverge, but not to 0.

As in the cases of sequences and series, we have a Cauchy Criterion for the convergence of products.

(7.24)   Cauchy Criterion   *An infinite product* $\prod\limits_{n=1}^{\infty} u_n$ *converges if and only if for each* $\epsilon > 0$ *there exists an* $N \in \mathbb{N}$ *such that*

(i) $$\left| 1 - \prod_{n=r}^{s} u_n \right| < \epsilon \quad \text{whenever } s \geq r \geq N.$$

Proof   Suppose the product converges. Choose $p \in \mathbb{N}$ such that $u_n \neq 0$ for $n \geq p$ and let

$$\lim_{q \to \infty} \prod_{n=p}^{q} u_n = L \in \mathbb{C} \setminus \{0\}.$$

Since $L \neq 0$ there is a $\delta > 0$ such that

$$\left| \prod_{n=p}^{q} u_n \right| > \delta \quad \text{for } q \geq p. \tag{1}$$

Let $\epsilon > 0$ be given. By Cauchy's Criterion for sequences, there is an $N \in \mathbb{N}$, $N > p$, such that

$$\left| \prod_{n=p}^{r-1} u_n - \prod_{n=p}^{s} u_n \right| < \epsilon \delta \tag{2}$$

whenever $s \geq r \geq N$. Now multiply both sides of (2) by $\left| \prod_{n=p}^{r-1} u_n \right|^{-1}$ and use (1) to obtain (i).

Conversely, suppose that (i) holds. Taking $\epsilon = 1/2$, let p be the corresponding N satisfying (i). Then

$$\frac{1}{2} < \left| \prod_{n=p}^{q} u_n \right| < \frac{3}{2} \quad \text{for } q \geq p. \tag{3}$$

Thus, $u_n \neq 0$ for $n \geq p$ and the sequence

$$\Pi_q = \prod_{n=p}^{q} u_n \quad (q \geq p) \tag{4}$$

cannot converge to 0. To see that it converges, let $\epsilon > 0$ be given and use our hypothesis to choose $N > p$ such that

$$|1 - \Pi_s \cdot \Pi_{r-1}^{-1}| = \left| 1 - \prod_{n=r}^{s} u_n \right| < \frac{2\epsilon}{3}$$

whenever $s \geq r \geq N$. Now multiply this inequality by $|\Pi_{r-1}|$ and use (3) to obtain

$$|\Pi_{r-1} - \Pi_s| < \frac{3}{2} \cdot \frac{2}{3} \epsilon = \epsilon.$$

Therefore, the sequence (4) is a Cauchy sequence, so it converges. □

(7.25) Corollary *If* $\displaystyle\prod_{n=1}^{\infty} u_n$ *converges, then* $u_n \to 1$ *as* $n \to \infty$.

Proof Take $r = s$ in (7.24.i). □

(7.26) Remarks (a) The converse of (7.25) is false, as is seen by Examples (7.23.a) and (7.23.c).

(b) In view of (7.24) we see that the convergence of an infinite product cannot be affected by changing a finite number of factors.

(c) With an eye to comparing products and series we often write the factors of a product in the form $u_n = 1 + a_n$. Then (7.25) takes the form $a_n \to 0$ if $\displaystyle\prod_{n=1}^{\infty} (1 + a_n)$ converges. When this is done we call a_n the nth *term* of the product.

Unsurprisingly, products are related to series via the logarithm, as we now prove.

(7.27)   Theorem   *Let* $(a_n)_{n=1}^{\infty} \subset C$. *Then* $\prod_{n=1}^{\infty} (1 + a_n)$ *converges if and only if for some* $p \in N$ *the series* $\sum_{n=p}^{\infty} \text{Log}(1 + a_n)$ *converges.** *[It is implicit that* $1 + a_n \neq 0$ *for* $n \geq p.$*] Moreover, if the sum of this series is* $S \in C$, *then*

$$\prod_{n=1}^{\infty} (1 + a_n) = e^S \cdot \prod_{n=1}^{p-1} (1 + a_n).$$

Proof   Suppose the series converges to the sum S. Since exp is continuous at S, it follows that

$$0 \neq e^S = \lim_{q \to \infty} \exp\left(\sum_{n=p}^{q} \text{Log}(1 + a_n) \right) = \lim_{q \to \infty} \prod_{n=p}^{q} (1 + a_n),$$

and so the product converges to the asserted value.

Conversely, suppose that the product converges and let $0 < \epsilon < 1$ be given. By (7.24), there is a $p \in N$ such that

$$\left| 1 - \prod_{n=r}^{s} (1 + a_n) \right| < \frac{\epsilon}{2} < \frac{1}{2} \tag{1}$$

whenever $s \geq r \geq p$. Using the Log Series (5.30), we see that

$$|\text{Log}(1 - z)| \leq \sum_{n=1}^{\infty} \frac{|z|^n}{n} \leq \sum_{n=1}^{\infty} |z|^n = \frac{|z|}{1 - |z|} \leq 2|z| \tag{2}$$

whenever $|z| < 1/2$. Now we combine (1) and (2) to get

$$\left| \text{Log} \prod_{n=r}^{s} (1 + a_n) \right| \leq \epsilon \tag{3}$$

whenever $s \geq r \geq p$. In particular,

$$|\text{Log}(1 + a_n)| \leq \epsilon \quad \text{for all } n \geq p. \tag{4}$$

We complete the proof by showing that

$$\left| \sum_{n=r}^{s} \text{Log}(1 + a_n) \right| \leq \epsilon \tag{5}$$

for $s \geq r \geq p$ and invoking Cauchy's Criterion for series. [It is not clear that (5) and (3) are equivalent, since $\text{Log}(zw)$ can differ from $\text{Log } z + \text{Log } w$.] To this end, fix r and proceed by induction on s. For $s = r$, (5) follows from (4). Now suppose that (5) holds for some fixed $s \geq r$. We can choose an integer k such that

$$\sum_{n=r}^{s+1} \text{Log}(1 + a_n) = \text{Log} \prod_{n=r}^{s+1} (1 + a_n) + 2k\pi i \tag{6}$$

**Here, as usual, Log denotes the principal branch of the complex logarithm [see (5.20)].

[Why?]. In view of (3), it suffices to show that $k = 0$ in order to obtain (5) for $s + 1$. But it follows from (6), (3), (4), and (5) for s that

$$2\pi|k| = |2k\pi i| \leq \left|\text{Log} \prod_{n=r}^{s+1} (1 + a_n)\right| + |\text{Log}(1 + a_{s+1})| + \left|\sum_{n=r}^{s} \text{Log}(1 + a_n)\right|$$

$$\leq \epsilon + \epsilon + \epsilon = 3\epsilon < 3 < 2\pi,$$

and so $k = 0$. □

For products having real terms of constant sign, the comparison with series takes a much simpler form, as we now see.

(7.28)   Theorem   *Let* $(a_n)_{n=1}^\infty \subset [0, \infty[$. *Then*

$$\prod_{n=1}^\infty (1 + a_n), \quad \prod_{n=1}^\infty (1 - a_n), \quad \sum_{n=1}^\infty a_n$$

either all three converge or all three diverge.

Proof   We may suppose that $a_n \to 0$ as $n \to \infty$, for otherwise all three must diverge. Summands 0 or factors 1 cannot affect convergence and neither can a finite number of changes of summands or factors; hence, we also suppose that $0 < a_n < 1$ for all n. Since

$$\lim_{x \downarrow 0} \frac{\log(1 + x)}{x} = \lim_{x \downarrow 0} \frac{-\log(1 - x)}{x} = \lim_{x \downarrow 0} \frac{-\log(1 - x)}{\log(1 + x)} = 1,$$

it follows from the Limit Comparison Test (7.6) that the three series

$$\sum_{n=1}^\infty \log(1 + a_n), \quad -\sum_{n=1}^\infty \log(1 - a_n), \quad \sum_{n=1}^\infty a_n$$

either all converge or all diverge. Therefore the proof is completed by invoking the preceding theorem. □

(7.29)   Examples   The hypothesis that $a_n \geq 0$ in (7.28) is crucial as the following two examples show.

(a) Let $a_n = (-1)^{n-1}/\sqrt{n}$ for $n \in \mathbb{N}$. Then $\sum a_n$ converges by Leibnitz's Test, but $\prod(1 + a_n)$ diverges. In fact, an easy computation shows that

$$(1 + a_{2k-1})(1 + a_{2k}) = \left(1 + \frac{1}{\sqrt{2k-1}}\right)\left(1 - \frac{1}{\sqrt{2k}}\right) = 1 - b_k$$

where $0 < b_k < 1$ and $kb_k \to 1/2$ as $k \to \infty$. Hence, (7.6) shows that $\sum b_k = \infty$, and so, by (7.28), $\prod(1 - b_k) = 0$. This proves that

$$\prod_{n=1}^{2m} (1 + a_n) = \prod_{k=1}^{m} (1 - b_k) \to 0$$

as $m \to \infty$ and

$$\prod_{n=1}^{2m+1} (1 + a_n) \to 0$$

as $m \to \infty$.

(b) Consider the product

$$\prod_{n=1}^{\infty} (1 + a_n) = \left(1 - \frac{1}{\sqrt{2}} \right)\left(1 + \frac{1}{\sqrt{2}} + \frac{1}{2} \right) \cdot \left(1 - \frac{1}{\sqrt{3}} \right)\left(1 + \frac{1}{\sqrt{3}} + \frac{1}{3} \right) \cdot \ldots ,$$

in which

$$a_{2n-1} = -1/\sqrt{n+1}, \qquad a_{2n} = 1/\sqrt{n+1} + 1/(n+1).$$

Plainly $a_n \to 0$,

$$a_{2n-1} + a_{2n} = 1/(n+1),$$
$$(1 + a_{2n-1})(1 + a_{2n}) = 1 - 1/(n+1)^{3/2}.$$

Thus $\sum a_n = \infty$ and $\prod(1 + a_n)$ converges. We leave the details as an exercise.

Despite the negative results of these examples, we do have the following useful fact.

(7.30)   Theorem   *Suppose that $(a_n)_{n=1}^{\infty} \subset \mathbb{C}$ satisfies*

(i) $$\sum_{n=1}^{\infty} |a_n|^2 < \infty.$$

Then $\displaystyle\sum_{n=1}^{\infty} a_n$ and $\displaystyle\prod_{n=1}^{\infty} (1 + a_n)$ either both converge or both diverge.

Proof   Choose $N \in \mathbb{N}$ so that $|a_n| < 1$ for $n \geq N$. By (5.30), we have

$$\text{Log}(1 + z) = z + z^2\left[-1/2 + z/3 - z^2/4 + - \ldots \right] \tag{1}$$

for $|z| < 1$. The power series in brackets represents a function continuous at $z = 0$ so, since $a_n \to 0$, the sequence $(b_n)_{n=N}^{\infty}$ defined by

$$b_n = \sum_{k=0}^{\infty} \frac{(-1)^{k+1} a_n^k}{k+2} \tag{2}$$

converges to $-1/2$ and is therefore bounded. This fact, (2.42), and (i) imply that

$$\sum_{n=N}^{\infty} a_n^2 b_n \quad \text{converges} \tag{3}$$

(in fact, absolutely). From (1) and (2) we have

$$\text{Log}(1 + a_n) = a_n + a_n^2 b_n \tag{4}$$

for $n \geqq N$. It follows from (3) and (4) that the two series

$$\sum_{n=N}^{\infty} \text{Log}(1 + a_n) \quad \text{and} \quad \sum_{n=N}^{\infty} a_n$$

either both converge or both diverge. This and (7.27) completes the proof. $\square$

(7.31) **Remark** With the notation and hypothesis of (7.30), write

$$s_k = \sum_{n=N}^{k} a_n, \qquad p_k = \prod_{n=N}^{k} (1 + a_n).$$

Using (7.30.4), we obtain

$$p_k e^{-s_k} = \exp\left(\sum_{n=N}^{k} a_n^2 b_n\right) \rightarrow e^c \quad \text{as } k \rightarrow \infty,$$

where

$$c = \sum_{n=N}^{\infty} a_n^2 b_n.$$

Just as for series there is a notion of absolute convergence for products which we define in the following theorem.

(7.32) **Theorem** If $(a_n)_{n=1}^{\infty} \subset \mathbb{C}$, then the following three assertions are equivalent:

(i) $$\sum_{n=1}^{\infty} |a_n| < \infty,$$

(ii) $$\prod_{n=1}^{\infty} (1 + |a_n|) < \infty,$$

(iii) $$\sum_{n=p}^{\infty} |\text{Log}(1 + a_n)| < \infty$$

for some $p \in \mathbb{N}$ $[a_n \neq -1$ for $n \geqq p]$.

If these three assertions obtain, we say that $\prod_{n=1}^{\infty} (1 + a_n)$ is *absolutely convergent*.

Proof The equivalence of (i) and (ii) follows from (7.28). If any of the three obtain, then $a_n \rightarrow 0$ $[\text{Log}(1 + a_n) \rightarrow 0$ implies $1 + a_n = \exp[\text{Log}(1 + a_n)] \rightarrow e^0 = 1]$, and so we may suppose that p is so large that $|a_n| < 1/2$ for $n \geqq p$. We may also suppose that $a_n \neq 0$ for all n. Then, using (5.30) again, we have

$$|a_n^{-1} \text{Log}(1 + a_n) - 1| = |-a_n/2 + a_n^2/3 - a_n^3/4 + \ldots|$$
$$< 1/2^2 + 1/2^3 + 1/2^4 + \ldots = 1/2,$$

and so

$$1/2 < \left| \left[\mathrm{Log}\,(1 + a_n) \right]/a_n \right| < 3/2$$

for $n \geq p$. It follows from (7.6) that (i) is equivalent to (iii).   □

(7.33)   Corollary   *If $\prod(1 + a_n)$ is absolutely convergent, then it is convergent.*

Proof   Apply (7.32) and (7.30).   □

(7.34)   Example   If $z \in \mathbb{C}\backslash\{0\}$, then the infinite product

$$\prod_{n=1}^{\infty} \frac{\sin(z/n)}{z/n}$$

is absolutely convergent. This is because, using the power series for sin, we can write

$$\frac{\sin(z/n)}{z/n} = 1 - b_n/n^2,$$

where $(b_n)_{n=1}^{\infty}$ is a bounded sequence; hence $\sum |b_n/n^2| < \infty$.

Exercises

1.   By computing partial products, prove that

(a) $\displaystyle\prod_{n=0}^{\infty} (1 + z^{2^n}) = \frac{1}{1 - z}$ $(|z| < 1)$,

(b) $\displaystyle\prod_{n=2}^{\infty} \left(1 - \frac{1}{n^2} \right) = \frac{1}{2}$,

(c) $\displaystyle\prod_{n=2}^{\infty} \left(1 - \frac{2}{n(n + 1)} \right) = \frac{1}{3}$,

(d) $\displaystyle\prod_{n=2}^{\infty} \frac{n^3 - 1}{n^3 + 1} = \frac{2}{3}$,

(e) $\displaystyle\prod_{n=2}^{\infty} \left(1 + \frac{2n + 1}{(n^2 - 1)(n + 1)^2} \right) = \frac{4}{3}$.

[Hints: In (a), multiply by $(1 - z)$. In the rest, express the nth factor as a single fraction, factor, write out the partial product, and look for cancellation.]

2.   Suppose that $\sum a_n$ converges with a nonzero sum and that no partial sum s_n is zero. Prove that

$$\sum_{n=1}^{\infty} a_n = a_1 \cdot \prod_{n=2}^{\infty} \left(1 + \frac{a_n}{s_{n-1}} \right).$$

[Hint: $s_k = s_1 \cdot \displaystyle\prod_{n=2}^{k} (s_n/s_{n-1})$.]

3. Use Exercise 2 to evaluate

(a) $\displaystyle\prod_{n=2}^{\infty}\left(1+\frac{1}{2^n-2}\right)$,

(b) $\displaystyle\prod_{n=2}^{\infty}\left(1+\frac{1}{n^2-1}\right)$.

[Hint: $a_n = 2^{-n}$; $a_n = n^{-1}(n+1)^{-1}$.]

4. Prove that $\displaystyle\prod_{n=1}^{\infty}(1+i/n)$ diverges and that $\displaystyle\prod_{n=1}^{\infty}|1+i/n|$ converges.

5. Find all $z \in \mathbb{C}$ for which the product

$$\prod_{n=1}^{\infty}\frac{n^2z-1}{n^2z+1}$$

converges. Is the convergence absolute?

6. Let p_n denote the nth prime number ($p_1 = 2$, $p_2 = 3$, $p_3 = 5$, $p_4 = 7$, $p_5 = 11, \ldots$).
Then

(a) $\displaystyle\prod_{n=1}^{\infty}\frac{1}{1-p_n^{-1}} = \infty$,

(b) $\displaystyle\prod_{n=1}^{\infty}\left(1-\frac{1}{p_n}\right) = 0$,

(c) $\displaystyle\sum_{n=1}^{\infty}\frac{1}{p_n} = \infty$.

[Hint: For (a), write $\prod_k = \displaystyle\prod_{n=1}^{k}$, expand each factor as a geometric series, multiply

these k series, and prove that $\prod_k \geq \displaystyle\sum_{m=1}^{k}\frac{1}{m}$.]

7. If $(a_n)_{n=1}^{\infty} \subset \mathbb{C}$ and $\displaystyle\sum_{n=1}^{\infty}|a_n|^2 < \infty$, then $\displaystyle\prod_{n=1}^{\infty}\cos a_n$ converges absolutely. [Hint:
Write $\cos z = 1 - 2\sin^2(z/2)$ and use (7.6).]

8. (a) If $z \in \mathbb{C}\backslash\{0\}$, then

$$\prod_{n-1}^{\infty}\cos\frac{z}{2^n} = \frac{\sin z}{z}.$$

(b) $\displaystyle\prod_{n=1}^{\infty}\cos\frac{\pi}{2^{n+1}} = \frac{2}{\pi}$.

(c) $\displaystyle\frac{2}{\pi} = \sqrt{\frac{1}{2}} \cdot \sqrt{\frac{1}{2}+\frac{1}{2}\sqrt{\frac{1}{2}}} \cdot \sqrt{\frac{1}{2}+\frac{1}{2}\sqrt{\frac{1}{2}+\frac{1}{2}\sqrt{\frac{1}{2}}}} \cdot \ldots$, that is,

$2/\pi = \displaystyle\prod_{n=1}^{\infty}u_n$ where $u_1 = \sqrt{1/2}$ and $u_{n+1} = \sqrt{(1+u_n)/2}$.

[Hints: For (a), use $2\cos w \sin w = \sin 2w$ to prove that $\prod_k \cdot \sin 2^{-k}, z = 2^{-k}\sin z$, where $\prod_k$ is the product of the first k factors. For (c), use (b) and $2\cos^2 w = 1 + \cos 2w$.]

Remark: The product in (c) is perhaps the first published infinite product. It appears in a 1646 book by F. Vieta.

9. Suppose that $(a_n)_{n=1}^\infty \subset \mathbb{R}$ and that $\sum a_n$ converges. Then $\prod(1 + a_n)$ converges if and only if $\sum a_n^2 < \infty$. [Hint: Use the ideas in the proof of (7.30) particularly noting (7.30.4).]

10. Which of the following products converge? Why?

(a) $\displaystyle\prod_{n=1}^\infty \left(1 + \frac{(-1)^n}{\sqrt{n}}\right)$

(b) $\displaystyle\prod_{n=2}^\infty \left(1 + \frac{(-1)^n}{\log n}\right)$,

(c) $\displaystyle\prod_{n=1}^\infty \frac{a+n}{b+n}$ $(a, b \in \mathbb{C})$,

(d) $\displaystyle\prod_{n=1}^\infty \sin(n\pi)$,

(e) $\displaystyle\prod_{n=1}^\infty \left(1 + \frac{(i)^n}{n}\right)$.

11. Let $0 \le a_n < 1$ for all $n \in \mathbb{N}$. For $k \in \mathbb{N}$, write

$$s_k = \sum_{n=1}^k a_n, \quad p_k = \prod_{n=1}^k (1 + a_n), \quad q_k = \prod_{n=1}^k (1 - a_n).$$

Use induction on k to prove the inequalities

(a) $p_k \ge 1 + s_k$,   (b) $q_k \ge 1 - s_k$,   (c) $p_k < 1/q_k$ for $k \ge 1$.

12. Use the preceding exercise to give an elementary proof of (7.28), that is, a proof that uses neither logarithms nor exponentials. [Hint: Supposing $\sum a_n < \infty$, you may wish to choose N so large that $\displaystyle\sum_{n=N}^\infty a_n < 1$.]

Some Theorems of Abel

A series $\sum a_n$ that converges but does not converge absolutely is said to *converge conditionally*. In this chapter and elsewhere, we have developed many tests for absolute convergence since this is just a question about series of nonnegative terms. However, to this point in the book, our only effective test for convergence of series of arbitrary terms is Cauchy's Criterion which, being both necessary and sufficient, is often difficult to apply. Then, too, our only general test for uniform convergence, other than the uniform Cauchy condition, is Weierstrass' M-test. In the present section, we partially rectify both of these matters by setting forth several tests that flow from the following simple, but ingenious, formula given by the Norwegian N. Abel in 1826. It is analogous to integration by parts.

(7.35)   **Summation by Parts**   *Let $(a_n)_{n=0}^\infty$ and $(b_n)_{n=0}^\infty$ be sequences in $\mathbb{C}$ and write*

(i) $$A_n = \sum_{k=0}^{n} a_k \quad \text{for } n \geqq 0, \quad A_{-1} = 0.$$

Then, for any integers $q > p \geqq -1$, we have

(ii) $$\sum_{n=p+1}^{q} a_n b_n = \sum_{n=p+1}^{q} A_n (b_n - b_{n+1}) + A_q b_{q+1} - A_p b_{p+1}.$$

Proof   For $n \geqq 0$, we have

$$a_n b_n = (A_n - A_{n-1}) b_n = A_n (b_n - b_{n+1}) + A_n b_{n+1} - A_{n-1} b_n.$$

Summing these equalities for $n = p + 1, p + 2, \ldots, q$ and taking account of a telescoping sum, we obtain (ii). $\square$

The following theorem, which is really four theorems, is variously attributed to Abel, Dirichlet, du Bois–Reymond, and Dedekind. Since it is essentially all due to Abel, we choose its present name.

(7.36)   **Abel's Tests**   *Let X be a nonvoid set and let $(a_n)_{n=0}^\infty$ and $(b_n)_{n=0}^\infty$ be sequences of bounded complex-valued functions defined on X. Define A_n as in (7.35.i). Then $\sum_{n=0}^{\infty} a_n b_n$ converges uniformly on X if any of the following four sets of hypotheses is satisfied.*

   (i)   *$(A_n)_{n=0}^\infty$ is uniformly bounded on X, $\sum_{n=0}^{\infty} |b_n - b_{n+1}|$ converges uniformly on X, $b_n \to 0$ uniformly on X.*
   (ii)   *$(A_n)_{n=0}^\infty$ is uniformly bounded on X, $(b_n)_{n=0}^\infty$ is a monotone sequence,* $b_n \to 0$ uniformly on X.*

   (iii)   *$(A_n)_{n=0}^\infty$ converges uniformly on X, $\left(\sum_{n=0}^{k} |b_n - b_{n+1}| \right)_{k=0}^\infty$ is uniformly bounded on X, $(b_n)_{n=0}^\infty$ is uniformly bounded on X.*
   (iv)   *$(A_n)_{n=0}^\infty$ converges uniformly on X, $(b_n)_{n=0}^\infty$ is a monotone sequence, $(b_n)_{n=0}^\infty$ is uniformly bounded on X.*

Proof   Regardless of which of the four alternative sets of hypotheses is given, there exist $\alpha, \beta \in \mathbb{R}$ such that

$$|A_n(x)| \leqq \alpha, \qquad |b_n(x)| \leqq \beta$$

for all $x \in X$ and $n \in \mathbb{N}$. Writing, as usual, $\| \cdot \|_u$ for the uniform norm over

---

*The hypothesis of monotoneness implicitly entails that the functions b_n are real-valued.

X, it follows from (7.35.ii) that

$$\left| \sum_{n=p+1}^{q} a_n(x)b_n(x) \right| \leq \sum_{n=p+1}^{q} |A_n(x)| \cdot |b_n(x) - b_{n+1}(x)|$$

$$+ |A_q(x)| \cdot |b_{q+1}(x)| + |A_p(x)| \cdot |b_{p+1}(x)|$$

$$\leq \alpha \left\| \sum_{n=p+1}^{q} |b_n - b_{n+1}| \right\|_u + \alpha(\|b_{q+1}\|_u + \|b_{p+1}\|_u) \quad (1)$$

for all $x \in X$ and $q > p > 0$. If (i) obtains, then each term on the right side of (1) has limit 0 as $p, q \to \infty$, and so the uniform Cauchy Criterion (3.105) assures that $\sum a_n b_n$ converges uniformly on X.

Supposing that (ii) obtains, we see that each function $b_n - b_{n+1}$ is of the same constant sign [each always nonnegative or each always nonpositive], and so

$$\sum_{n=p+1}^{q} |b_n - b_{n+1}| = \left| \sum_{n=p+1}^{q} (b_n - b_{n+1}) \right| = |b_{p+1} - b_{q+1}|$$

$$\leq \|b_{p+1}\|_u + \|b_{q+1}\|_u.$$

Then (1) yields

$$\left| \sum_{n=p+1}^{q} a_n(x)b_n(x) \right| \leq 2\alpha(\|b_{p+1}\|_u + \|b_{q+1}\|_u) \to 0 \quad \text{as } p, q \to \infty;$$

whence, again $\sum a_n b_n$ converges uniformly on X.

Suppose next that (iii) or (iv) obtains and let A be the uniform limit of $(A_n)_{n=1}^{\infty}$. Plainly $\|A\|_u \leq \alpha$ and

$$0 = \sum_{n=p+1}^{q} A(b_n - b_{n+1}) + Ab_{q+1} - Ab_{p+1}.$$

Subtracting this equality from (7.35.ii), we have

$$\sum_{n=p+1}^{q} a_n b_n = \sum_{n=p+1}^{q} (A_n - A)(b_n - b_{n+1})$$

$$+ (A_q - A)b_{q+1} - (A_p - A)b_{p+1}. \quad (2)$$

Writing $\delta_p = \sup_{n \geq p} \|A_n - A\|_u$, we have $\delta_p \to 0$ as $p \to \infty$.

Suppose that (iii) obtains. Using (2), we have

$$\left| \sum_{n=p+1}^{q} a_n b_n \right| \leq \sum_{n=p+1}^{q} \delta_p |b_n - b_{n+1}| + \delta_p \beta + \delta_p \beta \leq (M + 2\beta)\delta_p \to 0$$

as $q > p \to \infty$, where M is a uniform upper bound for the partial sums of $\sum |b_n - b_{n+1}|$. Thus, $\sum a_n b_n$ converges uniformly on X.

If (iv) obtains, then

$$\sum_{n=0}^{k} |b_n - b_{n+1}| = \left| \sum_{n=0}^{k} (b_n - b_{n+1}) \right| = |b_0 - b_{k+1}| \leq 2\beta$$

uniformly on X for all k, and so (iii) obtains. Thus, we need only apply the preceding paragraph. $\square$

(7.37) **Remarks** (a) Though we have phrased (7.36) in terms of two sequences of functions, one or both of these sequences may consist entirely of constants. Thus (7.36) applies to series of the forms $\sum c_n f_n$ and $\sum c_n d_n$ where $c_n, d_n \in \mathbb{C}$ and the f_n's are functions.

(b) Any series $\sum u_n$ can be expressed in the form $\sum a_n b_n$ in many different ways because each u_n can be factored in many ways. Thus (7.36) may potentially apply to *any* series. The success of its application depends upon judicious factorizations.

(7.38) **Corollary** *Let $(b_n)_{n=0}^{\infty} \subset \mathbb{R}$ be a nonincreasing sequence with limit 0. Then, for each $0 < \delta < 2$, the series*

(i)
$$\sum_{n=0}^{\infty} b_n z^n$$

converges uniformly on $X_\delta = \{z \in \mathbb{C} : |z| \leq 1, |z - 1| \geq \delta \}$. Moreover, for each $0 < \delta < \pi$, the series

(ii)
$$\sum_{n=0}^{\infty} b_n \cos(n\theta) \quad and \quad \sum_{n=1}^{\infty} b_n \sin(n\theta)$$

both converge uniformly on $[\delta, 2\pi - \delta]$, and the series

(iii)
$$\sum_{n=0}^{\infty} (-1)^n b_n \cos(n\theta) \quad and \quad \sum_{n=1}^{\infty} (-1)^{n+1} b_n \sin(n\theta)$$

both converge uniformly on $[-\pi + \delta, \pi - \delta]$. Also

(iv)
$$\left| \sum_{n=p}^{q} b_n e^{in\theta} \right| \leq \frac{2b_p}{\sin(\theta/2)}$$

whenever $q \geq p \geq 0$ and $0 < \theta < 2\pi$.

Proof The first assertion follows from (7.36.ii), for if we let $a_n(z) = z^n$, then

$$|A_n(z)| = \left| \sum_{k=0}^{n} z^k \right| = \left| \frac{z^{n+1} - 1}{z - 1} \right| \leq \frac{2}{|z - 1|} \leq \frac{2}{\delta} \tag{1}$$

for $n \geq 0$ and $z \in X_\delta$. If $z = e^{i\theta}$, then $|z - 1| = |e^{i\theta/2}| \cdot |e^{i\theta/2} - e^{-i\theta/2}|$

$= 2|\sin(\theta/2)|$, and so (iv) follows from (1), (7.35), and the estimates

$$\left| \sum_{n=p}^{q} b_n e^{in\theta} \right| \leq \sum_{n=p}^{q} |A_n(e^{i\theta})|(b_n - b_{n+1}) + |A_q(e^{i\theta})|b_{q+1}$$

$$+ |A_{p-1}(e^{i\theta})|b_p$$

$$\leq \frac{1}{\sin(\theta/2)}\left(\sum_{n=p}^{q}(b_n - b_{n+1}) + b_{q+1} + b_p \right) = \frac{2b_p}{\sin(\theta/2)}.$$

Again putting $z = e^{i\theta}$, the series (ii) are the real and imaginary parts of the series (i) and $\theta \in [\delta, 2\pi - \delta]$ implies $|z - 1| \geq 2\sin(\delta/2)$, so the series (ii) converge as asserted. Replacing θ in (ii) by $\pi - \theta$, we obtain the series (iii) and the interval $[\delta, 2\pi - \delta]$ gets replaced by $[-\pi + \delta, \pi - \delta]$. $\square$

(7.39)   Example   In (7.38), take $b_n = 1/n$ for $n \geq 1$. By (5.30), we have

$$\sum_{n=1}^{\infty} b_n z^n = \sum_{n=1}^{\infty} \frac{z^n}{n} = -\mathrm{Log}(1-z) \tag{1}$$

for $|z| < 1$. The right side of (1) is continuous on every X_δ and, by (7.38.i), so is the left side. Thus equality (1) obtains on every X_δ, and so it therefore obtains whenever $|z| \leq 1$, $z \neq 1$. In particular, for $z = -1$, we obtain

$$\sum_{n=1}^{\infty} \frac{(-1)^{n-1}}{n} = \log 2. \tag{2}$$

It is interesting to put $z = e^{i\theta}$ in (1) and to then find the sums of the trigonometric series in (7.38.ii). Recall (5.20) that

$$-\mathrm{Log}(1 - e^{i\theta}) = -\log|1 - e^{i\theta}| - i\,\mathrm{Arg}(1 - e^{i\theta}). \tag{3}$$

Now, multiplying by $|e^{-i\theta/2}| = 1$, we have

$$|1 - e^{i\theta}| = |e^{i\theta/2} - e^{-i\theta/2}| = 2|\sin(\theta/2)|.$$

Therefore

$$\sum_{n=1}^{\infty} \frac{\cos(n\theta)}{n} = -\log\left|2\sin\frac{\theta}{2}\right| \tag{4}$$

for $0 < \theta < 2\pi$. On the other hand, if $0 < \theta < 2\pi$, then $\sin(\theta/2) > 0$ and $1 - e^{i\theta} = (2\sin(\theta/2))e^{i(\theta-\pi)/2}$. Thus $-\mathrm{Arg}(1 - e^{i\theta}) = (\pi - \theta)/2$, and so

$$\sum_{n=1}^{\infty} \frac{\sin(n\theta)}{n} = \frac{\pi - \theta}{2} \tag{5}$$

for $0 < \theta < 2\pi$. Taking $\theta = \pi/2$, (5) yields the Leibnitz–Gregory Series

$$\pi/4 = 1 - 1/3 + 1/5 - 1/7 + - \dots . \tag{6}$$

As (7.38.ii) assures, the convergence in both (4) and (5) is uniform on every interval $[\delta, 2\pi - \delta]$ $(0 < \delta < \pi)$.

We can now give a simple proof of the following important result.

(7.40)   Abel's Limit Theorem   *Let $(a_n)_{n=0}^{\infty} \subset \mathbb{C}$ and suppose that $\sum_{n=0}^{\infty} a_n$ converges.*

Then the series

$$f(r) = \sum_{n=0}^{\infty} a_n r^n$$

converges uniformly on $[0, 1]$, *and so f is continuous on* $[0, 1]$. *In particular,*

(i) $$\lim_{r \uparrow 1} \sum_{n=0}^{\infty} a_n r^n = \lim_{r \uparrow 1} f(r) = f(1) = \sum_{n=0}^{\infty} a_n.$$

Proof   This follows at once from (7.36.iv) if we take $X = [0, 1]$ and $b_n(r) = r^n$ for $r \in X$. $\square$

(7.41)   Remarks   (a) Suppose that $f(z) = \sum_{n=0}^{\infty} c_n z^n$ is a power series with radius of convergence R, $0 < R < \infty$, where f is defined at those z for which the series converges. We know, of course (3.109), that f is continuous on the open disk $D = \{z \in \mathbb{C} : |z| < R\}$. It can happen, however, that $f(z_0)$ is defined for some $|z_0| = R$ and that f is *not* continuous on $D \cup \{z_0\}$ (see Exercise 13). It does follow from (7.40) that if $z_0 = R e^{i\theta_0}$ ($\theta_0 \in \mathbb{R}$) and the series converges at z_0, then

$$\lim_{s \uparrow R} \sum_{n=0}^{\infty} c_n (s e^{i\theta_0})^n = \sum_{n=0}^{\infty} c_n z_0^n$$

[we need only take $a_n = c_n z_0^n$ and $r = s/R$ is (7.40)]; hence, the restriction of f to the closed radius $\{r z_0 : 0 \leq r \leq 1\}$ is continuous. For larger subsets of $D \cup \{z_0\}$ on which f is surely continuous, see Exercise 12.

   (b) The limit (7.40.i) may very well exist even if the series is not convergent. For instance, let $a_n = (-1)^n$. Then

$$\lim_{r \uparrow 1} \sum_{n=0}^{\infty} a_n r^n = \lim_{r \uparrow 1} \frac{1}{1+r} = \frac{1}{2}.$$

   Any series $\sum a_n$ such that $\sum a_n z^n$ has radius of convergence $R \geq 1$ $\left(\overline{\lim}^n \sqrt{|a_n|} \leq 1 \right)$ for which

$$\lim_{r \uparrow 1} \sum_{n=0}^{\infty} a_n r^n = s$$

for some $s \in \mathbb{C}$ is said to be *Abel summable* to the *Abel sum s*. We write $A\text{-} \sum_{n=0}^{\infty} a_n$ $= s$. In this terminology, (7.40) says that every convergent series is Abel summable and that its Abel sum is its ordinary sum. The reader may check that

$$A\text{-} \sum_{n=0}^{\infty} (-1)^n n = -\frac{1}{4}.$$

We postpone further discussion of summability until later in this chapter.

Exercises

1. Suppose that $(b_n)_{n=0}^\infty \subset [0, \infty[$ is a nonincreasing sequence such that $\sum b_n < \infty$. Then

(a) $nb_n \to 0$ as $n \to \infty$,

(b) $\sum n(b_n - b_{n+1}) < \infty$,

(c) if $b_n + b_{n+2} \geq 2b_{n+1}$ for all n [such a sequence is called *convex*], then
$$n^2(b_n - b_{n+1}) \to 0 \text{ as } n \to \infty.$$

2. Suppose that $(b_n)_{n=0}^\infty \subset [0, \infty[$ is a nonincreasing sequence, $b_n \to 0$, and $\sum b_n = \infty$. Then $\sum b_n \cos(n\theta)$ and $\sum b_n \sin(n\theta)$ are conditionally [*not* absolutely] convergent for $\theta \neq k\pi$ $(k \in Z)$. [Hint: If $\sum b_n |\cos(n\theta)| < \infty$, then $\frac{1}{2}\sum b_n (1 + \cos(2n\theta)) = \sum b_n \cos^2(n\theta) < \infty$.]

3. We have, for $-\pi < \theta < \pi$,
$$\sum_{n=1}^\infty (-1)^n \frac{\cos(n\theta)}{n} = -\log\left(2\cos\frac{\theta}{2}\right)$$

and

$$\sum_{n=1}^\infty (-1)^{n+1} \frac{\sin(n\theta)}{n} = \frac{\theta}{2},$$

the convergence being uniform on $[-\pi + \delta, \pi - \delta]$ for any $0 < \delta < \pi$.

4. (a) $\sum_{n=1}^\infty (-1)^{n+1} \frac{1 - \cos(nx)}{n^2} = \frac{x^2}{4}$ for $-\pi \leq x \leq \pi$,

(b) $\sum_{n=1}^\infty \frac{(-1)^{n+1}}{n^2} = \frac{\pi^2}{12}$,

(c) $\sum_{n=1}^\infty (-1)^n \frac{\cos(nx)}{n^2} = \frac{3x^2 - \pi^2}{12}$ for $-\pi \leq x \leq \pi$,

(d) $\sum_{n=1}^\infty \frac{\cos(nx)}{n^2} = \frac{3x^2 - 6\pi x + 2\pi^2}{12}$ for $0 \leq x \leq 2\pi$,

(e) $\sum_{n=1}^\infty \frac{1}{n^2} = \frac{\pi^2}{6}$,

(f) $\sum_{n=1}^\infty \frac{1}{(2n - 1)^2} = \frac{\pi^2}{8}$.

[Hint: For $0 < x < \pi$, integrate the second equality of the preceding exercise over $[0, x]$ to obtain (a) for $-\pi < x < \pi$. Then note that both sides of (a) are continuous on $[-\pi, \pi]$. For (b), integrate (a) over $[-\pi, \pi]$.]

5. (a) For $0 < \theta < \pi$, we have
$$\sum_{n=0}^\infty \frac{\cos(2n + 1)\theta}{2n + 1} = \frac{1}{4}\log\left(\cot^2\frac{\theta}{2}\right),$$
$$\sum_{n=0}^\infty \frac{\sin((2n + 1)\theta)}{2n + 1} = \frac{\pi}{4},$$

the convergence being uniform on $[\delta, \pi - \delta]$ whenever $0 < \delta < \pi/2$. For which other real θ are these correct? [Hint: Use (7.39) and Exercise 3.]

(b) For $0 < \theta < \pi/2$, we have

$$\sum_{n=0}^{\infty} (-1)^n \frac{\sin((2n+1)\theta)}{2n+1} = \frac{1}{2} \log(\sec\theta + \tan\theta),$$

$$\sum_{n=0}^{\infty} (-1)^n \frac{\cos((2n+1)\theta)}{2n+1} = \frac{\pi}{4},$$

the convergence being uniform on $[\delta, \pi/2 - \delta]$ whenever $0 < \delta < \pi/4$. Sum these series for all other real θ for which they converge.

(c) If $\alpha, \beta \in \mathbb{R}$ and neither $\alpha + \beta$ nor $\alpha - \beta$ is an integral multiple of 2π, then

$$\sum_{n=1}^{\infty} \frac{\cos(n\alpha)\cos(n\beta)}{n} = -\frac{1}{2} \log|2(\cos\alpha + \cos\beta)|.$$

6. Find all $z \in \mathbb{C}$ for which the power series $\displaystyle\sum_{n=1}^{\infty} \left(\sum_{k=1}^{n} \frac{1}{k} \right) \frac{z^n}{n+1}$ converges.

7. $\displaystyle\sum_{n=2}^{\infty} \frac{\log(n+1) - \log n}{(\log n)^2} < \infty.$

8. Let $(a_n)_{n=0}^{\infty}$ and $(p_n)_{n=0}^{\infty}$ be sequences of nonnegative real numbers. Then

$$\lim_{r \uparrow 1} \sum_{n=0}^{\infty} a_n r^{p_n} = \sum_{n=0}^{\infty} a_n.$$

[Hint: For $\alpha < \displaystyle\sum_{n=0}^{\infty} a_n$, choose N such that $\displaystyle\sum_{n=0}^{N} a_n > \alpha$, then choose r_0 such that $\displaystyle\sum_{n=0}^{N} a_n r_0^{p_n} > \alpha$.]

9. If $\sum n a_n$ converges $(a_n \in \mathbb{C})$, then $\sum a_n$ converges and, writing $f(r) = \sum a_n r^n$, we have

$$\lim_{r \uparrow 1} \frac{f(1) - f(r)}{1 - r} = \sum n a_n.$$

10. Let $(b_n)_{n=0}^{\infty}$ be a nonincreasing sequence of positive real numbers. Then

(a) $\sum b_n \cos(nx)$ converges uniformly on $\mathbb{R}$ if and only if $\sum b_n < \infty$,

(b) $\sum b_n \sin(nx)$ converges uniformly on $\mathbb{R}$ if and only if $n b_n \to 0$.

[Hints: If $x = \pi/(4k)$, then

$$\sum_{n=k+1}^{2k} b_n \sin(nx) > k b_{2k} \sin(\pi/4).$$

Suppose $n b_n \to 0$. Write $\delta_k = \sup\{n b_n : n \geq k\}$. For $0 < x \leq \pi$, let $m = m(x) \in \mathbb{N}$ satisfy $\pi/(m+1) < x \leq \pi/m$. Note that $|A_k(x)| = \left| \displaystyle\sum_{n=1}^{k} \sin(nx) \right| = |[\cos(x/2) - \cos(k+1/2)x]/2\sin(x/2)| \leq \pi/x$ because $\pi \sin t \geq 2t$ for $0 \leq t \leq \pi/2$. Thus, for all $k \in \mathbb{N}$, $\left| \displaystyle\sum_{n=k+1}^{k+m} b_n \sin(nx) \right| \leq x \cdot \displaystyle\sum_{n=k+1}^{k+m} n b_n \leq x m \delta_k$

$\leq \pi \delta_k,$

$$\left|\sum_{n=k+m+1}^{\infty} b_n \sin(nx)\right| = |\Sigma A_n(x)(b_n - b_{n+1}) - A_{k+m}(x)b_{k+m+1}| \le 2b_{k+m+1}\pi/x$$

$$< 2(m+1)b_{k+m+1} \le 2\delta_k.]$$

(c) The series $\displaystyle\sum_{n=2}^{\infty} \frac{\sin(nx)}{n \log n}$ converges uniformly on $\mathbb{R}$, but it converges absolutely

only if x is an integral multiple of π. [Hint: Use Exercise 2.]

(d) If $(nb_n)_{n=1}^{\infty}$ is bounded, then there exists $C \in \mathbb{R}$ such that

$$\left|\sum_{n=1}^{p} b_n \sin nx\right| \le C$$

for all $p \in \mathbb{N}$ and all $x \in \mathbb{R}$. [Hint: If $nb_n \le M < \infty$ for all n, we can make
estimates similar to those in (b) to show that $C = \pi M + 2M$ will suffice.]

11.   Let $(c_n)_{n=0}^{\infty}$ be a sequence of complex numbers.

(a) If $\displaystyle\sum_{n=0}^{\infty} c_n$ and $\displaystyle\sum_{n=0}^{\infty} c_n^2$ both converge and if $\text{Re}(c_n) \ge 0$ for all n, then $\displaystyle\sum_{n=0}^{\infty} |c_n|^2$
$< \infty$. [Hint: $|c_n|^2 = 2(\text{Re}(c_n))^2 - \text{Re}(c_n^2)$.]

(b) It can happen that for every $k \in \mathbb{N}$ we have $\displaystyle\sum_{n=0}^{\infty} c_n^k$ converges and $\displaystyle\sum_{n=0}^{\infty} |c_n|^k$
$= \infty$. [Hint: Let θ be a fixed irrational number, let $c_n = e^{2\pi i n\theta}/(\log(n+2))$, and
use (7.36.ii).]

12.   (a) This is a generalization of Abel's Limit Theorem given by Otto Stolz (1875).
Suppose $\displaystyle\sum_{n=0}^{\infty} a_n$ converges $(a_n \in \mathbb{C})$ and $1 < \alpha < \infty$. Let $X_\alpha = \{z \in \mathbb{C} : |1 - z|$
$\le \alpha(1 - |z|)\}$. Then

$$\lim_{\substack{z \to 1 \\ z \in X_\alpha}} \sum_{n=0}^{\infty} a_n z^n = \sum_{n=0}^{\infty} a_n.$$

The power series converges uniformly on X_α.

(b) Geometrically, the set X_α consists of those z's for which $1 - z$ is inside or on the
inner loop of the limaçon whose polar equation is

$$r = \beta(\alpha \cos\theta - 1)$$

where $\beta = 2\alpha/(\alpha^2 - 1)$. This loop is inside the circle $|z| = 2$, between the
vertical lines $x = 0$ and $x = \beta(\alpha - 1)$, and between its two tangent lines at $r = 0$
whose slopes are $\pm\sqrt{\alpha^2 - 1}$. Sketch a graph of this limaçon and of the set X_α.

13.   Consider the power series

$$f(z) = \frac{z^3}{1} - \frac{z^{2\cdot3}}{1} + \frac{z^{3^2}}{2} - \frac{z^{2\cdot3^2}}{2} + \ldots + \frac{z^{3^n}}{n} - \frac{z^{2\cdot3^n}}{n} + - \ldots .$$

(a) The radius of convergence is $R = 1$.

(b) There are two dense subsets A and B of the unit circle $|z| = 1$ such that the
series converges at each point of A and diverges at each point of B. [Hint:
Consider the points $z = \exp(\pi i k/3^m)$ for the cases k even and k odd.]

(c) For each $m \in \mathbf{N}$ we have

$$\lim_{r \uparrow 1} |f(re^{\pi i/3^m})| = \infty.$$

[Hint: We have $\sum_{n=m}^{\infty} [r^{3^n}/n] \to \infty$ as $r \uparrow 1$ by Exercise 8.]

(d) There exists a sequence $(z_m)_{m=1}^{\infty}$ such that $|z_m| < 1$, $z_m \to 1$, and $|f(z_m)| \to \infty$ even though $f(1) = 0$. Compare this with the preceding exercise. [The trouble is that there is no fixed α such that $z_m \in X_\alpha$ for all m.]

14. We have

$$\text{Arctan } z = \frac{1}{2i} \text{Log} \frac{1 + iz}{1 - iz} = \sum_{n=0}^{\infty} (-1)^n \frac{z^{2n+1}}{2n+1}$$

whenever $|z| \leqq 1$ and $z \neq \pm i$. Moreover, if $0 < \delta < 1$, then the series converges uniformly on

$$X_\delta = \{ z \in \mathbf{C} : |z| \leqq 1, |z - i| \geqq \delta, |z + i| \geqq \delta \}.$$

[One can separate into real and imaginary parts to obtain the series of Exercise 5(b).]

15. If $\sum_{n=1}^{\infty} a_n$ converges ($a_n \in \mathbf{C}$), then

$$\lim_{x \downarrow 0} \sum_{n=1}^{\infty} [a_n/n^x] = \sum_{n=1}^{\infty} a_n.$$

16. Use (7.40) to give Abel's original proof of (2.72): If

$$\sum_{n=0}^{\infty} a_n, \quad \sum_{n=0}^{\infty} b_n, \quad \sum_{n=0}^{\infty} c_n$$

all converge and $c_n = a_0 b_n + a_1 b_{n-1} + \ldots + a_n b_0$ for all $n \geqq 0$, then

$$\sum_{n=0}^{\infty} c_n = \left(\sum_{n=0}^{\infty} a_n \right) \left(\sum_{n=0}^{\infty} b_n \right).$$

17. *Bernoulli Polynomials, Numbers, and Functions.*

(a) There exists a unique sequence $(\phi_n)_{n=0}^{\infty}$ of polynomials on $\mathbf{R}$ such that
   (1) $\phi_0(t) = 1$ for all $t \in \mathbf{R}$,
   (2) $\phi'_{n+1}(t) = \phi_n(t)$ for all $t \in \mathbf{R}$ and $n \geqq 0$,
   (3) $\int_0^1 \phi_n(t) \, dt = 0$ for all $n \geqq 1$.

These are called the *Bernoulli polynomials*.

(b) The degree of ϕ_n is n.

(c) Each ϕ_n has rational coefficients.

(d) $\phi_1(t) = t - 1/2$,
   $\phi_2(t) = (1/2)t^2 - (1/2)t + 1/12$,
   $\phi_3(t) = (1/6)t^3 - (1/4)t^2 + (1/12)t$,
   $\phi_4(t) = (1/24)t^4 - (1/12)t^3 + (1/24)t^2 - (1/720)$.

(e) $\phi_n(t) = (-1)^n \phi_n(1 - t)$ for all $t \in \mathbf{R}$ and $n \geqq 0$.
   [Hint: Write $P_n(t) = (-1)^n \phi_n(1 - t)$ and show that the P_n's satisfy the conditions of (a).]

(f) $\phi_n(0) = \phi_n(1)$ for all $n > 1$.

(g) $\phi_n(1) = \phi_n(1/2) = \phi_n(0) = 0$ for all odd $n > 1$.

(h) $\phi_n(t + 1) - \phi_n(t) = t^{n-1}/(n-1)!$ for all $t \in \mathbb{R}$ and $n \geq 1$.

[Hint: Show that $\phi_{n+1}(t + 1) - \phi_{n+1}(t) = \int_0^t \phi_n(u + 1)\, du - \int_0^t \phi_n(u)\, du$ and use

induction.]

(i) If $n, p \in \mathbb{N}$, then

$$\sum_{k=1}^p k^n = n!\,[\phi_{n+1}(p + 1) - \phi_{n+1}(1)].$$

The numbers $B_n = n!\,\phi_n(0)$ $(n \geq 0)$ are called the *Bernoulli numbers*. They appear quite naturally in many places in analysis. We shall see a few such instances below.

(j) For $t \in \mathbb{R}$ and $n \geq 0$, we have

$$\phi_n(t) = \frac{t^n}{n!} + \frac{B_1}{1!}\frac{t^{n-1}}{(n-1)!} + \frac{B_2}{2!}\frac{t^{n-2}}{(n-2)!} + \ldots + \frac{B_n}{n!} = \frac{1}{n!}\sum_{k=0}^n \binom{n}{k} B_k t^{n-k}.$$

Thus, if we expand purely formally according to the Binomial Theorem and interpret B^k to mean B_k, we may write

$$n!\,\phi_n(t) = (t + B)^n.$$

(k) Recalling (f), we have

$$B_n = -\frac{1}{n+1}\sum_{k=0}^{n-1}\binom{n+1}{k} B_k \quad \text{for } n > 0,$$

a recursion formula from which the Bernoulli numbers can be computed successively.

(l) $B_n = 0$ for odd $n > 1$, $B_1 = -1/2$, $B_2 = 1/6$, $B_4 = -1/30$, $B_6 = 1/42$, $B_8 = -1/30$, $B_{10} = 5/66$, $B_{12} = -691/2730$, $B_{14} = 7/6$.

(m) For $0 < t < 1$, we have

$$\phi_1(t) = -\frac{1}{\pi}\sum_{k=1}^\infty \frac{\sin(2\pi kt)}{k}.$$

For $0 \leq t \leq 1$ and $m \geq 1$, we have

$$\phi_{2m}(t) = (-1)^{m-1}\frac{2}{(2\pi)^{2m}}\sum_{k=1}^\infty \frac{\cos(2\pi kt)}{k^{2m}},$$

$$\phi_{2m+1}(t) = (-1)^{m-1}\frac{2}{(2\pi)^{2m+1}}\sum_{k=1}^\infty \frac{\sin(2\pi kt)}{k^{2m+1}}.$$

[Hints: Use (7.39.5), Exercise 4(d), (a), and (d).]

The periodic functions defined by the infinite series in (m) are called the *Bernoulli functions*.

(n) For $m \geq 1$, we have

$$\sum_{k=1}^\infty \frac{1}{k^{2m}} = \frac{(-1)^{m-1}B_{2m}(2\pi)^{2m}}{2\cdot(2m)!}.\qquad *$$

(o) The Bernoulli numbers with even subscripts alternate in sign: $(-1)^{m-1}B_{2m} > 0$ for $m \geq 1$.

---

*There is no known "finite expression" for $\displaystyle\sum_{k=1}^\infty \frac{1}{k^3}$ in terms of "known" numbers.

(p) $\displaystyle\sum_{k=1}^{\infty} \frac{1}{k^2} = \frac{\pi^2}{6}$, $\displaystyle\sum_{k=1}^{\infty} \frac{1}{k^4} = \frac{\pi^4}{90}$, $\displaystyle\sum_{k=1}^{\infty} \frac{1}{k^6} = \frac{\pi^6}{945}$, $\displaystyle\sum_{k=1}^{\infty} \frac{1}{k^8} = \frac{\pi^8}{9,450}$,

$\displaystyle\sum_{k=1}^{\infty} \frac{1}{k^{10}} = \frac{\pi^{10}}{93,555}$, $\displaystyle\sum_{k=1}^{\infty} \frac{1}{k^{12}} = \frac{691\pi^{12}}{3^6 \cdot 5^3 \cdot 7 \cdot 11 \cdot 91}$,

$\displaystyle\sum_{k=1}^{\infty} \frac{1}{k^{14}} = \frac{2\pi^{14}}{3^6 \cdot 5^2 \cdot 7 \cdot 11 \cdot 13}$.

(q)
$$\sum_{k=1}^{\infty} \frac{1}{(2k-1)^{2m}} = (-1)^{m-1} \frac{2^{2m}-1}{2 \cdot (2m)!} B_{2m}\pi^{2m}.$$

For $m = 1, 2, 3$, the sums are $\pi^2/8$, $\pi^4/96$, $\pi^6/960$.

(r)
$$\sum_{k=1}^{\infty} \frac{(-1)^{k-1}}{k^{2m}} = (-1)^{m-1} \frac{2^{2m-1}-1}{(2m)!} B_{2m}\pi^{2m}.$$

For $m = 1, 2, 3$, the sums are
$$\pi^2/12, \quad 7\pi^4/720, \quad 31\pi^6/30240.$$

(s)
$$\sum_{k=1}^{P} k^n = \frac{1}{n+1}[(p+1) + B]^{n+1} - \frac{1}{n+1} B_{n+1},$$

where we use the symbolic notation as in (j).

(t) For $m \geq 1$, we have

$$\frac{2 \cdot (2m)!}{(2\pi)^{2m}} < |B_{2m}| < \frac{4 \cdot (2m)!}{(2\pi)^{2m}},$$

$$\frac{(m+1)(2m+1)}{4\pi^2} < \left| \frac{B_{2m+2}}{B_{2m}} \right| < \frac{(m+1)(2m+\cdot1)}{\pi^2}.$$

(u) For $0 \leq t \leq 1$ and $0 < |z| < 2\pi$ $(z \in \mathbb{C})$, we have

$$\frac{ze^{tz}}{e^z - 1} = \sum_{n=0}^{\infty} \phi_n(t)z^n.$$

The Bernoulli polynomials are often defined by this equality. [Hints: Fix $|z| < 2\pi$. Let $f(t)$ denote the sum on the right when the series converges. Using (m), we have $|\phi_n(t)z^n| < 4(|z|/(2\pi))^n = M_n$ for $n \geq 2$ and $0 \leq t \leq 1$. Justify term-by-term differentiation and obtain $f'(t) = zf(t)$ for $0 \leq t \leq 1$. Differentiate $e^{-tz} f(t)$ and conclude that $f(t) = c(z)e^{tz}$ for $0 \leq t \leq 1$, where $c(z)$ does not depend on t. Justify termwise integration over $[0, 1]$ and obtain

$$c(z) = z/(e^z - 1).]$$

(v) If $0 < |z| < 2\pi$, then

$$\frac{z}{e^z - 1} = 1 - \frac{1}{2}z + \sum_{m=1}^{\infty} \frac{B_{2m}}{(2m)!} z^{2m},$$

$$\frac{z}{2} \cdot \frac{e^{z/2} + e^{-z/2}}{e^{z/2} - e^{-z/2}} = \sum_{m=0}^{\infty} \frac{B_{2m}}{(2m)!} z^{2m}.$$

(w) If $0 < |z| < \pi$, then

$$z \cot z = \sum_{m=0}^{\infty} (-1)^m \frac{2^{2m} B_{2m}}{(2m)!} z^{2m} = 1 - \frac{1}{3} z^2 - \frac{1}{45} z^4 - \frac{2}{945} z^6 - \cdots.$$

(x) If $|z| < \pi/2$, then

$$\tan z = \sum_{m=1}^{\infty} (-1)^{m-1} \frac{2^{2m}(2^{2m} - 1) B_{2m}}{(2m)!} z^{2m-1}$$

$$= z + \frac{1}{3} z^3 + \frac{2}{15} z^5 + \frac{17}{315} z^7 + \cdots.$$

[Hint: $\tan z = \cot z - 2 \cot(2z)$.]

(y) If $|z| < \pi$, then

$$z \csc z = \sum_{m=0}^{\infty} (-1)^{m-1} \frac{(2^{2m} - 2) B_{2m}}{(2m)!} z^{2m}$$

$$= 1 + (1/6)z^2 + (7/360)z^4 + (31/15,120)z^6 + \cdots.$$

[Hint: $\cot z + \tan(z/2) = \csc z$.]

(z) Use (t) to compute the radius of convergence of the power series given in (v)–(y).

18. *Euler's Summation Formula* (1738). In this excercise, f is a complex-valued function on $[0, \infty[$ for which we always suppose that all of the derivatives $f^{(j)}$ ($j \geq 0$) in question are continuous on the (closed) intervals in question. For brevity, we write $f_r = f(r)$ when r is an integer. We also use the notation of the preceding exercise. We write ψ_n for the nth Bernoulli function defined in Exercise 17(m). Thus, $\psi_n(t + j) = \psi_n(t)$ for $t \in \mathbb{R}$ and $j \in \mathbb{Z}$, $\psi_1(t) = t - [t] - 1/2$ for $t \in \mathbb{R}\backslash\mathbb{Z}$, $\psi_n = \phi_n$ on $[0, 1]$ for $n > 1$, and $\psi_{n+1} = \psi_n$ for $n \geq 1$ (except at integers when $n = 1$).

(a) For $r \geq 0$ an integer, we have

$$\frac{1}{2}(f_r + f_{r+1}) = \int_r^{r+1}\left(t - r - \frac{1}{2}\right)f'(t)\,dt + \int_r^{r+1} f(t)\,dt$$

$$= \int_r^{r+1} f(t)\,dt + \int_r^{r+1} \psi_1(t) f'(t)\,dt.$$

[Hint: Integrate by parts.]

(b) For $n \in \mathbb{N}$, we have

$$\sum_{r=0}^{n} f_r = \int_0^n f(t)\,dt + \frac{1}{2}(f_0 + f_n) + \int_0^n \psi_1(t) f'(t)\,dt.$$

(c) For $n, r \in \mathbb{N}$, we have

$$\int_0^n \psi_{2r-1}(t) f^{(2r-1)}(t)\, dt = \frac{B_{2r}}{(2r)!}\left(f_n^{(2r-1)} - f_0^{(2r-1)}\right) + \int_0^n \psi_{2r+1}(t) f^{(2r+1)}(t)\, dt.$$

(d) For $n, k \in \mathbb{N}$, we have *Euler's Summation Formula*:

$$\sum_{r=0}^n f_r = \int_0^n f(t)\, dt + \frac{1}{2}(f_0 + f_n)$$

$$+ \sum_{r=1}^k \frac{B_{2r}}{(2r)!}\left(f_n^{(2r-1)} - f_0^{(2r-1)}\right) + R_{k,n}$$

where

$$R_{k,n} = -\int_0^n \psi_{2k}(t) f^{(2k)}(t)\, dt = \int_0^n \psi_{2k+1}(t) f^{(2k+1)}(t)\, dt.$$

(e) If $a < b$ in $\mathbb{R}$, $n, k \in \mathbb{N}$, $h = (b-a)/n$, and $g : [a, b] \to \mathbb{C}$ has $g^{(2k)}$ continuous on $[a, b]$, then write a formula analogous to (d) for the sum $\sum_{r=0}^n g(a + rh)$.

[Hint: Put $f(t) = g(a + th)$.]
      The remainder $R_{k,n}$ in (d) need not tend to 0 as $k \to \infty$ or $n \to \infty$. However, even then, the formula is often very useful in estimating sums or integrals. The remaining parts illustrate its usefulness.

(f) Take $f(t) = t^p$ in (d) to give a proof of Exercise 17(s).

(g) If γ is Euler's constant (7.5.a), and $k \in \mathbb{N}$, then

$$\gamma = \frac{1}{2} + \sum_{r=1}^k \frac{B_{2r}}{2r} - (2k+1)! \int_1^\infty \frac{\psi_{2k+1}(t)\, dt}{t^{2k+2}}.$$

[Hint: Take $f(t) = (1 + t)^{-1}$ and replace n by $n - 1$ in (d), then let $n \to \infty$.]

(h) Take $k = 3$ in (g) and estimate $\left|\int_4^\infty \frac{\psi_{2k+1}(t)\, dt}{t^{2k+2}}\right| (|\psi_7| \le 4(2\pi)^{-7})$ to obtain

$$\gamma = \frac{1}{2} + \frac{1}{12} - \frac{1}{120} + \frac{1}{252} - 7! \int_1^4 \frac{\psi_7(t)\, dt}{t^8} + \epsilon$$

where $|\epsilon| < 10^{-6}$. One can also use (d) again to evaluate the last integral. This crude estimation yields $.5772146 < \gamma < .5772168$.

(i) For $k \in \mathbb{N}$ and $s \in \mathbb{C}$ with $\mathrm{Re}\,(s) > 1$, we have

$$\sum_{r=1}^\infty \frac{1}{r^s} = \frac{1}{s-1} + \frac{1}{2} + \sum_{r=1}^k \frac{B_{2r}}{2r}\binom{s+2r-2}{2r-1}$$

$$- (2k+1)! \binom{s+2k}{2k+1} \int_1^\infty \frac{\psi_{2k+1}(t)\, dt}{t^{s+2k+1}}$$

where the binomial coefficients $\binom{\alpha}{j}$ are defined in (7.46). [Hint: Put $f(t) = (1 + t)^{-s}$ and replace n by $n - 1$ in (d), then let $n \to \infty$.]

(j) For $k \in \mathbb{N}$, $s \in \mathbb{C}$, and $\mathrm{Re}\,(s) > -2k$ and $s \ne 1$ let $\zeta_k(s)$ denote the right side of the equality in (i). [It follows from (i) and analytic function theory that $\zeta_k(s) = \zeta_{k+1}(s) = \ldots$. Thus we write $\zeta(s) = \zeta_k(s)$ and call ζ *Riemann's* ζ-

function]. If $-s = p \in \mathbb{N}$, then $\zeta(0) = -1/2$ and

$$\zeta(-p) = -[1/(p + 1)](1 + B)^{p+1} = -B_{p+1}/(p + 1),$$

where we use the symbolic notation of Exercise 17(j).

19.   For which $\theta \in \mathbb{R}$ does $\sum_{n=1}^{\infty} \dfrac{\sin^2 (n\theta)}{n}$ converge?

20.   Suppose $(b_n)_{n=1}^{\infty} \subset \mathbb{C}$, $(nb_n)_{n=1}^{\infty}$ converges, and $\sum_{n=1}^{\infty} n(b_n - b_{n+1})$ converges. Then $\sum_{n=1}^{\infty} b_n$ converges.

21.   (a) $\lim\limits_{x\uparrow 1} \sum\limits_{n=1}^{\infty} \dfrac{(-1)^{n-1}}{n} \cdot \dfrac{x^n}{1 + x^n} = \dfrac{\log 2}{2}$.

(b) $\lim\limits_{x\uparrow 1}(1 - x) \sum\limits_{n=1}^{\infty} (-1)^{n-1} \dfrac{x^n}{1 - x^{2n}} = \dfrac{\log 2}{2}$.

22.   [Cesàro, 1888] Let $(a_n)_{n=1}^{\infty} \subset \,]0, \infty[$ be nonincreasing and let $(\epsilon_n)_{n=1}^{\infty} \subset \{-1, 1\}$. Suppose that $\sum a_n = \infty$ and $\sum \epsilon_n a_n$ converges. Then

$$\lim_{n\to\infty} \frac{\epsilon_1 + \epsilon_2 + \ldots + \epsilon_n}{n} \le 0 \le \overline{\lim_{n\to\infty}} \frac{\epsilon_1 + \epsilon_2 + \ldots + \epsilon_n}{n}.$$

That is, about half of the ϵ's are 1 and about half are -1. [Hints: Write $E_n = \epsilon_1 + \epsilon_2 + \ldots + \epsilon_n$. Assume $E_n/n > \alpha > 0$ for $n \ge N$. Then $\epsilon_1 a_1 + \ldots + \epsilon_n a_n = E_1(a_1 - a_2) + E_2(a_2 - a_3) + \ldots + E_n(a_n - a_{n+1}) + E_n \, a_{n+1} \ge C_N + N\alpha(a_N - a_{N+1}) + \ldots + n\alpha(a_n - a_{n+1}) + n\alpha a_{n+1} = C'_N + \alpha(a_{N+1} + \ldots + a_n) \to \infty$ as $n \to \infty$.]

23.   Let $\epsilon_n = 1$ if $2^{2k-1} \le n < 2^{2k}$ and $\epsilon_n = -1$ if $2^{2k} \le n < 2^{2k+1}$ for $k \in \mathbb{N}$. Prove that $\sum_{n=2}^{\infty} \dfrac{\epsilon_n}{n}$ diverges and $\sum_{n=2}^{\infty} \dfrac{\epsilon_n}{n \log n}$ converges. [Hint: Write $B_m = 1/m + 1/(m + 1) + \ldots + 1/(2m - 1)$. Compare with an integral to show that $\log 2 < B_m < \log[(2m - 1)/(m - 1)]$ for $m > 1$. Then show that

$$0 < B_m - B_{2m} < 1/(2m - 2)$$

by using $\log(1 + x) < x$. Show that the first series has bounded partial sums that fail to converge. Use (7.36).]

Another Ratio Test
and the Binomial Series

In this section, we present a generalization to complex series of Gauss' Test. This criterion, due to Weierstrass, is particularly useful in testing the boundary behavior of certain power series. We use it to give a complete description of the convergence of the Binomial Series.

(7.42)   **The Weierstrass Criterion** (1856)   *Consider a series* $\sum\limits_{n=0}^{\infty} a_n$ *of nonzero com-*

plex terms. Suppose that

(i) $\dfrac{a_{n+1}}{a_n} = 1 - \gamma/n - w_n/n^p$

for all $n \geq 1$, *where* $\gamma \in \mathbb{C}$, $1 < p < \infty$, *and* (w_n) *is a bounded sequence.* *Then the series converges absolutely if and only if* $\mathrm{Re}(\gamma) > 1$. *If* $\mathrm{Re}(\gamma) \leq 0$, *then the series diverges. If* $0 < \mathrm{Re}(\gamma) \leq 1$, *then the two series* $\displaystyle\sum_{n=0}^{\infty} (-1)^n a_n$ *and* $\displaystyle\sum_{n=0}^{\infty} |a_n - a_{n+1}|$ *both converge.*

Proof   Write $\gamma = \alpha + i\beta$, where $\alpha, \beta \in \mathbb{R}$, and choose $M \in \mathbb{R}$ such that $|w_n| \leq M$ for all n.

Suppose that $\alpha > 1$. Choose $1 < \alpha' < \alpha$. Since

$$\lim_{x \to \infty} (x - |x - i\beta|) = \lim_{x \to \infty} \frac{-\beta^2}{x + \sqrt{x^2 + \beta^2}} = 0,$$

we can take $x = n - \alpha$ to obtain

$$\lim_{n \to \infty} \left\{ n - |n - (\alpha + i\beta)| - M/n^{p-1} \right\} = \alpha.$$

Thus, for all sufficiently large n, the expression in braces is greater than α'. This implies that, for such n, we have

$$|a_{n+1}/a_n| \leq |1 - (\alpha + i\beta)/n| + M/n^p < 1 - \alpha'/n. \tag{1}$$

Therefore

$$\lim_{n \to \infty} n(1 - |a_{n+1}/a_n|) \geq \alpha' > 1,$$

and so Raabe's Test (7.16) shows that $\sum |a_n| < \infty$.

Next suppose that $\alpha \leq 1$. Then we have

$$|a_{n+1}/a_n| \geq |(1 - \alpha/n) - i\beta/n| - M/n^p \geq 1 - \alpha/n - M/n^p \tag{2}$$

for all $n \geq 1$. We will apply Bertrand's Test (7.19) to show that $\sum |a_n| = \infty$. We have

$$R_n = n\left(1 - \left|\frac{a_{n+1}}{a_n}\right|\right) \leq \alpha + M/n^{p-1}$$

and

$$B_n = (R_n - 1)\log n \leq (\alpha - 1)\log n + (M \log n)/n^{p-1};$$

whence $\overline{\lim} \, B_n \leq 0 < 1$. By (7.19), $\sum |a_n| = \infty$. Our first conclusion is therefore proved.

---

*This condition can be met only if $n(1 - a_{n+1}/a_n) \to \gamma$. Given this, it is then necessary that some $p > 1$ can be found such that the sequence (w_n), obtained by solving (i) for w_n, be *bounded.*

If $\alpha < 0$, it follows from (2) that $|a_{n+1}/a_n| > 1$ if $n^{p-1} > M/(-\alpha)$, and so $\sum a_n$ diverges because its terms do not tend to 0. Suppose $\alpha = 0$, then

$$a_{n+1}/a_n = 1 - i\beta/n - w_n/n^p,$$

and so, choosing $N \in \mathbb{N}$ such that $N > M^{1/p}$, we have

$$|a_{n+1}/a_n| \ge |1 - i\beta/n| - |w_n|/n^p \ge 1 - M/n^p > 0$$

for all $n \ge N$. Since $\sum M/n^p < \infty$, it follows from (7.28) that

$$\left| \frac{a_n}{a_N} \right| = \prod_{k=N}^{n-1} \left| \frac{a_{k+1}}{a_k} \right| \ge \prod_{k=N}^{n-1} \left(1 - \frac{M}{k^p} \right) \ge \prod_{k=N}^{\infty} \left(1 - \frac{M}{k^p} \right) = \delta > 0,$$

$$|a_n| \ge \delta |a_N|$$

for all $n > N$. Again a_n does not tend to 0. This proves that $\sum a_n$ diverges whenever $\alpha \le 0$.

Finally, suppose that $\alpha > 0$. Choosing $0 < \alpha' < \alpha$, we argue just as before that (1) holds for all large n, say $n \ge N$. This implies that $(|a_n|)_{n=N}^{\infty}$ is strictly decreasing, and so converges to some limit $a \ge 0$. Thus

$$\sum_{n=N}^{\infty} (|a_n| - |a_{n+1}|) = |a_N| - a \tag{3}$$

and all terms of this series are positive. Using (1), we have

$$\frac{|a_n - a_{n+1}|}{|a_n| - |a_{n+1}|} = \frac{|1 - a_{n+1}/a_n|}{1 - |a_{n+1}|/|a_n|} < \frac{|\gamma/n + w_n/n^p|}{\alpha'/n}$$

$$= \frac{|\gamma + w_n/n^{p-1}|}{\alpha'} \to \frac{|\gamma|}{\alpha'} < \infty.$$

Therefore, (3) and (7.6.i) show that

$$\sum |a_n - a_{n+1}| < \infty, \tag{4}$$

as was to be proved. Using (1) and (7.28), we have

$$\left| \frac{a_n}{a_N} \right| = \prod_{k=N}^{n-1} \left| \frac{a_{k+1}}{a_k} \right| < \prod_{k=N}^{n-1} \left(1 - \frac{\alpha'}{k} \right) \to 0 \quad \text{as } n \to \infty;$$

hence,

$$\lim_{n \to \infty} a_n = 0. \tag{5}$$

It follows from (4), (5), and (7.36.i) that $\sum (-1)^n a_n$ converges. $\square$

(7.43)   **Remark**   Regarding the preceding theorem, Weierstrass also proved that $\sum a_n$ diverges whenever $0 < \text{Re}(\gamma) \le 1$. Since we shall not need this fact, we merely sketch its proof in the copious hints to Exercise 12.

(7.44)   Corollary   *Let* $(a_n)_{n=0}^\infty$, $(w_n)_{n=1}^\infty$, γ, *and* p *be as in* (7.42). *Then the power series*

$$\sum_{n=0}^{\infty} a_n z^n$$

has radius of convergence $R = 1$. *Moreover,*
   (i) *if* $\mathrm{Re}(\gamma) > 1$, *the series converges absolutely for* $|z| \le 1$,
   (ii) *if* $0 < \mathrm{Re}(\gamma) \le 1$ *and* $0 < \delta < 2$, *then the series converges uniformly on* $X_\delta = \{ z \in \mathbb{C} : |z| \le 1, |1 - z| \ge \delta \}$, *but the convergence is not absolute if* $|z| = 1$,
   (iii) *if* $\mathrm{Re}(\gamma) \le 0$ *and* $|z| = 1$, *then the series diverges.*

Proof   Since $|(a_{n+1} z^{n+1})/(a_n z^n)| \to |z|$, we have $R = 1$ by the Ratio Test. Let $|z| = 1$. Then (i) and the nonabsolute part of (ii) follow from the first conclusion in (7.42). Moreover, (iii) obtains because, as proved in (7.42) for $\mathrm{Re}(\gamma) \le 0$,

$$\varlimsup |a_n z^n| = \varlimsup |a_n| > 0,$$

and so $(a_n z^n)$ cannot tend to 0. To prove the convergence part of (ii), we use (7.36.i). For $z \in X_\delta$ we have

$$\left| \sum_{k=0}^{n-1} z^k \right| = \left| \frac{z^n - 1}{z - 1} \right| \le \frac{2}{\delta}$$

for all $n \in \mathbb{N}$. For $0 < \mathrm{Re}(\gamma) \le 1$,   (7.42) yields

$$\sum_{n=0}^{\infty} |a_n - a_{n+1}| < \infty, \qquad a_n \to 0.$$

The proof is complete. $\square$

(7.45)   Example   Let α, β, and γ be complex numbers, none of which is 0 or a negative integer. Consider the *hypergeometric power series* $F(\alpha, \beta, \gamma, z) = \sum a_n z^n$, where a_n is as in (7.21). Then

$$a_{n+1}/a_n = 1 - [(\gamma + 1) - (\alpha + \beta)]/n - A_n/n^2$$

where $(A_n)_{n=1}^\infty$ is convergent. Thus the series has radius of convergence $R = 1$ and, writing

$$r = \mathrm{Re}(\gamma - \alpha - \beta),$$

we have for $|z| = 1$ that the series converges absolutely if $r > 0$, converges conditionally (not absolutely) if $-1 < r \le 0$ and $z \ne 1$, diverges if $r \le -1$.

We now turn to the main topic of this section.

(7.46)   Theorem   *Let* α *be any complex number that is not a nonnegative integer. Define the binomial coefficients*

$$\binom{\alpha}{0} = 1, \qquad \binom{\alpha}{n} = \frac{\alpha(\alpha - 1)(\alpha - 2) \ldots (\alpha - n + 1)}{n!}$$

for $n \in \mathbb{N}$ [*i.e.*, $\binom{\alpha}{n+1} = \binom{\alpha}{n} \cdot \frac{\alpha - n}{n+1}$ *for* $n \ge 0$]. *Then the binomial series*

$$\sum_{n=0}^{\infty} \binom{\alpha}{n} z^n \tag{*}$$

has radius of convergence $R = 1$: it converges absolutely if $|z| < 1$ and it diverges if $|z| > 1$. On the circle of convergence $|z| = 1$ we have the following.

(i)   If $\mathrm{Re}(\alpha) > 0$ and $|z| = 1$, then $(*)$ converges absolutely.

(ii)   If $-1 < \mathrm{Re}(\alpha) \le 0$ and $z = -1$, then $(*)$ diverges.

(iii)   If $-1 < \mathrm{Re}(\alpha) \le 0$ and $0 < \delta < 2$, then $(*)$ converges uniformly on $\{z \in \mathbb{C} : |z| \le 1, |1 + z| \ge \delta\}$. The convergence is not absolute if $|z| = 1$.

(iv)   If $\mathrm{Re}(\alpha) \le -1$ and $|z| = 1$, then $(*)$ diverges.

Finally, in all cases for which $(*)$ converges, we have

$$(1 + z)^{\alpha} = \sum_{n=0}^{\infty} \binom{\alpha}{n} z^n, \tag{**}$$

where $(1 + z)^{\alpha} = \exp[\alpha \, \mathrm{Log}(1 + z)]$ is the principal value if $z \ne -1$ and $0^{\alpha} = 0$ [recall: $\alpha \ne 0$].

Proof   Write $a_n = (-1)^n \binom{\alpha}{n}$. Then we have

$$\sum_{n=0}^{\infty} \binom{\alpha}{n} z^n = \sum_{n=0}^{\infty} a_n (-z)^n$$

and

$$a_{n+1}/a_n = (n - \alpha)/(n + 1) = 1 - (\alpha + 1)/n + (\alpha + 1)/(n(n + 1))$$

so, taking $\gamma = \alpha + 1$, $p = 2$, and $w_n = -(\alpha + 1)n^2/(n^2 + n)$, the fact that $R = 1$ and assertions (i), (iii), and (iv) all follow from (7.44).

Write

$$f_{\alpha}(z) = \sum_{n=0}^{\infty} \binom{\alpha}{n} z^n$$

whenever the series converges. By (4.63), we may differentiate term-by-term and get

$$f_{\alpha}'(z) = \alpha f_{\alpha - 1}(z) \tag{1}$$

for $|z| < 1$ and $\alpha \in \mathbb{C}$ [because $n \cdot \binom{\alpha}{n} = \alpha \cdot \binom{\alpha - 1}{n - 1}$]. We also have

$$(1 + z) f_{\alpha - 1}(z) = \sum_{n=0}^{\infty} \binom{\alpha - 1}{n} z^n + \sum_{n=0}^{\infty} \binom{\alpha - 1}{n} z^{n+1}$$

$$= 1 + \sum_{n=1}^{\infty} \left[\binom{\alpha - 1}{n} + \binom{\alpha - 1}{n - 1} \right] z^n$$

$$= \sum_{n=0}^{\infty} \binom{\alpha}{n} z^n = f_{\alpha}(z) \tag{2}$$

for $|z| < 1$ and $\alpha \in \mathbb{C}$. Combining (1) and (2), we get

$$(1 + z)f'_\alpha(z) - \alpha f_\alpha(z) = 0 \tag{3}$$

whenever $|z| < 1$. Writing $g_\alpha(z) = (1 + z)^\alpha = \exp[\alpha \operatorname{Log}(1 + z)]$ for $z \neq -1$, we use (5.20) and (5.21) to obtain

$$g'_\alpha(z) = \alpha(1 + z)^{\alpha - 1},$$

and so (3) yields

$$(f_\alpha / g_\alpha)'(z) = (1 + z)^{-\alpha - 1}[(1 + z)f'_\alpha(z) - \alpha f_\alpha(z)] = 0$$

whenever $|z| < 1$. It follows, with the use of (4.62), that f_α / g_α is a constant. Taking $z = 0$, we find that this constant is 1. Therefore

$$f_\alpha(z) = g_\alpha(z) = (1 + z)^\alpha \tag{4}$$

for $|z| < 1$. Now fix any $z \neq -1$ with $|z| = 1$ for which the series (*) converges. Then

$$(1 + z)^\alpha = \lim_{r \uparrow 1} g_\alpha(rz) = \lim_{r \uparrow 1} f_\alpha(rz) = \sum_{n=0}^{\infty} \binom{\alpha}{n} z^n, \tag{5}$$

where the first equality follows from the continuity of g_α at z, the second equality from (4), and the third from Abel's Limit Theorem (7.40). If $\operatorname{Re}(\alpha) = \beta > 0$, then (*) converges at $z = -1$, as already noted, and the argument for (5) still holds good except that the first equality is justified by

$$(1 - 1)^\alpha = 0 = \lim_{r \uparrow 1} e^{\beta \operatorname{Log}(1 - r)} = \lim_{r \uparrow 1} |g_\alpha(-r)|.$$

It remains only to prove assertion (ii). That is, we must show that

$$\sum_{n=0}^{\infty} (-1)^n \binom{\alpha}{n} \tag{6}$$

diverges if $\operatorname{Re}(\alpha) \leq 0$. Since

$$(-1)^{k+1}\binom{\alpha}{k+1} = -\frac{\alpha}{k+1} \cdot \frac{(1 - \alpha)(2 - \alpha) \dots (k - \alpha)}{1 \cdot 2 \cdot \dots \cdot k}$$

$$= -\frac{\alpha}{k+1} \prod_{n=1}^{k}\left(1 - \frac{\alpha}{n}\right),$$

it follows by induction that

$$\sum_{n=0}^{k} (-1)^n \binom{\alpha}{n} = \prod_{n=1}^{k}\left(1 - \frac{\alpha}{n}\right) \tag{7}$$

for all $k \geq 1$. Now $\sum |-\alpha/n|^2 < \infty$ and $\sum(-\alpha/n)$ diverges $(\alpha \neq 0)$, so, according to (7.30), the product $\prod_{n=1}^{\infty}\left(1 - \frac{\alpha}{n}\right)$ diverges. It follows from (7) that the series (6) diverges unless

$$\lim_{k \to \infty} p_k = 0, \tag{8}$$

where p_k denotes the right side of (7). Thus, we need only show that (8) fails whenever $\operatorname{Re}(\alpha) \leq 0$.

Suppose that $\text{Re}(\alpha) = -\beta < 0$. Then $|1 - \alpha/n| \geq 1 + \beta/n$ and so

$$|p_k| \geq \prod_{n=1}^{k}\left(1 + \frac{\beta}{n}\right) \to \infty$$

as $k \to \infty$ by (7.28); hence, (8) fails.

Suppose that $\text{Re}(\alpha) = 0$, $\alpha = iy$ $(y \in \mathbb{R}, y \neq 0)$. Writing

$$s_k = -\sum_{n=1}^{k}\frac{\alpha}{n} = -\sum_{n=1}^{k}\frac{iy}{n},$$

we have $|e^{-s_k}| = 1$ for all $k \geq 1$ and it follows from (7.31) that $\lim_k |p_k| = \left|\lim_k p_k e^{-s_k}\right|$ exists and is not 0. Thus (8) fails again. $\square$

Exercises

1. For which $\alpha \in \mathbb{C}$ do we have

$$\sum_{n=0}^{\infty}(-1)^n\binom{\alpha}{n} = 0?$$

2. For $-1 < \alpha < \infty$, we have

$$2^{\alpha/2}\cos\frac{\pi\alpha}{4} = \sum_{n=0}^{\infty}(-1)^n\binom{\alpha}{2n},$$

$$2^{\alpha/2}\sin\frac{\pi\alpha}{4} = \sum_{n=0}^{\infty}(-1)^n\binom{\alpha}{2n+1}.$$

[Hint: Put $z = i$ in the Binomial Series.]

3. Suppose that $-1 < \alpha < \infty$, $\theta \in \mathbb{R}$, and $\phi = \text{Arg}(e^{-i\theta}\cos\theta)$. Then

$$\sum_{n=0}^{\infty}\binom{\alpha}{n}\cos(\alpha - 2n)\theta = |2\cos\theta|^{\alpha}\cos\alpha(\theta + \phi),$$

$$\sum_{n=0}^{\infty}\binom{\alpha}{n}\sin(\alpha - 2n)\theta = |2\cos\theta|^{\alpha}\sin\alpha(\theta + \phi)$$

unless $\alpha < 0$ and $\cos\theta = 0$. (Here we write $0^{\alpha} = 0$ if $\alpha > 0$ and we define $\text{Arg}\,0 = 0$.) In particular, if $|\theta| < \pi/2$, then $\phi = -\theta$.

4. If $0 < \alpha < \infty$, we have

$$\sum_{n=0}^{\infty}\binom{\alpha}{n}^2 = \frac{2^{\alpha}}{\pi}\int_0^{\pi}(1 + \cos\theta)^{\alpha}\,d\theta,$$

$$\sum_{n=0}^{\infty}(-1)^n\binom{\alpha}{n}^2 = \frac{2^{\alpha}}{\pi}\cos\frac{\pi\alpha}{2}\int_0^{\pi}(\sin\theta)^{\alpha}\,d\theta.$$

[Hints: In each case multiply two absolutely convergent power series and then integrate over $[-\pi, \pi]$. Consider $(1 + e^{i\theta})^{\alpha}(1 + e^{-i\theta})^{\alpha}$ and $(1 + e^{i\theta})^{\alpha}(1 - e^{-i\theta})^{\alpha}$.]

5.   If $-\pi/4 < \theta < \pi/4$ and $\alpha \in \mathbb{R}$, then

$$\cos(\alpha\theta) = (\cos\theta)^{\alpha} \cdot \sum_{n=0}^{\infty} (-1)^n \binom{\alpha}{2n} \tan^{2n}\theta,$$

$$\sin(\alpha\theta) = (\cos\theta)^{\alpha} \cdot \sum_{n=0}^{\infty} (-1)^n \binom{\alpha}{2n+1} \tan^{2n+1}\theta.$$

[Hint: $e^{i\alpha\theta} = (\cos\theta)^{\alpha}(1 + i\tan\theta)^{\alpha}$.]

6.   The function $f(z) = (z-1) \cdot \sin[\mathrm{Log}(1-z)]$ (defined to equal 0 at $z = 1$) is the sum of a power series for $|z| \le 1$. [Hint: $2if(z) = (1-z)^{1-i} - (1-z)^{1+i}$.]

7.   (a) We have $\sup\{|(1+z)^{\alpha}| : |z| < 1\} < \infty$ if and only if $\mathrm{Re}(\alpha) \ge 0$.
        (b) Compute the supremum when $\mathrm{Re}(\alpha) = 0$.

8.   *The Inverse Sine.*   We want to solve the equation $\sin w = z$ for $w \in \mathbb{C}$ when $z \in \mathbb{C}$ is given and then select one such w for each z in a certain large domain $D \subset \mathbb{C}$ so that the map $z \to w$ of D into $\mathbb{C}$ is differentiable. Throughout this exercise we write $z = x + iy$ and $w = u + iv$, where $\{x, y, u, v\} \subset \mathbb{R}$.
        (a) Define the open quadrants

$$Q_1 = \{z : x > 0, \, y > 0\}, \quad Q_2 = \{z : x < 0, \, y > 0\},$$
$$Q_3 = \{z : x < 0, \, y < 0\}, \quad Q_4 = \{z : x > 0, \, y < 0\}$$

and the open strips

$$S_1 = \{w : 0 < u < \pi/2, \, v > 0\}, \quad S_2 = \{w : -\pi/2 < u < 0, \, v > 0\}$$
$$S_3 = \{w : -\pi/2 < u < 0, \, v < 0\}, \quad S_4 = \{w : 0 < u < \pi/2, \, v < 0\}.$$

Then, for $k = 1, 2, 3, 4$, sin is a one-to-one mapping of S_k onto Q_k. [Hint: First consider $k = 1$. Note that $\sin w = \sin u \cdot \cosh v + i \cos u \cdot \sinh v$. Given $z \in Q_1$, show that there is a unique $v > 0$ such that

$$\phi(v) = \frac{x^2}{\cosh^2 v} + \frac{y^2}{\sinh^2 v} = 1.$$

For the other k's, use $\sin(-w) = -\sin w$ and $\sin\overline{w} = \overline{\sin w}$.]
        (b) Let $I = \{w : u = 0\} = \{z : x = 0\}$ be the imaginary axis and define

$$E_1 = \{w : u = \pi/2, \, v \ge 0\}, \quad E_3 = \{w : u = -\pi/2, \, v \le 0\}.$$

Then the mappings

$$E_1 \to [1, \infty[, \quad E_3 \to]-\infty, -1], \quad I \to I, \quad]-\pi/2, \pi/2[\to]-1, 1[$$

by the function sin are each one-to-one and onto.
        (c) If we write $R = S_1 \cup S_2 \cup S_3 \cup S_4 \cup I \cup E_1 \cup E_3 \cup]-\pi/2, \pi/2[$, the function sin is one-to-one from R onto $\mathbb{C}$. We define the *principal branch of the inverse sine* to be the function Arcsin that is inverse to this mapping. Thus, Arcsin is a one-to-one mapping of $\mathbb{C}$ onto R and $\mathrm{Arcsin}(z) = w$ means $z \in \mathbb{C}$, $w \in R$, and $\sin w = z$.
        (d) Write $D = \mathbb{C} \setminus (]-\infty, -1] \cup [1, \infty[)$ and $S = \{w : |u| < \pi/2, \, v \in \mathbb{R}\} = R \setminus (E_1 \cup E_3)$. Then Arcsin is one-to-one and differentiable on D onto S, and

$$\mathrm{Arcsin}' z = (1 - z^2)^{-1/2}$$

for all $z \in D$, where the square root has its principal value. [Hint: Check that $\cos w = (1 - \sin^2 w)^{1/2}$ for $w \in S$ by showing that $\mathrm{Re}(\cos w) > 0$.]

(e) For $z \in D$, we have

$$\text{Arcsin } z = -i \operatorname{Log}\left(iz + \sqrt{1 - z^2}\right).$$

[Hints: Write $f(z)$ for the right side, show that f is differentiable on D by showing that the $1 - z^2 = -a \leq 0$ and $iz + \sqrt{1 - z^2} = -b \leq 0$ are both impossible for $z \in D$, and check that $f' = \text{Arcsin}'$ on D. Alternatively, solve a quadratic equation to show that $w = f(z)$ is a solution of $\sin w = z$ and show that $f(z) \in S$.]

(f) For $z \in D$, we have

$$\text{Arcsin } z = \int_0^1 \frac{z \, dt}{\sqrt{1 - z^2 t^2}}.$$

[Hint: For fixed z, apply the Fundamental Theorem of Calculus to $F(t) = \text{Arcsin}(zt)$.]

(g) $\dfrac{\pi}{2} = \displaystyle\int_0^1 \frac{dt}{\sqrt{1 - t^2}}.$

(h) For $|z| \leq 1$, we have

$$\text{Arcsin } z = z + \frac{1}{2} \cdot \frac{z^3}{3} + \frac{1 \cdot 3}{2 \cdot 4} \cdot \frac{z^5}{5} + \frac{1 \cdot 3 \cdot 5}{2 \cdot 4 \cdot 6} \cdot \frac{z^7}{7} + \cdots$$

$$= \sum_{n=0}^{\infty} \frac{(2n)!}{(2^n n!)^2} \cdot \frac{z^{2n+1}}{2n+1} = \sum_{n=0}^{\infty} (-1)^n \binom{-1/2}{n} \frac{z^{2n+1}}{2n+1},$$

and the series converges uniformly and absolutely for $|z| \leq 1$. [Hint: Use (d) or (f) and (7.44).]

(i) If $|z| < 1$, then $|\text{Arcsin } z| \leq \text{Arcsin } |z| < \pi/2$.

(j) If $z \in \mathbb{C}$ is given, then the set of $w \in \mathbb{C}$ such that $\sin w = z$ is given by

$$w = 2n\pi + \text{Arcsin } z, \qquad w = (2n+1)\pi - \text{Arcsin } z \quad \text{for } n \in \mathbb{Z}.$$

9. **The Inverse Cosine.** Since $\cos w = \sin(\pi/2 - w)$, we define the *principal branch of the inverse cosine* by the formula

$$\text{Arccos } z = (\pi/2) - \text{Arcsin } z.$$

We use the notation of the preceding exercise.

(a) The function Arccos is a one-to-one mapping of $\mathbb{C}$ onto $\pi/2 - R = \pi/2 + R$.

(b) Arccos is one-to-one and differentiable on D onto $\pi/2 - S = \pi/2 + S$ and

$$\text{Arccos}' z = -(1 - z^2)^{-1/2}$$

for all $z \in D$.

10. The formula

$$(1 + z)^{1/z} = 2 + \frac{1 - z}{2!} + \frac{(1 - z)(1 - 2z)}{3!} + \cdots$$

$$= 2 + \sum_{n=1}^{\infty} \frac{1}{(n+1)!} \prod_{k=1}^{n} (1 - kz)$$

is valid whenever $0 < |z| \leq 1$ and $z \neq -1$. For $z = -1$ and $|z| > 1$, the series diverges. For which values of z does the series converge absolutely?

11.   If $\gamma \in \mathbb{C}$ and $\text{Re}(\gamma) \leq 1$, then the series

$$\sum_{n=1}^{\infty} \frac{1}{n^{\gamma}}$$

diverges. [Hints: Write $\gamma = \alpha + i\beta$, where $\alpha, \beta \in \mathbb{R}$. Since $n^{-(1+i\beta)} = n^{-(1-\alpha)} \cdot n^{-\gamma}$, (7.36.iv) shows that we need only consider $\alpha = 1$, $\beta \neq 0$. Writing $c_n = n^{-i\beta} - (n-1)^{-i\beta}$, we have $\sum c_n$ diverges and

$$c_n = n^{-i\beta}\left[1 - (1 - n^{-1})^{-i\beta}\right] = n^{-i\beta}\left[-i\beta n^{-1} - w_n n^{-2}\right],$$

where $w_n \rightarrow \begin{pmatrix} -i\beta \\ 2 \end{pmatrix}$.]

12.   As in (7.42), suppose that

$$a_{n+1}/a_n = 1 - \gamma/n - w_n/n^p,$$

where (w_n) is bounded, $p > 1$, and $\text{Re}(\gamma) \leq 1$. Then $\sum a_n$ diverges. [Hints: First consider $\gamma \neq 1$. Writing $z_n = nw_n/(n - \gamma)$, we see that (z_n) is bounded and

$$\frac{a_{n+1}}{a_n} = (1 - \gamma/n)(1 - z_n/n^p),$$

$$a_{n+1} = a_1 \cdot \prod_{k=1}^{n} \frac{k - \gamma}{k} \cdot \prod_{k=1}^{n} (1 - z_k/k^p)$$

$$= a_1 \cdot (-1)^n \cdot \binom{\gamma - 1}{n} \cdot c_{n+1},$$

where the last equality defines c_{n+1}. Choose M and δ so that $0 < \delta \leq |c_n| \leq M$ and $|z_n| \leq M$ for all n. Then $|c_n^{-1} - c_{n+1}^{-1}| \leq M^2 \delta^{-2} n^{-p}$, so $\sum |c_n^{-1} - c_{n+1}^{-1}| < \infty$. If $\sum a_n$ converges, then $\sum a_n c_n^{-1}$ converges (7.36.iii), and so $\sum (-1)^n \binom{\gamma - 1}{n}$ converges, contrary to (7.46). If $\gamma = 1$, we can write

$$(n + 1)/n \cdot \frac{a_{n+1}}{a_n} = 1 + u_n/n^q,$$

where (u_n) is bounded and $q = \min\{p, 2\}$. Thus

$$a_n = a_1/n \cdot \prod_{k=1}^{n-1}\left(1 + \frac{u_k}{k^q}\right) = \frac{a_1}{n} \cdot d_n.$$

As before, $\sum |d_n^{-1} - d_{n+1}^{-1}| < \infty$, and $\sum a_n$ diverges.]

Rearrangements and Double Series

Unlike finite sums, the sum of an infinite series can sometimes be drastically affected by permuting the order of its terms.

(7.47)   Example   Consider the alternating harmonic series

$$1 - 1/2 + 1/3 - 1/4 + 1/5 - + \cdots = s$$

and its "rearrangement"
$$1 - 1/2 - 1/4 + 1/3 - 1/6 - 1/8 + 1/5 - - + \cdots$$
which is obtained by alternately taking one positive term and two negative terms of the original series. Let s_n and t_n denote the respective nth partial sums of these series. Then $s_n \to s$ [see (2.45)], where $0 < s < 1$ [$s = \log 2$ according to (7.39.2)] and

$$\begin{aligned}
t_{3n} &= (1 - 1/2) - 1/4 + (1/3 - 1/6) - 1/8 + \cdots \\
&\quad + (1/(2n - 1) - 1/(4n - 2)) - 1/(4n) \\
&= 1/2 - 1/4 + 1/6 - 1/8 + \cdots + 1/(4n - 2) - 1/(4n) \\
&= (1/2)(1 - 1/2 + 1/3 - 1/4 + \cdots + 1/(2n - 1) - 1/(2n)) \\
&= (1/2)s_{2n}.
\end{aligned}$$

Thus
$$\lim t_{3n} = (1/2) \lim s_{2n} = (1/2)s.$$

Moreover, one checks that
$$\lim t_{3n - 1} = \lim t_{3n - 2} = \lim t_{3n} = (1/2)s$$

and so $\lim t_n = (1/2)s$. Therefore the rearranged series has sum $(1/2)s \neq s$.

   This example is merely a special case of our next theorem. First we define "rearrangement" precisely.

(7.48)   Definition   Let $\sum\limits_{n=1}^{\infty} a_n$ be any given infinite series. An infinite series $\sum\limits_{k=1}^{\infty} b_k$ is called a *rearrangement* of the given series if there exists a one-to-one mapping ψ of $\mathbb{N}$ onto $\mathbb{N}$ such that $b_k = a_{\psi(k)}$ for all $k \in \mathbb{N}$. That is, each term of the given series appears exactly once in the second series and the second series has no other terms.

   Not only can the sum of a rearranged series differ from the original sum, but, under appropriate circumstances, the new sum can be preassigned. We have the following remarkable result.

(7.49)   Theorem   [Riemann]   *Let $\sum\limits_{n=1}^{\infty} a_n$ be a conditionally (not absolutely) convergent series of real terms and let $-\infty \leq \alpha \leq \beta \leq \infty$. Then there exists a rearrangement $\sum\limits_{n=1}^{\infty} a_n'$ of $\sum\limits_{n=1}^{\infty} a_n$ having partial sums s_n' such that $\underline{\lim} s_n' = \alpha$ and $\overline{\lim} s_n' = \beta$. In particular, the rearrangement can be chosen so as to diverge [$\alpha \neq \beta$ or $|\alpha| = \infty$] or to converge to any preassigned real number α [$\alpha = \beta$].*

   Proof   For each n write
$$a_n^+ = (1/2)(|a_n| + a_n), \qquad a_n^- = (1/2)(|a_n| - a_n).$$

Then $a_n^+ \geqq 0$, $a_n^- \geqq 0$, $a_n = a_n^+ - a_n^-$, and $|a_n| = a_n^+ + a_n^-$. If $\sum a_n^+ < \infty$, then $\sum |a_n| = 2 \sum a_n^+ - \sum a_n$ converges (2.45), contrary to hypothesis, and so $\sum a_n^+ = \infty$. Similarly, $\sum a_n^- = \infty$.

Let $p_1, p_2, \ldots$ denote the nonnegative terms of $\sum a_n$ in the order in which they occur and let $q_1, q_2, \ldots$ denote the absolute values of the negative terms of $\sum a_n$ in the order in which they occur. Then $\sum p_n = \sum q_n = \infty$, since these two series differ from $\sum a_n^+$ and $\sum a_n^-$ only in the absence of certain zero terms. Also, $p_n \to 0$ and $q_n \to 0$ because $a_n \to 0$.

We shall construct a series of the form

$$p_1 + \ldots + p_{k_1} - q_1 - \ldots - q_{m_1} + p_{k_1+1} + \ldots + p_{k_2} - q_{m_1+1}$$
$$- \ldots - q_{m_2} + \ldots, \tag{1}$$

where $0 < k_1 < k_2 < \ldots$ and $0 < m_1 < m_2 < \ldots$.

Any such series is evidently a rearrangement of $\sum a_n$. Let u_j and v_j denote those partial sums of (1) that end with p_{k_j} and q_{m_j}, respectively. We shall choose (k_j) and (m_j) so that $v_j \to \alpha$ and $u_j \to \beta$.

First, choose any two sequences (α_j) and (β_j) of real numbers [not $\pm \infty$] such that $\alpha_j \to \alpha$, $\beta_j \to \beta$, $\beta_1 > 0$, $\alpha_j < \beta_j$, and $\alpha_j < \beta_{j+1}$ ($j = 1, 2, \ldots$). Let k_1 be the smallest integer such that $u_1 > \beta_1 > 0$ $\left[\sum p_n = \infty\right]$. Then $0 < u_1 - \beta_1 \leqq p_{k_1}$. Next let m_1 be minimal such that $v_1 < \alpha_1 < \beta_2$. Then $0 < \alpha_1 - v_1 \leqq q_{m_1}$. Choose the minimal k_2 so that $u_2 > \beta_2 > \alpha_2$; hence $|u_2 - \beta_2| \leqq p_{k_2}$. Then choose the minimal m_2 such that $v_2 < \alpha_2$, etc. In this way, we choose strictly increasing sequences (k_j) and (m_j) in $\mathbb{N}$ in such a way that

$$|u_j - \beta_j| \leqq p_{k_j}, \qquad |v_j - \alpha_j| \leqq q_{m_j}.$$

Therefore $u_j \to \beta$ and $v_j \to \alpha$, and so α and β are cluster points of the sequence (s_n') of all partial sums of the series (1). Lastly, for any n there exists a $j = j(n)$ $[j(n) \to \infty$ as $n \to \infty]$ such that either $u_j \geqq s_n' \geqq v_j$ or $v_j \leqq s_n' \leqq u_{j+1}$ [just look at the position of the nth term of (1)]; hence all cluster points of (s_n') lie in $[\alpha, \beta]$. [Actually, each point of $[\alpha, \beta]$ is a cluster point of this sequence.] By (2.22), the proof is complete. $\square$

The following theorem and its corollary tell us that quite general rearrangements cannot alter sums in the presence of *absolute* convergence. Notice the (nonaccidental) analogy with Fubini's Theorem.

(7.50) Main Rearrangement Theorem *Suppose that* $c_{m,n} \in \mathbb{C}$ *for each* (m,n) $\in \mathbb{N} \times \mathbb{N}$ *and that* ϕ *is any one-to-one mapping of* $\mathbb{N}$ *onto* $\mathbb{N} \times \mathbb{N}$. *If any of the three* *sums*

$$\text{(i)} \qquad \sum_{m=1}^{\infty} \left(\sum_{n=1}^{\infty} |c_{m,n}| \right), \quad \sum_{n=1}^{\infty} \left(\sum_{m=1}^{\infty} |c_{m,n}| \right), \quad \sum_{k=1}^{\infty} |c_{\phi(k)}|$$

is finite, then all of the series

$$\text{(ii)} \qquad \sum_{n=1}^{\infty} c_{m,n} \qquad (m = 1, 2, \dots),$$

$$\text{(iii)} \qquad \sum_{m=1}^{\infty} c_{m,n} \qquad (n = 1, 2, \dots),$$

$$\text{(iv)} \qquad \sum_{m=1}^{\infty} \left(\sum_{n=1}^{\infty} c_{m,n} \right), \quad \sum_{n=1}^{\infty} \left(\sum_{m=1}^{\infty} c_{m,n} \right), \quad \sum_{k=1}^{\infty} c_{\phi(k)}$$

are absolutely convergent and the three series in (iv) *all have the same sum.*

Proof Theorem (2.50) shows that the three series in (i) all have the same sum and so they are all finite. Since no term of a convergent series of nonnegative terms can equal ∞, it follows that all the series in (ii) and (iii) are absolutely convergent—hence convergent.

Write

$$\sum_{n=1}^{\infty} c_{m,n} = b_m \qquad (m = 1, 2, \dots).$$

Since

$$|b_m| = \lim_{q \to \infty} \left| \sum_{n=1}^{q} c_{m,n} \right| \le \lim_{q \to \infty} \sum_{n=1}^{q} |c_{m,n}| = \sum_{n=1}^{\infty} |c_{m,n}|$$

for all m, the Comparison Test yields

$$\sum_{m=1}^{\infty} \left| \sum_{n=1}^{\infty} c_{m,n} \right| = \sum_{m=1}^{\infty} |b_m| \le \sum_{m=1}^{\infty} \left(\sum_{n=1}^{\infty} |c_{m,n}| \right) < \infty,$$

and so the first series in (iv) is absolutely convergent. Similarly, the second series in (iv) is absolutely convergent and, by the first sentence of this proof, so is the third. Let us write

$$\sum_{k=1}^{\infty} c_{\phi(k)} = s \in \mathbf{C}.$$

We shall next show that $\sum_{m=1}^{\infty} b_m = s$; i.e., the first and third series in (iv) have the same sum. That the second and third have the same sum can be proved in a similar manner.

Let $\epsilon > 0$ be given. Choose $k_0 \in \mathbf{N}$ such that

$$\sum_{k=k_0+1}^{\infty} |c_{\phi(k)}| < \frac{\epsilon}{3} \tag{1}$$

and

$$\left| s - \sum_{k=1}^{k_0} c_{\phi(k)} \right| < \frac{\epsilon}{3} . \tag{2}$$

Next choose $p_0, q_0 \in \mathbb{N}$ such that

$$\{\phi(k) : 1 \leq k \leq k_0\} \subset \{(m,n) : 1 \leq m \leq p_0, 1 \leq n \leq q_0\}.$$

Then, whenever $p \geq p_0$ and $q \geq q_0$, each term of the finite sum $\sum_{k=1}^{k_0} c_{\phi(k)}$

appears as a term $c_{m,n}$ in the finite sum $\sum_{m=1}^{p} \left(\sum_{n=1}^{q} c_{m,n} \right)$, and so, subtracting

those terms from the latter sum and using (1), we obtain

$$\left| \sum_{m=1}^{p} \left(\sum_{n=1}^{q} c_{m,n} \right) - \sum_{k=1}^{k_0} c_{\phi(k)} \right| \leq \sum_{k=k_0+1}^{\infty} |c_{\phi(k)}| < \frac{\epsilon}{3} . \tag{3}$$

We claim that

$$p \geq p_0 \quad \text{implies} \quad \left| s - \sum_{m=1}^{p} b_m \right| < \epsilon. \tag{4}$$

If (4) can be established, then $\sum_{m=1}^{\infty} b_m = s$, and our proof is complete. Fix

any $p \geq p_0$. Since

$$\lim_{q \to \infty} \sum_{n=1}^{q} c_{m,n} = b_m \qquad (m = 1, 2, \dots),$$

it follows from (2.8.i) [and induction] that

$$\lim_{q \to \infty} \sum_{m=1}^{p} \left(\sum_{n=1}^{q} c_{m,n} \right) = \sum_{m=1}^{p} b_m.$$

Thus we may choose some $q \geq q_0$ such that

$$\left| \sum_{m=1}^{p} \left(\sum_{n=1}^{q} c_{m,n} \right) - \sum_{m=1}^{p} b_m \right| < \frac{\epsilon}{3} . \tag{5}$$

Combining (2), (3), and (5), we obtain

$$\left| s - \sum_{m=1}^{p} b_m \right| < \frac{\epsilon}{3} + \frac{\epsilon}{3} + \frac{\epsilon}{3} = \epsilon.$$

Since $p \geq p_0$ was arbitrary, (4) has been proven. $\square$

(7.51) Remark In the notation of the preceding theorem, it can happen that all of the series in (ii) and (iii) and the first two series in (iv) are absolutely convergent, but that the first two series in (iv) do not have the same sum [cf. (2.53)].

(7.52) **Corollary** *Let $(a_n)_{n=1}^{\infty} \subset \mathbb{C}$. Then all rearrangements of $\displaystyle\sum_{n=1}^{\infty} a_n$ converge to the same sum [this property is called unconditional convergence] if and only if $\displaystyle\sum_{n=1}^{\infty} |a_n| < \infty$.*

Proof Suppose $\sum |a_n| = \infty$. If $\sum a_n$ diverges, there is nothing to prove, so we suppose that $\sum a_n$ converges. Write $a_n = b_n + ic_n$, where $b_n, c_n \in \mathbb{R}$. Since $|a_n| \leq |b_n| + |c_n|$, either $\sum |b_n| = \infty$ or $\sum |c_n| = \infty$; say $\sum |b_n| = \infty$. Then $\sum b_n$ is conditionally convergent, so according to Riemann's Theorem (7.49), there is a rearrangement $\sum b_{\psi(k)}$ that diverges. Then $\sum a_{\psi(k)}$ cannot converge.

Conversely, suppose that $\sum |a_n| < \infty$ and let ψ be any one-to-one mapping of $\mathbb{N}$ onto $\mathbb{N}$. For $m, n \in \mathbb{N}$, define $c_{m,n} = a_{\psi(m)}$ if $n = \psi(m)$ and $c_{m,n} = 0$ otherwise. Then

$$\sum_{n=1}^{\infty} c_{m,n} = a_{\psi(m)} \qquad (m = 1, 2, \dots), \tag{1}$$

$$\sum_{m=1}^{\infty} c_{m,n} = a_n \qquad (n = 1, 2, \dots), \tag{2}$$

because the only nonzero term in the second sum is $a_{\psi(m)}$, where m is such that $\psi(m) = n$. Since

$$\sum_{n=1}^{\infty} \left(\sum_{m=1}^{\infty} |c_{m,n}| \right) = \sum_{n=1}^{\infty} |a_n| < \infty,$$

it follows from (7.50), (1), and (2) that

$$\sum_{n=1}^{\infty} a_n = \sum_{m=1}^{\infty} a_{\psi(m)} . \qquad \square$$

(7.53) **Examples** (a) It can easily happen that

$$\sum_{m=1}^{\infty} \left(\sum_{n=1}^{\infty} c_{m,n} \right) = \sum_{n=1}^{\infty} \left(\sum_{m=1}^{\infty} c_{m,n} \right)$$

even though the hypothesis of (7.50) fails. As a simple example, let $c_{m,n} = (-1)^{m+n}/(mn)$. In this case, both sums equal $(\log 2)^2$.

(b) Let $\displaystyle\sum_{n=0}^{\infty} a_n$ be any absolutely convergent series. Write

$$2^{n+1}b_n = \sum_{k=0}^{n} 2^k a_k.$$

We can use (7.50) to prove that

$$\sum_{n=0}^{\infty} b_n = \sum_{n=0}^{\infty} a_n.$$

In fact, let $c_{m,n} = 2^m a_m / 2^{n+1}$ if $n \geq m$ and $c_{m,n} = 0$ otherwise. Then

$$\sum_{m=0}^{\infty} \left(\sum_{n=0}^{\infty} |c_{m,n}| \right) = \sum_{m=0}^{\infty} \left(\sum_{n=m}^{\infty} |a_m| 2^{m-n-1} \right) = \sum |a_m| < \infty,$$

$$\sum_{n=0}^{\infty} c_{m,n} = a_m, \quad \text{and} \quad \sum_{m=0}^{\infty} c_{m,n} = b_n.$$

(c) Let $|z| < 1$ and write $c_{m,n} = c_{m,n}(z) = z^{mn}$. Then

$$\sum_{n=1}^{\infty} \left(\sum_{m=1}^{\infty} c_{m,n} \right) = \sum_{n=1}^{\infty} \frac{z^n}{1 - z^n}$$

and, since $|z|^n/(1 - |z|^n) \leq |z|^n/(1 - |z|)$ for $n \geq 1$, we have $\sum \sum |c_{m,n}| < \infty$. Plainly we can arrange $\mathbb{N} \times \mathbb{N}$ into a single sequence [by some map ϕ as in (7.50)] such that (m, n) precedes (m', n') if $m \cdot n < m' \cdot n'$. It follows from (7.50) that

$$\sum_{n=1}^{\infty} \frac{z^n}{1 - z^n} = \sum_{k=1}^{\infty} \tau(k) z^k,$$

where $\tau(k)$ denotes the number of positive divisors of k.

Theorem (7.50) applies to the multiplication of series as follows.

(7.54) Theorem *Let $\sum_{n=1}^{\infty} a_n$ and $\sum_{n=1}^{\infty} b_n$ be any two absolutely convergent series and let ϕ be any one-to-one map of $\mathbb{N}$ onto $\mathbb{N} \times \mathbb{N}$. Define $p_k = a_m b_n$ if $\phi(k) = (m, n)$. Then $\sum |p_k| < \infty$ and*

$$\sum_{k=1}^{\infty} p_k = \left(\sum_{m=1}^{\infty} a_m \right) \left(\sum_{n=1}^{\infty} b_n \right).$$

Proof Let $c_{m,n} = a_m b_n$ so that $p_k = c_{\phi(k)}$. Then

$$\sum_{m=1}^{\infty} \left(\sum_{n=1}^{\infty} |c_{m,n}| \right) = \sum_{m=1}^{\infty} \left(|a_m| \sum_{n=1}^{\infty} |b_n| \right) = \left(\sum_{n=1}^{\infty} |b_n| \right) \left(\sum_{m=1}^{\infty} |a_m| \right) < \infty,$$

and so (7.50) applies to complete the proof. $\square$

(7.55) Note The preceding theorem says that the product of the sums of two absolutely convergent series $\sum a_m$ and $\sum b_n$ is obtained as the sum of any series whose terms are just the collection of all products $a_m b_n$ arranged in any way whatever into a single sequence (p_k).

If $1 = k_1 < k_2 < \dots$ and $g_j = p_{k_j} + \dots + p_{k_{j+1}-1}$, then $\sum_{j=1}^{\infty} g_j = \sum_{k=1}^{\infty} p_k$

since the partial sums of $\sum g_j$ form a subsequence of the partial sums of the *convergent* series $\sum p_k$ (this has nothing to do with absolute convergence). The Cauchy Product of two series is a special case of this "grouping of terms" or "insertion of parentheses." It is therefore a simple corollary of (7.54) that the Cauchy Product of two absolutely convergent series is absolutely convergent and the sum of the product is the product of the two sums. Mertens' Theorem (2.71) is different. It does not assert absolute convergence of the product, but it only requires one factor to be absolutely convergent. Take care, however, to realize that *Mertens' Theorem applies only to Cauchy Products.*

Our next goal is to prove a far-reaching theorem of Karl Weierstrass on series in which each term is a power series (compare (7.53.c)). First, we need the following important estimate for the coefficients in a power series.

(7.56) Cauchy's Estimate (1831)　*Let the power series* $f(z) = \sum_{n=0}^{\infty} c_n z^n$ *converge for* $|z| < R$. *Suppose that* $0 < r < R$ *and that* $|f(z)| \le M$ *whenever* $|z| = r$. *Then*
$$|c_n| \le M/r^n$$
for all $n \ge 0$.

Proof　Let $\epsilon > 0$ and an integer $p \ge 0$ be given. Choose an integer $N > p$ such that
$$\left| f(z) - \sum_{n=0}^{N} c_n z^n \right| < \epsilon$$
whenever $|z| = r$. Then
$$r^p \left| \sum_{n=0}^{N} c_n z^{n-p} \right| = \left| \sum_{n=0}^{N} c_n z^n \right| < M + \epsilon$$
for $|z| = r$. Writing $z = re^{i\theta}$ $(\theta \in \mathbb{R})$, we have
$$r^p |c_p| = r^p \left| \sum_{n=0}^{N} c_n \frac{1}{2\pi} \int_0^{2\pi} (re^{i\theta})^{n-p} \, d\theta \right|$$
$$\le \frac{1}{2\pi} \int_0^{2\pi} r^p \left| \sum_{n=0}^{N} c_n (re^{i\theta})^{n-p} \right| d\theta < \frac{1}{2\pi} \int_0^{2\pi} (M + \epsilon) \, d\theta = M + \epsilon.$$
Since ϵ and p are arbitrary, the proof is complete. $\square$

(7.57) Weierstrass' Double Series Theorem (1841)　*Let* $0 < R \le \infty$ *and let*
$$f_k(z) = \sum_{n=0}^{\infty} a_n^{(k)} z^n \qquad (k \ge 0)$$

be a given sequence of power series, all of which converge for $|z| < R$. *Suppose that*

$$F(z) = \sum_{k=0}^{\infty} f_k(z)$$

whenever $|z| < R$ *and that this convergence is uniform on every disk* $D_r = \{z \in \mathbb{C} : |z| \le r\}$, *where* $0 < r < R$. *Then the series*

(i) $$A_n = \sum_{k=0}^{\infty} a_n^{(k)} \qquad (n \ge 0)$$

all converge. For $|z| < R$, *we have*

(ii) $$F(z) = \sum_{n=0}^{\infty} A_n z^n,$$

the function F has derivatives of all orders at z, and

(iii) $$F^{(j)}(z) = \sum_{k=0}^{\infty} f_k^{(j)}(z) \qquad (j \ge 0).$$

Moreover, the series (ii) *and* (iii) *converge uniformly on every* D_r *with* $0 < r < R$.

Proof   Let $\epsilon > 0$ and $0 < r < s < R$ be arbitrary and fixed. For any integer $m \ge 0$, there exists a k_0 such that

$$\left| \sum_{n=0}^{\infty} \left(\sum_{k=p}^{q} a_n^{(k)} \right) z^n \right| = \left| \sum_{k=p}^{q} f_k(z) \right| < \epsilon_0 = \min\{1, s^m \cdot \epsilon\} \qquad (1)$$

for all $q \ge p \ge k_0$ and all $z \in D_s$. From (7.56), we infer that

$$\left| \sum_{k=p}^{q} a_m^{(k)} \right| \le \epsilon_0 s^{-m} \le \epsilon$$

whenever $q \ge p \ge k_0$. This proves that all of the series (i) converge. For any $q > k_0$ and $z \in D_s$, we have from (1) that

$$\left| \sum_{n=0}^{\infty} \left(\sum_{k=0}^{q} a_n^{(k)} \right) z^n \right| = \left| \sum_{k=0}^{q} f_k(z) \right|$$

$$\le \left| \sum_{k=k_0+1}^{q} f_k(z) \right| + \sup_{z \in D_s} \left| \sum_{k=0}^{k_0} f_k(z) \right| < 1 + C = M.$$

Another application of (7.56) yields

$$\left| \sum_{k=0}^{q} a_n^{(k)} \right| \le M/s^n$$

for all $n \ge 0$ and $q > k_0$. Letting $q \to \infty$, we have $|A_n| \le M/s^n$, and so $|z| \le r < s$ implies

$$\sum_{n=0}^{\infty} |A_n z^n| \le M \cdot \sum_{n=0}^{\infty} (r/s)^n < \infty.$$

Therefore $\sum A_n z^n$ converges (absolutely and uniformly) on D_r. Since r is arbitrary, this series converges (absolutely) for $|z| < R$. Let

$$G(z) = \sum_{n=0}^{\infty} A_n z^n \qquad (|z| < R).$$

We must show that $G = F$.

As above, there is some k_1 such that

$$\left| \sum_{n=0}^{\infty} \left(\sum_{k=p+1}^{q} a_n^{(k)} \right) z^n \right| = \left| \sum_{k=p+1}^{q} f_k(z) \right| < \epsilon_1 = \frac{\epsilon(s-r)}{s}$$

whenever $q > p > k_1$ and $z \in D_s$, and so

$$\left| A_n - \sum_{k=0}^{p} a_n^{(k)} \right| = \lim_{q \to \infty} \left| \sum_{k=p+1}^{q} a_n^{(k)} \right| \leq \frac{\epsilon_1}{s^n} \tag{2}$$

for all $n \geq 0$, $p > k_1$ by (7.56). Therefore, $|z| \leq r$ implies

$$\left| G(z) - \sum_{k=0}^{p} f_k(z) \right| = \left| \sum_{n=0}^{\infty} \left(A_n - \sum_{k=0}^{p} a_n^{(k)} \right) z^n \right|$$

$$\leq \sum_{n=0}^{\infty} \epsilon_1 (r/s)^n = \epsilon_1 \cdot \frac{s}{s-r} = \epsilon$$

for all $p > k_1$, and so

$$G(z) = \sum_{k=0}^{\infty} f_k(z) = F(z)$$

for $|z| \leq r$. Since r is arbitrary, $G(z) = F(z)$ for $|z| < R$.

Differentiating term-by-term according to (4.63), we have

$$f_k'(z) = \sum_{n=1}^{\infty} n a_n^{(k)} z^{n-1} \qquad (k = 0, 1, \dots),$$

$$F'(z) = \sum_{n=1}^{\infty} n A_n z^{n-1}$$

for $|z| < R$. As in the preceding paragraph, we can find k_2 such that (2) obtains, with ϵ_1 replaced by $\epsilon_2 = (s-r)^2 s^{-1} \epsilon$, for all $p > k_2$. Then

$$\left| F'(z) - \sum_{k=0}^{p} f_k'(z) \right| = \left| \sum_{n=1}^{\infty} n \left(A_n - \sum_{k=0}^{p} a_n^{(k)} \right) z^{n-1} \right|$$

$$\leq \sum_{n=1}^{\infty} n \epsilon_2 s^{-n} r^{n-1} = \frac{\epsilon_2}{s} \sum_{n=1}^{\infty} n (r/s)^{n-1} = \epsilon_2 \cdot s \cdot (s-r)^{-2} = \epsilon$$

whenever $p > k_2$ and $z \in D_r$. This proves that (iii) holds for $j = 1$ and the convergence is uniform on D_r for each $r < R$. The result for $j > 1$ now follows by repeated applications of what we have just proved. $\square$

As an instructive example of (7.57), we next obtain the power series for the secant function from its expansion in partial fractions.

(7.58) Example Recall (Exercise 10, page 255)

$$\frac{\pi}{4}\sec\frac{\pi z}{2} = \sum_{k=0}^{\infty}\frac{(-1)^k(2k+1)}{(2k+1)^2 - z^2} \tag{1}$$

if $z \in C$ is *not* an odd integer. Write

$$b_k(z) = \frac{2k+1}{(2k+1)^2 - z^2}.$$

We adopt the notation of (7.57). For all $z \in D_r$ $(0 < r < 1)$ and all $k \geq 0$, we have

$$|b_k(z)| \leq \frac{1}{2k+1-r^2}, \qquad |b_k(z) - b_{k+1}(z)| \leq \frac{2(2k+3)^2 + 2}{\left[(2k+1)^2 - r^2\right]^2},$$

and so it follows from (7.36.i) that the series (1) converges uniformly on D_r [it does not converge absolutely]. Write $F(z)$ for the left side of (1), $f_k(z) = (-1)^k b_k(z)$, and $R = 1$. Then

$$f_k(z) = \sum_{n=0}^{\infty}\frac{(-1)^k z^{2n}}{(2k+1)^{2n+1}}$$

for $|z| < 2k + 1$ (a geometric series).

If we write $A_{2n+1} = 0$ and

$$A_{2n} = \sum_{k=0}^{\infty}\frac{(-1)^k}{(2k+1)^{2n+1}}, \tag{2}$$

it follows from (7.57) that

$$\frac{\pi}{4}\sec\frac{\pi z}{2} = \sum_{n=0}^{\infty}A_{2n}z^{2n}$$

for $|z| < 1$. Just as with Bernoulli numbers, there is a simple way of computing these coefficients by a recurrence formula. First, define E_{2n} by the equation

$$(-1)^n\frac{E_{2n}}{(2n)!} = \frac{4}{\pi}\cdot\frac{2^{2n}}{\pi^{2n}}\cdot A_{2n}. \tag{3}$$

Then

$$\sec z = \sum_{n=0}^{\infty}(-1)^n\frac{E_{2n}}{(2n)!}z^{2n} \tag{4}$$

for $|z| < \pi/2$. If we multiply (4) by the cosine series, forming the Cauchy Product, and then invoke (4.63) to equate coefficients, we find that

$$E_0 = 1, \qquad \sum_{k=0}^{n}\binom{2n}{2k}E_{2k} = 0 \tag{5}$$

for $n > 0$. The numbers E_{2n} are called *Euler's numbers*. By (5) and induction, each is an integer. By (2) and (3), they alternate in sign and none are 0. We use (5) to compute them in succession:

$$E_0 = 1, \quad E_2 = -1, \quad E_4 = 5, \quad E_6 = -61, \ldots .$$

From (2) and (3), we obtain

$$\sum_{k=0}^{\infty} \frac{(-1)^k}{(2k+1)^{2n+1}} = \frac{(-1)^n E_{2n} \pi^{2n+1}}{2^{2n+2}(2n)!} .$$

Exercises

1.   If (a_n) is a sequence and (a_{n_k}) is a subsequence, then $\sum a_{n_k}$ is called a *subseries* of $\sum a_n$. A series of complex terms is absolutely convergent if and only if each of its subseries is convergent.

2.   If $\sum a_n$ is convergent, then parentheses can be inserted in it to form an absolutely convergent series. That is, there exists a sequence $1 = n_1 < n_2 < \ldots$ of integers such that $\sum g_k$ is absolutely convergent, where $g_k = a_{n_k} + \ldots + a_{n_{k+1}-1}$.

3.   Let $(a_n)_{n=1}^{\infty} \subset \,]0, \infty[$ be monotone and satisfy $\lim_{n\to\infty} na_n = L \in \mathbb{R}$. Given $p, q \in \mathbb{N}$, rearrange the convergent series

    (i) $\displaystyle\sum_{n=1}^{\infty} (-1)^{n-1} a_n = s$

by alternately taking p positive terms and q negative terms (in the order that they appear in (i)) to obtain the series

    (ii) $a_1 + a_3 + \ldots + a_{2p-1} - a_2 - a_4 - \ldots - a_{2q} + a_{2p+1} + \ldots = t.$

Then (ii) converges and

$$t = s + (L/2)\log(p/q).$$

[Hint: Write s_n and t_n for the partial sums. If $p > q$, then

$$t_{k(p+q)} = s_{2kp} + \sum_{j=kq+1}^{kp} a_{2j} = s_{2kp} + u_k.$$

If $\alpha < L < \beta$, then $\alpha < ja_j < \beta$ for all large j. Write

$$v_k = \sum_{j=kq+1}^{kp} \frac{1}{2j} .$$

Then $\alpha v_k < u_k < \beta v_k$ for large k. Comparing with an integral, one finds $v_k \to (1/2) \log(p/q)$.]

4.   (a)

$$\frac{1}{3} - \frac{1}{5} - \frac{1}{7} + \frac{1}{5} - \frac{1}{9} - \frac{1}{11} + \ldots + \frac{1}{2n+1} - \frac{1}{4n+1} - \frac{1}{4n+3} + \ldots$$

$$= \frac{1}{3} - \frac{1}{2} \log 2.$$

(b)

$$1 - \frac{1}{2} - \frac{1}{4} + \frac{1}{5} + \frac{1}{7} - \frac{1}{8} - \frac{1}{10} + \frac{1}{11} + \ldots + \frac{1}{6n-5} - \frac{1}{6n-4}$$
$$- \frac{1}{6n-2} + \frac{1}{6n-1} + \ldots = \frac{2}{3}\log 2.$$

5. Suppose that $a, b, c \in \mathbb{R}$, $a > 0$, $c > 0$, and $ac > b^2$ if $b < 0$. Then

$$\sum_{m=1}^{\infty} \sum_{n=1}^{\infty} (am^2 + 2b + bmn + cn^2)^{-\alpha}$$

converges if $\alpha > 1$ and diverges if $\alpha \leq 1$. [Hint: There exist positive constants M and N such that $Mmn \leq am^2 + 2b + bmn + cn^2 \leq N(m+n)^2$ for all m, n.]

6. Prove the following integral test for double series. Let f be a positive function defined on $\mathbb{N} \times \mathbb{N}$. Suppose that there exists a $t_0 > 0$ and two positive functions u and v defined on $[t_0, \infty[$ such that both $tu(t)$ and $tv(t)$ are monotone on $[t_0, \infty[$ and such that

$$u(k) \leq f(m, k - m) \leq v(k)$$

whenever $k, m \in \mathbb{N}$ satisfy $k > t_0$ and $1 \leq m \leq k - 1$. Then

$$\sum_{m=1}^{\infty} \sum_{n=1}^{\infty} f(m, n) = \infty$$

if $\int_{t_0}^{\infty} tu(t)\, dt = \infty$, and

$$\sum_{m=1}^{\infty} \sum_{n=1}^{\infty} f(m, n) < \infty$$

if $\int_{t_0}^{\infty} tv(t)\, dt < \infty$.

7. Let a, b, c be as in Exercise 5. Suppose that ϕ is a positive function on $]0, \infty[$ such that $\sqrt{t}\, \phi(t)$ is nonincreasing on $[t_0, \infty[$ for some $t_0 > 0$. Then

$$\sum_{m=1}^{\infty} \sum_{n=1}^{\infty} \phi(am^2 + 2bmn + cn^2) < \infty$$

if and only if $\int_{t_0}^{\infty} \phi(x)\, dx < \infty$. [Hint: There exist positive constants A and B such that

$$Bk^2 \leq am^2 + 2bm(k - m) + c(k - m)^2 \leq Ak^2$$

for $k \geq 2$ and $1 \leq m \leq k - 1$, so take $u(t) = \phi(At^2)$, $v(t) = \phi(Bt^2)$, and apply Exercise 6.]

8. Let $z \in \mathbb{C}$ with $\mathrm{Re}(z) > 0$. Then
(a)

$$\sum_{k=0}^{\infty} \frac{1}{(z+k)(z+k+1) \ldots (z+k+n)} = \frac{1}{nz(z+1) \ldots (z+n-1)}$$

for all $n \in \mathbb{N}$,
(b)

$$\sum_{n=1}^{\infty} \frac{(n-1)!}{z(z+1) \ldots (z+n)} = \frac{1}{z^2},$$

(c)

$$\sum_{k=0}^{\infty} \frac{1}{(z+k)^2} = \sum_{n=1}^{\infty} \frac{(n-1)!}{nz(z+1)\ldots(z+n-1)} .$$

[Hint: In (b), the difference between the sum of n terms and $1/z^2$ is

$$\frac{n!}{z^2(z+1)(z+2)\ldots(z+n)} ,$$

which has absolute value no larger than

$$\left[x^2 \prod_{k=1}^{n} \left(1 + \frac{x}{k} \right) \right]^{-1},$$

where $x = \text{Re}(z)$.]

9.   For $|z| < 1$, we have
(a)

$$\sum_{n=1}^{\infty} \frac{z^n}{1+z^{2n}} = \sum_{k=0}^{\infty} (-1)^k \frac{z^{2k+1}}{1-z^{2k+1}} ,$$

(b)

$$\sum_{n=0}^{\infty} \frac{z^{2n+1}}{1+z^{4n+2}} = \sum_{k=0}^{\infty} (-1)^k \frac{z^{2k+1}}{1-z^{4k+2}} ,$$

(c)

$$\sum_{n=1}^{\infty} (-1)^{n-1} \frac{z^n}{1+z^{2n}} = \sum_{k=0}^{\infty} (-1)^k \frac{z^{2k+1}}{1+z^{2k+1}} ,$$

(d)

$$\sum_{n=1}^{\infty} (-1)^{n-1} \frac{z^n}{(1+z^n)^2} = \sum_{k=1}^{\infty} (-1)^{k-1} \frac{kz^k}{1+z^k} ,$$

(e)

$$\sum_{n=0}^{\infty} \frac{(2n+1)z^{2n+1}}{1-z^{4n+2}} = \sum_{k=0}^{\infty} \frac{z^{2k+1}(1+z^{4k+2})}{(1-z^{4k+2})^2} .$$

10.   We have

$$\sum_{k=1}^{\infty} \frac{z^{2^{k-1}}}{1-z^{2^k}} = \frac{z}{1-z} \quad \text{or} \quad \frac{1}{1-z}$$

according as $|z| < 1$ or $|z| > 1$. This series diverges (or is undefined) whenever $|z| = 1$. [Hint: For $|z| > 1$, consider $|1/z| < 1$.]

11.   A *Lambert series* is a series of the form

$$\sum_{n=1}^{\infty} \frac{a_n z^n}{1-z^n} , \tag{*}$$

where $(a_n)_{n=1}^{\infty} \subset \mathbb{C}$.

(a) If $\sum a_n$ diverges and $|z| \neq 1$, then (*) converges if and only if $\sum a_n z^n$ converges. Moreover (*) converges uniformly on every disk $D_r = \{z \in \mathbb{C} : |z| \leq r\}$, where $r < R$, R being the radius of convergence of $\sum a_n z^n$. [Hint: For $|z| < 1$, we can apply (7.36.iii) with $b_n(z) = 1 - z^n$ and with $b_n(z) = (1 - z^n)^{-1}$. If (*) converges for some $|z_0| > 1$, use (7.36.iii) to show that $\sum a_n$ converges.]

(b) If $\sum a_n$ converges, then (*) converges uniformly on every disk D_r with $0 < r < 1$ and uniformly on $\mathbb{C} \backslash D_r$ for every $r > 1$. [Hint: $a_n z^n/(1 - z_n) = -a_n - a_n(1/z)^n/[1 - (1/z)^n]$.]

(c) Let $R_0 = \min\{R, 1\}$, where R is as in (a). If $|z| < R_0$, we have

$$\sum_{n=1}^{\infty} \frac{a_n z^n}{1 - z^n} = \sum_{n=1}^{\infty} A_n z^n,$$

where

$$A_n = \sum_{d \mid n} a_d.$$

That is, A_n is the sum of all a_d such that d is a divisor of n.

(d) The first equality in (c) can fail for $|z| > 1$ even when both series converge.

(e) Define $\mu : \mathbb{N} \to \{-1, 0, 1\}$ by $\mu(1) = 1$, $\mu(n) = (-1)^k$ if n is the product of k different prime numbers, and $\mu(n) = 0$ if n is divisible by the square of some prime. Then

$$\sum_{n=1}^{\infty} \frac{\mu(n) z^n}{1 - z^n} = z$$

whenever $|z| < 1$. [Hint: If $n > 1$ has just r different prime divisors ($r = 3$ if $n = 792 = 2^3 \cdot 3^2 \cdot 11$), then

$$\sum_{d \mid n} \mu(d) = \sum_{k=0}^{r} \binom{r}{k} (-1)^k = 0.]$$

(f) If $f(z)$ is the sum of (*) and $g(z) = \sum_{n=1}^{\infty} a_n z^n$, then $f(z) = \sum_{k=1}^{\infty} g(z^k)$ for $|z| < R_0$, where R_0 is as in (c).

(g) For $\alpha \in \mathbb{C}$ we have

$$\sum_{n=1}^{\infty} \frac{\alpha^n z^n}{1 - z^n} = \sum_{k=1}^{\infty} \frac{\alpha z^k}{1 - \alpha z^k}, \qquad \sum_{n=1}^{\infty} \frac{n \alpha^n z^n}{1 - z^n} = \sum_{k=1}^{\infty} \frac{\alpha z^k}{(1 - \alpha z^k)^2},$$

$$\sum_{n=1}^{\infty} \frac{\alpha^n z^n}{n(1 - z^n)} = -\sum_{k=1}^{\infty} \text{Log}(1 - \alpha z^k)$$

whenever $|z| < \min\{1, 1/|\alpha|\}$.

(h) For $|z| < 1$, we have

$$\sum_{n=1}^{\infty} \frac{z^n}{1 - z^n} = \sum_{n=1}^{\infty} z^{n^2} \cdot \frac{1 + z^n}{1 - z^n}.$$

(i) For $|z| < 1$, we have

$$\sum_{k=1}^{\infty} \frac{z^k}{(1 - z^k)^2} = \sum_{n=1}^{\infty} \sigma(n) z^n$$

where $\sigma(n)$ is the sum of the positive divisors of n.

(j) For $n > 1$, $n \in \mathbb{N}$, let $n = \prod_{j=1}^{n} p_j^{r_j}$, where the p_j are distinct primes and $r_j \in \mathbb{N}$.

Define $k(n) = r_1 + r_2 + \ldots + r_m$. That is, $k(n)$ is the number of factors, counting repetitions, in the unique representation of n as a product of primes. Define $k(1) = 0$. For example, $k(6) = 2$, $k(72) = 5$, $k(9) = 2$, $k(17) = 1$. Then

$$\sum_{d=1}^{\infty} (-1)^{k(d)} \frac{z^d}{1 - z^d} = \sum_{n=1}^{\infty} z^{n^2}$$

whenever $|z| < 1$. [Hint: $\sum_{d \mid n} (-1)^{k(d)} = \prod_{j=1}^{m} \sum_{s_j=0}^{r_j} (-1)^{s_j}$.]

12.   For $m, n \in \mathbb{N}$ define $c_{m,n} = (m^2 - n^2)^{-1}$ if $m \neq n$ and $c_{m,n} = 0$ if $m = n$. Then

$$\sum_{m=1}^{\infty} \sum_{n=1}^{\infty} c_{m,n} = -\frac{\pi^2}{8} = -\sum_{n=1}^{\infty} \sum_{m=1}^{\infty} c_{m,n}.$$

[Hint:

$$\sum_{n=1}^{\infty} c_{m,n} = \frac{1}{2m} \sum_{n=1}^{m-1} \left(\frac{1}{m-n} + \frac{1}{m+n} \right) - \frac{1}{2m} \sum_{k=1}^{\infty} \left(\frac{1}{k} - \frac{1}{2m+k} \right)$$

$$= \frac{1}{2m} \sum_{k=1}^{2m-1} \frac{1}{k} - \frac{1}{2m^2} - \frac{1}{2m} \sum_{k=1}^{2m} \frac{1}{k}.]$$

13.   Let the power series

$$f(z) = \sum_{n=0}^{\infty} a_n z^n, \qquad g(w) = \sum_{k=0}^{\infty} b_k w^k$$

have radii of convergence $R > 0$ and $S > 0$, respectively. For $k \geq 0$, write

$$[f(z)]^k = \sum_{n=0}^{\infty} a_n^{(k)} z^n,$$

where these series are obtained by a repeated application of the Cauchy Product rule if $k > 1$. Then these series all have radius of convergence R. Write

$$A_n = \sum_{k=0}^{\infty} b_k a_n^{(k)} \qquad\qquad (*)$$

whenever these series converge. We seek conditions on z which assure that the A_n's all exist and

$$g(f(z)) = \sum_{n=0}^{\infty} A_n z^n. \qquad\qquad (**)$$

(a) We have $a_0^{(0)} = 1$, $a_n^{(0)} = 0$ for $n > 0$, and

$$a_n^{(k+1)} = \sum_{j=0}^{n} a_j a_{n-j}^{(k)}$$

for $n \geq 0$, $k \geq 0$.

(b) If $|z| < R$, and $\sum_{n=0}^{\infty} |a_n z^n| < S$, then the series $(*)$ all converge and $(**)$ obtains.

[Hint: If the numbers $\alpha_n^{(k)}$ are formed from the $|a_n|$ just as the $a_n^{(k)}$ were formed from the a_n, then $|a_n^{(k)}| \leq \alpha_n^{(k)}$.]

(c) If $0 < r < R$, $M_r = \sup\{|f(z)| : |z| = r\}$, $|a_0| < S$, and $|z| < (S - |a_0|)r/$ $(M_r + S - |a_0|)$, then $\sum\limits_{n=0}^{\infty} |a_n z^n| < S$, and so (**) obtains. [Hint: Use Cauchy's Estimate.]

(d) If $|a_0| < S$, then there is some $T > 0$ such that (**) obtains whenever $|z| < T$.

(e) If $S = \infty$, then (**) obtains whenever $|z| < R$.

14. Let $h(z) = \sum\limits_{n=0}^{\infty} c_n z^n$ have radius of convergence $R > 0$ and suppose that $h(0) = c_0 \neq 0$. Then there exists a $T > 0$ and a power series expansion

$$\frac{1}{h(z)} = \sum_{n=0}^{\infty} A_n z^n$$

that is valid whenever $|z| < T$. Moreover, the largest such T satisfies

$$T \geq r|c_0|/(M_r + |c_0|)$$

whenever $0 < r < R$ and $M_r = \sup\{|h(z) - c_0| : |z| = r\}$. [Hint: Use the preceding exercise. Put $f(z) = h(z) - c_0$ and

$$g(w) = \frac{1}{c_0 + w} = \sum_{k=0}^{\infty} (-1)^k c_0^{-k-1} w^k.]$$

15. Suppose that $f(z) = \sum\limits_{n=0}^{\infty} c_n z^n$ has radius of convergence $R = \infty$. Suppose also that there exist an integer $p \geq 0$ and positive constants A and B such that

$$|f(z)| \leq A + B|z|^p$$

for all $z \in \mathbb{C}$. Then $c_n = 0$ for all $n > p$. That is, f is a polynomial of degree no larger than p. The case $p = 0$ is known as *Liouville's Theorem*. [Hint: Use Cauchy's Estimate.]

16. Let $f(z) = \sum\limits_{n=0}^{\infty} c_n z^n$ have radius of convergence $R > 0$ and let $|a| < R$. Let R_a be the radius of convergence of

$$f_a(z) = \sum_{k=0}^{\infty} \frac{f^{(k)}(a)}{k!} (z - a)^k.$$

(a) We have $R_a \geq R - |a|$ and $f(z) = f_a(z)$ whenever $|z - a| < R - |a|$. [Hint: Write $z = (z - a) + a$ in the first series, apply the Binomial Theorem, and rearrange a double series.]

(b) In fact, $R - |a| \leq R_a \leq R + |a|$. [Hint: Apply the same reasoning as in (a) to the second series by expanding $(z - a)^k$ in powers of z and conclude, as in (a), that $R \geq R_a - |a|$.]

(c) Either bound in (b) may be attained or both inequalites may be strict. Show this by proving that, in the case of the geometric series $\sum\limits_{n=0}^{\infty} z^n$, we have $R = 1$, $R_{1/2} = 1/2$, $R_{-1/2} = 3/2$, and $R_{3i/4} = 5/4$. The underlying key here is that the new open disks of convergence cannot include the "singular point" $z = 1$, a fact revealed in analytic function theory.

The Gamma Function

Aside from the so-called "elementary functions" studied earlier in this book [particularly Chapter 5], the special function that occurs most frequently in analysis is undoubtedly the Gamma Function. Because of its inescapable importance, and as a good application of many of our analytical techniques, we now make a brief study of this function.

The Gamma Function was first defined by Euler for $\mathrm{Re}(z) > 0$ by means of an integral [see (7.65)]. However, it is more convenient to follow Weierstrass in defining it by an infinite product. We begin with the following theorem.

(7.59) Theorem *The infinite product*

(i) $$F(z) = \prod_{n=1}^{\infty} \left[\left(1 + \frac{z}{n}\right) e^{-z/n} \right]$$

converges absolutely for all $z \in \mathbb{C}$. Thus, (i) defines a function $F : \mathbb{C} \to \mathbb{C}$ such that $F(z) = 0$ if and only if z is a negative integer. Moreover, there is a power series that converges to F at every $z \in \mathbb{C}$.

Proof Let R be any positive real number. Choose an integer $N > 2R$. Notice that $|z| \leq R$ and $n > N$ imply that $|z/n| < 1/2$; hence $|\mathrm{Arg}(1 + z/n) + \mathrm{Arg}\, e^{-z/n}| < \pi/6 + 1/2 < \pi$, and so

$$\mathrm{Log}\left[\left(1 + \frac{z}{n}\right) e^{-z/n}\right] = -\frac{z}{n} + \mathrm{Log}\left(1 + \frac{z}{n}\right)$$

$$= -\frac{z^2}{2n^2} + \frac{z^3}{3n^3} - \frac{z^4}{4n^4} + \cdots . \qquad (1)$$

This implies that

$$\left|\mathrm{Log}\left[\left(1 + \frac{z}{n}\right) e^{-z/n}\right]\right| \leq \left|\frac{z}{n}\right|^2 \left(1 + \left|\frac{z}{n}\right| + \left|\frac{z}{n}\right|^2 + \cdots\right)$$

$$< \frac{N^2}{4n^2}\left(1 + \frac{1}{2} + \frac{1}{2^2} + \cdots\right) = \frac{N^2}{2n^2}$$

whenever $|z| \leq R$ and $n > N$. Since $\sum 1/n^2 < \infty$, the Weierstrass M-test shows that the series

$$G_N(z) = \sum_{n=N+1}^{\infty} \mathrm{Log}\left[\left(1 + \frac{z}{n}\right) e^{-z/n}\right] \qquad (2)$$

converges absolutely and uniformly for $|z| \leq R$. We draw two conclusions. First, by (7.32) and (7.33), (i) converges absolutely at every $z \in \mathbb{C}$ [we could have chosen $R > |z|$], and so F is defined and $F(z) = 0$ only if some factor is 0. Second, since (1) shows that every summand in (2) has a power series expansion valid for $|z| \leq R$, the Weierstrass Double Series Theorem (7.57) entails that G_N is infinitely differentiable for $|z| < R$, all derivatives of G_N are obtained by termwise differentiation of (2), and G_N has a power series expansion valid for $|z| < R$. Now

$$F(z) = \exp\left[G_N(z) \right] \cdot \prod_{n=1}^{N} \left[\left(1 + \frac{z}{n} \right) e^{-z/n} \right]$$

for $|z| < R$; hence, upon substituting the power series for G_N, multiplying power series, and rearranging an absolutely convergent double series (7.50) [see Exercise 13(e) on p. 459], we obtain a power series that converges to $F(z)$ for all $|z| < R$. By (4.63), this latter series must be

$$F(z) = \sum_{n=0}^{\infty} \frac{F^{(n)}(0)}{n!} z^n. \tag{3}$$

But (3) does not depend on R [or N] and R is arbitrary, so (3) obtains for all $z \in C$. $\square$

(7.60)   Definition   Let F be as in (7.59). We define the *Gamma Function* Γ at each $z \in C$ that is neither 0 nor a negative integer by the formula

$$\Gamma(z) = \{ z e^{\gamma z} F(z) \}^{-1} = \left\{ z e^{\gamma z} \prod_{n=1}^{\infty} \left[\left(1 + \frac{z}{n} \right) e^{-z/n} \right] \right\}^{-1},$$

where

$$\gamma = \lim_{n \to \infty} \left(\sum_{k=1}^{n} \frac{1}{k} - \log n \right)$$

is Euler's constant as in (7.5.a).

(7.61)   Theorem   *The function Γ is well defined, never zero, and differentiable on $D = C\backslash\{0, -1, -2, \ldots \}$. The reciprocal of Γ, defined to equal 0 on $C\backslash D$, is given by*

$$\frac{1}{\Gamma(z)} = z e^{\gamma z} \prod_{n=1}^{\infty} \left[\left(1 + \frac{z}{n} \right) e^{-z/n} \right]$$

and is the sum of a power series that converges at each $z \in C$.

Proof   This is immediate from (7.60) and (7.59). $\square$

(7.62)   Theorem   [Euler, 1729]   *For each $n \in N$ and $z \in D = C\backslash\{0, -1, -2, \ldots \}$, define*

$$\Gamma_n(z) = \frac{n! \cdot n^z}{z(z+1)(z+2) \ldots (z+n)}.$$

Then

$$\lim_{n \to \infty} \Gamma_n(z) = \Gamma(z)$$

for all $z \in D$.

Proof   This is obvious from (7.61) and the computation

$$\frac{1}{\Gamma(z)} = z \cdot \lim_{n \to \infty} \left\{ e^{(1+1/2+\cdots+1/n-\log n)z} \cdot \prod_{k=1}^{n} \left[\left(1 + \frac{z}{k} \right) e^{-z/k} \right] \right\}$$

$$= z \cdot \lim_{n \to \infty} \left\{ n^{-z} \prod_{k=1}^{n} \left(\frac{z+k}{k} \right) \right\} = \lim_{n \to \infty} \frac{1}{\Gamma_n(z)} \cdot \quad \square$$

Our next theorem, though simple, reveals one of the most important properties of the Gamma Function; indeed, it shows that this function solves an interpolation problem that was the impetus for its invention.

(7.63)   Theorem   *For $z \in D$ [as in (7.62)], we have*

(i) $\Gamma(z + 1) = z\Gamma(z)$.

It follows that

(ii) $\Gamma(n + 1) = n!$

for each nonnegative integer n.

Proof   Let Γ_n be as in (7.62). Then $z \in D$ implies

$$\Gamma(z + 1)/\Gamma(z) = \lim_{n \to \infty} \left[\Gamma_n(z + 1)/\Gamma_n(z) \right]$$

$$= \lim_{n \to \infty} \frac{z(z+1)\ldots(z+n) \cdot n! \cdot n^{z+1}}{(z+1)\ldots(z+n)(z+1+n) \cdot n! \cdot n^z}$$

$$= \lim_{n \to \infty} \frac{z \cdot n}{z+n+1} = \lim_{n \to \infty} \frac{z}{1+(z+1)/n} = z,$$

and so (i) is proved. Since $\Gamma_n(1) = n! \cdot n/(n+1)! = n/(n+1)$ for all $n \in \mathbb{N}$, we have $\Gamma(0 + 1) = \Gamma(1) = 1 = 0!$, and so (ii) holds for $n = 0$. By (i), $\Gamma((n + 1) + 1) = (n + 1)\Gamma(n + 1)$. Therefore (ii) follows by induction. $\square$

The Gamma Function is connected to the Sine Function as follows.

(7.64)   Theorem   *For all $z \in \mathbb{C}$ we have*

$$\sin(\pi z) = \pi/(\Gamma(z)\Gamma(1 - z))$$

where, as in (7.61), the right side is 0 if z or $1 - z$ is a nonpositive integer. In particular, $\Gamma(1/2) = \sqrt{\pi}$.

Proof   If z is not an integer, then it follows from (7.63) and (7.61) that

$$\frac{1}{\Gamma(z)\Gamma(1 - z)} = \frac{1}{\Gamma(z)\Gamma(-z)(-z)}$$

$$= \left\{ ze^{\gamma z} \prod_{n=1}^{\infty} \left(1 + \frac{z}{n} \right) e^{-z/n} \right\} \cdot \left\{ e^{-\gamma z} \prod_{n=1}^{\infty} \left(1 - \frac{z}{n} \right) e^{z/n} \right\}$$

$$= z \prod_{n=1}^{\infty} \left(1 - \frac{z^2}{n^2} \right) = \frac{\sin(\pi z)}{\pi},$$

where the last equality follows from (5.40). If z is an integer, both sides are 0. For $z = 1/2$ we obtain $[\Gamma(1/2)]^2 = \pi$. Since it is clear from (7.61) that $\Gamma(x) > 0$ for $x > 0$, we obtain $\Gamma(1/2) = \sqrt{\pi}$. $\square$

Now we come to Euler's original definition of Γ via an integral.

(7.65) Theorem *For* $\mathrm{Re}(z) > 0$, *we have*

(i)
$$\Gamma(z) = \int_0^\infty e^{-t} t^{z-1}\, dt,$$

where, as usual, $t^{z-1} = \exp[(z-1)\log t]$ *is the principal value of the power.*

Proof Let $z \in \mathbb{C}$ with $\mathrm{Re}(z) = x > 0$ be given and fixed. Choose $a \geq 1$ such that $t^{2x} < e^t$ for all $t \geq a$. Then

$$\int_0^\infty |e^{-t} t^{z-1}|\, dt = \int_0^\infty e^{-t} t^{x-1}\, dt < \int_0^a t^{x-1}\, dt + \int_a^\infty e^{-t} t^x\, dt$$

$$< \frac{a^x}{x} + \int_a^\infty e^{-t/2}\, dt = \frac{a^x}{x} + 2e^{-a/2} < \infty,$$

and so the integrand in (i) is in $L_1([0, \infty[)$. Integration by parts shows that, for $n \in \mathbb{N}$, we have

$$\int_0^1 (1 - u)^n u^{z-1}\, du = \frac{n}{z} \int_0^1 (1 - u)^{n-1} u^z\, du$$

$$= \frac{n(n-1)}{z(z+1)} \int_0^1 (1 - u)^{n-2} u^{z+1}\, du$$

$$= \cdots = \frac{n(n-1)\cdots 2 \cdot 1}{z(z+1)\cdots(z+n-1)} \int_0^1 u^{z+n-1}\, du$$

$$= \Gamma_n(z) \cdot n^{-z},$$

where Γ_n is as in (7.62). The substitution $nu = t$ then shows that

$$\int_0^n (1 - t/n)^n t^{z-1}\, dt = \Gamma_n(z) \tag{1}$$

for all $n \in \mathbb{N}$. By (7.62), it suffices to show that the limit as $n \to \infty$ of the left side of (1) is the right side of (i). We claim that

$$0 \leq e^{-t} - (1 - t/n)^n \leq e^{-t} t^2/n \quad \text{for } n \geq 1 \quad \text{and} \quad 0 \leq t \leq n. \tag{2}$$

If (2) could be proved, we would have

$$\left| \int_0^\infty e^{-t} t^{z-1}\, dt - \int_0^n (1 - t/n)^n t^{z-1}\, dt \right|$$

$$\leq \left| \int_0^n [e^{-t} - (1 - t/n)^n] t^{z-1}\, dt \right| + \left| \int_n^\infty e^{-t} t^{z-1}\, dt \right|$$

$$\leq \frac{1}{n} \int_0^n e^{-t} t^{x+1}\, dt + \int_n^\infty e^{-t} t^{x-1}\, dt < \frac{1}{n} \int_0^\infty e^{-t} t^{x+1}\, dt + \int_n^\infty e^{-t} t^{x-1}\, dt$$

$$\to 0 \quad \text{as} \quad n \to \infty$$

and the proof would be complete. To prove (2), first notice that

$$1 + u \leqq e^u \leqq (1 - u)^{-1} \quad \text{for } 0 \leqq u < 1$$

[compare power series] so, taking $u = t/n$, $(1 + t/n)^{-n} \geqq e^{-t} \geqq (1 - t/n)^n$ for $0 \leqq t \leqq n$. Therefore

$$0 \leqq e^{-t} - (1 - t/n)^n = e^{-t}\left[1 - e^t(1 - t/n)^n\right]$$

$$\leqq e^{-t}\left[1 - (1 - t^2/n^2)^n\right] \leqq e^{-t}\left[1 - (1 - nt^2/n^2)\right]$$

$$= e^{-t}t^2/n$$

for $0 \leqq t \leqq n$, where the last inequality is an application of Bernoulli's Inequality (1.46): $(1 + v)^n \geqq 1 + nv$ for $v > -1$ and $n \in \mathbb{N}$. $\square$

(7.66)　Corollary　*For* $\text{Re}(z) > 0$, *we have*

$$\Gamma(z) = 2\int_0^\infty e^{-u^2}u^{2z-1}\, du.$$

In particular,

$$\int_0^\infty e^{-u^2}\, du = \frac{\sqrt{\pi}}{2}.$$

Proof　Substitute $t = u^2$ in (7.65.i). Then take $z = 1/2$ and use (7.64). $\square$

There is also a series representation of Γ as follows.

(7.67)　Theorem　*For* $\text{Re}(z) > 0$, *we have*

(i)
$$\Gamma(z) = \sum_{n=0}^\infty \frac{(-1)^n}{n!\,(z+n)} + \sum_{n=0}^\infty \frac{c_n}{n!}z^n$$

where

$$c_n = \int_1^\infty e^{-t}t^{-1}(\log t)^n\, dt.$$

The first of these series converges absolutely and uniformly on every compact subset of $\mathbb{C}\backslash\{0, -1, -2, \dots\}$ *and the second converges at every* $z \in \mathbb{C}$.*

Proof　For any $z \in \mathbb{C}$, we have

$$\int_1^\infty e^{-t}t^{z-1}\, dt = \int_1^\infty e^{-t}t^{-1}e^{z\log t}\, dt$$

$$= \int_1^\infty e^{-t}t^{-1}\sum_{n=0}^\infty \frac{1}{n!}(\log t)^n z^n\, dt$$

$$= \sum_{n=0}^\infty \frac{1}{n!}\int_1^\infty e^{-t}t^{-1}(\log t)^n\, dt\, z^n, \tag{1}$$

---

*Using the Identity Theorem of analytic function theory, it follows from (7.67) and (7.61) that equality (7.67.i) holds good for all $z \in \mathbb{C}\backslash\{0, -1, -2, \dots\}$.

where the interchange of $\int$ and $\sum$ is justified by Dominated Convergence (6.22). In fact,

$$\left| e^{-t}t^{-1} \sum_{n=0}^{N} \frac{1}{n!} (\log t)^n z^n \right| \le e^{-t} t^{|z|-1} = g(t)$$

for all $t \ge 1$ and all $N \ge 0$ and g has a finite integral over $[1, \infty[$.

If $\text{Re}(z) > 0$, then

$$\int_0^1 e^{-t} t^{z-1} dt = \int_0^1 \sum_{n=0}^{\infty} \frac{(-1)^n t^{n+z-1}}{n!} dt$$

$$= \sum_{n=0}^{\infty} \frac{(-1)^n}{n!} \int_0^1 t^{n+z-1} dt = \sum_{n=0}^{\infty} \frac{(-1)^n}{n!(z+n)}, \qquad (2)$$

where the interchange is justified just as above. In fact,

$$\left| \sum_{n=0}^{N} \frac{(-1)^n t^{n+z-1}}{n!} \right| \le e^t t^{\text{Re}(z-1)} = h(t)$$

whenever $0 \le t \le 1$ and $N \ge 0$, where h has a finite integral over $[0, 1]$. Now (i) follows if we add (1) and (2) and apply (7.65).

Finally, let $K \subset \mathbb{C} \setminus \{0, -1, -2, \dots\}$ be compact. Choose a positive number δ such that $|z + n| \ge \delta$ whenever $z \in K$ and n is a nonnegative integer. Then

$$\left| \frac{(-1)^n}{n!(z+n)} \right| \le \frac{1}{n!\,\delta}$$

for all such n and z. Thus, the last series in (2) converges absolutely and uniformly on K by the Weierstrass M-test. $\square$

We next define another function that is closely related to Γ. Its usefulness stems from Theorem (7.69).

(7.68) **Definition** For complex numbers p and q, neither of which is a nonpositive integer, define

$$B(p, q) = \Gamma(p)\Gamma(q)/\Gamma(p + q).$$

This function B is called the Beta Function.

(7.69) **Theorem** *For* $\text{Re}(p) > 0$ *and* $\text{Re}(q) > 0$, *we have*

(i) $\quad B(p, q) = \int_0^1 t^{p-1}(1 - t)^{q-1} dt = 2 \int_0^{\pi/2} (\cos^{2p-1}\theta) \cdot (\sin^{2q-1}\theta) \, d\theta.$

Proof Writing $a = \text{Re}(p) > 0$ and $b = \text{Re}(q) > 0$, we see that the first integrand in (i) is in $L_1([0, 1])$ because

$$\int_0^1 |t^{p-1}(1 - t)^{q-1}| \, dt = \int_0^1 t^{a-1}(1 - t)^{b-1} \, dt < \infty.$$

Now the second equality in (i) follows from the substitution $t = \cos^2\theta$. Use (7.66) and (6.121) to write

$$\Gamma(p)\Gamma(q) = 2\int_0^\infty e^{-x^2}x^{2p-1}\,dx \cdot 2\int_0^\infty e^{-y^2}y^{2q-1}\,dy$$

$$= 4\int_0^\infty\int_0^\infty e^{-(x^2+y^2)}x^{2p-1}y^{2q-1}\,dx\,dy$$

$$= 4\int_Q e^{-(x^2+y^2)}x^{2p-1}y^{2q-1}\,d(x,y),$$

where $Q = \{(x, y) \in \mathbb{R}^2:\ x > 0, y > 0\}$. The last integrand is in $L_1(Q)$ by (6.123), and so according to (6.148), we may transform that integral to polar coordinates to obtain

$$\Gamma(p)\Gamma(q) = 4\int_0^\infty\int_0^{\pi/2} e^{-r^2}(r\cos\theta)^{2p-1}(r\sin\theta)^{2q-1}r\,d\theta\,dr$$

$$= 4\int_0^\infty e^{-r^2}r^{2p+2q-1}\int_0^{\pi/2}(\cos^{2p-1}\theta)(\sin^{2q-1}\theta)\,d\theta\,dr$$

$$= 2\int_0^{\pi/2}(\cos^{2p-1}\theta)(\sin^{2q-1}\theta)\,d\theta\cdot\Gamma(p+q),$$

where the last equality uses (7.66) again. Since $\Gamma(p + q) \neq 0$, we may divide by it to complete the proof. $\square$

(7.70) Examples (a) Let $0 < \text{Re}(p) < 1$. Then we can apply (7.69) and (7.64) to write

$$\int_0^{\pi/2}(\cot\theta)^{2p-1}\,d\theta = \int_0^{\pi/2}(\cos^{2p-1}\theta)(\sin^{2(1-p)-1}\theta)\,d\theta$$

$$= (1/2)B(p, 1-p) = (1/2)\Gamma(p)\Gamma(1-p)$$

$$= \frac{\pi/2}{\sin(\pi p)}.$$

The substitution of $\pi/2 - \theta$ for θ yields

$$\int_0^{\pi/2}(\tan\theta)^{2p-1}\,d\theta = \int_0^{\pi/2}(\cot\theta)^{2p-1}\,d\theta = \frac{\pi/2}{\sin(\pi p)}$$

whenever $0 < \text{Re}(p) < 1$.

(b) Using (7.65), the substitution $t = x\log b$ shows that

$$\int_0^\infty x^a b^{-x}\,dx = \frac{\Gamma(a+1)}{(\log b)^{a+1}}$$

whenever $b > 1$ and $\text{Re}(a) > -1$. For instance, if $a = 1/2$ and $b = 2$, we get

$$\int_0^\infty \sqrt{x}\,/2^x\,dx = \frac{\sqrt{\pi}}{2}(\log 2)^{-3/2}$$

because $\Gamma(1/2 + 1) = (1/2)\Gamma(1/2) = \sqrt{\pi}\,/2$.

(c) If we make the substitution $t = \tan^2\theta$, we obtain

$$B(p,q) = B(q,p) = \int_0^\infty \frac{t^{p-1}\,dt}{(1+t)^{p+q}}$$

whenever $\text{Re}(p) > 0$ and $\text{Re}(q) > 0$.

The function Γ has a logarithm with some nice properties, as we now show.

(7.71) **Theorem** *The series*

$$L(z) = -\gamma z - \text{Log}\,z - \sum_{n=1}^\infty \left[\text{Log}\left(1 + \frac{z}{n}\right) - \frac{z}{n} \right] \qquad (*)$$

converges absolutely and uniformly on every bounded subset of $D = \mathbb{C}\backslash\{z \in \mathbb{R} : z \leq 0\}$.
The function L thereby defined is infinitely differentiable on D, and for all $z \in D$ we have

(i) $\exp[L(z)] = \Gamma(z)$,
(ii) $\Gamma'(z) = \Gamma(z)L'(z)$,

(iii) $L'(z) = -\gamma - \dfrac{1}{z} + z \cdot \displaystyle\sum_{n=1}^\infty \frac{1}{n(z+n)}$,

(iv) $L^{(k)}(z) = (-1)^k (k-1)! \cdot \displaystyle\sum_{n=0}^\infty (z+n)^{-k}$ *for* $k \geq 2$.

Here γ denotes Euler's constant.

Because of (i) and the fact that L is real-valued on $]0, \infty[$, we call L the principal branch of $\log \Gamma$.

Proof Let $S \subset D$ be bounded. Choose $R \in \mathbb{R}$ such that $|z| < R - 1$ for all $z \in S$ and then choose any integer $N > 2R$. Just as in the proof of (7.59), the function G_N given by

$$G_N(z) = \sum_{n=N+1}^\infty \left[\text{Log}\left(1 + \frac{z}{n}\right) - \frac{z}{n} \right] \qquad (1)$$

is infinitely differentiable for $|z| < R$, these derivatives may be obtained by termwise differentiation of (1), and the series (1) is absolutely and uniformly convergent on the disk $|z| \leq R$. Thus, the series $(*)$ is uniformly and absolutely convergent on S. Moreover,

$$L(z) = -\gamma z - \text{Log}\,z - \sum_{n=1}^N \left[\text{Log}\left(1 + \frac{z}{n}\right) - \frac{z}{n} \right] - G_N(z), \qquad (2)$$

so L is infinitely differentiable on $\{z \in D : |z| < R\}$, and these derivatives are obtainable by termwise differentiation of $(*)$. This proves (iii) and (iv) because R is arbitrary. By (2), the proof of (7.59), and (7.60), we have

$$\exp[L(z)] = e^{-\gamma z}z^{-1}e^{-G_N(z)}\left\{ \prod_{n=1}^N \left[\left(1 + \frac{z}{n}\right)e^{-z/n}\right] \right\}^{-1} = [e^{\gamma z}zF(z)]^{-1} = \Gamma(z)$$

for $z \in D$, $|z| < R$. Since R is arbitrary, this proves (i). Finally, (ii) follows from (i) by differentiation. $\square$

To estimate the growth of Γ as $|z| \to \infty$, we prove the following version of *Stirling's Formula*.

(7.72)   Theorem   *Let $0 < \delta < \pi$ and let $W(\delta) = \{z \in C : z \neq 0, -\pi + \delta \leq \operatorname{Arg} z \leq \pi - \delta\}$. Then*

$$\lim_{\substack{|z| \to \infty \\ z \in W(\delta)}} \left[\sqrt{2\pi}\, e^{-z} z^{z-1/2} / \Gamma(z) \right] = 1.$$

Proof   Fix any $z \in W(\delta)$ and write $f(t) = \operatorname{Log}(t + z)$ for $t \in \mathbb{R}$. As in Euler's Summation Formula, let $\psi_1(t) = t - [t] - 1/2$, where $[t]$ is the integer satisfying $[t] \leq t < [t] + 1$, be the first Bernoulli function. For an integer $n \geq 0$, integration by parts gives

$$\frac{1}{2}(f(n) + f(n+1)) = \int_n^{n+1} f(t)\, dt + \int_n^{n+1} (t - n - 1/2)f'(t)\, dt$$

$$= \int_n^{n+1} f(t)\, dt + \int_n^{n+1} \psi_1(t) f'(t)\, dt.$$

Therefore, summing for $n = 0, 1, \ldots, N - 1$, we have

$$\sum_{n=0}^{N} \operatorname{Log}(n + z) = \sum_{n=0}^{N} f(n) = \frac{1}{2} f(N) + \frac{1}{2} f(1) + \int_0^N f(t)\, dt + \int_0^N \psi_1(t) f'(t)\, dt.$$

Thus

$$\sum_{n=0}^{N} \operatorname{Log}(n + z) = -\left(z - \frac{1}{2}\right)\operatorname{Log} z + \frac{1}{2} \operatorname{Log}(z + N)$$

$$+ (z - 1)\operatorname{Log}(z + N) + (N + 1)\operatorname{Log}(z + N)$$

$$- N + \int_0^N \frac{\psi_1(t)}{z + t}\, dt \tag{1}$$

Denote this last integral by $I_N(z)$. Now put $z = 1$ in (1) and subtract the result from (1) to obtain

$$\sum_{n=0}^{N} \operatorname{Log} \frac{z + n}{1 + n} = -\left(z - \frac{1}{2}\right)\operatorname{Log} z + \frac{1}{2} \operatorname{Log} \frac{z + N}{1 + N} + (z - 1)\operatorname{Log}(z + N)$$

$$+ (N + 1)\operatorname{Log} \frac{z + N}{1 + N} + I_N(z) - I_N(1). \tag{2}$$

Now write

$$L_N(z) = (z - 1)\operatorname{Log} N - \sum_{n=0}^{N} \operatorname{Log} \frac{z + n}{1 + n}. \tag{3}$$

It follows from (2) that

$$L_N(z) = \left(z - \frac{1}{2}\right)\operatorname{Log} z - (N + 1)\operatorname{Log} \frac{z + N}{1 + N} - I_N(z) + I_N(1) + A_N(z),$$

$$\tag{4}$$

where $A_N(z) \to 0$ as $N \to \infty$. If we apply exp to both sides of (3), we find that

$$\lim_{N \to \infty} e^{L_N(z)} = \lim_{N \to \infty} \left[(1 + 1/N) \Gamma_N(z) \right] = \Gamma(z) \qquad (5)$$

by (7.62). Since $w^{-1} \text{Log}(1 + w) \to 1$ as $w \to 0$, we can write $w_N = (z - 1)/(1 + N)$ and obtain

$$\lim_{N \to \infty} (N + 1) \text{Log} \frac{z + N}{1 + N} = z - 1. \qquad (6)$$

If we write $\psi_2(t) = \int_0^t \psi_1(u) \, du$, it is clear that $\psi_2(m) = 0$ for all integers m and that $|\psi_2(t)| \le 1/8$ for all $t \in \mathbb{R}$. Integration by parts yields

$$I_N(z) = \int_0^N \frac{\psi_1(t)}{z + t} \, dt = \int_0^N \frac{\psi_2(t)}{(z + t)^2} \, dt,$$

and so

$$I(z) = \lim_{N \to \infty} I_N(z) \in \mathbb{C} \qquad (7)$$

because the last integrand is in $L_1([0, \infty[)$. Now apply exp to (4), let $N \to \infty$ and apply (5)–(7) to obtain

$$\Gamma(z) = z^{z-1/2} e^{-z} e^{-I(z)} \cdot C \qquad (8)$$

where $C = e^{1+I(1)}$ does not depend on z. If we write $z = re^{i\theta}$, where $r > 0$ and $-\pi + \delta \le \theta \le \pi - \delta$ [recall that $z \in W(\delta)$], we can make the substitution $t = ru$ and obtain

$$|I(z)| \le \frac{1}{8} \int_0^\infty |re^{i\theta} + t|^{-2} \, dt = \frac{1}{8r} \int_0^\infty |e^{i\theta} + u|^{-2} \, du$$

$$= \frac{1}{8|z|} \int_0^\infty \frac{du}{u^2 + 2u(\cos\theta) + 1}$$

$$\le \frac{1}{8|z|} \int_0^\infty \frac{du}{u^2 - 2u(\cos\delta) + 1} = \frac{\beta(\delta)}{|z|},$$

where $\beta(\delta) < \infty$ because the last integrand is bounded on $[0, \infty[$ and is less than $2/u^2$ for all large u. Therefore $I(z) \to 0$ as $|z| \to \infty$ with $z \in W(\delta)$. In view of (8), we need only show that $C = \sqrt{2\pi}$. But this follows from (5.44) since, if we take $z = n + 1$ in (8), we have

$$C = n! \, (n + 1)^{-(n+1/2)} e^{n+1} e^{I(n+1)}$$

$$= n! \cdot n^{-(n+1/2)} e^n \left[(1 + 1/n)^{-(n+1/2)} e^{1+I(n+1)} \right]$$

$$\to \sqrt{2\pi} \quad \text{as } n \to \infty$$

because $(1 + 1/n)^{n+1/2} \to e$. $\quad \square$

Exercises

1. If n is a nonnegative integer, then
   (a)
   $$\Gamma(z) = \Gamma(z + n + 1)/[z(z + 1)(z + 2) \ldots (z + n)]$$
   whenever $z \in C\backslash\{0, -1, -2, \ldots \}$,
   (b)
   $$\lim_{z \to -n} (z + n)\Gamma(z) = (-1)^n/n! ,$$
   (c)
   $$\Gamma(n + 1/2) = (2n)!\sqrt{\pi} /[2^{2n}n!],$$
   (d)
   $$\Gamma(-n + 1/2) = (-1)^n 2^{2n} n!\sqrt{\pi} /(2n)! .$$

2. Let $a_1, \ldots, a_k, b_1, \ldots, b_k$ be complex numbers such that $a_1 + \ldots + a_k = b_1 + \ldots + b_k$ and none is a nonnegative integer. Then
   (a)
   $$\prod_{n=1}^{\infty} \frac{(n - a_1) \ldots (n - a_k)}{(n - b_1) \ldots (n - b_k)} = \prod_{j=1}^{k} \frac{b_j\Gamma(-b_j)}{a_j\Gamma(-a_j)} .$$
   Moreover, if $a_1 + \ldots + a_k = 0$, then
   (b)
   $$\prod_{n=1}^{\infty} \left[\left(1 - \frac{a_1}{n}\right)\left(1 - \frac{a_2}{n}\right) \ldots \left(1 - \frac{a_k}{n}\right)\right] = (-1)^k \left[\prod_{j=1}^{k} a_j\Gamma(-a_j)\right]^{-1}.$$

   [Hints: Write the nth factor in (a) as
   $$\prod_{j=1}^{k} \left\{\left[\left(1 - \frac{a_j}{n}\right)e^{a_j/n}\right]\left[\left(1 - \frac{b_j}{n}\right)e^{b_j/n}\right]^{-1}\right\}.\right]$$

   (c) If $x \in C$, $k > 1$ is an integer, and $c = e^{2\pi i/k}$, then
   $$x^k \prod_{n=1}^{\infty} \left(1 - \frac{x^k}{n^k}\right) = -\left[\prod_{j=1}^{k} \Gamma(-xc^j)\right]^{-1}.$$

3. *Gauss's Multiplication Formula.* If $k \in \mathbb{N}$, then
   $$\prod_{n=0}^{k-1} \Gamma\left(z + \frac{n}{k}\right) = \sqrt{k} \cdot (2\pi)^{(k-1)/2} \cdot k^{-kz} \cdot \Gamma(kz)$$
   whenever $kz \in C\backslash\{0, -1, -2, \ldots \}$. The case $k = 2$
   $$\Gamma(z)\Gamma(z + 1/2) = \sqrt{\pi} \, 2^{1-2z}\Gamma(2z)$$
   is the *Duplication Formula* due to Legendre. [Hints: Define f by
   $$k\Gamma(kz)f(z) = k^{kz} \cdot \prod_{n=0}^{k-1} \Gamma\left(z + \frac{n}{k}\right).$$

Use (7.62) to show that f is a constant. Then $f(z) = f\left(\frac{1}{k}\right) = \prod_{n=1}^{k-1} \Gamma\left(\frac{n}{k}\right)$. Use (7.64)

to evaluate the square of this last product. Finally, prove that $\prod_{n=1}^{k-1} \sin \frac{n\pi}{k} = \frac{k}{2^{k-1}}$

by substituting $x = 1$ and $x = -1$ into the polynomial

$$\sum_{j=0}^{k-1} x^{2j} = \frac{(x^{2k} - 1)}{(x^2 - 1)} = \prod_{n=1}^{k-1}\left[x^2 - 2x \cos \frac{n\pi}{k} + 1\right]$$

and then multiplying the results.]

4. For $t \in \mathbb{R}$, $t \neq 0$, we have
    (a) $|\Gamma(it)|^2 = 2\pi/[t(e^{\pi t} - e^{-\pi t})]$
    (b) $|\Gamma(1/2 + it)|^2 = 2\pi/[e^{\pi t} + e^{-\pi t}]$
    [Hint: Use (7.64) and $\Gamma(\bar{z}) = \overline{\Gamma(z)}$.]

5. Let m and n be nonnegative integers. Express the value of the integral

    $$\int_0^{\pi/2}(\cos^m \theta)(\sin^n \theta)\, d\theta$$

    in terms of π and factorials.

6. Evaluate the following integrals in terms of Γ or B.
    (a)
    $$\int_0^1 (1 - x^a)^b\, dx \qquad (a > 0, \text{Re}\,(b) > -1),$$
    (b)
    $$\int_0^\infty e^{-x^a}\, dx \qquad (a > 0),$$
    (c)
    $$\int_0^1 \frac{dx}{\sqrt{1 - x^a}} \qquad (a > 0),$$
    (d)
    $$\int_0^1 \left(\log \frac{1}{x}\right)^{a-1} x^{b-1}\, dx \qquad (b > 0, \text{Re}(a) > 0),$$
    (e)
    $$\int_0^{\pi/2} \frac{dx}{\sqrt{\cos x}},$$
    (f)
    $$\int_0^\infty \frac{x^a\, dx}{1 + x^{2a}} \qquad (a > 1)$$

7. If $-\infty < a \leq x \leq b < \infty$ and $y \in \mathbb{R}$, then
    $$\lim_{|y| \to \infty} \left[|\Gamma(x + iy)|^2 |y|^{1 - 2x} e^{\pi|y|}\right] = 2\pi.$$

8. (a) If γ is Euler's constant, then
    $$\Gamma'(1) = -\gamma \quad \text{and} \quad \Gamma'(2) = 1 - \gamma.$$
    (b) If D and L are as in (7.71), then $\Gamma'' = [L'' + (L')^2] \cdot \Gamma$ on D.

(c) On $]0, \infty[$, Γ is always positive, Γ' is strictly increasing, and Γ attains an absolute minimum value at some $x \in]1, 2[$.

(d) If n is an odd negative integer, then, on $]n, n+1[$, Γ is always negative, Γ' is strictly decreasing, and

$$\lim_{x \downarrow n} \Gamma(x) = \lim_{x \uparrow n+1} \Gamma(x) = -\infty.$$

(e) If n is an even negative integer, then, on $]n, n+1[$, Γ is always positive, Γ' is strictly increasing, and $\lim_{x \downarrow n} \Gamma(x) = \lim_{x \uparrow n+1} \Gamma(x) = \infty.$

(f) $\lim_{x \downarrow 0} \Gamma(x) = \lim_{x \to \infty} \Gamma(x) = \infty.$

(g) On $]0, \infty[$, Γ is logarithmically convex and therefore convex.

(h) If $x_1, x_2, \ldots, x_n$ are positive real numbers, then

$$\left[\prod_{n=1}^{n} \Gamma(x_j) \right]^{1/n} \ge \Gamma\left(\frac{1}{n} \sum_{n=1}^{n} x_j \right)$$

and equality obtains if and only if $x_1 = x_2 = \ldots = x_n$.

9.   If $a > 0$, then

$$\lim_{x \uparrow 1} (1 - x)^{1/a} \cdot \sum_{n=0}^{\infty} x^{n^a} = \frac{1}{a} \Gamma\left(\frac{1}{a} \right).$$

[Hint: Write $f(t) = \exp(-t^a)$ and $x = f(h)$. Check that

$$\lim_{h \downarrow 0} h \sum_{n=0}^{\infty} f(nh) = \int_0^{\infty} f(t)\, dt.]$$

10.   If $a > 0$ and $\mathrm{Re}(s) > 1$, then

$$\sum_{n=1}^{\infty} (a + n)^{-s} = \frac{1}{\Gamma(s)} \int_0^{\infty} \frac{e^{-at} t^{s-1}\, dt}{e^t - 1}.$$

[Hint: Write the integrand as a geometric series.]

11.   Let γ be Euler's constant. Then

(a)

$$\gamma = \lim_{n \to \infty} \left\{ \int_0^1 \left[1 - \left(1 - \frac{t}{n} \right)^n \right] \frac{dt}{t} - \int_1^n \left(1 - \frac{t}{n} \right)^n \frac{dt}{t} \right\},$$

(b)

$$\gamma = \int_0^1 \frac{1 - e^{-t} - e^{-1/t}}{t}\, dt.$$

[Hints: For (a), check that

$$1 + \frac{1}{2} + \frac{1}{3} + \ldots + \frac{1}{n} = \int_0^1 \frac{1 - (1 - u)^n}{u}\, du.$$

For (b), use (a) and inequalities (7.65.2).]

12.   If $\mathrm{Re}(p) > 0$ and $\mathrm{Re}(q) > 0$, then

$$B(p, q) = \int_0^{\infty} \frac{x^{p-1}\, dx}{(1 + x)^{p+q}}.$$

13.　　Let $a \geq 0$ be a real number.

(a) A complex-valued function f defined on $]a, \infty[$ satisfies $f(x + 1) = xf(x)$ for all $x > a$ if and only if there is a complex-valued function g on $]a, \infty[$ such that $g(x + 1) = g(x)$ and $f(x) = g(x)\Gamma(x)$ for all $x > a$.

(b) If f is a complex-valued function on $]a, \infty[$ such that $f(x + 1) = xf(x)$ for all $x > a$ and

$$\lim_{n \to \infty} \frac{f(t + n + 1)}{n! \, n^t} = 1$$

for all $0 \leq t < 1$, then $f = \Gamma$ on $]a, \infty[$.

Divergent Series

We have seen via Cauchy's example (2.69) that the Cauchy Product of two convergent series may be divergent. However, the following theorem gives two methods of finding the "correct sum" of the product series whether it be divergent or not.

(7.73)　Theorem　*Let* $\displaystyle\sum_{n=0}^{\infty} a_n = A$ *and* $\displaystyle\sum_{n=0}^{\infty} b_n = B$ *be two convergent series of complex terms. Write*

$$c_n = \sum_{k=0}^{n} a_k b_{n-k}, \qquad C_n = \sum_{k=0}^{n} c_k.$$

Then

(i)
$$\lim_{r \uparrow 1} \sum_{n=0}^{\infty} c_n r^n = AB,$$

(ii)
$$\lim_{n \to \infty} \frac{1}{n+1} \sum_{k=0}^{n} C_k = AB.$$

Proof　Because the two power series

$$f(r) = \sum_{n=0}^{\infty} a_n r^n, \qquad g(r) = \sum_{n=0}^{\infty} b_n r^n$$

converge at $r = 1$, they are both absolutely convergent for $|r| < 1$　(2.64). Now (2.73) shows that

$$f(r)g(r) = \sum_{n=0}^{\infty} c_n r^n$$

for $|r| < 1$. This fact together with Abel's Limit Theorem (7.40) shows that

$$\lim_{r \uparrow 1} \sum_{n=0}^{\infty} c_n r^n = \left(\lim_{r \uparrow 1} f(r) \right)\left(\lim_{r \uparrow 1} g(r) \right) = AB,$$

which is (i). The proof of (ii) is the same as that of (2.72).　□

(7.74) Discussion Such early discoveries as (7.73.i) [Abel, 1826] and (7.73.ii) [Cesáro, 1890] made it clear that divergent series should not necessarily be discarded as meaningless or useless. Indeed, it is often possible, by some method or other, to assign a "sum" to a divergent series which is useful and "correct" according to the problem at hand. This has been emphatically demonstrated during this century in the theory of trigonometric series [see Chapter 8].

Very generally speaking, a *summation method* for series is a way of assigning a "sum" to each member of a certain set of series. Thus, it is a mapping of a set S_1 of series into a set S_2 of "sums." Ordinary convergence is a summation method: S_1 is the set of all convergent series of complex numbers, $S_2 = C$, and each member of S_1 is assigned the limit of its sequence of partial sums. A summation method is called *regular* if each convergent series in its domain S_1 is assigned its ordinary sum. In this section, we take a brief look at the two most important regular summation methods that go beyond ordinary convergence.

(7.75) Definition Let $(c_n)_{n=0}^{\infty}$ be a sequence of complex numbers. Write

(i) $$s_n = \sum_{k=0}^{n} c_k, \qquad \sigma_n = \frac{1}{n+1} \sum_{k=0}^{n} s_k.$$

If there exists a complex number s such that $\lim_{n \to \infty} \sigma_n = s$, then we say that the series $\sum_{n=0}^{\infty} c_n$ is C_1-*summable* [or $(C, 1)$ *summable*, or *summable by arithmetic means*] *to* s.

If $(c_n)_{n=0}^{\infty}$ is a sequence of complex-valued functions defined on some nonvoid set X, if σ_n is defined as in (i) [pointwise on X], and if there exists a complex-valued function s on X such that $\sigma_n \to s$ uniformly on X as $n \to \infty$, then we say that the series $\sum_{n=0}^{\infty} c_n$ is *uniformly C_1-summable to s on X.*

In these cases, σ_n is called the nth C_1-*mean* of $\sum_{n=0}^{\infty} c_n$ and of $(s_k)_{k=0}^{\infty}$ and we write

$$C_1\text{-} \sum_{n=0}^{\infty} c_n = C_1\text{-} \lim_{k \to \infty} s_k = s.$$

(7.76) Definition If $(c_n)_{n=0}^{\infty}$ is a sequence of complex numbers such that

(i) $$A(r) = \sum_{n=0}^{\infty} c_n r^n$$

converges whenever $0 \leq r < 1$ and if $\lim_{r \uparrow 1} A(r) = s \in C$, then we say that the series $\sum_{n=0}^{\infty} c_n$ is *Abel summable to s.*

If $(c_n)_{n=0}^{\infty}$ is a sequence of complex-valued functions defined on a nonvoid set X such that

(ii)
$$A(x,r) = \sum_{n=0}^{\infty} c_n(x) r^n$$

converges whenever $0 \leq r < 1$ and $x \in X$ and if there exists a complex-valued function s on X such that

$$\lim_{r \uparrow 1} A(x,r) = s(x)$$

uniformly for $x \in X$, then we say that the series $\sum_{n=0}^{\infty} c_n$ is *uniformly Abel summable* to s on X. This means that for each $\epsilon > 0$ there is some $r(\epsilon) \in [0, 1[$ such that

$$\sup_{x \in X} |s(x) - A(x,r)| < \epsilon$$

whenever $r(\epsilon) < r < 1$.

In these cases, we write

$$A\text{-}\sum_{n=0}^{\infty} c_n = s.$$

(7.77) **Remark** In each of the above two definitions, the first paragraph is the special case of the second in which each c_n is taken to be a constant function. For this reason, it is clear that the theorems which we phrase in terms of uniform summability of series of functions are also theorems about summability of numerical series.

(7.78) **Examples** (a) Consider the series

$$1 + \sum_{n=1}^{\infty} 2\cos(nx) \qquad (x \in \mathbb{R}) \tag{1}$$

$[c_0(x) = 1,\ c_n(x) = 2\cos(nx)$ for $n > 0]$. Then, according to (5.12),

$$\sigma_n(x) = K_n(x) = \frac{1}{n+1} \left(\frac{\sin\left[((n+1)/2)x\right]}{\sin\left[(1/2)x\right]} \right)^2$$

for $0 < x < 2\pi$.

For $0 < \delta < \pi$, put $X_\delta = [\delta, 2\pi - \delta]$. Then

$$\sup_{x \in X_\delta} |\sigma_n(x)| \leq \frac{1}{(n+1)\sin^2(\delta/2)},$$

and so $\sigma_n \to 0$ uniformly on X_δ. That is, (1) is uniformly C_1-summable to 0 on each X_δ ($0 < \delta < \pi$). In particular, taking $x = \pi$, the divergent series

$$1 - 2 + 2 - 2 + 2 - + \ldots \tag{2}$$

is C_1-summable to 0.

(b) Again consider series (1). Then

$$A(x,r) = 1 + \sum_{n=1}^{\infty} 2r^n \cos(nx) = 1 + 2\,\mathrm{Re}\left[\sum_{n=1}^{\infty} (re^{ix})^n\right]$$

$$= \mathrm{Re}\left[\frac{1 + re^{ix}}{1 - re^{ix}}\right] = \frac{1 - r^2}{1 - 2r\cos x + r^2}$$

whenever $0 \leqq r < 1$ and $x \in \mathbb{R}$. For fixed δ and r $(0 < \delta < \pi, 0 \leqq r < 1)$, we have

$$\sup_{x \in X_\delta} |A(x,r)| = \frac{1 - r^2}{1 - 2r\cos\delta + r^2},$$

which has limit 0 as $r\uparrow 1$. Therefore (1) is uniformly Abel summable to 0 on each X_δ and (2) is Abel summable to 0. This particular $A(x,r)$ is called *Poisson's Kernel* and is usually denoted $P_r(x)$.

(c) Consider the series

$$\sum_{n=1}^{\infty} n\sin(nx) \qquad (x \in \mathbb{R}) \tag{3}$$

obtained by differentiating series (1) term-by-term and then dividing by -2. Since, for fixed r $(0 \leqq r < 1)$, the series

$$A(x,r) = \sum_{n=1}^{\infty} nr^n \sin(nx)$$

converges uniformly on $\mathbb{R}$ [Weierstrass M-test], it follows from (4.56) that

$$A(x,r) = -(1/2)P_r'(x) = r(1 - r^2)\sin x / (1 - 2r\cos x + r^2)^2.$$

Thus (3) is Abel summable to 0 at every $x \in \mathbb{R}$. Moreover,

$$\sup_{x \in X_\delta} |A(x,r)| \leqq \frac{r(1 - r^2)}{(1 - 2r\cos\delta + r^2)^2}$$

for $0 < \delta < \pi$, and so (3) is uniformly Abel summable to 0 on X_δ for every such δ. This summability is *not* uniform on $[0, \pi/2]$; for if it were we would have

$$0 = \lim_{r\uparrow 1} \int_0^{\pi/2} A(x,r)dx = -\frac{1}{2}\lim_{r\uparrow 1}\left[P_r\left(\frac{\pi}{2}\right) - P_r(0)\right]$$

$$= -\frac{1}{2}\lim_{r\uparrow 1}\left[\frac{1 - r^2}{1 + r^2} - \frac{1 + r}{1 - r}\right] = \infty.$$

(d) If (3) is C_1-summable at some fixed $x \in \mathbb{R}$, then, using (7.79.i) below, we have $\lim_{n\to\infty} \sin(nx) = 0$, and this in turn implies that $x = k\pi$ for some integer k, as we now show. In fact, supposing that $\sin(nx) \to a$ and writing $u = e^{ix}$, we have $u^{2n} = (1 - 2\sin^2(nx)) + i\sin(2nx) \to (1 - 2a^2) + ia = c \neq 0$, $c = \lim_{n\to\infty} u^{2(n+1)}$ $= cu^2$, $u^2 = 1$, $x = k\pi$ for some integer k. We conclude that (3) is C_1-summable at $x \in \mathbb{R}$ if and only if x is an integral multiple of π.

(e) Taking $x = \pi/2$ in (3), we see from (c) and (d) that the divergent series

$$1 + 0 - 3 + 0 + 5 + 0 - 7 + \ldots$$

is Abel summable to 0 and is *not* C_1-summable.

The next theorem gives necessary, but not sufficient, conditions for summability.

(7.79) Theorem *Let* $(c_n)_{n=0}^{\infty} \subset C$ *and let* $s_n = c_0 + c_1 + \ldots + c_n$.

(i) *If* $\displaystyle\sum_{n=0}^{\infty} c_n$ *is* C_1-*summable, then* $\displaystyle\lim_{n \to \infty} (c_n/n) = \lim_{n \to \infty} (s_n/n) = 0$.

(ii) *If* $\displaystyle\sum_{n=0}^{\infty} c_n$ *is Abel summable, then* $1 < \beta < \infty$ *implies* $\displaystyle\lim_{n \to \infty} (c_n/\beta^n)$
$= \displaystyle\lim_{n \to \infty} (s_n/\beta^n) = 0$.

Analogous results obtain for uniform summability.

Proof (i) Let σ_n be as in (7.75) and suppose that $\sigma_n \to s \in C$. Then

$$s_n/n = \left[(n+1)\sigma_n - n\sigma_{n-1} \right]/n \to s - s = 0$$

and

$$\frac{c_n}{n} = \frac{s_n}{n} - \frac{n-1}{n} \cdot \frac{s_{n-1}}{n-1} \to 0.$$

(ii) Suppose $\sum c_n$ is Abel summable. This implies, by (7.76), that the power series $\sum c_n r^n$ has radius of convergence ≥ 1, and so, by (2.64), $\overline{\lim} \sqrt[n]{|c_n|} \leq 1$. Given $\beta > 1$, choose any α with $1 < \alpha < \beta$. Then $\sqrt[n]{|c_n|} < \alpha$, $|c_n| \leq \alpha^n$, for all sufficiently large n, and so

$$\overline{\lim_{n \to \infty}} \left(|c_n|/\beta^n \right) \leq \overline{\lim_{n \to \infty}} \left[(\alpha/\beta)^n \right] = 0.$$

For $|r| < 1$, we have

$$(1-r)^{-1} = \sum_{n=0}^{\infty} r^n, \qquad A(r) = \sum_{n=0}^{\infty} c_n r^n.$$

By (2.73), the Cauchy Product

$$(1-r)^{-1} \cdot A(r) = \sum_{n=0}^{\infty} s_n r^n$$

of these two power series converges for $|r| < 1$. The reasoning of the preceding paragraph now shows that $s_n/\beta^n \to 0$ for all $\beta > 1$. $\square$

The next theorem is simple but important. We leave its proof as an exercise for the reader.

(7.80)   Theorem   *Let $(a_n)_{n=0}^{\infty}$ and $(b_n)_{n=0}^{\infty}$ be sequences of complex-valued functions on a nonvoid set X and let α and β be complex numbers. Suppose $\sum_{n=0}^{\infty} a_n$ and $\sum_{n=0}^{\infty} b_n$ are uniformly C_1-summable on X to s and t, respectively. Then $\sum_{n=0}^{\infty} (\alpha a_n + \beta b_n)$ is uniformly C_1-summable on X to $\alpha s + \beta t$ and $\sum_{n=1}^{\infty} a_n$ is uniformly C_1-summable on X to $s - a_0$ where this latter series begins with 0th term equal to 0.*

The same assertion holds if "C_1-" is replaced by "Abel" throughout.

We now show the relations among our three summation methods: ordinary convergence, C_1-summability, and Abel summability.

(7.81)   Theorem   *Let $(c_n)_{n=0}^{\infty}$ be a sequence of* bounded *complex-valued functions on a nonvoid set X. Consider the series*

$$\sum_{n=0}^{\infty} c_n. \qquad (*)$$

(i) [Cauchy, 1821]. *If $(*)$ is uniformly convergent to s on X, then $(*)$ is uniformly C_1-summable to s on X.*

(ii) [Abel, 1826]. *If $(*)$ is uniformly convergent to s on X, then $(*)$ is uniformly Abel summable to s on X.*

(iii) [Frobenius, 1880]. *If $(*)$ is uniformly C_1-summable to s on X, then $(*)$ is uniformly Abel summable to s on X.*

Proof   Let s_n, σ_n, and A be the functions defined in (7.75) and (7.76) and write $\| \cdot \|_u$ for the uniform norm on X:

$$\| f \|_u = \sup\{ |f(x)| : x \in X \}.$$

Suppose that $(*)$ converges uniformly to s on X. This means that $\| s - s_n \|_u \to 0$ as $n \to \infty$. Given $\epsilon > 0$, choose $n_0 \in \mathbb{N}$ such that $\| s - s_k \|_u < \epsilon/2$ for all $k > n_0$. Since each c_n is bounded on X, we can choose $N > n_0$ such that

$$\frac{1}{N} \left\| \sum_{k=0}^{n_0} (s - s_k) \right\|_u < \frac{\epsilon}{2}.$$

Then $n > N$ implies

$$\| s - \sigma_n \|_u = \left\| \frac{1}{n+1} \sum_{k=0}^{n} (s - s_k) \right\|_u$$

$$\leq \frac{1}{n+1} \left\| \sum_{k=0}^{n_0} (s - s_k) \right\|_u + \frac{1}{n+1} \sum_{k=n_0+1}^{n} \| s - s_k \|_u$$

$$< \frac{\epsilon}{2} + \frac{n - n_0}{n+1} \cdot \frac{\epsilon}{2} < \epsilon.$$

This proves (i).

Clearly (ii) will follow from (i) and (iii). Alternatively, one may repeat the proof of (7.40).

Suppose that (∗) is uniformly C_1-summable to s on X. This means that $\|s - \sigma_n\|_u \to 0$ as $n \to \infty$. Let $\epsilon > 0$ be given. Choose $n_0 \in \mathbb{N}$ such that $\|s - \sigma_n\|_u < \epsilon/2$ for all $n > n_0$. Next choose $r_0 \in [0, 1[$ such that

$$(1 - r_0)^2 \cdot \sum_{n=0}^{n_0} (n + 1)\|s - \sigma_n\|_u < \frac{\epsilon}{2}.$$

By (7.79.i), $c_n(x)/n \to 0$ for all $x \in X$ [actually, the limit is uniform on X], from which it follows easily that the power series

$$A(x,r) = \sum_{n=0}^{\infty} c_n(x)r^n$$

converges whenever $x \in X$ and $|r| < 1$. Just as in the proof of (7.79.ii), we have

$$A(x,r) = (1 - r) \cdot \sum_{n=0}^{\infty} s_n(x)r^n,$$

and another application of the same reasoning shows that

$$A(x,r) = (1 - r)^2 \cdot \sum_{n=0}^{\infty} (n + 1)\sigma_n(x)r^n \qquad (1)$$

whenever $x \in X$ and $|r| < 1$. A special case of (1) [$c_0 = s$, $c_n = 0$ for $n > 0$] is

$$s(x) = (1 - r)^2 \cdot \sum_{n=0}^{\infty} (n + 1)s(x)r^n. \qquad (2)$$

Subtracting (1) from (2), we obtain

$$s(x) - A(x,r) = (1 - r)^2 \cdot \sum_{n=0}^{\infty} (n + 1)[s(x) - \sigma_n(x)]r^n$$

whenever $x \in X$ and $|r| < 1$. Therefore $x \in X$ and $r_0 \leq r < 1$ imply

$$|s(x) - A(x,r)| \leq (1 - r_0)^2 \cdot \sum_{n=0}^{n_0} (n + 1)\|s - \sigma_n\|_u$$

$$+ (1 - r)^2 \cdot \sum_{n=n_0+1}^{\infty} (n + 1)\|s - \sigma_n\|_u r^n$$

$$< \frac{\epsilon}{2} + (1 - r)^2 \cdot \sum_{n=n_0+1}^{\infty} (n + 1) \cdot \frac{\epsilon}{2} \cdot r^n$$

$$< \frac{\epsilon}{2} + \frac{\epsilon}{2} \cdot (1 - r)^2 \cdot \sum_{n=0}^{\infty} (n + 1)r^n = \epsilon.$$

Since r_0 is independent of $x \in X$, this proves (iii). ☐

(7.82) Remarks (a) The examples given in (7.78) show that each of the three assertions in (7.81) has a false converse, even for numerical series.

(b) The hypothesis in (7.81) that each c_n be bounded is important as the following examples show.

(7.83) **Examples** (a) Let $X = \mathbb{R}$ and write $c_0(x) = x$, $c_1(x) = -x$, and $c_n(x) = 0$ for $n > 1$. Then $s_0(x) = x$, $s_n(x) = 0$ $(n > 0)$, $\sigma_n(x) = x/(n + 1)$ $(n \geq 0)$, and $A(x, r) = (1 - r)x$. Therefore, the series $\sum_{n=0}^{\infty} c_n$ is uniformly convergent to 0 on $\mathbb{R}$, but it is neither uniformly C_1-summable nor uniformly Abel summable on $\mathbb{R}$.

(b) Let $X = \mathbb{R}$ and write $c_0(x) = x$, $c_1(x) = 1 - 2x$, $c_2(x) = x - 1$, and $c_n(x) = 0$ for $n > 2$. Then $s_0(x) = x$, $s_1(x) = 1 - x$, $s_n(x) = 0$ $(n > 1)$, $\sigma_0(x) = x$, $\sigma_n(x) = 1/(n + 1)$ $(n > 0)$, and $A(x, r) = r - r^2 + (1 - r)^2 x$. Therefore, the series $\sum_{n=0}^{\infty} c_n$ is uniformly convergent and uniformly C_1-summable to 0 on $\mathbb{R}$, but it is *not* uniformly Abel summable on $\mathbb{R}$.

We next give another example to show that, unlike the case of ordinary convergence, the insertion or deletion of zero terms in a series can affect the C_1- or Abel sums.

(7.84) **Example** Consider the two series

$$1 - 1 + 1 - 1 + 1 - + \ldots \tag{1}$$

$(c_n = (-1)^n$ for $n \geq 0)$ and

$$1 - 1 + 0 + 1 - 1 + 0 + \ldots \tag{2}$$

$(c_{3k} = 1$, $c_{3k+1} = -1$, $c_{3k+2} = 0$ for $k \geq 0)$.

For series (1), we have $s_{2k} = 1$, $s_{2k+1} = 0$, $\sigma_{2k} = (k + 1)/(2k + 1)$, $\sigma_{2k+1} = 1/2$ for $k \geq 0$ and $A(r) = 1/(1 + r)$. Thus (1) is both C_1- and Abel summable to $1/2$.

For series (2), we have $s_{3k} = 1$, $s_{3k+1} = s_{3k+2} = 0$, $\sigma_{3k} = (k + 1)/(3k + 1)$, $\sigma_{3k+1} = (k + 1)/(3k + 2)$, $\sigma_{3k+2} = 1/3$ for $k \geq 0$, and $A(r) = 1/(1 - r^3) - r/(1 - r^3) = 1/(1 + r + r^2)$. Thus (2) is both C_1- and Abel summable to $1/3$.

We next consider a far-reaching generalization of (7.81.i).

(7.85) **Theorem** [Toeplitz, 1911] *Consider an infinite matrix $T = (a_{j,k})_{j,k=0}^{\infty}$ of complex numbers. That is, $a_{j,k} \in \mathbb{C}$ for each ordered pair (j, k) of nonnegative integers. Suppose that T satisfies*

(i) $\lim_{j \to \infty} a_{j,k} = 0$ *for each $k \geq 0$,*

(ii) $\sum_{k=0}^{\infty} |a_{j,k}| \leq C < \infty$ *for each $j \geq 0$ $[C > 0$ does not depend on $j]$,*

(iii) $\sum_{k=0}^{\infty} a_{j,k} = A_j \to 1$ *as $j \to \infty$.*

Suppose that $(s_k)_{k=0}^{\infty}$ is a sequence of complex numbers such that the series

(iv) $\tau_j = \displaystyle\sum_{k=0}^{\infty} a_{j,k}s_k$

converge for each $j \geq 0$. If $s_k \to s \in \mathbb{C}$ as $k \to \infty$, then $\tau_j \to s$ as $j \to \infty$.

Proof   Write $s - s_k = \epsilon_k$ and

$$\tau_j' = \sum_{k=0}^{\infty} a_{j,k}\epsilon_k \qquad (j \geq 0)$$

[since (ϵ_k) is bounded ($\epsilon_k \to 0$), (ii) shows that these series all converge]. Then

$$\tau_j = sA_j - \tau_j' \qquad (j \geq 0)$$

so, since (iii) shows that $sA_j \to s$, it suffices to show that $\tau_j' \to 0$. Let $\epsilon > 0$ be given. Choose $k_0 \in \mathbb{N}$ such that $|\epsilon_k| < \epsilon/(2C)$ for all $k > k_0$. Next use (i) to obtain j_0 such that

$$\sum_{k=0}^{k_0} |a_{j,k}\epsilon_k| < \frac{\epsilon}{2} \quad \text{for } j > j_0.$$

Then, by (ii), $j > j_0$ implies

$$|\tau_j'| \leq \sum_{k=0}^{k_0} |a_{j,k}\epsilon_k| + \sum_{k=k_0+1}^{\infty} |a_{j,k}\epsilon_k| < \frac{\epsilon}{2} + \frac{\epsilon}{2C} \sum_{k=k_0+1}^{\infty} |a_{j,k}| \leq \epsilon.$$

Therefore $\tau_j' \to 0$. $\square$

(7.86)   Remarks   (a) If $s = 0$, then condition (7.85.iii) is unnecessary to prove (7.85) because $\tau_j = -\tau_j' \to 0$.

(b) A result analogous to (7.85) holds if the s_k are bounded complex-valued functions. We leave its formulation and proof as an exercise.

(c) Toeplitz also proved the converse of (7.85). Namely, if T is any infinite matrix such that the series (7.85.iv) all converge whenever $(s_k)_{k=0}^{\infty}$ is a convergent sequence, and if $s_k \to s$ implies $\tau_j \to s$, then T must satisfy (i), (ii), and (iii). The necessity of (i) and (iii) is obvious [consider $s_n = 0$ for $n \neq k$, $s_k = 1$ so that $\tau_j = a_{j,k}$ and $s_k = 1$ for all k so that $\tau_j = A_j$], while that of (ii) is more difficult [see Exercise 12(a)].

(7.87)   Definition   An infinite matrix T that satisfies (i), (ii), and (iii) of (7.85) is called a *Toeplitz matrix* [or a *regular summation matrix*]. If T is a Toeplitz matrix and $(s_k)_{k=0}^{\infty} \subset \mathbb{C}$ is such that all the series (7.85.iv) converge, then we say that $(s_k)_{k=0}^{\infty}$ [or the infinite series $\displaystyle\sum_{n=0}^{\infty} c_n$, where $c_0 = s_0$ and $c_n = s_n - s_{n-1}$ for $n > 0$] is *T-summable* to $s \in \mathbb{C}$ if $\tau_j \to s$ as $j \to \infty$. In this case, we call τ_j the *jth T-mean* of

$(s_k)_{k=0}^{\infty}$ and of $\sum_{n=0}^{\infty} c_n$, and we write

$$T\text{-}\sum_{n=0}^{\infty} c_n = T\text{-}\lim_{k\to\infty} s_k = s.$$

Finally, the infimum of the set of all C such that (7.85.ii) holds is called the *norm* of T and is denoted $\|T\|$. Thus $\sup_{j\geq 0}\sum_{k=0}^{\infty} |a_{j,k}| = \|T\|$.

(7.88) Examples (a) Write $a_{j,k} = 1/(j+1)$ for $0 \leq k \leq j$ and $a_{j,k} = 0$ for $0 \leq j < k$. Then $T = (a_{j,k})$ is a Toeplitz matrix and T-summability is the same as C_1-summability $[\tau_j = \sigma_j]$.

(b) Let $1 \leq C < \infty$ and let $(z_j)_{j=0}^{\infty} \subset C$ satisfy $|z_j| < 1$, $|1 - z_j| \leq C(1 - |z_j|)$, and $z_j \to 1$. Write $a_{j,k} = (1 - z_j)z_j^{k}$ $(j \geq 0, k \geq 0)$. Then $T = (a_{j,k})$ is a Toeplitz matrix. In the special case that $z_j = r_j$ with $0 \leq r_j < 1$ and $r_j \to 1$, T-summability is related to Abel summability by $\tau_j = A(r_j)$ [see the proof of (7.79.ii)].

(c) Write $a_{j,k} = 1/(k\log(j+1))$ for $1 \leq k \leq j$, $a_{j,k} = 0$ for $0 \leq j < k$, and $a_{j,0} = 0$ for $j \geq 0$. Then $T = (a_{j,k})$ is a Toeplitz matrix [see (7.5)]. In this case, we have $\tau_0 = 0$ and

$$\tau_j = \frac{1}{\log(j+1)} \sum_{k=1}^{j} \frac{s_k}{k} \qquad (j \geq 1)$$

and the τ_j's are called the *logarithmic means* of the sequence $(s_k)_{k=0}^{\infty}$, and we write $\tau_j = \lambda_j$.

(d) Theorem (7.85) can sometimes be used to show that one summation method is stronger than another. For example, let $(s_n)_{n=0}^{\infty} \subset C$, let $\sigma_k = (s_0 + \ldots + s_k)/(k+1)$ be the Cesáro means, and let λ_j be the logarithmic means as in (c). Since $s_k = (k+1)\sigma_k - k\sigma_{k-1}$, we find upon substitution that

$$\lambda_j = \frac{1}{\log(j+1)} \sum_{k=1}^{j} \frac{(k+1)\sigma_k - k\sigma_{k-1}}{k}$$

$$= \frac{1}{\log(j+1)} \left[-\sigma_0 + \sum_{k=1}^{j-1} \frac{\sigma_k}{k} + \frac{j+1}{j}\sigma_j \right].$$

One checks that the matrix that transforms $(\sigma_k)_{k=0}^{\infty}$ to $(\lambda_j)_{j=0}^{\infty}$ is a Toeplitz matrix, and so (7.85) shows that if $\sigma_k \to s$, then $\lambda_j \to s$. That is, every sequence $(s_n)_{n=0}^{\infty}$ [or series $\sum_{n=0}^{\infty} c_n$ with partial sums s_n] that is Cesáro summable to s is also summable by logarithmic means to s. The converse is false, because if $\sigma_k = (-1)^k$ $[s_k = (-1)^k(2k+1)]$, then $(s_k)_{k=0}^{\infty}$ is not Cesáro summable, but $\lambda_j \to 0$, so $(s_k)_{k=0}^{\infty}$ is summable by logarithmic means to $s = 0$.

(e) Let $\alpha \geq 0$ be a fixed real number. Define $a_{j,k} = 0$ if $k > j \geq 0$ and

$$a_{j,k} = \frac{\alpha^{j-k}}{(\alpha+1)^j}\binom{j}{k}$$

for $0 \leq k \leq j$. Then $a_{j,k} \geq 0$, $\displaystyle\sum_{k=0}^{\infty} a_{j,k} = 1$, and

$$(\alpha + 1)^k \cdot k! \cdot a_{j,k} \leq j^k [\alpha/(\alpha + 1)]^{j-k} \to 0$$

(as $j \to \infty$), so the matrix $E_\alpha = (a_{j,k})$ is a Toeplitz matrix. This matrix is called the *Euler–Knopp summation matrix of order* α. The E_α-means of a sequence $(s_k)_{k=0}^{\infty}$ are given by

$$\epsilon_j^{(\alpha)} = \tau_j = \frac{1}{(\alpha + 1)^j} \sum_{k=0}^{j} \binom{j}{k} s_k \alpha^{j-k}.$$

For example, let $z \in \mathbb{C}$, $z \neq 1$, and let

$$s_k = \sum_{n=0}^{k} z^n = \frac{1 - z^{k+1}}{1 - z}.$$

Then

$$\epsilon_j^{(\alpha)} = \frac{1}{1 - z} - \frac{z}{1 - z} \left(\frac{z + \alpha}{\alpha + 1} \right)^j,$$

and so

$$E_\alpha \cdot \sum_{n=0}^{\infty} z^n = \frac{1}{1 - z} \quad \text{if } |z + \alpha| < \alpha + 1$$

and for no other z. This gives a glimpse of the usefulness of the E_α-method in "analytic continuation" of functions defined by power series.

We next generalize (2.70.ii).

(7.89) **Theorem** [Knopp, 1911] *Let $(a_{j,k})_{j,k=0}^{\infty}$ be a Toeplitz matrix which is triangular ($a_{j,k} = 0$ for $k > j \geq 0$) and satisfies*

(i) $$\lim_{j \to \infty} a_{j,j-m} = 0 \qquad (m \geq 0).$$

If $(s_k)_{k=0}^{\infty}$ and $(t_k)_{k=0}^{\infty}$ are sequences of complex numbers such that $s_k \to s \in \mathbb{C}$ and $t_k \to t \in \mathbb{C}$ as $k \to \infty$, then

(ii) $$\lim_{j \to \infty} \sum_{k=0}^{j} a_{j,k} s_k t_{j-k} = st.$$

Proof For $j \geq 0$, we have

$$\sum_{k=0}^{j} a_{j,k} s_k t_{j-k} = \sum_{k=0}^{j} a_{j,k} (s_k - s) t_{j-k} + s \sum_{k=0}^{j} a_{j,k} t_{j-k}$$

$$= \sum_{k=0}^{j} b_{j,k} (s_k - s) + s \cdot \sum_{m=0}^{j} c_{j,m} t_m \qquad (1)$$

where $b_{j,k} = a_{j,k}t_{j-k}$ $(0 \leqq k \leqq j)$ and $c_{j,m} = a_{j,j-m}$ $(0 \leqq m \leqq j)$. Define $b_{j,k} = c_{j,k} = 0$ for $k > j \geqq 0$. Since $(t_n)_{n=0}^{\infty}$ is bounded, it is easy to see that the matrix $(b_{j,k})$ satisfies conditions (i) and (ii) of (7.85). Because $(s_k - s) \to 0$ as $k \to \infty$, it follows from (7.86.a) that

$$\lim_{j \to \infty} \sum_{k=0}^{j} b_{j,k}(s_k - s) = 0. \tag{2}$$

It is obvious, using hypothesis (i), that $(c_{j,m})$ is a Toeplitz matrix, and so

$$\lim_{j \to \infty} \sum_{m=0}^{j} c_{j,m}t_m = t. \tag{3}$$

Combining (1), (2), and (3), we obtain (ii). $\square$

Exercises

1.   (a) For $z \in \mathbb{C}$ with $|z| \leqq 1$ and $z \neq 1$, we have

$$C_1\text{-} \sum_{n=0}^{\infty} z^n = \frac{1}{1-z}, \qquad A\text{-} \sum_{n=0}^{\infty} (n+1)z^n = \frac{1}{(1-z)^2}.$$

(b) For $0 < \theta < 2\pi$, we have

$$C_1\text{-} \sum_{n=0}^{\infty} \cos(n\theta) = \frac{1}{2},$$

$$C_1\text{-} \sum_{n=1}^{\infty} \sin(n\theta) = \frac{1}{2}\cot\frac{\theta}{2},$$

$$A\text{-} \sum_{n=0}^{\infty} (n+1)\cos(n\theta) = \frac{1}{2} - \frac{1}{4}\csc^2\frac{\theta}{2},$$

$$A\text{-} \sum_{n=1}^{\infty} (n+1)\sin(n\theta) = \frac{1}{2}\cot\frac{\theta}{2},$$

$$A\text{-} \sum_{n=1}^{\infty} n\cos(n\theta) = -\frac{1}{4}\csc^2\frac{\theta}{2},$$

$$A\text{-} \sum_{n=1}^{\infty} n\sin(n\theta) = 0.$$

[Hint: Put $z = e^{i\theta}$ in (a) and justify separation into real and imaginary parts.]
(c) On which sets are the summations (limits) in (a) and (b) uniform?

2.   Let $T = (a_{j,k})$ be a Toeplitz matrix with $a_{j,k} \geqq 0$ for all $j \geqq 0$ and $k \geqq 0$.
(a) If $(s_k)_{k=0}^{\infty} \subset \mathbb{R}$ is such that the series (7.85.iv) all converge, then

$$\varliminf_{k \to \infty} s_k \leqq \varliminf_{j \to \infty} \tau_j \leqq \varlimsup_{j \to \infty} \tau_j \leqq \varlimsup_{k \to \infty} s_k.$$

[Hint: If $\alpha < \varliminf s_k$ and $s_k > \alpha$ for $k > N$, then

$$\tau_j \geqq \sum_{k=0}^{N} a_{j,k}(s_k - \alpha) + \alpha A_j,$$

so $\varliminf \tau_j \geqq \alpha$.]

(b) For real sequences $(s_k)_{k=0}^\infty$, Theorem (7.85) remains true for this T and the cases $s = \infty$ and $s = -\infty$.

(c) If a series $\sum_{n=0}^\infty c_n$, with $c_n \geq 0$ for all n, is T-summable to $s \leq \infty$, then $\sum_{n=0}^\infty c_n = s$.
[Hint: $s_0 \leq s_1 \leq \ldots .$]

3. (a) Let $(a_n)_{n=0}^\infty \subset \mathbb{R}$ be nonincreasing with limit 0 and let $b_n = a_0 + a_1 + \ldots + a_n$. Then

$$C_1 - \sum_{n=0}^\infty (-1)^n b_n = \frac{1}{2} \sum_{n=0}^\infty (-1)^n a_n.$$

[Hint: The partial sums s'_k and s_k of the respective series satisfy $s_k + (-1)^k b_k = 2s'_k$, so $2\sigma'_n = \sigma_n + s'_n/(n+1)$.]

(b) $C_1 = \sum_{n=0}^\infty (-1)^n \log(n+2) = \frac{1}{2} \log \frac{\pi}{2}$. [Hint: Use (a) and (5.40.i).]

(c) $C_1 = \sum_{n=0}^\infty (-1)^n \frac{n+1}{n+2} = \log 2 - \frac{1}{2}$. [Hint: Write $a_n = \frac{1}{(n+1)(n+2)}$.]

4. Let $(p_k)_{k=0}^\infty \subset \mathbb{C}$ satisfy

$$\sum_{k=0}^\infty |p_k| = \infty, \qquad \sum_{k=0}^j |p_k| \leq C \cdot \left| \sum_{k=0}^j p_k \right|$$

for all $j \geq 0$ and some real constant C, that does not depend on j.

(a) If $s_k \to s$ in $\mathbb{C}$, then

$$\lim_{j \to \infty} \frac{p_0 s_0 + p_1 s_1 + \ldots + p_j s_j}{p_0 + p_1 + \ldots + p_j} = s.$$

(b) If $(z_k)_{k=0}^\infty \subset \mathbb{C}$ and $z_k/p_k \to s \in \mathbb{C}$, then

$$\lim_{j \to \infty} \frac{z_0 + z_1 + \ldots + z_j}{p_0 + p_1 + \ldots + p_j} = s.$$

(c) If $n \in \mathbb{N}$, then

$$\lim_{j \to \infty} \left[\frac{1}{j^{n+1}} \sum_{k=1}^j k^n \right] = \frac{1}{n+1}.$$

[Hint: Take $p_k = k^{n+1} - (k-1)^{n+1}$ for $k > 0$, $p_0 = 0$.]

(d) $\lim_{j \to \infty} \frac{\log(j!)}{j \log j} = 1$.
[Hint: Take $z_k = \log k$ and $p_k = k \log k - (k-1)\log(k-1)$ for $k > 1$.]

5. Let $p \in \mathbb{N}$ be fixed. Then

$$A - \sum_{n=1}^\infty (-1)^{n-1} \cdot n^p = \frac{B_{2m}}{2m} \cdot (2^{2m} - 1)$$

if $p = 2m - 1$ is odd, while this Abel sum is 0 if p is even. Here, the B_{2m}'s are the Bernoulli numbers. In particular, we have the Abel sums

$$1 - 2 + 3 - 4 + - \ldots = 1/4,$$
$$1^3 - 2^3 + 3^3 - 4^3 + - \ldots = -1/8,$$
$$1^5 - 2^5 + 3^5 - 4^5 + - \ldots = 1/4,$$
$$1^7 - 2^7 + 3^7 - 4^7 + - \ldots = -17/16.$$

[Hints: Define $f(z)$ for $|z| < \pi$ by

$$f(z) = e^z/(1 + e^z) = 1/(e^{-z} - 1) - 2/(e^{-2z} - 1).$$

Use Exercise 17(v), p. 432, to obtain

$$f(z) = \frac{1}{2} + \sum_{m=1}^{\infty} \frac{B_{2m}}{(2m)!} (2^{2m} - 1)z^{2m-1}.$$

Thus $f^{(2m)}(0) = 0$ and $f^{(2m-1)}(0) = (B_{2m}/(2m)!)(2^{2m} - 1)$. Notice that $x < 0$ implies

$$f^{(p)}(x) = \sum_{n=1}^{\infty} (-1)^{n-1} n^p e^{nx}.]$$

6.   Let T be any Toeplitz matrix. Suppose that $T\text{-} \sum_{n=0}^{\infty} c_n = s$ and $T\text{-} \sum_{n=0}^{\infty} c'_n = s'$. If $\alpha, \beta \in \mathbb{C}$, then

$$T\text{-} \sum_{n=0}^{\infty} (\alpha c_n + \beta c'_n) = \alpha s + \beta s'.$$

In particular, the series

$$(c_0 - s) + c_1 + c_2 + c_3 + \ldots$$

is T-summable to 0. For which other infinite matrices T are these statements true?

7.   *Cesáro's Method (C, α).* Let $(c_n)_{n=0}^{\infty} \subset \mathbb{C}$ be such that $\sum_{n=0}^{\infty} c_n z^n$ converges for $|z| < 1$; i.e., $\varlimsup_{n \to \infty} |c_n|^{1/n} \leq 1$. Write $s_n = c_0 + c_1 + \ldots + c_n$ for the nth partial sum. For $\alpha \in \mathbb{R}$ and $n \geq 0$, define $A_n^{(\alpha)}$ and $S_n^{(\alpha)}$ by the requirements

$$\sum_{n=0}^{\infty} A_n^{(\alpha)} z^n = (1 - z)^{-\alpha - 1},$$

$$\sum_{n=0}^{\infty} S_n^{(\alpha)} z^n = (1 - z)^{-\alpha - 1} \cdot \sum_{n=0}^{\infty} c_n z^n = (1 - z)^{-\alpha} \cdot \sum_{n=0}^{\infty} s_n z^n$$

whenever $|z| < 1$ (cf. (7.46), (2.73), and (4.63)). Then we have the following:

(a) $A_0^{(\alpha)} = 1$.

(b) $A_n^{(\alpha)} = \binom{\alpha + n}{n} = \dfrac{(\alpha + 1)(\alpha + 2) \ldots (\alpha + n)}{n!}$ for $n \in \mathbb{N}$.

(c) $A_n^{(0)} = 1$, $S_n^{(0)} = s_n$, $S_n^{(-1)} = c_n$ for $n \geq 0$.

(d) $A_n^{(\alpha + \beta + 1)} = \sum_{k=0}^{n} A_k^{(\alpha)} A_{n-k}^{(\beta)}$ and $S_n^{(\alpha + \beta + 1)} = \sum_{k=0}^{n} S_k^{(\alpha)} A_{n-k}^{(\beta)}$ for $\alpha, \beta \in \mathbb{R}$ and $n \geq 0$.

(e) $A_n^{(\alpha + 1)} = \sum_{k=0}^{n} A_k^{(\alpha)}$, $S_n^{(\alpha + 1)} = \sum_{k=0}^{n} S_k^{(\alpha)}$ for $\alpha \in \mathbb{R}$ and $n \geq 0$.

(f) $S_n^{(\beta)} = \sum_{k=0}^{n} A_{n-k}^{(\beta)} c_k = \sum_{k=0}^{n} A_{n-k}^{(\beta - 1)} s_k$ for $\beta \in \mathbb{R}$ and $n \geq 0$.

(g) If $\alpha \in \mathbb{R}$ is not a negative integer, then

$$\lim_{n \to \infty} n^{-\alpha} \cdot A_n^{(\alpha)} = 1/[\Gamma(\alpha + 1)].$$

[Hint: Use (7.62) and (b).]

(h) For $\alpha \in \mathbb{R}$, $\sum_{n=0}^{\infty} |A_n^{(\alpha)}| = \infty$ if and only if $\alpha > -1$.

(i) If $\alpha > -1$, then $(A_n^{(\alpha)})_{n=0}^{\infty}$ is a monotone sequence of positive numbers (increasing if $\alpha > 0$, decreasing if $-1 < \alpha < 0$).

For $\alpha > -1$ and $n \geq 0$, define the *nth Cesáro mean of order* α of the series $\sum\limits_{n=0}^{\infty} c_n$ (and of the sequence $(s_n)_{n=0}^{\infty}$) to be the number $\sigma_n^{(\alpha)} = S_n^{(\alpha)}/A_n^{(\alpha)}$. We say that this series (or this sequence) is C_α-*summable* [or (C, α)-*summable*] to $s \in \mathbb{C}$ if $\lim\limits_{n \to \infty} \sigma_n^{(\alpha)} = s$. In this case, we write

$$C_\alpha \text{-} \sum_{n=0}^{\infty} c_n = s, \qquad C_\alpha \text{-} \lim_{n \to \infty} s_n = s.$$

We have the following:

(j) $\sigma_n^{(0)} = s_n$, $\sigma_n^{(1)} = \sigma_n = \dfrac{s_0 + s_1 + \ldots + s_n}{n + 1}$ for all $n \geq 0$.

(k) If $-1 < \alpha < \beta$ and $C_\alpha \text{-} \lim\limits_{n \to \infty} s_n = s \in \mathbb{C}$, then $C_\beta \text{-} \lim\limits_{n \to \infty} s_n = s$. [Hint: Use (d) to write

$$\sigma_n^{(\beta)} = \sum_{k=0}^{n} a_{n,k} \sigma_k^{(\alpha)},$$

where $a_{n,k} = A_{n-k}^{(\beta - \alpha - 1)} \cdot A_k^{(\alpha)}/A_n^{(\beta)}$, and check that $(a_{n,k})$ is a Toeplitz matrix.]

(l) If $(c_n)_{n=0}^{\infty} \subset \mathbb{R}$ and $-1 < \alpha < \beta$, then

$$\varliminf_{n \to \infty} \sigma_n^{(\alpha)} \leq \varliminf_{n \to \infty} \sigma_n^{(\beta)} \leq \varlimsup_{n \to \infty} \sigma_n^{(\beta)} \leq \varlimsup_{n \to \infty} \sigma_n^{(\alpha)}.$$

[Hint: Use Exercise 2(a).]

(m) If $\alpha > -1$ and $C_\alpha \text{-} \sum\limits_{n=0}^{\infty} c_n = s \in \mathbb{C}$, then $\lim\limits_{n \to \infty} n^{-\alpha} c_n = 0$. [Hints: Suppose $\sigma_n^{(\alpha)} \to 0$ [otherwise, replace c_0 by $c_0 - s$]. Write

$$c_n/A_n^{(\alpha)} = \sum_{k=0}^{n} a_{n,k} \sigma_k^{(\alpha)}$$

where $a_{n,k} = A_{n-k}^{(-\alpha-2)} \cdot A_k^{(\alpha)}/A_n^{(\alpha)}$ for $0 \leq k \leq n$ and $a_{n,k} = 0$ for $0 \leq n < k$. Check that $(a_{n,k})$ satisfies (i) and (ii) of (7.85) and then use (7.86.a). To prove (7.85.ii) it is helpful to use (i) above. If $-1 < \alpha < 0$, then $A_j^{(-\alpha-2)} < 0$ for $j > 0$ and one can take $C = 2$. If $\alpha \geq 0$, then

$$\sum_{k=0}^{\infty} |a_{n,k}| \leq \sum_{j=0}^{\infty} |A_j^{(-\alpha-2)}| = C < \infty \text{ by } (h).\Big]$$

(n) If $C_\alpha \text{-} \sum\limits_{n=0}^{\infty} c_n = s \in \mathbb{C}$ for some $\alpha > -1$ and if $(z_j)_{j=0}^{\infty}$ is any sequence as in (7.88.b), then $\lim\limits_{j \to \infty} \sum\limits_{n=0}^{\infty} c_n z_j^n = s$. [Hint:

$$\sum_{n=0}^{\infty} c_n z_j^n = (1 - z_j)^{\alpha+1} \cdot \sum_{n=0}^{\infty} z_j^n A_n^{(\alpha)} \sigma_n^{(\alpha)}$$

and (7.85) can be applied.]

(o) If $C_\alpha \text{-} \sum\limits_{n=0}^{\infty} c_n = s \in \mathbb{C}$ for some $\alpha > -1$, then $A \text{-} \sum\limits_{n=0}^{\infty} c_n = s$.

(p) If $\alpha > -1$ and $c_n = (-1)^n A_n^{(\alpha)}$, then

$$C_{\alpha+1}\text{-}\sum_{n=0}^{\infty} c_n = 2^{-\alpha-1}$$

and this series is *not* C_α-summable. [Hints: For $|z| < 1$,

$$\sum_{n=0}^{\infty} S_n^{(\alpha)} z^n = (1-z^2)^{-\alpha-1} = \sum_{j=0}^{\infty} A_j^{(\alpha)} z^{2j},$$

so $S_n^{(\alpha)} = A_{n/2}^{(\alpha)}$ or 0 according as n is even or odd. Therefore

$$S_n^{(\alpha+1)} = \sum_{j=0}^{[n/2]} A_j^{(\alpha)} = A_{[n/2]}^{(\alpha+1)}. \Big]^*$$

(q) If $c_0 = e$ and, for $n \geq 1$,

$$c_n = \sum_{k=1}^{\infty} \frac{1}{k!} \binom{n+k-1}{n},$$

then $A\text{-}\sum_{n=0}^{\infty} (-1)^n c_n = \sqrt{e}$, but there is no $\alpha > -1$ for which this series is C_α-summable. [Hints: Use (7.46) and (7.57) to see that

$$\sum_{n=0}^{\infty} c_n z^n = e^{1/(1-z)}$$

whenever $|z| < 1$. Given α, choose $p \in \mathbb{N}$ with $p \geq \alpha$. Then

$$c_n > \frac{1}{(p+2)!}\binom{n+p+1}{p+1} > \frac{n^{\alpha+1}}{(p+2)!\,(p+1)!} \quad \text{for all } n \geq 1. \Big]$$

(r) If σ_n is as in (j) and $\alpha > 0$, then $C_\alpha\text{-}\lim_{n\to\infty} s_n = s$ if and only if $C_{\alpha-1}\text{-}\lim_{n\to\infty} \sigma_n = s$.
[Hints: Let $\tau_n^{(\alpha)}$ be the nth C_α-mean of $(\sigma_n)_{n=0}^{\infty}$. That is, $\tau_n^{(\alpha)} = T_n^{(\alpha)}/A_n^{(\alpha)}$, where

$$T_n^{(\alpha)} = \sum_{k=0}^{n} A_{n-k}^{(\alpha-1)} \sigma_k \text{ as in (f). Since } T_n^{(\alpha-1)} = T_n^{(\alpha)} - T_{n-1}^{(\alpha)} \text{ and } (n+\alpha)$$

$$A_{n-k}^{(\alpha-2)} - (\alpha-1)A_{n-k}^{(\alpha-1)} = (k+1)A_{n-k}^{(\alpha-2)}, \text{ we have, using (f),}$$

$$(n+1)T_n^{(\alpha)} - (n+\alpha)T_{n-1}^{(\alpha)} = (n+\alpha)T_n^{(\alpha-1)} - (\alpha-1)T_n^{(\alpha)} = S_n^{(\alpha)}.$$

Division by $A_n^{(\alpha)}$ yields

$$(n+1)\tau_n^{(\alpha)} - n\tau_{n-1}^{(\alpha)} = \sigma_n^{(\alpha)} \tag{1}$$

and

$$\alpha\tau_n^{(\alpha-1)} - (\alpha-1)\tau_n^{(\alpha)} = \sigma_n^{(\alpha)}. \tag{2}$$

Using (1) to evaluate a telescoping sum, we have

$$\frac{1}{n+1}\sum_{k=0}^{n} \sigma_k^{(\alpha)} = \tau_n^{(\alpha)} \tag{3}$$

---

*This result is the special case $w = -1$, $\beta = \alpha + 1$ of the much deeper result that $\sum_{n=0}^{\infty} \binom{n+\alpha}{n} w^n$ is C_β-summable to $(1-w)^{-\alpha-1}$ whenever $\beta > \alpha > -1$, $|w| = 1$, and $w \neq 1$. A proof of this fact, using analytic function theory, can be found in G. H. Hardy's treatise, *Divergent Series*.

(this says that the C_1-means of the C_α-means are the C_α-means of the C_1-means). Substitution of (3) into (2) gives

$$\tau_n^{(\alpha-1)} = \frac{1}{\alpha}\sigma_n^{(\alpha)} + \left(1 - \frac{1}{\alpha}\right) \cdot \frac{1}{n+1} \sum_{k=0}^{n} \sigma_k^{(\alpha)}. \tag{4}$$

If $\tau_n^{(\alpha-1)} \to s$, then (2) shows that $\sigma_n^{(\alpha)} \to s$. If $\sigma_n^{(\alpha)} \to s$, then (4) shows that $\tau_n^{(\alpha-1)} \to s$.]

(s) If C_α- $\sum\limits_{n=0}^{\infty} u_n = U \in \mathbb{C}$, C_β- $\sum\limits_{n=0}^{\infty} v_n = V \in \mathbb{C}$, and $w_n = \sum\limits_{k=0}^{n} u_k v_{n-k}$, then

C_γ- $\sum\limits_{n=0}^{\infty} w_n = UV$ where $\gamma = \alpha + \beta + 1$ (here $\alpha > -1$, $\beta > -1$, $u_n \in \mathbb{C}$, $v_n \in \mathbb{C}$). [Hint: Let $U_n^{(\alpha)}$, $V_n^{(\beta)}$, and $W_n^{(\gamma)}$ be the Cesáro sums for these three series. That is,

$$\left(\sum_{n=0}^{\infty} U_n^{(\alpha)} z^n\right)\left(\sum_{n=0}^{\infty} V_n^{(\beta)} z^n\right)$$

$$= \left((1-z)^{-\alpha-1} \cdot \sum_{n=0}^{\infty} u_n z^n\right)\left((1-z)^{-\beta-1} \cdot \sum_{n=0}^{\infty} v_n z^n\right)$$

$$= (1-z)^{-\gamma-1} \cdot \sum_{n=0}^{\infty} w_n z^n = \sum_{n=0}^{\infty} W_n^{(\gamma)} z^n$$

for $|z| < 1$. Then

$$W_n^{(\gamma)}/A_n^{(\gamma)} = \sum_{k=0}^{n} a_{n,k}\left(U_k^{(\alpha)}/A_k^{(\alpha)}\right)\left(V_{n-k}^{(\beta)}/A_{n-k}^{(\beta)}\right)$$

where $a_{n,k} = A_k^{(\alpha)} A_{n-k}^{(\beta)}/A_n^{(\gamma)}$. Now check that (7.89) applies.]

(t) If $\alpha > 0$ and $\beta > 0$, then

$$C_\alpha\text{-}\sum_{n=0}^{\infty} (-1)^n A_n^{(\alpha-1)} = 2^{-\alpha}, \qquad C_\beta\text{-}\sum_{n=0}^{\infty} (-1)^n A_n^{(\beta-1)} = 2^{-\beta},$$

and the Cauchy Product of these two series is

$$\sum_{n=0}^{\infty} (-1)^n A_n^{(\alpha+\beta-1)}$$

which is $C_{\alpha+\beta}$-summable to $2^{-(\alpha+\beta)}$. Notice that $\alpha + \beta < \gamma = \alpha + \beta + 1$. ·

For nonnegative integers p, the *Hölder means of order p* of the sequence $(s_k)_{k=0}^{\infty}$ are defined recursively by

$$h_n^{(0)} = s_n, \qquad h_n^{(p+1)} = \frac{1}{n+1} \sum_{k=0}^{n} h_k^{(p)}.$$

We say that $(s_k)_{k=0}^{\infty}$ $\left(\text{or } \sum\limits_{n=0}^{\infty} c_n\right)$ is H_p-*summable* [or (H, p)-*summable*] to $s \in \mathbb{C}$ if $\lim\limits_{n \to \infty} h_n^{(p)} = s$. If this is so, we write

$$H_p\text{-}\sum_{n=0}^{\infty} c_n = H_p\text{-} \lim_{k \to \infty} s_n = s.$$

The Hölder means are not as easy to deal with as the Cesáro means, but we fortunately have the following fact.

(u) If $p \geq 0$ is an integer and $s \in \mathbb{C}$, then H_p- $\lim_{n\to\infty} s_n = s$ if and only if C_p- $\lim_{n\to\infty} s_n$ $= s$. [Hint: Use (r) to show that C_{p-j}- $\lim_{n\to\infty} h_n^{(j)} = s$ is equivalent to C_{p-j-1}- $\lim_{n\to\infty} h_n^{(j+1)} = s$ for all integers j such that $0 \leq j < p$.]

(v) If $s_{2n} = 2n(2n^2 + 1)$ and $s_{2n+1} = -2n(2n^2 + 3n + 1)$, then H_2- $\lim_{k\to\infty} s_k = \infty$ and C_2- $\lim_{k\to\infty} s_k$ does not exist. [Hint: $\sum_{n=1}^{k} n^2 = k(k + 1)(2k + 1)/6$.]

8. Let $A = (a_{j,n})$ and $B = (b_{n,k})$ $(j, n, k = 0, 1, 2, \ldots)$ be Toeplitz matrices.
(a) The matrix product $AB = C = (c_{j,k})$, where

$$c_{j,k} = \sum_{n=0}^{\infty} a_{j,n} b_{n,k},$$

is a well-defined Toeplitz matrix and $\|C\| \leq \|A\| \cdot \|B\|$.
(b) If A and B are triangular, i.e. if

$$(a_{j,n} = 0 \quad \text{if } n > j, \qquad b_{n,k} = 0 \quad \text{if } k > n),$$

then so is AB.
(c) If A and B are triangular and one of them satisfies (7.89.i), then AB also satisfies it.
(d) If $AB = BA$ and if a sequence (or series) is A-summable to s and B-summable to t, then $s = t$.

9. Let α, β, and γ be nonnegative real numbers. We consider the Euler–Knopp matrices as in (7.88.e). Let $(s_k)_{k=0}^{\infty} \subset \mathbb{C}$.
(a) The E_β-means of the E_α-means of $(s_k)_{k=0}^{\infty}$ are equal to the $E_{\alpha+\beta+\alpha\beta}$-means of $(s_k)_{k=0}^{\infty}$. That is, the matrices E satisfy

$$E_\beta E_\alpha = E_{\alpha+\beta+\alpha\beta}.$$

(b) If E_α- $\lim_{k\to\infty} s_k = s$ and $\gamma > \alpha$, then E_γ- $\lim_{k\to\infty} s_k = s$. [Hint: Take $\beta = (\gamma - \alpha)/(\alpha + 1)$ in (a).]
(c) The C_1-means of the E_1-means of $(s_k)_{k=0}^{\infty}$ are equal to the E_1-means of the C_1-means of $(s_k)_{k=0}^{\infty}$. That is, $E_1 C_1 = C_1 E_1$, where C_1 is the matrix T in (7.88.a). [Hint: For fixed k, use induction on $j \geq k$ to check that

$$\sum_{n=k}^{j} 2^{j-n}\binom{n}{k} = \sum_{n=k}^{j}\binom{j+1}{n+1}.]$$

(d) If E_1- $\lim_{k\to\infty} s_k = s$ and C_1- $\lim_{k\to\infty} s_k = t$, then $s = t$. [See Exercise 8(d).]
(e) If $s_k = (-2)^k$, then E_1- $\lim_{k\to\infty} s_k = 0$ and C_1- $\lim_{k\to\infty} s_k$ does not exist.
(f) If $s_k = \sqrt{k}$ when $k = n^2$ for some integer n and $s_k = 0$ for all other k, then C_1- $\lim_{k\to\infty} s_k = 1/2$ and E_1- $\lim_{k\to\infty} s_k$ does not exist. [Hint: Use (5.44.ii) and the inequality

$$\epsilon_{2j}^{(1)} = \frac{1}{2^{2j}} \sum_{k=0}^{2j}\binom{2j}{k} s_k \geq \frac{1}{2^{2j}}\binom{2j}{j}\sqrt{j} \quad \text{for } j = n^2$$

to see that $\overline{\lim_{j\to\infty}} \epsilon_j^{(1)} \geq 1/\sqrt{\pi} > 1/2$.]

(g) One can recapture the sequence $(s_k)_{k=0}^\infty$ from its E_α-means by the formula

$$s_n = \sum_{j=0}^n \binom{n}{j}(-\alpha)^{n-j}(1+\alpha)^j \epsilon_j.$$

[Hints: It is useful to prove that

$$\sum_{j=k}^n \binom{n}{j}\binom{j}{k}(-1)^j = 0$$

whenever $0 \le k < n$. To do this, apply the Binomial Theorem to $f(x) = (1-x)^n$ and evaluate $f^{(k)}(1)$.]

(h) If $E_\alpha\text{-}\lim_{n\to\infty} s_n = s \in \mathbb{C}$ and $\alpha > 0$, then $\lim_{n\to\infty}(1+2\alpha)^{-n}\cdot s_n = 0$. [Hint: It suffices to suppose $s = 0$. Apply (g) and (7.86.a).]

(i) In (h) the number $2\alpha + 1$ cannot be replaced by any smaller positive number.

10. Let $(s_k)_{k=0}^\infty \subset \mathbb{C}$ and $a > 0$ be given. Write $\sigma_n = (s_0 + s_1 + \ldots + s_n)/(n+1)$ and $t_n = as_n + (1-a)\sigma_n$. If $\lim_{n\to\infty} t_n = s \in \mathbb{C}$, then $\lim_{n\to\infty}\sigma_n = s$. [Hints: Write $b = 1/a$

and let $p_k = \binom{b-1+k}{k} = A_k^{(b-1)}$ as in Exercise 7. Then $p_0 + p_1 + \cdots$

$+ p_n = \binom{b+n}{n} = (na+1)p_n \to \infty$ as $n \to \infty$, $p_j t_j = p_j[(ja+1)\sigma_j - ja\sigma_{j-1}]$

$= p_j(ja+1)\sigma_j - p_{j-1}((j-1)a+1)\sigma_{j-1}$ for $j > 0$, and $p_0 t_0 = p_0 \sigma_0$. Therefore

$\sigma_n = \left(\sum_{j=0}^n p_j t_j\right)/\left(\sum_{j=0}^n p_j\right) \to s$ as $n \to \infty$ by Exercise 4(a).]

11. *Nörlund's Method N_p.* Let $p = (p_k)_{k=0}^\infty$ be a sequence of nonnegative real numbers

with $p_0 > 0$, and write $P_n = \sum_{k=0}^n p_k$ for $n \ge 0$. For $(s_k)_{k=0}^\infty \subset \mathbb{C}$, define the Nörlund

means of $(s_k)_{k=0}^\infty$ *corresponding to* p by

$$\nu_n^{(p)} = \frac{1}{P_n}\cdot\sum_{k=0}^n p_{n-k}s_k.$$

If $\nu_n^{(p)} \to s \in \mathbb{C}$ as $n \to \infty$, then we say that $(s_k)_{k=0}^\infty$ is N_p-*summable* to s and we write

$$N_p\text{-}\lim_{k\to\infty} s_k = N_p\text{-}\sum_{n=0}^\infty c_n = s$$

(here $c_0 = s_0$, $c_n = s_n - s_{n-1}$ for $n > 0$).

(a) If $\beta > 0$ and $p_k = \binom{\beta-1+k}{k}$, then $\nu_n^{(p)} = \sigma_n^{(\beta)}$, and so N_p-summability is the same as C_β-summability. [See Exercise 7.]

(b) The N_p-method is regular (i.e., $s_k \to s \in \mathbb{C}$ implies that $\nu_n^{(p)} \to s$) if and only if $p_n/P_n \to 0$ as $n \to \infty$. [Hint: Use (7.85). Notice that $p_{n-k}/P_n \le p_{n-k}/P_{n-k}$.]

(c) Any two regular Nörlund methods are consistent. That is, if the N_p-method and the N_q-method are both regular as in (b) and if $N_p\text{-}\lim_{k\to\infty} s_k = s \in \mathbb{C}$ and

$N_q\text{-}\lim_{k\to\infty} s_k = t \in \mathbb{C}$, then $s = t$. [Hint: Write $r_n = \sum_{k=0}^n p_k q_{n-k}$. Then $\nu_n^{(r)} =$

$\sum_{k=0}^n a_{n,k}\nu_k^{(p)}$, where $a_{n,k} = q_{n-k}P_k/\left(\sum_{j=0}^n q_{n-j}P_j\right)$. Write $a_{n,k} = 0$ if $0 \le n < k$

and check that $(a_{n,k})$ is a Toeplitz matrix. Thus, $\nu_n^{(r)} \to s$. Similarly, $\nu_n^{(r)} \to t$.]

(d) If $\alpha \in \mathbb{R}$ and $p_n = (n+1)^\alpha$, then the N_p-method is regular.

(e) If $\beta \in \mathbb{R}$, $\beta > 0$, and $p_n = \beta^n$, then the N_p-method is regular if and only if $0 < \beta \leq 1$.

12.　(a) Let $(a_{j,k})_{j,k=0}^{\infty}$ be an infinite matrix of complex numbers with the property that if $(s_k)_{k=0}^{\infty} \subset \mathbb{C}$ and $s_k \to 0$, then

(i) $\tau_j = \sum_{k=0}^{\infty} a_{j,k} s_k$ converges for all $j \geq 0$

and

(ii) $(\tau_j)_{j=0}^{\infty}$ is a bounded sequence.

Then

(iii) $M_j = \sum_{k=0}^{\infty} |a_{j,k}| < \infty$ for all $j \geq 0$

and

(iv) $(M_j)_{j=0}^{\infty}$ is a bounded sequence.

[Hints: Assume that (iii) fails for some fixed j. Let k_0 be the smallest k with

$$a_{j,k} \neq 0. \text{ Write } b_k = 0 \text{ for } 0 \leq k < k_0 \text{ and } b_k = \left(\sum_{n=k_0}^{k} |a_{j,n}| \right)^{-1} \text{ for } k \geq k_0.$$

Choose $s_k = b_k \cdot \operatorname{sgn} \bar{a}_{j,k}$. Then $s_k \to 0$ and, using (2.41.i), $\tau_j = \sum_{k=k_0}^{\infty} b_k |a_{j,k}| = \infty$,

contrary to (i).

　　For a fixed integer $p \geq 0$, let $s_p = 1$ and $s_k = 0$ for $k \neq p$. By (ii), $|a_{j,p}| = |\tau_j| \leq B_p < \infty$ for all $j \geq 0$. Write $B_k = \sum_{p=0}^{k} B_p$. Assume that (iv) is false. Let $j_0 = k_1 = 0$ and suppose that j_{r-1} and k_r have been chosen for some $r \in \mathbb{N}$. Choose $j_r > j_{r-1}$ such that

$$M_{j_r} > 2rB_{k_r} + r^2 + r + 1$$

and choose $k_{r+1} > k_r$ such that

$$\sum_{k=0}^{k_{r+1}} |a_{j_r,k}| > M_{j_r} - 1.$$

Then

$$\sum_{k=k_r+1}^{k_{r+1}} |a_{j_r,k}| > (M_{j_r} - 1) - B_{k_r} > rB_{k_r} + r^2 + r.$$

This defines $(j_r)_{r=0}^{\infty}$ and $(k_r)_{r=1}^{\infty}$. Now take $s_0 = 0$ and $s_k = r^{-1} \cdot \operatorname{sgn} \bar{a}_{j_r,k}$ for $k_r < k \leq k_{r+1}$. Then $s_k \to 0$ and

$$|\tau_{j_r}| \geq \sum_{k=k_r+1}^{k_{r+1}} r^{-1} \cdot |a_{j_r,k}| - \sum_{k=0}^{k_r} |a_{j_r,k}| - \sum_{k=k_{r+1}+1}^{\infty} |a_{j_r,k}|$$

$$> (B_{k_r} + r + 1) - B_{k_r} - 1 = r.$$

Thus, $\lim_{r \to \infty} |\tau_{j_r}| = \infty$, contrary to (ii).]*

---

*There is an elegant proof of this theorem based on the Banach–Steinhaus Theorem of Functional Analysis. However, we prefer to give this elementary proof because such constructions are useful elsewhere in analysis.

(b) Let $(b_n)_{n=0}^{\infty} \subset \mathbb{C}$. A necessary and sufficient condition that $\sum\limits_{n=0}^{\infty} b_n c_n$ be convergent for every convergent series $\sum\limits_{n=0}^{\infty} c_n$ of complex terms is that $\sum\limits_{n=0}^{\infty} |b_n - b_{n+1}| < \infty$. [Hints: For sufficiency, use (7.36.iii). For necessity, write $s_k = c_0 + c_1 + \ldots + c_k$, note that

$$\tau_j = \sum_{n=0}^{j} b_n c_n = \sum_{k=0}^{j-1} (b_k - b_{k+1}) s_k + b_j s_j = \sum_{k=0}^{\infty} a_{j,k} s_k,$$

and use (a).]

13. Let $\Phi = (\phi_k)_{k=0}^{\infty}$ be a sequence of complex-valued functions defined on some interval $]a, b[\subset \mathbb{R}$. Suppose that

(i) $\lim\limits_{t \uparrow b} \phi_k(t) = 0$ for all $k \geq 0$,

(ii) $\sum\limits_{k=0}^{\infty} |\phi_k(t)|$ is bounded on $]a, b[$,

(iii) $\lim\limits_{t \uparrow b} \sum\limits_{k=0}^{\infty} \phi_k(t) = 1$.

Let $(s_k)_{k=0}^{\infty} \subset \mathbb{C}$, $s \in \mathbb{C}$, $c_0 = s_0$, $c_n = s_n - s_{n-1}$ for $n > 0$. We say that $\sum\limits_{n=0}^{\infty} c_n$ and $(s_k)_{k=0}^{\infty}$ are Φ-*summable to* s, and we write

$$\Phi\text{-} \sum_{n=0}^{\infty} c_n = \Phi\text{-} \lim_{k \to \infty} s_k = s$$

if $S(t) = \sum\limits_{k=0}^{\infty} s_k \phi_k(t)$ converges for $a < t < b$ and $\lim\limits_{t \uparrow b} S(t) = s$.

(a) If $s_k \to s$, then $\Phi\text{-}\lim\limits_{k \to \infty} s_k = s$.

(b) If $a = 0$, $b = 1$, and $\phi_k(t) = (1 - t)t^k$, then Φ-summability is the same as Abel summability.

(c) If $a = 0$, $b = \infty$, and $\phi_k(t) = e^{-t} t^k / k!$, then (i), (ii), and (iii) are satisfied. In this case, Φ-summability is called *Borel summability*, and we use the prefix B- instead of Φ-. Thus, $B\text{-}\lim\limits_{k \to \infty} s_k = s$ means

$$\lim_{t \to \infty} e^{-t} \sum_{k=0}^{\infty} s_k t^k / k! = s.$$

(d) For $z \in \mathbb{C}$, $B\text{-} \sum\limits_{n=0}^{\infty} z^n = (1 - z)^{-1}$ if and only if $\mathrm{Re}(z) < 1$.

(e) If $E_\alpha\text{-}\lim\limits_{k \to \infty} s_k = s$ for some $\alpha \geq 0$ as in (7.88.e), then $B\text{-}\lim\limits_{k \to \infty} s_k = s$. [Hint: By computing a Cauchy Product, we have

$$e^{\alpha t} \cdot \sum_{k=0}^{\infty} \frac{s_k}{k!} t^k = \sum_{n=0}^{\infty} \frac{\epsilon_n^{(\alpha)}}{n!} [(1 + \alpha)t]^n.$$

Multiply by $e^{-(1+\alpha)t}$ and use (a).]

14. Let $p = (p_k)_{k=0}^{\infty}$ and $q = (q_k)_{k=0}^{\infty}$ be sequences of positive real numbers such that $P_n = p_0 + p_1 + \ldots + p_n \to \infty$ and $Q_n = q_0 + q_1 + \ldots + q_n \to \infty$ as $n \to \infty$. Let

$(s_k)_{k=0}^{\infty} \subset \mathbb{C}$ and $s \in \mathbb{C}$. We write A_p- $\lim_{k \to \infty} s_k = s$ if

$$t_n = \frac{1}{P_n} \sum_{k=0}^{n} p_k s_k \to s$$

as $n \to \infty$. We define A_q-$\lim s_k$ similarly.

(a) If $s_k \to s$, then A_p- $\lim_{k \to \infty} s_k = s$.

(b) If A_p- $\lim_{k \to \infty} s_k = s$, then $\lim_{n \to \infty} p_n(s_n - s)/P_n = 0$. [Hint: $p_n(s_n - s) = P_n(t_n - s) - P_{n-1}(t_{n-1} - s)$.]

(c) If $q_{k+1}/q_k \leqq p_{k+1}/p_k$ for all $k \geqq 0$ and if A_p- $\lim_{k \to \infty} s_k = s$, then A_q- $\lim_{k \to \infty} s_k = s$.

[Hint: Let $u_n = (q_0 s_0 + q_1 s_1 + \ldots + q_n s_n)/Q_n$ and let t_n be as above. Then

$$u_j = \sum_{n=0}^{\infty} a_{j,n} t_n, \text{ where}$$

$$a_{j,n} = \frac{P_n}{Q_j} \cdot \left(\frac{q_n}{p_n} - \frac{q_{n+1}}{p_{n+1}} \right)$$

for $0 \leqq n < j$, $a_{j,j} = (P_j q_j)/(Q_j p_j)$, and $a_{j,n} = 0$ for $0 \leqq j < n$. Use (7.85).]

(d) If $q_{k+1}/q_k \geqq p_{k+1}/p_k$ and $(P_k q_k)/(Q_k p_k) \leqq C < \infty$ for all $k \geqq 0$ and if A_p- $\lim_{k \to \infty} s_k = s$, then A_q- $\lim_{k \to \infty} s_k = s$.

(e) If there exists a number $\beta > 1$ such that $P_{n+1}/P_n \geqq \beta$ for all sufficiently large n and if A_p- $\lim_{k \to \infty} s_k = s$, then $s_k \to s$ as $k \to \infty$. [Hint: Use (7.85) and the fact that

$$s_j = \sum_{k=0}^{\infty} a_{j,k} t_k, \text{ where } a_{j,j} = P_j/p_j, a_{j,j-1} = -P_{j-1}/p_j, \text{ and all other } a_{j,k} \text{ are } 0.]$$

(f) If $\alpha > 1$ and $p_k = \alpha^k$ for all $k \geqq 0$, then A_p-summability is equivalent to ordinary convergence.

(g) If $\beta \geqq \alpha > -1$ and $p_k = (k + 1)^{\alpha}$, $q_k = (k + 1)^{\beta}$ for all $k \geqq 0$, then A_p-summability is equivalent to A_q-summability. [Hint: Use (c) and (d).]

(h) If $q_k \geqq q_{k+1}$ for all $k \geqq 0$ and if C_1- $\lim_{k \to \infty} s_k = s$, then A_q- $\lim_{k \to \infty} s_k = s$. [Hint: Take $p_k = 1$ in (c).]

(i) If $p_k \leqq p_{k+1}$ for all $k \geqq 0$ and if A_p- $\lim_{k \to \infty} s_k = s$, then C_1- $\lim_{k \to \infty} s_k = s$.

(j) If $q_k = (k + 1)^{-1}$ and $s_k = (-1)^k(k + 1)$ for all $k \geqq 0$, then A_q- $\lim_{k \to \infty} s_k = 0$ and C_1- $\lim_{k \to \infty} s_k$ does not exist. Here, A_q-summability is equivalent to summability by logarithmic means.

Tauberian Theorems

In the terminology of the preceding section, Abel's Limit Theorem (7.40) asserts that Abel summability is a regular method: $\sum_{n=0}^{\infty} c_n = s$ implies A- $\sum_{n=0}^{\infty} c_n = s$. For this reason, any theorem which asserts that some summation method M is regular [ordinary convergence to s implies M-summability to s] is called an *Abelian theorem*. Ordinarily, the converse of an Abelian theorem is false. However, such a false converse may admit correction by the addition of an extra hypothesis. In 1897, A. Tauber proved the first such corrected converse: if A- $\sum_{n=0}^{\infty} c_n = s$ *and if* $nc_n \to 0$, then

$\sum_{n=0}^{\infty} c_n = s$. Because of this, following G. H. Hardy, any theorem which asserts that ordinary convergence follows from some kind of summability *and* an additional hypothesis, is called a *Tauberian theorem*. The additional hypothesis is called a *Tauberian hypothesis*. Thus, a Tauberian theorem has the form: M-summability to s *and* a Tauberian hypothesis imply ordinary convergence to s. In this section we prove some celebrated Tauberian theorems.

(7.90) **Hardy–Littlewood Tauberian Theorem** [1914] *Let X be a nonvoid set and let $(a_n)_{n=0}^{\infty}$ be a sequence of real-valued functions on X. Suppose that there exists a positive real number C such that*

(i) $$na_n(x) \leq C$$

for all $x \in X$ and all $n \geq 0$. If the series $\sum_{n=0}^{\infty} a_n$ is uniformly Abel summable to a real-valued function s on X, then this series converges uniformly on X to the sum s.

Proof [H. Wielandt, 1952] The summability hypothesis means that for each $x \in X$, the series

$$A(x,r) = \sum_{n=0}^{\infty} a_n(x)r^n$$

converges for $0 \leq r < 1$ and $\lim_{r \uparrow 1} A(x,r) = s(x)$ uniformly for $x \in X$: i.e., given $\epsilon > 0$, there exists $0 < r_0 < 1$ such that $|s(x) - A(x,r)| < \epsilon$ whenever $x \in X$ and $r_0 < r < 1$. This clearly implies that if $k \in \mathbb{N}$, then $A(x,r^k) \to s(x)$ uniformly on X as $r \uparrow 1$ [$r_0^{1/k} < r < 1$ implies $r_0 < r^k < 1$]. Thus, if $P(r) = \sum_{k=1}^{m} c_k r^k$ is any real polynomial with $P(0) = 0$ and $P(1) = 1$, we have

$$\sum_{n=0}^{\infty} a_n(x)P(r^n) = \sum_{k=1}^{m} c_k \sum_{n=0}^{\infty} a_n(x)r^{nk} = \sum_{k=1}^{m} c_k A(x,r^k) \to \sum_{k=1}^{m} c_k s(x)$$

$$= s(x)P(1) = s(x) \qquad (1)$$

uniformly on X as $r \uparrow 1$.

Define a function ϕ on $[0,1]$ by $\phi(r) = 0$ if $0 \leq r < 1/2$ and $\phi(r) = 1$ if $1/2 \leq r \leq 1$. For $x \in X$ and $0 < r < 1$, define

$$\Phi(x,r) = \sum_{n=0}^{\infty} a_n(x)\phi(r^n) = \sum_{n=0}^{N(r)} a_n(x),$$

where $N(r)$ is the largest integer not exceeding the positive number $-(\log 2)/(\log r)$. Plainly, $N(r)$ is a nondecreasing function of $r \in [1/2, 1[$ *onto* $\mathbb{N}$, and so our proof will be complete as soon as we show that

$$\lim_{r \uparrow 1} \Phi(x, r) = s(x) \tag{2}$$

uniformly on X. To this end, we consider the function ψ on $[0, 1]$ defined by $\psi(r) = -1/(1 - r)$ if $0 \le r < 1/2$ and $\psi(r) = 1/r$ if $1/2 \le r \le 1$. Then

$$\phi(r) = r + r(1 - r)\psi(r) \tag{3}$$

for all $r \in [0, 1]$. We shall prove (2) by approximating ψ, and hence ϕ, by polynomials so as to take advantage of (1).

Let $0 < \epsilon < C$ be given. Write $\delta = \epsilon/(24C)$. Choose any two *continuous* real-valued functions f_1 and f_2 on $[0, 1]$ which satisfy

$$f_1(r) + \delta \le \psi(r) \le f_2(r) - \delta \tag{4}$$

for $0 \le r \le 1$ and

$$\int_0^1 [f_2(r) - f_1(r)] \, dr < 10\delta. \tag{5}$$

[For example, we can take

$$f_1 = \psi - \delta \quad \text{on } [0, 1/2[\cup [1/2 + \delta, 1]$$

and linear on $[1/2, 1/2 + \delta]$ and take $f_2 = \psi + \delta$ on $[0, 1/2 - \delta] \cup [1/2, 1]$ and linear on $[1/2 - \delta, 1/2]$.] Next apply Weierstrass's Approximation Theorem to obtain real polynomials Q_1 and Q_2 such that

$$|f_j(r) - Q_j(r)| < \delta \tag{6}$$

for $0 \le r \le 1$ and $j = 1, 2$. It is clear from (4) and (6) that

$$Q_1(r) < \psi(r) < Q_2(r) \tag{7}$$

for $0 \le r \le 1$ and, from (5) and (6), that

$$\int_0^1 [Q_2(r) - Q_1(r)] \, dr < 12\delta. \tag{8}$$

Now define polynomials P_1, P_2, and Q by

$$P_j(r) = r + r(1 - r) Q_j(r), \qquad Q(r) = Q_2(r) - Q_1(r).$$

Then

$$P_2(r) - P_1(r) = r(1 - r) Q(r)$$

and, using (3), (7), and (8),

$$P_1(r) \le \phi(r) \le P_2(r),$$

$$Q(r) > 0, \qquad \int_0^1 Q(r) \, dr < 12\delta = \frac{\epsilon}{2C},$$

$$P_1(0) = P_2(0) = 0, \qquad P_1(1) = P_2(1) = 1$$

for $0 \le r \le 1$. Notice that $0 \le r \le 1$ and $n \in \mathbb{N}$ imply that

$$1 - r^n = (1 - r)(1 + r + \ldots + r^{n-1}) \le n(1 - r).$$

The polynomial Q can be written $Q(r) = \sum_{k=0}^{q} b_k r^k$. Combining the preceding, we see that $0 < r < 1$ and $x \in X$ imply that

$$\Phi(x, r) - \sum_{n=0}^{\infty} a_n(x) P_1(r^n) = \sum_{n=1}^{\infty} a_n(x)\big[\phi(r^n) - P_1(r^n)\big]$$

$$\le \sum_{n=1}^{\infty} \frac{C}{n}\big[P_2(r^n) - P_1(r^n)\big]$$

$$= C \cdot \sum_{n=1}^{\infty} \frac{1}{n}(1 - r^n) r^n Q(r^n)$$

$$\le (1 - r)C \cdot \sum_{n=1}^{\infty} r^n Q(r^n)$$

$$= (1 - r)C \cdot \sum_{k=0}^{q}\Big[b_k \sum_{n=1}^{\infty} r^{n(k+1)}\Big]$$

$$= C \cdot \sum_{k=0}^{q} b_k \frac{(1 - r)r^{k+1}}{1 - r^{k+1}} = C \cdot g(r) \quad (9)$$

where the last equality defines g on $]0, 1[$. Similarly,

$$\sum_{n=0}^{\infty} a_n(x) P_2(r^n) - \Phi(x, r) \le C \cdot g(r) \tag{10}$$

for $0 < r < 1$ and all $x \in X$. Since $\lim_{r \to 1}[(1 - r^{k+1})/(1 - r)] = k + 1$, we have

$$\lim_{r \uparrow 1} g(r) = \sum_{k=0}^{q} \frac{b_k}{k + 1} = \int_0^1 Q(r)\, dr < \frac{\epsilon}{2C}$$

Consequently, there exists $0 < r_0 < 1$ such that $C \cdot g(r) < \epsilon/2$ whenever $r_0 < r < 1$. Therefore, (9) and (10) show that

$$\sum_{n=0}^{\infty} a_n(x) P_2(r^n) - \frac{\epsilon}{2} < \Phi(x, r) < \sum_{n=0}^{\infty} a_n(x) P_1(r^n) + \frac{\epsilon}{2} \tag{11}$$

if $r_0 < r < 1$ and $x \in X$. Now (1) applies to both P_1 and P_2, so for $j = 1, 2$, there exists $0 < r_j < 1$ such that

$$\Big|s(x) - \sum_{n=0}^{\infty} a_n(x) P_j(r^n)\Big| < \frac{\epsilon}{2} \tag{12}$$

whenever $r_j < r < 1$ and $x \in X$. Let $r_3 = \max\{r_0, r_1, r_2\}$. From (11) and (12), it follows that

$$s(x) - \epsilon < \Phi(x, r) < s(x) + \epsilon \quad \text{if } r_3 < r < 1 \quad \text{and} \quad x \in X.$$

This proves (2). $\square$

(7.91)   Littlewood's Tauberian Theorem [1911]   *Let X be a nonvoid set and let $(c_n)_{n=0}^{\infty}$ be a sequence of complex-valued functions on X. Suppose that there exists a positive real number C such that*

$$\text{(i)} \qquad\qquad |nc_n(x)| \le C$$

for all $x \in X$ and all $n \ge 0$. If the series $\sum\limits_{n=0}^{\infty} c_n$ is uniformly Abel summable to a complex-valued function s on X, then this series converges uniformly on X to the sum s.

> **Proof**   Write $c_n = a_n + ib_n$, where a_n and b_n are real-valued. Then $na_n(x)$ $\le |nc_n(x)| \le C$, $nb_n(x) \le |nc_n(x)| \le C$, and $\sum\limits_0^{\infty} a_n$ and $\sum\limits_0^{\infty} b_n$ are uniformly Abel summable to $\mathrm{Re}(s)$ and $\mathrm{Im}(s)$, respectively. Now apply (7.90). $\square$

(7.92)   Hardy's Tauberian Theorem [1909]   *Let X, $(c_n)_{n=0}^{\infty}$, and C be as in (7.91). Suppose that (7.91.i) obtains. If the series $\sum\limits_{n=0}^{\infty} c_n$ is uniformly C_1-summable to a complex-valued function s on X, then this series converges uniformly on X to the sum s.*

> **Proof**   By (7.80), $C_1\text{-}\sum\limits_{n=1}^{\infty} c_n = s - c_0$ uniformly on X. Since (7.91.i) shows that c_n is bounded on X for $n \ge 1$, we deduce from (7.81.iii) and (7.91) that $\sum\limits_{n=1}^{\infty} c_n = s - c_0$ uniformly on X. $\square$

(7.93)   Example   We have seen in Exercise 11, p. 443, that if $\beta \in \mathbb{R}$, then the series $\sum\limits_{n=1}^{\infty} n^{-(1+i\beta)}$ is divergent. Since $|n \cdot n^{-(1+i\beta)}| = 1$ for all $n \ge 1$, it follows from (7.91) and (7.92) that this series is neither Abel summable nor C_1-summable.

We close this section with a theorem in which the Tauberian hypothesis is different from that of the preceding three theorems.

(7.94)   Fejér's Tauberian Theorem   *Let X and $(c_n)_{n=0}^{\infty}$ be as in (7.91). Suppose that there is a positive real number M such that, for all $x \in X$,*

$$\text{(i)} \qquad\qquad \sum_{n=1}^{\infty} n|c_n(x)|^2 \le M^2$$

and that this series converges uniformly on X. If the series $\sum\limits_{n=0}^{\infty} c_n$ is uniformly Abel summable to a complex-valued function s on X, then it converges uniformly on X to s.

Proof For $0 < r < 1$ and $N \in \mathbb{N}$, write

$$A(x,r) = \sum_{n=0}^{\infty} c_n(x) r^n, \qquad s_N(x) = \sum_{n=0}^{N} c_n(x),$$

$$R_N = \sup_{x \in X} \left[\sum_{n=N+1}^{\infty} n |c_n(x)|^2 \right]^{1/2}.$$

We suppose that $R_N > 0$ for all N, for otherwise there is nothing to prove. For $0 < r < 1$, we have $1 - r^n \leq n(1 - r)$, so

$$|s_N(x) - A(x,r)| \leq \sum_{n=1}^{N} |c_n(x)|(1 - r^n) + \sum_{n=N+1}^{\infty} |c_n(x)| r^n$$

$$\leq (1 - r) \sum_{n=1}^{N} n|c_n(x)| + \sum_{n=N+1}^{\infty} |c_n(x)| r^n. \qquad (1)$$

Using Cauchy's Inequality and (i), we have

$$\sum_{n=1}^{N} n|c_n(x)| = \sum_{n=1}^{N} \sqrt{n} \cdot \sqrt{n} \, |c_n(x)| \leq \left[\sum_{n=1}^{N} n \right]^{1/2} \cdot \left[\sum_{n=1}^{N} n|c_n(x)|^2 \right]^{1/2}$$

$$\leq N \cdot M \qquad (2)$$

and

$$\sum_{n=N+1}^{\infty} |c_n(x)| r^n = \sum_{n=N+1}^{\infty} \left[\sqrt{n} \, |c_n(x)| \cdot \frac{r^n}{\sqrt{n}} \right]$$

$$\leq \left[\sum_{n=N+1}^{\infty} n|c_n(x)|^2 \right]^{1/2} \cdot \left[\sum_{n=N+1}^{\infty} \frac{r^{2n}}{n} \right]^{1/2}$$

$$\leq R_N \cdot \left[\frac{1}{N} \sum_{n=0}^{\infty} r^n \right]^{1/2} = \frac{R_N}{\sqrt{N(1 - r)}} \qquad (3)$$

for all $x \in X$, $N \in \mathbb{N}$, and $0 < r < 1$. Since the series (i) converges uniformly on X, we have $R_N \to 0$ as $N \to \infty$, and so $R_N < N$ for all large N. For each such N, define $r_N \in \,]0, 1[$ by $N(1 - r_N) = R_N$. Plainly $r_N \to 1$ as $N \to \infty$. Now (1), (2), and (3) imply that

$$\sup_{x \in X} |s_N(x) - A(x, r_N)| \leq MN(1 - r_N) + R_N^{1/2} = MR_N + R_N^{1/2} \to 0 \qquad (4)$$

as $N \to \infty$. By hypothesis,

$$\sup_{x \in X} |A(x, r_N) - s(x)| \to 0 \qquad (5)$$

as $N \to \infty$. Finally, (4) and (5) yield

$$\sup_{x \in X} |s_N(x) - s(x)| \to 0$$

as $N \to \infty$, which is the desired conclusion. $\square$

(7.95) Remark Neither of the Tauberian hypotheses (7.91.i) nor (7.94.i) is a consequence of the other. Indeed, if $n = k^4$ for some $k \in \mathbb{N}$, let $c_n = k^{-3}$ and take $c_n = 0$ for all other n. Then $\sum n|c_n|^2 = \sum k^{-2} < \infty$ and $\overline{\lim} |nc_n| = \infty$. On the other hand, if $c_n = n^{-1}(-1)^n$ for $n \in \mathbb{N}$, then $\sum n|c_n|^2 = \infty$ and $|nc_n| = 1$ for all n.

Exercises

1.   Without using the (more difficult) theorems of this section, give a simple proof of

Tauber's Theorem: If $(c_n)_{n=0}^{\infty} \subset \mathbb{C}$, $\lim\limits_{n \to \infty} nc_n = 0$, and the series $\sum\limits_{n=0}^{\infty} c_n$ is Abel summable to $s \in \mathbb{C}$, then this series converges to the sum s. [Hints: Write $s_n = c_0 + c_1 + \ldots + c_n$, $r_n = 1 - 1/n$, and $A(r) = \sum\limits_{j=0}^{\infty} c_j r^j$. Write

$$s_n - s = A(r_n) - s + \sum_{j=0}^{n} c_j(1 - r_n^j) - \sum_{j=n+1}^{\infty} c_j r_n^j.$$

Use the fact that $1 - r^j \leq j(1 - r)$ for $0 \leq r \leq 1$.]

2.   If a series $\sum\limits_{n=0}^{\infty} a_n$ having nonnegative real terms is Abel summable to $s \in \mathbb{R}$, then it converges to s. [Hint: $\sum\limits_{n=0}^{N} a_n r^n \leq s$ for $0 < r < 1$ and $N \in \mathbb{N}$.]

3.   Let $(c_n)_{n=0}^{\infty} \subset \mathbb{C}$ and let $s_n = c_0 + c_1 + \ldots + c_n$ for $n \geq 0$. If

$$A\text{-}\sum_{n=0}^{\infty} c_n = s \in \mathbb{C} \quad \text{and} \quad (s_n)_{n=0}^{\infty}$$

is bounded, then $C_1\text{-}\sum\limits_{n=0}^{\infty} c_n = s$. [Hints: Suppose $c_0 = 0$ and $c_n \in \mathbb{R}$. Write $u_n = c_1 + 2c_2 + 3c_3 + \ldots + nc_n = (n+1)(s_n - \sigma_n)$, $v_n = u_n/[n(n+1)]$, $A(r) = \sum\limits_{n=1}^{\infty} c_n r^n$, $f(r) = \sum\limits_{n=1}^{\infty} v_n r^{n+1}$. Then $f(r) + (1 - r)f'(r) = A(r) \to s$ as $r \uparrow 1$. Apply L'Hospital's Rule to ψ/ϕ, where $\psi(r) = f(r)/(1 - r)$ and $\phi(r) = 1/(1 - r)$, to infer that $f(r) \to s$ as $r \uparrow 1$. Since $nv_n = s_n - \sigma_n$ is bounded, (7.91) gives $\sum\limits_{n=1}^{\infty} v_n = s$. Now

$$\sigma_k = s_k - \frac{u_k}{k+1} = \sum_{n=1}^{k} \frac{(u_n - u_{n-1})}{n} - \frac{u_k}{k+1} = \sum_{n=1}^{k} v_n.]$$

4.   If $(c_n)_{n=0}^{\infty} \subset \mathbb{C}$, $A\text{-}\sum\limits_{n=0}^{\infty} c_n = s \in \mathbb{C}$, and $C_1\text{-}\lim\limits_{n \to \infty} nc_n = 0$, then $\sum\limits_{n=0}^{\infty} c_n = s$. [Hint: In the preceding hint, $nv_n \to 0$.]

5.   *Hardy's Tauberian Theorem*. This exercise outlines a simple direct proof of (7.92) that does not depend on the considerably more difficult theorems (7.90) and (7.91). Let the notation and the hypotheses be as in (7.92) and write $s_n = c_0 + c_1 + \ldots +$

c_n and $\sigma_n = (n+1)^{-1}(s_0 + s_1 + \ldots + s_n)$ just as in (7.75). Let $0 < \epsilon < 1$ be given and write $\|f\|_u = \sup\{|f(x)| : x \in X\}$ for any function $f : X \to \mathbb{C}$.

(a) There is an $N \in \mathbb{N}$ such that $\|s - \sigma_k\|_u < \epsilon^2$ whenever $k \geq N$.

(b) If $n > 2N + 1$ is fixed and k is the integer for which $k \leq (n - \epsilon)/(1 + \epsilon) < k + 1$, then we have $N \leq k < n$,

$$(k + 1)/(n - k) \leq 1/\epsilon \quad \text{and} \quad (n - j)/(j + 1) < \epsilon$$

for any integer $j > k$.

(c) If n and k are as in (b) and $k < j \leq n$, then $\|s_n - s_j\|_u < C\epsilon$.

(d) If n and k are as in (b), then

$$s_n - \sigma_n = \frac{k+1}{n-k}[\sigma_n - \sigma_k] + \frac{1}{n-k}\sum_{j=k+1}^{n}(s_n - s_j).$$

(e) If $n > 2N + 1$, then $\|s - s_n\|_u < \epsilon^2 + 2\epsilon + C\epsilon < (C + 3)\epsilon$.

(f) $\lim\limits_{n \to \infty} \|s - s_n\|_u = 0$.

8

TRIGONOMETRIC SERIES

In this final chapter we give a brief introduction to one of the most useful and fascinating topics in analysis. Here we find ample application of most of the material in the first seven chapters. Indeed, the reader may wish to regard this chapter as a dessert that rewards his hard labor expended in learning the fundamental principles of analysis. The author, at least, so regards it.

In very general terms, the program for analysis is to learn how to decompose (analyze) rather arbitrary functions into component parts that are well-known functions and how to put together (synthesize) these parts to rebuild the original function. For many compelling reasons [usefulness in physical applications; intrinsic mathematical properties], the most commonly chosen "parts" are the trigonometric functions

$$t \to a_n \cos(nt) + b_n \sin(nt).$$

Joseph Fourier himself seemed to believe that any 2π-periodic function at all can be "represented" as an infinite sum of such functions. Though this seems doubtful (particularly for nonmeasurable functions), we shall see here some very concrete ways in which this is true for Lebesgue integrable functions of period 2π. As we have seen, it is essentially hopeless to use the "parts" $t \to c_n t^n$ to represent reasonably arbitrary functions as sums of convergent power series, because any such sum must be infinitely differentiable. We shall see that many wildly discontinuous or continuous but nowhere differentiable functions can be obtained as the sums of convergent trigonometric series.

The reader should not get the impression that our concentration on 2π-periodic functions severely limits our scope. In fact, any function f defined on any bounded interval can be transformed to a function g defined on a proper subinterval of $]-\pi,\pi[$—say $]0,1[$—by an affine mapping $[x = at + b, g(t) = f(x)]$ and then g can be extended so as to be 2π-periodic.

We begin with precise definitions.

Trigonometric Series and Fourier Series

(8.1) **Definitions** A *trigonometric series* is any series of the form

$$\text{(i)} \qquad \frac{a_0}{2} + \sum_{k=1}^{\infty} (a_k \cos(kt) + b_k \sin(kt)),$$

where $(a_k)_{k=0}^{\infty}$ and $(b_k)_{k=1}^{\infty}$ are sequences of complex numbers and $t \in \mathbb{R}$. The nth *partial sum* of (i) is the function s_n defined on $\mathbb{R}$ by

$$\text{(ii)} \qquad s_n(t) = \frac{a_0}{2} + \sum_{k=1}^{n} (a_k \cos(kt) + b_k \sin(kt)).$$

Because of Euler's Formulas (5.5) we also have

$$\text{(ii')} \qquad s_n(t) = \sum_{k=-n}^{n} c_k e^{ikt}$$

where, if we write $b_0 = 0$,

$$c_k = (a_k - ib_k)/2, \qquad c_{-k} = (a_k + ib_k)/2,$$

$$\text{(iii)} \qquad a_k = c_k + c_{-k}, \qquad b_k = i(c_k - c_{-k}) \qquad \text{for } k \geq 0.$$

We also write $s_0(t) = c_0 = a_0/2$. For this reason, we also write (i) in the form

$$\text{(i')} \qquad \sum_{k=-\infty}^{\infty} c_k e^{ikt}$$

and we call s_n the nth partial sum of this series. To say that (i) or (i') converges in some sense [pointwise, a.e., uniformly, etc.] means that the sequence $(s_n)_{n=0}^{\infty}$ of functions converges in that sense. Any function of the form (ii) or (ii') is called a *trigonometric polynomial*.

If a function $f : \mathbb{R} \to \mathbb{C}$ is to be the pointwise sum of a trigonometric series or the pointwise limit of a sequence $(s_n)_{n=0}^{\infty}$ of trigonometric polynomials, then it must be 2π-*periodic*. That is, $f(t + 2\pi) = f(t)$ for all $t \in \mathbb{R}$. This is because every trigonometric polynomial is 2π-periodic. Thus, we isolate for study some classes of 2π-periodic functions.

(8.2) **Definitions** Let $f : \mathbb{R} \to \mathbb{C}$ be 2π-periodic. For a positive real number p, we write $f \in L_p(\mathbb{T})$ if f is Lebesgue measurable and

$$\frac{1}{2\pi} \int_{-\pi}^{\pi} |f(t)|^p \, dt < \infty.$$

In this case, we define the $L_p(\mathbb{T})$-*norm* of f to be the number

$$\|f\|_p = \|f\|_{L_p(\mathbb{T})} = \left(\frac{1}{2\pi} \int_{-\pi}^{\pi} |f(t)|^p \, dt \right)^{1/p}.$$

If f is continuous on $\mathbb{R}$ (and 2π-periodic), we write $f \in C(\mathbb{T})$ and we define the *uniform norm* of f to be the number

$$\|f\|_u = \|f\|_{C(\mathbb{T})} = \sup_{t \in \mathbb{R}} |f(t)| = \sup_{-\pi \le t \le \pi} |f(t)|.$$

We denote the set of all trigonometric polynomials by $TP(\mathbb{T})$.

Obviously, $TP(\mathbb{T}) \subset C(\mathbb{T}) \subset L_p(\mathbb{T})$ for all $p > 0$. Moreover, $TP(\mathbb{T})$ is dense in these spaces in the following sense.

(8.3)   **Theorem**   (i) *If $f \in C(\mathbb{T})$ and $\epsilon > 0$, then there exists some $P \in TP(\mathbb{T})$ such that $\|f - P\|_u < \epsilon$.*

(ii) *If $1 \le p < \infty$, $f \in L_p(\mathbb{T})$, and $\epsilon > 0$, then there exists some $P \in TP(\mathbb{T})$ such that $\|f - P\|_p < \epsilon$.*

Proof   (i) Let $f \in C(\mathbb{T})$ and $\epsilon > 0$ be given. Write $X = \{z \in \mathbb{C} : |z| = 1\}$ and define F on X by $F(z) = f(t)$, where $z = e^{it}$ [by the periodicity of f and (5.11), the definition of F is independent of the choice of $t \in \mathbb{R}$ such that $e^{it} = z$; we could take $t = \text{Arg}\, z$]. It is easy to see that F is continuous at every $z \in X$. In fact, $F(z) = f(\text{Arg}\, z)$, so continuity follows from (5.15) for $z \ne -1$, but F is also continuous at -1 because $f(\pi) = f(-\pi)$. Now apply (3.129) to obtain $(c_n)_{n=-N}^N \subset \mathbb{C}$ such that

$$\left| F(z) - \sum_{n=-N}^N c_n z^n \right| < \epsilon$$

for all $z \in X$. Then (i) follows by taking $P(t) = \sum_{n=-N}^N c_n e^{int}$.*

(ii) Let $1 \le p < \infty$, $f \in L_p(\mathbb{T})$, and $\epsilon > 0$ be given. Use (6.111) to obtain $g \in C([-\pi, \pi])$ such that

$$\left(\frac{1}{2\pi} \int_{-\pi}^{\pi} |f - g|^p \right)^{1/p} < \frac{\epsilon}{3}. \tag{1}$$

Choose $0 < \beta < \infty$ such that $|g(t)| \le \beta$ for $t \in [-\pi, \pi]$. Write $\delta = 2\pi(\epsilon/(6\beta))^p$. We may suppose $\delta < 2\pi$. Now alter g on $[\pi - \delta, \pi]$ to obtain $h \in C([-\pi, \pi])$ such that $|h| \le \beta$, $h = g$ on $[-\pi, \pi - \delta]$, and $h(\pi) = h(-\pi)$. For instance, one can define h on $[\pi - \delta, \pi]$ by the rule

$$h(t) = \delta^{-1}\left[(\pi - t)g(\pi - \delta) + (t - \pi + \delta)g(-\pi) \right].$$

Then

$$\frac{1}{2\pi} \int_{-\pi}^{\pi} |g - h|^p = \frac{1}{2\pi} \int_{\pi - \delta}^{\pi} |g - h|^p \le \frac{(2\beta)^p \delta}{2\pi} = \left(\frac{\epsilon}{3} \right)^p. \tag{2}$$

---

*A proof of (8.3.i) that does not depend on the Stone–Weierstrass Theorem (or on (8.3.i)) is given in (8.30).

Next, extend h to be 2π-periodic on $\mathbb{R}$ so that $h \in C(\mathbb{T})$. We can apply part (i) to obtain $P \in TP(\mathbb{T})$ such that $\|h - P\|_u < \epsilon/3$. Then

$$\left(\frac{1}{2\pi} \int_{-\pi}^{\pi} |h - P|^p\right)^{1/p} < \frac{\epsilon}{3}. \tag{3}$$

Finally, (1), (2), (3), and Minkowski's Inequality (6.107) complete the proof of (ii). $\square$

(8.4) **Remark** Restated, (8.3.i) says that if $f \in C(\mathbb{T})$, then there exists a sequence $(P_n)_{n=1}^{\infty} \subset TP(\mathbb{T})$ such that $P_n \to f$ uniformly on $\mathbb{R}$. However, it is not clear (or even true in general) that these P_n's can be taken to be the partial sums s_n of some fixed trigonometric series. If such a series could be found, then it is unique, as the following theorem shows.

(8.5) **Theorem** *Consider any trigonometric series* $\sum\limits_{k=-\infty}^{\infty} c_k e^{ikt}$ *with partial sums* s_n *as in (8.1.ii'). Suppose there exists some subsequence* $(s_{n_j})_{j=1}^{\infty}$ *and some function f such that either* (i) *$f \in C(\mathbb{T})$ and $\|f - s_{n_j}\|_u \to 0$ as $j \to \infty$ or* (ii) *$f \in L_p(\mathbb{T})$ for some $1 \leq p < \infty$ and $\|f - s_{n_j}\|_p \to 0$ as $j \to \infty$. Then, for every integer n, we have*

(iii) $$c_n = \frac{1}{2\pi} \int_{-\pi}^{\pi} f(t) e^{-int}\, dt.$$

Proof If (i) obtains, then $f \in L_p(\mathbb{T})$ for all p and $\|f - s_{n_j}\|_p \leq \|f - s_{n_j}\|_u$ for all j, and so (ii) also obtains. Now suppose that (ii) does obtain. Let n be any fixed integer. Choose $j_0 \in \mathbb{N}$ such that $n_j > |n|$ for $j > j_0$. The crucial, but obvious, fact here is that, for integers k,

$$\frac{1}{2\pi} \int_{-\pi}^{\pi} e^{ikt} e^{-int}\, dt = \frac{1}{2\pi} \int_{-\pi}^{\pi} e^{i(k-n)t}\, dt = 0 \text{ or } 1$$

according as $k \neq n$ or $k = n$. Thus, for all $j > j_0$ we have

$$\frac{1}{2\pi} \int_{-\pi}^{\pi} s_{n_j}(t) e^{-int}\, dt = \sum_{k=-n_j}^{n_j} c_k \cdot \frac{1}{2\pi} \int_{-\pi}^{\pi} e^{i(k-n)t}\, dt = c_n.$$

It follows that

$$\left| c_n - \frac{1}{2\pi} \int_{-\pi}^{\pi} f(t) e^{-int}\, dt \right| = \lim_{j \to \infty} \left| \frac{1}{2\pi} \int_{-\pi}^{\pi} [f(t) - s_{n_j}(t)] e^{-int}\, dt \right|$$

$$\leq \lim_{j \to \infty} \frac{1}{2\pi} \int_{-\pi}^{\pi} |f(t) - s_{n_j}(t)|\, dt$$

$$\leq \lim_{j \to \infty} \|f - s_{n_j}\|_p = 0,$$

where the second inequality follows from Hölder's Inequality (6.106). $\square$

(8.6)   **Remark**   Using the notation of (8.5), if we write $f_j = s_{n_j} - s_{n_{j-1}}$ for $j > 1$ and $f_1 = s_{n_1}$, then hypothesis (8.5.i) simply says that the series $\sum\limits_{j=1}^{\infty} f_j$, which is obtained from the given trigonometric series by grouping together disjoint blocks of terms, is uniformly convergent. This does *not* imply that the given series is even pointwise convergent [see (8.25) below].

With (8.5) as motivation, we single out one trigonometric series for each function $f \in L_1(\mathbb{T})$.

(8.7)   **Definition**   Let $f \in L_1(\mathbb{T})$. Define a function $\hat{f}$ on $\mathbb{Z}$ by

(i) $$\hat{f}(n) = \frac{1}{2\pi} \int_{-\pi}^{\pi} f(t) e^{-int} \, dt.$$

The trigonometric series

(ii) $$s(f, t) = \sum_{k=-\infty}^{\infty} \hat{f}(k) e^{ikt}$$

is called the *Fourier series of f* and, for each integer k, the complex number $\hat{f}(k)$ is called the kth *Fourier coefficient of f*. We denote the nth *partial sum* of (ii) $(n \geq 0)$ by

(iii) $$s_n(f, t) = \sum_{k=-n}^{n} \hat{f}(k) e^{ikt}.$$

If, as in (8.1.iii), we write

(iv) $a_n = \hat{f}(n) + \hat{f}(-n), \qquad b_n = i(\hat{f}(n) - \hat{f}(-n))$ for $n \geq 0$,

then we have

(v) $$a_n = \frac{1}{\pi} \int_{-\pi}^{\pi} f(t)\cos(nt)\,dt, \qquad b_n = \frac{1}{\pi} \int_{-\pi}^{\pi} f(t)\sin(nt)\,dt.$$

With this notation, we have

(vi) $$s_n(f, t) = \frac{a_0}{2} + \sum_{k=1}^{n} (a_k \cos(kt) + b_k \sin(kt))$$

and

(vii) $$s(f, t) = \frac{a_0}{2} + \sum_{k=1}^{\infty} (a_k \cos(kt) + b_k \sin(kt)).$$

When we write

(viii) $$f(t) \sim \sum_{k=-\infty}^{\infty} c_k e^{ikt},$$

we mean that $c_k = \hat{f}(k)$ for all integers k. When we write

(ix) $$f(t) \sim \frac{a_0}{2} + \sum_{k=1}^{\infty} (a_k \cos(kt) + b_k \sin(kt)),$$

we mean that the a_k and b_k are related to f via formulas (iv) and (v). Notice that in (viii) and (ix) we use the symbol $\sim$ and *not* the equality sign. There is no assurance that the Fourier series of f converges to f or to anything else [see (8.25)].

(8.8) **Remarks** (a) It follows from (8.5) that if a trigonometric series converges to f, either uniformly or in the L_p-metric ($p \geq 1$), then it must be the Fourier series of f. Take careful note, however, that there do exist trigonometric series which converge to 0 a.e. without having all coefficients equal to 0. The first known example of such a series, given by Menshov in 1916, has the coefficients

$$c_n = \prod_{j=1}^{\infty} \cos \frac{(j+2)n\pi}{j(j+1) \cdot 2^j}$$

for all integers n [see Exercise 23 at the end of this chapter]. A further exposition of such fascinating matters lies beyond the scope of this book. The reader is referred to the treatises of Bary or Zygmund cited in the bibliography.

(b) All integrals in (8.5) and (8.7) can be taken over any interval of length 2π without affecting their values. This is because the integrands are 2π-periodic. In fact, if $g \in L_1(\mathbb{T})$ and $a \in \mathbb{R}$, the substitution $u = t - 2\pi$ yields

$$\int_a^{a+2\pi} g(t)\,dt = \int_a^{\pi} g(t)\,dt + \int_{\pi}^{a+2\pi} g(t)\,dt = \int_a^{\pi} g(t)\,dt + \int_{-\pi}^{a} g(u)\,du = \int_{-\pi}^{\pi} g(t)\,dt.$$

We frequently apply this remark without further reference.

By way of an example, which we also find useful later, we have the following result.

(8.9) **Theorem** *Let* $f \in L_1(\mathbb{T})$ *be defined by* $f(t) = (\pi - t)/2$ *for* $0 < t < 2\pi$ *and* $f(0) = 0$ *[f is defined for all other* $t \in \mathbb{R}$ *to make* f 2π*-periodic]. Then*

(i) $$f(t) = \sum_{k=1}^{\infty} \frac{\sin(kt)}{k}$$

for all $t \in \mathbb{R}$, *the series* (i) *is the Fourier series of* f, *it converges uniformly on* $[\delta, 2\pi - \delta]$ *whenever* $0 < \delta < \pi$, *and its partial sums satisfy*

(ii) $$|s_n(f,t)| = \left| \sum_{k=1}^{n} \frac{\sin(kt)}{k} \right| < 2\pi + 1$$

for all $n \in \mathbb{N}$ *and all* $t \in \mathbb{R}$. [*Cf. Exercise 10(d), p. 428.*]

Proof The statements about convergence were proved in (7.39). It is easy to check that

$$\frac{2x}{\pi} < \sin x < x \tag{1}$$

for $0 < x < \pi/2$ [$(\sin x)/x$ is decreasing on this interval]. To prove (ii), we need only consider $0 < t < \pi$ [$\sin(-x) = -\sin x$], so fix any such t and let m be the integer such that $m \leq 1/t < m + 1$. If $m \geq 1$, then (1) shows that

$$0 < \sum_{k=1}^{m} \frac{\sin(kt)}{k} < \sum_{k=1}^{m} \frac{kt}{k} = mt \leqq 1 \qquad (2)$$

because $1 \leqq k \leqq m$ implies $0 < kt \leqq 1 < \pi/2$. If $n > m$, we can use (7.38.ii) and (1) to write

$$\left| \sum_{k=m+1}^{n} \frac{\sin(kt)}{k} \right| \leqq \frac{2}{(m+1)\sin(t/2)} < \frac{2}{(1/t)(2/\pi)(t/2)} = 2\pi. \qquad (3)$$

Now (ii) is a consequence of (2) and (3). To see that series (i) is the Fourier series of f, one can either use (i), (ii), Lebesgue's Dominated Convergence Theorem, and (8.5.ii) [with $p = 1$] or else compute directly as follows: for $k \neq 0$, we integrate by parts to obtain

$$\hat{f}(k) = \frac{1}{2\pi} \int_0^{2\pi} \frac{\pi - t}{2} e^{-ikt}\, dt$$

$$= \frac{1}{2\pi} \cdot \frac{\pi - t}{2} \cdot \frac{e^{-ikt}}{-ik} \Big|_0^{2\pi} - \frac{1}{2\pi} \int_0^{2\pi} \left(-\frac{1}{2} \right) \left(\frac{e^{-ikt}}{-ik} \right) dt = \frac{1}{2ki},$$

$$a_k = \hat{f}(k) + \hat{f}(-k) = 0, \quad b_k = i(\hat{f}(k) - \hat{f}(-k)) = 1/k, \quad a_0 = 2\hat{f}(0) = 0.$$

□

(8.10)   **Remark**   In computing Fourier coefficients, it is sometimes useful to use the following facts. If f is an odd function $[f(-t) = -f(t)]$, then

$$\int_{-\pi}^{0} f(t)\cos(kt)\, dt = - \int_0^{\pi} f(u)\cos(ku)\, du,$$

so that

$$a_k = \frac{1}{\pi} \int_{-\pi}^{\pi} f(t)\cos(kt)\, dt = 0.$$

Similarly, if f is an even function $[f(-t) = f(t)]$, then $b_k = 0$ for all $k \in \mathbb{N}$.

It is evident from (8.7.i) that the mapping $f \to \hat{f}$ from $L_1(\mathbb{T})$ into the space $\mathbb{C}^{\mathbb{Z}}$ of all complex-valued functions on $\mathbb{Z}$ is a linear mapping: $\widehat{\alpha f + \beta g}(n) = \alpha \hat{f}(n) + \beta \hat{g}(n)$ whenever $f, g \in L_1(\mathbb{T})$, $\alpha, \beta \in \mathbb{C}$, and $n \in \mathbb{Z}$. It is also clear that

$$\| \hat{f} \|_u = \sup_{n \in \mathbb{Z}} |\hat{f}(n)| \leqq \frac{1}{2\pi} \int_{-\pi}^{\pi} |f| = \| f \|_1$$

for all $f \in L_1(\mathbb{T})$, and so this mapping is into the space $l_\infty(\mathbb{Z})$ of all *bounded* complex-valued functions on $\mathbb{Z}$. We now show that this mapping is one-to-one in the sense of a.e. equality.

(8.11)   **Theorem**   *If $f, g \in L_1(\mathbb{T})$ and $\hat{f}(n) = \hat{g}(n)$ for all $n \in \mathbb{Z}$, then $f = g$ a.e. on $\mathbb{R}$.*

Proof Writing $h = f - g$, we have $\hat{h}(n) = \hat{f}(n) - \hat{g}(n) = 0$ for all n. Now define H on $\mathbb{R}$ by

$$H(x) = \int_0^x h(t)\,dt.$$

Then

$$H(x + 2\pi) - H(x) = \int_x^{x+2\pi} h(t)\,dt = 2\pi \cdot \hat{h}(0) = 0$$

so $H \in C(\mathbb{T})$. Define $F \in C(\mathbb{T})$ by $F(x) = H(x) - \hat{H}(0)$. Plainly

$$\hat{F}(0) = \frac{1}{2\pi} \int_0^{2\pi} [H(x) - \hat{H}(0)]\,dx = \hat{H}(0) - \hat{H}(0) = 0. \tag{1}$$

Since F is absolutely continuous on $[0, 2\pi]$ and $F'(x) = h(x)$ a.e. (6.84), we may apply integration by parts (6.90) to obtain, for $n \neq 0$,

$$\hat{\dot{F}}(n) = \frac{1}{2\pi} \int_0^{2\pi} F(t)e^{-int}\,dt = \frac{1}{2\pi in} \int_0^{2\pi} h(t)e^{-int}\,dt = \frac{1}{in}\hat{h}(n) = 0 \tag{2}$$

[the integrated part is 0 by periodicity]. If $P(t) = \sum_{n=-N}^{N} c_n e^{int}$ is any trigonometric polynomial, we have by (1) and (2) that

$$\frac{1}{2\pi} \int_0^{2\pi} F(t)\overline{P(t)}\,dt = \sum_{n=-N}^{N} \overline{c_n}\hat{F}(n) = 0.$$

Using (8.3.i), we choose a sequence $(P_k)_{k=1}^{\infty} \subset TP(\mathbb{T})$ such that $P_k \to F$ uniformly on $\mathbb{R}$. Then

$$\int_0^{2\pi} |F(t)|^2\,dt = \int_0^{2\pi} F(t)\overline{F(t)}\,dt = \lim_{k\to\infty} \int_0^{2\pi} F(t)\overline{P_k(t)}\,dt = 0.$$

Since $F \in C(\mathbb{T})$, we conclude that $F(t) = 0$ for all $t \in \mathbb{R}$. But then $f(t) - g(t) = h(t) = F'(t) = 0$ a.e. on $\mathbb{R}$. $\square$

(8.12)   Remarks   (a) The conclusion of (8.11) cannot be strengthened to the statement that $f = g$ everywhere. Indeed, the converse of (8.11) is true because if $f, g \in L_1(\mathbb{T})$ and $f = g$ a.e., then

$$\hat{f}(n) - \hat{g}(n) = \frac{1}{2\pi} \int_0^{2\pi} [f(t) - g(t)]e^{-int}\,dt = \frac{1}{2\pi} \int_0^{2\pi} 0\,dt = 0$$

for every integer n.

(b) It follows from (8.5) and (8.11) combined that if the Fourier series of a function $g \in L_1(\mathbb{T})$ converges uniformly on $\mathbb{R}$ to some function f [or only a subsequence $(s_{n_j}(g))_{j=1}^{\infty}$ of its partial sums converge uniformly], then $\hat{f}(n) = \hat{g}(n)$ for all n, and so $f = g$ a.e. Moreover, if $g \in C(\mathbb{T})$, then $f = g$ everywhere because $f \in C(\mathbb{T})$.

Which Trigonometric Series Are Fourier Series?

It is by no means true that every trigonometric series having a bounded sequence of coefficients is the Fourier series of a function in $L_1(\mathbb{T})$. In this section we prove this assertion by obtaining two further conditions that the coefficients in any Fourier series must satisfy. We begin with the following important fact.

(8.13) Riemann–Lebesgue Lemma *Let $f \in L_1(\mathbb{R})$. Then*

(i) $$\lim_{|x| \to \infty} \int_{-\infty}^{\infty} f(t)e^{-ixt}\,dt = 0 \qquad (x \in \mathbb{R}).$$

Proof First, suppose that $f = \xi_I$, where I is a bounded interval having endpoints $a \leq b$. Then $x \neq 0$ implies

$$\left| \int_{-\infty}^{\infty} f(t)e^{-ixt}\,dt \right| = \left| \int_a^b e^{-ixt}\,dt \right| = \left| -\frac{1}{ix}(e^{-ixb} - e^{-ixa}) \right| \leq \frac{2}{|x|},$$

and so (i) obtains for this f. It follows immediately that (i) also holds good for any step function $f \in M_0(\mathbb{R})$ (6.1).

Now let $f \in L_1'(\mathbb{R})$ and $\epsilon > 0$ be given. By (6.18) there is a function $\phi \in M_0(\mathbb{R})$ such that $\int |f - \phi| < \epsilon/2$. Now use the preceding paragraph to select $x_0 \in \mathbb{R}$ such that

$$\left| \int_{-\infty}^{\infty} \phi(t)e^{-ixt}\,dt \right| < \frac{\epsilon}{2}$$

whenever $|x| > x_0$. Then $|x| > x_0$ implies

$$\left| \int_{-\infty}^{\infty} f(t)e^{-ixt}\,dt \right| \leq \left| \int_{-\infty}^{\infty} [f(t) - \phi(t)]e^{-ixt}\,dt \right| + \left| \int_{-\infty}^{\infty} \phi(t)e^{-ixt}\,dt \right|$$

$$\leq \int_{-\infty}^{\infty} |f(t) - \phi(t)|\,dt + \frac{\epsilon}{2} < \epsilon.$$

This proves (i) for $f \in L_1'(\mathbb{R})$. The general case follows from this by separating f into real and imaginary parts. □

(8.14) Corollary *If $f \in L_1(\mathbb{T})$: then* $\lim_{|n| \to \infty} \hat{f}(n) = 0$. *That is, the mapping $f \to \hat{f}$ is into the space $c_0(\mathbb{Z})$ of all complex-valued functions ϕ on $\mathbb{Z}$ such that* $\lim_{|n| \to \infty} \phi(n) = 0$.

Proof Apply (8.13) to $f\xi_{[-\pi,\pi]}$. □

It follows from the next theorem that there are plenty of functions ϕ in $c_0(\mathbb{Z})$ that are *not* of the form $\hat{f}$ for any $f \in L_1(\mathbb{T})$. It also follows that the Fourier series of any $f \in L_1(\mathbb{T})$, even if divergent (see (8.26.b)), may be integrated term-by-term over any bounded interval to obtain a numerical series whose sum is the integral of f over that interval.

(8.15) **Theorem** *Let $f \in L_1(\mathbb{T})$ have the Fourier series*

$$f(t) \sim \frac{a_0}{2} + \sum_{k=1}^{\infty} (a_k \cos(kt) + b_k \sin(kt))$$

as in (8.7.ix). Then the series $\displaystyle\sum_{k=1}^{\infty} \frac{b_k}{k}$ *converges and we have*

(i)
$$\sum_{k=1}^{\infty} \frac{b_k}{k} = \frac{1}{2\pi} \int_0^{2\pi} f(t)(\pi - t) \, dt.$$

The function F defined on $\mathbb{R}$ by $F(x) = \int_0^x f(t) \, dt$ satisfies

(ii) $F(x) - \dfrac{a_0 x}{2} = \displaystyle\sum_{k=1}^{\infty} \frac{b_k}{k} + \sum_{k=1}^{\infty} \frac{1}{k}(a_k \sin(kx) - b_k \cos(kx))$ *for all $x \in \mathbb{R}$.*

The second series in (ii) converges uniformly on $\mathbb{R}$, and the right side of (ii) is the Fourier series of the function on the left side. If $\alpha \leq \beta$ in $\mathbb{R}$, then

(iii) $\displaystyle\int_\alpha^\beta f(t) \, dt = \int_\alpha^\beta \frac{a_0}{2} \, dt + \sum_{k=1}^{\infty} \int_\alpha^\beta (a_k \cos(kt) + b_k \sin(kt)) \, dt$

Proof We first observe that for any fixed $x \in \mathbb{R}$ the function $t \to f(x + t)$ has Fourier sine coefficients $b_k(x)$ given by

$$b_k(x) = \frac{1}{\pi} \int_{-\pi}^{\pi} f(x + t)\sin(kt) \, dt = \frac{1}{\pi} \int_{-\pi}^{\pi} f(u)\sin(k(u - x)) \, du$$

$$= b_k \cos(kx) - a_k \sin(kx). \tag{1}$$

Thus $b_k(x)/k$ is just the negative of the kth term of the second series in (ii). We claim that

$$\sum_{k=1}^{\infty} \frac{b_k(x)}{k} = \frac{1}{2\pi} \int_0^{2\pi} f(x + t)(\pi - t) \, dt \tag{2}$$

for all $x \in \mathbb{R}$ and that the series on the left converges uniformly on $\mathbb{R}$. Let $\epsilon > 0$ be given. Write $A = \dfrac{1}{\pi} \int_0^{2\pi} |f(t)| \, dt$ and $B = \pi/2 + (2\pi + 1)$. Use (6.79) to obtain $0 < \delta < \pi$ such that $\int_E |f| < \epsilon/(2B)$ whenever $E \subset [0, 4\pi]$ is measurable and $\lambda(E) \leq 2\delta$. Next use (8.9) to obtain $N \in \mathbb{N}$ such that

$$\left| \frac{\pi - t}{2} - \sum_{k=1}^{n} \frac{\sin(kt)}{k} \right| < \frac{\epsilon}{2A + 1} \tag{3}$$

whenever $n > N$ and $\delta \leq t \leq 2\pi - \delta$. Notice also that (8.9) shows that the left side of (3) is less than B for all $t \in [0, 2\pi]$ and all $n \in \mathbb{N}$. Write $D = [0, \delta] \cup [2\pi - \delta, 2\pi]$. For any $x \in [0, 2\pi]$ and any $n > N$, we have, using

(1) and (3), that

$$\left| \frac{1}{2\pi} \int_0^{2\pi} f(x+t)(\pi - t)\, dt - \sum_{k=1}^n \frac{b_k(x)}{k} \right|$$

$$= \left| \frac{1}{\pi} \int_0^{2\pi} f(x+t) \left[\frac{\pi - t}{2} - \sum_{k=1}^n \frac{\sin(kt)}{k} \right] dt \right|$$

$$\leq \frac{1}{\pi} \int_\delta^{2\pi - \delta} |f(x+t)| \cdot \frac{\epsilon}{2A + 1}\, dt + \frac{1}{\pi} \int_D |f(x+t)| B\, dt$$

$$\leq \frac{\epsilon}{2A + 1} \cdot A + \frac{B}{\pi} \int_{D+x} |f(u)|\, du$$

$$< \frac{\epsilon}{2} + \frac{\epsilon}{2\pi} < \epsilon.$$

Since both sides of (2) are 2π-periodic functions of x, this proves our claim made there. Since $b_k(0) = b_k$, (i) follows from (2). Next observe that, for any $x \in \mathbb{R}$,

$$\frac{1}{2\pi} \int_0^{2\pi} f(x+t)(\pi - t)\, dt = \frac{1}{2\pi} \int_x^{2\pi + x} f(u)(\pi - u + x)\, du$$

$$= \frac{1}{2\pi} \left\{ \int_0^{2\pi} + \int_{2\pi}^{2\pi + x} - \int_0^x \right\} f(u)(\pi - u + x)\, du$$

$$= \frac{1}{2\pi} \int_0^{2\pi} f(u)(\pi - u)\, du + \frac{x}{2\pi} \int_0^{2\pi} f(u)\, du$$

$$+ \frac{1}{2\pi} \int_0^x f(v + 2\pi)(x - v - \pi)\, dv$$

$$- \frac{1}{2\pi} \int_0^x f(v)(\pi - v + x)\, dv$$

$$= \sum_{k=1}^\infty \frac{b_k}{k} + \frac{a_0 x}{2} - F(x) \qquad (4)$$

because $f(v + 2\pi) = f(v)$. Substituting (1) and (4) into (2), we have (ii). Finally, (ii) with $x = \alpha$ subtracted from (ii) with $x = \beta$ yields (iii). □

(8.16) Example Since, by (7.3.b), $\displaystyle\sum_{n=2}^\infty \frac{1}{n \log n} = \infty$, it follows from (8.15) that the trigonometric series

$$(i) \qquad\qquad \sum_{n=2}^\infty \frac{\sin(nt)}{\log n}$$

is *not* the Fourier series of a function in $L_1(\mathbb{T})$. That is, the function $\phi \in c_0(\mathbb{Z})$ defined by $\phi(n) = (2i \log n)^{-1} = -\phi(-n)$ for $n \geq 2$ and $\phi(-1) = \phi(0) = \phi(1) = 0$ satisfies $\phi \neq \hat{f}$ for every $f \in L_1(\mathbb{T})$. Notice, however, that (7.38) implies that series (i) converges uniformly on $[\delta, 2\pi - \delta]$ whenever $0 < \delta < \pi$.

The problem of recognizing whether or not a trigonometric series is a Fourier series by inspecting its coefficients is unsolved and seems to be extremely difficult. Equivalently, the problem is to determine the range of the mapping $f \to \hat{f}$ of $L_1(\mathbb{T})$ into $c_0(\mathbb{Z})$. The preceding theorem provides a necessary condition. Theorem (8.5) and the following two theorems furnish sufficient conditions.

(8.17)   Riesz-Fischer Theorem   *Suppose that* $(c_n)_{n=-\infty}^{\infty} \subset \mathbb{C}$ *satisfies*

$$\sum_{n=-\infty}^{\infty} |c_n|^2 = \lim_{p \to \infty} \sum_{n=-p}^{p} |c_n|^2 < \infty.$$

Then there exists a function $f \in L_2(\mathbb{T})$ *such that* $\hat{f}(n) = c_n$ *for every integer* n. *Moreover,* $\|f - s_p(f)\|_2 \to 0$ *as* $p \to \infty$.

Proof   Write $s_p(t) = \sum_{n=-p}^{p} c_n e^{int}$ for $p \geq 0$. Then $q > p \geq 0$ implies

$$\|s_q - s_p\|_2^2 = \frac{1}{2\pi} \int_{-\pi}^{\pi} |s_q(t) - s_p(t)|^2 \, dt$$

$$= \frac{1}{2\pi} \int_{-\pi}^{\pi} \left(\sum_{p < |n| \leq q} c_n e^{int} \right) \left(\sum_{p < |k| \leq q} \overline{c_k} e^{-ikt} \right) dt$$

$$= \sum_{\substack{p < |n| \leq q \\ p < |k| \leq q}} c_n \overline{c_k} \frac{1}{2\pi} \int_{-\pi}^{\pi} e^{i(n-k)t} \, dt = \sum_{p < |n| \leq q} |c_n|^2.$$

By hypothesis, the last sum has limit 0 as $p, q \to \infty$. It follows from (6.110) that there exists $f \in L_2(\mathbb{T})$ such that $\|f - s_p\|_2 \to 0$ as $p \to \infty$. Now (8.5.ii) shows that $\hat{f}(n) = c_n$ for all n. Finally, $s_p(t) = \sum_{n=-p}^{p} \hat{f}(n) e^{int} = s_p(f, t)$. $\square$

The next theorem has very special hypotheses, but it is often useful for producing Fourier series. Also, its proof is instructive.

(8.18)   Theorem   *Let* $(a_k)_{k=0}^{\infty}$ *be a nonincreasing sequence of nonnegative real numbers having limit 0. Suppose also that this sequence is convex:* $2a_{k+1} \leq a_k + a_{k+2}$ *for all* $k \geq 0$. *Then the function* f *defined on* $\mathbb{R}$ *by*

(i) $$f(t) = \frac{a_0}{2} + \sum_{k=1}^{\infty} a_k \cos(kt)$$

is in $L_1^+(\mathbb{T})$ $[f \geq 0]$ *and the series* (i) *is the Fourier series of* f. *Moreover, this series converges uniformly on* $[\delta, 2\pi - \delta]$ *whenever* $0 < \delta < \pi$, *and so* f *is continuous except possibly at integral multiples of* 2π.

Proof   The last sentence, which does not require convexity, follows from (7.38), and so (i) defines a function f which is 2π-periodic and measurable on $\mathbb{R}$. Since $\sum_{k=0}^{\infty} (a_k - a_{k+1}) = a_0 < \infty$ and $(a_k - a_{k+1}) \geq (a_{k+1} - a_{k+2})$ for all k, it follows that

$$\lim_{n \to \infty} n(a_n - a_{n+1}) = 0. \tag{1}$$

Now let D_n and K_n be as in (5.12). Applying summation by parts (7.35) twice, we obtain

$$2s_n(t) = a_0 + \sum_{k=1}^{n} 2a_k \cos(kt)$$

$$= \sum_{k=0}^{n} (a_k - a_{k+1}) D_k(t) + a_{n+1} D_n(t)$$

$$= \sum_{k=0}^{n} (a_k - 2a_{k+1} + a_{k+2})(k+1) K_k(t)$$

$$+ (a_{n+1} - a_{n+2})(n+1) K_n(t) + a_{n+1} D_n(t).$$

If $t \neq 2m\pi$ $(m \in \mathbb{Z})$, it follows from (5.12) and the fact that $a_k \to 0$ that the last two terms have limit 0 as $n \to \infty$. This proves that

$$2f(t) = \sum_{k=0}^{\infty} (a_k - 2a_{k+1} + a_{k+2})(k+1) K_k(t) \tag{2}$$

for $t \neq 2m\pi$ $(m \in \mathbb{Z})$. By convexity and (5.12.ii), each term of this series is nonnegative, and so $f(t) \geq 0$ for all $t \in \mathbb{R}$. Notice that (5.12.ii) also implies that

$$\frac{1}{2\pi} \int_0^{2\pi} K_k(t) \cos(nt)\, dt = \begin{cases} 1 - \dfrac{n}{k+1} & (0 \leq n \leq k) \\ 0 & (0 \leq k < n) \end{cases} \tag{3}$$

using (7.35) again, we have

$$a_n = \sum_{k=n}^{q} (a_k - a_{k+1}) \cdot 1 + a_{q+1}$$

$$= \sum_{k=n}^{q} (a_k - 2a_{k+1} + a_{k+2})(k+1-n)$$

$$+ (a_{q+1} - a_{q+2})(q+1-n) + a_{q+1}$$

whenever $0 \leq n \leq q$. It follows from (1) that the last two terms have limit 0 as $q \to \infty$, and so

$$a_n = \sum_{k=n}^{\infty} (a_k - 2a_{k+1} + a_{k+2})(k+1-n). \tag{4}$$

Since all terms in (2) are nonnegative, (6.46) allows us to integrate term-by-term, and so, applying (3) and (4) (with $n = 0$), we obtain

$$\frac{1}{\pi} \int_0^{2\pi} f(t)\, dt = a_0 < \infty.$$

This proves that $f \in L_1^+(\mathbb{T})$.

Finally, multiply both sides of (2) by $\cos(nt)$, the right side term-by-term. It is evident that the right side of the resulting equality has its partial sums dominated by $2f$ on $]0, 2\pi[$, and so (6.55) permits term-by-term integration over that interval. Another application of (3) and (4) yields

$$\frac{1}{\pi} \int_0^{2\pi} f(t)\cos(nt)\,dt = a_n$$

for all $n \geq 0$. Since f is an even function, this proves that series (i) is the Fourier series of f. $\square$

We can now prove a corollary which shows that (8.14) is the best possible result on the behavior of $\hat{f}$ at infinity for arbitrary $f \in L_1(\mathbb{T})$.

(8.19) Corollary *Let $(c_n)_{n=0}^{\infty}$ be any sequence of nonnegative real numbers having limit 0. Then there exists a function $f \in L_1^+(\mathbb{T})$ such that $\hat{f}(-n) = \hat{f}(n) > c_n$ for every integer $n \geq 0$.*

Proof We need only find a convex sequence $(a_n)_{n=0}^{\infty}$ as in (8.18) such that $a_n > 2c_n$ for all n because, if this can be done, we can then define f as in (8.18.i) and notice that $\hat{f}(n) = \hat{f}(-n) = a_n/2$ for all $n \geq 0$.

To obtain the desired a_n's we first choose a sequence $(n_k)_{k=0}^{\infty}$ of nonnegative integers. Let $n_0 = 0$. If n_{k-1} has been chosen for some $k \geq 1$, then choose $n_k > 2n_{k-1}$ such that

$$c_n < 2^{-(k+2)} \quad \text{for all } n \geq n_k. \tag{1}$$

Clearly we have

$$2(n_{k+1} - n_k) > (n_k - n_{k-1}) \tag{2}$$

for all $k \geq 1$. Next let

$$M = 1 + 2 \cdot \max\{c_n : 0 \leq n < n_1\}. \tag{3}$$

Now we define a_n by

$$a_n = 1/2 + M(n_1 - n) \quad \text{if } 0 \leq n \leq n_1$$

and

$$a_n = (2n_{k+1} - n_k - n)/\left(2^{k+1}(n_{k+1} - n_k)\right) \quad \text{if } n_k \leq n \leq n_{k+1}$$

for some $k > 0$. Notice that these definitions are consistent when $n = n_k$ and that $a_{n_k} = 2^{-k}$ for $k \geq 1$. It is obvious that $(a_n)_{n=0}^{\infty}$ is a decreasing sequence having limit 0 and (1) and (3) show that $2c_n < a_n$ for all $n \geq 0$. As for the convexity condition $a_{n-1} + a_{n+1} \geq 2a_n$, it is clear that this holds with equality if $n_{k-1} < n < n_k$ for some $k > 0$. If $n = n_k$, we use (2) and the observation that $a_{n_1-1} > M \geq 1 \geq (n_1 + 1)/(2n_1)$ [for the case $k = 1$] to obtain

$$a_{n_k-1} + a_{n_k+1} \geq \frac{n_k - n_{k-1} + 1}{2^k(n_k - n_{k-1})} + \frac{n_{k+1} - n_k - 1/2}{2^k(n_{k+1} - n_k)}$$

$$= \frac{1}{2^k}\left[1 + \frac{1}{n_k - n_{k-1}} + 1 - \frac{1}{2(n_{k+1} - n_k)} \right]$$

$$= 2a_{n_k} + \frac{1}{2^k}\left[\frac{1}{n_k - n_{k-1}} - \frac{1}{2(n_{k+1} - n_k)} \right] > 2a_{n_k}. \quad \square$$

(8.20) Example   Let $(a_n)_{n=0}^{\infty}$ be any convex sequence with limit 0 as in (8.18) such that

$$\sum_{n=1}^{\infty} \frac{a_n}{n} = \infty.$$

Then (8.18) shows that the series

$$\sum_{n=1}^{\infty} a_n \cos(nt)$$

is a Fourier series while (8.15) shows that the related series

$$\sum_{n=1}^{\infty} a_n \sin(nt)$$

is *not* a Fourier series. As (8.19) implies, there are many such sequences. A specific example is $a_n = (\log(n+2))^{-1}$, [cf. (8.16)].

Despite the fact that (8.19) shows that sequences of Fourier coefficients may tend to 0 as slowly as we please, the next theorem shows that this statement fails for functions in $L_2(\mathbb{T})$.

(8.21) Lemma   *Let $f \in L_2(\mathbb{T})$. Then*

(i) $$\|s_p(f)\|_2 \leq \|f\|_2 \quad \text{for all } p \geq 0$$

and

(ii) $$\lim_{p \to \infty} \|f - s_p(f)\|_2 = 0.$$

[*Here $s_p(f)$ is as in (8.7.iii).*]

Proof   As abbreviations, we write

$$\langle g, h \rangle = \frac{1}{2\pi}\int_{-\pi}^{\pi} g(t)\overline{h(t)}\, dt, \quad e_n(t) = e^{int}, \quad \text{and} \quad s_p = s_p(f) = \sum_{n=-p}^{p} \hat{f}(n)e_n.$$

Then $\|f\|_2^2 = \langle f, f \rangle$, $\hat{f}(n) = \langle f, e_n \rangle = \overline{\langle e_n, f \rangle}$, $\langle e_n, e_k \rangle = 0$ or 1 according as $k \neq n$ or $k = n$, $\langle s_p, f \rangle = \sum_{n=-p}^{p} \hat{f}(n)\langle e_n, f \rangle = \sum_{n=-p}^{p} |\hat{f}(n)|^2$,

$$\langle f, s_p \rangle = \sum_{n=-p}^{p} \overline{\hat{f}(n)} \langle f, e_n \rangle = \sum_{n=-p}^{p} |\hat{f}(n)|^2$$

and

$$\langle s_p, s_p \rangle = \sum_{n=-p}^{p} \sum_{k=-p}^{p} \hat{f}(n) \overline{\hat{f}(k)} \langle e_n, e_k \rangle = \sum_{n=-p}^{p} |\hat{f}(n)|^2. \tag{1}$$

Therefore

$$0 \leqq \| f - s_p \|_2^2 = \langle f - s_p, f - s_p \rangle = \langle f, f \rangle - \langle f, s_p \rangle - \langle s_p, f \rangle + \langle s_p, s_p \rangle$$

$$= \| f \|_2^2 - \sum_{n=-p}^{p} |\hat{f}(n)|^2 \tag{2}$$

for all $p \geqq 0$, and so

$$\sum_{n=-\infty}^{\infty} |\hat{f}(n)|^2 \leqq \| f \|_2^2. \tag{3}$$

Now (i) is a consequence of (1) and (2). Inequality (3) is known as *Bessel's Inequality*. By (3) and (8.17) there is some $g \in L_2(\mathbb{T})$ such that $\hat{g}(n) = \hat{f}(n)$ for all n and $\| g - s_p \|_2 = \| g - s_p(g) \|_2 \to 0$ as $p \to \infty$. Invoking (8.11), we have $f = g$ a.e., and so $\| f - s_p \|_2 = \| g - s_p \|_2 \to 0$. ☐

(8.22) Parseval's Identities   *Let $f, g \in L_2(\mathbb{T})$. Then*

(i)
$$\sum_{n=-\infty}^{\infty} \hat{f}(n) \overline{\hat{g}(n)} = \frac{1}{2\pi} \int_{-\pi}^{\pi} f(t) \overline{g(t)} \, dt$$

and

(ii)
$$\sum_{n=-\infty}^{\infty} |\hat{f}(n)|^2 = \frac{1}{2\pi} \int_{-\pi}^{\pi} |f(t)|^2 \, dt.$$

In particular, these two series converge.

Proof   Using the notation of the preceding proof, we have

$$\langle s_p(f), s_p(g) \rangle = \sum_{n=-p}^{p} \sum_{k=-p}^{p} \hat{f}(n) \overline{\hat{g}(k)} \langle e_n, e_k \rangle = \sum_{n=-p}^{p} \hat{f}(n) \overline{\hat{g}(n)}$$

for all $p \geqq 0$, and so, using (8.21) and Schwarz's Inequality (6.106),

$$\left| \langle f, g \rangle - \sum_{n=-p}^{p} \hat{f}(n) \overline{\hat{g}(n)} \right|$$

$$= |\langle f, g \rangle - \langle s_p(f), s_p(g) \rangle|$$

$$\leqq |\langle f, g \rangle - \langle s_p(f), g \rangle| + |\langle s_p(f), g \rangle - \langle s_p(f), s_p(g) \rangle|$$

$$\leqq \| f - s_p(f) \|_2 \cdot \| g \|_2 + \| s_p(f) \|_2 \cdot \| g - s_p(g) \|_2$$

$$\leqq \| f - s_p(f) \|_2 \cdot \| g \|_2 + \| f \|_2 \cdot \| g - s_p(g) \|_2 \to 0 \quad \text{as } p \to \infty.$$

We conclude that

$$\lim_{p \to \infty} \sum_{n=-p}^{p} \hat{f}(n) \overline{\hat{g}(n)} = \langle f, g \rangle,$$

which is the same as (i). Finally, (ii) follows from (i) by taking $g = f$. ☐

(8.23)   Remarks   (a) Combining (8.22) and (8.17), we see that the mapping $f \to \hat{f}$, when restricted to $L_2(\mathbb{T})$, is a linear isometry of $L_2(\mathbb{T})$ *on*to the space $l_2(\mathbb{Z})$ of all complex-valued functions ϕ on $\mathbb{Z}$ such that

$$\|\phi\|_2 = \left(\sum_{n=-\infty}^{\infty} |\phi(n)|^2 \right)^{1/2} < \infty,$$

where this equality defines its left side. To say that it is an isometry means that $\|\hat{f}\|_2 = \|f\|_2$, and this is (8.22.ii).

(b) There is an imperfect version of (8.22.ii) known as the *Hausdorff–Young Inequality* for the case that $f \in L_p(\mathbb{T})$ and $1 < p < 2$. It states that

$$\left(\sum_{n=-\infty}^{\infty} |\hat{f}(n)|^{p'} \right)^{1/p'} \leq \|f\|_p,$$

where $p' = p/(p-1)$. There is no analogous result for $p > 2$. Proofs of these assertions lie beyond the scope of this book. As usual, the interested reader is referred to Zygmund's or Bary's treatises.

(c) One of the great triumphs of the Lebesgue integral is the Riesz–Fischer Theorem, which depends in the end on the completeness of $L_2(\mathbb{T})$. No such theorem is possible for Riemann integrable functions.

Exercises

1.   If $-\pi \leq x \leq \pi$, then

(a) $x^2 = \dfrac{\pi^2}{3} + 4 \cdot \displaystyle\sum_{n=1}^{\infty} \dfrac{(-1)^n}{n^2} \cos(nx)$,

(b) $|x| = \dfrac{\pi}{2} - \dfrac{4}{\pi} \cdot \displaystyle\sum_{n=1}^{\infty} \dfrac{\cos((2n-1)x)}{(2n-1)^2}$,

(c) $|\sin x| = \dfrac{2}{\pi} - \dfrac{4}{\pi} \cdot \displaystyle\sum_{n=1}^{\infty} \dfrac{\cos(2nx)}{4n^2 - 1}$,

(d) $\displaystyle\sum_{n=1}^{\infty} \dfrac{(-1)^{n-1}}{n^2} = \dfrac{\pi^2}{12}$,

(e) $\displaystyle\sum_{n=1}^{\infty} \dfrac{1}{(2n-1)^2} = \dfrac{\pi^2}{8}$,

(f) $\displaystyle\sum_{n=1}^{\infty} \dfrac{1}{n^2} = \dfrac{\pi^2}{6}$,

(g) $\displaystyle\sum_{n=1}^{\infty} \dfrac{1}{4n^2 - 1} = \dfrac{1}{2}$,

(h) $\displaystyle\sum_{n=1}^{\infty} \dfrac{(-1)^{n-1}}{4n^2 - 1} = \dfrac{\pi - 2}{4}$,

(i) $\displaystyle\sum_{n=1}^{\infty} \dfrac{1}{3 + 16n(n-1)} = \dfrac{\pi}{8}$,

[Hint: For (a), (b), and (c), compute the Fourier series of the g in $C(\mathbb{T})$ that equals the left side on $[-\pi, \pi]$ and show that the series converges to g by invoking (8.12.b).]

2. (a) For $z \in \mathbb{C}\backslash\mathbb{Z}$ and $-\pi \le t \le \pi$, we have

$$\cos(zt) = \frac{\sin(\pi z)}{\pi z} + \frac{2z\sin(\pi z)}{\pi} \cdot \sum_{n=1}^{\infty} \frac{(-1)^n \cos(nt)}{z^2 - n^2}.$$

[Hint: See the hint for Exercise 1.]

(b) Taking $t = \pi$ in (a), we obtain

$$\pi\cot(\pi z) - \frac{1}{z} = \sum_{n=1}^{\infty} \frac{2z}{z^2 - n^2}.$$

(c) For $0 < x < 1$, justify termwise integration of series (b) over $[0, x]$ and obtain

$$\log\left(\frac{\sin(\pi x)}{\pi x}\right) = \sum_{n=1}^{\infty} \log\left(1 - \frac{x^2}{n^2}\right) \quad \text{and} \quad \frac{\sin(\pi x)}{\pi x} = \prod_{n=1}^{\infty}\left(1 - \frac{x^2}{n^2}\right).$$

3. Let $f \in L_2(\mathbb{T})$, $n \in \mathbb{N}$, and $(c_k)_{k=-n}^n \subset \mathbb{C}$. Then

$$\frac{1}{2\pi}\int_{-\pi}^{\pi}\left| f(t) - \sum_{k=-n}^n c_k e^{ikt}\right|^2 dt = \|f\|_2^2 + \sum_{k=-n}^n |\hat f(k) - c_k|^2 - \sum_{k=-n}^n |\hat f(k)|^2,$$

and so this integral is smallest if and only if $c_k = \hat f(k)$ for $-n \le k \le n$. That is, there is a unique trigonometric polynomial of the above form that is nearest to f in the L_2-metric, namely $S_n(f)$. [Hint: As usual, use $|z|^2 = z\bar z$.]

4. (a) If $f(t) = \sum_{k=-n}^n c_k e^{ikt}$ is a trigonometric polynomial, then $c_k = \hat f(k)$ for $-n \le k \le n$ and $\hat f(k) = 0$ for $|k| > n$.

(b) If $f(t) = 2\sin^2 t$, then the Fourier series of f is $1 - \cos(2t)$ and $-(1/2)e^{-2it} + 1 - (1/2)e^{2it}$.

5. (a) Suppose that $f \in C(\mathbb{T})$ is absolutely continuous on $[-\pi, \pi]$ and that $f' \in L_2(\mathbb{T})$. Then

$$\sum_{n=-\infty}^{\infty} |\hat f(n)| \le |\hat f(0)| + \frac{\pi}{\sqrt3}\|f'\|_2.$$

[Hint: Use Exercise 6(a) to write $|\hat f(n)| = |\widehat{f'}(n)|n^{-1}$ for $n \ne 0$ and then use Cauchy's Inequality, Parseval's Identity, and Exercise 1(f).]

(b) If f is as in (a), then the Fourier series of f converges to f uniformly on $\mathbb{R}$.

6. Suppose that $F \in AC(\mathbb{T})$, by which we mean that $F \in C(\mathbb{T})$ and F is absolutely continuous on $[-\pi, \pi]$. Write $f = F'$ a.e.

(a) Then $f \in L_1(\mathbb{T})$ and $\hat f(n) = in\hat F(n)$ for all $n \in \mathbb{Z}$. [Hint: Integrate by parts.]

(b) $\lim_{|n|\to\infty} n\hat F(n) = 0$.

(c) The Fourier series $S(f)$ is obtained by formally differentiating the Fourier series $S(F)$ term-by-term.

(d) $S(F)$ converges to F uniformly on $\mathbb{R}$. [Hint: Use (8.15).]

(e) If, for some $p \in \mathbb{N}$, F is $p - 1$ times differentiable and $F^{(p-1)} \in AC(\mathbb{T})$, then $\lim_{|n|\to\infty} n^p \hat F(n) = 0$.

(f) There exist functions $F \in C(\mathbb{T})$ with $V_{-\pi}^\pi F < \infty$ such that (a) and (c)(and, by Exercise 8 following (8.58), even (b)) fail for F.

7.   (a) If a trigonometric series $\sum\limits_{n=-\infty}^{\infty} c_n e^{int}$ has the property that the formally differen-

tiated series $\sum\limits_{n=-\infty}^{\infty} in c_n e^{int}$ is the Fourier series of some $f \in L_1(\mathbb{T})$ ($\hat{f}(n) = inc_n$

for all n), then the original series is $S(F)$ for some $F \in AC(\mathbb{T})$ (see Exercise 6), it
converges uniformly on $\mathbb{R}$ to F, and $F' = f$ a.e. [Hints: Use (8.15). Check that

the sine coefficient b_n for f is $-n(c_{-n} + c_n)$ and take $F(x) = \sum\limits_{-\infty}^{\infty} c_n + \int_0^x f.$]

(b) The series $\sum\limits_{n=1}^{\infty} \dfrac{\sin(nt)}{n^2}$ is the Fourier series of the $AC(\mathbb{T})$ function F defined by

$F(t) = -\int_0^t \log|2\sin(\theta/2)|\, d\theta$. [Hint: See (7.39).]

(c) The series $\sum\limits_{n=2}^{\infty} \dfrac{\sin(nt)}{n \log n}$ is the Fourier series of a function $F \in AC(\mathbb{T})$. It

converges uniformly (but not absolutely) on $\mathbb{R}$. [Hint: Use (8.18).]

(d) The series $\sum\limits_{n=2}^{\infty} \dfrac{\cos(nt)}{n \log n}$ is the Fourier series of a function $f \in L_2(\mathbb{T})$ which is

continuous on $]0, 2\pi[$ but unbounded on every open interval containing 0.
[Hint: (8.18), (8.17) and (7.38.ii). We remark that there is a subtle question
raised here. If a series is known to be the Fourier series of a function f and is
also known to converge to a function g, can we conclude that $f = g$ a.e.? An
affirmative answer is given in (8.36).]

8.   Let $f \in C(\mathbb{T})$ be given. For $\delta \geq 0$ define $\omega(\delta) = \sup\{|f(s) - f(t)| : s, t \in \mathbb{R}, |s - t|$
$\leq \delta\}$. The (nondecreasing) function ω is called the *modulus of continuity* of f.
(a) For all $h \geq 0$, (8.22.ii) yields

$$4 \sum_{n=-\infty}^{\infty} |\hat{f}(n)\sin(nh)|^2 = \frac{1}{2\pi} \int_{-\pi}^{\pi} |f(t+h) - f(t-h)|^2\, dt \leq [\omega(2h)]^2.$$

(b) For all $j \in \mathbb{N}$, we have

$$2 \sum_{2^{j-i} \leq |n| < 2^j} |\hat{f}(n)|^2 \leq \left[\omega(\pi/2^j)\right]^2.$$

(c) For all $j \in \mathbb{N}$, we use Cauchy's Inequality to get

$$\sum_{2^{j-1} \leq |n| < 2^j} |\hat{f}(n)| \leq 2^{(j-1)/2}\omega(\pi/2^j) \leq 2^{1/2} \sum_{2^{j-1} < k \leq 2^j} k^{-1/2}\omega(\pi/k).$$

(d)

$$\sum_{n=-\infty}^{\infty} |\hat{f}(n)| - |\hat{f}(0)| \leq 2^{-1/2} \sum_{j=1}^{\infty} 2^{j/2}\omega(\pi/2^j) \leq 2^{1/2} \sum_{k=2}^{\infty} k^{-1/2}\omega(\pi/k)$$

$$\leq (2\pi)^{1/2} \int_0^\pi \delta^{-3/2}\omega(\delta)\, d\delta.$$

Given $\alpha > 0$, we say that f satisfies a *Lipschitz condition of order* α and
we write $f \in \mathrm{Lip}_\alpha(\mathbb{T})$ if there exists a constant $C < \infty$ such that

$$|f(s) - f(t)| \leq C|s - t|^\alpha \quad \text{for all } s, t \in \mathbb{R}. \tag{$*$}$$

(e) Condition ($*$) obtains if and only if $\omega(\delta) \leq C\delta^\alpha$ for all $\delta > 0$.

(f) [Bernstein, 1914] If $f \in \mathrm{Lip}_\alpha(\mathbb{T})$ for some $\alpha > 1/2$, then $\sum\limits_{n=-\infty}^{\infty} |\hat{f}(n)| < \infty$. [We remark that in Zygmund's treatise (vol. I, pp. 197–199) one finds Einar Hille's proof that, for $0 < \alpha < 1$, the series

$$\sum_{n=1}^{\infty} n^{-\alpha-1/2} e^{in \log n} e^{int}$$

converges uniformly on $\mathbb{R}$ to a function in $\text{Lip}_\alpha(\mathbb{T})$. Thus (f) fails for $\alpha = 1/2$.]

(g) Writing $V = V_0^{2\pi} f$, we have for each $N \in \mathbb{N}$ that

$$2N \int_{-\pi}^{\pi} \left| f\left(t + \frac{\pi}{2N}\right) - f\left(t - \frac{\pi}{2N}\right) \right|^2 dt = \sum_{k=1}^{2N} \int_{-\pi}^{\pi} \left| f\left(t + \frac{k\pi}{N}\right) - f\left(t + \frac{(k-1)\pi}{N}\right) \right|^2 dt$$

$$\leq 2\pi V \cdot \omega(\pi/N).$$

(h) From (g) and equality (a), we see that $j \in \mathbb{N}$ implies

$$2 \sum_{2^{j-1} \leq |n| < 2^j} |\hat{f}(n)|^2 \leq 2^{-j-1} V \cdot \omega(\pi/2^j).$$

(i) Once more Cauchy's Inequality yields

$$\sum_{2^{j-1} \leq |n| < 2^j} |\hat{f}(n)| \leq \frac{1}{2} V^{1/2} \left[\omega(\pi/2^j)\right]^{1/2} \leq V^{1/2} \cdot \sum_{2^{j-1} < k \leq 2^j} k^{-1} [\omega(\pi/k)]^{1/2}.$$

(j)

$$\sum_{n=-\infty}^{\infty} |\hat{f}(n)| - |\hat{f}(0)| \leq \frac{1}{2} V^{1/2} \cdot \sum_{j=1}^{\infty} \left[\omega(\pi/2^j)\right]^{1/2} \leq V^{1/2} \cdot \sum_{k=2}^{\infty} k^{-1} [\omega(\pi/k)]^{1/2}$$

$$\leq V^{1/2} \int_0^{\pi} \delta^{-1} [\omega(\delta)]^{1/2} \, d\delta.$$

(k) [Zygmund, 1928] If $f \in \text{Lip}_\alpha(\mathbb{T})$ for some $\alpha > 0$ and $V_0^{2\pi} f < \infty$, then

$$\sum_{n=-\infty}^{\infty} |\hat{f}(n)| < \infty.$$

(l) If $f \in AC(\mathbb{T})$ (see Exercise 6) and $f' \in L_p(\mathbb{T})$ for some $p > 1$, then $f \in \text{Lip}_\alpha(\mathbb{T})$, where $\alpha = (p-1)/p$, and so $\sum_{-\infty}^{\infty} |\hat{f}(n)| < \infty$. [Hint: Use Hölder's Inequality to show that (*) obtains with $C = \left(\int_0^{2\pi} |f'|^p\right)^{1/p}$.]

(m) The function F of Exercise 7(c) is in no $\text{Lip}_\alpha(\mathbb{T})$.

(n) [Lorentz, 1948] If there exist constants $A < \infty$ and $0 < \alpha < 1$ such that

$$\sum_{|n| \geq N} |\hat{f}(n)| \leq AN^{-\alpha} \quad \text{for all } N \in \mathbb{N},$$

then $f \in \text{Lip}_\alpha(\mathbb{T})$. [Hints: For fixed $0 < h < \pi$, choose $k \in \mathbb{N}$ such that $N = 2^k$ satisfies $\pi/N \leq h < 2\pi/N$. Write $f(t+h) - f(t-h)$ as the sum of its Fourier series (using (8.12.b)) and then estimate by using $|\sin(nh)| \leq |nh| < 2^j \cdot 2\pi/N$ for $2^{j-1} \leq |n| < 2^j$, $1 \leq j \leq k$, and using $|\sin(nh)| \leq 1$ for $|n| \geq N$.]

(o) If there exist constants $B < \infty$ and $0 < \alpha < 1$ such that $|n|^{1+\alpha} |\hat{f}(n)| \leq B$ for all $n \in \mathbb{Z}$, then $f \in \text{Lip}_\alpha(\mathbb{T})$. [Hint: Check the hypothesis of (n) by using an integral to estimate a sum.]

(p) Both (n) and (o) are false for $\alpha \geq 1$. [Hint: See Exercise 7(b).]

9.   Let $f \in L_1(\mathbb{T})$ have the Fourier coefficients a_n and b_n $(n \geq 0)$ as in (8.7.v), and let $\phi \in L_1(\mathbb{T})$ have the analogous Fourier coefficients α_n and $\beta_n : \alpha_n = \hat{\phi}(n) + \hat{\phi}(-n)$, $\beta_n = i\hat{\phi}(n) - i\hat{\phi}(-n)$.

(a) If both f and ϕ are in $L_2(\Pi)$, then

$$\frac{1}{\pi}\int_{-\pi}^{\pi} f(t)\overline{\phi(t)}\,dt = \frac{a_0\,\overline{a_0}}{2} + \sum_{n=1}^{\infty}(a_n\,\overline{\alpha_n} + b_n\,\overline{\beta_n}).$$

(b) Each of the following three statements implies the other two: (1) f is real-valued a.e. on $\mathbb{R}$; (2) a_n and b_n are real for all integers $n \geq 0$; (3) $\hat{f}(-n) = \hat{f}(n)$ for every integer n. [Hint: To show that (3) implies (1), use (8.11) to show that $f = \bar{f}$ a.e.].

10. Let $f \in C(\mathbb{T})$ be given.

(a) If $p \in \mathbb{N}$ and $\displaystyle\sum_{n=-\infty}^{\infty}|n^p\,\hat{f}(n)| < \infty$, then $f \in C^{(p)}(\mathbb{T})$: that is, f is p times differentiable on $\mathbb{R}$ and $f^{(p)} \in C(\mathbb{T})$. [Hint: For $p = 1$, the series $S'(f)$ obtained by (formal) term-by-term differentiation of $S(f)$ converges uniformly to some g, and one shows by term-by-term integration that $f(x) - \int_0^x g = \text{constant}$.]

(b) If $\alpha > 2$, $A < \infty$, and $|n|^\alpha|\hat{f}(n)| \leq A$ for every integer n, then $f \in C^{(p)}(\mathbb{T})$ where p is the largest integer less than $\alpha - 1$.

(c) Assertion (b) becomes false if "less than" is replaced by "not exceeding."

(d) If $p \in \mathbb{N}$, f is p times differentiable on $\mathbb{R}$, and $f^{(p)} \in L_1(\mathbb{T})$, then $|\hat{f}(n)| \leq \min\{|n|^{-k}\|f^{(k)}\|_1 : 0 \leq k \leq p\}$ for every integer $n \neq 0$. [Hint: The hypothesis implies $f^{(p-1)} \in AC(\mathbb{T})$ (see Exercise 6).]

11. For use in the next two exercises, the steps in this exercise outline Raphael Salem's elegant proof of the following theorem of J. E. Littlewood.

Let $n > 1$ be a natural number and let $r_1, r_2, \ldots, r_n$ be nonnegative real numbers. Write $s = (r_1^2 + r_2^2 + \ldots + r_n^2)^{1/2}$. Then there exists $\theta = (\theta_1, \theta_2, \ldots, \theta_n) \in \mathbb{R}^n$ such that the trigonometric polynomial

$$P_\theta(x) = \sum_{j=1}^{n} r_j\cos(jx - \theta_j)$$

satisfies

$$\|P_\theta\|_u = \sup\{|P_\theta(x)| : x \in \mathbb{R}\} \leq Cs(\log n)^{1/2}, \qquad (*)$$

where $C = 4\pi\sqrt{2}\,e^{3/2}$ is an absolute constant. First fix any $k \in \mathbb{N}$. We may suppose $s > 0$.

(a) For all $\theta \in \mathbb{R}^n$, we have $s^2/2 = \|P_\theta\|_2^2 \leq \|P_\theta\|_u^2$. [Hint: Check that $2\hat{P}_\theta(j) = 2\overline{\hat{P}_\theta(-j)} = r_j e^{-i\theta_j}$ for $1 \leq j \leq n$.]

(b) There is an $\alpha \in \mathbb{R}^n$ (depending on k) such that

$$\int_0^{2\pi}[P_\alpha(x)]^{2k-1}r_j\cos(jx - \alpha_j)\,dx \leq (2k-1)r_j^2\int_0^{2\pi}[P_\alpha(x)]^{2k-2}\,dx$$

for $1 \leq j \leq n$. [Big hint: Consider an α at which the function $f(\theta) = \int_0^{2\pi}[P_\theta(x)]^{2k}\,dx$ attains its absolute minimum value and then invoke the fact that all n of the partial derivatives $\partial^2 f/\partial\theta_j^2 = D_j^2 f$ are nonnegative at $\theta = \alpha$.]

(c) If α is as in (b), then

$$\left(\frac{1}{2\pi}\int_0^{2\pi}[P_\alpha(x)]^{2k}\,dx\right)^{1/k} \leq (2k-1)s^2.$$

[Hints: Sum inequalities (b). If $k > 1$, use Hölder's Inequality to obtain $\int_0^{2\pi} f^{2k-2} \leq (2\pi)^{1/k} \left(\int_0^{2\pi} f^{2k} \right)^{(k-1)/k}$. If $k = 1$, use (a).]

(d) If $a \leq c \leq b \leq a + 2\pi$ in $\mathbb{R}$ and $\theta \in \mathbb{R}^n$, then

$$(|P_\theta(c) - P_\theta(a)| + |P_\theta(b) - P_\theta(c)|)^2 \leq \left(\int_a^b |P_\theta'(x)| \, dx \right)^2 \leq (b-a) \int_a^b |P_\theta'(x)|^2 \, dx$$

$$\leq (b-a)\pi \sum_{j=1}^n (jr_j)^2 \leq \pi n^2 s^2 (b-a).$$

(e) If α is as in (c), $M = \|P_\alpha\|_u$, and $0 < t < 1$, then

$$2^k k^k s^{2k} \geq \frac{(tM)^{2k}}{2\pi} \cdot \frac{(2M - 2tM)^2}{\pi n^2 s^2} \geq \frac{t^{2k}(1-t)^2 M^{2k}}{n^2 \pi^2}.$$

[Big hint: Note that $\int_0^{2\pi} P_\alpha = 0$, so $P_\alpha(x) = 0$ for some x. Choose c so that $|P_\alpha(c)| = M$ and then choose a and b as in (d) so that $|P_\alpha(a)| = |P_\alpha(b)| = tM < |P_\alpha(x)|$ for $a < x < b$. Then the left end in (d) is $(2(1-t)M)^2$ and $\int_a^b [P_\alpha(x)]^{2k} \, dx > (tM)^{2k}(b-a)$. Also use (c) and (a).]

(f) If α is as in (c), then

$$\|P_\alpha\|_u < 2\sqrt{2} \, \pi e s k^{1/2} n^{1/k}.$$

[Hint: Choose $t = k/(1+k)$ and use $(1 + k^{-1})^k < e$ and $(k+1)^2 \leq 4^k$.]

(g) It is possible to choose k such that, when θ is the corresponding α, we have (*). [Hint: Let $2 \log n \leq k < 4 \log n$.]

12.   [R. Salem, 1940] Let $(r_j)_{j=1}^\infty$ be a sequence of nonnegative real numbers with the property that $\displaystyle\sum_{n=2}^\infty \frac{1}{n} \left(\frac{R_n}{\log n} \right)^{1/2} < \infty$, where $R_n = \displaystyle\sum_{j=n+1}^\infty r_j^2$. Then there exists a sequence $(\theta_j)_{j=1}^\infty \subset \mathbb{R}$ such that the series $\displaystyle\sum_{j=1}^\infty r_j \cos(jx - \theta_j)$ is the Fourier series of a function $f \in C(\mathbb{T})$. [Hints: Note that $R_1 \geq R_2 \geq \cdots$. Apply (7.1) twice to the given convergent series to see that if $n_k = 2^{2^k}$, then $\displaystyle\sum_{k=0}^\infty (R_{n_k} \log n_{k+1})^{1/2} < \infty$. For each k, apply Exercise 11 to obtain θ_j $(n_k < j \leq n_{k+1})$ so that $P_k(x) = \displaystyle\sum_{j=n_k+1}^{n_{k+1}} r_j \cos(jx - \theta_j)$ satisfies $\|P_k\|_u \leq C(R_{n_k} \log n_{k+1})^{1/2}$. Put $f(x) = \displaystyle\sum_{j=1}^{n_0} r_j \cos(jx) + \sum_{k=0}^\infty P_k$ and use (8.5.i).]

13.   (a) [T. Carleman, 1918] If $f \in L_2(\mathbb{T})$, then (8.22.ii) assures that $\displaystyle\sum_{j=-\infty}^\infty |\hat{f}(j)|^2 < \infty$.

However, there exist functions $f \in C(\mathbb{T})$ such that $\displaystyle\sum_{j=-\infty}^\infty |\hat{f}(j)|^p = \infty$ for every real number $p < 2$! [Hint: Use Exercise 12. Fix any $\alpha > 1$ and let $r_j = j^{-1/2}(\log j)^{-\alpha}$. Check by comparison with an integral that $R_n < (\log n)^{1-2\alpha}$.

Use (7.1) to see that $\sum r_j^p = \infty$ for all $p < 2$ and check that $|\hat{f}(j)| = |\hat{f}(-j)|$ $= r_j/2$.]

(b) [S. Banach, 1930] There is a function $f \in C(\mathbb{T})$ and a sequence $p_j \uparrow 2$ such that

$$\sum_{j=-\infty}^{\infty} |\hat{f}(j)|^{p_{|j|}} = \infty.$$ [Hint: Use the same f as in (a) and take $p_j = 2 - (\log j)^{-1/2}$. Show that $r_j^{p_j} > (j \log j)^{-1}$ for all sufficiently large j.]

14.   If g is the function defined by

$$g(x) = \sum_{n=2}^{\infty} \frac{\cos(nx)}{\log n}$$

for $x \in \mathbb{R} \setminus 2\pi\mathbb{Z}$ and $g(x) = 0$ for $x \in 2\pi\mathbb{Z}$, then the odd function f defined by $f(x) = g(x - \pi/3) - g(x + \pi/3)$ has the Fourier series

$$f(x) \sim \sum_{n=2}^{\infty} b_n \sin(nx) \quad \text{and} \quad \sum_{n=2}^{\infty} \frac{|b_n|}{n} = \infty.$$

In fact, $b_n = (2 \sin(n\pi/3))/\log n$, and this Fourier series converges uniformly to f on $[-\pi/3 + \delta, \pi/3 - \delta]$ and on $[\pi/3 + \delta, 5\pi/3 - \delta]$ whenever $0 < \delta < \pi/3$.

15.   *The Algebra $A(\mathbb{T})$.* We denote by $A(\mathbb{T})$ (or just A) the set of all $f \in C(\mathbb{T})$ for which

$$\|f\|_A = \sum_{n=-\infty}^{\infty} |\hat{f}(n)| < \infty.$$

We have seen in Exercises 5 and 8 (f) (k) (l) some important sufficient conditions for membership in $A(\mathbb{T})$. Now we consider more properties of $A(\mathbb{T})$.

(a) If $f \in A$, then the Fourier series of f converges to f uniformly on $\mathbb{R}$.

(b) If $(b_n)_{n=1}^{\infty} \subset \mathbb{R}$ satisfies $b_n \geq b_{n+1}$ for all n and $\lim_{n \to \infty} nb_n = 0$ then $\sum_{n=1}^{\infty} b_n \sin(nx)$ converges uniformly on $\mathbb{R}$. [See Exercise 10 just after (7.41).]

(c) The function f defined by

$$f(x) = \sum_{n=2}^{\infty} \frac{\sin(nx)}{n \log n}$$

has a uniformly convergent Fourier series, but $f \notin A$. However, $f \in AC(\mathbb{T})$ (see Exercise 7(c)).

(d) If $f, g \in A$ and $\alpha \in \mathbb{C}$, then αf, $f + g$, and fg are in A, $\|\alpha f\|_A = |\alpha| \cdot \|f\|_A$, $\|f + g\|_A \leq \|f\|_A + \|g\|_A$, and $\|fg\|_A \leq \|f\|_A \cdot \|g\|_A$. [Hint: $\widehat{fg}(n)$ $= \sum_{k=-\infty}^{\infty} \hat{f}(k)\hat{g}(n - k)$.]

(e) If $f \in A$, then $\|f\|_u \leq \|f\|_A$.

(f) If $\phi_n(x) = \sum_{k=1}^{n} k^{-1} \sin(kx)$, $c_n = \left(\sum_{k=1}^{n} k^{-1} \right)^{-1}$, and $f_n = c_n \phi_n$, then $(f_n)_{n=1}^{\infty}$ $\subset A$, $\|f_n\|_A = 1$ for all n, and $\|f_n\|_u \to 0$ as $n \to \infty$.

(g) If $(f_n)_{n=1}^{\infty} \subset A$, $\|f_n\|_A \leq \beta < \infty$ for all n, $f \in C(\mathbb{T})$, and $\lim_{n \to \infty} \|f - f_n\|_u = 0$, then $f \in A$ and $\|f\|_A \leq \beta$, but we *need not* have $\lim_{n \to \infty} \|f - f_n\|_A = 0$ even if $\lim_{n \to \infty} \|f_n\|_A$ exists. However, if we also have $\lim_{n \to \infty} \|f_n\|_A = \|f\|_A$, then $\lim_{n \to \infty} \|f - f_n\|_A = 0$. [Hints: write $\sum_{k=-p}^{p} |\hat{f}_n(k)| \leq \beta$ and check that $\lim_{n \to \infty} \hat{f}_n(k) = \hat{f}(k)$ for all

k. Also verify that $\|f - f_n\|_A \leqq [\|f_n\|_A - \|f\|_A] + 2 \sum\limits_{k=-p}^{p} |\hat{f}(k) - \hat{f}_n(k)| + 2 \sum\limits_{|k|>p}$

$|\hat{f}(k)|$ for all n and p. Consider part (f) with $f = 0$.]

(h) If $(f_n)_{n=1}^{\infty} \subset A$ and $\lim\limits_{m,n\to\infty} \|f_m - f_n\|_A = 0$, then $\lim\limits_{n\to\infty} \|f - f_n\|_A = 0$ for some
unique $f \in A$.

(i) If $f \in A$ and $\hat{f}(n) \geqq 0$ for all $n \in \mathbb{Z}$, then $\|f\|_A = f(0)$. [Hint: Use (a).]

(j) For fixed x with $0 < x \leqq \pi$, the *triangle function* $\Delta_x \in C(\mathbb{T})$ is defined by
$\Delta_x(t) = 1 - |t|/x$ if $|t| \leqq x$ and $\Delta_x(t) = 0$ if $x \leqq |t| \leqq \pi$. We have $\hat{\Delta}_x(n) \geqq 0$ for
all n, $\Delta_x \in A$, and $\|\Delta_x\|_A = 1$.

(k) Any function of the form $\phi = C + \sum\limits_{j=1}^{p} a_j \Delta_{x_j}$, where C is a nonnegative real

constant, $p \in \mathbb{N}$, $a_j \geqq 0$, and $0 < x_j \leqq \pi$ for all j satisfies $\phi \in A$, $\hat{\phi}(n) \geqq 0$ for
all n, and $\|\phi\|_A = \phi(0)$.

(l) If $f \in C(\mathbb{T})$, $f(-t) = f(t) \geqq 0$ for all $t \in \mathbb{R}$, and f is both convex and nonin-
creasing on $[0, \pi]$, then $f \in A$, $\hat{f}(n) \geqq 0$ for all n, and $\|f\|_A = f(0)$. [Hint: Show
that f is the uniform limit of some sequence of functions f_n of the form ϕ in (k)
with $C = f(\pi)$ and $f_n(0) = f(0)$. Indeed, if $0 = x_0 < x_1 < \cdots < x_n = \pi$, then the
even function in $C(\mathbb{T})$ that agrees with f at each x_k and is linear on each
$[x_{k-1}, x_k]$ has this form. Use (g) and (e).]

(m) If $0 \leqq a < b < \infty$, then

$$\left| \int_a^b \frac{\sin u}{u} \, du \right| \leqq \int_0^\pi \frac{\sin u}{u} \, du < \pi.$$

[Hint: First suppose $a = 0$ and $b \in \pi\mathbb{N}$.]

(n) If $f \in A$, then the function g defined for $x \in \mathbb{R}$ and $0 < a < b < \infty$ by $g(x, a, b)$
$$= \int_a^b \frac{f(x + t) - f(x - t)}{t} \, dt \text{ satisfies } |g(x, a, b)| \leqq 2\pi\|f\|_A \text{ and } \lim\limits_{a,b\to 0} g(x, a, b)$$
$= 0$ uniformly for $x \in \mathbb{R}$. [Hint: Write the integrand as a series that converges
uniformly for $x \in \mathbb{R}$ and $a \leqq t \leqq b$. Use (m).]

(o) If $f \in A$ is real and nondecreasing on an open interval I, then the oscillation
function $\omega(\delta, J) = \sup\{|f(u) - f(v)| : u, v \in J, |u - v| \leqq \delta\}$ satisfies $\lim\limits_{\delta \downarrow 0} \omega(\delta, J)$

$\log(1/\delta) = 0$ for each closed interval $J \subset I$. [Hint: For small enough $b > a > 0$,
we have $\sup\{g(x, a, b) : x \in J\} \geqq \omega(2a, \log(b/a))$, so take $2a = \delta$ and $2b = \sqrt{\delta}$.]

(p) If $f \in A$ is real-valued, odd, and nondecreasing on $[-b, b]$ for some $0 < b < \pi$,
then $\int_0^1 t^{-1} f(t) \, dt < \infty$. [Hint: $0 \leqq g(0, a, b) \leqq 2\pi\|f\|_A$ for all $0 < a < b$.]

(q) Let $f_1 \in C(\mathbb{T})$ be the odd function defined by $f_1(0) = f_1(\pi) = 0$ and $f_1(t)$
$= f_1(\pi - t) = [\log(\pi/t)]^{-1}$ if $0 < t \leqq \pi/2$ and let $f_2 = |f_1|$ be the even function
that agrees with f_1 on $[0, \pi]$. Then $f_1 \notin A$, $f_2 \in A$, and the conclusion of (o)
fails if $f = f_2$ and $J = [0, \pi/2]$. [Hints: Apply (o) or (p) to f_1. Apply (l) to
$f(t) = f_2(\pi/2) - f_2(t/2)$.]

(r) [Wiener–Lévy Theorem] Let $\Omega \subset \mathbb{C}$ be an open set and let $\phi : \Omega \to \mathbb{C}$ have the
property that if $a \in \Omega$, then ϕ is the sum of its Taylor series expansion about a
at every point z in the largest open disk that is centered at a and contained in

Ω.* (In particular, we suppose that ϕ is infinitely differentiable on Ω.) If $f \in A(\mathbb{T})$ and $f(\mathbb{R}) \subset \Omega$, then $\phi \circ f \in A(\mathbb{T})$, where, as usual, $\phi \circ f(t) = \phi(f(t))$. [Hints: We sketch a proof given by D. J. Newman in 1975. Fix $\delta > 0$ so that the compact set $E = \{z \in \mathbb{C} : |z - f(t)| \leq 3\delta$ for some $t \in \mathbb{R}\}$ is contained in Ω. Choose N so that $P = S_N(f)$ satisfies $\|f - P\|_A < \delta$. Use Exercise 5 to show that $\phi^{(n)} \circ P \in A$ and $\|\phi^{(n)} \circ P\|_A \leq \|\phi^{(n)} \circ P\|_u + 2\|\phi^{(n+1)} \circ P\|_u \cdot \|P'\|_u$. Write $M = \sup\{|\phi(z)| : z \in E\}$. For fixed $t \in \mathbb{R}$, note that the closed disk of radius 2δ centered at $P(t)$ is contained in E and apply Cauchy's Estimate (7.56) to the Taylor expansion of ϕ about $P(t)$ to obtain $|\phi^{(n)}(P(t))| \leq n! \, M(2\delta)^{-n}$. Use (d) to obtain $\|(f - P)^n\|_A < \delta^n$. Combine results to get

$$\left\| \frac{\phi^{(n)} \circ P}{n!} \cdot (f - P)^n \right\|_A \leq 2^{-n} M\big[1 + (n + 1)\delta^{-1}\|P'\|_u \big].$$

Conclude, using (h), that there is a $g \in A$ such that the partial sums

$$g_p = \sum_{n=0}^{p} \frac{\phi^{(n)} \circ P}{n!} \cdot (f - P)^n$$

satisfy $\lim_{p \to \infty} \|g - g_p\|_A = 0$. Finally note that $\lim_{p \to \infty} g_p(t) = \phi(f(t))$ for each $t \in \mathbb{R}$ and so $g = \phi \circ f$.]

(s) [Wiener, 1932] If $f \in A$ and $f(t) \neq 0$ for all $t \in \mathbb{R}$, then the reciprocal $1/f$ also has an absolutely convergent Fourier series: $1/f \in A$. [Hint: Take $\Omega = \mathbb{C}\setminus\{0\}$ and $\phi(z) = 1/z$.]

Divergent Fourier Series

Before turning to the problem of recapturing a function from its Fourier series by means of some sort of summation method or, hopefully, convergence, we prove a theorem which dramatically points out that this problem is by no means easy, even for continuous functions. Our exposition of this theorem is a detailed elaboration of the proof given in Katznelson's book (see the bibliography). Like all examples of divergent Fourier series, ours is built by using trigonometric polynomials having very special properties. We begin with a lemma that provides the polynomials that we use in proving the theorem.

(8.24) Lemma *Suppose that $V \subset [-\pi, \pi]$ is the union of some finite family of intervals of positive length and that $\delta = \lambda(V) > 0$. Then there exists a trigonometric polynomial Q such that*

(i) $\qquad\qquad \|Q\|_u = \sup\{|Q(t)| : t \in \mathbb{R}\} < 1$

and

(ii) $S^*(Q, t) = \sup\{|S_n(Q, t)| : n \geq 0\} > (1/\pi)\log\big[1/(3\delta)\big]$ *for every $t \in V$.*

[Here each of the three equalities defines its left side.]

---

*It is proved in analytic function theory that this property is implied by the apparently much simpler requirement that ϕ be differentiable on Ω.

Proof We first choose $0 < \epsilon < 1$ and a finite number of points $t_1, \ldots, t_p$ in V such that $p\epsilon < \delta$ and the intervals $[t_j - \epsilon, t_j + \epsilon]$ $(1 \leq j \leq p)$ cover V. To see that this can be done, let $\delta_1 \leq \delta_2 \leq \cdots \leq \delta_q$ be the lengths of finitely many *disjoint* intervals having V as union and choose any $\epsilon < 1$ with $0 < \epsilon < \delta_1$. For $1 \leq k \leq q$, choose $p_k \in \mathbb{N}$ with $p_k\epsilon < \delta_k \leq 2p_k\epsilon$. Now cover the kth interval by p_k closed intervals of length 2ϵ and let $p = p_1 + \cdots + p_q$.

For $z \in \mathbb{C}$, $z \neq 1$, write $\phi(z) = (1 - z)^{-1}$ and note that

$$\mathrm{Re}\left[\phi(re^{it})\right] = \frac{1}{2}\left[\phi(re^{it}) + \overline{\phi(re^{it})}\right] = \frac{1}{2}\left[1 + \frac{1 - r^2}{1 + r^2 - 2r\cos t}\right] \quad (1)$$

for $0 \leq r < 1$ and $t \in \mathbb{R}$. It is clear that $\phi(0) = 1$ and $\mathrm{Re}[\phi(z)] > 1/2$ for $|z| < 1$. Using (1) and the inequality $\cos\epsilon > 1 - \epsilon^2$ (valid for $0 < \epsilon < 1$), one easily checks that

$$\mathrm{Re}\left[\phi\left(\frac{1}{1 + \epsilon}e^{it}\right)\right] > \frac{1}{3\epsilon} \quad \text{if } -\epsilon \leq t \leq \epsilon. \quad (2)$$

It follows that the function f defined by

$$f(z) = \frac{1}{p}\sum_{j=1}^{p}\phi(e^{-it_j}z) \quad (3)$$

satisfies $f(0) = 1$, $\mathrm{Re}[f(z)] > 1/2$ for $|z| < 1$, and

$$\mathrm{Re}\left[f\left(\frac{1}{1 + \epsilon}e^{it}\right)\right] > \frac{1}{3p\epsilon} > \frac{1}{3\delta} \quad \text{if } t \in V^- \quad (4)$$

because if $|t - t_j| \leq \epsilon$ for some j, then (2) shows that at least one of the positive numbers $\mathrm{Re}[\phi((1 + \epsilon)^{-1}e^{i(t-t_j)})]$ exceeds $1/(3\epsilon)$. Writing $F(z) = \mathrm{Log}(f(z))$, where Log is as in Chapter 5, we have $F(0) = 0$,

$$\left|\mathrm{Im}\left[F(z)\right]\right| = \left|\mathrm{Arg}(f(z))\right| < \pi/2, \quad (5)$$

$$\left|F(z)\right| \geq \mathrm{Re}\left[F(z)\right] = \log|f(z)| \geq \log(\mathrm{Re}\left[f(z)\right]) \quad (6)$$

for $|z| < 1$.

We next show that F has a power series expansion

$$F(z) = \sum_{n=1}^{\infty}A_n z^n, \quad (7)$$

which converges uniformly on the circle $|z| = 1/(1 + \epsilon)$.* It is enough to obtain such an expansion which converges for $|z| < R = 1 - (2a)^{-1}$, where $a = 1 + \epsilon^{-1}$, for then $(1 + \epsilon)^{-1} = 1 - a^{-1} < R$. We have

$$|\phi(z) - a| = a\left|\frac{z - (1 - a^{-1})}{z - 1}\right| < a \quad \text{if } |z| < R$$

---

*Because F is differentiable for $|z| < 1$, this fact is well known in analytic function theory.

[indeed, $\alpha = 1 - a^{-1} < 1$ and $|z| < (1 + \alpha)/2 = R$ imply that $2\,\mathrm{Re}(z) < 1 + \alpha$, $|z - 1|^2 - |z - \alpha|^2 = (1 - \alpha) \cdot (1 + \alpha - 2\,\mathrm{Re}(z)) > 0$, and $|z - \alpha| < |z - 1|$]. This fact and (3) yield

$$|f(z) - a| < a \quad \text{if } |z| < R. \tag{8}$$

By (5.30.i), we have

$$\mathrm{Log}\, w = \mathrm{Log}\, a + \sum_{k=1}^{\infty} \frac{(-1)^{k-1}}{k} \left(\frac{w - a}{a} \right)^k \tag{9}$$

for $|w - a| < a$. Writing $f_k(z)$ for the kth summand in (9) after putting $w = f(z)$, it follows from (8) and (9) that

$$F(z) = \mathrm{Log}\, a + \sum_{k=1}^{\infty} f_k(z) \tag{10}$$

for $|z| < R$ and series (10) converges uniformly on any disk $|z| \leq r$, where $0 < r < R$ [because f maps such disks onto compact subsets of the disk $|w - a| < a$]. Like ϕ, the function f has a power series expansion convergent in the unit disk $|z| < 1$ and therefore, taking Cauchy Products, so does each f_k. Now (7.57) assures the existence of the sequence $(A_n)_{n=1}^{\infty} \subset \mathbb{C}$ such that (7) obtains for $|z| < R$. [$A_0 = 0$ since $F(0) = 0$.] For $t \in \mathbb{R}$, write

$$g(t) = \frac{2}{\pi} F\left(\frac{1}{1 + \epsilon} e^{it} \right) = \sum_{n=1}^{\infty} a_n e^{int} \tag{11}$$

where $a_n = (2/\pi)(1 + \epsilon)^{-n} A_n$. We see from (5) that $\|\mathrm{Im}(g)\|_u < 1$ and (4) and (6) combined reveal that the number $\beta = \inf\{|g(t)| : t \in V^-\}$ satisfies $\beta > (2/\pi)\log(1/(3\delta))$. The partial sums $g_N(t) = \sum_{n=1}^{N} a_n e^{int}$ of the *uniformly* convergent series (11) satisfy $\|\mathrm{Im}(g_N)\|_u \to \|\mathrm{Im}(g)\|_u$ and, for $t \in V^-$, $|g_N(t)| \geq \beta - \|g - g_N\|_u \to \beta$ as $N \to \infty$. Therefore, there exists some fixed N (any sufficiently large one will do) such that

$$\|\mathrm{Im}(g_N)\|_u < 1 \tag{12}$$

and

$$|g_N(t)| > (2/\pi)\log(1/(3\delta)) \quad \text{for all } t \in V. \tag{13}$$

At long last we define Q by

$$Q(t) = e^{-iNt} \mathrm{Im}(g_N(t)) = \frac{e^{-iNt}}{2i} \left[g_N(t) - \overline{g_N(t)} \right]$$

$$= \frac{1}{2i} \sum_{k=-N+1}^{0} a_{k+N} e^{ikt} - \frac{1}{2i} \sum_{k=-2N}^{-N-1} \overline{a}_{-k-N} e^{ikt}.$$

Now (i) clearly follows from (12) and, since

$$S^*(Q, t) \geq |S_N(Q, t)| = \left| \frac{e^{-iNt}}{2i} g_N(t) \right| = \frac{1}{2} |g_N(t)|,$$

(ii) follows from (13). $\quad\square$

(8.25)   **Theorem**   *If $E \subset [-\pi, \pi]$ and $\lambda(E) = 0$, then there exists a function $f \in C(\mathbb{T})$ whose Fourier series $S(f)$ diverges at every point of E and whose Fourier coefficients $\hat{f}(n) = 0$ if $n < 0$.*

Proof   First choose a family $\{I_n\}_{n=1}^{\infty}$ of subintervals of $[-\pi, \pi]$ such that $\sum_{n=1}^{\infty} |I_n| < \infty$ and, for each $t \in E$, the set $\{n \in \mathbb{N} : t \in I_n\}$ is infinite. [To do this select, for each $k \in \mathbb{N}$, a family $\{J_{k,j}\}_{j=1}^{\infty}$ of intervals covering E and having sum of lengths less than 2^{-k} and then let $I_n = J_{k,j}$, where $n = 2^{k-1}(2j - 1)$.] Next choose $1 \leq n_1 < n_2 < \ldots$ such that

$$3 \cdot \sum_{n=n_k}^{\infty} |I_n| < e^{-\pi k^3}.$$

Let $V_k = \bigcup_{n=n_k}^{n_{k+1}} I_n$ and let Q_k correspond to V_k as in (8.24). Then $3\delta_k = 3\lambda(V_k) < e^{-\pi k^3}$, so

$$S^*(Q_k, t) > k^3 \quad \text{for all } t \in V_k. \tag{1}$$

Our choice of the I_n's also assures that

$$\{k \in \mathbb{N} : t \in V_k\} \text{ is infinite} \quad \text{for all } t \in E. \tag{2}$$

Now let N_k be the degree of Q_k [this integer is defined as $\max\{|n| : n \in \mathbb{Z}, \hat{Q}_k(n) \neq 0\}$] and then choose integers $(p_k)_{k=1}^{\infty}$ such that $p_1 = N_1$ and

$$p_k - N_k > p_{k-1} + N_{k-1} \quad \text{for } k > 1. \tag{3}$$

Now define f on $\mathbb{R}$ by

$$f(t) = \sum_{k=1}^{\infty} k^{-2} e^{ip_k t} Q_k(t). \tag{4}$$

Since the kth term has absolute value less than k^{-2} for all $t \in \mathbb{R}$, the Weierstrass M-test shows that series (4) converges uniformly on $\mathbb{R}$, and so $f \in C(\mathbb{T})$. Also, this allows term-by-term integration when we apply (8.7.i), so for each $n \in \mathbb{Z}$ we have

$$\hat{f}(n) = \sum_{k=1}^{\infty} \frac{k^{-2}}{2\pi} \int_{-\pi}^{\pi} Q_k(t) e^{-i(n-p_k)t} \, dt = \sum_{k=1}^{\infty} k^{-2} \hat{Q}_k(n - p_k).$$

But $\hat{Q}_k(n - p_k) = 0$ unless $-N_k + p_k \leq n \leq N_k + p_k$ and (3) shows that (when n is fixed) these inequalities can obtain for at most one k. We therefore have

$$\hat{f}(n) = k^{-2} \hat{Q}_k(n - p_k) \quad \text{if } |n - p_k| \leq N_k \tag{5}$$

and $\hat{f}(n) = 0$ for all other n. In particular, $\hat{f}(n) = 0$ if $n < 0$ because $0 = -N_1 + p_1 \leq -N_k + p_k$ for all k. [One sees incidentally that the Fourier series of f is just the trigonometric series obtained by writing out the individual terms of the trigonometric polynomials $k^{-2} e^{ip_k t} Q_k(t)$ successively in the order of increasing frequencies.] For fixed $k \in \mathbb{N}$, $0 \leq N$

$\leqq N_k$, and $t \in \mathbb{R}$,   (5) yields

$$S_{p_k+N}(f,t) - S_{p_k-N-1}(f,t) = \sum_{n=p_k-N}^{p_k+N} \hat{f}(n)e^{int} = \sum_{j=-N}^{N} k^{-2}\hat{Q}_k(j)e^{i(p_k+j)t}$$

$$= k^{-2}e^{ip_k t}S_N(Q_k,t). \tag{6}$$

Now fix any $t \in E$. For any given $j \in \mathbb{N}$, we first use (3) and (2) to find $k = k(t, j)$ such that $k > j$, $p_k - N_k - 1 > j$, and $t \in V_k$ and then we use (1) to select N with $0 \leqq N \leqq N_k$ such that $S_N(Q_k, t) > k^3$. We next write $m_j = p_k - N - 1$, $n_j = p_k + N$, and infer from (6) that $|S_{n_j}(f,t) - S_{m_j}(f,t)| > k > j$. We conclude that the sequence $(S_n(f,t))_{n=0}^{\infty}$ is not a Cauchy sequence so it cannot converge. In fact, $\overline{\lim}_{n\to\infty} |S_n(f,t)| = \infty$. $\square$

(8.26) Remarks (a) The fact that the Fourier series of a continuous function may diverge at certain points was discovered by P. du Bois-Reymond in 1876. A particularly simple example was given by L. Fejér in 1911 (see Exercises 1–3 below).

(b) In 1926, the Russian mathematician A. N. Kolmogorov produced a function in $L_1(\mathbb{T})$ whose Fourier series diverges at *every* $t \in \mathbb{R}$.

(c) Until 1966 it was unknown whether or not there might exist a function in $C(\mathbb{T})$ whose Fourier series diverges everywhere. In that year the Swedish analyst Lennart Carleson proved that if $f \in L_2(\mathbb{T})$, then $S(f)$ converges to f a.e. Shortly after that the American R. A. Hunt used Carleson's methods to prove that this is also true for $f \in L_p(\mathbb{T})$ provided only that $p > 1$.

Exercises

1.   For $n \in \mathbb{N}$, define the *Fejér polynomials* Q_n and P_n by

$$Q_n(t) = \sum_{j=n}^{2n-1} \frac{\cos(jt)}{2n-j} - \sum_{j=2n+1}^{3n} \frac{\cos(jt)}{j-2n},$$

$$P_n(t) = S_{2n}(Q_n, t) = \sum_{j=n}^{2n-1} \frac{\cos(jt)}{2n-j}.$$

Also, define $\tilde{Q}_n$ and $\tilde{P}_n$, respectively, in exactly the same way except that $\cos(jt)$ is replaced by $\sin(jt)$.

(a) Combining terms for $j = 2n - k$ and $j = 2n + k$, we have

$$Q_n(t) = 2\sin(2nt) \cdot \sum_{k=1}^{n} \frac{\sin(kt)}{k}; \qquad \tilde{Q}_n(t) = -2\cos(2nt) \cdot \sum_{k=1}^{n} \frac{\sin(kt)}{k}.$$

(b) For all $n \in \mathbb{N}$ and all $t \in \mathbb{R}$, we have

$$|Q_n(t)| < 4\pi + 2, \qquad |\tilde{Q}_n(t)| < 4\pi + 2.$$

(c) For all $n, p \in \mathbb{N}$ and $t \in \mathbb{R}$, we have

$$|S_p(Q_n, t)| < 2 + 2\log n, \qquad |S_p(\tilde{Q}_n, t)| < 2 + 2\log n.$$

[Hint: $\sum_{k=1}^{n} k^{-1} \leqq 1 + \int_1^n \frac{dx}{x}$.]

(d) For all $n \in \mathbb{N}$, we have

$$P_n(0) > \log n, \qquad \tilde{P}_n(\pi/(4n)) > (\log n)/\sqrt{2} .$$

[Hint: $\sin(j\pi/(4n)) \geq \sin(\pi/4)$ if $n \leq j < 2n$.]

(e) If $0 \leq r < s$ are integers, $(b_j)_{j=r}^s \subset [-1, 1]$ is a monotone sequence, and $t \in \mathbb{R}$ is not a multiple of 2π, then

$$\left| \sum_{j=r}^s b_j e^{ijt} \right| \leq \frac{4}{|\sin(t/2)|} .$$

[Hint: Argue as in (7.38).]

(f) If $0 < \delta < \pi$, $M_\delta = 8/[\sin(\delta/2)]$, $\delta \leq t \leq 2\pi - \delta$, and $n, p \in \mathbb{N}$, then

$$|S_p(Q_n, t)| \leq M_\delta, \qquad |S_p(\tilde{Q}_n, t)| \leq M_\delta$$

[Hint: Each of these sums is the real or imaginary part of the sum of two sums of type (e).]

(g) For all $n \in \mathbb{N}$ and $t \in \mathbb{R}$, we have the derivatives

$$Q_n'(t) = 2 \sin(2nt) \cdot \sum_{k=1}^n \cos(kt) - 2n\tilde{Q}_n(t),$$

$$\tilde{Q}_n'(t) = -2 \cos(2nt) \cdot \sum_{k=1}^n \cos(kt) + 2nQ_n(t).$$

(h) For all $n \in \mathbb{N}$ and all $t, h \in \mathbb{R}$, we have

$$|Q_n(t + h) - Q_n(t)| \leq nC|h|, \qquad |\tilde{Q}_n(t + h) - \tilde{Q}_n(t)| \leq nC|h|$$

where $C = 8\pi + 6$. [Hint: Mean Value Theorem.]

2. Adopt the notation of Exercise 1. Suppose that $(n_k)_{k=1}^\infty \subset \mathbb{N}$ satisfies $n_{k+1} > 3n_k$ for all k. Define f and $\tilde{f}$ on $\mathbb{R}$ by

$$f(t) = \sum_{k=1}^\infty 2^{-k}Q_{n_k}(t) \qquad \tilde{f}(t) = \sum_{k=1}^\infty 2^{-k}\tilde{Q}_{n_k}(t). \qquad (*)$$

(a) These two series converge uniformly on $\mathbb{R}$ and f and $\tilde{f}$ are both in $C(\mathbb{T})$.

(b) The trigonometric series

$$\sum_{j=1}^\infty a_j \cos(jt) \qquad \sum_{j=1}^\infty a_j \sin(jt)$$

obtained by writing out the individual terms of the trigonometric polynomials Q_{n_k} and $\tilde{Q}_{n_k}$ as they occur in the above series (that is, $a_j = 2^{-k}(2n_k - j)^{-1}$ if $0 < |2n_k - j| \leq n_k$ for some k and $a_j = 0$ for all other j) are the Fourier series of f and $\tilde{f}$, respectively. [Hint: Use (8.5.i).]

(c) If $m, p \in \mathbb{N}$ and $n_m \leq p < n_{m+1}$, then

$$S_p(f) = \sum_{k=1}^{m-1} 2^{-k}Q_{n_k} + 2^{-m}S_p(Q_{n_m})$$

and the same is true if f and Q are replaced by $\tilde{f}$ and $\tilde{Q}$, respectively. (Here the first sum on the right is 0 if $m = 1$.)

(d) If $0 < \delta < \pi$, the Fourier series of f converges to f uniformly on $[\delta, 2\pi - \delta]$. The same is true for $\tilde{f}$. [Hint: Use (c) and Exercise 1 to obtain

$$|f - S_p(f)| < (4\pi + 2) \sum_{k=m}^{\infty} 2^{-k} + 2^{-m} M_\delta \quad \text{on } [\delta, 2\pi - \delta] \quad \text{if } n_m \leq p < n_{m+1}.]$$

(e) If $\varliminf_{m \to \infty} 2^{-m} \log n_m < \infty$ (for example, $n_m = 2^{2^m}$), then

$$\|S_p(f)\|_u < A, \cdot \quad \|S_p(\tilde{f})\|_u < A$$

for all $p \in \mathbb{N}$, where A is a real number not depending on p. [Hint: Choose $\beta \in \mathbb{R}$ such that $\log n_m < \beta 2^m$ for all $m \in \mathbb{N}$. Use (c) and Exercise 1.]

(f) If $\varlimsup_{m \to \infty} 2^{-m} \log n_m > 0$ (same example), then the Fourier series of f diverges at $t = 0$ and the Fourier series of $\tilde{f}$ is *not* uniformly convergent on $\mathbb{R}$ but it does converge at every $t \in \mathbb{R}$ including $t = 0$. [Hints: Choose $\alpha > 0$ so that $\log n_m > \alpha 2^m$ for infinitely many m. For such m, $\|S_{2n_m}(\tilde{f}) - S_{n_m - 1}(\tilde{f})\|_u \geq 2^{-m} \tilde{P}_{n_m}(\pi/(4n_m)) > \alpha/\sqrt{2}$.]

(g) If $\varlimsup_{m \to \infty} 2^{-m} \log n_m = \infty$ (for example, $n_m = 2^{3^m}$), then $\varlimsup_{n \to \infty} |S_n(f, 0)| = \infty$. [Hint: $S_{2n_m}(f, 0) - S_{n_m - 1}(f, 0) = 2^{-m} P_{n_m}(0)$.]

(h) Suppose that $0 < \varliminf_{m \to \infty} 2^{-m} \log n_m < \infty$ (say $n_m = 2^{2^m}$), and let $\{r_k\}_{k=1}^{\infty}$ be a dense subset of $\mathbb{R}$. Define g on $\mathbb{R}$ by

$$g(t) = \sum_{k=1}^{\infty} 2^{-k} \tilde{f}(t - r_k).$$

Then $g \in C(\mathbb{T})$ and the Fourier series of g converges to g at every $t \in \mathbb{R}$, but there is *no* nonvoid open interval of $\mathbb{R}$ on which $S(g)$ converges uniformly. [Hint: Using uniform convergence, show that $S_p(g, t) = \sum_{k=1}^{\infty} 2^{-k} S_p(\tilde{f}, t - r_k)$ and use (e) and (f).]

(i) Let $n_k = 2^{2^k}$ for all k. For all $t, h \in \mathbb{R}$ with $0 < |h| < 1$, we have

$$|f(t + h) - f(t)| < A \left[\log \frac{1}{|h|} \right]^{-1}$$

where $A = 16(3\pi + 2) \log 2$ is an absolute constant. The same inequality obtains with $\tilde{f}$ or g in place of f. [Hints: Consider a fixed h with $0 < |h| < 1/4 = 1/n_1$. Let $p \in \mathbb{N}$ satisfy $n_p |h| \leq 1 < n_{p+1} |h|$; hence, $2^{-p} \log |h|^{-1} < 2 \log 2$. Use (∗) to write $f(t + h) - f(t)$ as an infinite sum. Use Exercise 1(h) to estimate the first p summands and Exercise 1(b) to estimate the others. Note that $\sum_{1}^{p} 2^{-k} n_k < 2 \cdot 2^{-p} n_p$ because $x_{k+1} \geq 2x_k > 0$ for all k implies $x_1 + x_2 + \ldots + x_p < 2x_p$ for all p.] (We remark that (i) is relevant in connection with (8.44) and (8.51) below.)

3. Adopt the notation of Exercise 1. Let $(n_k)_{k=1}^{\infty} \subset \mathbb{N}$ satisfy $(k + 1)n_{k+1} > 3n_k$ and $\varlimsup_{k \to \infty} k^{-2} \log n_k = \infty$ (for example, $n_k = 2^{k^3}$). Define g on $\mathbb{R}$ by

$$g(t) = \sum_{k=1}^{\infty} k^{-2} Q_{n_k}(k! \, t).$$

(a) This series converges uniformly on $\mathbb{R}$ and $g \in C(\mathbb{T})$.

(b) The Fourier series of g is the trigonometric series

$$\sum_{m=1}^{\infty} c_m \cos(mt)$$

obtained from the above series by writing out the individual cosine terms of the Q's in the order that they appear there; that is, $c_m = k^{-2}(2n_k - j)^{-1}$ if $m = k!\,j$ and $0 < |2n_k - j| \leq n_k$ for some $j, k \in \mathbb{N}$ while $c_m = 0$ for all other m. [Hint: Since $k!\cdot 3n_k < (k+1)!\cdot n_{k+1}$, the $p!\cdot 2n_p$th partial sum of the latter series is the pth partial sum of the former, so apply (8.5.i).]

(c) Writing $p_k = k!\cdot n_k - 1$ and $q_k = k!\cdot 2n_k$, we have

$$S_{q_k}(g,t) - S_{p_k}(g,t) = k^{-2}P_{n_k}(k!\,t)$$

for all $k \in \mathbb{N}$ and all $t \in \mathbb{R}$.

(d) The set $E = \{t \in \mathbb{R} : \varlimsup_{n\to\infty} |S_n(g,t)| = \infty\}$ contains every rational multiple of 2π. [Hint: If $t = 2\pi a/b$, where $a \in \mathbb{Z}$ and $b \in \mathbb{N}$, then $P_{n_k}(k!\,t) = P_{n_k}(0) > \log n_k$ for $k \geq b$.]

(e) The sets $E_k = \{t \in \mathbb{R} : |S_n(g,t)| > k \text{ for some } n \in \mathbb{N}\}$ are open and $E = \bigcap_{k=1}^{\infty} E_k$.

(f) If V is a nonvoid open interval, then $E \cap V$ is uncountable.

(g) The set of t at which the Fourier series of g converges is of first Baire category in $\mathbb{R}$.

4.   The trigonometric series

$$\sum_{n=3}^{\infty} \frac{\cos[n(t - \alpha_n)]}{\log n}$$

where $\alpha_n = \log(\log n)$ has coefficients $a_n = (\cos(n\alpha_n))/(\log n)$ and $b_n = (\sin(n\alpha_n))/(\log n)$ tending to the limit 0 as $n \to \infty$, and yet it diverges at *every* $t \in \mathbb{R}$. This simple example was given by H. Steinhaus in 1929. The following steps outline a proof of the divergence. For integers $p \geq 3$, let r_p be the integer satisfying $r_p \leq \log p < r_p + 1$ and write

$$u_p(x) = \sum_{n=p+1}^{p+r_p} \frac{\cos[n(x - \alpha_n)]}{\log n}, \qquad v_p = \sum_{n=p+1}^{p+r_p} \frac{1}{\log n}.$$

(a) For $4 \leq p + 1 \leq n \leq p + r_p$ and $\alpha_p \leq x \leq \alpha_{p+1}$, we have

$$|x - \alpha_n| < \alpha_{p+r_p} - \alpha_p < \frac{r_p}{p \log p} \leq \frac{1}{p}.$$

[Hint: Apply the Mean Value Theorem for the second inequality.]

(b) For $p \geq 3$ and $\alpha_p \leq x \leq \alpha_{p+1}$, we have

$$0 \leq v_p - u_p(x) < \frac{r_p}{2\log(p+1)} \cdot \frac{(p+r_p)^2}{p^2} < \frac{1}{2}\left(1 + \frac{r_p}{p}\right)^2.$$

[Hint: Use $1 - \cos t = 2\sin^2(t/2) \leq t^2/2$ and (a).]

(c) $v_p \geq r_p/(\log(p + r_p)) \to 1$ as $p \to \infty$.

(d) There is a $p_0 > 2$ such that if $p > p_0$, then

$$v_p - (1/2)(1 + r_p/p)^2 > 1/3.$$

(e) If $p > p_0$ and $\alpha_p \leq x \leq \alpha_{p+1}$, then $u_p(x) > 1/3$.

(f) For any $t \in \mathbb{R}$, there are infinitely many $p \in \mathbb{N}$ such that $u_p(t) > 1/3$. [Hint: Each u_p is 2π-periodic.]

(g) The given series diverges at every $t \in \mathbb{R}$.

Summability of Fourier Series

In view of (8.26), it is a remarkable fact that for any $f \in C(\mathbb{T})$ the Fourier series of f is uniformly C_1-summable to f on $\mathbb{R}$. This result is a special case of (8.30) below. This then amply shows the usefulness of the C_1-method where convergence fails.

We begin our investigation of summability of Fourier series by showing how the partial sums, the C_1-means, and the Abel means can be expressed as integrals.

(8.27)   Theorem   *Let $f \in L_1(\mathbb{T})$, let D_n and K_n be as in (5.12):*

$$D_n(t) = \frac{\sin[(n + 1/2)t]}{\sin((1/2)t)}, \qquad K_n(t) = \frac{\sin^2[((n + 1)/2)t]}{(n + 1)\sin^2[(1/2)t]},$$

and let

$$P_r(t) = \sum_{k=-\infty}^{\infty} r^{|k|} e^{ikt}$$

for $0 \le r < 1$. Then

(i) $$P_r(t) = \frac{1 - r^2}{1 - 2r\cos t + r^2},$$

(ii) $$S_n(f,x) = \sum_{k=-n}^{n} \hat{f}(k) e^{ikx} = \frac{1}{2\pi} \int_{-\pi}^{\pi} f(x - t)D_n(t)\,dt,$$

(iii) $$\sigma_n(f,x) = \frac{1}{n+1} \sum_{k=0}^{n} S_k(f,x) = \frac{1}{2\pi} \int_{-\pi}^{\pi} f(x - t)K_n(t)\,dt,$$

(iv) $$\alpha_r(f,x) = \sum_{k=-\infty}^{\infty} r^{|k|} \hat{f}(k) e^{ikx} = \frac{1}{2\pi} \int_{-\pi}^{\pi} f(x - t)P_r(t)\,dt$$

for all $n \ge 0$, $0 \le r < 1$, and $t, x \in \mathbb{R}$.
[The first equalities in (ii), (iii), and (iv) are definitions. They define their left sides.]

Proof   To prove (i), fix $t \in \mathbb{R}$ and $0 \le r < 1$. The pth partial sum of the series defining $P_r(t)$ is

$$\sum_{k=-p}^{p} r^{|k|} e^{ikt} = \sum_{k=0}^{p} r^k e^{ikt} + \sum_{k=1}^{p} r^k e^{-ikt},$$

and as $p \to \infty$ this has the limit

$$\sum_{k=0}^{\infty} r^k e^{ikt} + \sum_{k=1}^{\infty} r^k e^{-ikt} = \frac{1}{1 - re^{it}} + \frac{re^{-it}}{1 - re^{-it}},$$

which is equal to the right side of (i).

To obtain (ii) and (iii), use the definitions of D_n and K_n given in

(5.12) to write

$$\sum_{k=-n}^{n} \hat{f}(k)e^{ikx} = \sum_{k=-n}^{n} \frac{1}{2\pi} \int_{-\pi}^{\pi} f(u)e^{-iku} \, du \, e^{ikx}$$

$$= \frac{1}{2\pi} \int_{-\pi}^{\pi} f(u) \left[\sum_{k=-n}^{n} e^{ik(x-u)} \right] du$$

$$= \frac{1}{2\pi} \int_{-\pi}^{\pi} f(u) D_n(x-u) \, du \tag{1}$$

and

$$\frac{1}{n+1} \sum_{k=0}^{n} S_k(f,x) = \frac{1}{n+1} \sum_{k=0}^{n} \frac{1}{2\pi} \int_{-\pi}^{\pi} f(u) D_k(x-u) \, du$$

$$= \frac{1}{2\pi} \int_{-\pi}^{\pi} f(u) \left[\frac{1}{n+1} \sum_{k=0}^{n} D_k(x-u) \right] du$$

$$= \frac{1}{2\pi} \int_{-\pi}^{\pi} f(u) K_n(x-u) \, du$$

and then make the substitution $u = x - t$. The proof of (iv) is almost the same as that of (1) except that interchange of the *infinite* sum and the integral is justified by the fact that, for fixed $x \in \mathbb{R}$ and $0 \le r < 1$, the series defining $P_r(x-u)$ converges *uniformly* for $u \in [-\pi, \pi]$. $\square$

(8.28)   Remark   The numbers $\alpha_r(f,x)$ defined in (8.27.iv) are the Abel means of the Fourier series of f at x as they were defined in (7.76.ii) since

$$\sum_{k=-\infty}^{\infty} r^{|k|} \hat{f}(k)e^{ikx} = \hat{f}(0)r^0 + \sum_{k=1}^{\infty} \left[\hat{f}(-k)e^{-ikx} + \hat{f}(k)e^{ikx} \right] r^k$$

where the square bracket is the kth term $S_k(f,x) - S_{k-1}(f,x)$ of $S(f,x)$. The function P_r defined in (8.27) is called the *Poisson Kernel.*

We next list some properties of the Fejér and the Poisson Kernels that allow us to prove the summability theorems that follow.

(8.29)   Theorem   Let K_n and P_r be as above. Then, for $n \ge 0$, $0 \le r < 1$, and $t \in \mathbb{R}$, we have

(i) $$\frac{1}{2\pi} \int_{-\pi}^{\pi} K_n(t) \, dt = \frac{1}{2\pi} \int_{-\pi}^{\pi} P_r(t) \, dt = 1,$$

(ii) $$K_n(t) = K_n(-t) \quad \text{and} \quad P_r(t) = P_r(-t),$$

(iii) $$0 \le K_n(t) \le n+1 \quad \text{and} \quad (1-r)/(1+r) \le P_r(t) \le (1+r)/(1-r),$$

(iv) $$0 \le K_n(t) \le \pi^2/((n+1)t^2) \quad \text{if } 0 < t \le \pi.$$

Proof   It is clear from (5.12) that $|D_k(t)| \le 2k + 1$, and so $K_n(t) \cdot (n+1)$

$$\leq \sum_{k=0}^{n} (2k + 1) = (n + 1)^2. \text{ Since } \sin(t/2) \geq t/\pi \text{ for } 0 \leq t \leq \pi, \text{ (iv) is evi-}$$

dent. The rest is obvious from (5.12) and (8.27). □

(8.30)   Fejér's Theorem   *Let $f \in L_1(\mathbb{T})$.*
  (i) *If $x \in \mathbb{R}$ is such that $f(x-) = \lim\limits_{t \uparrow x} f(t)$ and $f(x+) = \lim\limits_{t \downarrow x} f(t)$ exist and*

are finite, then

$$C_1\text{-} \sum_{k=-\infty}^{\infty} \hat{f}(k)e^{ikx} = \lim_{n \to \infty} \sigma_n(f, x) = \frac{f(x-) + f(x+)}{2}.$$

  (ii) *If f is continuous at every point of some closed interval $[a,b] \subset \mathbb{R}$*
[*two-sided continuity at a and b*], *then the Fourier series of f is uniformly C_1-summable on $[a,b]$ to f. That is, the trigonometric polynomials*

$$\sigma_n(f, x) = \sum_{k=-n}^{n} \left(1 - \frac{|k|}{n+1}\right)\hat{f}(k)e^{ikx}$$

converge uniformly to f on $[a,b]$.
  To honor Fejér, the polynomials $\sigma_n(f)$ are called the *Fejér sums* of f.

Proof   For fixed $x \in \mathbb{R}$, let $s(x) \in \mathbb{C}$ be arbitrary. Then (8.27.iii) and (8.29) imply that

$$\sigma_n(f, x) - s(x) = \frac{1}{2\pi} \int_{-\pi}^{\pi} \left[f(x - t) - s(x) \right] K_n(t)\, dt$$

$$= \frac{1}{2\pi} \int_{0}^{\pi} \left[f(x - t) + f(x + t) - 2s(x) \right] K_n(t)\, dt. \quad (1)$$

  Suppose the hypothesis of (i) is satisfied, write $s(x) = [f(x-) + f(x+)]/2$, and let $\epsilon > 0$ be given. Choose $\delta > 0$ such that $\delta < \pi$ and

$$|f(x - t) + f(x + t) - 2s(x)| < \epsilon/2 \quad (2)$$

whenever $0 < t < \delta$. Next choose $N \in \mathbb{N}$ such that

$$\left[2\pi^2/(N\delta^2)\right](\|f\|_1 + |s(x)|) < \epsilon/2. \quad (3)$$

Writing $\phi(x, t) = f(x - t) + f(x + t) - 2s(x)$, (1) yields

$$|\sigma_n(f, x) - s(x)| \leq \frac{1}{2\pi} \int_{0}^{\pi} |\phi(x, t) K_n(t)|\, dt.$$

From (2), (8.29.iii), and (8.29.i), we obtain

$$\frac{1}{2\pi} \int_{0}^{\delta} |\phi(x, t) K_n(t)|\, dt \leq \frac{1}{2\pi} \int_{0}^{\delta} \frac{\epsilon}{2} K_n(t)\, dt \leq \frac{\epsilon}{2}.$$

Similarly, (8.29.iv) and (3) show that if $n \geq N$, then

$$\int_\delta^\pi |\phi(x,t) K_n(t)| \, dt \leq \int_\delta^\pi |\phi(x,t)| \frac{\pi^2}{(n+1)t^2} \, dt$$

$$\leq \frac{\pi^2}{(n+1)\delta^2} \int_\delta^\pi \left[|f(x-t)| + |f(x+t)| + 2|s(x)| \right] dt$$

$$\leq \frac{\pi^2}{(n+1)\delta^2} \left(2\pi\|f\|_1 + 2\pi\|f\|_1 + 2\pi|s(x)| \right) < \pi\epsilon. \quad (4)$$

We conclude that

$$|\sigma_n(f,x) - s(x)| < \epsilon \tag{5}$$

whenever $n \geq N$, and so (i) is proven.

The proof of (ii) is almost the same as that above. Supposing that f is continuous on $[a,b]$ and at a and b, the only changes are as follows. We write $s(x) = f(x)$ for $x \in [a,b]$. Choose $0 < \delta < \pi$ independent of $x \in [a,b]$ so that (2) obtains whenever $x \in [a,b]$ and $0 < t < \delta$ [use uniform continuity on $[a,b]$ to select δ_1 so that $|f(u) - f(x)| < \epsilon/4$ if $|u - x| < \delta_1$ and $u, x \in [a,b]$, select δ_2 so that $|f(x) - f(u)| < \epsilon/8$ if $x = a$ or $b, u \in \mathbb{R}$ and $|x - u| < \delta_2$, then let $\delta = \min\{\delta_1, \delta_2\}$]. Choose N so that (3) obtains when $|s(x)|$ is replaced by $M = \sup\{|f(x)| : a \leq x \leq b\}$. Then N does not depend on any particular $x \in [a,b]$. Replace $|s(x)|$ by M in (4). Then $n \geq N$ implies that (5) holds for all $x \in [a,b]$. $\quad\square$

(8.31)   Remarks   (a) Theorem (8.30) holds good if C_1-summability is replaced by Abel summability [except that the $\alpha_r(f)$ are not trigonometric polynomials]. This follows immediately from Frobenius's Theorem (7.81.iii). Actually, a direct proof is almost the same as that given above.

(b) Theorem (8.30) fails of course if C_1-summability is replaced by ordinary convergence. This is shown by (8.25). The reason that an analogous proof won't work when K_n is replaced by D_n [and hence σ_n is replaced by s_n] is that

$$\overline{\lim_{n \to \infty}} \int_0^\delta |D_n(t)| \, dt = \infty$$

for all $\delta > 0$ [Exercise!].

(c) In case $f \in C(\mathbb{T})$, then (8.30.ii) (with $[a,b] = [0, 2\pi]$) assures that $\sigma_n(f) \to f$ uniformly on $\mathbb{R}$. This then is a simple proof of Weierstrass's theorem that any $f \in C(\mathbb{T})$ is a uniform limit of trigonometric polynomials. The proof, as one sees, requires very few prerequisites. In turn this theorem leads (via uniform convergence of the exponential series) to a simple proof of Weierstrass's Approximation Theorem (3.117).

Theorem (8.30) is closely bound to the continuity of the function f. Our next goal is to prove a deeper theorem (8.35) which assures that the Fourier series of *any* $f \in L_1(\mathbb{T})$ is C_1-summable to f a.e. on $\mathbb{R}$. This fact is surprising in view of Kolmogorov's example mentioned in (8.26.b). First, we prove the following important theorem.

(8.32)   Theorem [Lebesgue] *Let f be a complex-valued Lebesgue measurable function that is defined everywhere on $\mathbb{R}$ and has the property that*

$$\int_a^b |f(t)| \, dt < \infty$$

whenever $a < b$ are real numbers [such a function is said to be *locally integrable*]. *Then there exists a Lebesgue measurable set $E \subset \mathbb{R}$ such that $\lambda(\mathbb{R}\backslash E) = 0$ and*

(i) $\displaystyle\lim_{h \to 0} \frac{1}{h} \int_x^{x+h} |f(t) - c| \, dt = |f(x) - c|$

whenever $x \in E$ and $c \in \mathbb{C}$. In particular, for each $x \in E$ we have

(ii) $\displaystyle\lim_{h \to 0} \frac{1}{h} \int_0^h |f(x + t) + f(x - t) - 2f(x)| \, dt = 0.$

Moreover, E contains every x at which f is continuous.

Proof Let $\{c_n\}_{n=1}^\infty$ be a countable dense subset of $\mathbb{C}$. For each n, define F_n on $\mathbb{R}$ by

$$F_n(x) = \int_0^x |f(t) - c_n| \, dt,$$

and let $E_n = \{x \in \mathbb{R} : F_n'(x) = |f(x) - c_n|\}$. It follows from (6.84) that $\lambda(\mathbb{R}\backslash E_n) = 0$ and that E_n contains each x at which f is continuous. Now let $E = \bigcap_{n=1}^\infty E_n$. Then E is measurable and

$$\lambda(\mathbb{R}\backslash E) = \lambda\left(\bigcup_{n=1}^\infty (\mathbb{R}\backslash E_n) \right) \leq \sum_{n=1}^\infty \lambda(\mathbb{R}\backslash E_n) = 0.$$

Thus $\lambda(\mathbb{R}\backslash E) = 0$. To prove (i), let $x \in E$, $c \in \mathbb{C}$ and $\epsilon > 0$ be given. Fix an n such that $|c - c_n| < \epsilon/3$. Since $x \in E_n$, there exists a $\delta > 0$ such that

$$\left| \frac{1}{h} \int_x^{x+h} |f(t) - c_n| \, dt - |f(x) - c_n| \right| = \left| \frac{F_n(x+h) - F_n(x)}{h} - |f(x) - c_n| \right|$$

$$< \epsilon/3 \qquad\qquad\qquad (1)$$

whenever $0 < |h| < \delta$ ($h \in \mathbb{R}$). Obviously $\big| |f(t) - c| - |f(t) - c_n| \big| \leq |c - c_n| < \epsilon/3$ for all t, and so

$$\left| \frac{1}{h} \int_x^{x+h} |f(t) - c| \, dt - \frac{1}{h} \int_x^{x+h} |f(t) - c_n| \, dt \right| < \left| \frac{1}{h} \int_x^{x+h} \frac{\epsilon}{3} \, dt \right| = \frac{\epsilon}{3} \quad (2)$$

whenever $0 \neq h \in \mathbb{R}$. Also,

$$\big| |f(x) - c_n| - |f(x) - c| \big| < \epsilon/3. \qquad\qquad\qquad (3)$$

Combining (1), (2), and (3) gives

$$\left| \frac{1}{h} \int_x^{x+h} |f(t) - c| \, dt - |f(x) - c| \right| < \epsilon$$

whenever $0 < |h| < \delta$ $(h \in \mathbb{R})$. This proves (i).

To obtain (ii) fix any $x \in E$. Since $0 \neq h \in \mathbb{R}$ implies

$$0 \leq \frac{1}{h} \int_0^h |f(x + t) + f(x - t) - 2f(x)| \, dt$$

$$\leq \frac{1}{h} \int_0^h |f(x + t) - f(x)| \, dt + \frac{1}{h} \int_0^h |f(x - t) - f(x)| \, dt$$

$$= \frac{1}{h} \int_x^{x+h} |f(u) - f(x)| \, du + \frac{1}{-h} \int_x^{x-h} |f(v) - f(x)| \, dv,$$

(ii) follows from (i) by taking $c = f(x)$. $\square$

(8.33) Definition Let f be as in (8.32). A point $x \in \mathbb{R}$ is called a *Lebesgue point for f* if (8.32.ii) obtains. The set of all Lebesgue points for f is called the *Lebesgue set for f*.

(8.34) Example Theorem (8.32) implies that if f is locally integrable, then almost every real number is a Lebesgue point for f. However, (8.32.i) may fail at some Lebesgue points. For example, let $f(t) = 1 = -f(-t)$ for $t > 0$ and $f(0) = 0$. Then $x = 0$ is trivially seen to be a Lebesgue point for f, while, for $x = 0$, (8.32.i) fails for every $c \in \mathbb{C}$.

(8.35) Fejér–Lebesgue Theorem Let $f \in L_1(\mathbb{T})$. Then $\lim_{n \to \infty} \sigma_n(f, x) = f(x)$ whenever x is a Lebesgue point for f. In particular, the Fourier series of f is C_1-summable to f at almost every $x \in \mathbb{R}$.

Proof Let x be any fixed Lebesgue point for f. Write $\phi(t) = f(x + t) + f(x - t) - 2f(x)$ and

$$\Phi(h) = \int_0^h |\phi(t)| \, dt, \qquad \Phi(\pi) = a.$$

Let $\epsilon > 0$ be given. According to (8.33), there exists $0 < \delta < \pi$ such that

$$|(1/h)\Phi(h)| < \epsilon/13 \quad \text{for } 0 < |h| \leq \delta.$$

Next invoke (8.29.iv) to obtain $N > 1/\delta$ such that

$$0 \leq K_n(t) < \epsilon/(a + 1) \quad \text{if } n \geq N \quad \text{and} \quad \delta \leq t \leq \pi.$$

As in the proof of (8.30), we have

$$2\pi|\sigma_n(f, x) - f(x)| \leq \int_0^\pi |\phi(t)| K_n(t) \, dt = \int_0^{1/n} + \int_{1/n}^\delta + \int_\delta^\pi. \tag{1}$$

We now estimate these three integrals for an arbitrary $n \geq N$. First, we use (8.29.iii) to write

$$\int_0^{1/n} |\phi(t)| K_n(t)\, dt \leq \int_0^{1/n} |\phi(t)| (n+1)\, dt = (n+1)\Phi\left(\frac{1}{n}\right) \leq 2n\Phi\left(\frac{1}{n}\right) < \frac{2\epsilon}{13} \tag{2}$$

because $1/n < \delta$. Second, we have

$$\int_\delta^\pi |\phi(t)| K_n(t)\, dt \leq \int_\delta^\pi |\phi(t)| \frac{\epsilon}{a+1}\, dt \leq \frac{\epsilon}{a+1} \Phi(\pi) < \epsilon. \tag{3}$$

The integral over the middle interval is more difficult. It requires a more delicate use of (8.29.iv). We use integration by parts (6.90) and (6.85) to write

$$\int_{1/n}^\delta |\phi(t)| K_n(t)\, dt \leq \int_{1/n}^\delta |\phi(t)| \frac{\pi^2}{(n+1)t^2}\, dt$$

$$= \frac{\pi^2}{n+1}\left[\delta^{-2}\Phi(\delta) - n^2\Phi\left(\frac{1}{n}\right) + 2\int_{1/n}^\delta \Phi(t)t^{-3}\, dt \right]$$

$$< \frac{\pi^2}{n+1} \cdot \delta^{-1} \cdot \frac{\epsilon}{13} + \frac{2\pi^2}{n+1} \int_{1/n}^\delta \frac{\epsilon}{13} t^{-2}\, dt$$

$$< \pi^2\epsilon/13 + (2\pi^2\epsilon/13(n+1))(n - \delta^{-1}) < 3\pi^2\epsilon/13. \tag{4}$$

Combining (1), (2), (3), and (4) gives

$$|\sigma_n(f,x) - f(x)| < \frac{1}{2\pi}\left[\frac{2\epsilon}{13} + \frac{3\pi^2\epsilon}{13} + \epsilon \right] < \frac{\epsilon}{13} + \frac{6\epsilon}{13} + \frac{3\epsilon}{13} < \epsilon$$

for all $n \geq N$. $\square$

(8.36)   Corollary   *If $f \in L_1(\mathbb{T})$ and if its Fourier series $S(f)$ converges a.e. on $\mathbb{R}$ to some function g, then $g = f$ a.e.*

Proof   It follows from (7.81.i) and (8.35) that for almost every $x \in \mathbb{R}$ we have

$$g(x) = \lim_{n\to\infty} s_n(f,x) = \lim_{n\to\infty} \sigma_n(f,x) = f(x). \quad \square$$

Because of the polar representation $z = re^{i\theta}$, there is a close connection between complex variable theory and Fourier analysis. This is exemplified by the next theorem.

(8.37)   Theorem   *Let $f \in C(\mathbb{T})$. Then the function F defined on the closed unit disk $D = \{re^{i\theta} : \theta \in \mathbb{R}, 0 \leq r \leq 1\}$ by*

$$F(e^{i\theta}) = f(\theta) \quad and \quad F(re^{i\theta}) = \frac{1}{2\pi}\int_{-\pi}^\pi f(\theta - t) P_r(t)\, dt \quad if\ 0 \leq r < 1$$

is continuous on D. [Here P_r is the Poisson Kernel as in (8.27).]

Proof The periodicity of f and of exp show that F is well defined. For $|z| < 1$, write $z = re^{i\theta}$ and use (8.27.iv) to see that

$$F(z) = \sum_{k=-\infty}^{\infty} r^{|k|} \hat{f}(k) e^{ik\theta} = \sum_{k=0}^{\infty} \hat{f}(k) z^k + \sum_{k=1}^{\infty} \hat{f}(-k) \bar{z}^k.$$

Since $\hat{f}$ is bounded, these two power series converge for $|z| = |\bar{z}| < 1$, and so their sums are continuous for $|z| < 1$. This proves that F is continuous on the open disk $D^0 = \{z \in \mathbb{C} : |z| < 1\}$. By (8.30.ii) [with $[a, b] = [-\pi, \pi]$] and (8.31.a), we have

$$\lim_{r \uparrow 1} F(re^{i\theta}) = \lim_{r \uparrow 1} \alpha_r(f, \theta) = f(\theta) = F(e^{i\theta})$$

uniformly for $\theta \in \mathbb{R}$. That is, given $\epsilon > 0$, there exists $0 < \delta < 1$ such that $1 - \delta < r \leq 1$ implies

$$|F(re^{i\theta}) - F(e^{i\theta})| < \epsilon/2 \tag{1}$$

for all $\theta \in \mathbb{R}$. Since f is uniformly continuous on $\mathbb{R}$, there exists $0 < \eta \leq \delta$ such that

$$|F(e^{i\theta}) - F(e^{i\alpha})| = |f(\theta) - f(\alpha)| < \epsilon/2 \tag{2}$$

whenever $\alpha, \theta \in \mathbb{R}$ and $|\alpha - \theta| < \eta$. To check continuity of F at points of the boundary of D, let $\alpha \in \mathbb{R}$ be given. If ϵ is given and δ and η are as above, then (1) and (2) imply that $|F(e^{i\alpha}) - F(re^{i\theta})| < \epsilon$ whenever $1 - \delta < r \leq 1$ and $|\theta - \alpha| < \eta$. $\square$

As we have seen in (8.25), the partial sums $s_n(f)$ of a bounded (even continuous) function f need not be uniformly bounded. The following simple theorem shows that this pathology cannot occur for Fejér sums.

(8.38) Theorem *If $f \in L_1(\mathbb{T})$ is essentially bounded: $|f(x)| \leq M < \infty$ a.e. on $\mathbb{R}$, then $\|\sigma_n(f)\|_u \leq M$ for all $n \geq 0$.*

Proof By (8.27) and (8.29), we have

$$|\sigma_n(f, x)| \leq \frac{1}{2\pi} \int_{-\pi}^{\pi} |f(x - t)| \, K_n(t) \, dt \leq \frac{1}{2\pi} \int_{-\pi}^{\pi} M K_n(t) \, dt = M$$

for all x and n. $\square$

Riemann Localization
and Convergence Criteria

In this section we establish a few of the many classical theorems that are concerned with ordinary convergence of Fourier series at a point or on an interval. We want criteria that enable us to deduce convergence from some simple property of the function itself. We have already met (in the exercises) some such theorems which conclude absolute and uniform convergence from some differentiability or Lipschitz condition. We begin here with a useful lemma.

(8.39) Lemma *Let* $f, g \in L_1(\mathbb{T})$ *with* g *bounded on* $\mathbb{R}: |g(t)| \leq M < \infty$ *for all* $t \in \mathbb{R}$. *Then*

$$\lim_{|n| \to \infty} \int_{-\pi}^{\pi} f(x - t)g(t)e^{int} \, dt = 0$$

uniformly *for* $x \in \mathbb{R}$.

The point here is that the convergence to 0 is *uniform*. Since for any fixed x, the function $t \to f(x - t)g(t)$ is in $L_1(\mathbb{T})$, the pointwise convergence to 0 follows from (8.14).

Proof Let $\epsilon > 0$ be given. It is our job to find $N \in \mathbb{N}$ such that

$$\left| \int_{-\pi}^{\pi} f(x - t)g(t)e^{int} \, dt \right| < \epsilon \tag{1}$$

for all $x \in \mathbb{R}$ whenever $n \in \mathbb{Z}$ and $|n| \geq N$. As usual, we approximate.

First invoke (8.3.ii) to obtain a trigonometric polynomial

$$P(t) = \sum_{j=-p}^{p} c_j e^{ijt} \tag{2}$$

such that

$$\int_{-\pi}^{\pi} |f(u) - P(u)| \, du < \frac{\epsilon}{2M + 1}. \tag{3}$$

Write

$$\eta = \epsilon \bigg/ \left(1 + 4\pi \sum_{j=-p}^{p} |c_j| \right) \tag{4}$$

and then invoke (8.14) to obtain $N \in \mathbb{N}$ such that

$$|\hat{g}(k)| < \eta \quad \text{whenever } k \in \mathbb{Z} \quad \text{and} \quad |k| \geq N - p. \tag{5}$$

For $|n| > N$ and $-p \leq j \leq p$, we have $|j - n| \geq |n| - |j| \geq N - p$, so (5) yields

$$\left| \int_{-\pi}^{\pi} e^{ij(x-t)} g(t)e^{int} \, dt \right| = \left| e^{ijx} \int_{-\pi}^{\pi} g(t)e^{-i(j-n)t} \, dt \right| = |2\pi\hat{g}(j - n)| < 2\pi\eta.$$

Therefore, using (2) and (4), we have

$$\left| \int_{-\pi}^{\pi} P(x - t)g(t)e^{int} \, dt \right| \leq \sum_{j=-p}^{p} |c_j| \left| \int_{-\pi}^{\pi} e^{ij(x-t)} g(t)e^{int} \, dt \right|$$

$$\leq 2\pi\eta \cdot \sum_{j=-p}^{p} |c_j| < \frac{\epsilon}{2} \tag{6}$$

for all x whenever $|n| > N$. Plainly, (3) gives

$$\left| \int_{-\pi}^{\pi} f(x - t)g(t)e^{int}\, dt - \int_{-\pi}^{\pi} P(x - t)g(t)e^{int}\, dt \right|$$

$$\leq \int_{-\pi}^{\pi} |f(x - t) - P(x - t)| \, |g(t)e^{int}| \, dt$$

$$\leq M \int_{-\pi}^{\pi} |f(u) - P(u)| \, du < \frac{\epsilon}{2} \tag{7}$$

for all x and n (where we have used 2π-periodicity after the substitution $u = x - t$). Now add inequalities (6) and (7) to get (1). $\square$

The next theorem gives us a way of viewing the nth partial sum of a Fourier series as an integral that is simpler than the Dirichlet integral given in (8.27.ii). However, the simplification does introduce a uniformly small error term which is harmless for many purposes.

(8.40)   **Theorem** *Let $f \in L_1(\mathbb{T})$ and $0 < \delta \leq \pi$ be given. Then*

(i) $S_n(f, x) = \dfrac{1}{\pi} \displaystyle\int_0^\delta [f(x - t) + f(x + t)] \, \dfrac{\sin(nt)}{t} \, dt + \epsilon_n(x)$

where the functions ϵ_n (which are defined by (i) and also depend on the fixed f and δ) converge to 0 uniformly on $\mathbb{R}$ as $n \to \infty$. In particular,

(ii) $1 = \dfrac{2}{\pi} \displaystyle\int_0^\delta \dfrac{\sin(nt)}{t} \, dt + \epsilon_n'$,

where the numbers ϵ_n' (which depend on δ) have limit 0 as $n \to \infty$.

Proof   Since, by (8.27),

$$D_n(t) = \frac{\sin((n + 1/2)t)}{\sin((1/2)t)} = [\sin(nt)]\cot((1/2)t) + \cos(nt),$$

it follows from (8.27.ii) that

$$S_n(f, x) = \frac{1}{\pi} \int_{-\pi}^{\pi} f(x - t) \frac{1}{2} [\cot((1/2)t)] \sin(nt) \, dt + \alpha_n(x), \tag{1}$$

where

$$\alpha_n(x) = \frac{1}{2\pi} \int_{-\pi}^{\pi} f(x - t)\cos(nt) \, dt. \tag{2}$$

One checks by using l'Hospital's Rule or otherwise [e.g., (5.43) or via power series and $u^{-1} \sin u \to 1$ as $u \to 0$] that

$$\lim_{t \to 0} [(1/2)\cot((1/2)t) - 1/t] = 0, \tag{3}$$

so the 2π-periodic function g_1 such that $g_1(t)$ is the bracket in (3) for $0 < |t| < \pi$ and $g_1(0) = g_1(\pi) = 0$ is bounded and in $L_1(\mathbb{T})$ (since it is continuous except at odd multiples of π). By (1), we now have

$$S_n(f, x) = \frac{1}{\pi} \int_{-\pi}^{\pi} f(x - t) \frac{\sin(nt)}{t} \, dt + \beta_n(x) + \alpha_n(x), \qquad (4)$$

where

$$\beta_n(x) = \frac{1}{\pi} \int_{-\pi}^{\pi} f(x - t) g_1(t) \sin(nt) \, dt. \qquad (5)$$

Next, remembering our fixed δ, let g_2 be the 2π-periodic function such that $g_2(t) = t^{-1}$ if $\delta \leq |t| < \pi$ and $g_2(t) = 0$ if $|t| < \delta$ or $t = \pi$. Then g_2 is bounded and in $L_1(\mathbb{T})$. It follows from (4) that

$$S_n(f, x) = \frac{1}{\pi} \int_{-\delta}^{\delta} f(x - t) \frac{\sin(nt)}{t} \, dt + \gamma_n(x) + \beta_n(x) + \alpha_n(x), \qquad (6)$$

where

$$\gamma_n(x) = \frac{1}{\pi} \int_{-\pi}^{\pi} f(x - t) g_2(t) \sin(nt) \, dt. \qquad (7)$$

Splitting the interval of integration at 0, we see that the integral in (6) is equal to the integral in (i), and so $\epsilon_n(x) = \alpha_n(x) + \beta_n(x) + \gamma_n(x)$ for all $x \in \mathbb{R}$. But (2), (5), (7), and (8.39) show that the functions α_n, β_n, and γ_n each converge to 0 uniformly on $\mathbb{R}$ as $n \to \infty$, and hence so does ϵ_n. This proves our first assertion. The second is obtained by taking $f(t) = 1$ for all t and noticing that then $S_n(f, x) = 1$ for all n and x. $\square$

Since the Fourier coefficients of a function depend on the structure of the function over its entire domain, it seems surprising that the convergence or divergence behavior of the Fourier series at any particular point depends only on the structure of the function in an arbitrarily small neighborhood of that point. The next result makes this somewhat vague statement precise.

(8.41)   Riemann Localization Principle *Let $f_1, f_2 \in L_1(\mathbb{T})$ be given and suppose that $f_1(x) = f_2(x)$ for every x in some nonvoid open interval $I \subset \mathbb{R}$. Then*

$$\lim_{n \to \infty} |S_n(f_1, x) - S_n(f_2, x)| = 0$$

for all $x \in I$. Moreover, this convergence is uniform on every closed interval $J \subset I$.

Proof Given such a J, we can choose $0 < \delta < \pi$ such that $x + t$ and $x - t$ are in I whenever $x \in J$ and $0 \leq t \leq \delta$. Writing $f = f_1 - f_2$, we see that the integral in (8.40.i) is 0, and so $S_n(f_1, x) - S_n(f_2, x) = S_n(f, x) = \epsilon_n(x)$ for all $x \in J$. $\square$

Using (8.40), we can reformulate the question of convergence for Fourier series as an equivalent question about limits of integrals as follows.

(8.42)   Theorem *Let* $f \in L_1(\mathbb{T})$, $0 < \delta \leq \pi$, *and* $X \subset \mathbb{R}$ *be given. Write* $\phi(x, t)$ $= f(x + t) + f(x - t) - 2f(x)$. *Suppose that* f *is bounded on* X: $|f(x)| \leq M < \infty$ *for all* $x \in X$. *Then in order that*

(i) $\lim_{n \to \infty} S_n(f, x) = f(x)$ *uniformly for* $x \in X$

it is both necessary and sufficient that

(ii) $\lim_{n \to \infty} \dfrac{1}{\pi} \displaystyle\int_0^\delta \dfrac{\phi(x, t)}{t} \sin(nt)\, dt = 0$ *uniformly for* $x \in X$.

Proof Multiplying both sides of (8.40.ii) by $f(x)$ and then subtracting from (8.40.i), we find that

$$S_n(f, x) - f(x) = \frac{1}{\pi} \int_0^\delta \frac{\phi(x, t)}{t} \sin(nt)\, dt + \left[\epsilon_n(x) - \epsilon'_n f(x) \right].$$

Since f is bounded on X, the expression in brackets converges to 0 uniformly on X. ☐

We now obtain our first useful pointwise convergence criterion.

(8.43)   Dini's Test *Let* $f \in L_1(\mathbb{T})$ *and* $x_0 \in \mathbb{R}$ *be given. Suppose there exists* $0 < \delta \leq \pi$ *such that*

(i) $\displaystyle\int_0^\delta \frac{|\phi(x_0, t)|}{t}\, dt < \infty$,

where ϕ *is as in (8.42). Then* $\lim_{n \to \infty} S_n(f, x_0) = f(x_0)$. *That is, the Fourier series of* f *converges to* f *at* x_0.

Proof If we take $X = \{x_0\}$, the set whose only member is x_0, it follows from (i) and (8.13) (applied to the function g such that $g(t) = t^{-1}\phi(x_0, t)$ if $0 < t < \delta$ and $g(t) = 0$ elsewhere) that condition (8.42.ii) is fulfilled. ☐

(8.44)   Corollary *If* $f \in L_1(\mathbb{T})$ $x_0 \in \mathbb{R}$, *and*

(i) $|f(x_0 + t) - f(x_0)| \leq M|t|^\alpha$ *whenever* $0 < |t| < \delta$

for some positive real numbers α, δ, *and* M *not depending on* t, *then* $S_n(f, x_0)$ $\to f(x_0)$ *as* $n \to \infty$.

Proof Condition (8.43.i) is satisfied because $t^{-1}|\phi(x_0, t)| \leq 2Mt^{\alpha - 1}$ for $0 < t < \delta$ and $\displaystyle\int_0^\delta t^{\alpha - 1}\, dt = \delta^\alpha / \alpha < \infty$. ☐

(8.45)   Corollary *If* $f \in L_1(\mathbb{T})$, $x_0 \in \mathbb{R}$, *and* f *is both right and left differentiable at* x_0, *then* $S_n(f, x_0) \to f(x_0)$ *as* $n \to \infty$.

Proof   Condition (8.44.i) is satisfied with $\alpha = 1$ and M such that $\max\{|f'_+(x_0)|, |f'_-(x_0)|\} < M < \infty$ if we take δ sufficiently small. $\square$

We base the proof of our next convergence test on Fejér's Theorem and Hardy's Tauberian Theorem, so we need the following fact.

(8.46)   Theorem   Let $f \in L_1(\mathbb{T})$ satisfy $V_0^{2\pi} f < \infty$. In this case, we write $f \in BV(\mathbb{T})$. Then

   (i) $|n\hat{f}(n)| \leq (1/4)V_0^{2\pi} f$ for all $n \in \mathbb{Z}$.

Proof   Given $k, n \in \mathbb{Z}$ with $n \neq 0$, we have

$$\hat{f}(n) = \frac{1}{2\pi} \int_0^{2\pi} f(u)e^{-inu}\, du = \frac{(-1)^k}{2\pi} \int_0^{2\pi} f\left(t + \frac{k\pi}{n}\right)e^{-int}\, dt,$$

so

$$2\hat{f}(n) = \frac{(-1)^k}{2\pi} \int_0^{2\pi} \left[f\left(t + \frac{k\pi}{n}\right) - f\left(t + \frac{(k-1)\pi}{n}\right)\right]e^{-int}\, dt,$$

and therefore

$$4|n\hat{f}(n)| \leq \frac{1}{2\pi} \int_0^{2\pi} \sum_{k=1}^{2|n|} \left|f\left(t + \frac{k\pi}{n}\right) - f\left(t + \frac{(k-1)\pi}{n}\right)\right| dt.$$

But the last integrand does not exceed $V_0^{2\pi} f$, so (i) follows. $\square$

The next is probably the most useful test for convergence of Fourier series.

(8.47)   Jordan's Test   Suppose that $f \in L_1(\mathbb{T})$, $a < b$ in $\mathbb{R}$, and $V_a^b f < \infty$. If $a < x < b$, then

   (i) $\displaystyle\sum_{k=-\infty}^{\infty} \hat{f}(k)e^{ikx} = \lim_{n \to \infty} S_n(f, x) = \frac{f(x-) + f(x+)}{2}$.

If f is continuous on $]a, b[$ and $J \subset]a, b[$ is a closed interval, then

   (ii) $S_n(f) \to f$ as $n \to \infty$ uniformly on J.

Proof   Choose any $g \in L_1(\mathbb{T})$ such that $g = f$ on $[a, b]$ and $V_0^{2\pi} g < \infty$ [if $b - a \geq 2\pi$, take $g = f$; if $b - a < 2\pi$, take $g = 0$ on $]b, a + 2\pi[$]. We infer from (8.46) that

$$\left|k\left[\hat{g}(-k)e^{-ikx} + \hat{g}(k)e^{ikx}\right]\right| \leq \frac{1}{2} V_0^{2\pi} g = C < \infty$$

for all $k \in \mathbb{N}$ and all $x \in \mathbb{R}$. Thus, the Tauberian hypothesis (7.91.i) is satisfied by the Fourier series of g. Applying Fejér's Theorem (8.30) and Hardy's Tauberian Theorem (7.92) to g and its Fourier series $[X = \{x\}$ and

$X = J$, respectively] we see that the present assertions obtain with g in place of f. Since $f = g$ on $[a, b]$, we need only apply (8.41) with $I = \,]a, b[$ to obtain our conclusions for f instead of g. □

(8.48)   Corollary   *If $f \in C(\mathbb{T})$ and $V_0^{2\pi} f < \infty$, in which case we write $f \in CBV(\mathbb{T})$ then $S_n(f) \to f$ uniformly on $\mathbb{R}$ as $n \to \infty$.*

Proof   Take $[a, b] = [-\pi, 3\pi]$ and $J = [0, 2\pi]$. □

(8.49)   Remarks   (a) Before Jordan's time, Dirichlet had obtained (8.47.i) for all x in $\mathbb{R}$ under the hypotheses that f be bounded, real-valued, 2π-periodic, continuous except at a finite number of points in $[0, 2\pi]$, and have only a finite number of local maxima and minima in that interval. The set of functions having these properties is of course a very important subset of $BV(\mathbb{T})$. For this reason, (8.47) is often called the *Dirichlet–Jordan Theorem*.

(b) As our proof of (8.47) shows, its assertions are true of any $f \in L_1(\mathbb{T})$ for which $\sup\{|k\hat{f}(k)| : k \in \mathbb{Z}\} < \infty$ at any x where $f(x +\,)$ and $f(x -\,)$ both exist in $\mathbb{C}$. We simply take $g = f$.

The next theorem, which may be regarded as optional, is very powerful and has a correspondingly delicate proof. It is fair to say that nearly all of the known tests for convergence of Fourier series can be deduced from it. We include it here as a good example of the "divide and conquer" maxim of analysis.

(8.50)   Lebesgue–Gergen Test   *Let $f \in L_1(\mathbb{T})$, $X \subset \mathbb{R}$, and $0 < \delta' < \pi$ be given. Write $\phi(t) = \phi(x, t) = f(x + t) + f(x - t) - 2f(x)$. Suppose that f is bounded on X,*

(i) $$\lim_{h \downarrow 0} \frac{1}{h} \int_0^h \phi(t)\, dt = 0 \quad \text{uniformly for } x \in X,$$

and

(ii) $$\lim_{h \downarrow 0} \int_h^{\delta'} t^{-1} |\phi(t + h) - \phi(t)|\, dt = 0 \quad \text{uniformly for } x \in X.$$

Then $S_n(f, x) \to f(x)$ uniformly on X as $n \to \infty$.

Proof   With no harm done, we suppose that f is real-valued (which we use to get (11)). Let $\epsilon > 0$ be given. Write $\Phi(u) = \int_0^u \phi(t)\, dt$. Now use (i) and (ii) to obtain $0 < \delta < \delta'/2$ such that, for all $x \in X$, we have

$$|u^{-1}\Phi(u)| < \epsilon \quad \text{if } 0 < u \leq \delta \tag{1}$$

and

$$\int_h^{\delta'} t^{-1} |\phi(t + h) - \phi(t)|\, dt < \epsilon \quad \text{if } 0 < h < \delta. \tag{2}$$

Letting $g(s) = \int_0^s |f(u)|\,du$ and choosing $M < \infty$ such that $|f(x)| < M$ for all $x \in X$, we see that $h > 0$ and $x \in X$ imply that

$$\int_\delta^{\delta+h} t^{-1}|\phi(t)|\,dt \leq \delta^{-1}\int_\delta^{\delta+h}\big[|f(x+t)| + |f(x-t)| + 2|f(x)|\big]\,dt$$

$$\leq \delta^{-1}(2\omega_g(h) + 2Mh),$$

where $\omega_g(h) = \sup\{|g(u) - g(v)| : |u - v| \leq h\}$. Since g is uniformly continuous on $\mathbb{R}$, $\omega_g(h) \to 0$ as $h\downarrow 0$. Thus there exists $N \in \mathbb{N}$ with $N > 2\pi/\delta$ such that, for all $x \in X$,

$$\int_\delta^{\delta+h} t^{-1}|\phi(t)|\,dt < \epsilon \quad \text{if } h = \tfrac{\pi}{n} \quad \text{and} \quad n \geq N. \tag{3}$$

Now fix any $n > N$. In the remainder of this proof we take $h = \pi/n$. Note that $h < \delta/2$ and $\delta + h < \delta' \leq \pi$.

An integration by parts (6.90) yields

$$\int_a^b \phi(t)\,\frac{\sin(nt)}{t}\,dt = t^{-1}\Phi(t)\sin(nt)\Big|_a^b - \int_a^b \big[t^{-1}\Phi(t)\big]\{n\cos(nt) - t^{-1}\sin(nt)\}\,dt$$

so, since $|n\cos(nt) - t^{-1}\sin(nt)| \leq 2n$, (1) shows that, for all $x \in X$,

$$\left|\int_a^b \phi(t)\,\frac{\sin(nt)}{t}\,dt\right| < 2\epsilon + 2n\epsilon(b - a) \leq (4\pi + 2)\epsilon \quad \text{if } 0 \leq a < b \leq 2h. \tag{4}$$

Our plan is to now use only (1)–(4), which hold good for all $x \in X$, to show that, for our fixed arbitrary $n > N$, we have

$$\left|\int_0^\delta \frac{\phi(t)}{t}\sin(nt)\,dt\right| < 29\epsilon. \tag{5}$$

It will then follow from (8.42) that $S_n(f) \to f$ uniformly on X as desired. Plainly (4) yields

$$\left|\int_0^h \frac{\phi(t)}{t}\sin(nt)\,dt\right| < 15\epsilon. \tag{6}$$

Let us write

$$I = \int_h^\delta \frac{\phi(t)}{t}\sin(nt)\,dt = \int_{2h}^{\delta+h}\frac{\phi(t)}{t}\sin(nt)\,dt + \alpha, \tag{7}$$

where $\alpha = \int_h^\delta - \int_{2h}^{\delta+h} = \int_h^{2h} - \int_\delta^{\delta+h}$. Using (4) and (3), we have

$$|\alpha| < (4\pi + 2)\epsilon + \epsilon < 16\epsilon. \tag{8}$$

From (7) and $\sin(n(u + h)) = -\sin(nu)$, we have

$$2I - \alpha = \int_h^\delta \left[\frac{\phi(u)}{u} - \frac{\phi(u + h)}{u + h} \right] \sin(nu)\, du$$

$$= \int_h^\delta \frac{\phi(u) - \phi(u + h)}{u + h} \sin(nu)\, du$$

$$+ h \int_h^{2h} \frac{\phi(u)\sin(nu)}{u(u + h)}\, du + h \int_{2h}^\delta \frac{\phi(u)\sin(nu)}{u(u + h)}\, du$$

$$= I_1 + hI_2 + hI_3. \tag{9}$$

From (2), we have

$$|I_1| < \epsilon. \tag{10}$$

We now use (6.101) (with the decreasing function $(u + h)^{-1}$ redefined to be 0 at $2h$) and invoke (4) to obtain

$$|hI_2| = \left| \frac{1}{2} \int_h^\xi \frac{\phi(u)\sin(nu)}{u}\, du \right| < 8\epsilon. \tag{11}$$

Next we have

$$hI_3 = -h \int_h^{\delta - h} \frac{\phi(t + h)\sin(nt)}{(t + h)(t + 2h)}\, dt = -h \int_h^\delta \frac{\phi(t + h)\sin(nt)}{(t + h)(t + 2h)}\, dt + \beta \tag{12}$$

where, using (3), we have

$$|\beta| = \left| h \int_\delta^{\delta + h} \frac{\phi(u)\sin(nu)}{u(u + h)}\, du \right| \le \frac{h}{\delta + h} \int_\delta^{\delta + h} u^{-1}|\phi(u)|\, du < \epsilon. \tag{13}$$

Now (12) and (9) give

$$hI_2 + 2hI_3 - \beta = h \int_h^\delta \left[\frac{\phi(t)}{t(t + h)} - \frac{\phi(t + h)}{(t + h)(t + 2h)} \right] \sin(nt)\, dt = A + B, \tag{14}$$

where

$$A = h \int_h^\delta \frac{\phi(t) - \phi(t + h)}{(t + h)(t + 2h)} \sin(nt)\, dt, \qquad B = 2h^2 \int_h^\delta \frac{\phi(t)\sin(nt)}{t(t + h)(t + 2h)}\, dt.$$

Plainly (2) yields

$$|A| \le h \int_h^\delta \frac{|\phi(t) - \phi(t + h)|}{t \cdot 3h}\, dt < \frac{\epsilon}{3}. \tag{15}$$

Another integration by parts gives

$$B = \frac{2h^2 \Phi(\delta)\sin(n\delta)}{\delta(\delta + h)(\delta + 2h)} - 2h^2 \int_h^\delta \Phi(t) \frac{d}{dt} \left[\frac{\sin(nt)}{t(t + h)(t + 2h)} \right] dt.$$

One checks that the indicated derivative here has absolute value less than $nt^{-3} + 3t^{-4}$, and so we obtain from (1) that

$$|B| < \epsilon + 2h^2 \int_h^\delta \epsilon \cdot (nt^{-2} + 3t^{-3})\, dt < \epsilon + 2\pi\epsilon + 3\epsilon < 11\epsilon. \qquad (16)$$

We combine (9) and (14) to see that

$$4I = 2\alpha + 2I_1 + hI_2 + \beta + A + B$$

and then we use (8), (10), (11), (13), (15), and (16) in that order to get

$$|4I| < 32\epsilon + 2\epsilon + 8\epsilon + \epsilon + \epsilon + 11\epsilon < 56\epsilon.$$

This result, (7), and (6) imply (5). $\square$

After Jordan's Test, the next test is probably the most useful test for *uniform* convergence. We deduce it easily from (8.50), even though a direct proof would be less difficult than proving (8.50) first.

(8.51)   Dini–Lipschitz Test   *Let $f \in L_1(\mathbb{T})$ and let $I \subset \mathbb{R}$ be a nonvoid open interval. Suppose that*

(i) $$\lim_{t \to 0}\sup_{x \in I} (|f(x + t) - f(x)|\log(1/|t|)) = 0.$$

Then $S_n(f) \to f$ as $n \to \infty$ uniformly on each closed interval $J \subset I$.

Proof   Let such a J be given and select $0 < \delta' < 1$ such that $x \pm t \in I$ whenever $x \in J$ and $0 < t \leq \delta'$. We check the hypotheses of (8.50) with $X = J$. Clearly it follows from (i) that f is continuous on I, so f is bounded on X. Given $\epsilon > 0$, use (i) to obtain $0 < \delta < \min\{\delta', e^{-1}\}$ such that $|f(u + v) - f(u)| < (\log(1/|v|))^{-1}(\epsilon/2)$ if $u \in I$ and $0 < |v| < \delta$. Then $0 < t < \delta$ and $x \in I$ imply that

$$|\phi(t)| \leq |f(x + t) - f(x)| + |f(x - t) - f(x)| < (\log t^{-1})^{-1}\epsilon < \epsilon$$

because $\delta < 1/e$. In particular,

$$\left|\frac{1}{h}\int_0^h \phi(t)\, dt\right| < \epsilon \quad \text{if } x \in X \quad \text{and} \quad 0 < h < \delta,$$

which establishes (8.50.i). If $x \in X$, $0 < t < \delta'$, and $0 < h < \delta$, then $|\phi(t + h) - \phi(t)| \leq |f(x + t + h) - f(x + t)| + |f(x - t - h) - f(x - t)| < (\log h^{-1})^{-1}\epsilon$ because $x \pm t \in I$. Therefore, since $\log \delta' < 0$ and $-\log h = \log h^{-1}$,

$$\int_h^{\delta'} t^{-1}|\phi(t + h) - \phi(t)|\, dt < (\log h^{-1})^{-1}\epsilon \cdot \int_h^{\delta'} t^{-1}\, dt < \epsilon$$

whenever $x \in X$ and $0 < h < \delta$, which establishes (8.50.ii). $\square$

(8.52)   Remarks   (a) As is clear from the proof of (8.50), we only need (8.50.ii) for h of the form π/n with $n \in \mathbb{N}$.

(b) The weaker version of (8.50) in which $|\phi(t)|$ replaces $\phi(t)$ in (8.50.i) is due to Lebesgue and is known as Lebesgue's Test. The three most important tests, (8.43), (8.47), and (8.51), can all be deduced from it. The drawback is that (8.50.ii) is often hard to check. The improved version presented here is due to J. J. Gergen (1930).

(c) Hypothesis (8.50.i) is very often satisfied. It is equivalent to the requirement that the indefinite integral $F(u) = \int_0^u f$ satisfy

$$\lim_{h\to 0} \frac{F(x+h)-F(x-h)}{2h} = f(x) \quad \text{uniformly on } X.$$

This certainly obtains at any x where $F'(x) = f(x)$ and so a.e. The above limit, if it exists, is known as the *first symmetric derivative* of F at x. Lebesgue's stronger hypothesis mentioned in (b) is, when $X = \{x\}$, just the requirement that x be a Lebesgue point for f.

(d) The Dini–Lipschitz Test is *not* a test for convergence at individual points. Indeed, there exist functions $f \in C(\mathbb{T})$ whose Fourier series diverge at $x = 0$ and yet (8.51.i) is satisfied for $I = \{0\}$ (see Exercise 10(d) on page 559).

(e) Hypothesis (8.51.i) is "best possible" in the sense that there exist functions $f \in C(\mathbb{T})$ such that

$$\sup_{0<|t|<1}\left\{\sup_{x\in\mathbb{R}}\left(|f(x+t)-f(x)|\log\frac{1}{|t|}\right)\right\} < \infty$$

and yet $S(f)$ converges uniformly on *no* nonvoid open interval or even diverges at some points (see Exercise 2(i) on page 532).

Growth Rate of Partial Sums

(8.53) Definition   The *Lebesgue constants* are the numbers

$$L_n = \frac{1}{\pi}\int_0^\pi |D_n(t)|\,dt = \frac{1}{2\pi}\int_{-\pi}^\pi |D_n(t)|\,dt = \|D_n\|_1,$$

where, as usual, D_n is the Dirichlet kernel.

(8.54) Theorem   *The Lebesgue constants satisfy*

(i) $\lim_{n\to\infty}\left[L_n/(\log n)\right] = 4/\pi^2.$

In fact, writing $A = 1 + 2/\pi + 4/\pi^2$, *we have*

(ii) $(4/\pi^2)\log n - A < L_n < (4/\pi^2)\log n + A + 2 \quad \text{for all } n \geq 1.$

Proof   As in the proof of (8.40), we have

$$D_n(t) = \frac{2\sin(nt)}{t} + \left[\cot((1/2)t) - 2/t\right]\sin(nt) + \cos(nt)$$

and the expression in brackets decreases from 0 to $-2/\pi$ as t increases from 0 to π. Therefore

$$2t^{-1}|\sin(nt)| - (2/\pi + 1) \leq |D_n(t)| \leq 2t^{-1}|\sin(nt)| + (2/\pi + 1)$$

for $0 < t \leq \pi$ and so, writing $C = 1 + \pi/2$, we have

$$I_n - C \leq \frac{\pi}{2} L_n \leq I_n + C, \quad \text{where } I_n = \int_0^\pi \frac{|\sin(nt)|}{t} \, dt. \tag{1}$$

It is easier to estimate I_n than L_n. Indeed, we have

$$I_n = \int_0^{n\pi} \frac{|\sin x|}{x} \, dx, \qquad I_{n+1} - I_n = \int_{n\pi}^{(n+1)\pi} \frac{|\sin x|}{x} \, dx = \int_0^\pi \frac{\sin u}{u + n\pi} \, du,$$

and

$$\frac{2}{(n+1)\pi} = \int_0^\pi \frac{\sin u}{\pi + n\pi} \, du < I_{n+1} - I_n < \int_0^\pi \frac{\sin u}{0 + n\pi} \, du = \frac{2}{n\pi}.$$

Adding these inequalities for $n = 1, 2, \ldots, p - 1$, we obtain

$$\frac{2}{\pi} \sum_{n=2}^p \frac{1}{n} = \frac{2}{\pi} \sum_{n=1}^{p-1} \frac{1}{n+1} < I_p - I_1 < \frac{2}{\pi} \sum_{n=1}^{p-1} \frac{1}{n}.$$

But, using (7.4) and (7.5.a) if necessary, we get

$$\log p - 1 < \frac{1}{p} + \int_2^p \frac{dx}{x} \leq \sum_{n=2}^p \frac{1}{n},$$

$$\sum_{n=1}^{p-1} \frac{1}{n} \leq 1 + \int_1^{p-1} \frac{dx}{x} < 1 + \log p,$$

and hence

$$\frac{2}{\pi} \log p - \frac{2}{\pi} < I_p < \frac{2}{\pi} + \frac{2}{\pi} \log p + I_1 \tag{2}$$

for all $p \geq 2$. Comparing (1) and (2) and noting that $I_1 < \pi$ because $\sin t < t$ $(t > 0)$, we obtain (ii) and thence (i).   □

(8.55) Corollary *If* $f \in L_1(\mathbb{T})$ *is bounded:* $|f(x)| \leq M < \infty$ *for all* x, *then* $|S_n(f, x)| \leq ((4/\pi^2)\log n + A + 2)M$ *for all* $x \in \mathbb{R}$ *and* $n \geq 1$.

Proof Apply (8.27.ii) and (8.54).   □

The next result is a vast a.e. improvement on (8.55). Though Fourier series can diverge unboundedly at every point, this divergence is rather slow.

(8.56) Theorem [Lebesgue, 1909] *Let* $f \in L_1(\mathbb{T})$ *be arbitrary. Then*

(i) $$\lim_{n \to \infty} \left[S_n(f, x)/(\log n) \right] = 0$$

for almost every $x \in \mathbb{R}$. Indeed, (i) *obtains at every Lebesgue point of f (see* (8.33)). *Moreover, if f is continuous on some open interval I, then the convergence in* (i) *is uniform on every closed interval $J \subset I$.*

Proof   Write $\phi(x, t) = \phi(t) = f(x + t) + f(x - t) - 2f(x)$ and $\Phi(x, h) = \Phi(h) = \int_0^h |\phi(t)| \, dt$. Let x, a Lebesgue point for f, and $\epsilon > 0$ be given. By (8.33), there exists $0 < \delta < 1$ such that

$$h^{-1}\Phi(h) < \epsilon \quad \text{if } 0 < h \leq \delta. \tag{1}$$

[If f is continuous on I, then uniform continuity on J provides δ such that (1) obtains for all $x \in J$.] By (8.40) we have

$$S_n(f, x) - f(x) = \frac{1}{\pi} \int_0^\delta \phi(t) t^{-1} \sin(nt) \, dt + \epsilon_n(x) - f(x)\epsilon_n', \tag{2}$$

where the last two terms tend to 0 uniformly on any set where f is bounded. Now choose $N > \delta^{-1} > 1$ such that

$$2\epsilon/\pi + |\epsilon_n(x) - \epsilon_n' f(x)| + |f(x)| < (\epsilon/2)\log n \tag{3}$$

if $n > N$ [if f is continuous on I, then f is bounded on J, so N can be chosen so that (3) obtains for all $x \in J$ and $n > N$]. For any $n > N$, we use (1) and integration by parts to write

$$\left| \int_0^\delta \phi(t) t^{-1} \sin(nt) \, dt \right| \leq \int_0^{1/n} |\phi(t)| n \, dt + \int_{1/n}^\delta |\phi(t)| t^{-1} \, dt$$

$$= n\Phi\left(\frac{1}{n}\right) + \left[t^{-1}\Phi(t) \right]\Big|_{1/n}^\delta + \int_{1/n}^\delta t^{-2}\Phi(t) \, dt$$

$$< \epsilon + \delta^{-1}\Phi(\delta) + \int_{1/n}^\delta t^{-1}\epsilon \, dt < 2\epsilon + \epsilon \log n.$$

Thus, $n > N$ implies

$$\left| \frac{1}{\pi} \int_0^\delta \phi(t) t^{-1} \sin(nt) \, dt \right| < \frac{2\epsilon}{\pi} + \frac{\epsilon}{\pi} \log n. \tag{4}$$

Combining (2), (4), and (3), we have

$$|S_n(f, x)| < (\epsilon/\pi)\log n + (\epsilon/2)\log n < \epsilon \log n$$

whenever $n > N$ [and this holds for all $x \in J$ if f is continuous on I]. $\square$

For functions of bounded variation, the partial sums (which we know to converge pointwise by Jordan's Test) are uniformly bounded as we now show.

(8.57)   **Theorem**   *If $f \in BV(\mathbb{T})$ (see* (8.46)), *then*

$$|S_n(f, x)| \leq (1/2)V + \|f\|_u$$

for all $x \in \mathbb{R}$ and all $n \geq 0$, where $V = V_0^{2\pi} f$.

Proof Using (8.27) and (8.46), we have

$$|S_n(f,x) - \sigma_n(f,x)| = \frac{1}{n+1}\left|\sum_{k=-n}^{n} |k|\hat{f}(k)e^{ikx}\right| \le \frac{1}{n+1}\sum_{k=-n}^{n}|k\hat{f}(k)|$$

$$\le \frac{2}{n+1}\sum_{k=1}^{n}\frac{1}{4}V \le \frac{1}{2}V.$$

Now invoke (8.38). □

(8.58)   Remarks It is not possible to replace the term $(1/2)V$ in (8.57) with a term that tends to 0 as $n\to\infty$ and does not depend on x. In fact, let us consider the function $f \in BV(\mathbb{T})$ given by $f(0) = 0$ and $f(t) = (\pi - t)/2$ for $0 < t < 2\pi$. By (8.9), we have

$$S_n(f,x) = \int_0^x \sum_{k=1}^{n} \cos(kt)\,dt = \frac{1}{2}\int_0^x [D_n(t) - 1]\,dt = -\frac{x}{2} + \frac{1}{2}\int_0^x D_n(t)\,dt.$$

As in (8.54), we have

$$(1/2)D_n(t) = [\sin(nt)]/t + g(t)\sin(nt) + (1/2)\cos(nt),$$

where g decreases from 0 to $-1/\pi$ on $[0,\pi]$. Thus, for $0 < x \le \pi$, (6.101) yields

$$\left|S_n(f,x) - \int_0^x \frac{\sin(nt)}{t}\,dt\right| = \left|-\frac{x}{2} + g(x)\int_\xi^x \sin(nt)\,dt + \frac{1}{2}\int_0^x \cos(nt)\,dt\right|$$

$$\le x/2 + 2/(n\pi) + |\sin(nx)|/(2n)$$

and so, taking $x = \pi/n$, we get

$$\left|S_n\left(f,\frac{\pi}{n}\right) - \int_0^\pi \frac{\sin u}{u}\,du\right| \le \frac{1}{n}\left(\frac{\pi}{2} + \frac{2}{\pi}\right).$$

It follows that

$$\lim_{n\to\infty} S_n\left(f,\frac{\pi}{n}\right) = \int_0^\pi \frac{\sin u}{u}\,du = y_0.$$

One checks that $u^{-1}\sin u > 1 - u/\pi$ for $0 < u < \pi$; hence $y_0 > \pi/2 = \|f\|_u$. Computation gives $y_0 = (1.179\ldots)\pi/2$.

This example is an illustration of the *Gibbs Phenomenon*: the functions $S_n(f)$ converge to f everywhere on $]0,\pi[$ but they "overshoot the mark near 0" in the sense that the union of the graphs $\bigcup_{n=1}^{\infty} \{(x, S_n(f,x))\colon 0 < x < \pi\}$ has in its closure (in $\mathbb{R}^2$) a point $(0, y_0)$ where $y_0 > f(0+) = \pi/2$. In fact, they overshoot by nearly 18% and the entire segment $[0, y_0]$ on the y-axis is in this closure.

Exercises

1.   (a) If $f \in L_1(\mathbb{T})$ is even $(f(-t) = f(t))$ and $f_1 \in L_1(\mathbb{T})$ is such that $f_1(t) = f(t)$ for $-\pi < t < 0$ and $f_1(t) = 0$ for $0 < t < \pi$, then $S_n(f,0) = 2S_n(f_1,0)$ for all $n \ge 0$.

(b) There exists a function $f_1 \in C(\mathbb{T})$ such that $f_1(t) = 0$ for $0 \leq t \leq \pi$ and the Fourier series of f_1 diverges at $x = 0$. [Hint: Let f be as in Exercise 2(f) following (8.26).]

(c) Any function f_1 as in (b) has Fourier series *uniformly* convergent to 0 on every closed interval $J \subset]0, \pi[$ but *not* on $]0, \pi[$ itself. Thus (8.41) can fail for $J = I$. [Hint: Take $f_2 = 0$.]

2.   Consider any trigonometric series and its C_1-means (Fejér sums):

$$S = \sum_{k=-\infty}^{\infty} c_k e^{ikt}, \qquad \sigma_n(t) = \sum_{k=-n}^{n} \left(1 - \frac{|k|}{n+1}\right) c_k e^{ikt}.$$

(a) Series S is the Fourier series of some $f \in C(\mathbb{T})$ if and only if $(\sigma_n)_{n=0}^{\infty}$ converges uniformly on $\mathbb{R}$. [Hint: If $\sigma_n \to f$ uniformly, we may interchange $\int_{-\pi}^{\pi}$ and $\lim_{n\to\infty}$ when computing $\hat{f}(k)$.]

(b) Series S is the Fourier series of some bounded $f \in L_1(\mathbb{T})$ if and only if $\sup\{\|\sigma_n\|_u : n \in \mathbb{N}\} < \infty$. [Hint: If the sup is $\beta < \infty$, obtain

$$\beta^2 \geq \|\sigma_n\|_2^2 \geq \sum_{k=-p}^{p} \left(1 - \frac{|k|}{n+1}\right)^2 |c_k|^2$$

whenever $n \geq p \geq 0$, let $n \to \infty$, and find $f \in L_2(\mathbb{T})$ with $\hat{f}(k) = c_k$ for all k.]

(c) Series S is the Fourier series of some $f \in L_2(\mathbb{T})$ if and only if $\sup\{\|\sigma_n\|_2 : n \in \mathbb{N}\} < \infty$.

(d) If $c_k = 1$ for all k, then $\sigma_n = K_n$ for all n, so $\sup\{\|\sigma_n\|_1 : n \in \mathbb{N}\} < \infty$, but S is *not* the Fourier series of any $f \in L_1(\mathbb{T})$.

(e) If $f \in L_p(\mathbb{T})$, where $1 \leq p < \infty$, then

$$\lim_{t \to 0} \int_{-\pi}^{\pi} |f(x - t) - f(x)|^p \, dx = 0.$$

[Hint: First consider $f \in C(\mathbb{T})$ and then approximate.]

(f) Series S is the Fourier series of some $f \in L_1(\mathbb{T})$ if and only if $\lim_{m,n\to\infty} \|\sigma_m - \sigma_n\|_1 = 0$. [Hints: If $f \in L_1(\mathbb{T})$, use (e), Fubini's Theorem and (8.29.iv) to show that $\|f - \sigma_n(f)\|_1 \to 0$. Conversely, use (6.56) to produce f and then compute $\hat{f}(k)$ as in (a).]

(g) Series S is the Fourier series of some $f \in BV(\mathbb{T})$ if and only if the derivatives σ_n' satisfy $\sup\{\|\sigma_n'\|_1 : n \in \mathbb{N}\} < \infty$. [Hints: Use (6.103) to see that $2\pi\|\sigma_n'\|_1 = V_0^{2\pi}\sigma_n$. If these are bounded by β, use (8.46) to obtain $|kc_k| \leq \beta/4$, and so invoke (8.17) to obtain $g \in L_2(\mathbb{T})$ such that $\sigma_n = \sigma_n(g)$ for all n. If $0 \leq x_0 < \cdots < x_p \leq 2\pi$ are Lebesgue points for g, then (8.35) yields $\sum_{j=1}^{p} |g(x_j) - g(x_{j-1})| \leq \beta$. Take $f = g$ on the Lebesgue set of g and define f on the remaining null set so that $f \in BV(\mathbb{T})$. Conversely, if $f \in BV(\mathbb{T})$ and $\hat{f}(k) = c_k$ for all k, one uses (8.27.iii) to check directly that $V_0^{2\pi}\sigma_n \leq V_0^{2\pi} f$ for all n.]

3.   Let $f \in L_1(\mathbb{T})$, $x_0 \in \mathbb{R}$, $0 < \alpha \leq 1$, and $0 < C < \infty$ be given. Suppose that, for all $t \in \mathbb{R}$, we have

$$|f(x_0 - t) - f(x_0)| \leq C|t|^\alpha.$$

(a) If $\alpha < 1$ and $n \geq 0$, then

$$|\sigma_n(f, x_0) - f(x_0)| < 2C\pi^\alpha / \left[(1 - \alpha^2)(n + 1)^\alpha\right].$$

(b) If $\alpha = 1$ and $n \geq 0$, then

$$|\sigma_n(f, x_0) - f(x_0)| < [C\pi/(n+1)][1/2 + \log(n+1)].$$

[Hint: In both cases, write the difference as an integral over $[0, \pi]$ (see (8.30.1)), split this interval at $\pi/(n+1)$, and estimate K_n by using (8.29).]

4. (a) If $f \in \text{Lip}_\alpha(\mathbb{T})$ (that is, $f \in C(\mathbb{T})$, $\alpha > 0$, $C < \infty$, and $|f(s) - f(t)| \leq C|s - t|^\alpha$ for all $s, t \in \mathbb{R}$), then $2|n|^\alpha|\hat{f}(n)| \leq C\pi^\alpha$ for all $n \in \mathbb{Z}$. [Hint: Use the idea in (8.46).]

(b) If $0 < \alpha < 1$, then the function f defined by

$$f(x) = \sum_{n=1}^\infty \frac{\sin(2^n x)}{2^{\alpha n}}$$

is in $\text{Lip}_\alpha(\mathbb{T})$. [See Exercise 8(n), on p. 521.]

(c) The conclusion in (a) cannot be changed to $\lim_{|n| \to \infty} |n|^\alpha |\hat{f}(n)| = 0$ when $0 < \alpha < 1$.

(d) If $\alpha = 1$, then the function in (b) is *not* in $\text{Lip}_\alpha(\mathbb{T})$. [Hint: $D^+ f(0) = \infty$.]

5. If P is a trigonometric polynomial of degree $n \geq 1$ ($\hat{P}(k) = 0$ if $|k| > n$, $|\hat{P}(n)| + |\hat{P}(-n)| > 0$), then the derivative P' satisfies

$$P'(x) = -\frac{n}{\pi} \int_{-\pi}^\pi P(x - t) K_{n-1}(t) \sin(nt) \, dt$$

for all $x \in \mathbb{R}$. Thus $\|P'\|_u \leq 2n\|P\|_u$ and $\|P'\|_1 \leq 2n\|P\|_1$.

6. *Divergence in $L_1(\mathbb{T})$.* This exercise outlines an example due to F. Riesz of a function $f \in L_1(\mathbb{T})$ whose Fourier series does not converge in the L_1-metric. We remark that there is a theorem of M. Riesz lying beyond the scope of this book which assures that $\lim_{n \to \infty} \|f - S_n(f)\|_p = 0$ if $f \in L_p(\mathbb{T})$ *and* $1 < p < \infty$. For $p = 2$ this is trivial (see (8.21)) and for $p = 1$ it is false, which is the point of this exercise.

Consider polynomials P_n and Q_n defined for $n = 0, 1, 2, \ldots$ by

$$P_n(z) = \frac{1}{n+1} \left[\sum_{k=0}^n z^k \right]^2, \qquad Q_n(z) = \frac{1}{n+1} \sum_{k=0}^n (k+1) z^k.$$

(a) For all $n \geq 0$, we have

$$P_n(z) = \frac{1}{n+1} \left[\frac{1 - z^{n+1}}{1 - z} \right]^2 = Q_n(z) + \frac{1}{n+1} \sum_{k=n+1}^{2n} (2n + 1 - k) z^k$$

(b) For $t \in \mathbb{R}$ and $n \geq 0$, we have

$$2\text{Re}\left[Q_n(e^{it}) \right] = \frac{n+2}{n+1} D_n(t) + \frac{1}{n+1} - K_n(t).$$

(c) For $n \geq 0$, we have

$$2\|Q_n(e^{it})\|_1 = \frac{1}{\pi} \int_{-\pi}^\pi |Q_n(e^{it})| \, dt > L_n - 2 > \frac{4}{\pi^2} \log n - 5,$$

where the equality defines its left side. [Hint: See (8.54).]

(d) For all $t \in \mathbb{R}$ and all $n \geq 0$, we have

$$|P_n(e^{it})| = K_n(t).$$

Now fix sequences $(c_k)_{k=1}^\infty \subset \mathbb{C}$, $(m_k)_{k=1}^\infty \subset \mathbb{N}$, and $(n_k)_{k=1}^\infty \subset \mathbb{N}$ such that

$$\sum_{k=1}^\infty |c_k| < \infty, \quad \lim_{k \to \infty} |c_k \log n_k| = \infty, \quad m_{k+1} > m_k + 2n_k \quad \text{for all } k.$$

(e) The choices $c_k = 2^{-k}$ and $m_k = n_k = 2^{3^k}$ have these properties.

(f) The formula

$$f(t) = \sum_{k=1}^{\infty} c_k e^{im_k t} P_{n_k}(e^{it})$$

defines a function $f \in L_1(\mathbb{T})$, this series is absolutely and uniformly convergent on $[\delta, 2\pi - \delta]$ whenever $0 < \delta < \pi$, its pth partial sums f_p satisfy $\lim_{p \to \infty} \|f - f_p\|_1$

$= 0$, and $\|f\|_1 \leq \sum_{k=1}^{\infty} |c_k|$. [Hint: Use (d) and (8.29).]

(g) For all $p \in \mathbb{N}$, $p > 1$, we have

$$S_{n_p + m_p}(f, t) = f_{p-1}(t) + c_p e^{im_p t} Q_{n_p}(e^{it}),$$
$$S_{m_{p-1} + 2n_{p-1}}(f, t) = f_{p-1}(t).$$

(h) It is *false* that $\lim_{m,n \to \infty} \|S_m(f) - S_n(f)\|_1 = 0$.

(i) It is *false* that $\lim_{n \to \infty} \|f - S_n(f)\|_1 = 0$.

(j) The Fourier series of f diverges at $x = 0$.

7. Here is an example due to Lebesgue of an $f \in C(\mathbb{T})$ whose Fourier series diverges at 0 and converges uniformly in $[\delta, 2\pi - \delta]$ whenever $0 < \delta < \pi$. This example is perhaps more illuminating geometrically than those given above: one can "see" the graph of f.

Suppose the fixed sequences $(n_k)_{k=1}^{\infty} \subset \mathbb{N}$, $(c_k)_{k=1}^{\infty} \subset \mathbb{R}$, and $a_k = \prod_{j=1}^{k} n_j$ satisfy $n_k > 1$ for all k, $c_k \downarrow 0$, $c_k \log n_k \to \infty$, and $a_k / n_{k+1} \to 0$ as $k \to \infty$. Write $a_0 = 1$.

(a) The choices $n_k = 2^{k!}$ and $c_k = 1/k$ meet these requirements.

Now define f on $\mathbb{R}$ by $f(t) = c_k \sin(a_k t)$ if $\pi/a_k \leq t \leq \pi/a_{k-1}$ and $k \in \mathbb{N}$, $f(0) = 0$, and $f(-t) = f(t) = f(t + 2\pi)$ for all t.

(b) The function f is in $C(\mathbb{T})$.

(c) $$S_n(f, 0) = \frac{2}{\pi} \int_0^{\pi} f(t) \frac{\sin(nt)}{t} dt + \epsilon_n \quad \text{where} \quad \epsilon_n \to 0 \text{ as } n \to \infty.$$

(d) $$\left| \int_0^{\pi/a_k} f(t) \frac{\sin(a_k t)}{t} dt \right| < \pi c_{k+1}.$$

(e) $$2 \int_{\pi/a_k}^{\pi/a_{k-1}} f(t) \frac{\sin(a_k t)}{t} dt = c_k \log n_k - c_k \int_{\pi/a_k}^{\pi/a_{k-1}} \frac{\cos(2a_k t)}{t} dt.$$

(f) The last term in (e) has absolute value not exceeding $c_k/(2\pi)$. [Hint: Use (6.101) with $\alpha(\pi/a_{k-1}) = 0$.]

(g) $$\left| \int_{\pi/a_{k-1}}^{\pi} f(t) \frac{\sin(a_k t)}{t} dt \right| < \frac{c_1}{n_k} \left(\frac{1}{\pi} + a_{k-1} \right).$$ [Hint: Apply (6.101) again and then use (6.90). Estimate $|f'|$ by $c_1 a_{k-1}$.]

(h) $\lim_{k \to \infty} S_{a_k}(f, 0) = \infty$.

(i) If $0 < \delta < \pi$, then $S_n(f) \to f$ as $n \to \infty$ uniformly on $[\delta, 2\pi - \delta]$.

(j) $V_0^{\delta} f = \infty$ for all $\delta > 0$.

(k) $\overline{\lim_{t \to 0}} |f(t)| \cdot |t|^{-\alpha} = \infty$ for all $\alpha > 0$.

(1) $$\varlimsup_{h\downarrow 0}\int_h^\delta\frac{|f(t+h)-f(t)|}{t}\,dt>0\quad\text{for every }\delta>0.$$

8. This exercise shows that the conclusion of (8.46) cannot be replaced by $\lim\limits_{|n|\to\infty}|n\hat f(n)|$
= 0 even when f is continuous.

Let ψ be the Lebesgue singular function defined in terms of Cantor's ternary set P just as in (3.92). Now define f on $\mathbb{R}$ by $f(t)=f(-t)=\psi(t/\pi)$ for $0\le t\le\pi$ and $f(t+2\pi)=f(t)$ for all $t\in\mathbb{R}$. Then we have

(a) $f\in C(\mathbb{T})$ and $V_0^{2\pi}f=2$;

(b) $\psi(x)=2\psi(x/3)$ if $0\le x\le 1$;

(c) $\psi(x+2/3)=\psi(x)+1/2$ if $0\le x\le 1/3$;

(d) $\hat f(n)=\int_0^1\psi(x)\cos(n\pi x)\,dx$ for all $n\in\mathbb{Z}$;

(e) $\hat f(n)=6\int_0^{1/3}\psi(u)\cos(3n\pi u)\,du$ for all $n\in\mathbb{Z}$;

(f) $\hat f(n)=6\int_{2/3}^1\psi(t)\cos(3n\pi t)\,dt$ for all $n\neq 0$ in $\mathbb{Z}$;

(g) $\int_{1/3}^{2/3}\psi(t)\cos(3n\pi t)\,dt=0$ for all $n\neq 0$ in $\mathbb{Z}$;

(h) $\hat f(n)=3^k\,\hat f(3^k n)$ for all $k\in\mathbb{N}$ and $n\neq 0$ in $\mathbb{Z}$;

(i) $\psi(x)+\psi(1-x)=1$ if $0\le x\le 1$;

(j) $3^k\,\hat f(3^k)=\hat f(1)=-(1/\pi)+2\int_0^{1/2}\psi(x)\cos(\pi x)\,dx<-(1/\pi)+\int_0^{1/2}\cos(\pi x)\,dx$
= 0.

(k) The Fourier series of f converges to f uniformly on $\mathbb{R}$.

9. Suppose that $a<b$ in $\mathbb{R}$ and $(u_n)_{n=1}^\infty\subset L_1([a,b])$ are given. Write

$$T_n=\int_a^b|u_n(t)|\,dt\quad\text{and}\quad\tau_n(f)=\int_a^b f(t)u_n(t)\,dt$$

if $f:[a,b]\to\mathbb{C}$ is bounded and measurable. Also write $\|f\|_u=\sup\{|f(t)|:a\le t\le b\}$ for such f.

(a) For each n, there exists $f_n\in C([a,b])$ such that $\|f_n\|_u\le 1$, $f_n(a)=f_n(b)=0$, and $|\tau_n(f_n)|\ge(3/4)T_n$. [Hints: Let $v_n=\operatorname{sgn}\bar u_n$; that is, $v_n(t)=|u_n(t)|/u_n(t)$ if $u_n(t)\neq 0$ and $v_n(t)=0$ if $u_n(t)=0$. Then $\tau_n(v_n)=T_n$. Use (6.76) to find functions g_j meeting the first three conditions on f_n with $g_j:\to v_n$ a.e. and then use (6.55) to find j so that $f_n=g_j$ meets the fourth.]

(b) If $\sup\{T_n:n\in\mathbb{N}\}=\infty$, then there exists $f\in C([a,b])$ such that $\|f\|_u\le 1$, $f(a)=f(b)=0$, and $\sup\{|\tau_n(f)|:n\in\mathbb{N}\}=\infty$. [Hints: Suppose $\beta_n=\sup\{|\tau_j(f_n)|$ $:j\in\mathbb{N}\}<\infty$ for all n. Choose $(n_k)_{k=1}^\infty\subset\mathbb{N}$ such that $n_1=1$ and $5^{-p}T_{n_p}>4\cdot$
$\max\{p,\sum_{k=1}^{p-1}5^{-k}\beta_{n_k}\}$ for $p>1$. Write $f=\sum_{k=1}^\infty 5^{-k}f_{n_k}$. Then $p>1$ implies

$$|\tau_{n_p}(f)|\ge\frac{3}{4}5^{-p}T_{n_p}-\sum_{k=1}^{p-1}5^{-k}\beta_{n_k}-\sum_{k=p+1}^\infty 5^{-k}T_{n_p}>p.]$$

(c) If $[a,b]=[-\pi,\pi]$ and $u_n=D_n$, then $T_n=2\pi L_n$ (Lebesgue's constants) and $\tau_n(f)=2\pi S_n(f,0)$ so (b) yields an $f\in C(\mathbb{T})$ whose Fourier series diverges at $x=0$.

10. Suppose that $\phi \in L_1^r(]0, \pi])$ and $0 < \delta < \pi$ satisfy $\phi(t) \geq 0$ for all $t \in]0, \pi]$, $\int_0^\delta t^{-1}\phi(t)\,dt = \infty$, and the function $t \to t^{-1}\phi(t)$ is nonincreasing on $]0, \delta]$.

(a) It follows that

$$\lim_{n \to \infty} \int_0^\pi t^{-1}\phi(t)|\sin(nt)|\,dt = \infty.$$

[Hint: For fixed n, consider the intervals $[a_k, b_k]$ with $\pi/(4n) = a_0 < a_1 < \cdots < a_p < \delta$ and $a_p \geq \delta/2$ on which $|\sin(nt)| > 2^{-1/2}$. Check that

$$2\int_{a_k}^{b_k} t^{-1}\phi(t)|\sin nt|\,dt \geq 2^{-1/2}\int_{a_k}^{a_{k+1}} t^{-1}\phi(t)\,dt.$$

Sum for $k = 0, 1, \ldots, p - 1$.]

(b) There exists an even $f \in C(\mathbb{T})$ with $\|f\|_u \leq 1$ and $f(0) = f(\pi) = 0$ such that

$$\overline{\lim_{n \to \infty}} \left| \int_0^\pi f(t)\phi(t)\frac{\sin(nt)}{t}\,dt \right| = \infty.$$

[Hint: Use the preceding exercise.]

(c) If, in addition, $\phi \in C([0, \pi])$, then there exists an even function $g \in C(\mathbb{T})$ with $g(0) = g(\pi) = 0$ such that $|g(t)| \leq \phi(t)$ for $0 \leq t \leq \pi$ and $\overline{\lim_{n \to \infty}} |S_n(g, 0)| = \infty$. [Hint: Let $g(-t) = g(t) = f(t)\phi(t)$ for $0 \leq t \leq \pi$.]

(d) There exists a function $g \in C(\mathbb{T})$ whose Fourier series diverges at $x = 0$ while $\lim_{t \to 0}(|g(t) - g(0)|\log(1/|t|)) = 0$. (Compare (8.52.d) and (8.44).) [Hint: Check that $\phi(t) = [\log(\pi^2/t) \cdot \log\log(\pi^2/t)]^{-1}$, $\phi(0) = 0$ satisfies our hypotheses for some δ.]

11. Define functions $f, g \in BV(\mathbb{T})$ by $f(x) = g(x) = 1$ for $0 < x < \pi$, $f(x) = -1$ for $-\pi < x < 0$, $f(-\pi) = f(0) = 0$, and $g(x) = (4/\pi)|2x + \pi| - 3$ for $-\pi \leq x \leq 0$. Then, for all $x \in \mathbb{R}$, we have

(a)
$$f(x) = \frac{4}{\pi} \cdot \sum_{k=0}^\infty \frac{\sin(2k+1)x}{2k+1}$$

and

(b)
$$g(x) = \frac{16}{\pi^2} \cdot \sum_{k=0}^\infty \frac{1}{(2k+1)^2}\left[\frac{1}{2}\cos(4k+2)x + (-1)^k \sin(2k+1)x.\right]$$

Moreover, series (a) converges uniformly on every closed interval not containing an integral multiple of π and series (b) converges uniformly on $\mathbb{R}$. Notice that both series converge to 1 at every x in $]0, \pi[$.

12. *Poisson's Summation Formula.* Let $\phi : \mathbb{R} \to \mathbb{C}$ be in $L_1(\mathbb{R})$ and write

$$g_n(x) = \sum_{k=-n}^n \phi(x + 2\pi k), \qquad G(x) = \sum_{k=-\infty}^\infty |\phi(x + 2\pi k)|.$$

(a) We have $G(x) < \infty$ a.e. on $\mathbb{R}$ and $G \in L_1(\mathbb{T})$. [Hint: Use (6.46).]

(b) For $n, j \in \mathbb{Z}$ with $n \geq 0$, we have

$$\int_{-\pi}^\pi g_n(t)e^{-ijt}\,dt = \int_{-(2n+1)\pi}^{(2n+1)\pi} \phi(x)e^{-ijx}\,dx.$$

(c) The limit $g(x) = \lim_{n\to\infty} g_n(x)$ exists in $\mathbb{C}$ at almost every $x \in \mathbb{R}$, $g \in L_1(\mathbb{T})$, and

$$\hat{g}(j) = \frac{1}{2\pi} \int_{-\infty}^{\infty} \phi(x)e^{-ijx}\,dx = \hat{\phi}(j)$$

for all $j \in \mathbb{Z}$. (The last equality defines the *Fourier transform* $\hat{\phi}$ of ϕ at every $j \in \mathbb{R}$.) [Hint: Use (6.55).]

Now suppose *in addition* that $\phi \in BV(\mathbb{R})$, by which we mean that $\lim_{b\to\infty} V_{-b}^{b}\phi < \infty$. We write this limit as $V_{-\infty}^{\infty}\phi$.

(d) The series

$$g(x) = \sum_{k=-\infty}^{\infty} \phi(x + 2k\pi)$$

converges uniformly (and absolutely) on $[0, 2\pi]$, $g \in BV(\mathbb{T})$, and $V_0^{2\pi} g \le V_{-\infty}^{\infty}\phi$. [Hint: Use the Weierstrass M-test: if $x_0 \in [0, 2\pi]$ with $G(x_0) < \infty$, then $|\phi(x + 2k\pi)| \le |\phi(x_0 + 2k\pi)| + V_{2k\pi}^{2(k+1)\pi}\phi$ for all $x \in [0, 2\pi]$ and all $k \in \mathbb{Z}$.]

(e) If $\phi(x-) + \phi(x+) = 2\phi(x)$ for all $x \in \mathbb{R}$, then

$$\sum_{k=-\infty}^{\infty} \phi(x + 2k\pi) = \sum_{j=-\infty}^{\infty} \hat{\phi}(j)e^{ijx}$$

for all $x \in \mathbb{R}$ (where these sums are the limits of the symmetric partial sums $\sum_{-n}^{n}$).

(f) If $\phi \in C(\mathbb{R}) \cap L_1(\mathbb{R}) \cap BV(\mathbb{R})$, then formula (e) obtains for all $x \in \mathbb{R}$, the series on the left converges uniformly on $[0, 2\pi]$, and the series on the right converges uniformly on $\mathbb{R}$.

(g) If $a > 0$ and $\phi(x) = e^{-a|x|}$, then $\hat{\phi}(u) = a\pi^{-1}(a^2 + u^2)^{-1}$ and we obtain

$$\frac{1}{2} + \frac{1}{e^{2\pi a} - 1} = \frac{1}{2\pi a} + \frac{a}{\pi} \cdot \sum_{j=1}^{\infty} \frac{1}{a^2 + j^2}.$$

Consider what happens as $a \to \infty$.

(h) If $\psi \in L_1(\mathbb{R}) \cap C(\mathbb{R}) \cap BV(\mathbb{R})$ and

$$\Psi(u) = \frac{1}{\sqrt{2\pi}} \int_{-\infty}^{\infty} \psi(x)e^{-iux}\,dx = \sqrt{2\pi}\,\hat{\psi}(u),$$

then, for any positive real numbers a and b with $ab = 2\pi$, we have

$$\sqrt{a} \sum_{k=-\infty}^{\infty} \psi(ka) = \sqrt{b} \sum_{j=-\infty}^{\infty} \Psi(jb).$$

This is *Poisson's Summation Formula*. [Hint: Define $\phi(x) = \psi(x/b)$ and use (e) with $x = 0$.]

13.   *More Parseval Formulas*. Let $f \in L_1(\mathbb{T})$ be given.

(a) If $g \in L_1(\mathbb{T})$ and $f\bar{g} \in L_1(\mathbb{T})$, then

$$\frac{1}{2\pi} \int_{-\pi}^{\pi} f(x)\overline{g(x)}\,dx - \sum_{k=-n}^{n} \hat{f}(k)\overline{\hat{g}(k)} = \frac{1}{2\pi} \int_{-\pi}^{\pi} f(x)\big[\,\overline{g(x)} - \overline{S_n(g, x)}\,\big]dx$$

and

$$\frac{1}{2\pi} \int_{-\pi}^{\pi} f(x)\overline{g(x)}\,dx - \sum_{k=-n}^{n} \Big(1 - \frac{|k|}{n+1}\Big)\hat{f}(k)\overline{\hat{g}(k)}$$

$$= \frac{1}{2\pi} \int_{-\pi}^{\pi} f(x)\big[\,\overline{g(x)} - \overline{\sigma_n(g, x)}\,\big]dx.$$

(b) If $g \in L_1(\mathbb{T})$ is bounded on $\mathbb{R}$, then

$$\frac{1}{2\pi} \int_{-\pi}^{\pi} f(x) \overline{g(x)} \, dx = \lim_{n \to \infty} \sum_{k=-n}^{n} \left(1 - \frac{|k|}{n+1}\right) \hat{f}(k) \overline{\hat{g}(k)} = C_1\text{-} \sum_{k=-\infty}^{\infty} \hat{f}(k) \overline{\hat{g}(k)},$$

where the second equality defines its right side. [Hint: Use (a), (8.35), (8.38), and (6.55).]

(c) If $g \in BV(\mathbb{T})$, then

$$\frac{1}{2\pi} \int_{-\pi}^{\pi} f(x) \overline{g(x)} \, dx = \lim_{n \to \infty} \sum_{k=-n}^{n} \hat{f}(k) \overline{\hat{g}(k)} = \sum_{k=-\infty}^{\infty} \hat{f}(k) \overline{\hat{g}(k)},$$

where the second equality defines its right side. [Hint: Use (a), (8.47), (8.57), and (6.55).]

14. *Term-by-term Integration.* Let $f \in L_1(\mathbb{T})$ and $\phi \in L_1(\mathbb{R}) \cap BV(\mathbb{R})$ be given (see Exercise 12). Then

$$\int_{-\infty}^{\infty} f(x) \phi(x) \, dx = \sum_{k=-\infty}^{\infty} \hat{f}(k) \int_{-\infty}^{\infty} \phi(x) e^{ikx} \, dx.$$

In particular, the numerical series on the right converges whether or not $S(f)$ converges. [Hints: Use Exercise 12(d) to obtain $\int_{-\pi}^{\pi} fg = \int_{-\infty}^{\infty} f\phi$ and $2\pi\hat{g}(-k) = $ the kth integral in the above series. Then apply Exercise 13(c) with $\bar{g}$ in place of g.]

15. Let $\alpha \in \mathbb{C}\backslash\mathbb{Z}$ be fixed. By considering the Fourier series of the function $f \in BV(\mathbb{T})$ defined by $f(x) = e^{-i\alpha x}$ for $0 < x < 2\pi$ and $f(0) = e^{-i\pi\alpha}\cos(\pi\alpha)$, prove that the following equalities obtain whenever $0 < x < 2\pi$.

(a) $$\frac{\pi e^{i\pi\alpha}}{\sin(\pi\alpha)} = \sum_{n=-\infty}^{\infty} \frac{e^{i(n+\alpha)x}}{n+\alpha}.$$

(b) $$\frac{\pi e^{-i\pi\alpha}}{\sin(\pi\alpha)} = \sum_{n=-\infty}^{\infty} \frac{e^{-i(n+\alpha)x}}{n+\alpha}.$$

(c) $$\pi = \sum_{n=-\infty}^{\infty} \frac{\sin[(n+\alpha)x]}{n+\alpha}.$$

(d) $$\pi \cot(\pi\alpha) = \sum_{n=-\infty}^{\infty} \frac{\cos[(n+\alpha)x]}{n+\alpha}.$$

(e) $$\frac{\pi}{\sin(\pi\alpha)} = \sum_{n=-\infty}^{\infty} \frac{(-1)^n}{n+\alpha}.$$

(f) If $x = 0$, then (d) is true but the first three are all false.

(g) If $x = 2\pi$, then the first four are all false with the exception that (c) is true if 4α is an odd integer. Of course, $\sum_{-\infty}^{\infty}$ means $\lim_{p \to \infty} \sum_{-p}^{p}$ throughout.

16. (a) Suppose that $f: [0, 2\pi] \to \mathbb{C}$ is continuous with $V_0^{2\pi} f < \infty$ and that $\alpha \in \mathbb{C}$ is such that $f(2\pi) = f(0)e^{2\pi i\alpha}$ (for example, $\alpha = (1/(2\pi i))\text{Log}(f(2\pi)/f(0))$ if $f(0) \neq 0 \neq f(2\pi)$). Write

$$c_n = \frac{1}{2\pi} \int_0^{2\pi} f(t) e^{-i(n+\alpha)t} \, dt$$

for $n \in \mathbb{Z}$. Then $f(x) = \lim\limits_{p \to \infty} \sum\limits_{n=-p}^{p} c_n e^{i(n+\alpha)x}$ uniformly for $0 \le x \le 2\pi$. [Hint:
Consider the function $g \in C(\mathbb{T})$ such that $g(x) = f(x)e^{-i\alpha x}$ for $0 \le x \le 2\pi$.]

(b) Compute the expansion (a) for the case that $f(x) = (\pi - x)/2$ and $\alpha = 1/2$ and then compare your result with (8.9.i).

(c) Given a fixed $\alpha \in \mathbb{C}$, find a simple description of the set of functions on $[0, 2\pi]$ that can be obtained as uniform limits of functions that are finite linear combinations of the functions

$$e^{i(\alpha+n)x} \qquad (n = 0, \pm 1, \pm 2, \ldots).$$

17.   The Lebesgue constants L_n satisfy

$$L_n = \frac{16}{\pi^2} \sum_{k=1}^{\infty} \frac{a_k(n)}{4k^2 - 1},$$

where

$$a_k(n) = \sum_{j=1}^{k(2n+1)} \frac{1}{2j - 1}.$$

In particular, $L_n < L_{n+1}$ for all $n \ge 0$. [Hints: Use the Fourier series of $|\sin x|$ and the identity $[\sin(kx)]^2/(\sin x) = \sin x + \sin(3x) + \ldots + \sin[(2k - 1)x]$.]

18.   Suppose that $f \in L_1(\mathbb{T})$, $a < b$, and $V_a^b f < \infty$. Then

$$\sup\{|S_n(f, x)| : n \in \mathbb{N}, a + \delta \le x \le b - \delta\} < \infty$$

whenever $0 < \delta < (b - a)/2$, but not necessarily for $\delta = 0$. [Hints: Use (8.41) and (8.57). See Exercise 1 for the case $\delta = 0$.]

19.   If $0 < \delta < \pi$ and D_n is Dirichlet's kernel, then

$$\lim_{n \to \infty} \frac{\pi}{4 \log n} \int_0^{\delta} |D_n(t)| \, dt = 1.$$

20.   *Weierstrass's Nowhere Differentiable Functions.* This exercise outlines a very simple proof (adapted from Katznelson's book) that certain functions with absolutely convergent Fourier series are nowhere differentiable.

Let $\alpha > 1$ be fixed and let $(n_k)_{k=1}^{\infty} \subset \mathbb{N}$ satisfy $n_{k+1} \ge \alpha n_k$ for all k. Suppose that $f \in L_1(\mathbb{T})$ is a given function that is differentiable at some $x_0 \in \mathbb{R}$ and satisfies $\hat{f}(n) = 0$ if $n \in \mathbb{N}$ and $n_k < n < n_{k+1}$ for some k. Define g on $\mathbb{R}$ by $g(t) = f(x_0 + t) - f(x_0)\cos t - f'(x_0)\sin t$. Let p_k be the integer satisfying $2p_k < (1 - \alpha^{-1})n_k \le 2p_k + 2$ and let Q_k be the trigonometric polynomial $Q_k = \|K_{p_k}\|_2^{-2} K_{p_k}^2$, where K_n is Fejér's Kernel. Then we have the following.

(a) $g(0) = 0$ and $\lim\limits_{t \to 0} t^{-1} g(t) = 0$.

(b) $\hat{g}(n) = 0$ if $n \in \mathbb{N}$ and $n_k < n < n_{k+1}$ for some k.

(c) If $j \in \mathbb{Z}$ and $1 \le |j| \le 2p_k$ for some $k > 1$, then $n_{k-1} < n_k + j < n_{k+1}$, and so $\hat{g}(n_k + j) = 0$. [Hint: $2 - \alpha^{-1} < \alpha$.]

(d) $\hat{Q}_k(0) = 1$ and $\hat{Q}_k(j) = 0$ if $|j| > 2p_k$.

(e) $$\hat{f}(n_k) = \hat{g}(n_k) = \frac{1}{2\pi} \int_{-\pi}^{\pi} g(t)Q_k(t)e^{-in_k t} \, dt \text{ if } k > 1.$$

[Hint: Q_k has the form $\sum\limits_{j=-2p_k}^{2p_k} c_j e^{ijt}$ with $c_0 = 1$.]

(f)
$$\| K_{p_k} \|_2^2 = \sum_{j=-p_k}^{p_k} \left(1 - \frac{|j|}{p_k + 1} \right)^2 > \frac{2(p_k + 1)}{3}.$$

(g)
$$0 \le Q_k(t) \le \frac{3\pi^4}{2(p_k + 1)^3 t^4} \quad \text{if } 0 < |t| \le \pi.$$

[Hint: See (8.29) and (f).]

(h) If $k > 1$ and $\epsilon_k = \sup\{|t^{-1} g(t)| : 0 < |t| \le (p_k + 1)^{-1/4}\}$, then

$$|\hat{f}(n_k)| \le \frac{3\pi^4 + 2}{2p_k + 2} \left(\epsilon_k + \frac{\|g\|_1}{p_k + 1} \right).$$

[Hints: Use (e). Divide the interval $[-\pi, \pi]$ into the sets E_1 where $|t| \le (p_k + 1)^{-1}$, E_2 where $(p_k + 1)^{-1} \le |t| \le (p_k + 1)^{-1/4}$, and E_3 for the rest. Write $I_j = (2\pi)^{-1} \int_{E_j} |g| Q_k$. In I_1, estimate $|g|$ by $(p_k + 1)^{-1} \epsilon_k$ and use $\| Q_k \|_1 = 1$. In I_2, use (g) to estimate Q_k and $|t|\epsilon_k$ to estimate $|g(t)|$. In I_3, replace Q_k by its sup over E_3 and use (g).]

(i) $\lim_{k \to \infty} n_k \hat{f}(n_k) = 0$.

From these results we deduce the following.

(j) If $(n_k)_{k=1}^\infty \subset \mathbb{N}$ satisfies $\inf\{n_{k+1}/n_k : k \in \mathbb{N}\} > 1$ and $f \in L_1(\mathbb{T})$ satisfies $\hat{f}(n) = 0$ whenever $n \in \mathbb{N}$ and $n_k < n < n_{k+1}$ for some k but $\overline{\lim}_{k \to \infty} n_k |\hat{f}(n_k)| > 0$, then f is differentiable at *no* point of $\mathbb{R}$.

(k) If $a \in \mathbb{N}$ and $b \in \mathbb{R}$ satisfy $0 < b < 1$ and $ab \ge 1$ and if $(\theta_k)_{k=1}^\infty \subset \mathbb{R}$ is arbitrary, then the function f defined by

$$f(t) = \sum_{k=1}^\infty b^k \cos(a^k t + \theta_k)$$

is in $C(\mathbb{T})$ but has a finite derivative *nowhere* on $\mathbb{R}$. [Hint: $2|\hat{f}(a^k)| = b^k$.]
    We remark that over a century ago Karl Weierstrass used different methods to prove (k) under the additional hypotheses that a is odd, $2ab > 3\pi + 2$, and $\theta_k = 0$ for all k. For example, $a = 7$, $b = 6/7$.

(l) The function f defined by $f(x) = \sum_{n=1}^\infty 2^{-n} \sin(2^n x)$ satisfies $\sup\{|n\hat{f}(n)| : n \in \mathbb{Z}\} < \infty$ and $f \in C(\mathbb{T})$ but $f \notin BV(\mathbb{T})$. Compare (8.46).

21.   (a) Use (8.40) to prove that

$$\lim_{b \to \infty} \int_0^b \frac{\sin x}{x} dx = \frac{\pi}{2}.$$

(b) Use $\sin(2x) = 2(\sin x)\cos x$ to obtain

$$\lim_{b \to \infty} \int_0^b \frac{(\sin x)\cos x}{x} dx = \frac{\pi}{4}.$$

(c) Integrate by parts to get

$$\int_0^\infty \left(\frac{\sin x}{x} \right)^2 dx = \frac{\pi}{2}.$$

22.   *A Localization Theorem.* Let $f \in L_1(\mathbb{T})$ be constant on some *open* interval $I \subset \mathbb{R}$ and suppose that $\lim_{|n| \to \infty} n\hat{f}(n) = 0$. Then the series

$$\sum_{n=-\infty}^{\infty} in\hat{f}(n)e^{inx} \tag{*}$$

(obtained by formally differentiating the Fourier series of f term-by-term) converges to 0 uniformly on every closed subinterval of I. This result is needed for the next exercise. Its proof can be obtained by the following steps. Let $J \subset I$ be a closed interval and let $0 < \delta < \pi$ be such that $x - t \in I$ if $x \in J$ and $-\delta \le t \le \delta$. Choose any $g \in C(\mathbb{T})$ which has a continuous third derivative and satisfies $g(t) = \cot(t/2)$ if $\delta \le |t| \le \pi$.

(a) Such a function g exists and we also have $\sum_{k=-\infty}^{\infty} |k\hat{g}(k)| < \infty$. [Hints: Take $g(t) = \phi(t)\cot(t/2)$, where $\phi(t) = 1$ for $\delta \le |t| \le \pi$, $\phi(t) = 0$ for $|t| < \delta/2$, and $\phi \in C^{(3)}$ ((4.31) may help). Use integration by parts to show $|k^3\hat{g}(k)| \le \|g^{(3)}\|_1$.]

(b) The partial sums of series (*) are given by

$$S_p(x) = \sum_{n=-p}^{p} in\hat{f}(n)e^{inx} = \frac{1}{2\pi}\int_{-\pi}^{\pi} f(x-t)D_p'(t)\, dt,$$

where D_p is Dirichlet's Kernel.

(c) If $\delta \le |t| \le \pi$, and $p \ge 0$, then $D_p(t) = g(t)\sin(pt) + \cos(pt)$.

(d) If $x \in J$ and $p \ge 0$, then

$$S_p(x) = \frac{1}{2\pi}\int_{-\pi}^{\pi} f(x-t)\{pg(t)\cos(pt) + g'(t)\sin(pt) - p\sin(pt)\}\, dt.$$

[Hint: $f(x-t) = f(x)$ if $-\delta \le t \le \delta$ and $D_p(\delta) = D_p(-\delta)$.]

(e) For all $p \in \mathbb{Z}$ and $x \in \mathbb{R}$ we have

$$\frac{1}{2\pi}\int_{-\pi}^{\pi} f(x-t)g(t)e^{ipt}\, dt = \sum_{k=-\infty}^{\infty} \hat{f}(p+k)\hat{g}(k)e^{i(p+k)x}.$$

[Hint: Replace g by its Fourier series and integrate term-by-term.]

(f) $$\lim_{|p| \to \infty} |p| \cdot \sum_{k=-\infty}^{\infty} |\hat{f}(p+k)\hat{g}(k)| = 0.$$

[Hints: Write $|p| \le |p+k| + |k|$ and then dominate by

$$A \cdot \sum_{|k|>N} (|k|+1)|\hat{g}(k)| + B \cdot \sum_{n=p-N}^{p+N} (|n|+1)|\hat{f}(n)|,$$

where $A = |\hat{f}(0)| + \sup\{|n\hat{f}(n)| : n \in \mathbb{Z}\}$ and $B = |\hat{g}(0)| + \sup\{|k\hat{g}(k)| : k \in \mathbb{Z}\}$. First fix N to make the first term small and then let $|p| \to \infty$.]

(g) The right side of (d) has limit 0 uniformly on $\mathbb{R}$ as $p \to \infty$. [Hint: Don't forget (8.39).]

(h) $\lim_{p \to \infty} S_p(x) = 0$ uniformly on J.

(i) The Cantor–Lebesgue Theorem shows that (*) cannot converge on a nontrivial interval if $\overline{\lim}_{|n| \to \infty} |n\hat{f}(n)| > 0$. [See Exercise 24(d), p. 313.]

23. *Nonuniqueness.* Can a trigonometric series converge a.e. on $\mathbb{R}$ to a function ϕ $\in L_1(\mathbb{T})$ and yet *not* be the Fourier series of ϕ? This question was first answered in the affirmative with $\phi = 0$ in 1916 by the Russian analyst D. E. Menshow.* This exercise leads us through (a slightly more general version of) his example.

Let P be the Cantor set constructed from a sequence $(a_n)_{n=0}^{\infty}$ as in (2.81) and let ψ be the corresponding function as in (3.92). Adopt all notation of (2.81). For each $n \in \mathbb{N}$, let $\psi_n \in C([0, 1])$ be the function that agrees with ψ on $[0, 1] \setminus P_n^0$ and is linear on each $J_{n,k}$ ($1 \leq k \leq 2^n$). That is, $\psi_n = 2^{-m}(2k - 1)$ on $I_{m,k} (1 \leq m \leq n,$ $1 \leq k \leq 2^{m-1})$, $\psi_n(0) = 0$, $\psi_n(1) = 1$, and ψ_n is linear on each component interval of P_n.

(a) Sketch graphs of ψ_1, ψ_2, and ψ_3.

(b) For each $n \in \mathbb{N}$, ψ_n is nondecreasing on $[0, 1]$, $\psi_n' = 2^{-n} a_n^{-1}$ on P_n^0, $\psi_n' = 0$ on $[0, 1] \setminus P_n$, and $|\psi_n(x) - \psi_{n+1}(x)| \leq \epsilon_{n+1} 2^{-(n+1)}$ for all $x \in [0, 1]$, where ϵ_{n+1} $= 1 - 2a_{n+1}/a_n$. [Hint: For the inequality, sketch the portions of the graphs of the two functions that lie above one $J_{n,k}$ and note that we have equality at the endpoints of the middle interval $I_{n+1,k}$.]

(c) The sequence $(\psi_n)_{n=1}^{\infty}$ converges to ψ uniformly on $[0, 1]$.

(d) For all $n \in \mathbb{N}$ and $x \in [0, 1]$ we have $|\psi(x) - \psi_n(x)| \leq 2^{-n} \delta_n$, where $\delta_n = \sup\{\epsilon_j :$

$$j \in \mathbb{N}, j > n\}. \ [\text{Hint: } \psi - \psi_n = \sum_{j=n}^{\infty} (\psi_{j+1} - \psi_j).]$$

(e) If $k, n \in \mathbb{N}$, then

$$\int_0^1 |\psi(x) - \psi_n(x)| \, dx \leq 2^{-n} \delta_n \lambda(P_n) = \delta_n a_n,$$

$$\left| \int_0^1 \psi_n(x) \cos(k\pi x) \, dx \right| = \left| \frac{1}{k\pi} \int_{P_n} \psi_n'(x) \sin(k\pi x) \, dx \right| \leq \frac{2}{k^2 \pi^2 a_n},$$

$$\left| \int_0^1 \psi(x) \cos(k\pi x) \, dx \right| \leq \delta_n a_n + \frac{2}{k^2 \pi^2 a_n}.$$

Now define $f \in C(\mathbb{T})$ by $f(t) = f(-t) = \psi(t/\pi)$ if $0 \leq t \leq \pi$.

(f) For all $k \in \mathbb{Z}$ we have

$$\hat{f}(k) = \int_0^1 \psi(x) \cos(k\pi x) \, dx.$$

(g) If $k \in \mathbb{Z}$, $p \in \mathbb{N}$, and $|k| \geq a_p^{-1} \delta_p^{-1/2}$, then

$$|k\hat{f}(k)| \leq \delta_p^{1/2} [1 + 4\pi^{-2}(1 - \delta_p)^{-1}].$$

[Hints: Note that $1 \geq \delta_1 \geq \delta_2 \geq \ldots$, $a_n \leq 2^{-n}$, $a_n^{-1} \delta_n^{-1/2} \geq 2^n$, $a_{n-1} a_n^{-1}$ $= 2(1 - \epsilon_n)^{-1}$, and $\delta_p \geq \epsilon_n$ if $p < n$. Consider the $n > p$ for which $a_{n-1}^{-1} \delta_{n-1}^{-1/2}$ $\leq |k| < a_n^{-1} \delta_n^{-1/2}$ and use (f) and (e).]

---

*It had already been proved in 1908 by the British analyst J. W. Young that if the question is changed by replacing "a.e. on $\mathbb{R}$" by "everywhere on $\mathbb{R} \setminus E$, where E is countable," then the answer is negative. The German analyst Georg Cantor was led to his invention and investigation of transfinite numbers by just such questions. Indeed, Cantor had earlier proved Young's theorem for the case that the exceptional set E has countable closure. Time and space do not allow us to include more of the fascinating story of uniqueness in this volume.

Now we suppose that $2^n a_n \to 0$ and $\epsilon_n \to 0$ as $n \to \infty$. (Menshov's example was $a_n = 2^{-n}(n+1)^{-1}$, so $\epsilon_n = (n+1)^{-1}$.)

(h) We have $\lim_{p \to \infty} \delta_p = 0$, $\lim_{|k| \to \infty} |k\hat{f}(k)| = 0$, and $\lambda(P) = 0$.

(i) If $\alpha_k = -k\hat{f}(k)$ for $k \in \mathbb{N}$, then the trigonometric series

$$\sum_{k=1}^{\infty} \alpha_k \sin(kt)$$

converges to 0 at almost every $t \in \mathbb{R}$ but not all α_k are 0. In fact, $\alpha_1 > 0$ and this series converges to 0 uniformly on any closed interval $J \subset \{t : -\pi \leq t \leq \pi, |t/\pi| \notin P\}$. [Hints: Use Exercise 22, (8.11), and (f).]

(j) The series in (i) is not the Fourier series of any function in $L_1(\mathbb{T})$. [Hint: (8.36).]

The truly heroic reader may wish to verify the remaining assertions in order to compute $\hat{f}$ in the general case. Here $r_j = a_{j-1} - a_j$ and $F_n = \{ \sum_{j=1}^{n} \xi_j r_j : \text{each } \xi_j \text{ is 0 or 1} \}$ is the set of left endpoints of the 2^n intervals $J_{n,k}$ just as in the proof of (2.83).

(k) If $n \in \mathbb{N}$ and $0 \neq k \in \mathbb{Z}$, then

$$\int_0^1 \psi_n(x) e^{ik\pi x} \, dx = \frac{(-1)^k}{ik\pi} + \frac{e^{ik\pi a_n} - 1}{2^n a_n k^2 \pi^2} \cdot \sum_{t \in F_n} e^{ik\pi t}.$$

(l) For $n \in \mathbb{N}$ and $k \in \mathbb{Z}$, we have

$$\frac{1}{2^n} \sum_{t \in F_n} e^{ik\pi t} = e^{ik\pi s_n/2} \cdot \prod_{j=1}^{n} \cos(k\pi r_j/2),$$

where $s_n = r_1 + r_2 + \ldots + r_n = 1 - a_n$.

(m) If $n \in \mathbb{N}$ and $0 \neq k \in \mathbb{Z}$, then

$$\int_0^1 \psi_n(x) \cos(k\pi x) \, dx = \frac{[-2\sin(k\pi/2)]\sin(k\pi a_n/2)}{a_n k^2 \pi^2} \cdot \prod_{j=1}^{n} \cos(k\pi r_j/2)$$

and

$$\int_0^1 \psi(x) \cos(k\pi x) \, dx = -\frac{\sin(k\pi/2)}{k\pi} \cdot \prod_{j=1}^{\infty} \cos(k\pi r_j/2).$$

(n) If $k \in \mathbb{N}$, then

$$\alpha_k = \frac{\sin(k\pi/2)}{\pi} \cdot \prod_{j=1}^{\infty} \cos(k\pi r_j/2).$$

(o) If $a_n = 3^{-n}$, so that P is Cantor's ternary set, and $0 \leq k \in \mathbb{Z}$, then $3^k \hat{f}(3^k) = -\frac{1}{\pi} \prod_{j=1}^{\infty} \cos(3^{-j}\pi) < 0$. (Compare Exercise 8(j).)

BIBLIOGRAPHY

1.   Apostol, T.M. *Mathematical Analysis.* 2d ed. Reading, Mass.: Addison-Wesley, 1974.

2.   Bartle, R.G. *The Elements of Real Analysis.* New York: John Wiley & Sons, Inc., 1964.

3.   Bary, N.K. *A Treatise on Trigonometric Series.* 2 vols. New York: Macmillan, 1964.

4.   Boas, R.P. *A Primer of Real Functions.* Carus Mathematical Monograph No. 13. New York: John Wiley & Sons, Inc., 1960.

5.   Burckel, R.B. *An Introduction to Classical Complex Analysis.* New York: Birkhäuser Verlag, 1979.

6.   Fleming, W. *Functions of Several Variables.* New York: Springer-Verlag, 1977.

7.   Halmos, P.R. *Naive Set Theory.* New York: Springer-Verlag, 1975.

8.   Hardy, G.H. *Divergent Series.* New York: Oxford University Press, 1956.

9.   Hardy, G.H. *A Course of Pure Mathematics.* 10th ed. New York: Cambridge University Press, 1952.

10.   Hardy, G.H.; Littlewood, J.E.; and Polya, G. *Inequalities.* New York: Cambridge University Press, 1952.

11.   Hardy, G.H. and Rogosinski, W.W. *Fourier Series.* 3d ed. New York: Cambridge University Press, 1956.

12. Hewitt, Edwin and Stromberg, Karl. *Real and Abstract Analysis*. New York: Springer-Verlag, 1975.

13. Katznelson, Y. *An Introduction to Harmonic Analysis*. New York: John Wiley & Sons, Inc., 1968.

14. Knopp, K. *Theory and Application of Infinite Series*. 2d ed. New York: Hafner, 1951.

15. Landau, E. *Foundations of Analysis*. New York: Chelsea, 1951.

16. Nagy, B. *Real Functions and Orthagonal Expansions*. New York: Oxford University Press, 1965.

17. Natanson, I.P. *Theory of Functions of a Real Variable*. New York: Ungar, 1955.

18. Niven, Ivan. *Irrational Numbers*. Carus Mathematical Monograph No. 11. New York: John Wiley & Sons, Inc., 1956.

19. Ross, K.A. *Elementary Analysis: The Theory of Calculus*. New York: Springer-Verlag, 1980.

20. Rubin, J.E. *Set Theory for the Mathematician*. San Francisco: Holden-Day, 1967.

21. Rudin, W. *Principles of Mathematical Analysis*. 3d ed. New York: McGraw-Hill, 1976.

22. Saks, S. and Zygmund, A. *Analytic Functions*. New York: Hafner, 1952.

23. Smiley, M.F. *Algebra of Matrices*. Boston: Allyn & Bacon, 1965.

24. Zygmund, A. *Trigonometric Series*. 2 vols. New York: Cambridge University Press, 1959.

INDEX